RECURSING ADA

Figure 1: Ada Lovelace.

Charles Babbage, Byron's Daughter and the Genesis of a Deception ...

Recursing Ada

Charles Babbage, Byron's Daughter and the Genesis of a Deception ...

Gavan Tredoux

galton.org

ISBN: 978-1-963225-33-4

Contents.

Figures.

By the time she [Ada] was fifteen she was showing her extraordinary talent for mathematics, which she inherited from her mother, the former Miss Annabella Milbanke, and had already taught herself part of Paisley's *Geometry*. ... All Charles's tender anxiety seems at this time [1842-3] to have been focused on Lady Lovelace, for he was undoubtedly astonished by the rapid progress she was making in her studies. In the last year or two she had not only mastered the details of the analytical engine but also those of symbolic logic,[1] of which Thomas Hobbes of Malmesbury was an early exponent. ... The Byron Vault ... was reopened ... on 15th June 1938. There the coffins of father and daughter, both thirty-six years of age, were found to lie side by side, within a foot of each other, the poet and the mathematician, geniuses both.

—Maboth Moseley.M. Moseley 1964, 156, 168-9, 215.

I am in good spirits; for I hope another year will make me really something of an Analyst. The more I study, the more irresistible do I feel my genius for it to be. I do not believe that my father was (or ever could have been) such a Poet as I shall be an Analyst; (& Metaphysician); for with me the two go together indissolubly.

— Ada Lovelace, letter to Babbage, 1843-07-30.

Here will be a fine field for my clear, logical & accurate, mind, to work its powers upon; & to develop its metaphysical genius, which is not least amongst its qualifications & characteristics.

—Ada Lovelace, letter to Babbage, 1843-07-02.

Both Lovelace and Wheatstone were probably better informed about the Difference and Analytical Engines than anyone other than Babbage himself and possibly Somerville, and they had more knowledge than Menabrea, who had met Babbage only briefly in Turin.

—Fuegi and Francis, 2003.

[Babbage] had genius, he was the affectionate friend of one of the most beautiful and talented women of his time—no less a person than Lord Byron's daughter—who was one of the very few people who understood him, but I think he was born to be miserable!

—B. V. Bowden. Foreword to M. Moseley 1964, 16.

Was she indeed seeing "coming events"? Ada's name has now come to represent a computer language developed by the U.S. Department of Defense and named in her honor.

— Toole 1992, remarking on Ada's whimsical description of herself in a letter as a General.

Sir: In your otherwise beautiful poem 'The Vision of Sin' there is a verse which reads – 'Every moment dies a man, Every moment one is born.' It must be manifest that if this were true, the population of the world would be at a standstill ... I would suggest that in the next edition of your poem you have it read – 'Every moment dies a man, Every moment $1\frac{1}{16}$ is born.' ... The actual figure is so long I cannot get it onto a line, but I believe the figure $1\frac{1}{16}$ will be sufficiently accurate for poetry. I am, Sir, yours, etc., Charles Babbage

—Babbage to Tennyson, 1851.

There is one subject ancillary to Babbage on which far too much has been written, and that is the contributions of Ada Lovelace. Some combination of the connection to Byron, the appeal

[1] There is no evidence that Ada ever worked on symbolic logic.

of the myth that an aristocratic woman was "the world's first programmer," and the even stranger mythology concerning Lovelace and Babbage circulated by Maboth Moseley have led to a popular fascination with her. I am somewhat sorry I did not debunk her role more vigorously in my thesis. It would be only a slight exaggeration to say that Babbage wrote the "Notes" to Menabrea's paper, but for reasons of his own encouraged the illusion in the minds of Ada and the public that they were authored by her. It is no exaggeration to say that she was a manic depressive with the most amazing delusions about her own talents, and rather shallow understanding of both Charles Babbage and the Analytical Engine. As Maboth Moseley laid some claim to technical competence, it is hard to understand her generating so much misinformation about Ada in her book *Irascible Genius* (London, 1964). In *Ada, Countess of Lovelace, Byron's Legitimate Daughter* (New York, 1977), Doris Langley Moore explicitly disavows any intent or competence to judge Ada's technical contributions (and the book is almost more about Ada's mother), yet she may still have inadvertently reinforced popular misconceptions about them. However, she did not leave any mistaken impression that Ada had a charming personality, or played a large role in Babbage's life. In their paper "Lady Lovelace and Charles Babbage" (*Annals* 2(4): 299-329), Velma R. Huskey and Harry D. Huskey quote at great length from the correspondence between Ada and Babbage (and some other sources). To me, this familiar material seems to make obvious once again that Ada was as mad as a hatter, and contributed little more to the "Notes" than trouble. Yet the Huskeys seem to feel otherwise, for they largely quote without comment. Most recently, and, one may hope, finally, we have Dorothy Stein's *Ada, A Life and Legacy* (Cambridge, Massachusetts, 1985). This is, to my tastes, another work of tedious length, but it does finally seem to get the facts straight. Stein enthusiastically debunks Ada's talents and contributions, I believe convincingly and accurately (quoting the very same materials provided by the Huskeys). I will retain an open mind on whether Ada was crazy because of her substance abuse (as Moore would have it) or despite it (as Stein believes). I hope nobody feels compelled to write another book on the subject. But, then, I guess someone has to be the most overrated figure in the history of computing.

—Bruce Collier 1990

Preface.

In 2018 the Senate (115th congress) of the United States passed S. Res. 592 designating October 9, 2018, as 'National Ada Lovelace Day' and 'honoring the life and legacy' of Byron's daughter Ada Lovelace (1815–1852) as 'the first computer programmer'. The resolution determined among other things that

1. 'from a young age, Lovelace displayed a gift for mathematics, languages, and the sciences';
2. 'in 1833, Lovelace was introduced to inventor and mechanical engineer, Charles Babbage, and began to study his designs for the Analytical Engine';
3. she was 'the first person to recognize that the Analytical Engine could be used to manipulate symbols and letters and was the first person to theorize that the Analytical Engine could be used to create music and graphics';
4. 'Lovelace published step-by-step instructions for using the Analytical Engine to calculate Bernoulli numbers';
5. she died 'leaving behind a legacy of poetic science and reasoning, in which the arts and sciences are woven together to find new insights';
6. she was 'a leading woman in science and mathematics and the first computer programmer'.

This is Ada Lovelace mythology in list form, a cult of personality now extending beyond popular science books to novels, documentaries, dolls, stationery, and figurines—office-desk, classroom and hallway novelty items. Ephemera multiply at a compounding rate. All this aside from a programming language named for her, with which she bears no connection, and now even a Graphical Processing Unit (GPU) micro-architecture, even more remote from anything she ever did.[2]

All the claims in the list above are false, as shown here in careful detail. For an immediate example, the Analytical Engine, a conceptual cogwheel computer, was not even invented in 1833, unless the Senate is prepared to attribute clairvoyance to Ada, a possibility one cannot rule out. More fatuous resolutions have surely been passed by legislators—the Indiana General Assembly once tried to legislate a fractional value for π, failing all round—but this may be the only such attempt to impose findings on the history of science, to legislate the past.

It is certainly *not* true that Ada Lovelace was the world's first computer programmer, since that was Charles Babbage himself. Consider how Babbage could possibly invent a programmable computer (on paper) without programming it (on paper). We still have those first programs. Nor was Ada ever a 'leading scientist' or mathematician—she left no published results, no unpublished manuscripts containing anything scientifically worthwhile, no mathematical advances, no new applications. Her general knowledge of mathematics was rudimentary at best.

[2] Nvidia.

It *is* true that Ada helped to translate, with annotations, a paper in French by L. F. Menabrea describing Babbage's Analytical Engine, a paper for which Babbage himself had supplied most of the content. For those annotations, Babbage provided outlines of the mathematical workings for all the examples they contain, and beyond that much of the written content, but not all, is due directly to him. The notes are a pastiche. Where Ada attempted to explicate the material her exposition is riddled with elementary mathematical errors, which forces us to conclude that Babbage did not read her mathematics closely. The programming examples in her annotations are all simplifications of exercises prepared by himself. In particular, the program to calculate the Bernoulli numbers was due to him—he had been working on that idea since 1835. Where Ada did make a definite contribution it was to deliberately obfuscate a critical deficiency, a deception in which Babbage himself became complicit.

Determining Ada's role requires knowing a lot about Babbage—his personal history and peculiar personality—and the saga of the Difference and Analytical Engines. Much of this material has been inadequately excavated in the past. A rich collection of primary sources is provided here so that the (properly skeptical) reader can examine the details. Many letters are included, along with a new translation of the famous paper by Menabrea, really co-written with Babbage, to extricate it from Ada's marked-up and subtly altered translation.

Gavan Tredoux
2025.

Chronology.

1791. Birth of Charles Babbage (CB), 12-26, in Walworth, Surrey, to Benjamin Babbage and Elizabeth Plumleigh Teape. His father was a well-off banker from a family of goldsmiths, his mother from an influential Totnes family.

1810. CB enters Trinity College, Cambridge, 04-21 as a pensioner (ordinary fee-paying student). Befriends J.F.W. Herschel, St John's College, son of the astronomer.

1812. CB helps to found the Analytical Society with Herschel and others, to reform Cambridge mathematics along continental lines, 03-12. Migrates to Peterhouse, 04-07.

1813. *Memoirs of the Analytical Society* is published. Herschel is Senior Wrangler, George Peacock second.

1814. CB is 'clasped' during his oral examination (Act), excluding him from honours; leaves Cambridge via the ordinary poll examination, BA. Marries Georgiana Whitmore 07-25[3] against his father's wishes. Commences serious mathematical research. Receives a comfortable allowance from his father.

1815. CB settles at 5, Devonshire Street, Portland Place, lectures on Astronomy at the Royal Institution. Benjamin Herschel Babbage is born 08-06. Birth of Augusta Ada Byron (AB) [later Lovelace], to Lord Byron the poet and Anne Isabella Milbanke, 12-10.

1816. CB is elected FRS after being proposed by William and JFW Herschel. Continues serious mathematical research. Applies but is turned down for several positions. Georgiana's father dies, leaving a considerable legacy to his daughter.

1817. A son, Charles Whitmore Babbage, is born. Takes MA from Cambridge (a formality).

1818. A daughter, Georgiana Whitmore Babbage, is born.

1819. CB visits Paris with JFW Herschel.

1820. CB helps to found the Royal Astronomical Society and serves as its secretary.

1821. Herschel and CB visit Northern Italy and Switzerland in late summer. CB conceives the Difference Engine, a mechanical cogwheel computer for calculating tables of values by addition, using the method of differences on polynomials.

1822. CB writes a letter to Sir Humphry Davy, 07-03, announcing plans for the Difference Engine. First demonstration model of a small Difference Engine is constructed in wood.

1823. The Treasury refers the Difference Engine plans to the Royal Society, 04-01. Given a favourable report on 05-01, the Chancellor of the Exchequer agrees a month later to fund the construction of the full engine, starting with £1,000. CB receives Gold Medal of the Astronomical Society for the following year (on the advice of Herschel), 06. CB engages the established artisan Joseph Clement (1779–1844) to help construct it. Dugald Bromhead Babbage is born.

[3] Snyder 2011, 376 fn. 34.

1824. CB is readmitted at Trinity College, 04-07. Lord Byron dies in Greece from an infection, 04-19. Henry Prevost Babbage is born. CB visits Italy with Herschel.

1826. CB visits Paris with Georgiana, meets Laplace and Quetelet. Applies and canvasses for the Lucasian professorship at Cambridge but George Airy is preferred.

1827. Benjamin Babbage, father of Charles, dies at Teignmouth, 02-27, leaving his son CB financially independent on £100,000. Charles Whitmore Babbage dies at Devonshire street, aged 10, on 07-20. Georgiana, wife of CB, dies at Boughton House in Worcester, aged 35, as a result of childbirth, 09-01.[4] The child expired a short time after.[5] CB has nervous breakdown.

1828. CB travels in France, Italy and Germany. Appointed Lucasian Professor at Cambridge vacated by Airy, 03-06, due to Whewell's influence prompted by Herschel, but will never lecture.[6] Herschel superintends the Difference Engine construction in his absence. CB attends the scientific congress in Berlin 09-18, after visiting von Humboldt. CB leases Dorset House in London on his return. His soirées there begin.

1829. CB travels in Italy. Herschel marries Margaret Brodie Stewart, robust daughter of a Gaelic scholar. Disputes begin with Clement about charges and ownership of tools, which are ultimately settled by arbitration in favour of Clement.

1830. CB publishes *The Decline of Science in England*, aimed at the Royal Society.[7]

1831. The British Association for the Advancement of Science is founded with support from CB.

1832. A working portion of the Difference Engine is assembled. CB publishes *Economy of Machines and Manufactures*.

1833. Joseph Clement and CB fall out and work halts on the Difference Engine, never to resume. CB has spent £17,000 so far but the government takes some time to reimburse it all. AB accompanies her mother to see the Difference Engine in operation.

1834. The Royal Statistical Society is founded with CB as member. Dionysius Lardner publishes a description of the Difference Engine,[8] with extensive collaboration by CB. Georgiana, daughter of CB, dies suddenly. CB conceives of the Analytical Engine, a general programmable computer, and makes rapid progress on paper.

1835. CB, chairman of the Royal Statistical Society that year, writes to Quetelet (Belgium), Bowditch (USA) and Plana (Turin) describing the Analytical Engine. Resolves against ever physically constructing the engine, in correspondence with friends. AB marries Charles King (soon to be made Lord Lovelace).[9]

1836. 05-12, AL gives birth to a son Byron Noel.

1837. CB describes the 'Mathematical Capabilities of the Analytical Engine' at length but does not publish. 09-22, AL gives birth to a daughter, Anne Isabella King.

1838. Plana enlists the help of CB to equip the Sardinian Navy steam warship with state of the art engines. 06-30, Lord King becomes Earl of Lovelace. Murchison rescinds his offer of the British Association presidency to Babbage, in favour of Herschel, prompting CB to storm out for good.

---

4 Collier 1970, 46, revises this date to October, but it is confirmed by multiple newspaper reports, e.g. *Sun* (London), Wednesday 05 September 1827, 4.

5 M. Moseley 1964, 83.

6 He was listed as 'Trinity College' from 1824 onward, but not as a fellow, which he could not be as he was married when readmitted.

7 Babbage 1830.

8 Lardner 1834.

9 AL below.

1839. CB resigns the Lucasian professorship. AL gives birth to a son, Ralph Gordon Noel King, 07-02.

1840. CB attends the scientific congress in Turin and lectures privately on the Analytical Engine, which he brings several descriptions and worked examples for, along with a wooden model of the addition mechanism with 'anticipatory carry'. Ada Lovelace corresponds with Augustus de Morgan, seeking help with basic mathematics.

1842. Menabrea publishes the first paper on the Analytical Engine, in French. Ada Lovelace commences translating Menabrea to English. British Government formally terminates financial support for the moribund Difference Engine project.

1843. Ada, assisted by prepared examples from CB and his continual help, annotates Menabrea, which is published in Taylor's *Scientific Memoirs*.[10] An additional note by CB is published in the *Philosophical Magazine*.[11]

1846. CB begins to design a simplified Difference Engine.

1851. The Exhibition of 1851 at the Crystal Palace omits the Difference Engine, provoking CB to write a scathing pamphlet.[12]

1852. Death of Ada Lovelace, 11-27.

1862. The Difference Engine No. 1 is included at the Exhibition of 1862 but obscured.

1871. Death of Charles Babbage, 10-18.

10 Menabrea 1843b.
11 Babbage 1843.
12 Babbage 1851.

Chapter 1

Précis of a Deception.

What really happened, reconstructed. All these conclusions are extensively supported in the analysis of primary sources that follows.

1. When Charles Babbage finally had his moribund Difference Engine project cancelled by the British government in late 1842, he wanted to state his side of the dispute in print, both to justify his role and to exert pressure in official circles.
2. He had recently ghost-written a description of his improved Analytical Engine, published that October in Switzerland, but it was in French, in the name of Luigi Menabrea.
3. The Menabrea piece now had to be translated to English with dispatch and published at home, so Babbage arranged to use Ada, Lady Lovelace, to do the work and add explanatory notes, since she had considerable cachet as a daughter of Byron, her mother was a cousin of Lord Melbourne, a recent prime minister, and her husband the Earl of Lovelace was similarly well-connected.
4. Babbage supplied the same worked programming examples he had constructed as early as 1836/7 and had given to Menabrea when he visited the Scientific Congress at Turin in 1840, copies of which are in the Turin archives today. Now he supplied a further program for calculating the Bernoulli numbers using 'course of values' recursion, an idea he had mentioned to multiple correspondents since 1835. This makes him the world's first computer programmer in the modern sense, and Menabrea his first follower outside his own family.
5. Ada reworked the examples to simplify them slightly less than Menabrea had, without adding anything original to them, and supplied explanatory notes with extensive contributions by Babbage which became famous when they were rediscovered in the modern era.
6. The notes show that Ada had an extremely tenuous grasp of mathematical ideas leading her to invent nonsensical and pretentious sigma notation for counting repetitions of operations. This nonsense could not have been produced by Babbage.
7. Parts of the notes were written directly by Babbage himself, including learned remarks about the history of the study of functional equations, and references to the 'three-body problem' and errors in lunar tables, especial interests of his far beyond Ada's sphere of knowledge.
8. The Bernoulli number example proved especially baffling to Ada, so Babbage supplied an outline of its mathematical justification which Ada attempted to explicate in her notes. He admitted this much later in his autobiography but was not fully forthcoming

about the extensive help he had given for *all* the worked examples and the portions he had written directly himself.

9. Ada did not understand the mathematics involved in the Bernoulli example and produced paragraphs of outright gibberish while trying to expand the outline that Babbage had given her. Many other mistakes mar her treatment.

   (a) The wrong formula for the Bernoulli numbers is used, causing them to have signs. The (now old-style) definition she used requires them to always be positive.

   (b) An entirely bogus 'alternative formula' for the Bernoullis is displayed but never used.

   (c) After deriving the course-of-values recursion relation

   $$0 = (A_0 + A_1 B_1 + \cdots + A_{2n-3} B_{2n-3}) + B_{2n-1}$$

   an elementary mathematical mistake is introduced when using it to deduce the next Bernoulli number from a combination of the previous ones. This means all the later values calculated will be (disastrously) wrong.

   (d) Ada's 'explanation' of how the above recursion is derived contains gross mathematical blunders concerning elementary concepts like linear dependence and independence, showing that she did not know what these meant.

   (e) Ada knew that the numbers her formulae produced were wrong, since they could be compared to the enumeration in De Morgan's *Cyclopedia* article (1840), and the few she checked had the wrong sign. She attempted to cover this up by claiming that she cared only for their magnitude and promised that the machine would get the sign right when it executed the corresponding program, which is false. If she had checked a few more of them by hand, a simple exercise requiring only the recursion relation and arithmetic, she would also have seen that subsequent numbers are wrong in magnitude too and rapidly lose all resemblance to De Morgan's (correct) enumeration.

   (f) The program that Babbage had constructed to illustrate the calculation of the Bernoulli numbers was annotated by Ada in a deliberately misleading form, since she had discovered when working through it that Babbage had not provided addressable memory in the engine, which is needed to store previously calculated Bernoullis and retrieve them during the calculation of subsequent numbers. This appears to be the mysterious programming 'error' that she shared with Babbage and he comments on in his autobiography but does not claim had been fixed, merely altered.

   (g) Ada covered up the lack of addressable memory by fudging the critical portion of the code, which she represented as a 'program trace' to avoid clearly exposing the loop structure and handling of stored values. She boasted about this evasion in a letter to Babbage so he was aware of it too.

   (h) Executable code can easily be produced from the pseudo-'program trace' listing once the addressable memory deception is recognized and variable zeroing is done by toggling their loading instructions. Babbage's code is very nearly correct in its looping structure, easily inferred from the 'trace' using modern knowledge, but the lack of addressable memory completely precludes it from working as a general example for all Bernoulli numbers.

   (i) The program repeats the mathematical errors and introduces additional coding errors of a more familiar, less interesting, kind. Since it is inconceivable that

Babbage, Lucasian professor of mathematics at Cambridge for a decade, made the mathematical blunders, Ada probably modified the example code provided to her in order to incorporate those mistakes in the obfuscated program. In any event she was, as the only identified author of the notes, responsible for them.

(j) Once the critical mathematical and minor programming errors are fixed the program inferred from the trace works perfectly, showing that its non-trivial structure had been carefully worked by Babbage over time (given that he had no way to execute the program even in simulation).

(k) In subsequent years, Babbage fixed the lack of addressable memory by introducing variable cards which can be moved *both backward and forward* by a dynamically variable number of positions. It can be shown (though he did not explain this) that modular packs of these variable cards can be used to provide workable blocks of addressable memory. With this extension, Babbage's Analytical Engine is Turing complete (if the need for infinite memory is waived as per usual).

10. Babbage could not have read Ada's mathematics with any care (the same may be said for subsequent commentators) otherwise he would have corrected the obvious blunders and gibberish she perpetrated in her notes.

11. Babbage had a compelling reason to misleadingly boost Ada's supposed mathematical powers: she had written the most detailed description in English of his analytical engine, so suggesting otherwise would have diminished his own reputation and involvement with her. This matches his deliberately misleading description of his role in the construction of the notes.

12. Dorothy Stein's doubts in her biography of Ada (1985) about her subject's mathematical abilities, largely based on Ada's homework exercises with De Morgan, were prescient and understated—they can be rendered unanswerable by a close reading of the mathematics in her notes to Menabrea. Subsequent attempts to resurrect her reputation on the basis of the homework are easily refuted. Ada Lovelace was not a mathematician at all. Nor was she the world's first programmer let alone computer scientist.

Chapter 2

The Genesis of Charles Babbage.

Charles Babbage was born in 1791 on Boxing Day, at Walworth, then in Surrey, to Benjamin Babbage, a prosperous banker, and Elizabeth Plumleigh Teape of Totnes.[1] His father's family worked themselves up from the yeomanry of the west country and were probably nonconformists before they assimilated to the mainstream. They were known as goldsmiths before Benjamin made the transition to banking, a profession in which he accumulated enough fortune to make his son independently wealthy. We know little about his mother's family, except that they were prominent in the civic life of Totnes. The fact that Charles valued her advice all his life suggests that she was a person of mental substance.

As a child Babbage already displayed strong tastes for mathematics and all things mechanical: dissecting toys, studying on his own and outpacing his peers by a long distance. He was schooled locally under a clergyman in Alphington, Devon; at Holmwood Academy in Enfield, outside London, under Stephen Freeman; and finally tutored under an 'excellent clergyman' near Cambridge. There is little independent information to be had about this period, except for an amusing report by a later acquaintance Mrs Crosse: 'I once learnt, not many years ago, from an old man who had been a boy at the same class with him at Dartmouth, that "Babbage was the stupidest boy in the whole school in arithmetic." I asked if he remembered anything remarkable about the great calculator in his boyhood. "No, nothing—we used to call him 'Barley Cabbage,' and he didn't like it"'.[2] Forced to endure that sort of company, he was largely self-educated. By the time he got to Trinity College, Cambridge, in April of 1810, he was already reading French textbooks on mathematics, prompted by an early English textbook by Woodhouse.[3] Continental mathematicians used more expressive notation and embraced a broader conception of the subject than Britain had become becalmed in ever since Isaac Newton invented a useful version of calculus.

At the university Babbage soon befriended JFW Herschel (1792–1871), a son of the celebrated astronomer William Herschel and destined to become a noted scientist in his own right. Herschel was slightly his senior, having started at St John's the previous term in Oct. 1809. Together they rebelled—Herschel covertly, Babbage overtly—against the fusty atmosphere of the university, which they considered, with some exaggeration, to be moribund mathematically-speaking. They helped to form the Analytical Society in 1812 (first mooted on May 7th), to promote continental methods. Initial members included Michael Slegg (1791?—1889?), Edward Bromhead (1789–1855), George Peacock (1791–1858), Richard Gwatkin (1791–1870), John Whittaker (1790–1854) and Henry Wilkinson (1791?–1838).

---

1 Hyman 1982, 5, citing the baptism record in St Mary Newington, County Hall, London.

2 Crosse 1890.

3 Woodhouse 1803. Some of the early mathematical history of Babbage has been disputed by Fisch 2014, who argues that he left a deliberately misleading private account of it in his papers, calculating that nobody would be around to dispute it when it was unearthed. This complicated argument, against which many objections can be made, will not be evaluated here.

The 'Analyticals', propelled by Herschel and Babbage, proved to be influential, publishing a volume of *Memoirs* and eventually some textbooks of exercises with solutions. However they were not the only ones working to modernize the subject in Britain, and it was some time before their efforts had any effect.[4] It was George Peacock who first made controversial inroads for their programme at the university in the 1817 Tripos, only triumphing by the early 1820s when William Whewell added his weight. The idea that expressive notation was an important facilitator of mathematical and technical understanding remained a theme for Babbage throughout his career.

In 1812 Babbage transferred from Trinity to St. Peter's (Peterhouse), much smaller and less well-endowed than either Trinity or Herschel's college St John's. It has been suggested the move was because taking religious orders was not essential at Peterhouse for a fellowship.[5] A likelier explanation is that in a smaller less distinguished college Babbage would have a better chance at a fellowship if he only placed moderately well in the crucial Senate House 'Tripos' Examination, which was dominated, but not monopolized, by St Johns and Trinity.

The Tripos was an unenlightening marathon of calculation and artificial problem-solving. By 1813/4 it had stretched to five days in January, with three full days of written whole-day examination, a fourth for posting ranked results, and a fifth for oral interrogation to confirm the rankings. (The entire examination was originally held orally, with the subject sitting on a three-cornered stool at bay to the examiners, hence 'Tripos'.) It was a test of sheer endurance, speed, and accuracy under pressure—a quest for marks in bulk rather than insight. For almost all candidates, securing a good place required years of concerted cramming under tutors and even private coaches. The subject matter was mostly 'mixed mathematics', now called 'applied mathematics', rather than theory. There were four tiers: wranglers, senior and junior optimes, and ordinary pass (*hoi polloi*), into which candidates were pre-sorted. (JMF Wright, who was excluded from writing the tripos when he missed his preliminary 'act', having been gored through glass shop-window panes by a rampaging bull—apparently not an adequate excuse—published two volumes of solutions to these exams a few years later.[6])

Unlike Babbage, who preferred to continue with self-education, Herschel took the tripos very seriously and applied all his effort to it from the beginning, emerging as the coveted Senior Wrangler in 1813. He also secured a Smith's prize, an additional competition which few but the very best entered, and was rewarded with a fellowship, no small consideration to his frugal parents. But even he found the whole experience frustrating and stultifying. Babbage, a term behind him, had already delayed a year. As it turned out, he was barred from the honours tier for 1814 and had to compete in the ordinary pass category instead. Some mystery still surrounds this event, which placed Babbage at a distinct disadvantage early in his career, no matter how little he affected to care.

It seems Babbage ran into trouble in his first 'Act', an oral disputation among small groups of disputants—firing syllogisms and ethical declarations at each other in dog Latin—conducted by a more senior (MA) 'moderator' over multiple weeks. There was a second 'Act' for those who passed the first, held just before the Tripos itself. Before they were abolished in 1840, these rituals were used to pre-sort and (less commonly) exclude candidates before they took the Tripos itself.

> it was the invariable custom to have ... three opponents to each respondent. Of these the first, who took the lead in the discussion, was expected to urge five objections against the first of

4 Babbage, Herschel, and Peacock 1813, Philip C. Enros 1983; Philip Charles Enros 1979; Grattan-Guinness 2000; Peacock, Herschel, and Babbage 1820.

5 Becher 1995.

6 J. M. F. Wright 1825.

> the propositions laid down by the respondent, three against the second, and one against the third. The respondent replied to each in turn, and when an argument had been disposed of, the moderator called for the next by saying *Probes aliter*. When the disputation had continued long enough the opponent was dismissed with some such phrase as *Bene disputasti*. The second opponent followed, and urged three objections against the first proposition and one against each of the others. His place was then taken by the third opponent, of whom but one argument against each question was required. If a candidate failed utterly he was dismissed with the order *Descendas*, which was equivalent to a modern pluck. Such cases were extremely rare. Finally, the respondent was examined by the moderator, and according as he acquitted himself was released with some suitable remark.[7]

Babbage was 'clasped' by the moderator, Rev. Thomas Jephson (1784?–1864?), a fourth wrangler in 1806 from St. John's, who told him to 'go down'.[8] The student gossip was that 'Babbage ... wishing most devoutly to prove that God was material, Jephson's Piety received such a violent shock that *'Descendas'* thundered from his lips ... All Peterhouse was in an uproar, when the direful news came that their crack man had got a *descendas*'.[9]

This was not a chance outcome from one indiscreet remark. Babbage had already decided to marry Georgiana Whitworth—they had met at the seaside and apparently became secretly engaged in 1812.[10] Fellows could not be married, so it is likely that by January 1813 he had ceased to take the process at all seriously. His banker-father was fairly wealthy, and there were few opportunities then for practising mathematics as a living anyway. The leading men—who had excelled in the 'Tripos' honours that Babbage had in effect declined to compete in—usually served out their fellowship for a while, then took up the law as barristers, burgeoned the ranks of the clergy, or became private coaches for modest pay. Even Herschel, who was immediately elected to the Royal Society, briefly took up the law at Lincoln's Inn. That would remain the case for most of the century.

Babbage had progressively riled up Jephson with pointed assertions during his 'Act', probably playing to the audience. Writing to Herschel in February, he crowed 'I made a thesis for my act on equations at which Jephson muttered something which no mortal could hear. Higman[11] says it contained something like gravitas. Now this in one sense may signify a stinking smell. I had but finished until past 6 o'clock and performed my part of the farce very well. My first opp. did not understand his arguments and puzzled the Mod.; my second understood his arguments but could not find words to express them; and Jephson who had words could not understand them. My first opp. [honestly] was very ludicrous. The Mod. did not ask me one question, which was lucky, as I could not have answered one; in the moral argument I quoted St Luke in English and in Greek; and spoke half-Latin half-English. This together with the singularity of the arguments made Macfarlan and every one in the schools laugh most heartily ... I have been rather idle with my act but shall read a good deal I hope in the next fortnight.'[12]

Whittaker of the Analytical Society, who was delivering an act at the same time, was quizzed by an already-nostalgic Herschel: 'I want much to know how you were satisfied with your act. Has anything more absurd than usual (Babbage's act excluded, of course) taken place in the schools? Did the moderator make his first appearance drunk or has any

[7] Rouse Ball 1889, 165-6. However the descendas was distinct from a 'pluck', as the correspondence between Babbage's contemporaries shows. The claim by Maurice Wilkes 2002 that the descendas had nothing to do with Babbage not taking honours is nonsensical.

[8] Pollock (Trinity), Walter (St John's) and Hustler (Trinity) were ahead of him, *York Herald*. Saturday 01 February 1806, 3.

[9] Cambridge University Library (CUL), Add. MS 8194/1, T. G. [Thomas Greenwood] to J. W. [Whittaker], 14 April 1813; cited in Becher 1995.

[10] Snyder 2011, 58, but no source is cited for this claim.

[11] See section J.5, p. 442.

[12] 1813-02-19, C. Babbage to JFW Herschel, Royal Society, section I.3, p. 319.

one called him fool yet? There are two momentous questions, which I beg you will duly answer.'[13] Whittaker did not exclude their friend: 'Babbage had a very crowded audience. He kept ye freshmen in a roar of Laughter, you can judge what at. Jephson asked him if he recollected some formula — & Babbage coolly answered 'Non recolligi illud expressio'. Abominable Quantities — cosinus & methodus'.[14]

After the *descendas* from Jephson that inevitably followed, a confrontation between the friends ensued in the flesh at the rooms of Thomas Greenwood[15] when Herschel—now decompressing with his parents at Slough—visited the university and Babbage stood his ground. 'Babbage's fortitude [?]:- Oh' twas e.e.exemplary!! the only symptoms of impatience under his afflicting trial which I saw was to call Herschel a Blockhead for telling him he deserved all he got. Such a return for his noble endeavours in the cause of Infidelity or as some call it, in the cause of Reason versus Fanaticism he could not bear: he then rushed out with the memorable words: "Herschel you are as great an Ass as myself" '.[16]

Whittaker in turn quizzed Herschel, 'I want to have your opinion as to ye justice of Jephson's behaviour to Babbage. Having publickly reproved him in ye schools, he refuses to hear every thing said in his favour, or in extenuation of his offences. I very strongly suspect that Babbage deserved ye treatment he has received'.[17]

Herschel's considered opinion was duly pronounced a few weeks later. 'A certain acquaintance of ours, who got roughly handled by Jephson in the schools, you well know was very wroth about it. He richly deserved his punishment however in my opinion, if I have heard the story right. Upon my soul it is a pity the man is so intemperate. He is like a hot poker, which you may play with till you burn your fingers unless you take a devilish deal of care — but this *entre nous*. However if you should happen to meet with him, I would thank you just to mention that I am in hopes of hearing something of him before he leaves Cambridge. If he does not write to me, I must to him, which I do not want to do'.[18]

(Things would take a curious turn in future years. Jephson, ensconced at St John's, was tried in July of 1823 for an *act* of his own: 'an assault upon James Welsh, and with making an attempt to commit an unnatural crime; the 2d with having solicited the same James Welsh to commit, &c.; and the third count charged him with having indecently exposed his person, with the view of exciting the passions, &c. of the same James Welsh'. While crossing a field Jephson—saying that he did not like girls—solicited 'an unnatural act' from Welsh. The twenty year-old labourer accepted a later assignation. Accomplices were enlisted to entrap Jephson, aroused and untrousered in a ditch, and blackmail him. Paid off, they confronted him at St John's for more. The trial had its moments. One witness 'Being asked by Mr. Denman if he ever accommodated a friend with the loan of his wife' answered 'no more than (the learned counsel) did'. Jephson was acquitted, perhaps assisted by the fact that the extortioners appear to have done it before to 'the Bishop', but widespread doubt remained about his complicity.[19])

In January 1814 Babbage sat the ordinary examination in the same unheated Senate House building as the Tripos men, who were examined in the gallery while he shivered in the 'gulph' below. This is contrary to many commentators. For example, Doron Swade states that Babbage 'either disdained to compete in the Senate House examinations, or was disqualified from doing so'. They may have been misled by the memoirs of JMF Wright, who actually noted that those who had *already qualified* for the honours exam by passing

13 1813-02-22, Herschel to Whittaker, section I.4, p. 319. For Whittaker, see section J.7, p. 443.

14 'I don't remember that expression'. 1813-02-26, JWW Whittaker to JFW Herschel, section I.6, p. 321.

15 See section J.3, p. 442.

16 CUL, Add. MS 8194/2, Thomas Greenwood to JW Whittaker, 22 April 1813, cited in Becher 1995.

17 1813-05-04, JW Whittaker to JFW Herschel, Royal Society, section I.9, p. 325.

18 Herschel to Whittaker, May 22, 1813, St. John's College, section I.10, p. 325.

19 This drew a squib from a radical pamphleteer, Cobbett 1823. See also *Jackson's Oxford Journal*, 1823-07-26; *The Examiner*, 1823-07-27; *Morning Chronicle*, 1823-12-23; reproduced in Norton 2015.

their acts might, if they had a *valid excuse* such as illness for not writing the exam, qualify instead for an ordinary degree without further examination.[20] Babbage was (definitely) disqualified from writing the *honours* exam but he (certainly) wrote the ordinary exam.

A letter to Herschel implies that Babbage had been paying little attention to the curriculum, 'Pet. Coll. Evening before Execution ... I have been Chemicising a good Deal and have met with much success in cutting burning and defacing my own hands. Such things must be endured. I hope you will soon come and assist me. I have been endeavouring to put a stop to my plans, and to the continual evolution of mad schemes from my brain'. 'I hope you will soon revisit these denizens of knavery, pomposity and bigotry. Ten good men would have saved Sodom and Gomorrah. Quere: how many would this place require?' 'I don't know what but let us eat and drink for tomorrow we may be plucked. Quere the odds!'[21]

Whittaker wrote the tripos at the same time (he would be ranked 13th wrangler). 'In the course of the examination two things were asked out of the *Analytical Memoirs*. Babbage, who is in the gulph, but would without doubt have been plucked if he had not been clasped, worked one of them, & report says it was the only thing he did. ... We had most severe frost during the examination: surely nothing is more abominable than those morning problems, kneeling down in the gallery, & writing with frozen fingers & hearing the distant hum of friends shilling & pen-problems below. I never hated an occupation so much in my life.' Herschel had speculated that Babbage might fail the exam entirely, as Babbage would himself, even though he had already been 'clasped' down to the *hoi polloi*. 'Let me know whether Babbage is pluckt'.[22] That January the Thames was freezing over, and frigid fairs would be held on it.

If Babbage only attempted one question, it was enough. Herschel promptly gibed his friend, 'The Revd C. Babbage, Bachelor of Arts of St. Peter's College in the University of Cambridge', 'So I am told you are Captain of the Gulph with all your blushing honours thick upon you. I am sorry to hear of Whittaker's ill success, which I am sure could not have been deserved. I want to hear what sort of examination was given & who got the Spoon?'[23] (The spoon was last place in the 'junior optimes', the next class above the 'gulph'; the newspapers did not report the rankings of the poll men.)

Against the wishes of his father, the 'Captain of the Gulph' promptly married Georgiana in July. Even his closest friend Herschel found out only after the fact. Subsisting on a comfortable allowance from his father, Charles continued to pursue the mathematical researches he had started at Cambridge, not in the least deterred by having 'gulphed it'. Functional equations—known then as 'the calculus of operations'—were his main interest and would form the basis of much of his later algorithmic thinking about computing.[24] He applied for multiple positions—the Professorship of Mathematics at Haileybury in 1816, the chair at Edinburgh in 1819—but was turned down repeatedly. He attributed this to a lack of social connections, and his biographer Tony Hyman was prepared to put it down, rather implausibly, without any evidence, to his liberal politics. The fact that he had not taken honours is a more likely explanation.[25]

In 1816 Herschel proposed his friend for the Royal Society and he was duly elected a fellow. Co-sponsors included Herschel's father William and an old schoolfriend, Peter Roget, later famous for his thesaurus. Babbage had by then settled down in 5, Devonshire Street,

20 Dubbey 1978; Snyder 2011; D. Swade 2003, 19-20, John Martin Frederick Wright 1827, Vol. 1, 60.

21 1814-01-21, C. Babbage to JFW Herschel, Royal Society, section I.16, p. 332.

22 1814-01-16, Herschel to Whittaker; 1814-01-21, C. Babbage to JFW Herschel.

23 Royal Society Herschel Papers, JFW Herschel to C. Babbage, 1814-01-26. See also Herschel, Hardcastle, et al. 1932.

24 Grattan-Guinness 1992.

25 Babbage 1864, Hyman 1982, 39. Note the remarks Whewell made (below) about Babbage's application in 1826-8 for the Lucasian chair at Cambridge.

Portland Place, and had lectured at the Royal Institution on Astronomy the previous year. That August, writing to another Analytical Society member, Henry Wilkinson, Herschel was bemused by the latest effusions from their former colleague. 'I am on my way at present to Dawlish with my father – I expect to see Babbage. He is I hear intent upon nothing in the world but sailing, and functions. He has just sent me a copy of his second paper in the Phil. Transactions, which is abstract enough to choke any reasonable man and to turn the brains of any analytical one who has contrived to escape madness hitherto'.[26] Babbage continued to publish papers on 'the calculus of operations' up until the early 1820s. In 1819 he travelled to Paris with Herschel, meeting the senior mathematicians, including Fourier and Laplace. In 1821 they went to Northern Italy and Switzerland in late summer.

It is not possible now to determine exactly how Babbage first got the idea of constructing a mechanical Difference Engine. He left varying descriptions of how it came about. The immediate stimulus was the Astronomical Society and the presence of errors in printed mathematical tables of logarithms (and other functions). These tables allowed multiplication and division to be performed as addition or subtraction, a much easier operation. Exponentiation maps the values back to the desired result. The entries were calculated laboriously by hand using approximations. As it happened, both Herschel and Babbage were founding members of the Astronomical Society, formed in 1820 to bypass the then-sclerotic Royal Society. Appointed to a committee for analyzing tables, crucial to astronomy, the friends were struck by the vagaries of human error in the formation of these and other tables. From manual calculation to manual transcription to manual typesetting, every stage introduced potential errors.

> My friend Herschel calling upon me, brought with him the calculations of the computers,[27] and we commenced the tedious process of verification. After a time many discrepancies occurred, and at one point these discordances were so numerous, that I exclaimed "I wish to God these calculations had been executed by steam", to which Herschel replied "It is quite possible." We continued our tedious comparison without further comment. The labor and uncertainty of reducing formulae to numbers, and then of calculating tables by means of differences, had however been strongly impressed on my mind, and the uncertainty in the result of such computations with the pen, induced me to enquire whether one part at least, namely, that depending on the method of differences, might not be accomplished by mechanism. ... The idea continually presented itself during the few succeeding days, and soon after, having a leisure evening, I resolved on devoting it to the preliminary enquiry.
>
> During the next few days the possibility of calculating by machinery, (which I should not for a moment have doubted had I ever proposed it to myself seriously, as a question to be worked out practically,) recurred constantly to my imagination. The idea appeared to possess a species of novelty which affords so much pleasure, and makes so strong an impression on the mind, when, for the first time we express in words some principle of, or precept to which we have long tacitly assented. When we have clothed it in language we appear to have given permanent existence to that which was before only transient, and we admit what is frequently only a step in the process of generalization as the creation of our own intellect.
>
> In a few hours I satisfied myself that it was possible to make machinery compute tables by differences, and even to print them when computed.
>
> This however was but a general view and was very far indeed from the subsequent realization of the conclusion. The importance of the subject urged me on,

[26] 1816-08-14, JFW Herschel to Henry Wilkinson, section I.20, p. 336. See Babbage 1815, 1816.

[27] Then a term for human calculators.

> whilst on the other hand a reasonable diffidence made me doubt whether I could command the means necessary for accomplishing the object.
>
> The excitement of the enquiry had an unfavorable effect upon my bodily health; and I was recommended to abstain entirely for a time from all thought of the "Calculating Engine". I accordingly went to visit my friend at Slough and continued to obey strictly the injunctions of my medical adviser. I had however been taking certain general views of the art of mechanical invention, and one day when Mr. Herschel made a visit to London, to attend the Board of Longitude,[28] I resolved on applying my new views, and on indulging in their application to my engine.
>
> On the return of my friend late in the evening, I briefly explained to him the new form the engine had assumed; and I remember well the remark he made, expressive of his conviction, that it was an entirely new machine. Such it must necessarily have appeared to any one except the inventor, who had watched the gradual transitions which its parts underwent, when they were submitted to the principles he had laid down.[29]

In an earlier manuscript of 1822, closer to the events, 'one of us' suggested mechanization, and the reply that it was 'quite possible' was left to the other.[30] Though some have used these conflicts to suggest that Babbage was being deceptive, letters of that period between Herschel and Babbage make it clear that Babbage was the inventor, a priority that Herschel endorsed.[31]

The above incident must have taken place sometime in the summer of 1821, after June 7. By year-end, various plans had been considered and discarded. Then Babbage wrote to his friend shortly before Christmas about a new breakthrough, which he was already calling the Arithmetical Engine:[32] 'Can you come to me in the evening as early as you like. I want to explain my Arithmetical engine and to open to you sundry vast schemes which promise to reach the third and fourth generation. You have witnessed the wave-like succession of my plans and I can laugh with you at the rapidity and evanescence of the past, but there are reasons which incline me to reject the strong probability of failure derived from ~~past~~ experience and I am confident that when you hear them you will allow their weight. Do let me see you for I cannot rest until I have communicated to you a world of new thought'.

A comment published in December 1822 confirms the proximate source of Babbage's interests here. 'Astronomical tables were ... the first objects on which I turned my attention, when attempting to improve the power of the engine, *as they had formed the first motive for constructing it*'.[33] Yet Babbage would write much later in his autobiography that glimmerings of this idea had occurred to him years earlier, while still at Cambridge:

> The earliest idea that I can trace in my own mind of calculating arithmetical Tables by machinery arose in this manner :—
>
> One evening I was sitting in the rooms of the Analytical Society at Cambridge, my head leaning forward on the Table in a kind of dreamy mood, with a Table of logarithms lying open before me. Another member, coming into the room, and seeing me half asleep, called out, "Well, Babbage, what are you dreaming about?"

---

28 Herschel was appointed to the Board of Longitude on June 7, 1821, at which time he appears to have been living at Slough with his parents.

29 Buxton 1988, 46-7, from an MS which Buxton dates to 1834.

30 Babbage 1989c.

31 For an example, consult Snyder 2011. Compare the letter of 1828-05-09, C. Babbage to JFW Herschel, 'I most heartily wish I had never thought of the infernal machine', section I.33, p. 346.

32 1821-12-20, C. Babbage to JFW Herschel, section I.22, p. 338.

33 Babbage 1822c, emphasis added.

> to which I replied, "I am thinking that all these Tables (pointing to the logarithms) might be calculated by machinery."
>
> I am indebted to my friend, the Rev. Dr. Robinson, the Master of the Temple, for this anecdote.[34] The event must have happened either in 1812 or 1813.

The industrial revolution, with its mechanical steam engines and precision instruments, was already gathering pace, but Babbage did nothing about it at the time. This account from 1864 entirely left out the table-checking episode with Herschel. Instead Babbage described how he had consulted William Hyde Wollaston[35] in 'about 1819' concerning some plans for astronomical instruments (not novel) and the idea of the engine (a more promising subject). This gives, by omission, the impression of a simple transition from the Cambridge idea to practice.[36] Wollaston certainly was consulted, but only three years later, probably only once an initial model was built, certainly only after the omitted Herschel episode. Confusions about dates are common in autobiographies written by aged men, but it is much harder to account for the omission of the very interesting detail above, such as the debilitating obsession that the excitement of invention had brought on.

The first target of Babbage's machine was to be the production of the tables he had been checking when the idea had surfaced again. The era was awash in tomes of ready-to-hand lookups constructed by *human* computers. Products, squares, cubes and even higher powers for numbers up to 1,000; square roots and cube roots; reciprocals; sines, cosines, tangents and cotangents; logarithms of whole numbers up to 108,000; logarithms of sines, cosines, tangents, and cotangents; hyperbolic logarithms up to 10,000; logarithms of sums and differences (those logarithms being known); squares of sines, cosines etc; specialist astronomical tables for lunar distances ... ad infinitum. A mechanical device for assisting in the computation of these found many willing takers, like Francis Baily of the Astronomical Society.[37] Accuracy was a thorny problem. Copious errata for published tables led to errata of the errata, even to errata of the errata of the errata and so on. Improving this process was a proximate motivator, but not the only projected application, as we will see. Babbage developed an abiding interest in these tables and their errata, producing his own (apparently widely-used) table of logarithms, followed by an ultra-rare 151-volume version in multiple ink colors. He eventually amassed many hundreds of volumes of tables from various sources in his personal library. All this has contributed to the misconception that tables exhausted the interest of his Difference Engine.[38]

The method of differences was already in use for manual production of tabled values, so using exactly that in the mechanical version was an obvious step, and not in itself any great advance. A polynomial with highest term an n-th power will reduce in power since $a(x+1)^n - ax^n$ cancels the n-th power. Repeated differences will produce a constant (0-th power) term. Reversing the process, starting from given initial values, a table of values can be produced by repeated addition of the differences, starting with the known constant difference. A very broad range of well-behaved mathematical functions can be approximated by power series, so that this approach can be used to calculate tables of (approximate) values for them, as long as care is taken to control the accumulation of approximation errors by periodically calculating direct (initial interval) values afresh, which had to be done by hand for the difference engine.

Babbage was well aware of this himself. When Herschel wrote to him, sometime early in 1822, that he had heard tell of a mechanical calculator invented in Paris, he was not perturbed, because that sort of thing was, to him, relatively easy. He already had more

[34] See section J.6, p. 442.
[35] See section J.8, p. 443.
[36] Babbage 1864, 42. This account is repeated naively by some secondary sources e.g. Dubbey 1978.
[37] Baily 1823.
[38] Babbage 1827a,b, 1831; Sotheby, Wilkinson, and Hodge 1872.

elevated visions of where the credit would really lie. 'The same cause (namely want of sufficient information on the subject) which will overrate its difficulty in the opinion of the multitude will in our scientific world underrate the value of those other machines I have contrived for reducing different parts of mathematics to the dominion of gravity: every moment I devote to this enquiry convinces me that although at present they are few and in an infant state yet that there is a vast field of discovery open and that the reputation to be derived from these ought deservedly to be greater and more permanent'. Whatever it was, the Parisian invention did not materialize, though Babbage asked Herschel to make further inquiries about it without disclosing the English invention just yet.[39]

It has been claimed that at least one other person had thought of constructing a calculator to use finite differences for the approximation of functions.[40] A German, Philipp Engel Klipstein, edited a brief booklet in 1786 containing a design of a simple watch-like calculator (addition, multiplication, subtraction and division). This was really due to his compatriot Johann Helfrich Müller (1746–1830), as the title discloses: *J. H. Müller's Description of his newly-invented Calculating Machine according to its form, its use and utility.*[41] Herschel translated a portion of Müller's text for Babbage at some unknown date, and it has been said that this is 'a seemingly compromising document', suggesting that Babbage really got his ideas there.[42] But this is a simple mistake. In a suffix, Müller promised to deliver 'other calculating machines manufactured that perform even more than the present one' if the simpler one found a buyer. As an example he mentions calculating and automatically printing tables derived from sequences. The sequence can be constructed using a difference, but the resulting elements are just operated on as usual to produce squares or cubes, e.g. by direct multiplication of the sequence element. There are no *higher-order* differences, no polynomials mentioned, no Taylor-series approximations of any well-behaved function on a given interval, and no mention of the ability to do all this with addition only. Thus Müller's contemplated device is *not* a difference engine at all, just a slightly more capable calculator, for which he provides no concrete design.[43]

> Should a suitable buyer and promoter of this useful invention soon emerge, I will not hesitate both to make known the internal arrangement of this machine as clearly as possible, and also to have other calculating machines manufactured that perform even more than the present one; and among others, also deliver whole series of numbers that progress in composite arithmetic proportions, each number just by 2 to 3 turns of the crank (thus even much faster than the present machine).
>
> But because the copying of many numbers, growing into large tables, not only takes up a lot of time, but can also become erroneous itself: I have thought about having the machine, if desired, print the resulting numbers along with the necessary naturally progressing numbers and intermediate lines on underlying paper, without having to set type, or blacken them with printer's balls, as all this must be done by the machine. Thus, if once the first number along with the associated difference numbers (be they common or named) are set on the number discs and the necessary printer's ink is put into a certain vessel in the machine, also the printing paper is placed in the suitable frame: then the worker has nothing more to do than to constantly turn the crank, and to turn the paper, after a page or the entire sheet is fully printed, or in the latter case to exchange it with another, for which the

[39] 1822-04-09, C. Babbage to JFW Herschel, section I.24, p. 339. This letter refutes the claim invented by Lindgren 1990, 2023 that Babbage was dismayed by Herschel's news.

[40] Lindgren 1990, 64-70.

[41] Müller 1786.

[42] British Library Add.Ms. 37198, 189-191, cited in Lindgren 1990, 64-70.

[43] Müller 1786.

crank, which then fixes itself, gives him a sign.

The advantage here is particularly this: that one obtains the numbers calculated by the machine and coming out one after another in much less time than copying requires and quite correctly printed.

Thus, according to a preliminary estimate, an arithmetic series of 60 numbers, common or named, (such as stand on p. 24 for example) each number having many or few digits, can be calculated and brought to paper merely by a common day laborer in one minute, thus in about the twentieth part of the time otherwise required for this, and with the 60th part of the costs.

Thus, for example, completely correct cubic tables from the roots 1 to 100000 could be delivered in about $10^1/_2$ days, even if only 3 hours were worked per day. One would obtain only one copy in this time, but in cases where only a few copies are necessary, the work could still be repeated with advantage; for at least 10 copies could be correctly printed in the same time as would be required for one and perhaps erroneous copy.

This much I dare to promise bindingly on request; but I hope perhaps to bring it to the point that by means of the machine, with not much more effort, a desired number of copies can be delivered just as correctly.

After completion of the above described and manufactured machine, I intended to have one or a few lesser ones manufactured just for adding and subtracting; but because through them, according to the very simple arrangement that I intended to give them, addition could indeed be conveniently done, but progressions could not be produced with advantage: I have so far refrained from doing so.

Anyone can dream of constructing devices which (are said to) achieve stated outcomes, the point is to 'reduce to practice' and provide plausible paths to their realization. Some preliminary investigation by Babbage led to a wooden model of a 'Difference Engine' in which 10-place figured cogwheels were physically moved between stopped positions. The details of his design would mature in coming years but the principle did not vary much. The machine had multiple columns, each of which was a spindle containing stacked figure-wheels. These columns represented decimal numbers, with each figure wheel a decimal digit. Each column had duplicate machinery for addition and the needed carry operations, so they could in principle operate in parallel. Addition of a digit from another wheel was accomplished by accumulation as the source and receiving figure-wheels were rotated while connected by an interlocking mechanism (the source wheel would automatically restore since it received a full rotation). The overflow would be carried up. The decimal point is fixed and implicit, as it makes no difference to the machine. The figure wheels had appropriate knobs to trigger carrying once the wheel passed the appropriate point. This has been described as an attempt by Babbage to reconcile Newton and Leibniz, through which he 'was led to regard motion as an expression of the accretion of quantity; and to consider the possibility of making actual motion, under certain conditions, the index of the quantities generated, in arithmetical operations'.[44] But mechanical clocks already did that. Making it happen efficiently and reliably, with redundant machinery to prevent accidental movements, was the major challenge. The finished device appears to avoid lubrication—it was not explicitly built-in, which may be considered less than ideal for metal-on-metal parts in extended use—and was powered by a hand crank only. Delicate engineering questions were posed by the need to prevent component breakage through excessive transmission of force. The number of mechanical operations required had to be minimized. To calculate tables, certain columns held the differences. Once these were set up manually, the machine was set in motion to

[44] Buxton 1988, 46.

calculate the table from them automatically.[45]

From the start Babbage also aimed to add automatic printing of the results, since transfer to type was a major source of error. He toyed with various ways of doing this, from composing type blocks automatically to hammering letters into moulds for manufacturing type. He was never able to bring these plans to fruition, as they constantly limped behind his other technical problems. Still, the plan was always to construct more than a mere calculator, as the Astronomical Society would later recognize.[46]

Babbage diarized 'May 10 [1822] My calculating machine is nearly finished. Of those who I have made acquainted with the principles [too?] many think too little of it, of those ignorant of them the greater part think far too much'. Wollaston represented the former class. 'Monday June 8th [1822] ... Dr. W examined it and I explained all its parts, and he then worked it and after about an hour and a half the result of his opinion was expressed in these words. "All this is very pretty but I do not see how it can be rendered productive" '. Earlier that day his old friend George Peacock from the Analytical Society had dropped by. 'Maule and Peacock called and examined the Engine, admired it considerably. P[eacock] candidly confessed that since Herschel first told him of it he had ~~thought~~ considered the subject and thought it impossible. Maule appeared to take considerable interest in its success but rather doubted [it]'. Other visitors included Wilkins, Barrow, Murdoch and (surely) William Whewell.[47]

The Board of Longitude was suggested by several of these visitors as a sponsor, but its influential Dr. Thomas Young was known to be skeptical, drawing a catty remark from the mathematical treasurer of the Royal Society: 'Thursday [11th June 1822] This morning Davies Gilbert breakfasted with me. He examined the machine and we had much discourse on mathematical subjects. He admired it much and was particularly pleased at the contrivance for dividing the root of eq[uations]. He advised that it should be brought before the Board of Longitude. I observed that Dr. Y[oung] had already expressed a decided opinion of the mathematical part and that if he was a bad judge of the mechanical part his opinion was not of much value, and that from what I knew of the Board I did not think it had power or will to do any thing. With respect to the government he observed that they were perfectly unacquainted with the nature of the thing and that the only chance was that Mr Peel might take it up. He still recommended me to go on with the [printing?] part. He remarked that Dr Y[oung] disliked every thing that did not originate with himself'.[48] Subsequent events would disprove the last remark.

The model implemented the calculus of finite differences to calculate power series through addition only. When constructing tables of values, regular intervals are a natural choice, so that finite differences are sufficient. The first wooden engine was limited to three figure-wheels, with two wheels for the first order difference and one for the second order. As noted above, using finite differences here was a natural step given that human computers of tables already used the technique, which was also useful for error checking: errors would cumulate and be revealed by periodic checks.[49]

Babbage began promoting the invention, drawing up an open letter to Sir Humphry Davy, the president of the Royal Society, 'pointing out the probable extent and important consequences' to which his investigations 'appear to lead'. 'I wrote it at his suggestion' Babbage told Herschel.[50] This was printed as a pamphlet and widely circulated within

[45] For further mechanical details, which are not our concern here, see Collier 1970, Hyman 1982, Bromley 1982b, Bromley 1990, D. Swade 2002, D. Swade 2003.

[46] Colebrooke 1824.

[47] Babbage 1822–1825.

[48] Babbage 1822–1825.

[49] Errors found would have to be reworked back to the last checkpoint. Of course this is only useful, given the rework required, if errors are rare.

[50] 1822-08-03, CB to JFW Herschel.

government and broader scientific circles.[51] It was soon apparent to Babbage, as he worked on implementing the Difference Engine, that he could generalize its powers in unexpected ways, and this was already suggested in the letter to Davy, which projects the power to 'calculate tables governed by laws which have not been hitherto shown to be explicitly determinable' and to 'solve equations for which analytical methods of solution have not yet been contrived'.

This provoked a caution from Herschel, 'I question whether the mention of those Engines of which you have only sketches & general conceptions, especially those destined to the Solution of Equations of differences and those which work series reducible to no law but that which the mechanism itself determines, will tend so much to advance your speculations in the eye of the world as the more tangible and practical results to which you have already come. But as I know you think me a thingy fellow, this is only by the bye'.[52]

Babbage was not deterred by Herschel's prudence. Computed results could be fed back into the differences used, e.g. the units part of the last computed result could be used as the second order difference, rather than a fixed value. Or a multiple of any digit of the computed result could be used for any order of difference.[53] Although multiples never seem to have been implemented in the real machine fragment that was eventually produced, some feedback of results was realized, amounting to a kind of recursion.[54] Babbage would later refer to this as the engine 'eating its own tail'. Though he would take some time to fully comprehend its far-reaching implications, in the form of the successor Analytical Engine, the germs of the idea were already present in 1822. He saw at once that new kinds of difference series would follow which would require novel mathematical treatments where they were even tractable analytically.[55]

The Board of Longitude was never involved in the project. At the suggestion of Davies Gilbert, seconded by Humphry Davy, the government (represented by Robert Peel and the exchequer Robinson) asked the Royal Society to report on the engine, for which they appointed an ad-hoc committee. Since the future Astronomer Royal George Biddell Airy would assert in jaundiced retrospect that the committee was stacked with partisans of Babbage[56]—incorrectly recalling its composition and methods, in which he was not involved being a student at the time—it is worth examining their deliberations in more detail. The committee was formed by the President, Humphry Davy, and two Secretaries, Brande and Combe, along with Gilbert, JFW Herschel, Wollaston, Young, the Astronomer Royal John Pond, the secretary of the Astronomical Society Francis Baily, the preeminent civil engineer Marc Isambard Brunel, Major Thomas Colby (1784–1852), who would soon superintend the Ordnance Survey of Ireland, and Captain Henry Kater (1777–1835), the Anglo-German Copley medalist.[57]

The committee worked briskly, formally inspecting the engine model at Babbage's house. 'All agreed as to the practicability, even Dr. Young but he denied the utility of the tables when formed. Dr Wollaston observed that there was only one objector and no objection', 'Baily, Herschel, the President and one other member met and agreed that their report was to be that Mr. B's invention is highly ingenious, founded on just scientific principles and worthy of encouragement'. Charles Weld, historian of the Royal Society, would later report that 'I am informed upon good authority, that Dr.Young differed in opinion from his colleagues. Without doubting that an engine could be made, he conceived that it would be

[51] Babbage 1822a.
[52] JFW Herschel to CB, 1822-07-31, section I.25, p. 340.
[53] Babbage 1824b.
[54] Franksen 1981.
[55] Babbage 1824a, incorrectly dated 1826 in Babbage 1989b following Van Sinderen 1980; the offprint explicitly states 1824.
[56] D. Swade 2003, 201-3, citing Airy to Goulburn, 1842-09-15.
[57] Weld 1848, 371-2.

far more useful to invest the probable cost of constructing such a calculating machine as was proposed, in the funds, and apply the dividends to paying calculators'.[58] But as it happened, two favourable reports were drawn up, and 'Dr. Young who was the only [objector] last time said he still did not see the utility but he had drawn up a report which he would read. It was the strongest of the three and was adopted with little variation. I was at Sir H. Davy on Saturday evening. He mentioned the conduct of Dr. Y[oung]. with great suprize and said he thought the report too strong but as Dr. Y[oung] had proposed it he agreed'.[59]

In writing the report Young may have been motivated by the same sentiments as Herschel, who told Babbage that he had 'not the least hesitation in signing, as however strongly it is worded, (I mean when the Degree of perfection to which the Engine may be brought is spoken of) you have done so much that I can easily believe you have it in your power to go as far as is said and I have no scruple to declare the faith that is in me to all the world'.[60] Which is to bank on the capabilities of Babbage and what he might unlock by working on the project, considerations that go far beyond the utility of the tables the engine as conservatively projected could produce. Airy's later claim that Babbage positively hated Young because of all this makes no sense given that Young wrote the highly favourable committee report and Babbage knew that. We have no reason to believe the decision was not honestly reached. It is not as if the highly-eminent members were relatives of Babbage. Instead, it is Airy we must suspect of animus, obtusely describing the Difference Engine in his autobiography as 'worthless'.[61]

Through the influence of Herschel, the Astronomical Society awarded its gold medal to Babbage in 1823 'for his ingenious invention of an engine for computing and printing astronomical and mathematical tables', though the medal would only be *struck and presented* the following year—it had to be done at the anniversary meeting, already past that year—so that it bears the date 1824.[62] Henry Colebrooke, the president for 1824, recognized further uses of the 'singular and pregnant discovery' beyond mere table creation, to extract roots and solve simultaneous equations.[63]

At around the same time, promotion of the invention in official circles had finally borne fruit. In mid-July Babbage met with PJ Robinson, the Chancellor of the Exchequer. As he wrote Herschel, the chancellor 'acquired the knowledge of the Astronomical Medal two days since from [Davies] Gilbert so that you got it for me just at the fortunate moment', 'the manner in which he spoke of it was more gratifying than the grant itself as I have liberal people to deal with I shall not be annoyed about pence and the particular mode in which I may think it right to distribute them'.[64] An initial payment to Babbage for £1,500 was authorized. His (misconceived) optimism that he had 'liberal people to deal with' and need not 'be annoyed about pence' is striking in retrospect.

At the suggestion of Marc Brunel, Babbage engaged the noted engineer Joseph Clement (1779–1844) to construct the engine to his specifications.[65] Clement, who would play an im-

58 Weld 1848, 372.

59 Babbage 1822–1825.

60 1823-07-05, JFW Herschel to C. Babbage, section I.29, p. 342.

61 Airy to Goulburn, 1842-09-15, D. Swade 2003, 200-7. See also Airy 1896, 152. Lindgren 1990, 51, repeats the specious idea that the committee was packed with Babbage's 'good friends'.

62 Contrary to Campbell-Kelly in Babbage 1989b, 57. Multiple newspaper reports confirm this—e.g. *Saint James's Chronicle*, Tuesday 24 June 1823, 2; *Morning Post*, Saturday 21 June 1823, 3—along with a letter to Babbage from their secretary John Millington (who had just replaced Babbage himself), 1823-06-13, section I.26, p. 341; also cited in Collier 1970, 41. There is also the letter from Babbage to Herschel from 1823-06-27, quoted below.

63 Colebrooke 1824.

64 C. Babbage to JFW Herschel, 1823-06-27, section I.28, p. 341. This contradicts Collier 1970, 41-2. Gilbert (1767-1839) was a mathematician and the treasurer of the Royal Society at the time. He would later serve as its president, to the consternation of Babbage.

65 Williams 1992, who cites a letter of 1863 to George Clowes.

portant role in these events, was a self-made man, originally from the north country. Unembarrassed by much schooling—he received some brief tertiary education later, at Marischal College in Aberdeen—he had worked his own way up from thatcher and slater to a power loom manufacturer. He would become the inventor and operator of a lucrative precision lathe, picking up expert draughtsmanship along the way. Drawn to Glasgow and then London in 1814, he had served the engineers Joseph Bramah and Henry Maudsley as a draughtsman. After setting out on his own in 1817, he showed considerable business acumen. The Society of Arts would award him several medals for his work (1818, 1827, 1828). Babbage certainly never had any cause to complain about the technical quality of Clement's work, and always took pains to note so. However their interests were not fully aligned, with critical consequences later.

The partnership with Clement was ultimately very productive. The engine would need many copies of identical parts, so new tools for making the parts to consistent (indeed remarkably high) tolerances were proposed by Babbage, with Clement devising the most practicable mechanisms for working them. Clement also attracted skilled draughtsmen and engineers to assist him, expanding his workshop accordingly. But the work went slowly—we have little record of its detailed progress until years later—largely as a consequence of the demands by Babbage, cleverly exploited by Clement, for those extraordinarily high levels of precision.

In order to simplify the design of the engine, always enthusiastic about the conceptual affordances of expressive notation, Babbage invented a new system of 'signs' for specifying the dynamic behavior of a complex machine—'the state of motion or rest of any individual part at any given instant of time'.[66] Timing and direction of motion and velocity of parts could be expressed. He would later credit this system of named parts abbreviated by letters—marked up with a thicket of accents, with lines of multiple types decorated with arrows and strokes—with a large role in his success, evolving the notation progressively. He was disappointed that nobody else ever took it up seriously, though the *Encyclopedia Britannica* would describe it succinctly, with a more comprehensible example than he had offered, in 1840.[67]

> The first step in this method consists in drawing, on a sheet of paper, as many vertical lines as there are pieces or moving parts in the machine; to each of these pieces its name must be affixed, and the letter which marks it out in the drawing. These vertical lines must then be crossed with horizontal lines, upon which we must write,
>
> 1. The nature of the motion of each piece, indicated by a conventional sign.
> 2. The number of teeth in the wheels, pinions, or sectors, or the number of pins or studs on any revolving barrel.
> 3. The velocities, whether absolute or relative, of the several parts of the machine.
> 4. The mode in which the motion is communicated or transmitted.
> 5. The modes of adjustment, some of which are permanent and made by the mechanist, while others depend on the nature of the work they are intended to perform, such as the distance between the mill-stones in a flour-mill, on which the fineness of the flour depends, or such as the winding up of the weight or spring, which may be called a periodic adjustment.
> 6. The times or durations of the movements.
>
> Mr Babbage employs two classes of signs, viz. those for indicating the peculiar nature of the means by which the motion is transmitted, and those which denote the state of motion or rest of every particular part of the machine.

[66] Babbage 1826.
[67] Anonymous 1840.

The lines pack in a lot of information, e.g. unbroken lines indicate motion, parallel straight lines denote uniform motion, curved lines denote a variable velocity, dotted lines imply rest and so on (see Figure 2.1, p. 20). The notation would find echoes in his later work on the Analytical Engine, where a complex (and as we will see, not particularly useful) notation about dynamic states is introduced.[68]

In 1824 Babbage had himself readmitted to Trinity College in Cambridge—more prestigious than Peterhouse—a move that few have noticed.[69] It was the start of a campaign to secure himself a post at his alma mater, which he would disguise later.[70] He had already written to Herschel in 1821, suggesting that they both apply for the Plumian professorship vacated by the death of Samuel Vince (1749–1821). Herschel wryly declined, noting that he had made three unsuccessful passes already at a position there without any support from the dons. 'I have no longer the keen relish for abstract mathematical studies I formerly felt, & except excited to their application by something wh. interests me more, am perfectly content to relinquish the distant fame & equivocal enjoyment they afford to others', 'as far as mere science is concerned I had rather pass my days among those who are advancing eagerly & rapidly & running a race with ardour — than in goading up hill the sluggish paces of any established institution under the sun'.[71]

The Lucasian professorship—originally endowed by a land bequest from Henry Lucas, a wealthy Cambridge MP who died unmarried—once held by no less than Isaac Newton—became vacant late in 1826. The position was high in external prestige, its low-pay and encumbrances offset by nominal duties, and according to Whewell 'for a mathematician far the most desirable we have'. Babbage energetically lobbied the electors—heads of the colleges—to claim it. Whewell reported to friends in November that 'Babbage is making application, and has written to people here on the subject. He has no chance whatever, and it is mere extravagance, at least as appears to me, his taking up the thing. I do undoubtedly believe that he would be a good Professor now, but it is too much to expect that our Heads should understand not only his merits, but the varying shape of them as time and circumstances may have modified it'. The problem was that he had not taken honours, but nevertheless 'I wish Babbage had any chance. He would be an admirable person'. The following month Whewell wrote to another friend that 'Babbage came here to canvass, and is now here. He has effected a complete conquest of your Master's good graces and is staying at Caius Lodge.[72] Herschel has also been here to support him'. His rivals were George Airy and Rev. Dr William French (1786–1849). Babbage took legal advice as to the eligibility of French for the position, given that he was then Master of Jesus College—one could not hold both the Lucasian and other positions. At the suggestion of Davy, 'French withdrew, finding the office untenable with his present one'.[73]

Airy was Senior Wrangler in 1823 and like Herschel a Smith's prizeman, but less distinguished as yet for publications and actual accomplishment than Babbage, his senior. Airy was well-aware of Babbage's work early on, as William Whewell informed Herchel in 1822. 'You have of course heard from [George] Peacock of Airy, a pupil of his, and certainly a man of very extraordinary talents. The reports about Babbage's machine have, it seems, excited him to attempt something of the same kind. He and another man have made a machine to solve cubic equations, but besides this he has, so far as I can make out, invented something a good deal in the way of Babbage's contrivance. He is not here at present, but his friend

68 Menabrea 1843b.
69 See section J.1, p. 441.
70 Babbage 1864.
71 1821-12-02, JFW Herschel to CB, section I.21, p. 337.
72 The Master of Gonville and Caius College in 1826 was Martin Davy (1763–1839).
73 1826-11-18, Whewell to Rev. R. Jones; 1826-11-19, Whewell to Rose; 1826-12-10, Whewell to Rev. R. Jones, Todhunter 1876, Vol. 2, 74–80. Thomas Turton (1780–1864) was compelled to surrender the Lucasian in 1826 when he took up the Regius Professorship of Divinity.

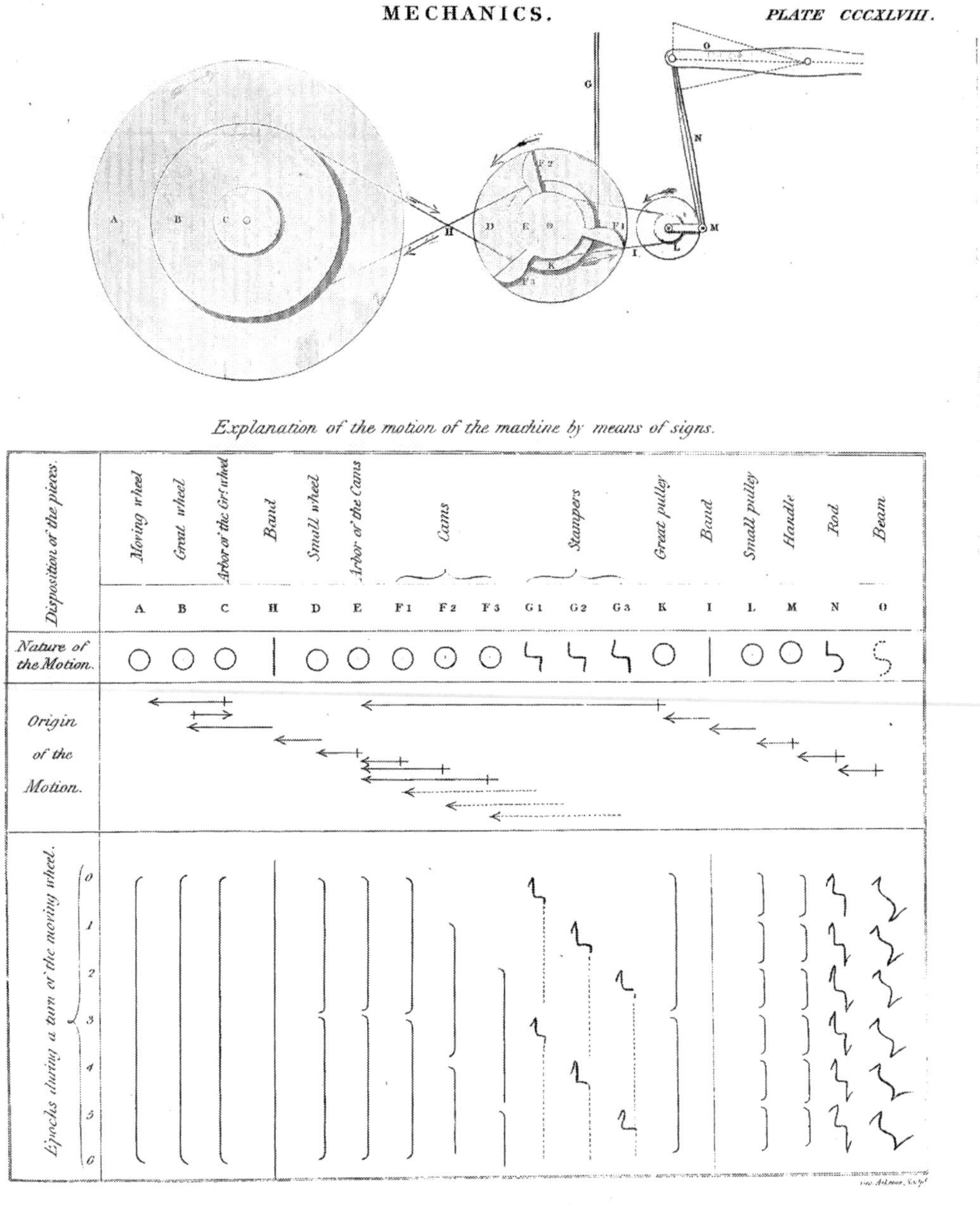

Figure 2.1: *Babbage's mechanical notation applied to 'the case of an ideal but simple machine, which consists of a large wheel driving a system of levers and stampers, also a large beam driven by a crank'. From the* Encyclopedia Britannica, *1840.*

tells me that it has got toothed wheels, working in one way for the differences and in another for the digits. If it be a similar invention to that, it is probably an independent one; for I do not know any way by which he has got any lights about Babbage's affair. I will enquire into it as soon as he returns'.[74] But Airy's machine appears to have failed.

In the end Babbage's campaigning and even Herschel's support came to nought. Airy easily prevailed among the heads, as he crowed in his autobiography. 'I was elected (I think unanimously) and admitted. I believe that this gave great satisfaction to the University in general'.[75] But it proved a mixed blessing to the striving Scotsman. 'So far as income is concerned, the Lucasian is rather a starving matter, for they compelled him [Airy] in taking it to give up all his College employments and all emoluments except his fellowship'.[76]

This was the start of a calamitous sequence of events for Babbage. In late February of 1827 his father Benjamin died—one doubts that they were sufficiently estranged for this not to affect the inventor. In late July, his ten year-old son Charles Whitmore Babbage died at Devonshire street. Then his wife Georgiana died at Boughton House in Worcester, as a result of childbirth, on the first of September. She was only 35 years old. The son she bore also died shortly after. The result was a complete breakdown for Babbage, who was granted permission to place Herschel temporarily in charge of the Difference Engine project and travel abroad on the continent for his health.

Babbage's luck would change for a while at least. Early in 1828—just over a year after his failed bid for the Lucasian professorship—the Plumian professor died, opening up a more lucrative post for Airy. 'We have another professorship vacant by Woodhouse's death. Airy will probably succeed him, and the vacant one will be the Lucasian for which Babbage was a candidate. Some people here hope that Herschel will take it, which I much doubt. I should rejoice to have Babbage, but I am not so sure that he (one of the *polloi*) would succeed, and not at all certain that he would now offer himself. He is at Naples and therefore long to get at'.[77] (As noted above, Babbage had not taken honours but rather an ordinary poll degree in 1814.)

In early March Herschel, who had tried but failed to reach Babbage in time, wrote to Whewell, giving the go-ahead anyway. It was a calculated gamble that he would accept, based on the enthusiasm with which he had campaigned for it a year earlier. The principal rival this time was Watkin Maddy (1798–1857) the astronomer, second wrangler in 1820 and author of *The Elements of the Theory of Plane Astronomy* (1826). Whewell replied to Herschel that it 'was entirely in consequence of your letter, which I got yesterday, that I bestirred myself in the matter, for before that I did not consider myself sufficiently authorized to press his claims upon the electors'. John Philips Higman and George Peacock lent their considerable weight too, having 'considered that a private proposition of his [Babbage's] friends without his knowledge could not in any way compromise his character or his delicacy'. They worked the heads of colleges, who acted as electors, with vigour and dispatch.[78] As a result, Babbage was elected with overwhelming (but not unanimous) support on March 6th.

Babbage told the story somewhat misleadingly in his autobiography many years later: 'at Rome, I accidentally read in Galignani's newspaper the following paragraph, dated Cambridge :– "Yesterday the bells of St. Mary rang on the election of Mr. Babbage as Lucasian Professor of Mathematics."'.[79] Memory plays tricks. On 1828-03-15, Galignani merely reported: 'CAMBRIDGE, FEB. 29.[sic] — Charles Babbage, Esq., M.A., of Trinity Coll., is

74 Todhunter 1876, Vol. 2, 50-1.
75 Airy 1896, 71.
76 Todhunter 1876, Vol. 2, 86.
77 Whewell to Jones, 1828-01-06, Todhunter 1876, Vol. 2, 89-90.
78 1828-03-09, Herschel to Babbage, section I.31, p. 343.
79 Babbage 1864.

elected Lucasian Professor of Mathematics, vacant by the election of G. B. Airy, Esq., to the Plumian Professorship'. (No trace of this report mentioning the bells of St. Mary could be found in any contemporary newspaper, suggesting that Babbage mentally transferred it from another source, e.g. a private letter to him.) We are left with the impression that this was a position he had not cared all that much for, had not even applied for (technically true, but he had applied for it with litigious zest just over a year earlier). He even states that he seriously considered refusing it, only changing his mind when pressed by friends in Rome who had also read the announcement and persuaded him that he had to avoid disappointing those who had exerted themselves on his behalf, which would include Herschel and Whewell. But in a letter to Herschel, Babbage assigns the discovery to a letter from his mother, and immediately sees the hands of his friends. 'Nothing ever surprised me so completely as my mother's letter announcing my election to the Lucasian professorship. The unceasing exertions of your friendship have become so habitual to me through life that I could not for a moment doubt the prime mover in an event which I look upon with feelings of the highest pride and gratification. I do not yet feel strong enough in mind or body yet to return to England nor do I think such a course immediately necessary. I shall however shorten my tour and shall return with far different feelings from those which I anticipated: as far as Cambridge is concerned at least. I shall hold the post and shall try to do credit to the recommendation of my friends'.[80]

Herschel did his best to look after the engine, but was perplexed by the glacial progress made by its frail artisan, who kept demanding further payments at an average of £130 a month, or else, he threatened, all work would stop. '[Joseph] Clement's health continues good — Machine gets on slowly and more is done on paper than in brass'.[81] In the months to come the astronomer would strongly suspect sharp practice on the part of Clement.

Babbage was frustrated too, but not because of the slow progress by Clement, who he expected was still making the necessary tools. He was personally out of pocket for the escalating expenses entailed by this exacting-precision toolmaking, but the government had changed hands and now were not eager to acknowledge any definite commitments to cover the overrun. 'I have not thought about the machine except [when] some one has enquired and whenever I have thought about it, it has been to wish I had never undertaken it. Now however a new and most unexpected difficulty has arisen on finding that Lord Goderich ceased to be at the head of the government. I wrote to W.W. Whitmore[82] to get some communication from Lord Goderich to his successor. It seems (I hope it arises from my imperfect instructions to my brother in-law) that Lord G. contends that there never was any understanding that I should have more money'. A certain despair is evident. 'The result will be that I shall finish the machine and be more than half ruined by it. But I can scarcely bring myself to believe that Lord Goderich will not recall to his memory the bearing of that conversation. If he do not I will repay the money the government advanced. However I must think of these things when I return'.[83]

This was to be a repeated theme in coming years. When the project started, it was a gentleman's (verbal) agreement and no formal commitment was ever made by the government addressing the future conduct of it. Ambiguity persisted for so long that it seems likely it was positively in the interest of successive governments *not* to clarify it. They could cover further costs if they chose to, at their discretion, or they could change their minds without embarrassment.

Babbage had no experience in project management, and naively supposed that the government would simply cover all overruns as long as he deemed them necessary. At no stage

[80] 1828-04-17, C. Babbage to JFW Herschel, section I.32, p. 345.
[81] 1828-03-09, Herschel to Babbage, section I.31, p. 343.
[82] Babbage's brother-in-law.
[83] 1828-04-17, C. Babbage to JFW Herschel, section I.32, p. 345.

did he try to curtail expenses pragmatically by limiting goals. He would find out that no government could have inexhaustible tolerance for that.

A few months later, writing from Naples in response to Herschel's doubts about Clement's progress, Babbage raised another concern that would prove even more important than government funding. 'Clement is spending much time in making tools. Now this is to a certain extent necessary ~~but~~ and requires considerable supplies of money but I should wish you incidentally if possible to find out whether it is not Clement's intention to make me pay for the construction of these tools and then to keep them as his own property—from the multitude he is making it looks so'. At this stage Babbage was prepared to underwrite it all himself if needed, but would rethink that in years to come. And bitterness was setting in.

> 1st I most heartily wish I had never thought of the infernal machine.
>
> 2nd I must finish it at any expense and as I am in an isolated way I feel little uneasiness at the prospect of sacrificing a large portion of my fortune.
>
> 3rd Of the machine itself I think the drawings advance more rapidly then they did before, and the work on the whole does make considerable progress, more indeed in your hands then it did in mine. The money expended is large with regard to the labor part but this I cannot help and as I must complete it I would not have its progress checked on any account indeed if any sacrifice could terminate it instantly I would gladly make it. It is the wear and tear of moderate annoyances that I most dread. It would be impossible to pay for it out of my own income. I have therefore decided upon consuming capital and have made arrangements at Messrs Praeds[84] so that you may continue to draw for money. This has all along been the case: I thought I was advancing instead of consuming capital. I left Clement so well acquainted with all the mechanical actions of the machine that I have not the least fear of his making such changes us he may think necessary.[85]

Less insouciant about Clement, Herschel was increasingly dubious about the material progress he could see, and the incessant demands for money. Soon he was addressing the engineer curtly as 'Sir' and asking him to keep expenses down. But the astronomer's oversight was limited by his knowledge of the design—'I find it impossible to master all its details'—and the opaque path from detailed design and tooling to manufacture of parts.[86]

Babbage lingered abroad through 1828, visiting Alexander von Humboldt in Germany and attending the scientific congress there in September. He only returned to England late in the year, acquiring the spacious old house of the late William Hyde Wollaston (see above) at 1 Dorset Street in London. This address would form part of the lore surrounding him in years to come. Presently, correspondence with Clement turned increasingly frigid and snippy, as Babbage finally began to query the amounts spent and whether Clement really could claim to own the machining tools. A friend in the Colonial Office, Richard Penn, advised him that 'as Mr Clement has chosen by his offensive conduct to forfeit all claim to that liberality on your part which disposed you at first not to question the amount of his charges but merely to require the particulars of them' so that 'your best course would now be to force the matter to a more hostile arbitration', anticipating that 'Mr Clement's low cunning will be defeated'.[87] At Cambridge the developing tension was noticed. 'Babbage has been here in his capacity of professor, and I have really enjoyed his society much, having seen him more closely than I had done before. But his anxiety about the success and fame of his machine is quite devouring and unhappy'.[88]

The prickly artisan, much more experienced in such matters, welcomed arbitration by a panel of two professionals in April 1829. In practice such a panel would only have limited

[84] Praeds & Co., London, his bankers. They would later merge with Lloyds.
[85] 1828-05-09, C. Babbage to Herschel, section I.33, p. 346.
[86] 1828-07-24, JFW Herschel to J. Clement, section I.34, p. 347.
[87] 1829? Richard Penn to C. Babbage, section I.35, p. 348.
[88] William Whewell to Rev. R. Jones, 1829-02-04, Todhunter 1876, Vol. 2, 97.

access to information about the workings of Clement's shop on an entirely novel design. Information asymmetry was heavily in Clement's favour, as he knew from past encounters, so that the outcome was the opposite of Penn's anticipation. Babbage now learned, evidently to his surprise, that Clement claimed all ownership of the tools that had been made, declaring that they had been made with his own money. When asked if he would undertake to not make further copies of the machine himself, 'Mr Clement declines doing so'. Clement was to keep the tools, and his expenses were deemed reasonable.[89]

At this stage any further dispute between the two men could only terminate the project. The large amount of money already spent made it increasingly difficult to obtain more, and at least a large proportion of it would have to be spent all over again if Clement decamped with the essential machining tools that so much of the effort had gone into creating. The dispute was already brewing. The engineer considered himself to be employed by Babbage, rather than the government, which was hard to argue with, so Babbage's protests that he could not advance more money without government support were not accepted.

The inventor had been forced to advance his own money for quite some time, and appealed for reimbursement and further funds. Babbage wrote to the Duke of Wellington to ask for clarification. Another committee of the Royal Society—Herschel featured once more—was appointed, this time to assess only the progress of the project toward its stated goals. It found that progress was reasonable given the novelty and complexity of the subject-matter, and further funds were approved. 'My Lords are of opinion that they are fully justified in directing a further payment of £1500 to Mr Babbage to enable him to complete the machine by which such important benefit to Science may be expected to result'.[90]

In a letter to the arbitrator Bryan Donkin, Babbage himself confessed to having been uncertain up until then about the exact financial arrangement between himself, the government and Clement, which sits uneasily with his other recollections. 'I have hitherto in conversation with Mr Clement always treated the work as my own private property for two reasons: first I thought his charges might be more moderate than if they were paid by government, and secondly I did not feel myself quite authorised to state the contrary. Since I have been acquainted with the result of my letter to the Duke of Wellington and the report of the Council of the R.S. I can scarcely continue to consider this to be the case'.[91] (The idea that Clement would have charged less to Babbage—who thought, like Clement himself, that Clement deserved to be 'well-paid'—is preposterous.)

With more funds, a working version of the manufactured engine—it still works today—could be assembled. It was intended to be housed in some fire-proofed buildings added at government expense to 1 Dorset Street. This version included a less ambitious number of differences and more limited precision than the full device called for, and did not include the printer, but was still a complete demonstration of the computational core. A larger engine would have been more of the same: the machining tools would be used to manufacture more copies of the pieces used, to add more columns and more wheels. Arguments that the device was a technical failure, because the assembled portion was not the *full* specification and did not include an automatic printer,[92] are wrong-headed. Although Babbage had often talked of automatic printing, and had conceived several designs, that portion of the engine never seems to have gone much beyond the drawing board.

Relations with Clement reached a climax at this point. Babbage wanted the engineer to move his shop into 1 Dorset place to facilitate further work. The assembly of the working core had taken many months, at least partly because of the need to transport pieces from the shop to its secure location. Clement's shop was about six miles away at 21 Prospect

[89] 1829-04-22, Bryan Donkin to C. Babbage, section I.36, p. 349.
[90] 1829-04-28, Treasury Minute, section I.38, p. 350.
[91] 1829-04-11, C. Babbage to Bryan Donkin, section I.37, p. 350.
[92] E.g. Lindgren 1990.

Place, no small distance in that period to travel to and transport material from. Further work there would have required additions since there was no room at his shop. Brunel and the Royal Society supported the need for the move. The designs of the new buildings were approved by Decimus Burton. Clement insisted that making such a move would be costly to him and submitted a steep estimate, which now had to be transmitted directly to the government for approval. This demand was greeted with 'suprize' by the treasury and found to be 'unreasonable and inadmissable' in claiming an annuity for himself and repayment of interest on his expenses, which they held to be the usual business expenses. Their analysis—transmitted to Babbage, who largely agreed with it—was brutal.[93]

> Mr Clement desires that a Lease of the Premises may be given to him for three or four years, and estimates the expense of removal at the sum of £350 in which sum is included the expense of a second removal and fixing of tools etc. at the expiration of the time specified in the lease. He presses also a personal claim for £130 for interest on Money laid out in his furniture, and for wear and tear of his furniture and requires various alterations to be made in the House in East Street for his accommodation. He also claims to himself an annuity from the Public of £660 per Annum with additions for rent, rates and taxes in East Street. ... Mr Clement must recollect that £12,340 of Public Money has been expended on the Engine, the greater part, if not the whole, of which sum has passed through his hands, from which a very large amount of profit has accrued to him, and that his employment by you has constituted, and still continues to constitute, as My Lords are informed, the principal part of his business, and that he has not been restricted from using for his general business any tools for which the public have paid, and that the further expense of completing the engine will yield to him a considerable profit.

Unbowed, impervious to attempts by Babbage to arrange a compromise, Clement refused to proceed. When payment for his outstanding invoices was not immediately forthcoming—Babbage insisted that the Treasury would have to pay them—he dismissed the staff working on the project. When the Treasury agreed to pay in due course, he repeated his previous demands for compensation to move, or permission to resume work at his own premises. He added further invoices for work already done: if these were not paid he would not transfer the drawings or equipment to the care of Babbage.

This deadlock, reached in June of 1833, would prove to be the undoing of the engine. It was a pitfall present since its inception. With Clement owning the special machine tools, work would have to start again from scratch even if his dismissed craftsmen were re-employed directly by Babbage. The diagrams were by arbitration the property of Babbage and were eventually transferred to 1 Dorset street in early 1834, but it appears to have taken an astounding 18 months to secure them, as Clement was in no mood to cooperate until his invoices were settled.[94] The whole project was now in limbo. To make things even more problematic, in August 1834 Babbage had already started work on an incomparably more capable machine, news of which only perplexed the government.

Babbage never published a written description of the Difference Engine after his prospective articles from the 1820s, preferring to let others introduce its mature state to the wider world while he waited in the wings. It was a technique he would use more than once. Dionysius Lardner (1793–1859), Anglo-Irish editor of the *Cabinet Cyclopedia*, author of mid-level textbooks and a prodigious popularizer of science, had become interested by 1833. Lardner gave public lectures on the device and other calculators, up and down the country, to 'Mechanics' Institutes', the Royal Institution and other hives of autodidacts. He quickly enlisted the aid of Babbage and his son Herschel to understand all the details, soliciting 'interesting anecdotes which you may know respecting the history of tables and more especially any which relate to the invention of the machine. State the circumstances which first directed

93 1832-09-14, J. Stewart to C. Babbage., section I.40, p. 352.
94 Hyman 1982, 164.

Amount (£-s-d)	Minuted	Paid to	Received
1,547-0-0	11th Jul 1823	Babbage	7th Aug 1823
1,500-0-0	29th Apr 1829	Babbage	1st May 1829
3,000-0-0	24th Nov 1829	Babbage	3rd Dec 1829
600-0-0	24th Dec 1830	Babbage	31st Dec 1830
1,000-0-0	12th Jul 1831	Babbage	14th Dec 1831
1,000-0-0	30th Sep 1831	Babbage	post 4th Oct 1831
1,701-10-7	29th May 1832	Babbage	post 2nd Jun 1832
1,956-05-11	1st Sep 1832	Babbage	post 7th Sep 1832
1,782-11-4	13th Aug 1833	Clement	13th Aug 1833
1,200-13-6	12th Aug 1834	Clement	post 12th Aug 1834
TOTAL: 15,288-1-4			

Table 2.1: Difference Engine Expenditure. Add '£2,190-13s-6d on the construction of buildings to house the completed engine, workshops and the adaptations to a house for the superintendent', Roberts 2014b.

your attention to the subject and any other matters that may occur to your recollection. It would also be very interesting if you could state some of the difficulties moral, mechanical or mathematical which you had to overcome. Also ... examples of ingenious methods of evading difficulties'.[95] An extensive correspondence followed.

A sixty five page article emerged in July of the following year, published in the staid *Edinburgh Review*.[96] At the suggestion of Augustus de Morgan this was framed as a review of Babbage's earlier articles, including the Royal Society report and the sign notation, which Lardner had taken a liking to. Much of this should be treated as *de facto* Babbage's own work, 'written partially, if not entirely, at the house of Mr. Babbage, where Dr. Lardner had not only the advantage of the inspection of the plans, drawings, and model of the Difference Engine, but had furthermore the benefit of the advice and assistance of Mr. Babbage' (which is a little fast and loose, since the correspondence shows Lardner working on the paper on his own for some time).[97] Some pointed closing remarks about the suspension of the project demonstrate that Babbage was not entirely in control of the content. It includes a comprehensive overview of mathematical tables, their history and manifold sources of error, the method of differences, with illustrations, and an exhausting history of lesser calculating devices. All this was especially well-suited to Lardner's talent as a general-purpose scientific amanuensis and still holds up today. The British Association invited him to reprise his lectures shortly after. Lardner would often find himself on the wrong side of scientific disputes, and fell foul of public mores when he took to philandering. Yet the project which inspired his sure feel for publicity, where he was for once squarely on the side of posterity, had ground to a dead stop and could not be helped by the considerable attention he generated for it.

At the root of this suspension of work lay Babbage's own naïveté. He was a scientist, not a businessman, and appears to have had no talent for managing practical projects involving other parties. The contracts that he should have had in place to secure the tools—even prevent loss of intellectual property—were only thought of when it was far too late and enormous sums of money had already been spent. He wrote to one of his supporters, Lord Somerset, that it was 'singular that I should not have perceived the full difficulty

95 British Library Add Ms 37188 f140, D. Lardner to C. Babbage, Roberts 2014a.
96 Lardner 1834.
97 Buxton 1988, 288.

of my position until the great mechanical difficulty was overcome'.[98] Later events would show that the engineering tolerances demanded were grossly over-specified, ballooning the cost, a situation that did not inconvenience Clement.[99] Here the engineer pulled where his client leaned, as more than one of Clement's former employees would later inform Babbage. Charles Godfrey Jarvis wrote that 'the greater part of the time charged was expended in considering the propriety of adding a "hair breadth" to the head of that screw, or taking a "thirty-second" off that boss ... things were of no consequence whatever to the efficiency of the thing itself' (at the time Jarvis was still employed by Clement and some allowance must be made for hostility between the two men).[100] Nevertheless, the inventor was disinclined to enter into public recriminations with his artisan, and always took care to note that Clement was 'a most excellent workman and draftsman and ought to be well paid'.[101] Laudable sentiments, perhaps, for someone with unlimited funds. Baffling notions—how much is well?—for anyone with a finite budget, who must pay no more than the work actually demands. Not even comprehending that there was some limit to the spending that any government could authorize was fatal.

When communicating with the government *in private*, Babbage told a different story, ascribing the delay in part to 'Clement slow', with 'interest to cause delay' along with 'obstinacy and refusal to allow me to have the drawings'. He described Clement's 'Excessive demands in order to retain work at his own place'. He noted that 'Everybody in [Communication] with him discontented'.[102] When communicating with others, Babbage swung around again, placing most of the blame on the failure of the government to reimburse him timeously, writing to Lord Somerset, 'amongst all civilised countries England is that which there exist the greatest number of practical engineers who can appreciate the mechanical part whilst at the same time it is of all others that country in which the governing powers are most incompetent to understand the merit either of the mechanical or mathematical', 'I am deeply indignant at the treatment I have experienced from the country'.[103] That attitude would grow to consume him. Yet the public expenditure ultimately granted to Babbage's project, over £17,000, was generous, the cost of at least 14 steam locomotives (see Table 2.1). Granted, he was initially out-of-pocket between 1824 and 1829, but the costs had ballooned, and he was fully reimbursed in the end.[104] He had been inconvenienced and (arguably) inconsiderately treated, but that is not why the project failed to complete. Though he would complain that he had received no recognition in his own country, in 1831 he was offered a knighthood, in the Hanoverian Order of the Guelph. He turned it down because it was in the Third Class and not the First, perhaps because it would not carry the honorific 'Sir'.[105] It would not be the last time he turned down honours hoping for more.

---

98 1832? Babbage to Lord Somerset, British Library Add Ms 37187 f302.

99 There have been some attempts to revive Clement's reputation, such as Lindgren 1990 and Williams 1992, infected by the enthusiasm of Smiles 1881.

100 1831? C.G. Jarvis to C. Babbage, section I.39, p. 351.

101 1829-04-11, C. Babbage to Bryan Donkin, section I.37, p. 350.

102 British Library Add Ms 37189 f112, 26 May 1835. Notes made by Babbage of his interview with Spring Rice. Based on Roberts 2014a.

103 1832, C. Babbage to Lord Somerset, section I.41, p. 353.

104 Roberts 2014b.

105 Babbage to Brougham, 1831-09-11, British Library Add Ms 37186 f97.

Chapter 3

The Analytical Engine.

After the Difference Engine ground to an inconclusive halt, Babbage re-examined the drawings he had finally obtained from his foot-dragging engineer Clement. He had seen some tantalizing glimpses of more general capabilities from the beginning. The engine could break free from simple difference calculations to change its rules, producing results for which no recurrence relation was known. But he had pragmatically postponed further work on more advanced computational ideas. A central problem was how to simplify the manual calculations using the power series expansion that were still necessary—the engine only filled in the gaps through differences.

In a letter written a year later to the American autodidact Nathaniel Bowditch (1773-1838)—his son Henry had visited London in 1834 and come to see the working core of the Difference Engine at Dorset Street—Babbage communicated a major breakthrough made that September.[1]

> I can scarcely write to you on a topic of greater interest nor will I attempt to conceal my own opinion of the importance it now possesses ~~in my own estimation~~. ... I speak of an entirely different Engine from that which your son is acquainted with. Shortly after he left England in September last I got possession of the drawings of the old Engine which I had not seen for two years. In explaining these to my son many new views struck me and a totally new and more extensive arrangement occurred to me which appeared of such importance that I thought it worthwhile before I proceeded to re-examine every part of the old Engine. I soon found many new modes of adding, of carrying, and of performing other processes which were mechanically much more simple than the ancient methods. These new simplifications enable me to add new powers to the Engine and yet to make it upon the whole less complicated than the former one. It is somewhat curious that in the Engine I am now drawing there scarcely occurs a single contrivance which has been used in the former one. ... I have had a draftsman constantly at work since September and I have given up all other pursuits for the sake of this and the progress I have already made surpasses my expectations. It is indeed more than I ever before made in several years. This arises partly from a more enlarged experience, partly by having a better tempered draftsman, partly by having all my drawings in my own house instead of at a distance of four miles.

He went on to explain that the new design implemented all the arithmetical operations—not just addition but also subtraction, multiplication and division—so that it could calculate any polynomial, and thus any power series approximation of more general functions. Key to this was the ability to temporarily store and feed intermediate results to subsequent calculations. It was from this thinking that *programmability* for combining complex computations was born. Later the machine would become generally programmable through conditional branches and loops to support an even broader range of calculations. At this early stage

[1] 1835-08-02, Charles Babbage to Nathaniel Bowditch, section I.51, p. 359.

the programming was to be performed mechanically by a drum mechanism on which knobs for driving levers were adjusted, somewhat like a player piano. This anticipated modern microcode—the primitive instructions from which seemingly simple, but in reality more complex, operations can be built—but Babbage didn't share these details with Bowditch. He already envisioned accuracy of up to thirty places, and automated printing of both inputs and outputs. He was also planning a particular program which we will return to at great length later. 'Of course it can calculate the values of π, e, the numbers of Bernoulli etc'. A complaint followed that we now know correctly estimates the importance of his own work but must have been hard at the time to separate from the conceit that usually surrounds inventors wrapped up in their own causes.

> You will be able to appreciate the influence of such an engine on the future progress of science. I live in a country which is incapable of estimating it and I am much displeased with the treatment I have experienced. You who have yourself made great pecuniary sacrifices for the production of a great and important work will understand my feelings. I have made large sacrifices and am prepared to make still larger. There is however a limit at which my duty to my children demands any forbearance and I place that at the completion of the drawings which is in fact the completion [of] the invention. These will remain when I am gone and when some more enlightened country than my own shall at some future period become sensible of the value of the engine or ambitious of the fame of calling it into real existence, those drawings will undoubtedly be sought and from them a machine will be made which shall give to all nations and to all after [times] tables at once perfect and enduring and which shall execute calculations from which all human agents would shrink in hopeless despair. I have thus given you an outline of the work I am employed in, the views that it has opened and which it continually suggests are amongst the most interesting and important I have met with.

As we will see, Babbage would excite Giovanni Plana, the renowned astronomer based in Turin, with similar details in letters probably written later in 1835.[2] An earlier notice with few concrete details had already been sent privately to the pioneering statistician Alphonse Quetelet in May of 1835, and published in extract by the Belgian, apparently without the consent and contrary to the intentions of Babbage.[3] Although he describes his new ideas in all of these communications as a breakthrough, the germs for his thinking had been present since at least 1822. He had written to Sir Humphry Davy then, after describing the difference engine prototype, of another 'machine, whose plans are much more advanced than several of those just named', 'for constructing tables which have no order of differences constant'.[4] This more advanced device would cover the transcendental functions like sin and cos which, unlike polynomials, do not have a constant order of difference. Writing to the Royal Astronomical Society in December of the same year, he had enlarged on a method for treating functions like $A \sin \theta$, which recur often in astronomical tables. 'In examining the analytical theory of the various differences of the sine of an arc, I noticed the property which it possesses of having any of its even orders of differences equal to the sine of the same arc increased by some multiple of its increment multiplied by a constant quantity. With the aid of this principle an engine might be formed which would require but little attendance, and I believe that it might in some cases compute a table of the form $A \sin \theta$ from the 1st value of $\theta = 0$ up to $\theta = 90°$ with only one set of figures being placed in it'.[5] This is a special case for sin but, as Babbage noted, it would have wide application to astronomical tables.

A more expansive vision of the mechanization of mathematics was present, as we have seen, in a letter to Herschel in 1822, when he had talked of 'other machines I have contrived

2 See 1835, C. Babbage to Plana, section I.54, p. 363.

3 Babbage 1835. The complete letter is reproduced in section I.49, p. 357. See also Babbage 1989b.

4 Babbage 1822a.

5 Babbage 1822c reprinted in Babbage 1889. For the trigonometric recurrence $\Delta^2 \sin x = K \sin(x + 1)$ see Appendix B, p. 131.

for reducing different parts of mathematics to the dominion of gravity'.[6] He had kept this in the back of his mind for 12 years while he concentrated on developing the (much simpler) engine for constant orders of difference. In the 1831/2 prototype engine he had introduced an echo of these ideas by showing how previous results could be fed forward into subsequent ones, though in a way that was only theoretically interesting (one digit of any order could be transferred from the previous result, just for fun). He referred to this as 'the engine eating its own tail' (we call it recursion). Re-examining the whole process now, he realized that the ability to 'eat' that 'tail' flexibly, in ways that could be preset, was key, assuming that missing arithmetical operations like multiplication and division could be provided.

This was the foundation of programmability for his engine, where prior results could be stored and then operated on again in future calculations, using very general formulae. Lower-level programming was needed to even implement an operation like division, which required branching conditions to realize. He worked hard to achieve efficient implementations of multiplication and addition, in which digit carrying was a thorny issue. Eventually he devised an ingenious highly-efficient 'anticipatory carry' mechanism, after trying many alternatives, the mechanical details of which will not concern us here. From a modern point of view, this treats addition as a family of functions, with the first argument to the sum selecting a slightly different function to apply to the second, so that the digit carry is 'anticipated' efficiently. In the next century, this higher-order sort of concept would be generalized in the 'lambda calculus'.[7]

Babbage quickly noticed that once the algebraic operations involved in power series could be supported by the engine, a broad range of possibilities opened up. These were the series used to approximate many functions in the first place, to get the values that the difference engine had to reset to at intervals, to overcome the accumulation of approximation errors—but they need no longer be calculated by hand. Programming those operations through a revolving drum on which knobs, struck by levers, are set to enable composition of primitive operations into more complex ones was his first attempt at a solution, anticipating 'microcode' by a century. This advance was closely tied to an earlier decision, made sometime in 1835, to concentrate processing within the new engine into a distinct 'mill', which would operate on values transmitted to it and yield results which could be stored in the columns for reuse. The difference engine had supplied an addition operation only, duplicated at each column and not centralized at all. That duplication would now be impractical given the new operations to be supported. Once operations were centralized in the mill, connected to the storage columns by means of racks and pinions, it became feasible to think of programming its operations, using the drum to combine the sequences of primitive mechanical manipulations making up the arithmetical ones. At the same time, the columns used to store values were simplified. (The mill also had temporary internal storage analogous to modern registers, for intermediate results.)

The drum worked and would be retained but was complicated to use in practice, complications which grew as more sophisticated combinations were needed. The real breakthrough came sometime in July of 1836 when Babbage noticed that cardboard cards with holes punched in them, already used to program brocading patterns in powered looms by J. M. Jacquard, could also be used to program his 'analytical engine', as the more advanced device came to be known. Now the levers would be driven by the absence or presence of holes in the cards, rather than by knobs they struck on the barrel. The cards were of distinct types, encoding constant values and instructions, with separate readers. It was a major conceptual advance, the equivalent of modern software, which would also be realized a century later using punch cards. The device could be reprogrammed in one fell swoop using a prepared set of cards, taped together and fed onto a spooling device to read them. Card programs

6 1822-04-09. C. Babbage to JFW Herschel, section I.24, p. 339.
7 For anticipatory carry see Hyman 1982.

could be independently duplicated and combined *as instructions*. Those instructions had been abstracted as information, what we call 'software'.

Even Jacquard had not started from scratch, improving his pattern cards from earlier ideas due to Jean-Baptiste Falcon (circa 1728, for a hand operated loom) and Jacques de Vaucanson. It is not exactly clear how Babbage made the leap from pattern weaving to his own machine, with its radically different purpose. But since he had already made a special project in the earlier 1830s, as he was grinding to a halt with Clement, of the study of machines in general and their economies—looms included— he was already primed to make it.[8] Though he would never publish it under his own name, in 1837 Babbage drew up a long report on the mathematical capabilities of his new engine which gives a good idea of his intentions.[9]

Piqued by what he believed to be a failure on the part of the government to support him, Babbage had already resolved in 1834 not to attempt to build the new device, only to design it—at least so he wrote to Nathaniel Bowditch, in which some retrospective tidying-up is possible. The technical drawings would be 'in fact the completion [of] the invention'.[10] The mechanical details would be worked out in those drawings, aided by what amounted to thought experiments, and the elaborate 'Method of Expressing by Signs the Action of Machinery' he had invented.[11] At considerable expense he employed a full-time draftsman, C. G. Jarvis, who had previously worked under Clement, to facilitate capturing and evolving his ideas. Though pieces would later be constructed, Babbage ultimately stuck to this resolution. As he told it, the diagrams would be comprehensive enough to permit anyone who cared to build the physical realization. This proved to be false: the full details of the engine, though its outlines and general intent are well-known, died with him. But for some time it seems to have been unclear whether the machine was expected to be constructed or not.

At the end of December 1834 Babbage informed the prime minister, then the Duke of Wellington, in writing that, just as the Difference Engine had ground to a halt, he had invented 'a totally new engine possessing much more extensive powers, and capable of calculations of a nature far more complicated'.[12] He did not explicitly say that he was abandoning the old one, and did not conclude that it was now redundant, but it was a reasonable question. The Duke asked for clarification which met with an inconclusive response. Letters from November of the following year suggest that Babbage himself was unsure about the role he expected the authorities to play. Jarvis, his draftsman, peremptorily proposed leaving for higher pay elsewhere, pushing Babbage into a state of panic. He felt that this was important enough to inform the exchequer Spring Rice about, and he even attempted to deliver the letter by hand.[13] Unable to do so, leaving it to be forwarded, he settled with Jarvis, agreeing to pay the draftsman what he asked. (One suspects that Jarvis had learned more from his former employer Clement than he cared to admit.) If Babbage did not expect the government to fund the Analytical Engine, even if it was never to be constructed and only a set of diagrams, this panic makes no sense.

Spring Rice was alive to these ambiguities, and wrote to Babbage in the new year asking once again for clarification: did he expect them to pay or not, and was the old to be abandoned for the new?[14] The reply from Babbage, accompanied by a report on the current state of the engines, was that he now considered the old engine redundant—he had changed

---

8 Babbage 1832.

9 See Appendix D, p. 137.

10 1835-08-02, Charles Babbage to Nathaniel Bowditch, section I.51, p. 359.

11 Babbage 1826.

12 1834-12-23, C. Babbage to Wellington, British Library Add Ms 37188 f525, Roberts 2014a.

13 British Library Add Ms 37189 f201, 20th November 1835, Babbage to Spring Rice. British Library Add Ms 37189 f202, 20th November 1835, note written by Babbage. See Roberts 2014a.

14 See section I.58, p. 368.

his mind after new developments—but that he left the decision up to the government! Nor was he asking for funds! 'I certainly never intended to apply for what might be considered a favor when I have been treated ~~which~~ with what I ~~consider~~ look upon as injustice'.[15] It seems he desired supplicants instead. He continued to sulk in his report: 'I do not entertain the slightest doubt of the success of the first Engine nor do I intend it as any application to finish the one or ~~make~~ to construct the other'.[16] This elaborate game would continue for years, with Babbage protesting personal injury and demanding a decision about the Difference Engine but refusing to make a recommendation on a topic that he, and not an increasingly-bored government, was the expert on. If the Difference Engine project failed entirely, it seems he wanted the blame to lie with the government for cancelling it, and thus wasting the investment they had made. But not desiring new funding does not square with his eagerness to inform them of Jarvis' threatened defection. And the waste of earlier funding was by then moot, since everyone realized that work would simply have to start over if the Difference Engine was to be completed, given the loss of Clement, his tools and his expertise, *regardless of the new invention.* Babbage blithely informed the government that modern inventions were typically superseded before they were completed anyway, a politically-impossible argument regardless of its technical truth or not. It was surely a matter of wounded pride now. (Read as a whole, this correspondence does not support the argument by his biographer Tony Hyman that Babbage was the victim of 'The Circumlocution Office' of Dickens' *Little Dorrit (1855-7)*, a sclerotic government, in all this.[17] The circumlocution was more his than theirs.)

The design of the Analytical Engine changed over the years, as Babbage experimented with ideas at every level and Jarvis thrived on the diagrams, but its core features remained comparable.[18] One feature he carried over from the Difference Engine was the ability to automatically print inputs and outputs, to reduce typesetting errors, but he made no more progress than before in figuring out exactly how this would work in practice. Many possibilities were immediately evident, but the theoretical complications posed by printing did not approach the level of complexity demanded by the computational facilities and their mechanical realization, which took up most of his attention. We will return to these details when we examine a famous example program for the engine and reveal the limitations of the design as it stood in 1843. We will also see how these limitations were at first obfuscated and then later quietly overcome.[19]

As he continued to flesh out his new engine, Babbage's reputation extended, despite his failure to complete the implementation of its precursor. A steady stream of visitors, foreign and domestic—Chevalier Binkhausen the Russian consul, Signor Ribus and Hollier lately arrived from Spain, M Scharling of Copenhagen, General Fullmudge, a distinguished citizen of New York, and his family, Joseph Hume, Mr Henry Siddons, Grandson of *the* Mrs Siddons, the Prussian Mr Thiede, Thomas Bradley, Revd Mr Pyne, Count Mensdorf, Lord Lansdowne, Lord Brougham, Viscount Melbourne, Sir Robert Peel, Lord John Russell, and many others—sought and were granted personal demonstrations of that device, sometimes on any day of the week that suited them. Babbage would tire of this by late 1836 and often refuse individual showings after then.

Since the death of his wife and subsequent return to England the inventor had held regular at-home gatherings on Saturday evenings, to which many scientific and literary personalities of London society were drawn. At first this was to make it easier for his mother, frail and housebound, to meet his distinguished friends (or so a friend recalled him

15 1836-02-01, C. Babbage to Spring Rice, section I.59, p. 369.
16 1836-02-01, C. Babbage to the Exchequer, section I.60, p. 369.
17 Hyman 1982.
18 Bromley 1982a, 1987, 1990, 1991, 2000.
19 See chapter 6, p. 59.

saying).[20] After 1834 they were treated to a showing of the 'silver lady'—a twelve-inch high automaton made by John Joseph Merlin (1735–1803), capable of dancing, coquetting her eyes, and making graceful gestures while a fluttering bird perched on her finger—which had impressed Babbage as a child. The showman Thomas Weeks had bought Merlin's gadgets, originally exhibited at Merlin's *Mechanical Museum* in Hanover Square, on the inventor's death. After Weeks' own death in 1834, his museum long closed, they were up for auction. Babbage had rescued, repaired and then clothed the 'lady'[21]. The assembled portion of the Difference Engine, the 'Leviathan', stood on a stand in the middle of one room. Talk about new developments swirled, though this could be 'an endless subject of monologue' to those who had come rather for the society.[22] Along with this, there was general immersion in the 'progressive' thinking that Babbage favoured in those days. Thomas Jefferson's granddaughter Ellen Wayles Coolidge attended while on her European tour and left a perceptive account of 'Babbage's Menagerie'.[23]

> Saturday morning we breakfasted with Mr Rogers and Saturday evening at half past nine, went to Mr Babbage's house in Dorset St. Manchester Sq. Here was a gathering of the elect, a "re-union" of literary & scientific men, artists, authors, celebrities of both sexes. Those who like myself had no claim of learning or letters for admittance into so choice an assembly, could only rejoice in the opportunity of seeing so many Lions in one cage. We had, Mr Babbage himself the inventor of the famous calculating machine, Mr Hallam, the author of the "Middle Ages," the Revd Mr Whewell, whose Bridgewater Treatise I had read in the U.S. and whose Architectural Notes on German Churches I have been studying this summer, a propos to my recent acquaintance with English Cathedrals; Mr Morier, whose Hadji Baba in Persia & England, I have cried & laughed over; Judge Haliburton, alias Sam Slick; Mr Rogers, the poet of Memory & of Italy; Mr Hume, the political calculator; Mr Grote, the friend of the Analytical lot & of Dugald Stewart; Mr Taylor, the author of Philip Van Artevelde; Mr Justice Maule, who possessed of talents, knowledge, every thing which fits him to do honour to the place he holds, might have remained a simple barrister for all his life, had he not bought himself into Parliament—he gave two thousand pounds for a seat in the House of Commons, and a seat upon the Bench soon paid him his money with interest. This is the way they manage things in England it appears. Then came Sir Francis Chauntrey; Sir David Wilkie; Turner, a very favorite painter here though rather slightingly mentioned by Dr Waagen, who allows him great talent it is true, but reproves him for the abuse of it; Sir Charles & Lady Morgan; Mrs Austen, whose translation of the Letters of a German Prince (Puckler-Muskau) the best translation ever made into the English language, has put her on a footing with original writers; Miss Martineau, to whom there is no denying the praise of talent, though accompanied by self-conceit & prejudice—such was the character of the company. There were a few titled persons present, among whom I heard the name of Lady Anna Maria Donkin, sister of the Earl of Minto, and the Earl of Selkirk, whose family history connects him with the more famous Paul Jones, who was the son of a gardener in the service of Lord Selkirk's father. The Americans of the party were Mr & Mrs Stevenson, Mr John Van Buren, Mr Charles Sumner, Mr C.[24] & myself. It was a pleasure to me to see so many persons whose names I had seen in print, and the assembly of the evening was literally what Carlyle calls "that crowning phenomenon, and summary of modern civilisation, a soirée of Lions." There were no doubt "many blue men and women, dear Mrs Rigmaroles, great Mr Rigmaroles" in this company of distinguished persons—many "pewter tankards scoured bright" and mistaken by "dim-eyed animals" like myself, for "solar luminaries," but there were also some rightful objects of that hero worship which even Carlyle applauds. "Veneration of great men is perennial in the nature of man; this in all times, especially in these, is one of the blessedest facts predicable of him."

[20] Lloyd 1880, 47-69.

[21] Babbage 1864, 365.

[22] Such as Mrs Crosse, Crosse 1890. See also Lloyd 1880, 47-69.

[23] Travel Diary of Ellen Wayles Coolidge, Mon. 1839-02-18, Birle and Francavilla 2011, 244-7, 250-1.

[24] Her husband Joseph Coolidge (1798–1879).

> The rooms, after a while, became crowded and warm, and I went below with Mrs Stevenson to get a cup of tea. Refreshments of various kinds were served up on tables behind which servants were stationed to attend upon the guests; there was tea, coffee, lemonade, wine, fruit, cake, thin bread & butter, and I believe sandwiches. Enough of every thing and good, but no affectation of style or shew. Returning up stairs I encountered Miss Rogers whom I like better & better. She is not only a sensible but a good-natured person. Now good nature is not a common virtue among the English. If they happen to possess they are ashamed to acknowledge it, and when you do find it, it has the charm of rarity. I was presented to some few persons, exchanged a few words with Mr & Mrs Hallam, Mr Whewell & Judge Haliburton, & held something approaching to a conversation with Mr Morier & Mr Babbage—but as to becoming acquainted with any one in such a crowd, and where all talk is necessarily desultory, I do not understand the possibility of it. Miss Martineau whom I knew in America, says, I am told, that she should be pleased to renew the acquaintance. If so let her make the first advance. She is on her own ground now and I am the stranger. I certainly shall not do homage to her. Lady Morgan was rather cold. One of my friends told me that I had committed an unpardonable mistake in speaking to her of any thing but herself & her own writings. That Glorvina expected to hear of nothing but her own perfections, and that when I talked to her on other subjects I had shewn a want of taste & feeling, at which she was naturally disgusted. Alas!—The little Goody was dressed in scarlet velvet with a black hat & feathers and a band of small diamonds over her brow, but the Irish witch prevails over all her finery. Mr Babbage invited us to join his saturday evening parties whenever we were not otherwise engaged—a permission of which we shall gladly avail ourselves. ... The persons, on the other hand, whom I meet in society have all, more or less, the same style of manners and of dress, and their ordinary conversation is pitched nearly in the same key. They vary because Nature has put it out of their power to conform in all things to a given standard, but they vary as little as they can. This, in general society, produces a certain amount of insipidity, a want of heartiness, or earnestness, of any sort of warmth or glow. At "Babbage's Menagerie," as Lady Anna Maria Donkin calls his Saturday evening parties, where so many political, literary, scientific & artistic characters assemble, I should say that the distinguishing mark was want of all character for good or evil. This is, I suppose, unavoidable, at least to a certain extent. If each individual carried into the company of others too much of his individual self there might be many petty shocks & some rude collisions. Yet it seems a pity that Babbage, Hallam, Whewell, Wilkie &c &c should move about requiring, as Mr Morier said, to have labels pinned to their backs, in order to tell one from another. Yet it is better to have them simply ticketed with their names than be obliged to add as Mr William Coke did when he rode a vicious mare, "she kicks," which might be the case were these gentlemen to come into society without leaving their individualities behind.

Even the sober *Quarterly Review* was enthused. 'Mr. Babbage's Saturday evening-parties should be mentioned to show what science can effect for itself. They were amongst the most brilliant of the last season, and presented an interesting *coup-d'oeil* of beauty and celebrity. As the Sheridan sisters entered the room, where Mrs. Leicester Stanhope, Mrs. Austin, Mrs. Lister, Lady Vincent, the present Lady King [soon to be Lady Lovelace] ('Ada, sole daughter of my house and heart'), Lady Morgan's nieces, &c. &c., were already clustered, M. de Beaumont exclaimed half aloud to M. de Tocqueville—*Ah, cette beauté Anglaise! c'est vraiment étonnante!* [Ah, this English beauty! It's really astonishing!]'.[25]

Curiously, the up-and-coming naturalist Charles Darwin attended in the early years of his return from the *Beagle*, before he retreated to Down House and social isolation. On his way home Darwin had met and been impressed by Herschel at the Cape, where the astronomer had already spent two years surveying the firmament through the 20-foot telescope he had transported from Slough, when he was not drawing the exquisite *camera lucida* sketches of the landscape that he left to posterity.[26] Herschel would not return to London for another

[25] *Quarterly Review*, 1836-02, 476.

[26] Herschel 2007; Herschel, Evans, et al. 1969.

year when Darwin briefly entered the orbit of Dorset Street in 1837. Darwin's ineffectual brother Erasmus also attended, and Charles recalled later dinners with Thomas Carlyle:[27]

> The last man whom I will mention is Carlyle, seen by me several times at my brother's house, and two or three times at my own house. His talk was very racy and interesting, just like his writings, but he sometimes went on too long on the same subject. I remember a funny dinner at my brother's, where, amongst a few others, were Babbage and Lyell, both of whom liked to talk. Carlyle, however, silenced every one by haranguing during the whole dinner on the advantages of silence. After dinner Babbage, in his grimmest manner, thanked Carlyle for his very interesting lecture on silence.[28]

Carlyle reciprocated this disdain. He had first met Babbage on Feb. 1 of 1839, at a dinner given by Lady Harriet Baring. Writing to his brother John he described canvassing for subscriptions to a London lending library: 'I spoke also one night to Babbage about it: did you ever see him? A mixture of craven terror and venomous-looking vehemence; with no chin too: "cross between a frog and a viper," as somebody called him; forever loud on the "wrongs of literary men," tho' he has his £20,000 snug! I advise men to keep off that man'.[29] In a letter from the following year the theme repeats, and Babbage was lucky not to have provoked more thorough treatment in the historian's sensational (posthumously-published) *Reminiscences*. 'Babbage continues eminently unpleasant to me, with his frog mouth and viper eyes, with his hidebound wooden irony, and the acridest egoism looking thro' it'.[30]

Darwin had also found Babbage's pride and desire for honours ludicrous. When Herschel had returned from the Cape in 1838, Sir Roderick Murchison, keenly attuned to the effects of publicity, had rescinded an offer he had made to the engineer to head the coming British Association, pivoting to the astronomer instead. This had infuriated Babbage, always jealous of honours. Darwin, who had no desire for honours himself, wondered as the engineer resigned and refused to deal with the British Association again. 'I have been much amused with an account I have received of the wars of Don Roderick [Murchison] & Babbage—what a grievous pity it is that the latter should be so implacable, & if one might so call the calculating machine, so very silly'.[31]

Harriet Martineau recalled that Babbage had a reputation for paying unusual attention to slights. 'I was told repeatedly that he spent all his days in gloating and grumbling over what people said of him, having got it all down in a book, which he was perpetually poring over. People who so represented him had little idea what a domestic tenderness is in him,—though to me his singular face seemed to show it,—nor how much that was really interesting might be found in him by those who viewed him naturally and kindly'. She preferred to believe that he recorded positive as well as negative opinions, as a kind of experiment, and offered a more generous interpretation of his character. 'All were eager to go to his glorious soirées; and I always thought he appeared to great advantage as a host. His patience in explaining his machine in those days was really exemplary. I felt it so, the first time I saw the miracle, as it appeared to me; but I thought so much more, a year or two after, when a lady, to whom he had sacrificed some very precious time, on the supposition that she understood as much as she assumed to do, finished by saying "Now, Mr. Babbage, there is

---

27 Charles Darwin, *Autobiography*,Darwin 1887, 77.

28 John Morley made a similar remark in 1870: 'The new library edition of Mr. Carlyle's works may be taken for the final presentation of all that the author has to say to his contemporaries', 'The canon is definitely made up. The golden Gospel of Silence is effectively compressed in thirty fine volumes', Morley 1870, Morley 1871, 195.

29 Thomas Carlyle to John A. Carlyle, 1839-02-04, `https://carlyleletters.dukeupress.edu/volume/11/lt-18390204-TC-JAC-01`.

30 Thomas Carlyle to John A. Carlyle, 1840-11-24, `https://carlyleletters.dukeupress.edu/volume/12/lt-18401124-TC-JAC-01`.

31 Charles Darwin, Letter to Charles Lyell, 1838-09-14, Darwin Correspondence, DCP-LETT-428.

only one thing more that I want to know. If you put the question in wrong, will the answer come out right?" All time and attention devoted to lady examiners of his machine, from that time forward, I regarded as sacrifices of genuine good nature'.[32] Yet Babbage certainly did attach, contrary to his magnanimous friend, obsessive weight to supposed sleights—this emerged clearly in notes he kept of an interview in November 1842 with the prime minister Robert Peel (see below).[33]

Friends like Marie Lloyd also recalled his 'almost painfully sensitive feelings'. The famous soirées eventually came to an end, Babbage told her, when he was no longer able to recognize 'more than half of the guests in my room'.[34] He gave the impression in his autobiography that he published nothing at this time because, having resolved to give every waking hour to the new engine, he preferred to solve its technical complexities (the future) rather than write about it (merely the past). Here the inventor misled his readers. He had actually written at length about the new engine, but the manuscripts lay gathering dust among his burgeoning 'scribbling books', and can be read today. It was just that he preferred not to publish the new descriptions. He had the time and opportunity, since he published other material of a far slighter nature—including a curious booklet against Hume, arguing unconvincingly that the miracles the Scotsman scorned might have been programmed into the nature of things by a divine being, like a difference engine altering its rule using feedback.[35] The fact was that Babbage did want the invention publicized, but by others, to add lustre; ideally abroad, led by his fancy (as he had written to Bowditch) that he was more likely to be properly appreciated there. All that may be interpreted instead as a desire to spite the British government and the society he thought had let him down. In 1840 the opportunity presented itself.

[32] Martineau 1877, 201, 265-8.

[33] 1842-11-11, CB notes on meeting Peel, section I.100, p. 397.

[34] Lloyd 1880, 47-69.

[35] Babbage 1837.

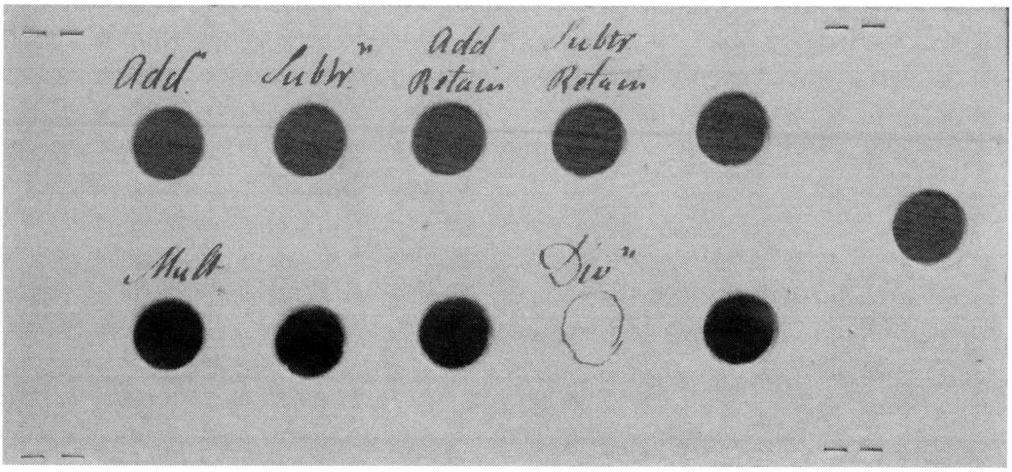

Figure 3.1: Analytical Engine operation card, Turin Archives.

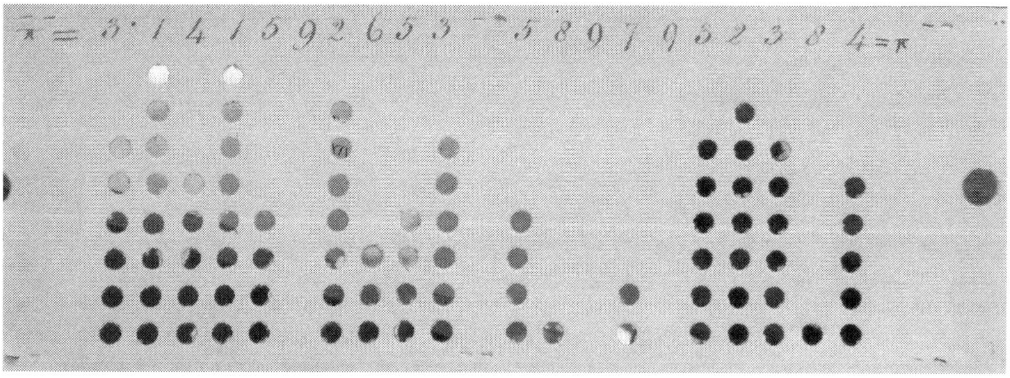

Figure 3.2: Analytical Engine card for loading the constant π, Turin Archives.

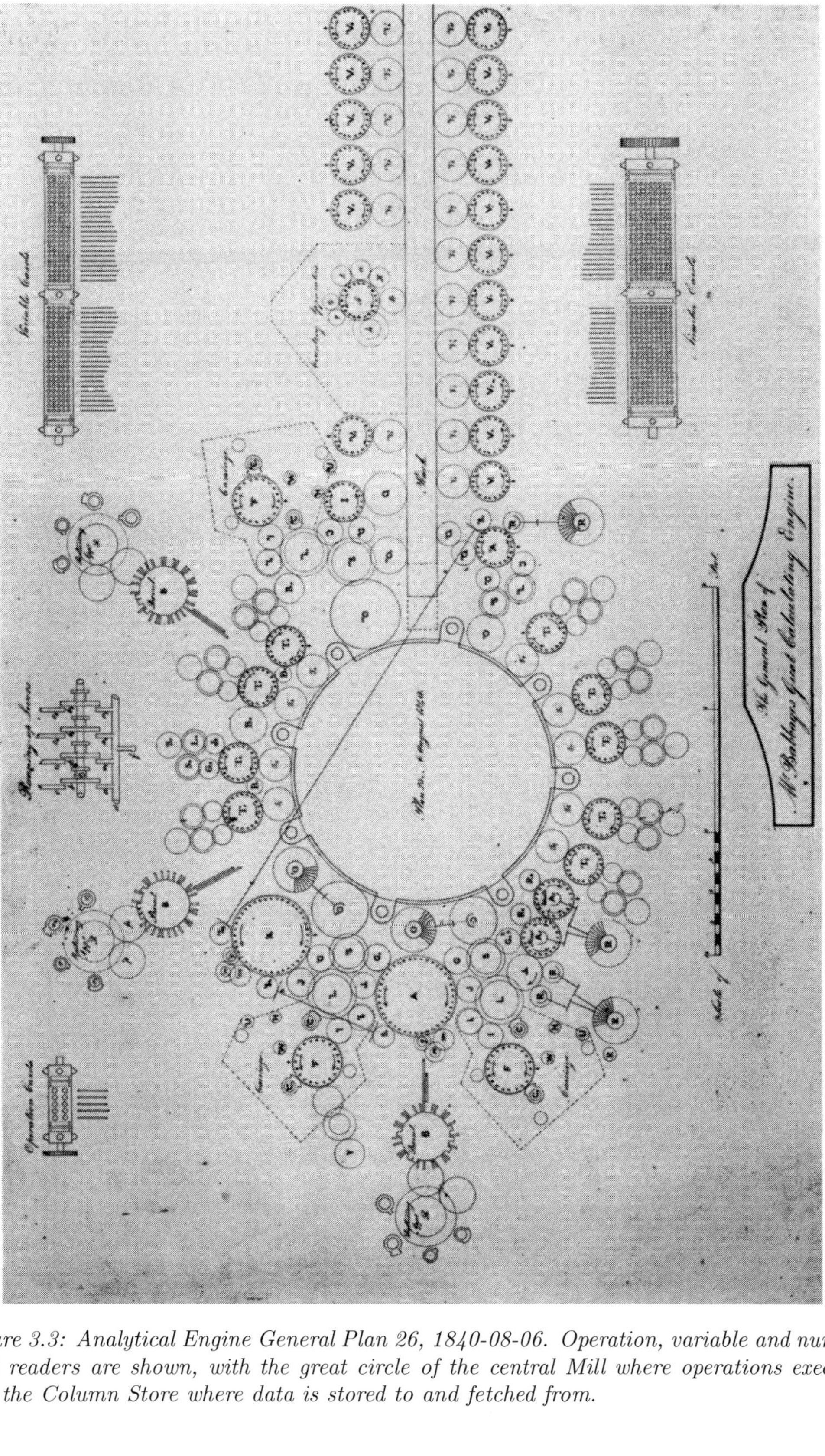

Figure 3.3: Analytical Engine General Plan 26, 1840-08-06. Operation, variable and number card readers are shown, with the great circle of the central Mill where operations execute, and the Column Store where data is stored to and fetched from.

Chapter 4

Ada and the Byron Mystique.

It is now difficult, perhaps impossible, to wholly comprehend the influence that Byron had on Western culture in the first part of the 19th century. As his only legitimate daughter, Ada carried a great deal of the poet's mystique, enhanced by his reputation as a sublimely talented roué. That mystique runs throughout Ada's association with Babbage and her subsequent reputation.

To take just one of countless examples of Byron mania, Charles Darwin's cousin Francis Galton, later a distinguished and very sober scientist in his own right, developed an 'unlimited admiration' for Byron and all his associations.[1] Both his uncles, Francis Darwin and Theodore Galton, had met the poet in Athens and Smyrna in 1809-10, before his full flowering. As a nine year-old Frank was cribbing *Childe Harold* to his sisters. In 1840, shortly to go up to Cambridge and mathematics, he had set off for Germany, ostensibly to study chemistry but really in the grip of wanderlust and mental fag. On the boat over he had seen Byron's widow, by then styling herself Lady Noel, and wrote home with some excitement. 'Please tell me in your Trieste letter who Lady Noel is as regards Lord Byron & by the peerage, what her age is & her sisters'.[2] En route in the Drachenfels and Baden Bergen he wrote home that 'the scenery is infinitely superior to any thing I had any conception of. Byron's description in the 3rd or 4th canto of the Childe Harold (I forget which) & in his letter to Countess Guicioli which you will find in his life is most exact. The mixture of scenery is quite curious, passing from the wild copse coloured or bare mountains to vineyards, corn fields &c.' Within a few days of arriving in Germany he met up with a local professor of literature, Adrian, who had translated Byron and now enthused the young man with the idea of travelling down the Danube to Constantinople and Greece, via a newly opened penultimate shortcut, to find Byron's scent there. It was a remarkable journey for an 18 year old.[3] His *Memories* show that the veneration persisted into old age. 'The mementoes of Lord Byron at Newstead Abbey were well cared for, and most touching to me, for I had in my youth an unlimited admiration of his works; so I drank greedily with my eyes all that I saw connected with him'.[4]

Those who claim Ada as 'the world's first programmer' derive considerable energy from the fact that she was Byron's daughter. They never fail to mention it, and how could they? Yet she never wrote any poetry, preferring to think of herself as 'mathematical' and 'musical' rather than 'literary', a habit she seems to have acquired from her impossible mother. Nor did she ever write about her father. Lady Byron (Anne Isabella Milbanke, familiarly 'Annabella') had separated from her husband in 1816, less than a year after they

---

1 Galton 1908, 49.

2 Tredoux 2023b. She was actually on her way to locate Byron's other (illegitimate) daughter Medora Leigh and (supposedly) to inform her that she was a product of incest between her father and his half-sister.

3 Tredoux 2023b, 429–438, Tredoux 2023c.

4 Galton 1908, 170.

were married, and the poet had stalked off to Europe when the break was formalized, never to return. He would see nothing of the daughter who had been born to him in December 1815, about a month before the separation.

Ada was raised by her mother. Byron had other interests entirely and did not seek the custody he may well have been entitled to by law. She was educated at home by a series of hapless tutors forced to follow Annabella's idiosyncratic theories about education reinforced by punishments and rewards. When her father died in Greece—romantically attracted to the war against the Turks without ever seeing any action, possibly bled to death by his own doctors while feverish—she was only going on nine years old.

Though much has been written about Ada's 'mathematical childhood', the records which have survived in the Lovelace papers, including the diaries of her tutors and other letters, do not bear that idea out. She certainly had well above average intelligence and from her early teens already styled herself as 'mathematical', but her progress was indifferent at best. She had seen the Difference Engine in 1833 with her mother, but Lady Byron's description to Dr. King suggests that her mother at least had misunderstood the effect of Babbage's difference-altering feedback mechanism. 'I am only certain that the numbers were no longer successive ones, and that their differences were neither in arithmetical nor Geometrical ratio, as far as I could apprehend. If this occult principle of change existed in the law according to which the machine was constructed, (for Babbage discovered it to be latent in the mathematical formula originally applied by him) it may be consistent with the general laws of our solar system that the sun shall not rise tomorrow'.[5] (It is not latent in the 'formula', it alters the 'formula' by arrangement of the engineer, which Babbage likened to a miracle altering the 'laws of nature'.)

A Cambridge man, the defrocked radical and family acquaintance Dr. William Frend (1757–1841), had also provided some ineffectual advice over the years. Now Ada agitated for more. Aspirations overwhelmed her. In 1834, at the age of eighteen, she sought advice from Dr King, the correspondent above, who was one of her mother's Brighton doctors. Though he too had once studied mathematics at Cambridge, King proved to be little more than a beginner himself.[6] Writing to him, Ada confessed that she needed to master 'elementary' mathematics in order to read Astronomy. She was inspired, she said, by a meeting with Babbage the previous year, when she had 'come out' in London society (though without landing a prize).

Help was soon present in the form of Mary Somerville, an unpretentious textbook writer noted for translating Laplace.[7] Somerville secured welcome invitations to meet Babbage more intimately (he had already invited Ada to his Saturday evening soirées, which she would often attend in future years). 'We are to see the drawings of his new machine and to have a few hours conversation with him'.[8] However this was mostly going through the motions, since she had none of the background required to grasp the principles of the machine—dependent as they are on power series and their use to approximate transcendental functions—or the mechanical niceties of the new design. And even though she took up Dion Lardner's mammoth article that year on its simpler precursor, the Difference Engine, it is unlikely that, without a knowledge of difference equations and the approximation of functions, she penetrated any further than the median reader of the *Edinburgh Review* (that is, not very far).[9]

A nervous breakdown, one of many mysterious illnesses, derailed Ada early in 1835.

[5] 1833-06-21, Lady Byron to Dr. William King (1786-1865), section I.42, p. 354.

[6] For King, a cooperative enthusiast, one might consult Corina 1994 but note that King's relationship with Ada is grossly misunderstood there.

[7] Laplace 1831.

[8] 1834-11-25, Somerville to Ada Byron, section I.48, p. 357.

[9] Lardner 1834.

There had been a more serious interruption a year or two prior, when she had inconclusively eloped with a tutor.[10] That July she married William, Lord King, after only a month-long engagement. She would rapidly give birth to a series of offspring—a son, in 1836, a daughter in 1837, another son in 1839—despite confessing that like her mother she did not care overly much for children. Surviving correspondence with Somerville shows her struggling through the basics of trigonometry, having tried her hand at Euclid's geometry in previous years. The addition formulae troubled her.[11] Abstract algebraic symbolism seems to have baffled her, a barrier that stubbornly persisted as she tried to proceed into more advanced material. Nonetheless, Ada always received every encouragement and accommodation from her husband, family and other scholars to pursue her persistent desire to become known as a mathematician and scientist. When Somerville left for Italy in 1840 her communication with Ada appears to have already petered out (as far as one can tell). This meant she was now on the hunt for a new adviser.

Counter-intuitively, the search for help was all the more pressing because Ada suspected that she had a real *genius* for mathematics, or so she wrote to Babbage in late 1839. 'Do not reckon me conceited, for I am sure I am the very last person to think over-highly of myself; but I believe I have the power of going just as far as I like in such pursuits, & where there is so very decided a taste, I should almost say a passion, as I have for them, I question if there is not always some portion of natural genius even'.[12] Though Babbage promised to search for a suitable candidate, he seems to have drawn a blank, since the guide who did turn up in the middle of the following year was Augustus De Morgan (1806–1871), a consequential figure in British mathematics who was based at University College London.[13]

De Morgan was a student and later a close friend of George Airy, anathema to Babbage, so it is more likely that he was introduced to Ada through his wife Sophia Frend, daughter of the ineffectual Dr. Frend above. He was a productive worker, having already written some textbooks on mathematics, and was issuing a new textbook on calculus in parts when he became involved with Ada sometime in mid-1840. At the same time he wrote a long series of encyclopedia articles (Sophia counted over 700) on a very broad range of scientific subjects, not limited to mathematics.[14] He was forced to supplement his income by tutoring, but no money was ever exchanged in his extensive relationship with Ada, even though it was often of the most exasperating kind. Since it provides some clues about her abilities, and has been the subject of some controversy, it is necessary to examine it in some detail.

That need is odd in itself. With genuinely accomplished mathematicians, it is not necessary to examine their schoolwork to determine their level of competence. They have left results, theorems, proofs—conceptual and notational innovations—for which they are known. Exceptions may be made for feats at exceptionally young ages. Yet through a combination of circumstances, historians have been placed in the position of having to do exactly that for Ada Lovelace: examine her elementary homework, even though all of what survives was done at a mature age.

In 1985, Dorothy Stein's biography of Ada contained some surprising information about the mathematical abilities of her subject.[15] Ada had a widespread reputation as a mathematician, even (some said) a mathematical *genius*. Her correspondence with Augustus de Morgan had flummoxed her previous biographer, Doris Langley-Moore. 'She wrote often, enquiring, speculating, arguing, filling pages with equations, problems, solutions, algebraic

[10] For these sorts of biographical details, which are of minor interest here and sometimes thinly-sourced, see Moore 1977 and more accurately Stein 1985.

[11] 1835-11-18, AL to Mary Somerville, section I.56, p. 366.

[12] 1839-11, Ada Lovelace to C. Babbage, section I.64, p. 372.

[13] Attar, Rice, and Stray 2024.

[14] A. De Morgan 1835, 1842b; S. E. De Morgan 1882.

[15] Stein 1985, 89-90.

formulae, like a magician's cabalistic symbols'.[16]

It was with some consternation that Stein found this material, on inspection, to be nothing more than calculus homework of the most elementary kind. Ada had been working through some introductory texts—principally *Elements of Algebra* (1835) and *Differential Calculus* (1842, issued previously in parts) — conscripting De Morgan, their author, as a guide. She issued a flood of exasperated queries in his direction: that she was not convinced by what she read, that her workings (included, and about which she *was* convinced) came out differently; that she would not accept a previous clarification; that this or that basic idea was illogical; or, much less often, that she was baffled and owned up to it. The reader expecting to find any mathematical discoveries or insight in all this will be disappointed.

De Morgan had to argue Ada through each step, pointing out her stubborn conceptual confusions, repeated missteps and algebraic errors, a process which often required multiple iterations for a single point. Aside from the occasional pheasant from the Lovelace estates, the poor man, by no means wealthy, seems to have received no payment at all for this barren task, which required no small effort and almost saintly patience. It is not easy to determine how long it all lasted, since the letters are often only partially dated, and sometimes misdated or devoid of any date at all, even a weekday. Later, guessed dates were annotated by other hands. The sequence is uncertain.

A combination of techniques must be used to assign the incomplete dates. Since the notes must have been exchanged between 1840 and 1843, those with month, numerical day, and weekday can be placed in a year using the calendar for 1840-3. Sometimes these *prima facie* combinations are unlikely or even impossible, as a result of transcription errors, or errors original to the correspondent. To supplement this inference, and sometimes supersede it, the mathematical content must be matched to form clusters in which one or more contiguous letter has an unambiguous date. Some uncertainty must be accepted.[17]

The errors perpetrated repeatedly by Ada in this exchange led Stein to conclude that her mathematical capabilities had been greatly overrated. Not because those errors exist but because they are so elementary and recur so often, placing her at every stage far behind where one would expect a person with genuine mathematical ability to be. Recall that she had received lower-level coaching from Mary Somerville and others. On examination, that work also turns out to be full of errors and confusions, far more than would be expected for someone mathematically talented. We will only concern ourselves further with the later work, since it contains more than enough evidence at a much higher level, when she was more mature. It is enough to note that the idea that Ada Lovelace was ever a 'mathematical child' has no basis and is disproved by the modest level at which the work commencing in 1840 was conducted, when she was well over 25 years old.

Some basis for evaluating this homework—running from mid-1840 to the close of 1841, say a year and a half—needs to be established first. Charles Spearman and Alfred Binet were the first to clearly understand and account for the fact that comparison of students within an age group is necessary to account for maturation.[18] This is complicated by the fact that Ada was already in her mid to late twenties when the correspondence was started, as opposed to her mid to late teens, which is when a modern student would start. The work must be compared to what one would expect of any mature student of median ability within a similar timeframe, starting from a similar point. That way we can try to continuously judge at each stage if she was above or below that level—bearing in mind that she has acquired the reputation of a *mathematical genius*, which would place her *vastly* above the median student. No great precision is possible, but useful conclusions can still be drawn.

One must bear in mind that Ada and her 'tutor' met in person regularly. She also

[16] Moore 1977, 99.

[17] See the calendar reconstructed in section G.1, p. 306.

[18] Binet and Simon 1916; Spearman 1904.

contacted De Morgan by letter with questions, peculiar ideas (usually mistaken), concerns and things she simply got stuck on. Those are the records which survive. It is necessary to pay close attention to the nature of the content, not to the mere fact that an interaction occurred, and to always bear in mind that mathematical skill and knowledge is coherent and not uncorrelated. We must ask: are these topics that we would reasonably expect a mature student to get stuck on? What percentage of students would have such problems, make similar mistakes, require extended help to deal with, and not be able to resolve on their own?

A detailed consideration of Ada's homework is presented in Appendix G, for the more mathematically inclined reader, along with a critical evaluation of some more recent attempts to rebut Stein's initial treatment and rehabilitate Ada through a kind of 'gilt by association'.[19] It goes far beyond Stein and confirms her preliminary conclusions. *What it all strongly suggests is well-below-average ability for a student of mathematics at every stage of Ada's development, over the course of a year and a half.* The particular examples given there may be replaced or supplemented with equivalent ones from the letters, the details are not at issue here: almost none of them involve advanced concepts. As stated before, similar observations can be made of her earlier more elementary work with Mary Somerville, thus to all of her mathematical development prior to 1843 that we have evidence for.

Ada was not in the least discouraged by her ongoing tangles with mathematics, which she thought showed herself flourishing. Writing to Babbage in early 1841, she generously offered her services to him. 'I am very anxious to talk to you. I will give you a hint on what. It strikes me that at some future time (it might be even within 3 or 4 years, or it might be many years hence), my head may be made by you subservient to some of your purposes & plans. If so, if ever I could be worthy or capable of being used by you, my head will be yours'.[20] A month later she pressed him again, 'I think much of the possible (I believe I may say the probable) future commission between us; and it is an anticipation I increasingly like to dwell on. I think great good may be the result to both of us; and I suspect that the idea, (which by the bye is one that I believe I have long entertained, in a vague & crude form), was one of those happy instincts'.[21] Letters survive between them until August of 1842, after which something appears to have happened before they pick up again in February of the next year. Without that new development, Ada would be completely forgotten today, receding into the general background of London society, outshone by the multitude of brilliant figures who had streamed through the Saturday evening soirées at Dorset Street.

---

[19] Consult Appendix G, p. 297.
[20] 1841-01-12, Ada Lovelace to C. Babbage, section I.77, p. 381.
[21] 1841-02-22, Ada Lovelace to C. Babbage, section I.78, p. 381.

Chapter 5

Turin and Menabrea's Sketch.

On the 10th of September, 1840, Charles Babbage stepped off the coach from Lyon to Turin, then part of the Kingdom of Sardinia. Not fluent in Italian, he was accompanied by an interpreter and general factotum that he had engaged in London—or so he thought—a talented Italian exile called Fortunato Prandi. Officially the engineer was there to present his Analytical Engine concept to the 2nd Scientific Congress, which would run from the 16th to the 30th later that month. After lodging overnight at the Swiss Pension, the companions took two furnished second floor rooms in 22 Po Street, the old 'Hospital of the Poor' building. They were closely watched by the Turin Police, who had been waiting for them at the coach stop and would keep minutely detailed records of all their movements and contacts in the coming weeks. Prandi was well-known to the Police. However, nothing here is quite what it seems.

It was as an engineer and the author of a treatise *On the Economy of Machinery and Manufactures* (1832)—a subject on which he had developed exceptionally comprehensive knowledge—that Babbage had first made his name in Italy, rather than as an inventor of calculators. The continent was eager to emulate the British industrial success that Babbage briefly came to represent. A French translation of the book had been quickly followed by a German edition. An Italian edition came out in 1834.[1]

The invitation to the Scientific Congress had come from Giovanni Plana (1781–1864), or at least that's how Babbage remembered it. Plana was an astronomer and a disciple of Lagrange—he would marry Lagrange's niece. He had when he was not fraternizing with Henri Beyle, the future 'Stendhal'—won the Copley Medal of the Royal Society of 1834 for his *Theorie du Mouvement de la Lune* (1832). This was a coveted honour, open to international competition, the highest the society could award. Past winners included Michael Faraday, G.B Airy and JFW Herschel. Future winners would include CF Gauss, Charles Darwin, Karl Weierstrass, TH Huxley and Francis Galton; a roll call of 19th century science. Major international figures like Amedeo Avogadro (1776–1856) would also be present at the conference. But as we will see, Babbage would only attend the meeting ambiguously, at one remove.

Plana, something of an Anglophile, was well-known to Babbage, and had been in touch with him as early as 1834. His colleague at Turin, Giuseppe Francesco Baruffi (1801–1875) bore a letter of introduction from the astronomer when he visited London that October. Babbage was mourning the shocking loss of his daughter Georgiana[2] and at first refused to see any callers at all. But on reading Plana's letter he rushed a reply to Baruffi at his hotel in Leicester Square, 'among the many things he told me, I noted with patriotic pride, his desire to go to Turin to consult in person our supreme geometer about the possible uses

[1] Babbage 1832, 1833a,b.

[2] She died at age 17 on Sep. 26th at Boughton House in Worcestershire.

and applications of his machine to the progress of mathematical sciences. After various reciprocal visits, he kindly gave me the gift of some papers and drawings concerning said machine'.[3]

The Englishman missed meeting Plana when the Italian was prevented by family matters from traveling to England the next year. But he had followed up with a comprehensive letter to the astronomer to 'explain to you the nature of the calculating machine, of which you have shown me so much desire to be informed'. In 1831, Plana had himself devised a perpetual calendar, still operating today at the Chapel of Merchants and Bankers in Turin. Now he learned that using the new Engine 'by simply turning the handle (without making the least effort of mind) it will give the numerical value of all the algebraic functions'. Babbage had not yet introduced programming through cards (he did not describe the drum mechanism he was then using to choreograph operations), but already saw expansive possibilities. 'It will calculate logarithms, or their differences of any order whatsoever, without passing through any of the antecedents. It will calculate the values of π, of e and of Bernoulli numbers, up to 30 places, and perhaps up to 50, by giving it more time. It will calculate according to a given date, tables of annuities (life annuities) on the life or any sort of combination between the life of one, two or three individuals of any age whatsoever'.

However, the engineer was already gloomy about the reception of his advances. 'I have spent a great deal of money in my researches, and have not yet received any encouragement from my country. I am not in a position to have a machine made according to the designs, and I do not care to do so; for when the advantage of such a machine shall be appreciated, whether in my lifetime or not, recourse will always be had to my designs. The subject is indeed too far beyond the understanding of the persons who first occupied themselves with it; and the men at the head of our affairs, who would know its value, are too busy keeping their places, to pay the least attention to subjects of high science'.[4]

Plana enthused over what he had read. 'How can one remain indifferent to the sight of a machine capable of calculating the ratio of circumference to diameter, the base of Napier logarithms, and Bernoulli numbers, with 30 figures? Doubtless you can also calculate the tables on elliptic transcendents that you see in Legendre's work, and the numerical coefficients that one obtains by the return of a given sequence. Now all this is astonishing, for anyone who is aware of the strength of the legislative power and the weakness of the executive power of algebraic analysis'.[5] Those 'executive' and 'legislative' powers would often be invoked by the Englishman in the future.

From a follow-up letter from Babbage that year it appears that Plana may have promptly written a description of the new engine, which can only have been based on the material sent to him through Baruffi. 'I have distributed almost all the copies you sent of your valuable paper and together with my own thanks I am commissioned by many of your friends in England to return their best acknowledgements. You seem to have seized on the mathematical principles which the Calculating Engine embraces most completely. It will as you observe revert any given series and give the numerical coefficients and it will also prepare the Numbers and differences of any particular function at such intervals as will enable the former engine to fill up the whole table by ordinary differences. It is impossible by description to convey to you any idea of the mechanical contrivances but I hope you will yourself be able to examine them next summer and I assure you nothing will give me more delight than to give you every opportunity of judging of it for yourself and of discussing with you the varied uses to which it can be applied'.[6] If there really was such a paper about the Analytical Engine it would be among the first published descriptions of it, but no

[3] Baruffi 1841, Vol. 1, 279-81. See also Maquet 1965, 161-2.
[4] Babbage to Plana, 1835, see section I.54, p. 363.
[5] 1835-10-03. Plana to Babbage, section I.55, p. 366.
[6] Babbage to Plana, 1835-11-30, see section I.57, p. 367.

such article by Plana could be located. It is possible that the paper first referred to in the letter is not connected with the follow-on sentence, which may refer quite separately and confusingly to Plana's previous *letters* to Babbage, a clumsy juxtaposition.

In late 1838 Plana was in touch again. The King of Sardinia had asked him through an intermediary to help construct a new warship, the *Tripoli*, 'a steamboat of the strength of 160 or 180 horsepower, capable of transporting a battalion of 400 and even 500 men across the Mediterranean with the usual rapidity inherent in these machines'. The few steamships the Kingdom possessed had crippling problems with limestone deposits in their equipment. Plana sought Babbage's advice about technical details and the best choice of English firms to help with the project. 'Render me this important service, my dear Mr. Babbage, and my gratitude will be all the greater because, through this, I would have the satisfaction of having contributed, as far as it depended on me, to supporting a grand vision, worthy of the Sovereign who ordered me to study it, and in keeping with the progressive march of current civilization, which harnesses at will the expansive force of steam, to multiply its astonishing influence and make it serve to realize conceptions inaccessible to ancient modes of navigation'.[7]

The engineer was naturally happy to help 'realize conceptions inaccessible', recommending the leading English firm of marine engineers Maudsley, Sons & Field for construction and fitting of the engines, and the purchase of precision navigation instruments. Babbage took great 'pains' as he later put it, to make sure the right specifications were used. The collaboration was still in progress when he stepped off the coach in Turin that September of 1840. By then the *Tripoli*, built locally from premium Sardinian lumber, had already launched from the Genoa shipyard (on May 25th) and sailed to England to be fitted with its advanced 180 horsepower steam engines (on July 14th).[8] Plana's invitation to the engineer to attend the Congress was surely motivated by interest in more engines than one. The involvement of Fortunato Prandi also appears to have been cleverly arranged.

Prandi (1799–1868), a former Sardinian Guard turned postman, had been exiled from the Kingdom ever since the farcical 1821 'Piedmont Insurrection', in which he had played a desultory part. Sentenced to death in absentia, hanged in effigy, he had fled first to France, then to Spain, then back to France and finally to England. In London he had established himself among a network of Italian radicals and exiles, earning a precarious living as a translator. Since Babbage liked to mingle with 'advanced' circles, he had probably already met Prandi in passing before the exile asked him to show the Difference Engine to some friends of his, from the '*entourage* of the duke of Lucca', in early March of 1840. Soon Babbage introduced the exile to Ada Lovelace, who wrote a few days later to ask for his address.[9] A friendship quickly developed. Prandi let Babbage have a copy of his recent translation of Andryane's *Mémoires d'un Prisonnier d'Etat*.[10] When the invitation to Turin was finalized, Babbage appointed Prandi, by all accounts an accomplished linguist, to be his interpreter.

The fact that Prandi had been sentenced to death in Sardinia might have been problematic. However, the King of Sardinia had already granted the exile permission to return for three months and commuted his sentence. The Turin police, who kept their own counsel, believed that Prandi was by then *in the service of Carlo Alberto*, and that the King had awarded him a medal a few years previously for services rendered (none of which meant that the police trusted Prandi one bit).[11] Having someone to keep tabs on the radical Italian exiles of London had obvious uses, but Carlo Alberto was also thinking of the Austrians.

[7] Plana to CB, 1838-12-08, cited in Bulferetti 1970.
[8] Bulferetti 1970.
[9] AL to CB, 1840-03-14, section I.68, p. 374.
[10] Alexander Andryane 1838; Alexandre Andryane 1837; Bulferetti 1968b.
[11] See section E.4, p. 176.

The King could see his way to using controlled radicals against a common enemy. It is no coincidence that Andryane's prison *Memoir*, which Prandi had translated to English, was anti-Austrian.

Nor could it be a mere coincidence that an agent already in the service of Carlo Alberto just happened to accompany Babbage, whose proven technical expertise was of ongoing service to the Sardinian state. Perhaps it was really the cicerone Prandi who had 'engaged' Babbage. It is likely that the invitation to the 1840 Congress was not uninformed by all this. Carlo Alberto had been warned by his advisers not to approve the congress of scientists, widely believed to be a troublesome coven of radicals, but the King had brushed those objections aside. Babbage himself may not have been fully aware of the intentions of his hosts, but he took care not to mention any of the work on the *Tripoli* warship in his memoirs.[12] There is every indication that Babbage expected the Italians to take his calculating machine ideas seriously, but he was inevitably disappointed. The Police took at least a cursory interest, procuring a diagram printed in Paris.[13]

After setting up in their new quarters in Po Street, Babbage and Prandi circulated widely in the town, dining out enthusiastically, joined at times by Prandi's priest brother, frequenting the booksellers and calling on the professionals, the Ministry of War, the minor nobility of the area, and other relatives of Prandi, many of whom seem to have had politics sympathetic to the exile's tempered form of radicalism. All these movements were carefully recorded by the police, who had agents gamely trying, but not always succeeding, to keep up with the energetic movements of the pair. Their reports suggest, perhaps not reliably, that neither of the pair directly attended the congress, held later that month—certainly, the published proceedings only seem to contain references to Babbage by others, though he may have attended the socials. Instead, the Englishman held some private sessions in his rooms, to which a select group of scientists was invited.

The Police were focused on Prandi, recording that 'on the morning of the 18th around 10 o'clock, he went to Mr. Giuseppe Pomba's printing shop from where he exported various books, returned home and received a visit around two in the afternoon from 14 or 15 scientists, including Professor S. Martin Director of the Royal Botanical Establishment of S. Salvario who came expressly with others to view the Englishman Babbage's machine'. (A misapprehension: there were only models and plans on display.) Other attendees included Plana and his former student Luigi Federico Menabrea (1809–96), the radical physicist Ottaviano Fabrizio Mossotti (1791–1863), and the astronomer Emile Plantamour (1815–82). By then a friend of Babbage, James MacCullagh (1809–47), a professor of mathematics from Dublin and an enthusiastic supporter of his ideas, had interrupted a climbing trip to the Tyrol (as Babbage recalled) to join the party (he was listed in the conference program as an attendee).[14] It is probable that the geologist Angelo Sismonda (1807–1878) also attended.

Babbage had brought boxes of material, arranged for him by Prandi, to illustrate the design of his new engine, including operation, variable and number punch cards, a wooden model of the addition component (possibly with the sophisticated 'anticipatory carry' mechanism), detailed technical drawings, several worked examples, and some handwritten summaries. The latter include: 'Notes Respecting the Analytical Engine', which summarizes the history of his research to that date; 'Calculating Engine', which describes some of its capabilities at a high level; and an untitled document describing its individual components (this material survives today in the archives of the Turin Academy of Sciences, where Plana deposited it after Babbage departed).[15] Babbage later recalled that these sessions went on

---

[12] Babbage 1864.

[13] Bulferetti 1968b.

[14] Babbage 1864, 130. MacCullagh would commit suicide in October 1847, after a period of mental instability.

[15] See Appendix E, p. 169, section E.1, p. 171, section E.2, p. 172, section E.3, p. 174.

for 'several mornings'. 'Around the room were hung the formula, the drawings, notations, and other illustrations which I had brought with me. I began on the first day to give a short outline of the idea. My friends asked from time to time further explanations of parts I had not made sufficiently clear'.

Of all the attendees at the Po Street rooms, it was the engineer and former student of Plana, Menabrea—by then resident member of the Class of Physical and Mathematical Sciences of the Academy of Turin—who would prove most significant. In his (posthumously published) memoirs the Italian left some little-noticed remarks about these sessions.[16]

> I became particularly attached to Charles Babbage, who had come to Italy to talk about his analytical engine with some mathematicians who had the patience to listen to him: in fact he expressed himself with difficulty and he himself admitted that he had had to use his mind so hard to solve the serious problem he had posed, and that he had tired his brain so much, to the point of not being able to clearly explain his system in a way that would make it understandable to the public. He had in fact taken on a problem that was difficult to solve, because it was no longer a question of an arithmetic machine to do the usual calculations; he himself had already built one and with it he had calculated the logarithmic tables that were later published.[17] Now he proposed instead to do analytical or algebraic calculations and, given an analytical formula, to transform it, develop it and then calculate it by introducing numbers to obtain the numerical results. The engine had to do all these operations taking into account all the elements, especially the signs,[18] which was particularly difficult. ... For better or worse, Babbage explained to me his machine, partly already constructed, which had required him to use all the resources of kinematics. I thought I had understood the basic idea, which was based on the use of Jacquard cards which, in the manufacture of brocades, regulate the different threads that must form those designs. Babbage used similar cards to perform the various operations indicated by the formula.

Menabrea was more prepared than most to appreciate the labour savings that such a device could bring. One of his tasks as a student of Plana had been to check the conclusions of his mentor's treatise on irregularities of lunar orbits, *Théorie du mouvement de la lune* (1832).[19] 'This gigantic undertaking required immense calculations, especially numerical computations. Since I had shown a certain aptitude for this activity, he asked me to review the calculations as the proofs of his book came out of the printers. This work lasted almost two years and, at the beginning, I found it so hard that at certain moments I despaired of being able to get to the bottom of it'.

Aside from Plana, to whom we will return presently, the other scientists present at these sessions left little evidence of their considered thoughts. Babbage spent the rest of his time in Turin in the company of Prandi, leading the police tails, over the course of days, *inter alia* from Count Balbo's country house, to a certain 'Casa Chanaz', which Prandi did not enter, on to Gianolio, a merchant in Rivoli, then back to Count Balbo. The Englishman was able to obtain, doubtless through his cicerone, more than one audience with various government officials and even Carlo Alberto himself. If he had expected the King or his ministers to quiz him about calculating engines, or even to spark some interest in financing its construction, he was disappointed. Most likely it was steam power and his new battleship that enthused the monarch the most, but Babbage kept quiet about those conversations, recalling only casual talk about his friend Charles Wheatstone's electric telegraph and its usefulness for advance warning of developing storms.

The police maintained their close surveillance of Prandi, who stayed on for a while after

[16] Menabrea 1971, 36-7.

[17] The machine was *not* used when these tables were published in Babbage 1827b and again in Babbage 1831, as it was not yet assembled.

[18] Plus (positive) or minus (negative).

[19] Plana 1832.

Babbage left on October 8th, taking the coach to Genoa.[20] They noticed that the exile changed his clothes several times a day and suspected an attempt to slip his tails. They also wondered where an underemployed exile got the money to eat out so frequently and generally live so well. All his contacts seemed to be of a similar, dangerous, political complexion. It was not apparent to them that Prandi, tired of hardscrabble life and (perhaps) minor espionage, was mostly in business for himself now. He was in the process of parleying his English engineering contacts into commercial advantage, and Babbage would play a crucial role there by introducing him to Isambard Kingdom Brunel, the leading civil engineer. Carlo Alberto instructed the police to let the exile travel unhindered. The now-prospering Prandi would be back in England by the end of the year, leaving Turin on October 31, but would return to Turin for good in 1842, with the permission of Carlo Alberto. With Philip Taylor (1786-1870) as a partner, he would found the railway engineering firm Taylor and Prandi (later sold to Giovanni Ansaldo) in Sampierdarena.[21]

Babbage may well have expected the Sardinians to reciprocate his help with their steam engine by showing more interest in his Analytical Engine. Instead they offered him an honour from the King, apparently through Prandi's suggestions to Angelo Sismonda and Cesare Saluzo, who had presided over the recent conference. Writing to Sismonda, who had chaired the Geological Section, Babbage feigned surprise, 'such a favor becomes the more acceptable from its being entirely unsought and unexpected on my part. The pause I took to arrange the contract and to see its faithful execution in the construction of the steam-engines for one of your vessels of war were justly due to a Sovereign in alliance with my own—to a King whose appreciation of science had already excited my admiration long before I was honored by those delicate and refined attentions which his Majesty condescended to bestow on me during my short residence in his capital'. He only did so because the honour would please his aged mother (this is hardly convincing) and raise the social status of science (which may well be true).

Unfortunately, as a result of some court intrigue which painted the engineer as a less worthy figure, the offered knight's cross, once bit into, turned out to be too middling to accept. A draft survives of a reply to an unknown recipient, in which the engineer regretted that he had hoped for an honour which 'would have undoubtedly made much effect in my country, and it is even probable that it would have opened the path to the construction of this machine which, as Plana wrote, encompasses all the executive of analysis ... It would have also helped to elevate science in England; and I hope to be able to add that H.M. himself [Carlo Alberto] would have in some way added to his glory, being the first to appreciate and worthily reward a discovery that is destined to change the aspect of modern ~~society~~ analysis. But it was perhaps much more than Sardinian forms and customs allowed to do for a simple private individual like me'.[22] Taking this hint, an upgraded honour would soon be bestowed, which Babbage could accept and proudly display on the title page of his autobiography: Commander of the Order St. Maurice and St. Lazare.[23]

What was most on Babbage's mind now was persuading Plana to assist his cause, but the Italian's health was reportedly failing. The Englishman hoped he could be assisted, and so wrote to Sismonda. 'The great object of my visit to Turin was to convey to Plana and to some of the analysts of Italy the principles on which I had contrived an engine to perform as he has beautifully expressed it "The whole Executive of Analysis". The discovery is so much in advance of my own country and I fear even of the age, that it is very important for

20 Hyman 1982 has Babbage leaving in early November but this does not match the police surveillance report, though that may be incomplete, see section E.4, p. 176.

21 It was Taylor's naturalist-publisher brother Richard (1781-1858) who would publish the English translation of Menabrea 1842, in his journal for circulating foreign research, *Scientific Memoirs selected from the Transactions of foreign Academies of Science.*

22 CD to unknown, draft circa January 1841.

23 CB to Sismonda, 1840-12-01, 1841-03-24, section I.73, p. 378, section I.81, p. 383.

its success that the fact should not rest upon my own unsupported authority. I therefore selected the meeting at Turin as the time of publication partly from the celebrity of its Academy and partly from my high estimation of Plana, and I had hoped that a report on the principles on which the Calculating Engine is founded could have been already made to the Royal Academy. I am aware of the difficulty of the subject and of Plana's ill health; but you have amongst you several others who favored me with their attention who are quite able and I hope not unwilling to assist him. If you could in any way accelerate this report it would be of great importance. I have already troubled you too much with my affairs but I must rely once more on your kindness. ... PS. I have just received a letter from Plana but I fear no Report has been made to the Academy'.[24]

Plana pled pressure of work, and the need to concentrate, 'the immensity of science and the short duration of an individual's life increase the difficulties in properly directing the use of our intellectual faculties. I must therefore limit myself to special research suggested by my previous studies: and from then on I must ignore a multitude of other truths contained within an immense sphere of which I can barely see the surface.' He had, he said, deposited in the archives of the Academy, 'the model of your calculating machine with the designs and papers that you entrusted to me for such a destination. Since then, nothing has taken place that relates to this communication. You alone have an intimate knowledge of this invention and its details. Everyone else recoils before the difficulties of execution'.

The astronomer included a barbed closing remark. 'Let us hope that a fortunate combination will better enable you to overcome all obstacles. Furthermore, I hope that complete justice will be done to you, and that you will not have to declare acts of ingratitude comparable to what I read on page 188 of your work *The Ninth Bridgewater Treatise*'.[25] Perhaps Plana meant the following statement by Babbage: 'Here I should willingly have left the subject; but the public having erroneously imagined; that the sums of money paid to the workmen for the construction of the engine, were the remuneration of my own services, for inventing and directing its progress; and a Committee of the House of Commons having incidentally led the public to believe that a sum of money was voted to me for that purpose, I think it right to give to that report the most direct and unequivocal contradiction'.[26]

Some degree of bewilderment on the part of the Italians is understandable here. The best-placed person to write a memoir on the Analytical Engine was Babbage himself. He invented it. Only he really knew how it would work. He was an internationally-known scientist, from a country then at the forefront of the industrial revolution. It was no accident that Plana and his colleagues were definite Anglophiles, envious of British success (all their writing shows this). Yet here was Babbage asking *them* to write about *his* invention?

It was not pressure of work. Babbage had written about the engine several times both privately and semi-publically: in lengthy letters of 1835 to Quetelet, Plana, and Bowditch; in an 1837 manuscript 'On the Mathematical Powers of the Calculating Engine'; in the three manuscripts he brought with him to Turin.[27] Of these only the letter to Quetelet was published, contrary to his own intentions (some doubt must exist about that). There was also a passing description published in his *Ninth Bridgewater Treatise* (1837), and a year later several copies of plans printed from wood-engravings were given to colleagues at the British Association meeting at Newcastle. A grandson of Benjamin Franklin, Alexander Dallas Bache (1806–1867) had attended the meeting and wrote to Babbage early in 1840 to ask what happened. 'Why have you not published on the subject of the new Calculating Machine? Henry and I have been looking with impatience for the appearance of the sketch

---

[24] CB to Sismonda, 1841-03-24, section I.81, p. 383.
[25] Plana to CB, 1841-03, cited in Bulferetti 1970.
[26] Babbage 1837, 171-2.
[27] Reproduced in section I.49, p. 357; section I.51, p. 359; section I.54, p. 363; Babbage 1973; section E.1, p. 171, section E.2, p. 172, section E.3, p. 174.

which was begun and partly in shape when you were at Newcastle'.[28] Later that year Babbage had one of the general plans of the engine lithographed at Paris, and took copies to Turin.[29]

If the invention was too far in advance of the country driving technical innovation internationally, what hope was there for the rest of the world? What reason to suppose that those following in its wake would succeed where Britain had failed? Babbage certainly had some arguments in his broadside *Reflections on the Decline of Science in England, and some of its Causes* (1830),[30] but in retrospect these appear peculiarly blind to contemporary cross-national comparisons, as if he was wearing spectacles which only became rose-coloured abroad. But then again he wrote to Plana, in October of 1841, that if 'you had made a report on the subject to the Academy of Turin during last year it might have been of essential service to me in the discussion of the question with my own Government'.[31] So was he seeking government support for the Analytical Engine or not? Was he proudly hoping they would come to him as supplicants, cup in hand? If international support would bolster his case, or remove any doubt that this was the eccentric fancy of one individual, that could be achieved by *co-authoring* a full description of it. Why did Babbage want to retreat so completely into the background?

Around the middle of 1841, Babbage turned to his old friend Alexander von Humboldt to see if he could rouse some interest in other parts of the continent. His noticeably desperate letter contains references to the computation of the Bernoulli numbers, using a recurrence relation, which would prove significant later. 'I am very desirous to make you acquainted with the object that I have proposed to accomplish and to convince you that the means exist of constructing an [analytical] engine ... The greater part of the drawings exist by which an engine might be made which shall reduce to numbers any explicit function whose laws of formation can be assigned. For instance, the equation of finite differences expressing the law of formation of the numbers of Bernoulli being given to it, it will work out and print the successive numbers. This engine is unfortunately far too much in advance of my own country to meet with the least support. I have at an expense of many thousand pounds caused the drawings to be executed, and I have carried on experiments for its perfection. Unless however some country more enlightened than my own should take up the subject, there is no chance of that machine ever being executed during my own life, and I am even doubtful how to dispose of these drawings after its termination'.[32]

A similar sentiment was present as early as 1835 in his letter to Bowditch, where the engineer had anticipated 'when some more enlightenend country than my own shall at some future period become sensible of the value of the engine'.[33] Yet there is no evidence that Babbage ever directly sought any funding for, or support of, the Analytical Engine within his own country. He certainly did not do so *explicitly* from the British government itself, with which he was still in a state of limbo over its precursor, the Difference Engine. Later, when he was directly asked if he wanted them to fund the successor rather than the predecessor, he made a great point of leaving the decision up to them without ever arguing the merits, perhaps to escape any responsibility for the eventual cancellation of the first engine.[34] In all these dramatic complaints—denials, resignations and disavowals—there are strong signs of spite.

---

28 1840-03-25, Alexander Dallas Bache to C. Babbage, British Library Add Ms 37191 f354. Based on Roberts 2014a. Joseph Henry (1797–1878), later at the Smithsonian, had toured Europe in 1837, receiving Babbage and Wheatstone in his London rooms.

29 Babbage 1837,Weld 1848, 386.

30 Babbage 1830.

31 1841-10-03, C. Babbage to Plana, section I.87, p. 387.

32 CB to von Humboldt, circa 1841-07, section I.83, p. 384.

33 See section I.51, p. 359.

34 D. D. Swade 2000.

As it happened, Babbage's appeal to Sismonda for some other scientist to take up his cause, in lieu of Plana, was ultimately answered by the astronomer's former student Luigi F. Menabrea, who had paid close attention to the Po Street sessions in Turin. Mosotti seems to have been involved too, as Babbage wrote to Antinori from London 'I have found my presence here very necessary for the progress of the Engine and am now working hard at the descriptions which Mossotti is to translate'.[35] His role may only have involved translating material Babbage was supplying to Menabrea, but it demonstrates how closely the inventor involved himself with the writeup.

A member of the Savoyard and Aosta Valley nobility, Menabrea had pursued his mathematical studies against the wishes of his family, who did not approve of a career so close to industry (his elder brother Leon, 1804–1857, was an historian and diplomat, while his father Octave, ?–1842, was a lawyer). He had entered the University of Turin in 1828 anyway, to study under Plana. When the university was closed for a year in 1830-1, due to student riots—what has changed?—he secretly continued to work under his mentor. On graduation he proceeded immediately to the Royal Engineers, as a second lieutenant, which he recalled to be an 'unusual appointment' resulting from the reforms of King Carlo Alberto.[36] In 1839 he was promoted to Captain in General Staff of the Military Engineers. Plana inducted him into the Turin Academy of Science, in the division of Physical and Mathematical Sciences, in the same year. He was not of Plana's scientific stature, but he was certainly no lightweight either, publishing an influential paper two years later in the *Philosophical Magazine* of London, 'On Cavendish's Experiment' (for determining the mean density of the earth).[37]

Menabrea took up the Analytical Engine in earnest early in 1841, producing a 'sketch' by May 22.[38] In September he regretted that a letter to the inventor containing 'my musings on your astonishing machine' had been delayed—it seems that a foot-dragging returning British diplomat had promised to convey it by hand. But reading the reply from Babbage he was 'happy to see, sir, that I have managed to grasp the principle of your admirable invention. Certainly, I am far from claiming to have understood all its secrets, and since I read your letter, it appears even more significant to me than before. Indeed, you have mechanically realized the conception that could pass as one of the most metaphysical of all analysis, I mean the idea that one must be content with the passage of one quantity by zero to infinity. This is due to the simplicity and exactitude of the principle on which your machine is founded and which, as you think, seems to be the most general of mathematical language'.

The Italian hoped to meet Babbage in order to clear up any misconceptions about the operation of the device, 'I would still essentially need to confer with you for the drafting of the article that I wish to dedicate to you. What I had written in my letter was only a first sketch that I proposed to modify according to your observations'. If a meeting in person was not possible, 'I would ask you to kindly write to me with some detail concerning the points of view under which you wish that your invention should be presented and the remarks that you consider most essential. On that information, I will draft the article in question and before delivering it for publication I will have it reach you in England through the Embassy'.[39]

Since Plana had already directly refused to write about the engine, Fortunato Prandi, who was back in Turin by January of 1842, could only add some paranoid notes while

[35] C. Babbage to Chevalier Antinori, British Library Add Ms 37191 f489, 15 November 184[0], Roberts 2014a. However the date here is not certain.

[36] Menabrea 1971.

[37] Menabrea 1841.

[38] 1841-08-26, F. Prandi to C. Babbage, section I.84, p. 385.

[39] Menabrea to CB, 1841-09-18, section I.86, p. 386.

pretending to have privileged information about the intentions of the astronomer. 'Plana will not write anything about the engine. He seems to think that you delude yourself, that the engine, if ever executed, will be a great curiosity, but perfectly useless. He expressed great friendship and regard for you, but this is in substance what he told me concerning the engine. I suspect that the report made of it by Menabrea has disgusted him. He says he will write to you if you write to him. Pray say nothing of this. I tell you all I hear without restraint, but you must take care that you do not compromise me'.[40]

Babbage was already aware that Prandi's suspicions about Menabrea's findings were nonsense. The engineers continued to exchange communications as they converged to a shared understanding and a first draft of a report. In a letter just a few days after Prandi's presumptive confidences, Menabrea mentions that he had been delayed by the illness and subsequent death of his father. More importantly, 'I have given much thought to the analytical machine and to the clarifications you were kind enough to give me regarding it, which allowed me to expand considerably the field of ideas on this matter. I therefore believe I must entirely rewrite the article that I had written and of which I had already given you a communication. This is the main reason why it has not yet been published'. Some examples were still in progress. 'I am studying a question that consists of explaining in a clear manner, for less informed minds, the solution to numerical equations. I believe this is an important point to persuade of the utility of this invention and to demonstrate the rationality of the principle upon which it is founded'.[41]

MacCullagh was also aware of Menabrea's work-in-progress, and increasingly impatient at Babbage's failure to publish. The Dublin mathematician wrote a month later to thank Babbage for his help in winning special election—thus bypassing the decade-long queue—to the prestigious Athenaeum Club in Pall Mall, which boasted many mountaineers among its literati and scientists. Anticipating more frequent visits to London, he added 'when shall we have something *in print* about the Machine? For you know, *that* has always been the burden of my song. Has Menabrea published anything yet? I live quite out of the world here, and never hear a word of news—scientific or political'.[42]

Plana wrote to encourage the Englishman, showing an acute understanding that this was necessarily joint work, not a solo effort by Menabrea. 'I presume that you are ceaselessly occupied with your calculating machine; I am assured that you are composing a work that will contain a complete explanation of what concerns it as well as details of its construction. *You alone can properly create such a work.* By thus delivering your ideas to others' meditation, one will better appreciate the goal and the means: then, many doubts will cease, and the consideration of the expense to construct this machine will be balanced by the advantages it presents. And if difficulties arise against your project, they will need to be clearly stated; and you will know how to put them to profit'.[43]

It was October of 1842 before the article was published. It appeared in French, as 'Notions Sur La Machine Analytique De M. Charles Babbage', through a Geneva journal, *Bibliothèque Universelle de Génève.*[44] This was a natural choice, since the journal was edited by one of the attendees at the Turin conference, Augusto de la Rive (1801–1873), who may also have attended the sessions in Po Street. Menabrea was misleadingly listed as the sole author. Later the Italian recalled a curious fact, 'I published a memoir in a Geneva magazine in which I explained the idea I had formed of the analytical engine; indeed, *I added the calculation of signs, which Babbage instead seemed to want to keep secret.* I sent my

40 Prandi to CB, 1842-01-04. Cited in D. D. Swade 2000, 133 and D. Swade 2003, 223.

41 Menabrea to CB, 1842-01-27, section I.89, p. 387.

42 MacCullagh to CB, 1842-02-25, section I.92, p. 389.

43 Plana to CB, 1842-05-25, cited in Bulferetti 1970, emphasis added.

44 Menabrea 1842. See section F.1, p. 185 for the French text, and section F.2, p. 212 for a fresh English translation of this paper with the original notation preserved.

memoir to him in England and, some time later, I received a letter from Babbage in which he congratulated me because I had understood his machine and, above all, the method of signs'.[45] This omits to mention the continuous communication between the two. In the article itself, Menabrea qualified his explanation of the 'method of signs'. 'Not having had the leisure to discuss with Mr. Babbage the manner of introducing into his machine the combination of algebraic signs, I do not claim to expose here the method which he uses for this purpose; but I believed it necessary to supplement it'. But there had been leisure enough, the will was lacking.

Not everyone swallowed the story that Menabrea was the sole author. Over a decade later, Lord Rosse even went so far as to casually refer, in his presidential address to the Royal Society, to 'the ideas of Mr. Babbage, as presented in the small essay published by him under the title Menabrea'. The latter rushed to refute this, though without being entirely forthcoming about the full extent of the collaboration between himself and Babbage.[46]

The Menabrea article provided the first comprehensive description of the Analytical Engine in print, succinctly covering all its major features from the programmable point of view, while glossing over the mechanical implementation details, of less interest to the public. It concentrates on the use of the machine for the evaluation of algebraic expressions, providing several detailed examples, without overcomplicating these. Programs are shown in table format to align the required operation and variable cards, which required separate readers but may logically be read together. The contents may be summarized as follows:

- Physical instruments to aid calculation currently inadequate, subsection F.2.1, p. 212.
- Log Tables, subsection F.2.2, p. 212.
- Difference Engine, subsection F.2.3, p. 213.
- Analytic Expressions and their evaluation. subsection F.2.4, p. 215.
- Store and Mill, subsection F.2.5, p. 215.
- Jacquard-like operation and variable cards for programs, subsection F.2.6, p. 216.
- Solving two simultaneous equations, subsection F.2.8, p. 217.
- Table of operation and variable cards showing multiplication, addition and division, subsection F.2.8, p. 217.
- Table expanded with annotations for variable zeroing, and constant loading, subsection F.2.8, p. 218.
- Propagating signs, subsection F.2.9, p. 219.
- Determining the coefficients of multiplied polynomials, subsection F.2.10, p. 220.
- Table of operations for calculating coefficients, subsection F.2.10, p. 221.
- Calculating ax^n using a loop, subsection F.2.11, p. 222.
- Looking up fixed values for infinite sums like π, subsection F.2.12, p. 223.
- Constants can be loaded from constant cards or tabled values but may be easier to just calculate directly, subsection F.2.13, p. 223.

The tables given are straightforward programs showing the operation and variable cards required in order, aligned. They resemble execution traces—Ada would greatly expand on this theme—because of the clumsy annotations that Menabrea used to indicate which variable cards zeroed out their registers and which did not, rather than marking up the relevant cards as the zeroing or non-zeroing types, which is how it is actually programmed.

45 Menabrea 1971, emphasis added.
46 Menabrea 1855.

The example in which simultaneous equations are solved—all the algebra is first done manually—is a simplified version of a worked example, dated 1840-07-23, that Babbage had brought with him and survives in the Turin archives today.[47]

Another example shows a primitive kind of symbolic computing on multiplied polynomial expressions, where the programmer must do all the algebraic work gathering (trigonometric) coefficients, and the engine merely evaluates the resulting expressions coded by the cards. It is again a simplified version of one provided by Babbage, dated 1837-09-07 (the original, more complicated, example would also be selected later by Ada Lovelace for simplification and has been mistaken for her own work).[48]

Instructions for performing loops and branches are only covered in the text, not in any of these card programming tables. Menabrea's vague hand-waving explanation of how ab^n would be programmed is, moreover, not entirely accurate.[49] In fact Babbage would never nail down exactly what the instruction set was for looping and conditional branching, except at a high level, though one can work out what it must have looked like.[50]

The overall result of this exercise, two years in the making, was a surprisingly well-written and informed article—not just because of Menabrea's considerable personal merits, but also because it had undergone several undeclared revisions by Babbage himself. Lord Rosse did not have to be clairvoyant to notice the hidden hand. Nevertheless, the strategy of having it published abroad in French, under the name of a relatively obscure Italian engineer, backfired. Far from lending some kind of enhanced prestige, it resulted only in temporary oblivion, regardless of the technical merits of the invention or the very clear write-up. Nobody of any consequence showed any sign of even noticing it. Something had to be done.

---

47 See Table E.2, p. 171.

48 See Table E.2, p. 171, section 262, p. 67.

49 See subsection F.2.11, p. 222, footnote 9. In Ada Lovelace's later tabled examples, given to supplement this, deliberately leaving out the loop instructions results in what appear to be program traces but are not really. See for example section A.4, p. 112.

50 Details are given in subsection A.4.2, p. 120.

Chapter 6

Translating and Annotating.

Once Menabrea's paper was published in French, the issue of its translation to English immediately arose. There is considerable irony in this fact, given that its chief subject and disguised contributor was English in the first place.

The British Association (BA) at Newcastle had already established a committee in 1838 to promote the translation and republication of important papers first published in foreign languages abroad. For this they allocated £100 a year. The naturalist Richard Taylor (1781–1858), whose brother Phillip would go into business with Fortunato Prandi, had started a journal *Scientific Memoirs, Selected from the Transactions of Foreign Academies of Science and Learned Societies, and from Foreign Journals* in June of the previous year, expressly for the purpose of republishing foreign papers, not all of which required translation. In the first volume, among those thanked by Taylor—who served as both editor and publisher—was none other than Charles Babbage. Taylor also credited Professor Charles Wheatstone 'for his valuable contributions'.

Wheatstone (1802–1875) was the inventor of the electric telegraph, and an old friend of Babbage, sometimes to be seen at the Saturday evening soirées the engineer held at his house in Dorset street. He served on the British Association translations committee (the other members were Dr. Robinson, Sir John Herschel and Sir D. Brewster, 'with the power to add to their number').[1] Additional volumes of Taylor's journal would only appear sporadically while he ran it—1837, 1843, 1846 and 1852—as demand was not overwhelming. The £100 grant was intended to help make up for this lack of financial viability, but was spent sparingly. The first two volumes contained translations of papers by Gauss, Cauchy, Jacobi and many other leading figures on the continent.[2]

Recommendations of papers to translate and publish came from influential scientists within the Taylor and broader British Association orbit. It is not known now who proposed the Menabrea paper, but it was selected directly on its publication in French that October of 1842. Wheatstone would later write that 'I made no stipulations whatever with Mr. Taylor when he sent Menabrea's paper for translation'.[3] Taylor may have *sent* the paper for translation, but that does not entail that he selected it in the first place, or that the selection was not influenced by Babbage at one remove.

As the first draft was already done by the end of December, it may have been started even earlier than that from an offprint or manuscript, given that it was such an arcane subject. It was certainly not from a field that Wheatstone had ever shown a definite interest in or knowledge of—he was taken up with electricity and telegraphy, along with the transmission of sound and musical instruments, even inventing a concertina. Why choose it and so quickly?

1 Bowers 2001, 102; British Association for the Advancement of Science 1839.
2 Sabine 1842.
3 Wheatstone to AL, 1843-08-25, section I.147, p. 425.

It was common knowledge in the inner circle of Babbage that he had worked with Menabrea for nearly two years to get the paper out. Like Taylor, Wheatstone was a member of that coterie for quite some time. It appears that the visiting Danish physicist Hans Christian Oersted (1777–1851) had introduced him to Babbage and his friend JFW Herschel circa 1823.[4] They were both active in the British Association together, though Babbage would later feud with the organization. They also shared a passion for cryptography.

On April 22nd and 24th of 1837, Babbage and Wheatstone were together in the rooms of Joseph Henry (1797-1878) witnessing electrical experiments.[5] The geologist Charles Lyell wrote to Herschel on May 26, 'At Babbage's yesterday, there was much talk of Wheatstone's new plan for telegraphing information by five wires conducting electricity, and carried through an india-rubber tube, or rather rope, each wire representing a letter of the alphabet, or several letters, I believe by different intensities of the charge. The wires isolated by the india-rubber'.[6] Two years later Babbage wrote in his notebook for 1839-05-18. 'Yesterday saw Wheatstone's model for telegraph & his drawings for Multi[plication] Engine', 'It would therefore require seventeen times as many units of electrical time as it does of engine time—But the unit of Elec[trical] must really be limited by the unit of escapement time'.[7] (Amusingly two of Ada's latter-day champions interpreted this as 'Wheatstone had also worked on designs for calculating by machinery', 'Considering the date, 1839, the idea is breathtaking, coming almost a century ahead of Howard Aiken making his first advanced calculator proposals to IBM'. But here 'multiplier' is really an electrical force multiplier to overcome the problem that the power declines as it is transmitted over distance, and not a calculator at all. Johann Schweigger invented a version of this back in 1820. Wheatstone worked extensively on similar 'multipliers'.[8])

The thespian William Charles Macready recorded a *conversazione* in April of 1840 at Dorset street attended by Wheatstone, Sydney Smith, and others; then another party that May attended by Wheatstone, Charles Dickens, and the banker-poet Samuel Rogers.[9] In August of 1841, Wheatstone even directed a stereoscopic photograph that Henry Collen (1800–1875) took of Babbage — 'successful' in Collen's opinion but too protracted in the taking—a process that the telegraphist pioneered.[10] In June of 1842, Wheatstone had written to Babbage about Brunel's 'atmospheric railway', which proved to be a dud (a hollow pipe between the rails was pumped remotely to produce a vacuum that would power a piston in an engine flexibly coupled to it).[11]

All this intimacy suggests that it is wildly implausible that Babbage was unaware of the translation before it was completed early in 1843, even though many have repeated this idea, which he had encouraged. The explanation is sometimes given that the engineer was ill towards the end of 1842, hence did not hear of it, but illness would not preclude the exchange of letters or brief notes. We know that Babbage was well enough to confront the prime minister Robert Peel in early November of 1842.[12] He had also corresponded that

[4] Appleyard 1926. 'For Charles Wheatstone, Oersted had considerable regard. The two men had much in common. It is stated by Mr. M. C. Harding that, in 1823, on the occasion of Oersted's visit to London, Wheatstone had just opened a musical-instrument shop where he carried out acoustic experiments so much appreciated by the Danish physicist that Oersted himself introduced Wheatstone to Sir John Herschel and to Babbage'.

[5] Henry 1899.

[6] Lyell 1881, 13.

[7] Hyman 1982, 227.

[8] Füegi and Francis 2003, Bowers 2001.

[9] Hyman 1982, 178; Macready 1875, 159, 289.

[10] Henfrey 1854.

[11] Hyman 1982, 200.

[12] Hyman 1982; D. D. Swade 2000; D. Swade 2002; Toole 1992, 1998; Woolley 1999, Stein 1985, 88. Bizarrely, Essinger 2014 omits Wheatstone completely, making it appear as if the translation was Ada's own idea. For Peel see section 242, p. 62.

December with an old family friend, the iron founder James Maclaren, about publishing what he had already written about the engines, receiving sound advice which he would resolutely ignore, 'I have looked over your papers and should have returned them sooner had I not waited for an opportunity to send them to you; you cannot do better than complete and publish them; I am competent to advise in this matter only on one point; I advise you to be magnanimous and publish without any allusion at all as to the treatment you have received from Government'.[13]

Wheatstone, Taylor and Babbage were all located within easy reach of each other. What could be more interesting to Babbage than the translation of a paper he had moved cubic yards of earth to get published abroad? That the engineer was explicitly promised a reward for his involvement in its writing at an *unspecified* stage was later confirmed by Wheatstone himself, in the letter to Ada Lovelace mentioned above. 'I made no stipulations whatever with Mr. Taylor when he sent Menabrea's paper for translation, and the only arrangement made since has been that you should be allowed to have for the mere expense of paper and print 100 copies and Mr. Babbage 50'.[14]

Similar questions may also be asked of the choice Wheatstone made for the translator: Ada Lovelace. They were already known to each other on a visiting basis. The scientist had attended experiments in mesmerism at her house in London the previous year.[15] Yet Ada had no proven experience taking on a scientific translation. She had neither been educated in, nor worked in, the field, nor even in any allied field. Was the choice purely the result of Wheatstone's own deliberation, or was Babbage involved again? One would expect Wheatstone to, at the very least, consult with his friend about the best choice of translator, especially for a paper *on a topic that only Babbage really had mastery of*, a fact that Wheatstone was well aware of. They both lived in London and notes could be sent back and forth even when in-person visits could not be made. Babbage also knew a lot about Ada Lovelace, having met her on many occasions and having exchanged not a few letters with her. Recall that it was exactly Ada who had more than once offered to place her 'mind' at his disposal just over a year earlier. And while Ada offered no scientific credentials, her mother was a first cousin of William Lamb, Lord Melbourne (1779–1848), prime minister as recently as August that year. Lord Melbourne's mother Elizabeth Milbanke was an aunt of Annabella Milbanke.[16] Her husband was similarly well-connected in official circles as Earl Lovelace and Lord Lieutenant of Surrey.

Babbage told the story himself in his memoirs, written twenty years later, more than a decade after Ada's death. His telling requires careful parsing.[17]

> Some time after the appearance of his memoir on the subject in the "Bibliothèque Universelle de Geneve," the late Countess of Lovelace informed me that she had translated the memoir of Menabrea. I asked why she had not herself written an original paper on a subject with which she was so intimately acquainted? To this Lady Lovelace replied that the thought had not occurred to her. I then suggested that she should add some notes to Menabrea's memoir; an idea which was immediately adopted.

It is an ambiguous recollection, raising more questions than it answers.[18] It does not exclude the possibility that Babbage already knew that she was working on it, then learned that she

[13] 1842-12-24, J. Maclaren to C. Babbage, section I.101, p. 400. Hyman 1982 describes Maclaren as an iron founder and old friend of Babbage.

[14] Wheatstone to AL, 1843-08-25, section I.147, p. 425.

[15] Woolley 1999, 233-4, citing the MS diary of Joseph Kay held at the John Rylands Library, Manchester University.

[16] See also Belanger and Stein 2005.

[17] Babbage 1864, 136.

[18] The original MS is still extant in the Wanganui Regional Museum in Auckland, New Zealand, see Baum 1986, and is accurately transcribed.

had *completed* the task. It does not disclose what we already know from others, that she had been explicitly commissioned by Wheatstone for Taylor, and had not taken the job on her own initiative.

Just before the translation project had started, a significant event had also happened in Babbage's long-standing impasse with the government over his Difference Engine and its funding (he persisted in refusing to ask for any funding for the Analytical Engine). Robert Peel and his exchequer Goulburn had finally given him the resolution he had asked for. After consulting with Airy, who had always myopically scorned the Difference Engine, and Herschel, who continued to support its long-term benefits, the government definitely killed funding for the project in early November of 1842.[19] There would be no new funding and Babbage could keep the engine core built so far (he preferred to forever lodge the product with its original funders, hoping perhaps to embarrass them).

Babbage promptly asked for a personal audience with the prime minister Sir Robert Peel, which was granted. A remarkable document survives with his summary of the interview, during which he complained that he had not received the recognition or the remuneration that had been accorded to others, naming Airy, Peacock, Herschel, Whewell and the geologists Henry Buckland and Adam Sedgwick. Peel was annoyed by this jarring *faux pas* and swept it aside brusquely, as he did Babbage's complaints that his reputation had been impugned by scurrilous rumours—surely, Peel said, men of sense ignored that sort of thing?[20]

This was no isolated incident. Lyon Playfair would recollect that 'Babbage always considered himself a badly treated man, and this feeling at last produced an egotism which restricted the numbers of his friends'. When Prince Albert wanted to confer honours on some men of science he 'asked what I would recommend. I suggested that it would produce a favourable impression if one or two men of undoubted position were made privy councillors, mentioning Faraday and Babbage as two men entitled to this honour. Shortly after this conversation I was commissioned to sound the philosophers and ascertain whether they would like to be appointed to the Privy Council. Unfortunately I first went to Babbage, who was delighted with the suggestion, but made it a condition that he alone should be appointed, as a reparation for all the neglect of the Government towards his inventions. Even the association of such a distinguished man as Faraday would take away from the recognition which was due to him. This condition was naturally disagreeable to the Prince Consort, and no further steps were taken to open the Privy Council to men of science'.[21]

The meeting with Peel and the Playfair-Faraday incident underscore how acutely status and honour played on the mind of the scientist. It is likely that cancellation of the Difference Engine stung Babbage into moving publication at home of a defence, but after first considering doing so directly (see the letter from James Maclaren above), preferred to arrange one at a discreet distance. That is certainly how he thought until August of the following year about the annotations to be added to the translation. He revealed in this a desire to have others act as his public champion—always more convincing than a personal statement in his own interest—giving him opportunity to smuggle his self-defence on board. Rather than an afterthought, this may have been the impetus for the whole project—facilitated for him by his friends Taylor and Wheatstone, and disguised in the retelling as something like a pleasant surprise to the inventor.

Babbage also left a note in his papers that suggests more was going on between himself and Lady Lovelace.

---

[19] Babbage 1843; Hyman 1982; D. D. Swade 2000; D. Swade 2003. See also Herschel's draft of his cautiously positive evaluation of the project, 1842-09, Herschel to H. Goulburn, section I.98, p. 391.

[20] 1842-11-11, CB notes on meeting Peel, section I.100, p. 397. His papers contain several drafts before the meeting listing the specific complaints and status comparisons he would raise.

[21] Playfair 1899, 155-6.

CB to Lady Lovelace
Tuesday 7th Feb 43
accept invitation
long since we met,
begin under new circumstances

Why would they need to 'begin under new circumstances', and why was it 'long since we met'? The extant correspondence between them disappears between August of 1842 and sometime early in 1843, a six month or so gap. It is possible—one can only speculate—that they had quarrelled, hence a need to 'begin under new circumstances'. This might also explain the annotation 'accept invitation': was he refusing them previously?

Ada's translation more than doubles the length of the original paper, because of the long, repetitive, symbol-adorned, frequently vague notes appended to it. Few people ever seem to have worked through these in sufficiently close detail, or their problematic nature would be more well-known. It is not straightforward to attribute the ideas in them to their sources. They were initialled 'A.A.L.', easily decoded to 'Augusta Ada Lovelace', and she certainly wrote them up. However they were extensively discussed with Babbage, in person and by letter. He revised them multiple times and supplied much, though as we will see definitely not all, of the content himself. He had given her his standard list of polished programming examples to select from and explain, together with (we may infer) whatever other written materials he had to hand, which were at least as extensive as those he had brought to Turin in 1840, and had already been used by Menabrea.

> We discussed together the various illustrations that might be introduced: I suggested several, but the selection was entirely her own. So also was the algebraic working out of the different problems, except, indeed, that relating to the numbers of Bernouilli [sic], which I had offered to do to save Lady Lovelace the trouble. This she sent back to me for an amendment, having detected a grave mistake which I had made in the process.

Their private correspondence shows, contrary to Babbage's earlier claim, that this was definitely *not* a subject with which Ada was 'intimately acquainted'.[22] That correspondence shows her asking about basics. 'The day I called on you, you wrote off on a scrap of paper (which I have unluckily lost), that the Difference Engine would do ... (something or other), but that the Analytical Engine would do ... (something else that is absolutely general). Be kind enough to write this out properly for me'.[23] The examination below of her notes emphatically confirms that. Babbage had his own motives for saying that she was. There was the need he felt for affirmation by prestigious supporters, related to the use of the article as a tactic in his quarrel with the government. Familiarity of the supporter with the material and skill in the art is necessary to carry conviction. Self-deception is an ever-present danger. We will also see that Babbage really supplied 'algebraic working out' for *all* the examples, which were simplified by Ada, just as they had been by Menabrea, from his more complex ones. The defensive assurance that 'the selection was entirely her own' really reveals that it was *no more* than a selection from his own material.

A general word of caution is needed. For the longest time this material has acted as a sort of Rorschach inkblot for projection by commentators, who seize on vague and ambiguous statements within them, in the full light of subsequent technological history, to find supposed anticipations of many of the major ideas of computer science.[24] More self-discipline is needed to avoid emulating the cult of Nostradamus. The context of the notes is explicitly and only the Analytical Engine, as invented by Charles Babbage, to which its

[22] Babbage 1864, 136.
[23] 1843. Ada Lovelace to C. Babbage. Not dated but mention of the notes to Menabrea places it sometime in 1843 after February.
[24] See especially Toole 1992, 1998.

statements are meant to apply, either as already realized capabilities, or as straightforward extensions of mechanisms already present. Wherever some anticipation is said to occur, it must be shown that the idea is unambiguously present, that the implications are properly understood and some plausible realization is at least outlined, so that it is not just idle talk.

There is a lot of technical and mathematical detail stemming from the notes—we will follow them using Ada's numbering for easy reference—including original and corrected programs, which the general reader may not care for. Here we will only summarize the background and main conclusions. The motivated reader can verify the full, sometimes gory, details at leisure in Appendix A.[25] Highlights are given below.

It was not a straightforward exercise in translation, because Ada (or rather Babbage) felt free to deliberately alter the meaning of the original text, an unusual decision in a translation. Menabrea's notation—showing simple programs using operation, variable and number cards, presented compactly in table format—was altered to use novel superscripts with new rules. This was the notation Ada used in her notes too. The reader may discover this by consulting the original paper with care.[26] Ada does declare within her Note B that she has altered the original, rationalizing that she has merely clarified the text. This change is dealt with in greater detail in section A.2 where we will show that her 'improved' notation conflates ideas and leads to logical inconsistencies.[27] Taking such a liberty suggests that Babbage (for how could Ada presume so?) viewed Menabrea's paper as more-or-less his own to alter.

As it happens, Babbage had sent the Italian previews of the translation, though not in enough time for anything like revision or approval, and apparently without mentioning Ada. 'You will probably have received the rough proofs I sent to Turin by my son'.[28] In 1884, Menabrea published a little-noticed recollection of the translation-with-notes in *Comptes Rendu*. 'At first, I had believed that Babbage himself was the author; but, in a letter of August 28, 1843 ... he undeceived me by giving me the name of my mysterious translator, who was none other than a very noble and very beautiful English lady, whose name will be transmitted to posterity on the wings of one of the greatest poets of our century' (notice that this is confined to the *translator*).[29] In his unpublished *Memorie*, written a few years later, Menabrea recalled that the publication added 'very important analytical notes—*probably supplied by Babbage himself*' and that Babbage hoped 'that the memoir would serve to spread the idea of his inventions'.[30] The suspicion was understandable and shared by others. Babbage's admission quoted above, written twenty years later, that he had indeed played an undeclared role in writing the notes, as he had in writing the Menabrea paper in the first place, complicates any interpretation.

The question of authorship can be clarified by internal evidence from the notes, laid out below in detail. Often these deal with technical topics in the history of mathematics, such as the three-body problem, which we may be sure that Ada knew little about. These could only have come directly from Babbage, since they lie squarely within his research interests and were often mentioned by him in correspondence and in print, from which sources they are lightly reworded if at all. They presuppose advanced technical knowledge, fully up to date, extensive familiarity with the history of astronomy and mathematics. Glaring errors in the description of the engine and its programming, which Babbage would almost certainly not have made, identify Ada as the likely author. Likewise, elementary mathematical blunders and confabulation are highly unlikely to have come from a distinguished mathematician like

[25] See Appendix A, p. 107.

[26] See section F.3, p. 226, page 681 in the original, and section F.2, p. 219, where the original notation is restored.

[27] See section A.2, p. 108.

[28] CB to Menabrea, 1843-08-28, section I.148, p. 425.

[29] Menabrea 1884.

[30] Menabrea 1971, quoted in Losano 2014. Emphasis added.

Babbage, and are best interpreted as Ada's attempts to explain material, working from an outline provided by him, that she did not understand.

Consider this from Note F: 'The calculus of operations is likewise in itself a topic of so much interest, and has of late years been so much more written on and thought on than formerly, that any bearing which that [Analytical] engine, from its mode of constitution, may possess upon the illustration of this branch of mathematical science, should not be overlooked'. A statement which could only be founded on extensive experience. Babbage had that, Ada did not.[31]

Or notice this especial interest of Babbage: 'The methods in Arbogast's *Calcul des Derivations* are peculiarly fitted for the notation and the processes of the engine. Likewise the whole of the Combinatorial Analysis, which consists first in a purely numerical calculation of indices, and secondly in the distribution and combination of the quantities according to laws prescribed by these indices'.[32] (Louis François Antoine Arbogast (1759–1803) separated operations from the quantities they operate on, so that they are first-class manipulable entities in their own right. He was often cited by Babbage, who made much of the idea.)

A striking example is given at the close of Note F when Ada talks of using the engine to obtain solutions, using numerical methods given initial conditions, to the sets of differential equations which describe the intractable three-body gravitational problem.[33]

> In the solution of the famous problem of the Three Bodies, there are, out of about 295 coefficients of lunar perturbations given by M. Clausen (Astro Nachrichten, No. 406)[34] as the result of the calculations by Burg, of two by Damoiseau, and of one by Burckhardt, fourteen coefficients that differ in the nature of their algebraic sign; and out of the remainder there are only 101 (or about one-third) that agree precisely both in signs and in amount. These discordances, which are generally small in individual magnitude, may arise either from an erroneous determination of the abstract coefficients in the development of the problem, or from discrepancies in the data deduced from observation, or from both causes combined. The former is the most ordinary source of error in astronomical computations, and this the engine would entirely obviate.

All this was right up Babbage's street, not Ada's. He was a connoisseur of errors in tables, a problem ideal for the application of computational machinery, and *au fait* with all the literature. There is even direct evidence. In a letter to Ada from early September that year, *after* the notes had been completed, he wrote 'I shall bring some books about that horrible problem—the three bodies which is almost as obscure as the existence of the celebrated book "De Tribus Impostoribus"'.[35] We may be sure then that the example and its description, all of the wording, come straight from him.

The notes start with a refutation of the idea that Babbage abandoned his Difference Engine because he invented the Analytical Engine. These definite assertions imply knowledge of events and motivations which could only come from the inventor himself, who is transmitting corrections through her from behind the scenes, as any alert reader would immediately deduce. It is argued that this cannot be because the latter invention was much more advanced than the former (surely all the more reason to abandon the former). This

---

31 Menabrea 1843b, 694.

32 Menabrea 1843b, 724), Babbage 1864, 136-7.

33 Menabrea 1843b, 722.

34 'It will presumably be of interest to many to see a comparative compilation of the coefficients of the newest lunar tables. To this end, I have carefully reduced the expressions of the periodic equations from Bürg's and Burckhardt's tables to mean arguments, and retained all terms greater than 1/10 second; likewise the latitude equations from Damoiseau's prize essay and his lunar tables, ... Since all three authors follow the parallax theory, the deviations in Burckhardt's coefficients appear significant ... These can be fully explained by assuming that he neglected by mistake to transform them according to his longitude arguments. ... // Altona, June 26th 1840 / Thomas Clausen', Clausen 1840.

35 1843-09-09, C. Babbage to AL, section I.149, p. 425.

argument from Babbage need not be taken seriously. As we have already seen, Babbage had explained exactly how one engine led to the other in a letter to Nathaniel Bowditch in August 1835, 'in September last I got possession of the drawings of the old Engine which I had not seen for two years. In explaining these to my son many new views struck me and a totally new and more extensive arrangement occurred to me which appeared of such importance that I thought it worthwhile before I proceeded to reexamine every part of the old Engine. I soon found many new modes of adding, of carrying, and of performing other processes which were mechanically much more simple than the ancient methods'.[36] The Difference Engine project had *stalled* for prosaic reasons, but it was not resumed exactly because the Analytical Engine was much more interesting (though much later the inventor would return to a simpler design for the Difference Engine).

Note A also contains a phrase that has received an inordinate amount of attention ever since. 'We may say most aptly that the Analytical Engine weaves algebraical patterns just as the Jacquard-loom weaves flowers and leaves'. A trite flourish and strictly inaccurate. It is no more penetrating than saying that the cogwheel device *cog*itates. Insofar as there are 'algebraical patterns' here they are 'woven' in the brain of the programmer and then submitted to the (passive) engine in code. Given values for the symbols, the engine actively and unromantically 'develops' the coded expression—that is, reduces it—to a numerical result. As we will see, the engine in this realization does *no algebraic manipulation itself*, indeed could not do so without design changes, and should not be confused with systems in the far-future which have finally been able to do that.

It is here the first clear indication emerges that Ada doesn't understand how the engine actually works. When discussing operations (by which she just means functions in general) she offers the thought that in mathematics 'the symbols of operation are frequently also the symbols of the results of operations. We may say that these symbols are apt to have both a retrospective and a prospective signification ... figures, the symbols of numerical magnitude, are frequently also the symbols of operations, as when they are the indices of powers'. Thus she believes that for say the function $f(x) = x^3$ the symbol for the number 3 may instead refer to the cubing operation itself and not the number, and that 2^3 may mean either the number 2 *about* to be cubed or one that has *already been* cubed(!) Entering into the spirit of this, Ada has passed over the *intermediate* stages, as when 2^3 is merely a middle-aged 2^2 on the way to becoming full-fledged 2^3. Spare a thought for $\sqrt{2}$, forever thwarted, condemned to remain irrationally prospective, passing through endless Siberian train stations on a journey to a destination it will never reach, only glimpse snowbound in the distance.

This eccentric line of thinking, a kind of machine-mathematics double-vision, gets her into real trouble when she decides that the analytical engine eliminates these supposed ambiguities by enforcing segregation in its column store: 'whenever numbers meaning operations and not quantities (such as the indices of powers), are inscribed on any column or set of columns, those columns immediately act in a wholly separate and independent manner, becoming connected with the operating mechanism exclusively, and re-acting upon this. They never come into combination with numbers upon any other columns meaning quantities though, of course, if there are numbers meaning operations upon n columns, these may combine amongst each other, and will often be required to do so, just as numbers meaning quantities combine with each other in any variety'.[37] Which is to say, x^{2+4} is allowed, but 2 and 4 still supposedly refer to operations. It is hard to explain why this was noticed by just one other commentator as a 'rather brief passage in Note A that Lady LOVELACE appended to her translation of Menabrea's article. This passage refers rather obscurely to numbers representing operations rather than quantities appearing on the same portion of

[36] Babbage to Bowditch, 1835-08-02, section I.51, p. 359.
[37] Menabrea 1843b, 693-4.

the Analytical Engine as numbers representing quantities'.[38] It is really a badly-garbled version of the idea that operations are distinct in the engine from the objects that they operate on.

There is no segregation of *numbers* in the analytical engine, which has no concept of indices being themselves operators. When x^k is evaluated, k is of course just a number. This is the power function $f(x,k)$ where both arguments are mere numbers. And of course k may be combined with any other number in any other context, as a number, e.g. $k+9$ can be punched onto an output card without complaint. When k is an integer it is treated exactly as that, where x^k is evaluated by a multiplicative loop with a counter which decrements down from k until exhausted, as described by Menabrea.[39] For a real value of k a power series approximation of the exponential function $e^{k\,log x}$ would be used. Neither function is a built-in operator in any of Babbage's descriptions, though they could be, or they could be subroutines, but even if they were their arguments would just be numbers as above.

Aside from commentary, the notes contain two examples extending those already provided by Menabrea: multiplication of more complicated polynomials, and exponentiation. In truth these are both taken directly from the worked examples provided by Babbage himself, which are more general. Ada 'simplifies' them slightly less than Menabrea had. These are followed by a third: the recursive computation of the Bernoulli numbers, which as we will see was also provided in outline form by Babbage.

Note E. expands the polynomial multiplication which Menabrea had already discussed via a simple example of

$$(ax^0 + bx^1)(A\,cos^0 x + B\,cos^1 x).$$

However the programmer has to do all the symbolic work, expanding the product, gathering coefficients of terms, and mentally tagging columns which will hold their values. In Note E, Ada chooses to supplement this with a more complicated exercise in the gathering of coefficients, involving an infinite series and requiring a familiar trigonometric identity. But it is not original.

That this extended example was already worked out by Babbage himself might be deduced from its deliberately neat internal structure, of the classroom exercise kind. There is more direct evidence though: it is one of multiple examples that he took with him to Turin, where the archives still contain the large-format three-page write-up. The date on it shows that it was originally composed in September of 1837.[40] Nevertheless, it proved very heavy going for Ada. 'You cannot conceive the trouble I have had with the trigonometrical Note E! In fact no one but me, I really believe, would have doggedly stuck to it, as I have been doing, in all its [wearying] minutiae'.[41] Her 'dogged' treatment fails to account for the fact that the infinite series involved, which she proposes to approximate for concrete values of its parameters, might not converge, depending on those values.

The remaining examples make use of loops, at first simply in the case of exponentiation, then more complicatedly in the case of the Bernoulli numbers. It is significant that Ada, verbose about all else, never explains the loops which Menabrea had described in detail and which are implemented via 'combinatorial cards' to be mixed in with the operator cards. These are somewhat vaguely described by Babbage, but included the equivalent of conditional GOTO operations for backing and advancing cards, powerful enough to implement any loop. Operation cards and variable cards could be moved separately. She mentions repetitions many times, but never explains how they would really work, leading her to miscount how many operations they require (she must count the loop operations including integer decrements and increments, and conditional branches). It is possible that they were

38 Randell 1973, 349.
39 See section F.2, p. 222.
40 See no. 23 in the list of materials in Turin, Table E.2, p. 171.
41 AL to CB, 1843-08-02, section I.136, p. 416.

deliberately omitted here to be consistent with Note G, where they are also crucial but are obscured to conceal memory addressing problems, as we will see.[42]

In note B, Ada considers a simple program to calculate ax^n for arbitrary a, x and n, a key component of a power series and already the subject of Menabrea's slightly simpler loop.[43] Although $a = 5$, $x = 98$ and $n = 7$ are assigned to variables V_1, V_3, V_2 respectively, no indication is given of how V_2, that is n, is actually used and by which operations! Instead the reader is blithely informed that $V_3 = 98$ will be multiplied by itself $n - 1 = 6$ times, then by $V_1 = 5$. What is described amounts to the fixed program $a * x * x * x * x * x * x * x$. The arbitrary power $V_2 = n$ is just an illusion unless it is shown how it is used. If Ada deliberately intended to omit the details of the necessary loop for implementing this she could easily have said so in no more than one or two sentences, with a reference to Menabrea. The Italian *does* explicitly describe how loops, which are not first-class operations in themselves, have to be implemented by the programmer, e.g. 'a card will order a counter to *mark* $(n - 1)$ and at the same time will execute the multiplication of b by b'. These loops lie at the core of the engine's capabilities. Without them it is not much more than an inflexible calculator.

This is closely related to Note B where we encounter some remarks by Ada about symbols and 'developed' formulae, which have been taken by some to suggest the use of the engine for 'symbolic computation', rather than numerical evaluation. In fact she is talking about something closer to logging during numerical evaluation: 'it must be easy by means of a few simple provisions and additions in arranging the mechanism, to bring out a double set of results., viz.—1st, the numerical magnitude which are the results of operations performed on numerical data ... 2ndly, the symbolical results to be attached to those numerical results, which symbolical results are not less the necessary and logical consequences of operations performed upon symbolical data, than are numerical results when the data are numerical'.[44] Her meaning is clarified by a footnote referring to Menabrea. 'In fact such an extension as we allude to, would merely constitute a further and more perfected development of any system introduced for making the proper combinations of the signs plus and minus'. Supposedly this has not been done 'because the necessary arrangements for effecting it would increase the complexity and extent of the mechanism to a degree that would not be commensurate with the advantages'.[45] (We will soon see that Babbage had really put this question to one side without resolving it.)

Signs are handled in the engine by an extra digit on each variable, and some simple rules for correctly propagating them through operations, e.g. so that $-2 \times -2 = +4$. This sign digit *tags* the variable in the column store and Ada is proposing that symbols might tag variables in much the same way. Thus if V_1 is tagged with x and V_2 with y, and we set $V_3 = V_1 + V_2$ then V_3 would be tagged as '$x + y$'. Similarly for the other basic operations $\times, -, \div$.

This cannot just be a 'more perfected development' of a sign, since the signs do not carry their history with them, being continually reduced when combined, and never occupy more than one digit. Nowhere did Babbage, let alone Ada, discuss the encoding of arbitrary symbols by numbers (as in a modern ASCII table), or the formation of growing strings of them, occupying chunks of memory somewhere, to make up formulae as variables are combined. No simple mechanisms existed here for any data structures at all, since memory in the engine is *named* ('V_{132}') and not *addressed* (unlike modern arrays in say the C programming language, '$p[i]$' where $i = 132$, or the pointer de-reference '$*p$' where the variable p holds the address value 132). Aggregation of storage into dynamically addressable

[42] See section 276, p. 71. Modern O(n) notation doesn't count fixed overhead per repetition, rolling them all into one fixed cost, but Ada does not comprehend that focus on iterative complexity only.

[43] Menabrea 1843b, 703; also subsection F.2.11, p. 222.

[44] Menabrea 1843b, 695.

[45] Menabrea 1843b, 713.

units is not straightforward.[46]

But let's assume that somehow these issues, which belong to a much later phase of the history of computing, are overcome, and that programmers supply initial symbolic tags in their code, to propagate and compose as growing strings. The engine does *much more* than just compute expressions like $3x + 4y + 5$. It has conditional branches and loops. A program might find the smallest prime number larger than a given n, but only if the 143245th digit of π is odd. The products of such programs cannot in general be represented as familiar mathematical formulae. Even well-known operations are not built-in and will be approximated, as Babbage explained, either by a few terms of their power-series expansions or any other numerical approximation algorithm. The formula tag for a variable holding the calculation of $x \cos x$ will not be '$x \cos x$' at all, but rather a mess of terms combining x with some approximation of $\cos x$.

Babbage had already given thought to something like this idea himself a decade earlier, but not as a side-effect of numerical calculation. A 'scribbling book' entry, which shows signs of being written after the fact, suggests that he had soon realized the implications for the engine as it stood and rejected it. 'On machinery for the Algebraic Development of Functions. About this date the idea of making a development engine arose with considerable distinctness. It is obvious that if the Calculating Engine could print the substitutions which it makes in an Algebraic form we should arrive at the algebraic development "it can print all those substitutions which are noted in the composition in some of the notations"—it will however be better to construct a new engine for such purposes.'[47]

Even earlier he had toyed with what look like more general ideas, probably as a result of considering the automatic solution of simultaneous equations by elimination. A 'scribbling book' entry from 1836-07-10, records: 'This day I had for the first time a general but very indistinct conception of the possibility of making an engine work out *algebraic* developments. I mean without *any* reference to the *value* of the letters. My notion is that as the cards (Jacquards) of the Calc. engine direct a series of operations and then recommence with the first so it might perhaps be possible to cause the same cards to punch others equivalent to any given number of repetitions. But [their holes] might perhaps be small pieces of formulae previously made by the first cards and possibly some mode might be found for arranging such detached parts according to the powers of nine numbers and of collecting similar ones'.[48]

Ada's confused communication of these formulae 'development' ideas, which we must assume sprang from discussions with Babbage, do not describe how symbolic *computation* would really work. The case of $\sqrt{(x^2)}$ is instructive. The formula tag 'development', as a side-effect from its numerical calculation, would be an inscrutable mess of terms produced by the approximation of square roots. This is *not* useful symbolic computation, which would instead immediately reduce $\sqrt{(x^2)}$ to x. Likewise, real symbolic computation would reduce $xy - yx$ to nothing, and $x \times x \times y$ to x^2y. It would be capable *at a minimum* of collecting terms, factoring, distributing and all those other features inherent in the reduction of numerical computations but real work for any implementation of symbolic computation.

Extending the Analytical Engine even to do what amounts to no more than a kind of unhelpful logging would have required major advances in its memory addressing. As Belanger and Stein remark, the claim that this or that simple calculator might encompass general 'thinking' is all too easily made, as say by Pascal's sister Giilberte (her brother having invented no more than an elementary device).[49] None of those who made these facile

[46] See subsection A.4.5 for the elements of a solution for this, not disclosed by Babbage.

[47] Babbage, vol. 3, D4, 3/9, 1837-12-13. Quoted in Hyman 1982, 244.

[48] Babbage, vol. 2, D3, 3/5, 150. Quoted in varying degrees of completeness by Collier 1990, 139, Randell 1973, 349, Stein 1985, 102 and Belanger and Stein 2005, 89.

[49] Belanger and Stein 2005.

claims actually showed how to do that until the modern era.

In 1851, Babbage would refer to an 'apparent impossibility' he had investigated in the late 1830s, which 'seemed to present far greater difficulty' than numerical computation. 'Fortunately it was not one of immediate practical importance, although as a question of philosophical inquiry it possessed the highest interest I had frequently discussed with Mrs. Somerville and my highly gifted friend the late Professor M'Cullagh of Dublin, the question whether it was possible that we should be able to treat algebraic formula by means of machinery. The result of many inquiries led to the conclusion, that if not really impossible, it was almost hopeless'.[50] But he chose to break out of this impasse by addressing a question that is not really about symbolic computing but rather the subject we now call *computability.* He construed 'algebra' as the programming of the engine using symbols (via variable names) to be bound later to concrete arguments (numbers, or functions) when the expressions they represent are 'developed' (reduced to number), not what we call symbolic computing.

When thinking through the idea that the engine could calculate formulae like x^2, Babbage asked: how far is this really true? Could x be absolutely *any* (say real or complex) number, as a mathematician would require? Would the engine ever be able to handle that? This is nothing to do with rewriting of formulae in which the variables remain symbols and are *not* bound to values. It is about which values those variables really could be bound to in practice, to evaluate the result. At first glance it might seem that since the variables of any real engine would always be of a finite size, arbitrary values could not be supported. If one wants to calculate π^2 for example, even representing π would require infinitely many digits, and only an approximation could be calculated in a fixed machine architecture.

The solution conceived by Babbage really does anticipate later developments by nearly a century. While the variables in the engine certainly are limited in size, its stream of punch cards is not. If we assume it has much time as desired, and an infinite stream of them to consume, 'it was found possible to evade the difficulty, by connecting indefinite number with the infinite in time instead of with the infinite in space'.[51] (But it is left unsaid that having unlimited cards requires that the space surrounding the machine is itself infinite.) Then the engine can compute π^2 to *absolutely any degree of precision required* by using as many cards and as much time as needed. This is equivalent to being able to compute the value π^2 itself. Babbage adds an enigmatic principle, 'the nature of a function might be indicated by its position', to fully mechanize evaluation of algebraic expressions. It may be conjectured that he meant a primitive form of type-inference from context. In order to 'develop' the functional equations he was so fond of, substitutions of concrete executable functions for abstract symbols representing functions only make sense when limited to functions which fit, i.e. which operate on the right number of operands and return results which may appear in the surrounding expression.

Babbage was also much taken with the idea of operations being kept distinct in the engine from the objects that they operate on (named variables), an idea that Ada frequently repeats. This appealed to him because of its analogue in analysis, where an algebra of operations had proven very powerful.[52] This would be useful in the engine itself if there was a way of doing symbolic computation in the first place. Even if $V_1 - V_1$ puts the operator '$-$' at one remove from V_1, it cannot be manipulated without first binding V_1 to a particular value. The engine cannot resolve it to 0 *irrespective of the value of* V_1, without rewriting the expression algebraically.

---

[50] Babbage 1851, 184, Babbage 1864, 124, Belanger and Stein 2005.

[51] Babbage 1851, 185.

[52] The basic idea is simple. Consider $f(x) + g(h(x)) = y$, This can be written as $(f + gh)(x) = y$ where what looks like multiplication gh is really composition of operators. Then the operator expression can be manipulated algebraically, which may lead to say $x = (f + gh)^{-1}y$ where $(...)^{-1}$ is an inverse operator. The algebra of operators is a lot like the algebra of numbers, but a little more complicated, since composition may not be commutative, $gh \neq hg$ but obey more complex laws instead.

The closest that Babbage got to realizing this 'separation of operator from operand' is through the separate card readers for operation and variable cards, which could be moved independently. Thus for any fixed sequence of operation cards, a different result can be achieved by advancing or retreating the variable cards, which specify both which operands are used and how stored results feed back into subsequent operations. This is not an idea that ever caught on in other systems, perhaps because of the difficulties it creates when others try to understand the program.[53]

Having counted, in Note D, the number of operations involved in solving the simultaneous equations that Menabrea had tackled, Note E proposes some peculiar notation for counting the number of *repeated* operations involved in 'reducing' expressions using loops. The logician Robin Gandy described Ada's symbolism as 'rather inappropriate', but that is an understatement.[54] Ada repeatedly confused the indexes of the 'sum' expressions she tried to repurpose with their starting and ending limits, a common mistake made by tyros. The repetition shows that these were not mere typos. Bizarre elements of her notation like $(+1)^n$, which might appear profound to the unwary or the merely busy, were really produced by a muddle-headed attempt to instantiate a general example that Babbage must have provided. Once her examples are reconstructed from that basis the eccentric notation collapses. This is not harmless: she uses this notation later in Note G to attempt to 'count' instructions.

A much more sophisticated illustration of loops is the calculation of the Bernoulli numbers, using course-of-values recursion. This takes up all of Note G. It is the most remarked on of all the notes, and has been mentioned by many commentators as evidence of Ada's mathematical capabilities.[55] That impression must be based on the profusion of formulae and purported program traces on offer, a poor guide. It seems few have ever worked through its technical details—or if they have, they have kept silence (see section A.4 for the not especially complex but lengthy details[56]). Some have even suggested that this whole example was her own idea, which is certainly false.

Before proceeding, the reader should recognize that the Bernoulli numbers, which arise in sums of certain power series,[57] are not treated in the note in any interesting light, mathematically speaking. They were described 130 years earlier by Jacob Bernoulli. They appear only as an illustration of the computational powers of the Analytical Engine. It is a gross error to suppose that invoking various standard formulae (sometimes given incorrectly) and manipulating them (inaccurately, as it happens) shows any sort of advanced mathematical ability.

Babbage had been considering the Bernoulli numbers as an illustration for his engine from the very beginning. In August of 1835 he had written about the proposed machine to Nathaniel Bowditch, 'Of course it can calculate the values of π, e, the numbers of Bernoulli'.[58] A few months later he had written the same to Giovanni Plana, adding 'up to 30 places, and perhaps up to 50, by giving it more time'.[59] However if it was among the set of examples which he brought with him to Turin in August 1840, it has not survived there.[60]

A year later, in 1841, Babbage referred to the Bernoulli numbers once again when writing to Alexander von Humboldt. He even called out the recurrence relation that would be used in Note G. 'The greater part of the drawings exist by which an engine might be made which

[53] But see subsection A.4.5 for an important application of this for addressable memory in this design.
[54] Gandy 1995, 54.
[55] For example by Hyman 1988, 243, with its veiled swipe at Stein 1985, not named or cited.
[56] Turn to section A.4, p. 112.
[57] Particularly in power series expansions of $x/(e^x - 1)$.
[58] Reproduced in section I.51, p. 359.
[59] Reproduced in section I.54, p. 363.
[60] See Appendix E, p. 169

shall reduce to numbers any explicit function whose laws of formation can be assigned. *For instance, the equation of finite differences expressing the law of formation of the numbers of Bernoulli being given to it, it will work out and print the successive numbers*'.[61] On top of all this, Menabrea had himself mentioned the Bernoulli numbers in the text that Ada had already translated.

Thus when Babbage solicited notes from Ada for the translation of Menabrea's paper, as we have seen, it is no surprise that the Bernoulli example came up. However, Babbage's confession that he had helped Ada with the example is ambiguous and worth re-examining: 'the selection was entirely her own. So also was the algebraic working out of the different problems, except, indeed, that relating to the numbers of Bernouilli [sic], which I had offered to do to save Lady Lovelace the trouble. This she sent back to me for an amendment, having detected a grave mistake which I had made in the process'. Exactly how much of the written-up version, the most complex topic dealt with in the notes, did he supply? Was it really 'to save Lady Lovelace the trouble'? Or had she already stuck fast in 'trouble'?

There is a sense of embarrassment in this belated admission by Babbage. The published notes did not contain any admission of his role, though many, like Menabrea, privately suspected his hand in it all. The general public and even specialists have credited Ada with all of it, and this has formed the basis of the idea that she was a mathematician, down to the present day. The claim that 'the algebraic working out' of the non-Bernoulli problems was entirely hers is, as we have already seen, not forthcoming—Babbage had explicitly provided the worked examples for those, she had merely simplified them a little less than Menabrea had. Appending 'she sent back to me for an amendment, having detected a grave mistake which I had made' positively invites the suspicion that Babbage was a little overeager to sugar the pill two decades later. And what was this mysterious 'grave mistake' anyway? It is never identified, but some excellent clues exist.

Ada raised the Bernoullis in an undated letter, possibly circa May 1843, undoubtedly prompted by Babbage's repeated talk about them. 'I want to put in something about Bernoulli's Numbers, in one of my Notes, as an example of how an implicit function may be worked out by the engine, without having been worked out by human head & hands first. Give me the necessary data & formulae'. By July 2 Babbage was writing 'I like much the improved form of the Bernoulli Note but can judge of it better when I have the diagram and Notation'. A few days later on the 5th, Ada reported that she was researching. 'I am doggedly attacking & sifting to the very bottom, all the ways of deducing the Bernoulli Numbers. In the manner I am grappling with this subject, & connecting it with others, I shall be some days upon it'.

A note in Babbage's papers shows that he visited Ada on the 11th to supply much-needed help. 'C. B. to Lady L / Tuesday 11 July 43 / Bernouillis [sic] numbers'. *Over a week later*, on the 21st, Ada was still lost. 'I am in much dismay at having got into so amazing a quagmire & botheration with these Numbers, that I cannot possibly get the thing done today ... I am in a charming state of confusion'. Yet Brooke's *Formulae*, which she had earlier requested from Babbage, explicitly covers the Bernoulli numbers. De Morgan's *Differential and Integral Calculus*, published the previous year, in which she had worked for some time, has a full explanation of the methods Babbage used. There was also De Morgan's *Penny Cyclopaedia* article of 1840 on the Bernoulli Numbers.[62]

An undated letter from this period has Ada responding to Babbage pointing out some errors, perhaps in this material. 'I really cannot believe it to be incorrect, for nothing can exceed the care with which I have gone over it. But the fact is I am plagued out of my life,

---

[61] CB to von Humboldt, circa 1841-07, cited in Collier 1970, 1990, 174-5. See section I.83, p. 384. Emphasis added.

[62] See Lovelace to Babbage, circa 1843, section I.108, p. 402; Brooke 1829; A. De Morgan 1840b,A. De Morgan 1842b, 248, 553-4.

just now'.[63] By the 28th the material was already being revised and Ada was complaining. 'The beginning of Note G (by which I mean the Table & all that precedes it) never has been returned into my hands; a small part of the remainder was, but that I speedily gave you back, & thus it is, now printed. The missing part must be either at your house or at the printers'. Babbage recorded the same day 'Note G with its beautiful Table is missing' but by the next day 'Sat 29 July. Missing Mss found'. After this we hear only of non-specific proofing and revising of the notes into August.

This leaves us with little positive information about the genesis of Note G, aside from the fact that Babbage certainly supplied technical assistance, but in response to Ada's distress, not 'to save her the trouble', a dissimulation. This included an outline of the deduction of the recurrence relation needed, and a well-honed sample program for the Analytical Engine (Babbage already had related examples to hand). So we will need to pay close attention to the details of the note to learn more. We will only summarize here what is formally demonstrated in section A.4, but motivated readers are encouraged to work through the details there.

The first step is to choose which Bernoulli numbers to work with. The definition chosen yields what we now call the old-style Bernoulli numbers, which are always positive.[64] However the modern definition was already known then, and is covered in Augustus de Morgan's *Differential and Integral Calculus* (1842, see above), which Ada owned and had worked from extensively when he attempted to tutor her. The problem with these definitions though is that they involve infinite series and are difficult to work with computationally. So a more tractable formula must be chosen. Curiously a complicated alternative formula is displayed but never used.[65] This is just as well, since it is completely bogus and can never yield the Bernoulli numbers! A fact that could have been verified by Ada by trying a few values and computing the results by hand, an exercise of a few minutes.

The formula which *is* used also happens to be incorrect.[66] It gets the *signs* of the Bernoulli numbers wrong. Recall that those are always positive (on this 'old-style' definition). On her definition they will sometimes be positive, sometimes negative. That is significant, since if one consults (as she may have done) De Morgan's list of 49 computed Bernoulli numbers in say the *Penny Cyclopaedia* (1840)

$$\frac{1}{6}, \frac{1}{30}, \frac{1}{42}, \frac{1}{30}, \frac{5}{66}, \frac{691}{2730}, \frac{7}{6}, \frac{3617}{510}, \frac{43867}{798}, \frac{174611}{330}, \frac{854513}{138}, \ldots$$

one sees that they are all positive.[67] She certainly realized there was a problem. A little later in the note, Ada states 'In the above table and diagram we are not considering the signs of any of the B's, merely their numerical magnitude. The engine would bring out the sign for each of them correctly of course, but we cannot enter on every additional detail of this kind, as we might wish to do'.

Here it is absurd to claim that 'engine would bring out the sign for each of them correctly' since the formulae used and the consequent program are wrong. (One is reminded of the anecdote told by Harriett Martineau in which a lady, when shown the Difference Engine, asks if it will produce correct results given wrong inputs. Ada would have us believe that the Analytical Engine would produce correct results given incorrect programs.) The Bernoulli numbers are their own magnitudes, given that they must be positive, and implying otherwise is a bad blunder. It is certainly *not* the case that Ada 'cannot enter on every additional detail of this kind, as we might wish to do'. Rather it is the case that she could not identify

63 1843, Ada Lovelace to C. Babbage. See section I.125, p. 411.
64 Equation 2, p. 112.
65 See subsection A.4.7, p. 130.
66 Equation 4, p. 113.
67 A. De Morgan 1840b.

what was wrong, and tried to brush over it. Yet fixing the formula to ensure that they are positive, as they must be on this definition, is trivial.[68]

From that formula a recurrence relation can be deduced using some simple algebraic manipulations and known facts about polynomials, as the appendix shows. The recurrence relates the next Bernoulli number to a linear combination of the previous ones. As many of these numbers as desired can be calculated, provided that one can store all the ones so far and calculate the dependent factors needed for their combination. We can be sure that Ada got this from Babbage himself since she asked him for it in her letters and he, as we have seen, admitted doing so, though supposedly just 'to save her the trouble'.

Ada attempts to work through his (typically terse) derivation line-by-line, supplying her own interpretations of (apparently missing) steps, but comes completely unstuck at the point where a linear dependence is established involving the (incorrectly signed) Bernoulli numbers B_k.

$$0 = (A_0 + A_1B_1 + \cdots + A_{2n-3}B_{2n-3}) + B_{2n-1}$$

She fails to see that simple arithmetic will yield the required recurrence relation for the next Bernoulli number B_{2n-1} in terms of $-(A_0 + A_1B_1 + \cdots + A_{2n-3}B_{2n-3})$, where each A_k coefficient really depends on n and should be written $A_{k,n}$. Her 'explanation' goes completely awry, leading her to insert several breathtakingly nonsensical arguments about 'linear dependence' and 'independence', conditions which she must have heard were somehow involved but did not understand.

She persuades herself that the 'linear dependence' here refers to a property of the index n, so that for D_n 'n is then not arbitrary, but is always a certain function of the distance of that D from the beginning'. Perhaps because *lines* may measure distance? But curiously this applies only in earlier steps of the derivation of the above equation, not in the equation itself. She is led to ruminate on n, displaying two empty expressions for it in terms of distance.

$$2n + 1 = d, \text{ and } n = \frac{d-1}{2} \text{ (for any \textit{even} power of } x\text{)}$$
$$2n = d, \text{ and } n = \frac{d}{2} \text{ (for any \textit{odd} power of } x\text{)}$$

Still in search of B_{2n-1}, leaving the former expressions dangling helplessly, she adds more useless truisms, while persuading herself that the equation above is now linearly independent since n is an independent variable (its supposed dependency on 'distance' has evaporated for obscure reasons).

> This circumstance, combined with the fact (which we may easily perceive) that whatever n is, every term ... after the $(n + 1)$th is $= 0$, and that the $(n + 1)$th term itself is always $= B_{2n-1} \cdot \frac{1}{1} = B_{2n-1}$, enables us to find the value (either numerical or algebraical) of any nth Number of Bernoulli B_{2n-1}, *in terms of all the preceding ones*, if we but know the values of $B_1, B_3 \ldots B_{2n-3}$.

There is no point in resurrecting phantom terms in the recurrence relation for

$$A_{2n+1}B_{2n+1}, A_{2n+3}B_{2n+3}, A_{2n+5}B_{2n+5}, \ldots$$

nor is there a need for A_{2n-1}, which is here mysteriously $\frac{1}{1}$ instead of just 1. The attempt to do so suggests confusion with the infinite series formulation of what we now call a generating function. These pseudo-explanations show decisively that Ada did not understand what the

[68] See Equation 4*, p. 113.

concepts meant here and had not learnt how to use mathematical analysis to make an argument.[69]

More important than these bogus explanations is the failure to deduce the correct recurrence relation. In all the subsequent working, including the program 'trace' that is eventually shown, an incorrect relation is used without the subtraction: $A_0 + A_1B_1 + \cdots + A_{2n-3}B_{2n-3}$. Thus there are two errors stemming from the signs of the B_k and the missing subtraction (which are here *not* the same thing). These are mathematical errors, not the slips we call 'programming bugs' (though there are a handful of those too). That they are incorporated into the program suggests that Ada edited the version she was given by Babbage to match the mathematical mistakes she had made.

More complicated scenarios seem far less likely. Say it is Babbage who makes the mistakes in the code that he gave her (not impossible, since they could just be common-or-garden bugs) but not in his mathematical outline, then Ada allows them to pervert her mathematical treatment, even unto the gibberish about linear (in)dependence, without noticing that they are simple mistakes. Failing to check them with Babbage himself she constructs nonsense on their backs. But if so, her grasp of the mathematics involved had to be very poor indeed, and her methods even worse. She must take responsibility for her treatment regardless of its provenance: she was exceptionally eager to claim it as her own.

Why did Babbage, a practicing mathematician, not catch and correct these howlers before publication? It is inconceivable that he would not have noticed them if he had worked through the details. If he had noticed them, it would have been in his own interest to correct them. The surviving correspondence (certainly not complete) suggests he was most interested in the use of the article in his dispute with the British government over funding. It is likely that, once he had given her the gist of what she needed for Note G, he paid little close attention to the mathematics in it afterwards. Interested self-deception[70] about Ada's ability to follow his terse explanations, always simpler to their author than to a beginner, may well have played a role.

There is also the possibility that, by then, haggling with Ada had passed the point of rescue. Her letters show her furiously refusing corrections from Babbage. 'I am much annoyed at your having altered my Note. You know I am always willing to make any alterations myself, but that I cannot endure another person to meddle with my sentences', 'I wish you were as accurate, & as much to be relied on, as I am myself. You might often save me much trouble, if you were; whereas you in reality add to my trouble not infrequently, and there is at any rate always the anxiety of doubting if you will not get me into a scrape; even when you don't'.[71] He may simply have given up and left it up to her. After all, he wasn't going to put his name to any of it. Be that as it may, these blunders have stood mute and unnoticed in the text ever since.

The recurrence relation leads naturally then to the aforementioned program to calculate the 'Bernoulli numbers' (but the wrong ones) using the Analytical Engine. Specifically the case $n = 4$ is considered but that is said to be arbitrary. It is important to note though that if Ada had simply calculated the first few numbers by hand using her incorrect recurrence relation she would have verified with a few minutes of elementary arithmetic that they were, after the first two, wholly incorrect, and not just in their signs. The program that is shown also contains a remarkable *deception* introduced by Ada. Calculating the n-th Bernoulli number, if n is to be truly arbitrary, requires loops. These are not explicitly shown. Instead vague comments are inserted in the code about previous instructions being 'repeated'.

The loops are not fully described in order to cover up a major defect: the engine did not, at that stage, support addressable memory, for example an array or memory pointer.

69 See section A.4, p. 112 for further analysis of these confused tautologies.

70 Trivers 2011.

71 Two undated letters, circa July 1843. See Appendix I, p. 317.

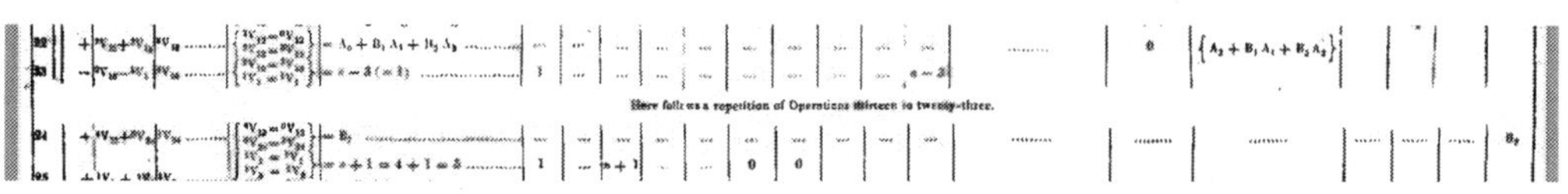

Figure 6.1: Ada's Addressable Memory Deception. 'Here follows a repetition', Menabrea 1843a, Note G.

To calculate the n-th number, where n is arbitrary, you need the first $n-1$ numbers already stored—and a place to put the n-th if you want to get the $(n+1)$-th number next. With each iteration of the loop you need a new dynamic location to store the number you calculate, and Babbage only had fixed named variables. Those names cannot be dynamically altered as the loop iterates without special measures: that would require altering the program (but hold that thought).

Letters exchanged between Ada and Babbage show that she stumbled on the fixed variable name problem while trying to work out her explication step-by-step, and was positively proud that she was able to subtly disguise the shortcoming. She boasted to Babbage about this evasion. 'I have (I think very judiciously & easily) touched on the only departure from perfect identity which could exist during the repetitions of (13 ... 23), & yet have not committed myself by saying if these departures would require to be met by the introduction of one or more new cards, or not; but have simply indicated that as the variations follow a regular rule, they would be easily provided for. I think I have done it admirably & diplomatically. Here comes in the intrigante & the politician!'[72]

Leaving out the loop instructions requires showing only an 'execution trace', for the case $n = 4$, and the language used by Ada implies 'instructions' being somehow 'repeated' below, where they are omitted, but this cannot be so without changes. However, if the loop operations are restored, the addressable memory issue is fixed, the signs are calculated correctly, the recurrence relation is multiplied by the right factor, and one or two other minor bugs are corrected, the resulting card program *executes perfectly*, as a card program, and produces the correct sequence of genuine Bernoulli numbers for any n (as far as storage permits). The correct detailed structure of the program once the loops are restored clarifies that Babbage did, in fact, mean to provide those facilities, for the intricate details of the segments to be looped would not make sense otherwise. The reader may verify this from the executable code provided in the appendix.[73] The program therefore has almost exactly the right structure for looping, bar the fixes noted above, and must have been carefully honed by Babbage over time, with considerable skill, *as a program* (recall that there was no device to actually execute it on). The appearance of a 'trace' here is a red herring, which led some to suppose wrongly that programs as such were only dubiously being shown.[74] Yet they were, since Menabrea had already shown them, and this was just a disguised program once variable zeroing is correctly notated and the loops restored.

Most likely the addressable memory problem was the mysterious 'grave mistake' that Babbage remembered Ada had 'detected'. If his reminiscence is read carefully, it is striking that he does not claim the problem was actually *fixed*, only that the program was sent back to him 'for an amendment' (which in itself confirms that he wrote its prototype, and modified it). However that may be, no other problem could be more serious: it is fatal to the Bernoulli program and any other program requiring such memory, and cannot be worked around without conceptual changes to the instruction set. It is not a mere 'bug' to be ironed out. It had to be papered over at this late stage of the translation production,

[72] Dated only 1843. See section I.117, p. 407.
[73] See subsection A.4.3, p. 121. The uncertainty that Bromley 1982a expresses over this is overwrought.
[74] E.g. Hyman 1997.

there wasn't time enough left for a proper solution. The response by Ada—hiding it in the 'trace'—was a crafty obfuscation, which Babbage played along with, rather than a correction. It was not the program that was amended, but rather the supposed 'trace' of it. When read with his history of invective against scientific malpractice, this placed Babbage in a compromised position. Alan Bromley always suspected that something was fishy here. 'In the notes to the Menabrea paper, Charles Babbage seems to me to studiously avoid explaining how the control is to be implemented, and I sense the same from the AAL [Lovelace] correspondence'.[75]

Given this limitation for something so simple as $n = 4$, and the above confession, it is hard to fathom Ada's brazen claim in a footnote that 'during the processes for the computation of millions of these Numbers, no other arbitrary modification would be requisite in the arrangements, excepting the above simple and uniform provision for causing one of the data periodically to receive the finite increment unity'. Since all the previous Bernoullis have to be stored for the recurrence relation to work, calculating the millionth number using the previous 999,999 stored values would be pretty difficult in a machine like the Analytical Engine, a locomotive-sized object with just 1,000 variables. This was also noticed by Robin Gandy. 'Lady Lovelace uses a course-of-values recursion. She claims, quite falsely, that the program will only need a fixed number of registers. (Falsely because the formula expresses the next nonvanishing Bernoulli number in terms of all the previous ones)'.[76]

An earlier claim in the notes about the recurrence relation, 'we may derive from it the numerical value of every Number of Bernoulli in succession, from the very beginning, *ad infinitum*'[77] illustrates an amusing duality in Babbage's approach to these questions. On the one hand he filled thousands upon thousands of engineering drawings with painstaking, minute, now-enigmatic detail about the concrete design of parts of the engine and their interconnection. On the other, by a seamless transition, we are transported to the world of infinite possibility, the theoretical computer in which as many cards as needed are always available, stretching off through infinite space and time, in which arbitrarily large numbers can be split over those cards and computed with, using just as many past results as needed, stored on their limitless expanse. It is a vexed Isambard Kingdom Brunel, the practical engineer, crossed with Alan Turing, the limitless mathematician.

All this may also help to explain why Babbage was not writing or signing this article himself. The engine was still a work in progress. He knew that critical issues remained unresolved. Not signing it gave him plausible deniability. He would not be committed to particular readings of how the engine worked, or ought to work.

The noted Babbage expert, Alan Bromley, left some poignant comments about this addressable memory problem, which he feared downgraded the Analytical Engine to no more than a versatile calculator. 'It is a little difficult to admit this conclusion after expending so many years studying Babbage's designs. Perhaps it is why I "took a break from Babbage" in the late 1980s and never really came back, and still find it somewhat difficult to focus on completing the work'.[78] But Bromley, who was not convinced that Babbage understood this limitation, was mistaken.

One would expect Babbage to quickly devise a solution for the issue once he faced up to the consequences of not having it—which the interaction with Ada and the use of an obfuscation shows he was definitely aware of—rather than trying to hide it. For example, the variable cards could be extended to express ideas like 'load (or store to) the variable V_k where k is in V_3'. No doubt Babbage could have rustled up a mechanical implementation in no time. He actually chose a simpler route leading to a logically workable solution, adding

[75] Babbage 1830, Grier 2004.
[76] Menabrea 1843b, 729, Gandy 1995.
[77] Menabrea 1843b, 726.
[78] Grier 2004, 67.

a new instruction for independently moving *variable* cards a dynamic number of places.

It is shown in subsection A.4.5 how a 'card pack', combined with Babbage's new instruction for moving the variable cards by a number of positions, can enable the equivalent of relatively efficient dynamic memory arrays by mapping an index to a named variable at *execution* time.[79] This method would physically expand any card program containing it, but the length of the program is not at issue here, only its computational capabilities. The packs can be made as small or as large as the memory range needed, and pre-manufactured copies of packs may be inserted as often as required in a modular fashion. Such packs enable the successful calculation of the Bernoulli numbers in a completely general way, as they do all programs with similar memory needs.

Babbage nowhere explains this card-pack technique, applied to the addressing problem, but the structure of the extended variable card operation shows that he understood its principle. (Actually executing programs containing such packs requires an Analytical Engine emulator to implement the moving of variable cards in their reader.) All this means that his Analytical Engine was indeed, as far as is practically possible, 'Turing Complete', since it possessed the equivalent of a (theoretically infinite) movable tape to read from and store to. That lays to rest the contentions otherwise by Maurice Wilkes, Martin Davis and others.[80]

[79] See subsection A.4.5.
[80] Grier 2004; Herken 1995.

Chapter 7

Aftermath.

Babbage's principal motivation in (probably) suggesting the translation of Menabrea, and (certainly) annotating it with notes, was to put his side of the dispute he had with the British Government over continued funding of his engine production. When a decision was finally reached to terminate this support in November 1842, the process of soliciting the English version—probably suggested to Taylor by Babbage—became urgent. It was typical of Babbage's methods that he wanted it to appear under somebody else's name, as he had done with both Dionysius Lardner and Menabrea.

Ada was, in effect, being used by Babbage. As she completed writing up the annotations in late July 1843, he promptly produced a pointed note of his own explaining how the project had come to grief (apparently through no fault of his own) and sketching the progression of ideas from the first engine to the next. He wanted this self-justification inserted, just as he had inserted so much of the other material in the notes. Ada, no doubt acting on the advice of Earl Lovelace, baulked at this manipulation. She had no desire to become involved retroactively in a dispute with the government, to which her husband had close ties, a dispute from which she could hope to gain nothing.[1] She protested that the note was out of place in a venue for scientific translations.

Babbage complicated things by proposing instead that the paper be withdrawn and published elsewhere. The result was an exchange leading to a condescending, astonishingly high-handed letter from Ada running to sixteen pages — enumerating his personal defects while extolling 'a mind of my particular class' — eventually terminating with point blank demands. Further cooperation between them would only be possible if he let her handle practical matters—the whole conduct of the engine development—to remedy his defective judgment: 'can you undertake to give your mind wholly & undividedly, as a primary object that no engagement is to interfere with, to the consideration of all those matters in which I shall at times require your intellectual assistance & supervision: & can you promise not to slur & hurry things over; or to mislay, & allow confusion & mistakes to enter into documents'. But she feared that, unlike herself, he loved fame and glory more than he loved God.[2]

Whether or not Babbage loved God as much as Ada vowed she did—whether or not he hoped to be His Prophet as she proposed to be—he refused to agree to these feverish conditions, and after some negotiation with Taylor and Wheatstone, the translation was published as had been agreed earlier, while Babbage's self-justification was published separately in the *Philosophical Magazine.*[3]

Though they restored sociable terms after this and continued to correspond, their relationship never really recovered. He continued to work on the Analytical Engine, but no

[1] 1843-08-06. Ada Lovelace to C. Babbage, section I.139, p. 417.

[2] 1843-08-08. C. Babbage to Ada Lovelace, 1843-08-14; Ada Lovelace to C. Babbage, section I.143, p. 420.

[3] Babbage 1843.

trace survives of any concrete work by Ada on anything related to it, though there are hints of unrealized work on unspecified topics which she seemed to have progressively lost interest in, and some other unoriginal dabbling.[4] The following year she began to correspond with Michael Faraday, hoping to work with him as a scientific assistant—flattering him, elevating herself and slighting Babbage *en passant*.[5]

> I know I can be a minister in your service, in a thousand ways; & that without interfering with my own views & avocations, but so as to make all my services to you co-operate with my pursuits & progress.
>
> One thing I am now writing is again in the service of my friend Babbage; but of a very different description from the last. How I wish by the bye that his mind were (in some points) more akin to that which in you I so admire. There is in him too much self, & too little of what I would term divine love. He is capable of great devotion to individuals now & then, but there is not in him the Scripture the Christian love which I so adore. He is a great man, & has a great intellect; but he would have been a yet greater, & his intellect would have been yet higher & wider, had his moral feelings corresponded.
>
> This, by the way, is an illustration of some of the feelings & principles I slightly sketched in my first letter to you. Babbage has I think in him the capacity however, for a high moral & religious development, tho' I doubt if it will be in this world. I am much attached to & interested in him, up to a certain point; but then comes something which creates a bar to that degree & kind of affection I feel for several others.

Wisely, the physicist cried off, saying that he no longer had the vim required. Mounting use of opiates, combined with reckless love affairs and a ruinous turf-betting scheme, dragged both Ada and her husband into serious debt, which she attempted to evade by secretly pawning her jewels. She was increasingly ill and late in 1852 fell to cancer, just short of 37 years old. Her mother soldiered on, persecuting Lord Lovelace for the whole fiasco.

In the meantime Babbage had temporarily switched his attention to designing a simplified Difference Engine which he hoped would be easier to manufacture, perhaps in the hope of redeeming his reputation. (This would form the basis for a successful project in the 1980s and 1990s to construct that machine from scratch at the Science Museum in London, using only techniques available to Babbage—though not without some improvisation.[6]) The Great Exhibition of 1851 passed his work over, reinforcing his sense of alienation and calling forth a defiant pamphlet.[7]

In parallel, Georg and Edvard Scheutz in Sweden had since 1834 been working on their own difference engine, based on Lardner's description of Babbage's design (which we have seen was as much the work of Babbage himself as it was of Lardner).[8] By the mid 1850s they had a working version, pragmatically manufactured to much lower tolerances than he had called for. Babbage did not resent this in the least and helped to promote their device when it was presented to him. Their machine met with limited adoption, suggesting that the market at that time for Difference Engines was hard to realize. Later in the 1850s Babbage resumed work on the Analytical Engine, building a working portion of the Mill (later tinkered with by his son), experimenting with cheaper ways to cast its parts, but making no more conceptual progress. The exhibition of 1862 did feature Babbage's original

---

4 Stein 1985, 121ff.
5 1844-11-08, Ada Lovelace to Faraday, section I.163, p. 435.
6 D. D. Swade 2000; D. Swade 1991.
7 Babbage 1851; Hyman 1982.
8 Lardner 1834; Lindgren 2023.

portion of the Difference Engine, constructed 30 years earlier, but he complained that it was obscured in a corner out of public view.[9]

Towards the end Babbage's powers began to fail. Street organists outside his windows in Dorset Street drove him to distraction, delighting in pestering him all the more because he pursued a protracted public campaign against them. 'He told me with distress one day that he had forgotten my name, and my Father's name, when he came to see me. He had also forgotten his cards, so he took a small brass cog-wheel out of his waistcoat pocket and scratched his name on it! and he did the same at Professor Owen's door'.[10] After his death from kidney disease in 1871, he took most of the knowledge needed to interpret his thousands of diagrams, insofar as he was still competent, with him (though his brain was physically preserved and one half is still on display).

Nobody has ever built a working Analytical Engine in physical form, though it seems clear enough that a machine of that type would work if somebody ever rises to the challenge. It would require an expansive interpretation of Babbage's intentions to do this, rather than a slavish adherence to the letter of his surviving plans, given the inherent ambiguity of what remains. Since it was Turing-complete, capable of addressing memory using the technique described here, it would be an interesting project.

9 Babbage 1889.
10 Lloyd 1880, 47-69.

Chapter 8

Ada's Reputation.

The Menabrea translation with its mathematical notes was, in a sense, a deception by Babbage. He had arranged for its publication and probably suggested its translation, definitely its annotation. Yet his name did not appear anywhere as a co-author even though he was responsible for the core of the routine mathematics, the analytical engine conception and its programming. He had been pursuing all three for years, yet published hardly anything of substance despite having written a very full (if technical) description in English, as well as distributing detailed technical diagrams.[1]

The tactic worked so well that even Menabrea disappeared into the background as enthusiasts of Ada thrust her to the fore. The English translation of Menabrea appeared in 1989 in the eleven volume *Works of Charles Babbage* as 'Sketch of the Analytical Engine, by A. A. Lovelace (1843)'—only the original French version appears under Menabrea's own name.[2] This is strictly contrary to precedent, as the editor acknowledges in a curious footnote, without giving reasons. We do not credit Constance Garnett as the author of *Crime and Punishment* merely because she translated it from Dostoevsky. Nor do we credit Alexander Pope as the author of the *Illiad* and the *Odyssey* merely because he translated it from Homer. If clarification of authorship is intended, we should list *Babbage-squared* as the primary author, given his compounded role from the conception of the paper through its annotation, with Menabrea a secondary and Ada a minor contributor.

Ada had long suspected that she was much more than a minor contributor. Recall that in 1839 she had written to Babbage regarding her mathematical abilities 'I believe I have the power of going just as far as I like in such pursuits, & where there is so very decided a taste, I should almost say a passion, as I have for them, I question if there is not always some portion of natural genius even'.[3] Before the Menabrea translation and annotation was complete she was already casting about for another 'fine field for my clear, logical & accurate, mind, to work its powers upon; & to develop its metaphysical genius, which is not least amongst its qualifications & characteristics'.[4] Not long after this, she wrote 'I am in good spirits; for I hope another year will make me really something of an Analyst. The more I study, the more insatiable do I feel my genius for it to be. I do not believe that my father was (or ever could have been) such a Poet as I shall be an Analyst, (& Metaphysician), for with me the two go together indissolubly'.[5] Of the notes she had already written 'I am rather amazed at them, & cannot help being struck quite malgre moi [if I don't say so myself], with the really masterly nature of the style, & its superiority to that of the Memoir itself'. Indeed, 'I do not think you possess half my forethoughts, & powers of foreseeing all

[1] Babbage 1973, Appendix D, p. 137.
[2] Babbage 1989a.
[3] 1839-11. Ada Lovelace to C. Babbage, section I.64, p. 372.
[4] 1843-07-02. Ada Lovelace to C. Babbage, section I.112, p. 404.
[5] 1843-07-30. Ada Lovelace to C. Babbage, section I.131, p. 413.

possible contingencies (probable & improbable, just alike)'.[6]

As soon as the paper was published, Ada began to forward copies—she ordered 250 offprints, but the publisher experienced persistent difficulty collecting her payment for them—to friends and acquaintances.[7] In letters she proclaimed her talents, which she now estimated rather superior to those of Babbage in multiple respects. As she told her mother, 'He is beyond measure careless & desultory at times. I shall be willing to be his Whipper-in during the next 3 years, if I see fair prospect of success', moreover 'I suppose many hundreds & thousands of ... formulae will come forth from my pen, in one way or another'. (Not a single correct formula original to Ada appears in the notes.) She supposed Babbage would seize 'the advantage of having my pen as his servant' and that her 'strong coercive influence over him' was his only hope of ever getting the engine built.[8] Perhaps she felt free to adopt this sort of tone (with and about him) because she had come to consider him within her power now that he was complicit in their deception about addressable memory.[9]

In the same month she wrote to her mother identifying 'much latent power' and 'a most indomitable industry which despises no even apparently trivial minutia', and which had produced 'clear, forcible & logical documents'.[10] To Babbage she felt compelled to write once more sharing her self-enthusiasm. 'The pithy & vigorous nature of the style seem to me to be most striking; and there is at times a half-satirical & humorous dryness, which would I suspect make me a most formidable reviewer. I am quite thunder-struck at the power of the writing. It is especially unlike a woman's style surely; but neither can I compare it with any man's exactly'.[11]

Contemporaries who did not directly know her soon heard by word of mouth that the 'A.A.L.' of the notes was Countess Lovelace. Most could not have known the undeclared central role played by Babbage himself, though not a few of the well-informed shrewdly suspected his hand in it all. Those who accepted the paper at face value saw mathematical symbolism strange to them, even though there are only elementary mathematical ideas there once one gets beyond the notation. The public just assumed that Lady Lovelace must be an accomplished mathematician to write all *that.*

Even Mary Somerville, evidently not one of those who suspected Babbage's undeclared role, was impressed and wrote back to 'most sincerely congratulate you on the proficiency you have made in the highest branches of mathematics'.[12] Likewise William Farr, in a tribute to Babbage, would state of Ada that 'Her notes display a surprising knowledge of analysis'.[13]

De Morgan received a copy and was studiously polite, calling it only 'a pretty thing' without much enthusiasm. If he had paid close attention to it he would surely have spotted the confusions we have identified here, but if he did then politeness restrained him. (Privately, De Morgan and Babbage detested each other, as we have seen.) To her mother, Ada wrote 'I am very much afraid as yet of exerting the powers I know I have over others'.[14]

An attempt to interest Faraday ran into the obstacle that he, an experimentalist only, was the last great physicist to know no mathematics. Ada sent him an offprint anonymously, provoking him to write to Babbage, to thank him for the gift, only to be surprised by the information that the real author of the notes was Ada, a further deception by Babbage.[15]

[6] 1843-07-27. Ada Lovelace to C. Babbage, section I.128, p. 412.
[7] Baum 1986, 122-3.
[8] 1843-08-15, Ada Lovelace to Lady Byron, section I.145, p. 424.
[9] See section 276, p. 71.
[10] 1843-08, Ada Lovelace to Lady Byron, section I.133, p. 414.
[11] See section I.134, p. 415.
[12] 1843, Mary Somerville to Ada Lovelace, section I.150, p. 426.
[13] Farr 1871.
[14] 1843-09-15, Ada Lovelace to Lady Byron, section I.155, p. 428.
[15] 1843-09-01, Faraday to C. Babbage, section I.151, p. 426.

Faraday wrote her a flattering note in turn, but had not comprehended the paper at all. Nevertheless Ada got the impression from his letters that he believed she was 'the rising star of Science'. This provoked her imagination. 'You know I am not a bit my own agent as to my scientific progress & objects. I am simply the instrument for the divine purposes to act on & thro': happening to be appropriate for that object. Like the Prophets of old, I shall but speak the voice I am inspired with'. 'I am a Prophetess born into the world; & this conviction fills me with humility, with fear & trembling!' In time these apprehensions of grandeur would mushroom under the influence of opiates.[16]

In Ada's subsequent reputation, Babbage himself played up her abilities. He was committed to this by the peculiar ghostwriting arrangement he had disguised himself behind, both for her translation and Menabrea's original paper, not to mention Lardner's. For the papers to play the role he wanted, promoting the engines and his own reputation, furthering his dispute with the government, the competence of 'A.A.L.' had to be backed—to be backed convincingly, it had to be believed, inviting self-deception. The fact that Ada had no formal mathematical education and had published nothing else required special stress on sheer *natural talent*, supposedly displayed here for the first time. In his memoirs Babbage was noticeably keen to claim, after finally disclosing some, though not all, of his covert assistance, that she 'entered fully into almost all the very difficult and abstract questions connected with the subject'. Not *all* of them then, but *almost all*. This has a definitely defensive air.

Here some have seized on a letter Babbage wrote to Ada that September, drawing far-reaching conclusions from it.[17]

> I find it quite in vain to wait until I have leisure so I have resolved that I will leave all other things undone and set out for Ashley [her country estate] *taking with me papers enough to enable me to forget this world and all its troubles and if possible its multitudinous charlatans—everything in short but the Enchantress of Number.*
>
> My only impediment would be my Mother's health which is not at the moment quite as good as I could wish. Are you at Ashley? and is it still consistent with all your other arrangements that I should join you there? and will next Wednesday or Thursday or any other day suit you? and shall I leave the [...] road at [Taunton] or at Bridgwater—and have you got Arbogast *Du Calcul Des Derivations* with you there (i.e.at Ashley)?
>
> I shall bring some books about that horrible problem—the three bodies: which is almost as obscure as the existence of the celebrated book *De Tribes Importoribus*. So if you have Arbogast I will bring something else.
>
> Farewell my dear and much admired Interpretress.

It is this letter which gave rise to the idea, apparently first due to Maboth Moseley, that Babbage called Ada an 'Enchantress of Number', later corrupted by the Huskeys, Joan Baum and Betty Toole to 'Enchantress of *Numbers*'.[18] From there it spread far and wide, and can be found now in many places.[19] On a close reading it may be a simple mistake, but this is hard to settle definitely.

Ada flirted relentlessly in her letters, often referring to herself as a 'fairy', not just to Babbage but to others too. In response Babbage played on 'fairy' as 'enchantress', perhaps

[16] Stein 1985.

[17] 1843-09-09. C. Babbage to Ada Lovelace, section I.149, p. 425.

[18] Baum 1986; V. R. Huskey and H. D. Huskey 1980; M. Moseley 1964; Toole 1992. Toole blithely includes a facsimile of this letter in which 'Number' is clearly stated. Subsequent editions did not correct the error, Toole 1998, 2010. Baum unconsciously uses both forms.

[19] Adams 2023; Chiaverini 2017; Essinger 2014.

elicited by her flirting. On the same day, he had written to Faraday of 'that Enchantress who has thrown her magical spell around the most abstract of Sciences'.[20] But Ada was in no sense a great calculator, and he surely used this common word for more than one purpose. A closer examination of the letter suggests he may not have meant 'enchantress of *number*' here to apply to her, but was introducing a further play on the word.

In the letter Babbage is proposing to *travel* to stay *with* Ada, at the Lovelace retreat Ashley Coombe—in Porlock, on the coast of Somerset—provided that she will be there. He states that he will bring certain things, 'papers enough to enable me to forget this world and all its troubles and if possible its multitudinous charlatans'. But he will either not be *taking* 'the Enchantress of Number' with him or, what is more likely, the papers he does take will let him forget every *thing* unpleasant except the 'Enchantress' while he is there.

He is unlikely to have meant Ada herself. It would be nonsensical to take Ada to Ashley, where she would already be. Again, if he meant that he would not forget Ada while he was there, he would have said 'everybody' and, besides, the idea of forgetting his hostess would too silly to need denying—he would also need to deny that he would forget his host, her husband, and everyone else there. Moreover, the papers that he would take could have no connection with not forgetting either one of them, present before his eyes. Neither could his hosts have possibly supposed that they ranked among those 'charlatans' and 'troubles', or have expected the dubious reassurance that they were not thought of as such.

Babbage probably meant something else by this 'enchantress' left behind and/or not forgotten. Most likely he just meant one of his calculating devices, which had absorbed all his thoughts, as a refuge from the world, and would be alive in the papers that he would take: either the (first) Difference Engine (of which a working part had long been completed), or the more recent Analytical Engine, which existed on paper only. And it makes perfect sense that one would call either of these an 'Enchantress of Number', and a 'thing' to think about. They had been endlessly, delightedly, thought about by him. And they really did do things with numbers that most found mysterious, almost magical. Reusing 'enchantress' would be an additional play on the word that she would have recognized.

If, instead, the phrase really did apply to Ada herself, and bearing in mind that Babbage had also stretched to 'that Enchantress who has thrown her magical spell around the most abstract of Sciences', one must assume it was no more than calculated flattery. We already know Ada's *demonstrated* mathematical abilities only too well—not just in her homework exercises, but in the notes themselves, and in her frustrated interaction with Babbage during their creation, not disclosed by him.

Ada's reputation soon became intertwined with that of Babbage himself. As a result both sank from sight for nearly a century, with only occasional glimpses.

Babbage himself was widely rediscovered in the late 1940s, as modern computing emerged and gathered pace, though he had never been completely forgotten after his death. Some pioneers of electronic computing knew all about him while others did not.[21] Ada Lovelace was suddenly prominent too. An influential collection of essays edited by B.V. Bowden, *Faster than the Speed of Thought* (1952) enthused about Ada's role. A portrait of her is the frontispiece, and she is repeatedly credited, quite inaccurately, with key ideas concerning the Analytical Engine:[22]

- 'Lady Lovelace ... seems in a way to be foreshadowing the use of the drum' (94);
- 'As our last example we show how the B-tube can now be used to help the programmer in other ways which Lady Lovelace suggested' (73);

[20] 1843-09-09. C. Babbage to Faraday, section I.152, p. 426,

[21] Comrie 1946; Ludgate 1909, 1914; Mullett 1948. For many more early references see Metropolis and Worlton 1980; Randell 1973.

[22] Bowden 1953.

- 'Lady Lovelace realized that a machine would be able to guide itself through a calculation of great complexity by following out in detail a routine which had been prescribed for it, and that it could reach decisions by following a mathematical version of the game of "Twenty Questions"' (318);
- 'Harvard Mark I machine ... was probably the first machine to have a conditional transfer of control instruction in the sense that Babbage and Lady Lovelace recommended' (175);
- 'It was Lady Lovelace who made it possible for us to appreciate Babbage's genius' (22);
- 'A remarkable understanding both of mathematics and of horses seems to have persisted in Lady Lovelace's family for many generations, much as did pure mathematics among the Bernoullis or music among the Bachs' (21).

Many more examples could be given. Ada's granddaughter, Lady Wentworth, gave Bowden access to her correspondence, a few bits of which are reproduced. Bowden included a full transcription of Ada's Menabrea paper, silently correcting some of the errors. The collection could be said to have set the tone for subsequent developments.[23]

The physicist Philip Morrison and his then wife Emily wrote a popular article on Babbage for *Scientific American* in 1952.[24] This was expanded as part of a collection of Babbage's writings they edited in 1961, issued in an affordable reprint format. They included their own transcription of Ada's translation, which left the errors unaltered.[25] This popular collection was no doubt responsible for spreading the word far and wide about Ada too.

Maboth Moseley's lax popular biography of Babbage, *Irascible Genius* (1964), devotes inordinate attention to Ada. 'By the time she was fifteen she was showing her extraordinary talent for mathematics, which she inherited from her mother, the former Miss Annabella Milbanke, and had already taught herself part of Paisley's *Geometry*'. References back to her father were irresistible. 'The Byron Vault ... was reopened ... on 15th June 1938. There the coffins of father and daughter, both thirty-six years of age, were found to lie side by side, within a foot of each other, the poet and the mathematician, geniuses both'. Bowden, soon to be a life peer, weighed in again from the foreword, '[Babbage] had genius, he was the affectionate friend of one of the most beautiful and talented women of his time—no less a person than Lord Byron's daughter—who was one of the very few people who understood him, but I think he was born to be miserable!'.[26]

In the *Dictionary of Scientific Biography* (1970) Ada is mentioned only briefly, in a closing remark to the entry for Babbage. 'Lady Lovelace, Lord Byron's mathematically gifted daughter, and her detailed annotations (especially a sketch of how Bernoulli numbers could be computed by the engine) are excellent'.[27]

A 1973 collection of papers on *The Origins of Digital Computers*, edited by Brian Randell with commentary, discusses Lovelace enthusiastically, but not always accurately. It refers to her as 'a gifted young mathematician' whose annotation of Menabrea is 'particularly valuable for its commentary on the capabilities and limitations of the engine, even though it gives comparatively few details of its mechanical design'. However 'one must be careful to avoid reading too much into these passages by simply supplying a modern interpretation, coloured

[23] See also Bowden 1960.

[24] P. Morrison and E. Morrison 1952. Morrison (1915–2005) had worked on the Manhattan project and had been a member of the Communist Party, to which he was recruited at Berkeley. Later he was accused of spying for the Soviets.

[25] Babbage 1961.

[26] M. Moseley 1964. In their letters Ada shows no sign of understanding Babbage at all.

[27] Gridgeman 1970.

by knowledge of present-day computers and their usage, and forgetting the unavoidable limitations, particularly as regards speed, of an entirely mechanical device'.[28]

Malcolm Elwin's posthumously-issued book on *Lord Byron's Family* (1975), based on extensive manuscript material, had some coverage of Ada's childhood, including her arithmetic lessons, but revealed nothing about her mature interests.[29]

The first full-length biography of Ada herself only appeared in 1977, the work of Doris Langley Moore, who had previously published several studies of Byron and was steeped in the Byronalia.[30] Since Moore was neither a computer scientist nor a mathematician herself, even innocent of any background at all in either field, she concentrated on the Byronic drama of Ada's life, especially the role of Lady Byron. Moore's treatment of the Analytical Engine is perfunctory and follows precedent, echoing praise from others for Ada's abilities and contributions. In her preface she allows herself to refer to Ada's 'genius for mathematics'. Some mention is made of her subject's interests in studying mathematics, and of her mother's reputed interests when younger (though these seem to have gone no further than examining a book on geometry and perhaps others). As we have already noted, Ada's correspondence with Augustus De Morgan looked like magic to Moore.[31]

By now Ada's reputation was spreading rapidly, with the growth of computing in general, and several popular histories promoted wild misconceptions about her, including the idea that she had invented binary numbers![32] All this provoked Dorothy Stein, who had some background in computing and mathematics, to re-examine the extant manuscript material.[33] She suspected that Ada's mathematical abilities in particular had been greatly exaggerated, based on an obvious solecism in her translation of Menabrea. Ada had blithely translated a misprint in Menabrea's original text, asserting that 'for *cos* $n = \infty$', which should have been corrected to read 'for the case $n = \infty$'. This was not a typo by Ada, contrary to some commentators, but rather a literal and accurate translation of the misprint in the French original, 'le cos de n = ∞' (cos should be cas), oblivious to its mathematical absurdity. Later reprints of her translation sometimes corrected, sometimes missed this error, but they were not compelled to pay word-for-word attention to the text and its meaning, unlike the translator.[34]

When Stein examined the surviving correspondence with Augustus de Morgan, which Moore had mistaken for magic, it revealed nothing more than a series of elementary calculus homework exercises. De Morgan was acting as her tutor. In these letters Ada shows only rudimentary mathematical ability, puzzling over curves, basic trigonometric formulae, constants of integration, and simple substitute-rearrange-simplify algebraic operations. This led Stein to doubt whether Ada had the required mathematical ability to write the most complex note to Menabrea, Note G about the Bernoulli numbers, herself. She argued that Ada's role in the meat of the notes had been greatly exaggerated. (This material is examined in greater mathematical detail in Appendix G which confirms and extends the substance of Stein's conclusions.[35])

Reaction to Stein's extensively detailed investigation was peculiar. Few confronted her argument directly. Some found the idea risible, even personally insulting, that Ada might not be a pioneering mathematician—a trend that remains strong today. Many preferred not to cite Stein at all or mention her by name. It is *The Case of Disappearing Dorothy*. In his collection of Babbage writings published just three years later, his biographer Anthony

28 Randell 1973, 11, 349.
29 Elwin 1975.
30 Moore 1977.
31 Moore 1977, 99.
32 Rosenberg 1969, 71.
33 Stein 1985.
34 See Toole 1992 for the typo defence; Bowden 1953 silently corrected the error; Babbage 1961 missed it.
35 See chapter 6, p. 59 and Appendix G, p. 297.

Hyman would only refer to Ada's 'far better grasp of mathematics than some commentators have allowed' (that nameless 'commentator' had called out Hyman's careless treatment of Ada more than once).[36] Betty Toole's collection of Ada letters, published in 1992/8, lists an extensive set of Ada sources, including the very slightest, in which Stein doesn't appear at all. The text of the book never mentions her, but refers darkly at the close to 'rumors that are responsible for creating a new myth—that Ada was an incompetent mathematician and did not write the Notes'. This is a topic Toole undertook to investigate in the future, but the results seem to have been decided by her in advance of that, as a 'myth'. Along the way she casually introduces an astonishing non sequitur. 'I have deleted all mathematical formulae from Ada's letters to Mrs Somerville and later to her tutor Augustus De Morgan. The literary remains of Ada's mathematical correspondence are a skewed sample since they reflect what she did not understand'.[37] Mathematical knowledge is not independent and talk of 'sampling' it is wildly inappropriate.

Doron Swade, who helped to coordinate the successful project to construct Babbage's simplified Difference Engine II at the Science Museum in London in the 1980s and 1990s, took care to list Stein's book as 'controversial' in the accompanying booklet's bibliography, but made no other mention of it.[38] Later Swade wrote a popular account about Babbage and the project, *The Cogwheel Brain* (2000), which again lists Stein, not marked now as 'controversial', but does not discuss her findings in the text.[39]

By this date popular science publications on Ada began to multiply quickly, often consisting of the same recycled material, with little new research. As one example consider James Essinger's not especially careful but astutely marketed *A Female Genius: How Ada Lovelace Lord Byron's daughter started the computer age* (2014).[40] This includes the claim that Ada was the world's first programmer, of which more below, and represents her as a consummate mathematician. Stein is listed in the bibliography but not discussed in the text, nor are her findings confronted even indirectly.[41] There are many more treatments like this, usually targeted, with a very low level of fidelity.

Sticking to ostensibly more serious material, the comprehensive *Dictionary of Nineteenth-Century British Scientists* (2004) bulges with marginal female figures of questionable relevance, including obscure botanical illustrators, writers of religious tracts, and wives of leading men like James Clerk Maxwell, who (it is stated) *must* have contributed something, even if those things are not known. No marginal male figures are included.[42] A fulsome entry on Ada refers to her as a 'mathematician' and gives her almost as much space as eminent figures like Augustus de Morgan, Arthur Cayley and Karl Pearson. The authors do not clarify what Ada's mathematical accomplishments were, exactly. The bibliography there of Ada references entirely omits Stein, though it finds space for items like Sadie Plant's *Zeroes + ones: digital women + the new techoculture* (1997). Neither Stein nor her arguments are mentioned in the text either.[43]

An early exception to this boycott of Stein was *The Calculating Passion of Ada Byron* by Joan Baum, a professor of English, in which Stein receives extensive notice, but mostly in the footnotes. The book was written as a brief for the defense of Ada's abilities and importance. However it presents no new evidence or analysis, resting on repeated denial of Stein's interpretations, without the benefit of technical expertise on the subject matter.

[36] Hyman 1988, 243, Stein 1985, xiv, x.

[37] Toole 1992, 1998.

[38] D. Swade 1991.

[39] D. D. Swade 2000; D. Swade 2002.

[40] Essinger 2014.

[41] It seems the same author has also produced a libretto with lyrics, *Ada's Algorithm: the Ada Lovelace Musical: a new two-act musical* The Conrad Press, S.l, 2020.

[42] Lightman 2004.

[43] Füegi and Francis 2004.

A belief has also caught on far and wide that Ada, aside from being a 'mathematician', was the first programmer of the Analytical Engine.[44] Variations on this theme have her as the first *real* programmer, or the first to realize its *general potential* in the modern sense as a universal device. Between 1975 and 1979 the US Department of Defense ran a competition for a new programming language on which to standardize its systems, renaming the winning entry from 'Green' to 'Ada', though it was never widely adopted elsewhere.[45]

None of these claims could possibly be true. It is not in dispute that Babbage invented the Analytical Engine nearly a decade earlier, starting in late 1834. It is logically impossible to invent a programmable machine without devising a way to program it—programs which would necessarily co-evolve with the capabilities of the machine. Babbage's 'scribbling books', which still survive today, contain many such programs. He had taken detailed examples with him to Turin in 1840 where they are still in the archives. These examples were given to Ada, as they were given to Menabrea before her. The Italian's paper of 1842 contains straightforward programs, a fact that appears to get lost in its general assimilation with the translation a year later by Lovelace. Few people, if any, ever read Menabrea's original version, shorn of Lovelace.[46]

Experts have refuted this myth for decades, with little effect beyond their specialist circles. The first substantial study of Babbage, a PhD. thesis by Bruce Collier (1945–2006), appeared in 1970.[47] Collier reacted to Maboth Moseley in particular:

> the ideas were indeed Babbage's and not [Ada's]; indeed, she never made any claim to the contrary. She made a considerable contribution to publicizing the Analytical Engine, but there is no evidence that she advanced the design or theory of it in anyway. And she did not even express an interest in learning about the machine until January 5, 1841, even as late as June 30, 1843, she apparently knew quite little about the mechanical details of the Engine. All of this is said not to belittle Lady Lovelace, but because a very exaggerated view has been formed by some recent writers of the significance of her contribution to the Engine or of her role in Babbage's life.

Robert Anthony Hyman (1928–2011), who wrote the first substantial biography of Babbage was annoyed by the attention given to the simple programs outlined in Ada's notes.[48]

> Babbage, probably some of his assistants, and possibly Herschel Babbage, had carried out similar exercises about five years before, and Babbage carried out the calculations for the Bernoulli numbers to save Ada the work. Ada was probably the third or fourth person in the world to write simple programs, if that is the right word for them. She worked under Babbage's careful guidance and they are student exercises rather than original work. Indeed there is not a scrap of evidence that Ada ever attempted original mathematical work ...

Hyman was nevertheless careful to show some balance. 'Ada's importance was as Babbage's interpretress. As such her achievement was remarkable'. Later he would be more direct and downgrade her precedence. 'It is often suggested that Ada was the world's first programmer. This is nonsense: Babbage was, if programmer is the right term. After Babbage came a mathematical assistant of his, Babbage's eldest son, Herschel, and possibly Babbage's two younger sons. Ada was probably the fourth, fifth or six person to write the programmes'. Inexplicably, Hyman does not mention Menabrea here.[49]

---

[44] See for instance Adams 2023; Baum 1986; Chiaverini 2017; Essinger 2014; Woolley 1999.

[45] Toole 1992 is packed with distracting material about this programming language.

[46] Menabrea 1842. See section F.2, p. 212 for a fresh translation with no Lovelace interventions.

[47] Collier 1970.

[48] Hyman 1982, 198.

[49] Hyman 1997. Hyman's skepticism about the programs being programs is unwarranted. That is exactly what they are as the code in subsection A.4.3, p. 121 shows.

When Collier republished his thesis in 1990 he became much more outspoken in his new preface about Ada's role.[50]

> There is one subject ancillary to Babbage on which far too much has been written, and that is the contributions of Ada Lovelace. Some combination of the connection to Byron, the appeal of the myth that an aristocratic woman was "the world's first programmer," and the even stranger mythology concerning Lovelace and Babbage circulated by Maboth Moseley have led to a popular fascination with her. I am somewhat sorry I did not debunk her role more vigorously in my thesis. It would be only a slight exaggeration to say that Babbage wrote the "Notes" to Menabrea's paper, but for reasons of his own encouraged the illusion in the minds of Ada and the public that they were authored by her. It is no exaggeration to say that she was a manic depressive with the most amazing delusions about her own talents, and rather shallow understanding of both Charles Babbage and the Analytical Engine. ... Ada was as mad as a hatter, and contributed little more to the "Notes" than trouble. ... I guess someone has to be the most overrated figure in the history of computing.

The Australian Allan Bromley (1947–2002) spent many years working on the Babbage papers, of which he had microfilms, leading to a string of technical reports and papers on the technical design of the Analytical Engine. His conclusions support Hyman and Collier.[51]

> ... there are several dozen sample programs prepared between 1837 and 1840 (all, incidentally, substantially predating the preparation of Ada Lovelace's notes, which incorporate several of them). ... Strictly speaking, they are not programs in the modern sense, but are walkthroughs of particular execution instances of programs. ... One group of programs deals with the tabulation of polynomials by difference techniques. ... A second group of programs deals with the tabulation of iterative formulas of varying complexity. Frequently these programs involve the tabulation of a polynomial by difference techniques as a subfunction. The most complex program of this style is that for the Bernoulli numbers included in Ada Lovelace's notes to Menabrea's paper on the Analytical Engine. This program requires a nested-loop structure with a varying range for the inner loop.
>
> A third group of programs deals with the solution of simultaneous equations by variants of Gaussian elimination. Part of the group explores various methods of doing the row reductions, with a view to minimizing the computing time. The remainder of the programs explore ways in which the coefficients of the equations can be arranged in the store. Babbage is clearly seeking simple looping structures for the operation and variable cards. The cycle of operation cards is easily found; it is of similar complexity to that required for the Bernoulli-numbers program.

Bromley subsequently dismissed Ada's supposed role as a 'programmer', with some force.[52]

> Ada Lovelace has sometimes been acclaimed as the "world's first programmer" on the strength of her authorship of the notes to the Menabrea paper. This romantically appealing image is without foundation. All but one of the programs cited in her notes had been prepared by Babbage from three to seven years earlier. The exception was prepared by Babbage for her, although she did detect a "bug" in it. Not only is there no evidence that Ada Lovelace ever prepared a program for the Analytical Engine but her correspondence with Babbage shows that she did not have the knowledge to do so. Babbage seems to have deliberately employed independent persons to convey knowledge of the Analytical Engine to the wider public in exactly the same manner as, a decade earlier, he had used the well-known popularizer of science

50 Collier 1990.
51 Bromley 1980a,b,c, 1982a,b, 1987, 1990, 1991, 2000.
52 Bromley 1990, 88-9. There are very similar remarks in Bromley 1982a, xv.

Dionysius Lardner to convey into print a detailed account of the purpose of the Difference Engine.

In an email exchange with the noted computer pioneer Maurice Wilkes, in the year 2000, Bromley commented, 'I am much concerned that it is so easy to build and spread a myth, as we saw happen with Ada Lovelace'.[53]

The force of this is borne out by remarks made by Doron Swade as late as 2015—at the 200th anniversary of Ada's birth, held at Oxford—again correcting the myth. 'I confirm that the manuscript evidence clearly shows that Babbage wrote "programs" for his Analytical Engine in 1836-7 i.e. 6-7 years before the publication of Lovelace's article in 1843. There are about 24 of such "programs" and they have the identical features of the Lovelace's famous "program".' Swade considered these facts 'unarguable' and that they 'do not support, indeed they contradict the claim that Lovelace was the "first programmer".'[54]

Swade hedges by claiming that Lovelace did contribute 'an original understanding of where the power and potential of computers lay'. Here he is echoing Betty Toole, 'she saw what Babbage did not see: the machine might work on other things beside numbers, e.g. symbols'.[55] There are no grounds whatsoever for this 'good in parts' fallback. The idea that the founder of computing—after nearly two decades of work on the subject, immersed among some of the best minds of his generation at home and on the continent, including Herschel, Wheatstone, Somerville, McCullagh, Jacobi, Plana, Menabrea and von Humboldt, and many others—had not already speculated how mechanized computing might be used in the future and needed *Ada* to raise it, is preposterous. It was the other way around. For years Babbage's Saturday evening soirées swirled with fanciful talk about his automata, his engines and what they might do or not do. 'Reasoning' naturally featured prominently, as it already had in the memoir of Pascal and his calculator by Pascal's sister Gilberte.[56] The *Morning Herald* was getting carried away by this sort of thing as early as 1837, writing of 'Doctor Babbage's accountant machine, into one end of which if you tumble all a bankrupt's books, in five minutes thereafter a clean balance sheet will come out at the other'.[57]

As we have already seen, there are discussions in Babbage's own scribbling books from 1837 on the subject of symbolic computing, a topic closely allied to the genesis of his interest in algorithmic procedures, functional equations. This topic was not convincingly addressed by Ada's offhand annotations. She offered no solution at all beyond the symbol-logging we have shown would not work, even after assuming the solution of stiff technical problems that lay far in the future of computing. Though much has been made of remarks she made in Note A about possible use of the device to 'compose elaborate and scientific pieces of music', she really means the expression of sounds through operations encoding 'the science of harmony and of musical composition' i.e. *playing* music not inventing it, and then only *supposing it is possible*, which she does not commit to. Playing music using a device programmed by a drum is just a flexible 'music box', invented in the previous century. The cards already extended the programmability of the drum in the Analytical engine. Sounds, like ringing a bell, were well within scope of both engines. Programmable printing extended the device to control additional machinery, as did the notion of manufacturing punch cards using programs.

Then there was the highly-sophisticated drum microprogramming of the Analytical Engine, where vertical rows of studs representing instructions come into contact with between 50 and 100 levers as the barrel rotated, and were used to control the mechanical and logical implementation of more complex operations within the engine itself, which Babbage

53 Grier 2004.

54 Quoted in Yanes 2015.

55 Quoted in Yanes 2015. See also D. Swade 2002; Toole 1992, 1998, `https://podcasts.ox.ac.uk/series/ada-lovelace-symposium-celebrating-200-years-computer-visionary`.

56 Belanger and Stein 2005. See the analysis of Note A, section 258, p. 65.

57 1837-02-24, 3. Spreadsheets!

had already placed at the core of it. These instructions also conditionally controlled the movement of the barrel itself, to activate subsequent instructions, with multi-way branching possible.[58] Babbage already comprehended applications of programming beyond mere numbers and had described them in his detailed designs for the Analytical Engine itself. He did not necessarily have solutions for all the technical problems this posed in general, but he understood, as Ada did not, that those problems would not be brushed aside by blithe poorly-reasoned remarks, void of real substance, effortlessly 'extending' this or that feature of the engine.

The analysis by Collier, Hyman, Stein, Bromley and Swade is supported by the investigation here, which shows how the Bernoulli example and all the other examples in the notes are originally taken from prototypes by Babbage, diluted and often mangled by Ada. The role that she played in the notes—really a pastiche of Babbage himself and Babbage misconstrued by Ada—is that of a technical writer, confused explicator and temporary assistant. It extended to stumbling on a major problem in the Bernoulli program, due to an *engine* defect, but also to its successful *concealment*, which has led almost all readers astray even up until the present, along with the insertion of much mathematical gibberish and some fatal errors, also curiously undetected until now. She was not the 'world's first programmer', nor was she any sort of mathematician. Far from being neglected, from the very beginning she received concerted encouragement she was not able to live up to, followed by greatly exaggerated credit, for which Babbage himself was initially responsible.

[58] Bromley 1982a.

Chapter 9

How Feasible was the Analytical Engine?

The Analytical Engine was never built. It existed only as a set of plans, though some models and key components were made for demonstration purposes. Babbage had announced his intention from the start not to build it, but his interactions with the British government and the Kingdom of Sardinia suggested that this was not sincere. He wanted them to ask for it, so that he wasn't seen to be pushing it on them. That was also insurance in the event of failure. But could it have been built and would it have worked?

To answer the question we must agree what 'the Analytical Engine' was. If one is forced to limit the engine to exactly the plans Babbage had on paper we can be confident that changes would have been needed as construction proceeded. Even the Difference Engine II, which was finally built in the 1980s and was a far simpler affair, needed some improvising when it was constructed.[1] The plans on paper were not enough. The Analytical Engine would have raised many more issues. The question is whether an engine built in the *spirit* of the plans, using only engineering techniques already available then, or extensions congruent with contemporary knowledge, would have worked as a general purpose computer.

One objection has already been disposed of here: the lack of addressable memory in the 1843 design. We have shown how a simple change introduced circa 1845 enabled 'card packs' which would have solved the deficiency, without any further changes. More efficient solutions would surely have been found by Babbage without much trouble, but a working solution was already to hand without more advanced engineering. We can conclude with confidence now that the engine was Turing Complete, at least insofar as any machine constrained in practice by finite memory is so.

The question can be analyzed with the above concessions. It would probably be possible today to build an Analytical Engine on the lines intended by Babbage, using his methods, provided pragmatic leeway is granted. But we can say more. The technology to massively *scale* the engine, a locomotive-sized mechanical object in Babbage's conception, was already at hand in 1849, though he was not aware of it. A graduate of Trinity College, Cambridge, slowly recovering from a mental breakdown, had devised the means—therapeutic invention—to control machinery remotely using minute electrical signals which conveyed information rather than power. His name was Francis Galton, and he was a cousin of Charles Darwin. He would later go on to a fabulous scientific career of his own, but his first scientific work would remain almost completely unknown.[2]

Galton's focus, working solo in his rooms in London, was on a printing electric telegraph, which he described in two editions of a now very scarce pamphlet, *The Telotype* (1849).[3] The early telegraphs required hand-coding and decoding to signal. This remained true for

[1] Grier 2004; D. Swade 1991, 2002.

[2] For a comprehensive description of Galton's genesis and interests see Tredoux 2023a,b and `http://galton.org`.

[3] Galton 1849, 1850.

many years, because human operators proved to be remarkably adept at the task. However, Galton could see the enormous potential of automating the entire process, with a (piano-like) keyboard for input at the message source, so that the system was, from a remote distance, 'capable of printing what letters we wish simply as a result of our touching certain keys'. Multiple copies could be generated, making it a duplication system too. His critical insight was that the 'subtle agent' electricity—he experimented with recently-invented batteries housed in his study table—could be used to convey *information*, to effect control, rather than motive power. The apparatus at either end would be powered by its own source, e.g. steam or water-power. Over long distances, electrical power would decline rapidly, especially in those days of inadequate insulation. The signals, feeble in force, 'multiplied' as needed, would only direct the equipment. Like Babbage he built only portions of the device, not the whole, but published comprehensive descriptions of its mechanics and actively experimented with different methods for realizing it.

Galton specified a ternary information-encoding scheme for information, which included not just letters and numerals but also equipment control. 'The Telotype can do more than print letters, it can direct electric currents through different channels', completing remote electrical circuits. Thus, a signal could ring a bell, move the remote paper carrier, mechanically insert a new typeface, or any feasible mechanical action. This scheme was something like a modern ASCII table. Ternary because he was thinking of a needle kept neutral or moved left or right—three states (o,l,r) rather than the two used in binary encoding. This followed Wheatstone's original telegraph design— Galton had met Babbage's friend at Richard Partridge's house when he was still a medical student in 1839/40—but now without involving humans to encode and decode the needle positions. Three 'trits' (carried on three wires) of three states each would give 27 characters, covering the alphabet.

Trit 3	Trit 2	Trit 1	Decimal	
0	0	0	1	'a'
0	0	1	2	'b'
0	0	2	3	'c'
0	1	0	4	'd'
0	1	1	5	'e'
0	1	2	6	'f'
...				
2	2	0	25	'y'
2	2	1	26	'z'
2	2	2	27	CTRL

More expressive systems were also contemplated by Galton. Words could be formed from successive triples. Or physical wires, which were a nuisance, could be reduced by using different currents, as Wheatstone himself had already experimented with. 'Besides sending a positive or a negative current, we may also send a current of greater or less electro-motive power, and we may send any current at perceptibly differing intervals of time'. Remarkably, a method for manufacturing ribbon cables of insulated parallel wires is described. In the second edition of his pamphlet he even designed an automatic encryption module for scrambling the signals at source and descrambling them on receipt, using a codeword key and a faceted rotating drum.

The inventor also envisaged hub-and-spoke networking of the devices to exchange data at scale. 'If telegraphs that printed satisfactorily were once found practicable, most large houses, public and private, would soon become supplied with them. The communication being so immediate, answer following question as fast as it is put, affords much more nearly the advantages of a personal interview then the best regulated Post Office ever could. Any scheme to introduce such telegraphs generally, would probably be first confined to London. There would be central offices, and from these bundles of wires would radiate to numerous branch offices. From the branch offices again wires would pass along adjacent streets, supplying houses as they passed. ... As an example how one house would communicate with another, let C be the central station, A a house in wire connection with the branch office B and A1 with B1. The A signals to B to connect its wire with C, then to C to connect it with B1, next to B1, to connect it to A1, and the chain is complete'.[4]

Galton described how mechanical means could be used to interpret the signals in the case of printing apparatus, releasing an amplifying cascade of detained cogwheels by means of raised detaining arms. 'The signal is taken up by each arm in its turn, and each having forwarded it on to the next, drops back to its place'. Receipt of a complete signal is automatically signalled back by the receiver to the transmitter. Using his scheme, generalizing from the special case of printing messages to equipment control, a cluster of Analytical Engines could be harnessed to exchange information, including initiation and intermediate results of complex computations conducted in parallel. This would have solved a glaring defect in the Analytical Engine. There were mechanical limitations to how large the devices could feasibly be built. More storage, more significant digits, required additional physical columns of digit cogs. Communication within the device required rods and levers in direct mechanical contact, subject to constraints on the physical force used to avoid equipment jamming and breakage. These posed major challenges even at the room-sized scale Babbage was considering. Now the devices could be split over multiple locations. Once that was mastered it would have been obvious that the room-sized devices could be split up into smaller parts which could communicate with each other using electrical signals rather than rods and levers.

The combination of these two inventions would have supplied what the other lacked. Galton's 'telotype' was not a general purpose computer, stopping at automatic encryption of its messages. Babbage's device was a general purpose computer but could only communicate and compute through mechanical movements, actuating cogwheels through pinions and rods. Combined they would soon have evolved to their modern form, once it was realized that the mechanical clockwork mechanisms only got in the way of information manipulation and exchange, physically limiting the power of the devices. But the opportunity was missed.

In 1850 Galton departed to explore South-West Africa for two years and never resumed work on the *Telotype*. It is possible that he had overworked himself on the problem and had broken down again, a recurring tendency in those years. Babbage continued to tinker with the Analytical Engine for a decade more but made no more advances. The two men would meet in later years through the Royal Geographical Society. Galton sent Babbage a copy of his *Art of Travel: or Shifts and Contrivances Available in Wild Countries* (1855, always in print),[5] which a grateful Babbage declared one of the few books he could afford to 'read through' and sent a copy of to one of his sons in South Australia.[6] Galton's papers contain a faded photograph of Babbage,[7] curiously and incorrectly annotated 'Babbage. Given me by Sir Wm Grove. He never was photographed except on this occasion & by him, July / 82'. Unfortunately, neither man appears to have left any record of ever discussing computing and electrical communication with the other.

4 Galton 1849.
5 Galton 1855, 1856.
6 Babbage to Galton, 1857-04-17, UCL GALTON/3/2/2.
7 UCL, GALTON/1/5/4.

Chapter 10

Portraits.

Figure 10.1: Georgiana Whitmore and Charles Babbage, around the time of their marriage. Georgiana is by Isaac Wane Slater, Soho, London, 1813. The Babbage miniature is not attributed. Powerhouse collection, object ID: 413046.

Figure 10.2: J. F. W. Herschel (1792–1871) by Alfred Edward Chalon, around 1829.

Figure 10.3: Charles Babbage (1792—1871) by Samuel Laurence, likely around 1840 but sometimes dated 1845.

Figure 10.4: Charles Babbage, likely around 1852. Claudet daguerreotype.

Figure 10.5: Ada Lovelace in 1836, by Margaret Sarah Carpenter.

Figure 10.6: Giovanni Antonio Amedeo Plana (1781-1864), professor of astronomy at Turin, around 1850.

Figure 10.7: Luigi Federico Menabrea (1809–1892).

Appendix A

Ada's Notes: Technical Details.

Those notes which require more involved technical treatment, with some light mathematics, are dealt with here for the more motivated reader. After the first note we will use more compact notation, e.g. $V_{12} \leftarrow V_{22} * V_{11}$ $[B_3 * A_{3,4}$ since it is logically equivalent to the longform, e.g. 'MULT ...' and much easier to read. The machine itself uses punch cards.

A.1 Note B. Exponentiation and Symbolization.

The explanation of exponentiation given in the note is flawed, since it doesn't exhibit the crucial loop and show that the operation can handle *any* positive power in principle. Nothing more than a hard-coded fixed multiplication is described there.[1] A general implementation of exponentiation in the Analytical Engine instruction set is given below. An IF 'combination card' has been assumed, which executes the next card if the given variable is > 0, and skips over it otherwise. The GOTO operator (a special case of the conditional GOTOIF statement) skips the given number of cards, forward or backward, where for simplicity the number to skip is encoded directly on the card (operations and corresponding variable triples). A little work is needed to make up for not having logical operators. This program only works correctly for non-negative integer values of $V_2 = n$ but will always terminate.

Operation Cards	Variable Cards	Constant Cards	
		$V_1 \leftarrow 5$	
		$V_2 \leftarrow 7$	
		$V_3 \leftarrow 98$	
		$V_4 \leftarrow 1$	$t \leftarrow 1$
		$V_6 \leftarrow 1$	
MINUS	V_6 V_2 V_8		$1 - n$
IF	V_8		$(1 - n) > 0 =$ $(n <= 0)$
GOTO +4			
MULT	V_4 V_3 V_4		$t \leftarrow t \times x$

1 See chapter 6, 68.

MINUS	V_2 V_6 V_2		$n \leftarrow n-1$
GOTO -5			
MULT	V_4 V_1 V_4		V_4 is ax^n

A.2 Note D. Annotations for Zeroed Variables.

Note D explicates the superscript notation Ada (or perhaps Babbage, here we cannot be sure) introduces in an attempt to clarify the behavior of these programs, which replaces that used by Menabrea in his own text (a remarkable liberty). One problem the Italian had to overcome was a side-effect of loading variables from the store. To allow the programmer to keep things neat and tidy, a variable card may instruct the engine, by a special mark on it, to clear the value from the store once the operation using it is complete. Then the columns can be used to display results without the confusion of intermediate values, which are erased. But if the variable has multiple uses, the programmer can choose to retain its value instead of clearing it.

Ada and/or Babbage introduce their own ideas about this using superscripts for 'before' and 'after'. Their extensive use of these annotations creates the impression that Menabrea's programs are really execution traces, which they are not (their own example is more complicated, see below). The result is unhappy (really, a mess) and plagued with logical inconsistencies.

Menabrea had chosen to indicate variable zeroing by an additional descriptive column in his tables showing the program instructions. Ada (or perhaps it is Babbage) believes that she (or it may be he) can improve this by specifying, using a superscript, how many changes the variable has gone through as it is used in operations. Thus ${}^0V_1$ is the initial value of zero, ${}^nV_1$ is its n-th non-zero value. One might think, and Ada suggests in her note, that this may be used to show how operations themselves (and not the variable cards) have an effect on stored variables, but that merely restates the operation involved, as in ${}^2V_1 = {}^1V_1 \times {}^1V_2$, which is not very helpful. Moreover it supposes that one can even tell whether V_1 changes in value (so that we must increment its superscript) which it will not if $V_2 = 1$, or if that is the 341234-th decimal digit of π calculated by the program as V_2, we simply do not know. This is not in any event how the notation is actually used in the Menabrea tables, rewritten by Ada to contain a blizzard of superscripts, because the only concern there is to disambiguate what the variable cards requested as a side-effect: were the variables to be reset to zero or not? Now ${}^1V_1 = {}^1V_1$ specifies no zeroing, whereas ${}^1V_1 = {}^0V_1$ specifies that it has been zeroed.

This just exacerbates the awkwardness. A non-sequitur ${}^1V_1 = {}^0V_1$ has been added to the truism ${}^1V_1 = {}^1V_1$. A non-sequitur since ${}^1V_1$ is the 1-st non-zero value of V_1. Ada has committed herself to the idea that zero is a special value and she does not want to label duplicates of zero as new states of the variable. This led her to the rule that when the n-th value ${}^nV_1$ is zeroed by a variable card, it reverts to ${}^0V_1$, but then if it is changed again, it progresses to ${}^{n+1}V_1$.

Putting the above problem aside, letting ${}^nV_k$ just be whatever value V_k has after n changes to it, the more fundamental trouble is that this notation confuses changes in a variable as the result of the operation itself with changes as a side-effect of the variable card used in the operation. Consider the program $V_1 \leftarrow V_2 \times V_1$ where ${}^1V_1$ is initialized by a constant card but ${}^nV_2$ is the result of prior calculations, and assume that the variable card for V_1 does *not* request zeroing. If we annotate this ${}^1V_1 = {}^1V_1$ (as Ada does for all such situations in which zeroing was not requested) then it may be false, because the *operation* $\times$ may have altered V_1. Since we do not know the value of ${}^nV_2$ (suppose it was calculated earlier as 0 if the 231423523-th decimal place of π is 2, and 1 otherwise) we cannot say if ${}^2V_1$ is non zero or really back at square one, ${}^0V_1$. So was zeroing requested by the card for V_1 or not?

The notation creates more problems than it solves. One obvious solution is to just mark up the variable (exactly as it would be in practice on the card) where it appears so that the requested side-effect is visible. Using yet another superscript 0 we might write $V_2 \leftarrow V_1^0 + V_1$ in the *program*, not in a descriptive column, to note unambiguously that zeroing was requested for V_1, otherwise it was not, irrespective of the results of the operation. This appears in its natural place as an instruction. Any other markup might be used—this is just a choice about how programs should be presented in written form in explanatory documents, the card themselves already have the required facility. With that the superscript notation may simply be discarded, along with the tables of values, and the presentation simplified.

A.3 Note E. Multiplying Expressions.

As noted previously, this is a modest simplification of an example prepared by Babbage.[2] Regularizing notation a little, Ada multiplies into an infinite series

$$(B_0 + B_1 \cos\theta)\,(\sum_0^\infty A_k \cos k\theta)$$

Approximation is not mentioned, but it emerges later that she only proposes to calculate finitely many terms of this (four in her example). Much rearrangement and use of simple facts like

$$\cos P \cos Q = \frac{1}{2}\cos(P+Q) + \frac{1}{2}\cos(P-Q)$$

give a laborious analogue of the same methods Menabrea had already shown adequately.[3] However Ada naively fails to account for the fact that the series might *diverge*, in which case approximation is not possible. Set $B_0 = 1, B_1 = 0, A_0 = 0, \theta = 0$ and for $k > 0$, $A_k = 1$ and then we get $\sum_1^\infty 1 = \infty$.

The conveniently manipulable terms are chosen to be so. The omission of a crucial detail like the assumed convergence of the series is understandable in rough notes by Babbage, which Ada failed to comprehend fully.

The fact that coefficients for arbitrarily many terms of the expansion above can be calculated, using the easily-derived general term, where $n \geq 1$

$$C_n \cos n\theta = \{B_0 A_n + \frac{1}{2}B_1(A_{n-1} + A_{n+1})\}\cos n\theta$$

This again floats the *repetition of instructions* theme, which could easily be demonstrated by just showing the programming of a loop for producing all the first n terms. But that is not shown, just as it was not for ax^n or anywhere else in these notes (loops are deliberately obscured in the Bernoulli number example too). Instead much is made of counting saving of overall operation cards by executing cycles of them (again without counting the looping operations themselves). This counting of operations angle was previously covered extensively

2 See section 262, p. 67.

3 Note that equation (3) on p. 714 of the note should be $\cos n\theta \cdot \cos\theta = \frac{1}{2}\cos(n+1)\theta + \frac{1}{2}\cos(n-1)\theta$.

by Babbage in his unpublished description of the engine from 1837, and in the materials he brought to Turin.[4]

There is also vague mention in note E of some ideas from the calculus of functions. This was a Babbage speciality and can only have come directly from him, but it is sketched by Ada with the sort of vagueness that implies incomprehension. She dilates at great length on her inability to assume the reader has enough background to understand what this is about—why mention it then?—but it can be explained sufficiently in a single paragraph, less space than she allocates to doubting the competence of her readers. Babbage had already shown in 1820 how recursive expressions can produce (what amount to) cycles of cycles, and cycles of cycles of cycles, and so on.[5] Define the n-th function f^n of f by $f^0 = I$ (the identity function $I(x) = x$) and $f^{n+1} = f(f^n)$ which we can simply write as ff^n. Thus what appears to be multiplication is composition, and a power is really repeated composition, or recursion as that came to be known. However, algebraically the indices behave as one would hope, so that $f^m f^n = f^{n+m}$. If we define f^{-1} to be the inverse of f then $f^{-1}f^1 = f^{-1+1} = f^0 = I = ff^{-1}$. Then if $f(x)$ contains a cycle of instructions using x (produced by a loop) then $f^2(x)$ will contain a cycle of a cycle. One can make this as complicated as one likes by, instead of just composing f with itself, substituting it into another expression D, the 'law of development', and then recursing the result back into f to get more cycles of the 'development' as fDf. And so on with other expressions, interweaved to taste. This will produce a series of 'cycles within cycles' of arbitrary complexity.

Rather than explain this, which took less than a paragraph, Ada absurdly pleads lack of space—at great length. 'We should premise that we suppose the reader to understand what is meant by an nth function. We suppose him likewise to comprehend distinctly the difference between developing an nth function algebraically, and merely calculating an nth function arithmetically. If he does not, the following will be by no means very intelligible; but we have not space to give any preliminary explanations'. The last argument is a repetition of the idea already noticed, that merely because the engine program names operations, assuming that functions can be referenced similarly, and only binds them to specific variables when expressions are 'developed' into numerical results, then this enables algebraic manipulation of expressions involving those functions without any variable binding. Of course an engine *can* be built to do that, but the Analytical Engine is not that engine. Ada's repeated assurances that only minor changes are needed is nonsense. Babbage himself had already abandoned the idea of building that kind of engine. The only way the program fDf above will ever progress in the Analytical Engine is when $(fDf)(x)$ is evaluated, where x is a concrete value.

Ada finishes her note E with an attempt to find notation for expressing the structure of cycles of operators succinctly.[6] This notation has to be carefully reconstructed from the statements she has provided, since these are somewhat confused. She starts by considering expressions which have a mixture of operators, as in $a + bx + cx^2$, from which a (bracketed) list of operators can be formed $\{+, \times, +, \times\}$, assuming that x^2 is already calculated. Group this as $\{2(+, \times)\}$. Similarly we may extract from

$$(a + bx + cx^2)/(r + sx)$$

the list $\{\div, 3(+, \times)\}$ (two from the numerator and one from the denominator).

Next she introduces what she describes as 'a technique from the Integral Calculus' to achieve more expressiveness. This puffs Euler's humble symbol Σ, here always used in the finite sense as shorthand for plain old addition, and therefore nothing to do with the 'Integral Calculus' (a phrase repeated three times in quick succession by Ada). It is necessary to work

4 See subsection D.4.4, p. 163 and subsection E.2.6, p. 173.

5 Babbage 1820.

6 Menabrea 1843b, 719-20.

backward from her equations (9) and (10), which show that Σ counts elements *concatenated* onto the list of operators, just as $n(+,\times)$ expands into n repetitions of $+,\times$. However (10) must be corrected because Ada continually confuses the (variable) indexes of terms with their constant limits, a tyro's error we will return to below. Numbering each operator uniquely as $1, 2, 3, ..., m$, discarding their symbols, we may reduce the list of operators, where p is a constant

$$p(1,2,\ldots m),(p-1)(1,2,\ldots m),\ldots,(p-n)(1,2,\ldots m) \tag{9}$$

to the more compact form

$$\sum_{i=p-n}^{p} i(1,2,...,m) \tag{10}$$

Attempting to generalize expressions of this sort, Ada wants a flexible multiplier function of the variable index i, call it $\phi(i)$ for reasons which will emerge below. She does not explain this, but it would look in the case of (10) like

$$\sum_{i=p-n}^{p} \phi(i)(1,2,3,...,m) \tag{10*}$$

She applies this technique in equations (6) and (7), but the confused description she gives only mystifies. In both these equations she conflates the variable index with its constant limits. Thus in (6) below p is mangled, 'p stands for the variable; $(+1)^p$ for the function of the variable, that is, for ϕ p; and the limits are from 1 to p, or from 0 to $p-1$, each increment being equal to unity'. Again, in (7) n is supposed to be the name of the index *and* its upper limit, with the 'limits of n being from 1 to n'.

$$(\div),\sum(+1)^p(\times,-) \tag{6}$$

$$\sum(+1)^n\{(\div),\sum(+1)^p(\times,-)\} \tag{7}$$

We must correct these misunderstandings, and will use i or k for the indexes in (6*) and (7*) below.

But Ada also wants to illustrate the idea in (10*) above of using a function $\phi(i)$ for the repetition, which she garbled without any explanation as 'ϕ p'. In this case she uses a degenerate one $\phi(i) = (+1)^i = 1$ just to maintain that form. (Using $+1$ rather than just 1 was rare but not completely unknown in those times.) This she enigmatically describes as 'the function of the variable' (i.e. index) and doesn't bother to explain why it is there even though in this case it does nothing whatsoever, looking like some mysterious operator, which it is not. We may correct her (6) and (7) then to

$$(\div),\sum_{i=1}^{p}(+1)^i(\times,-) \tag{6*}$$

$$\sum_{k=1}^{n}(+1)^k\{(\div),\sum_{i=1}^{p}(+1)^i(\times,-)\} \tag{7*}$$

As Ada now sheepishly points out, this means *exactly the same thing*, i.e. produces the same lists in these two cases as the far simpler notation,

$$(\div),p(\times,-) \quad \text{and} \quad n\{(\div),p(\times,-)\}.$$

Her point is that the more complex notation handles more sophisticated cases, as in (9).

This mess of potage jars with Babbage's advocacy of the importance of good notation in mathematics, which he argued encouraged productive analysis. Lack of comprehension combines with overt pretension ('Integral Calculus'). The underlying ideas were probably based on notes provided to Ada by Babbage, poorly understood and garbled. That is why $\phi\ p$ survives in her unhelpful description of equation (6), as symbolic debris. Babbage had been interested in the analogous problem of counting how many symbols recursion produces since at least 1820, and had already developed applicable techniques.[7]

The ability of the engine to repeat operation cards in (possibly nested) loops leads Ada to attempt an analysis of how extensive a savings in sheer number of cards this brings. The idea is not to reduce the number of operations themselves, only the physical cards corresponding to them. Repetition of cards does enable fewer taped-together cards to be used, but that does not reduce the number of operations executed. Nor does it even reduce the number of times the cards are read, since they must be read anyway whether they are reused or not, and moving back to them is no cheaper operationally than advancing to new ones. The introduction of combination cards to control the loops actually increases the total number of operations needed, though not by much. *This should not be confused with modern analysis of computational complexity*, which seeks to reduce the number of executed operations by analyzing the logical structure of the program (here, the loops) from which the number of operations is deduced, not permitting the constant-time operations to obscure the dynamic complexity, which is the real interest.

For the solution of a system of 10 simultaneous equations in 10 variables (ignoring subsequent substitutions to actually find the unknowns) she counts 330 total calculations on the coefficients, taken individually (a mixture of multiplication and subtraction). Grouping the operations, she notices repeated applications of the triple $(\times, \times, +)$. If the variable and result cards are arranged in the right order, looping around this operation card triple will reduce the card count, since the operand and result variable cards are repeatedly consumed. This leads her to conclude that 'three Operation-cards would perform the office of 330 such cards'. Here she is comparing the engine to another (also hypothetical) system where each coefficient calculation above gets its own cards. Notice that her total omits the combination cards and index decrements needed to program the loop, and it doesn't count the variable cards, all of which are needed to actually produce the 330 calculations she is making this comparison for.

A.4 Note G. Bernoulli Numbers.

The reader should beware that the note uses an archaic definition of the Bernoulli numbers.[8]

$$
\begin{aligned}
B_{2n-1} &= 2 \cdot \frac{1 \cdot 2 \cdot 3 \cdots 2n}{(2\pi)^{2n}} \cdot \left\{ 1 + \frac{1}{2^{2n}} + \frac{1}{3^{2n}} + \cdots \right\} \qquad (2) \\
&= \frac{2(2n)!}{(2\pi)^{2n}} \sum_{k=1}^{\infty} k^{-2n} > 0
\end{aligned}
$$

Here B_{2n-1} is the n-th Bernoulli number, i.e. it ranges over all of them, so that $\{B_{2n-1}\}$ is the complete set. This is a completely arbitrary label, and $B^{\dagger}_{2n}$, or simply B^{*}_{n}, or other indexes could be used instead without changing anything of importance below.

The direct formula (2) is inconvenient to calculate with, though it does tell you that every Bernoulli number (under this definition) is positive (hold that thought). Ada also

7 Babbage 1820.

8 For the genesis of Note G, see section 276, p. 71.

lists another direct formula, which she labels (3) and does not use—which is just as well, since it doesn't work.[9] Instead a recurrence relation can be used to calculate B_{2n+1} from the preceding $\{B_1, \ldots, B_{2n-1}\}$, provided that these values were previously calculated and stored, which seems like a suitable task for an Analytical Engine program, but as we shall see imposes the onerous requirement to remember all of the prior calculations. The values of the Bernoulli numbers can then be tabulated, provided that the signs are always corrected, after B_1 is calculated directly and the rest $B_2, B_3...$ accumulated. Several such (manual) tabulations would appear in the 19th century, and there is one by De Morgan in the *Penny Cyclopaedia.*[10] The following equation is used in the note to deduce a recurrence relation for B_{2n+1} (factorial notation, known at the time but not used by Ada, neatens things). The recurrence relation is obtained by manipulating the following expansion:

$$\frac{x}{e^x - 1} = 1 - \frac{x}{2} + B_1\frac{x^2}{2!} + B_3\frac{x^4}{4!} + B_5\frac{x^6}{6!} + \cdots \tag{4}$$

This is the start of a great deal of trouble. The formula (4) is wrong, and will lead to every other supposed 'Bernoulli number' B_{2n-1} having the incorrect sign in what follows, a mistake has been blithely copied many times.[11] Relabelling the actual Bernoullis as B_n^* to obviate confusion, the formula should be something like:

$$\begin{aligned}\frac{x}{e^x - 1} &= 1 - \frac{x}{2} + (-1)^0 B_1^* \frac{x^2}{2!} + (-1)^1 B_2^* \frac{x^4}{4!} + \cdots \\ &= 1 - \frac{x}{2} + \sum_{n=1}^{\infty} (-1)^{n-1} B_n^* \frac{x^{2n}}{(2n)!}\end{aligned} \tag{4*}$$

It is just barely possible that Ada was led astray here by puzzle-patedly copying De Morgan's text.[12] On page 248, De Morgan has signs but there he is using the 'modern definition' of Bernoulli numbers we have denoted $B_n^\dagger$.[13] Multiple confusions are needed to copy this. On pages 553-4 De Morgan switches to the 'old definition', which are all positive, and clearly notes the change of form. In the *Penny Cyclopaedia* 'Numbers of Bernoulli' entry, Ada would have seen the form she was using with the numbers correctly enumerated all the way up to B_{49}^* (De Morgan personally contributed over 700 articles to this encyclopedia, about 1/6 of its total length, including this entry, according to his wife).[14]

We will follow Ada's working using (4) as she did. If we multiply both sides by $\frac{e^x - 1}{x}$, expand and simplify (described elliptically by Ada as 'expand the denominator of the first side in powers of x, and then divide both numerator and denominator by x') we get

$$1 = (1 - \frac{x}{2} + B_1\frac{x^2}{2!} + B_3\frac{x^4}{4!} + \ldots) \times (1 + \frac{x^1}{2!} + \frac{x^2}{3!} + \frac{x^3}{4!} + \ldots) \tag{5}$$

$$= 1 + \sum_{k=1}^{\infty} D_k x^k \tag{6}$$

where the D_k can be found by gathering all the coefficients of x^k in the product. However from (6) $0 = \sum_{k=1}^{\infty} D_k x^k$ so we must have $D_k = 0$ for all k.

9 See subsection A.4.7, p. 130.
10 A. De Morgan 1840b.
11 E.g. Stein 1985, 89, Hollings, Martin, and Rice 2018, 77-8.
12 A. De Morgan 1842b, 248, 553-4.
13 See subsection A.4.6, p. 129.
14 A. De Morgan 1840b, S. E. De Morgan 1882, 411.

We care about the coefficient D_{2n} of x^{2n} because it involves the Bernoulli numbers. Equation (6), considering only terms which can maybe produce x^{2n}, can be written as

$$[-\frac{1}{2}x^1 + \sum_{k=0}^{n} \frac{B_{2k-1}}{(2k)!}x^{2k}][\sum_{0}^{2n} \frac{1}{(k+1)!}x^k] \tag{A.1}$$

To get a power $2n$ we must combine the only pair of odd powers $(1, 2n-1)$ or any of the pairs of even powers $(2n-2k, 2k)$ where $k = 0 \ldots n$. Let's define $B_{-1} = 1$ and collect these as

$$\begin{aligned} D_{2n} = & -\frac{1}{2}\frac{1}{(2n)!} + \sum_{k=n}^{0} \frac{B_{2n-2k-1}}{(2n-2k)!}\frac{1}{(2k+1)!} \\ = & \frac{1}{(2n+1)!} - \frac{1}{2}\frac{1}{(2n)!} + \frac{B_1}{2!}\frac{1}{(2n-1)!} + \frac{B_3}{4!}\frac{1}{(2n-3)!} \\ & + \frac{B_5}{6!}\frac{1}{(2n-5)!} + \cdots + \frac{B_{2n-1}}{(2n)!} \\ = & \, 0 \quad \text{since all} \quad D_k \quad \text{are zero} \end{aligned} \tag{7}$$

Multiplying through by $(2n)!$ and simplifying a little gives us a linear dependency between $B_1, \ldots, B_{2n-1}$:

$$\begin{aligned} 0 = & (-\frac{1}{2})(\frac{2n-1}{2n+1}) + B_1\frac{1}{2}2n + B_3\frac{1}{4!}\frac{(2n)!}{(2n-3)!} \\ & + B_5\frac{1}{6!}\frac{(2n)!}{(2n-5)!} + \cdots + B_{2n-1} \end{aligned} \tag{8}$$

Introducing constants for the coefficients in (8) is convenient. Ada does not take appropriate care in the notation: the coefficients A_k must all be recalculated afresh for each n.

$$0 = (A_0 + A_1B_1 + \cdots + A_{2n-3}B_{2n-3}) + B_{2n-1} \tag{9}$$

For example $A_5 = \frac{1}{6!}\frac{(2n)!}{(2n-5)!}$. She notices in the text that these constants are actually functions of n but adding subscripts would have made that much clearer and harder to overlook subsequently, when forgetting this gets her into trouble.

$$0 = A_{0,n} + A_{1,n}B_1 + A_{3,n}B_3 + \cdots + B_{2n-1} \cdots \tag{9*}$$

The next step is clearly to restate (9) to give an expression for B_{2n-1} which can then be calculated if the previous B_{2k-1} are known, say because we calculated them earlier and saved the results, and the appropriate $A_{2k-1,n}$ are calculated as per (9*).

$$B_{2n-1} = -(A_0 + A_1B_1 + \cdots + A_{2n-3}B_{2n-3}) \tag{9b}$$
$$= -(A_{0,n} + A_{1,n}B_1 + \cdots + A_{2n-3,n}B_{2n-3}) \tag{9*b}$$

Bizarrely Ada does not take this elementary step. Instead, though she does not state this explicitly, when she tries to calculate B_{2n-1} subsequently she uses the bracketed part of (9). This is stated repeatedly, in both the table at the end of the note and in the text of the note itself, as she considers how B_1, B_3 *would* be stored, and how B_7 *is* stored. We will refer to this implicit incorrect form as (9c) though she does not.

$$B_{2n-1} = A_0 + A_1B_1 + A_3B_3 + \cdots + A_{2n-3}B_{2n-3} \tag{9c}$$

Even if the right formula (9b) is used, she is missing the $(-1)^{n-1}$ multiplier for each B_{2n-1} because she has the incorrect formula in (4). These multipliers ensure that the Bernoulli numbers are positive as they must be by the definition in (2). This can easily be fixed by setting the n-th Bernoulli number $B_n^* = (-1)^{n-1}B_{2n-1}$. For example, a little arithmetic shows that $B_1 = \frac{1}{6}$ which is indeed B_1^*, but $B_3 = -\frac{1}{30}$, which has the incorrect sign, so the Bernoulli is really $B_2^* = \frac{1}{30}$.

As noticed before, Ada realized there was a problem with the signs and tried to brush over it. The correct formula (4*) and the right expression for B_{2n-1} from (9b), elementary alterations, would fix that, but she was unable to deduce those corrections. The recurrence relation requires that signs consistent with it be used. If instead of calculating the preceding 'Bernoulli numbers' explicitly they are loaded from a known enumeration, in which they are positive, Ada's recurrence will produce not just 'Bernoulli numbers' with the wrong signs, but entirely the wrong numbers. The loaded numbers need their signs adjusted appropriately to become the pseudo-Bernoulli numbers mistakenly solved for.

It is a lot harder to explain how Ada failed to derive formula (9b) than how she got the wrong definition in (4). Deriving (9b) is simple arithmetic. Silently using the incorrect formula (9c) has a domino effect as subsequent B_{2k-1} are calculated from earlier incorrect ones. We must either suppose that she failed to do the subtraction, because of sheer incomprehension, or we can look for an explanation in the trouble that the signs caused her. Even though (9b) would not fix that problem per se, it is possible that she was first driven to abandon (9b) and silently use (9c) in the *hope* that her sign problem would be resolved. When later calculations showed that it did not, she may, thoroughly perplexed, perhaps embarrassed, simply have left things as they were. A failed 'correction'. Of course, writing out, as we may deduce that she did, the calculated Bernoullis for the first few values and comparing them to known results (by hand or from published enumerations) should soon have showed her that she needed a factor like $(-1)^{n-1}$ to correct them. But evidently she could not see that. Which leaves no good explanation apparent. But what follows is even more interesting.

Unable to deduce (9b) directly Ada still feels compelled to explain how she gets the Bernoulli numbers from (9c), which is what she will use in her 'program trace'. It seems that the term 'linear dependence' (which unknown to her is exactly what (7), (8) and (9) express) had lodged in her memory, perhaps from conversations with Babbage, or the worked out details he had given her that she was expanding on. She persuaded herself that this 'linear dependency' was a statement about the index n of D_n in (6).

> On examining (7.) and (8.), we perceive that, when these formulae are isolated from (6.) whence they are derived, and considered in themselves separately and independently, n may be any whole number whatever; although when (7.) occurs at one of the D's in (6.), it is obvious that n is then not arbitrary, but is always a certain function of the distance of that D from the beginning.

It is certainly (uselessly) true that the n in D_n expresses its 'distance from the beginning' of the equation, since it just denotes the $(n+1)$-th term. It is possible that Ada is led to this by 'linear' part of 'linear dependence' because a line can be used to measure distance. Lacking a mathematical repertoire, she uses her 'verbal intelligence' instead. However that may be, she then proceeds to derive some more completely useless expressions of n in terms of this supposed distance d.

$$2n + 1 = d, \text{ and } n = \frac{d-1}{2} \text{ (for any \textit{even} power of } x\text{)}$$

$$2n = d, \text{ and } n = \frac{d}{2} \text{ (for any \textit{odd} power of } x\text{)}$$

Deriving useless tautologies is a familiar experience for mathematicians. The sheet of paper is crumpled up and deposited in a waste paper basket. Instead Ada states this empty 'result' and leaves it hanging there. It plays no further part in her discussion, nor could it. Done with the 'linear dependence' she found in (6), she resolves instead to move on to what she believes is *linearly independent*, that is equations (8). Independent of n that is.

> This circumstance, combined with the fact (which we may easily perceive) that whatever n is, every term of (8.) after the $(n+1)$th is $= 0$, and that the $(n+1)$th term itself is always $= B_{2n-1} \cdot \frac{1}{1} = B_{2n-1}$, enables us to find the value (either numerical or algebraical) of any nth Number of Bernoulli B_{2n-1}, *in terms of all the preceding ones*, if we but know the values of $B_1, B_3 \ldots B_{2n-3}$.

A quicksand of nonsense. One can only speculate how she floundered into it. The equation is certainly dependent on n, not independent of it, which is why it appears there. She may have imagined phantom terms $0 = A_{n+1}B_{2n+1} = A_{n+2}B_{2n+3} = \ldots$ omitted from the linear dependency (8), confused by the infinite series form of (what we now call) a generating function. Those can be inserted but they are quite useless. Notice that the *coefficients* $A_{n+1}, A_{n+2}, \ldots$ should really be written $A_{n+1,n}, A_{n+2,n}, \ldots$ since they depend on n, and only then set to zero. All of which tells us nothing interesting about the Bernoulli numbers $B_{n-1}, B_{n+1}, \ldots$ which are definitely not zero.

It is also anybody's guess what Ada meant by $B_{2n-1} \cdot \frac{1}{1} = B_{2n-1}$ and what she thought that implied. It may merely be that the omitted coefficient $A_{2n-1}1$ is 1, with $\frac{1}{1}$ used to resemble the result of work, or from some other confusion. Again, no conclusions follow from this useless tautology, nothing is explained by it.

All this symbol shuffling is used as a spell to summon the conclusion 'enables us to find the value (either numerical or algebraical) of any nth Number of Bernoulli B_{2n-1}, *in terms of all the preceding ones*, if we but know the values of $B_1, B_3 \ldots B_{2n-3}$'. But nowhere does she explicitly state that she will (wrongly) just use (9c) as her formula.

Returning to the real world, it is the *terms* in (7) and (8) that are linearly dependent on each other, and both equations depend on the index n, which may in turn take any positive integer value. None of which matters since (9b), which is what she needs, follows directly from (9) by simple arithmetic. Both 'linear dependence' and 'linear independence' are just phrases which have assumed the status of incantations in her thinking.

Next Ada turns to a description of the program Babbage had given her to calculate the 4-th Bernoulli number, not by hand but automatically. She shows not the program itself but rather a purported trace of the program listing specific operations performed and their effects on variables, in table format. (Recall that all of Menabrea's examples may be interpreted, by contrast, as straightforward programs on cards with annotations.) From this one must attempt to reconstruct what the program would have to really look like to produce such a trace. Notice here that Ada states that the 'signs' of the Bernoullis are 'intentionally left blank in the diagram' but of course, as we have already seen, the B_{2n-1} are not the Bernoullis B_n^* but rather $(-1)^{n-1}B_n^*$, which we will continue to notate 'Bernoulli numbers'.

A fundamental problem arises in this exercise. To have a *completely general* program to calculate the n-th 'Bernoulli number', the previous 'Bernoulli numbers' must have been stored, say by prior invocations of that program (a process that Ada purports to describe), or loaded from constant number cards.[15] Where would these previously calculated numbers be stored? How would they be used? The index n is dynamic and only determined when the program is run. A loop would be used to implement that. However the Analytical Engine only has *named* variables to store results in. Memory is not addressable and there are no arrays. The program must use those explicit names, say for simplicity V_1 to V_n. Variable cards stating those names literally must be placed in the program.

[15] These constants would have to be $B_{2k-1} = (-1)^{n-1}B_k^*$ and *not* literally tabulated values of B_k^*.

What is shown in the note is not, as it stands, a general program at all, but rather a *hard-coded* example for the case of the 4th 'Bernoulli number' $B_{2(4)-1} = B_7$. The code simply assumes that the previous B_1, B_3, B_5 have been explicitly stored under known names, and the result will also explicitly be stored in a known variable name. The trace purports to come from nested loops, 'we suppose the engine to be *in the course of* computing the Numbers to an indefinite extent, from the very beginning'. Operations are shown incrementing a counter for n up, and for each value of n decrementing a counter from $n-1$ down to 0, though no 'combination cards' are actually mentioned, even parenthetically, *using* those counters.

All this doesn't really work as it stands, though it comes frustratingly close.[16] So close—recall that this code was never executed on a real machine—that it was surely honed over many years, as Babbage presented and explained it to his collaborators. The trace that is given *deliberately glosses over how the loop would work* with the misleading comment 'Here follows a repetition of operations 13 to 23'. This omission is not for textual convenience, though it is represented as such. The named variable that would be required here *cannot* be altered, as it must be, by 'repeating' operations 13 to 23. As we have already seen from her correspondence with Babbage, this gloss was no oversight but rather a deliberate deception by Ada, in which Babbage himself participated.[17]

Ada shows some sly awareness that the named variables would not work even for this hard-coded special case trying to pass as a general solution. As n increments, a new named variable is need at each increment to store B_{2n-1}, and in future that fixed name will also have to be used to calculate further results. 'The only exception to a perfect identity in all the processes and columns used, for every repetition of Operations (13...23) is, that Operation 21 [multiplying $A_{k,n}$ by B_k] always requires one of its factors from a new column, and Operation 24 [the calculated result for this iteration B_{2n-1}] always puts its result on a new column.' A deliberate evasion follows. 'But as these variations follow the same law at each repetition, (Operation 21 always requiring its factor from a column one in advance of that which it used the previous time, and Operation 24 always putting its result on the column one in advance of that which received the previous result), they are easily provided for in arranging the recurring group (or cycle) of Variable-cards.'[18]

What does this 'arranging the recurring group' really mean? From the concise listing below (see subsection A.4.1, p. 119)

$$\begin{aligned} 21 &: V_{12} \leftarrow V_{22} * V_{11} \quad [B_3 * A_{3,4}] \\ 22 &: V_{13} \leftarrow V_{12} + V_{13} \quad [T \leftarrow B_3 * A_{3,n} + T] \end{aligned}$$

this 'arrangement' would have to do what a 'repetition' could never do, literally substitute V_{23} for V_{22}.

$$\begin{aligned} 21 &: V_{12} \leftarrow V_{23} * V_{11} \quad [B_5 * A_{5,4}] \\ 22 &: V_{13} \leftarrow V_{12} + V_{13} \quad [T \leftarrow B_5 * A_{5,n} + T] \end{aligned}$$

If each value of n requires a different 'arrangement' to use the 'one in advance' named variables, that is really manually rewriting the program for every single Bernoulli number calculation. The use of the Analytical Engine for constructing *any* sort of tabulation would be crippled by this limitation (but see variable storage blocks below).

The idea of 'operation cycles' draws the note to a close with a corresponding analysis of the number of operations those repetitions would lead to. This is ponderously expressed once again in the bizarre "$\Sigma(+1)^n \ldots$" notation we have already shown should really be written "$\sum_1^n \ldots$" and merely means 'expands n times'.[19]

[16] See how close the executable version is at subsection A.4.3, p. 121

[17] See section 276, p. 71.

[18] Menabrea 1843b, 729-30.

[19] See section 262, p. 110.

A penultimate remark puzzles. We are told that the A_k can be tabulated too! 'Supposing that it was desired not only to tabulate B_1, B_3, &c., but A_0, A_1, &c.; we have only then to appoint another series of Variables, V_{41}, V_{42}, &c., for receiving these latter results as they are successively produced upon V_n'. It is a silly idea, given that the A_k are really $A_{k,n}$ so that tabulating them would not be much use to anyone. This contradictory remark appears to be an addition by Ada, since the calculation of $A_{k,n}$ in the program was clearly shown to depend on the n, and the program itself was due initially to Babbage, then adumbrated by Ada. That explains why the note speaks with two minds.

As it stands the program that the 'trace' in Note G represents would not calculate the 4th Bernoulli number correctly, or any other except the very first. The two-fold error of using the wrong definition, thereby confusing the signs, and not using the correct subtraction (9b) along with some other minor programming 'bugs', leads to wholly incorrect results. A simplified version of the program appears below, without the distracting table of superscripted variables that Ada added but does not really clarify how it works. It keeps as closely as possible to the original semantics. The program is followed by a corrected version of the inferred executable code (here realized in R but any other language such as Javascript would do), with loops which work. Direct markup of the variable cards indicates zeroing as needed.

Dynamic storage via arrays is used in the inferred code to keep things simple. However the critical drawback of non-addressable memory in the engine could in principle be overcome, using facilities later added by Babbage, as shown below (see subsection A.4.5, p. 126).

The note ends with a terse statement the can only have come verbatim from Babbage himself. It cannot be ascribed to Ada. (A confusing typo gives z_n as z^n in the original.)

> The formula (8.) is interesting in another point of view. It is one particular case of the general Integral of the following Equation of Mixed Differences:—
>
> $$\frac{d^2}{dx^2}\left(z_{n+1}x^{2n+2}\right) = (2n+1)(2n+2)z_n x^{2n}$$
>
> for certain special suppositions respecting z, x and n. The *general* integral itself is of the form,
>
> $$z_n = f(n)\cdot x + f_1(n) + f_2(n)\cdot x^{-1} + f_3(n)\cdot x^{-3} + \cdots$$
>
> and it is worthy of remark, that the engine might (in a manner more or less similar to the preceding) calculate the value of this formula upon most *other* hypotheses for the functions in the integral, with as much, or (in many cases) with more, ease than it can formula (8.).

Here the double differentiation in the 'mixed differences' equation would eliminate any even power of x when it is reflected in the general integral, with those coefficients being gathered as some function of n not dependent on x. Thus only the odd powers remain. The 'special suppositions' required to get the Bernoulli recurrence from the extremely flexible form of the 'general integral', an exercise of dubious value, include truncating the infinite series beyond $2n$, and zeroing the coefficients in terms involving even powers. By the claim that evaluating the infinite series—which could in general only be an approximation—would be 'easier' than calculating Bernoulli numbers via the recurrence in (8), Babbage may have meant that the course-of-values recursion requires storage of all previous Bernoulli values, whereas approximating the more general infinite series to an arbitrary degree of precision requires only cumulating the results so far. And recall that Babbage argued that x^n really can be evaluated for any real number x assuming boundless time instead of boundless storage, treating an arbitrary degree of precision as equivalent to algebraic calculation.

A.4.1 Concise Original Pseudo-Trace.

The pseudo-trace in the table contains a number of minor errors of the kind we now readily recognize as 'bugs'. These are in addition to and should be clearly distinguished from the *conceptual errors* discussed above, which stem from confused mathematics, e.g. about the signs of the Bernoulli numbers (necessarily positive on this definition) and the incorrect recurrence relation. The problems with the inner loop structure for the $A_{k,n}$ have already been covered above too.

The bugs are inherently of lesser interest—all coders make inadvertent mistakes, though some far more than others—and a list of them should be kept in mind by anyone who wants to slavishly emulate the sort of code that could be inferred from the obfuscated 'trace' in the table, as far as possible. Most of these concern upper indices to specify when intermediate variables should be zeroed out by the cards that load them and other annotations.[20] The zeroing of intermediate variables is shown correctly in the next section, subsection A.4.3, p. 121.

1. Op. 4, should be $V_4 \div V_5$ i.e. $(2n-1) \div (2n+1)$ and not inverted as $V_5 \div V_4$.
2. Op. 21, In the change annotation, beware that ${}^2V_{12} = {}^0V_{12}$ should be ${}^0V_{12} = {}^2V_{12}$. It was zero, now it is not. Moreover annotations usually do not record (obvious) changes in the *stored* variable.
3. Op. 24, V_{13} definitely must be zeroed out for the next iteration of the loop.
4. Op. 8. Prior to this V_7 must be zeroed out from any previous iteration of the loop. A conditional test for $V[10] > 0$ is also needed to handle calculating $B[1]$ properly from scratch.
5. B_{2k-1} needs to be dynamically indexed or addressed but fixed names $B_1 = V_{21}$ $B_3 = V_{22}$ and $B_5 = V_{23}$ were used. This will not work for n in general, or even for $n = 4$ when the loop is implemented fully.

From number (data) cards: $V_1 = 1, V_2 = 2, V_3 = n$.

$$1 : V_4, V_5, V_6 \leftarrow V_2 \times V_3 \quad [2n]$$
$$2 : V_4 \leftarrow V_4 - V_1 \quad [2n-1]$$
$$3 : V_5 \leftarrow V_5 + V_1 \quad [2n+1]$$
$$4 : V_{11} \leftarrow V_4 / V_5 \quad [(2n-1)/(2n+1)]$$
$$5 : V_{11} \leftarrow V_{11} / V_2 \quad [(1/2)(2n-1)/(2n+1)]$$
$$6 : V_{13} \leftarrow V_{13} - V_{11} \quad [T \leftarrow A_{0,n} = -(1/2)(2n-1)/(2n+1)]$$
$$7 : V_{10} \leftarrow V_3 - V_1 \quad [j \leftarrow n-1]$$

$$8 : V_7 \leftarrow V_2 + V_7 \quad [V_7 \leftarrow 2]$$
$$9 : V_{11} \leftarrow V_6 / V_7 \quad [A_{2k-1,n} \leftarrow 2n/2]$$
$$10 : V_{12} \leftarrow V_{21} * V_{11} \quad [B_{2k-1} * A_{2k-1,n}]$$
$$11 : V_{13} \leftarrow V_{12} + V_{13} \quad [T \leftarrow A_{0,n} + B_{2k-1} * A_{2k-1,n}]$$
$$12 : V_{10} \leftarrow V_{10} - V_1 \quad [j \leftarrow j-1]$$

[20] See Glaschick 2016, whose valuable paper proved useful to confirm findings already made by the author before it was discovered. However like Ada the paper gets the signs of the Bernoulli numbers wrong and fails to detect her other mathematical errors, and the title is misleading in giving no credit to Babbage for this program. He certainly wrote the basis for it, and said as much.

$13 : V_6 \leftarrow V_6 - V_1 \quad [m \leftarrow m - 1 \quad \text{starts at 2n-1 counts down}]$
$14 : V_7 \leftarrow V_1 + V_7 \quad [k \leftarrow 1 + k \text{ starts at 1+2 counts up}]$
$15 : V_8 \leftarrow V_6/V_7 \quad [m/k]$
$16 : V_{11} \leftarrow V_8 * V_{11} \quad [A_{2k-1,n} \leftarrow (m/k) * A_{2k-1,n}]$
$17 : V_6 \leftarrow V_6 - V_1 \quad [m \leftarrow m - 1]$
$18 : V_7 \leftarrow V_1 + V_7 \quad [k \leftarrow k + 1]$
$19 : V_9 \leftarrow V_6/V_7 \quad [m/k]$
$20 : V_{11} \leftarrow V_9 * V_{11} \quad [A_{2k-1,n} \leftarrow (m/k) * A_{2k-1,n}]$
$21 : V_{12} \leftarrow V_{22} * V_{11} \quad [B_{2k-1} * A_{2k-1,n}]$
$22 : V_{13} \leftarrow V_{12} + V_{13} \quad [T \leftarrow B_{2k-1} * A_{2k-1,n} + T]$
$23 : V_{10} \leftarrow V_{10} - V_1 \quad [j \leftarrow j - 1]$
$: \text{IF } V_{10} > 0 \text{ GOTO 13 [while } j > 0]$

$$24 : V_{24} \leftarrow V_{24} - V_{13} \quad [-B_{2n-1} \leftarrow A_{0,n} + \sum_{1}^{n-1} A_{2k-1,n} * B_{2k-1}]$$

$25 : V_3 \leftarrow V_1 + V_3 \quad [n \leftarrow n + 1]$

A.4.2 Conditional Branches.

Babbage was somewhat vague about the 'Combinatorial Cards' in his (unpublished MS) Mathematical Powers of the Analytical Engine[21]

> One or more peculiar cards may be inserted among the operation cards of certain formulae. They are called Combinatorial Cards. The object of these Cards is: To govern the Repeating Apparatus of the Operation and of the Variable Cards and thus to direct at certain intervals the return of those Cards to given places and to direct the number and nature of the repetitions which are to be made by those cards. Whenever Combinatorial Cards are used other cards called index cards must occur amongst those of the formulae. The use of these cards is to compute the numbers which are to serve successively for the indices of the combinatorial cards. At what time the Combinatorial Cards shall act depends on the number of repetitions the last of those cards appointed. What orders each Combinatorial Card shall give depends on the nature of each individual Card.

Allan Bromley dug up more details from later notebooks and plans (1842-6), in which an operator one might call a conditional GOTOIF is described.[22] This tests whether an index is > 0 and if it is moves the operation cards a fixed number of places, specified on the conditional GOTOIF card itself, and the variable cards a flexible number of places, loaded from a variable card. Otherwise the operation completes and the next statement will be executed. This certainly enables a "do .... while($i > 0$)" idea, if the fixed number backtracks the operation cards. Can it go forward though? The direction could be specified by the sign of the number of places for both operations and variables: say negative is back, positive is forward. Bromley found this unspecified and opted to only commit to the backward direction, since this is needed by inter alia the loops in the Bernoulli number program.

But close inspection of the Bernoulli program will show that the forward direction is definitely needed there as well, to enable an IF statement. There is no other way the B_{2n-1} can be accumulated from $n = 1$ onward; the code needs to skip over the section that follows that. The direction cannot go backward, the counter for n must be incremented

[21] Babbage 1973, Appendix D, p. 137.
[22] Bromley 1982a, 2000.

below it. Moreover, Babbage in his autobiography specifically speaks of the cards being moveable forward as well as backward. 'Mechanical means have been provided for backing or advancing the operation cards to any extent'.[23] In a handwritten document that he left at Turin in 1840 he had also stated that 'The mill has also the power of turning backward or forward any number of cards according to any laws or even arbitrary numbers which may be prescribed for it'.[24] One may assume this also applies to the variable cards since these have the same mechanism.

This can be used as an IF statement if one just inverts the condition. Say GOTOIF n V1 V2 where n specifies how many operation cards to advance, V1 is tested for $V1 > 0$, V2 holds the number of variable cards to advance. Abusing the syntax, GOTOIF n (-1*V1) V2 is the same as IF $(V1 > 0)$ ....

A conditional GOTOIF also yields an unconditional GOTO (just make the condition always true), e.g. GOTOIF n 1 V2.

We assume that GOTOIF 1 1 V2 moves the operation cards to the next one, and moves the variable cards by the contents of V2. When it completes the current operation card (which was the very next) is executed.

A.4.3 Implied Executable Code.

The code below implements the algorithm implied by the trace, in R code. There are no GOTOs in R, so simple while loops are used (it is trivial to do these using IF conditions and GOTOs in some suitable language but it would clarify nothing to do so here). It contains the fixes needed to make it work for any N, within practical limits, but only checks the first four Bernoulli numbers calculated. The reader may execute the code and verify that it works. A simple way of demonstrating zeroing of variables on load is included, and is needed for it to work properly. This is not a full-scale emulator but it is close enough for our purposes here.

Rather than the mangled B_{2k-1} of the note, the B_k^* (see subsection A.4.6, p. 129) are stored dynamically and reused in later calculations. The signs are corrected so that they really are Bernoullis. The right recurrence relation is used, and the minor programming bugs noted above are also fixed. The code does not check if variables being stored are properly zero before being written to. That would clutter the syntax with functions for operations though it is easy enough to implement that and the reader is welcome to do so. The code does however correctly set variables to zero in each case, using the Z notation shown instead of V.

```
###### SETUP CODE
V = rep(0,1000)

Z = function(i) {
  val <- V[i]
  V[i] <<- 0 # set in global scope
  return (val)
}

# To verify the results only
Bern <- c(1:4)
Bern[1] = 1/6
Bern[2] = 1/30
Bern[3] = 1/42
```

---

[23] Babbage 1864, 135.

[24] See section E.3, p. 174.

```
Bern[4] = 1/30

N = 4

# To store the recalculated prior Bernoullis
# Always with the right signs
# We label it B[n] rather than B[2n-1] for convenience
B <- rep(0,N)

##### REAL CODE STARTS HERE

V[1]<-1; V[2]<-2

V[3] <- 1 # increment n from 1 to N to calc all B
V3_signer <- 1

while (V[3] <= N) {

  # 1
  V[4] <- V[2] * V[3] # 2n
  V[5] <- V[2] * V[3] # 2n
  V[6] <- V[2] * V[3] # 2n

  # 2
  V[4] <- Z(4) - V[1] # 2n-1

  # 3
  V[5] <- Z(5) + V[1] # 2n+1

  # 4
  V[11] <- Z(4) / Z(5) # (2n-1) / (2n+1)

  # 5
  V[11] <- Z(11) / V[2] # (1/2) (2n-1)/(2n+1)

  # 6
  V[13] <- Z(13) - Z(11) # T = A0 = - (1/2) (2n-1)/(2n+1)

  # 7
  V[10] <- V[3] - V[1] # j <- n-1

  # Bugfix. Must reset V[7] to 0.
  # V[7] <- 0  A Trick. V[30] not set so always 0
  V[7]  <- Z(30) + Z(30)

  #     IF j==0  GOTO END
  if (V[10] > 0) {
    V10_signer = 1

    # 8
    V[7] <- V[2] + Z(7) # r <- 2
```

```
# 9
V[11] <- V[6] / V[7] # A1

# 10
V[12] <- (V10_signer*B[1]) * V[11] # (was V21)  B1 * A1

# 11
V[13] <- Z(12) + Z(13) # A0 + B1 * A1

# 12
V[10] <- Z(10) - V[1] # j = j-1

while (V[10] > 0) {
  # 13
  V[6] <- Z(6) - V[1] # m <- m-1

  # 14
  V[7] <- V[1] + V[7] # r <- 1+r

  # 15
  V[8] <- V[6] / V[7] # m/r

  # 16  A_(2k-1) <- (m/k) * A_(2k-1)
  V[11] <- Z(8) * Z(11)

  # 17 repeat 13
  V[6] <- Z(6) - V[1] # m <- m-1

  # 18 repeat 14
  V[7] <- V[1] + V[7] # r <- 1+r

  # 19 repeat 15
  V[9] <- V[6] / V[7] # m/r

  # 20 repeat 16
  # A_(2k-1) <- (m/k) * A_(2k-1)
  V[11] <- Z(9) * Z(11)

  # 21 B[k] * A_(2k-1)  note k = n-j
  # was V22 (V7=3) and V23 (V7=5)
  V10_signer <- (-1)*V10_signer
  V[12] <- (V10_signer*B[V[3]-V[10]]) * V[11]

  # 22 T <- T + B[k] * A_(2k-1)
  V[13] <- Z(12) + Z(13)

  # 23
  V[10] <- Z(10) - V[1] # j <- j-1
 } # GOTO 13
}
```

```
  # 24 DOUBLY FIXED, ADA HAS IN EFFECT Z(24) + Z(13)
  B[V[3]] <- V3_signer*(Z(24) - Z(13))

  # 25
  V[3] <- V[1] + Z(3) # n <-n+1
  V3_signer <- V3_signer * (-1)
}

# END

#### ADDITIONAL CODE TO CHECK RESULTS

checkN = function(N) {
  print(paste("N=", N, ": ", B[N], " =?=", Bern[N]))
  print(paste("Diff is ", round(B[N] - Bern[N],10)))
}

checkSome = function(i) {
  for (i in 1:4) {
    checkN(i)
  }
}

checkSome(N)
```

A.4.4 General Code.

The following simple example in R gives a general way of generating correct Bernoulli numbers using the recurrence relation in (9b), and demonstrates clearly that the code implied by Note G produces incorrect results from the second number onward if it is not fixed.

```
# Calculate coefficient of B
A = function(j,n) {
  if (j == 0) {
    r = (-1/2)*(2*n-1)/(2*n+1)
    return (r)
  }

  k = 2*j - 1

  perm = 1
  fact = 1
  for (i in 0: (k-1)) {
     perm = perm * (2*n - i)
     fact = fact * (2+i)
  }
  return (perm/fact)
}

B = function(n) {
```

```
  t = A(0,n)
  if (n == 1) {
    return (-t)
  }

  for (j in 1: (n-1) ) {
    t = t + A(j,n) * B(j)
  }

  return (-t)
}

Bern = function(n) {
  b = B(n)
  return (b*((-1)^(n-1)))
}

B_ada = function(n) {
  t = A(0,n)
  if (n == 1) {
    return (t)
  }

  for (j in 1: (n-1) ) {
    t = t + A(j,n) * B_ada(j)
  }

  return (t)
}

Bern1 = 1/6
Bern2 = 1/30
Bern3 = 1/42
Bern4 = 1/30
Bern5 = 5/66
Bern6 = 691/2730
Bern7 = 7/6

A(0,1)
Bern(1)
B_ada(1)

d = 10
round(Bern(1) - Bern1, digits=d)

Bern(2)
round(Bern(2) - Bern2, digits=d)

B_ada(2)
round(B_ada(2) - Bern2, digits=d)

Bern(3)
round(Bern(3) - Bern3, digits=d)

B_ada(3)
```

```
round(B_ada(3) - Bern3, digits=d)

Bern(4)
round(Bern(4) - Bern4, digits=d)

B_ada(4)
round(B_ada(4) - Bern4, digits=d)

Bern(5)
round(Bern(5) - Bern5, digits=d)

B_ada(5)
round(B_ada(5) - Bern5, digits=d)

Bern(6)
round(Bern(6) - Bern6, digits=d)

Bern(7)
Bern7
round(Bern(7) - Bern7, digits=d)
```

A.4.5 Dynamically Indexed Storage Blocks.

It is possible to overcome Babbage's lack of addressable memory without changing any of the other features of the Analytical Engine. This establishes that the Analytical Engine is indeed Turing-complete, contrary to Martin Davis. 'If the Analytical Engine could have been built, it would certainly have been a very useful tool. But it would not have been an all-purpose computer'.[25]

First, recall that the operation and variable cards may move independently, as they have separate card readers. We are not concerned primarily with efficiency here, or with the size of programs, let alone elegance, but only with the logical capabilities of the machine. We also assume here that any constants have been loaded already, and that for convenience we can just name them by their value rather than the (pedantic) name of the variable they are really in.

Babbage added an operation circa 1845 which allows the variable cards to be moved a *dynamic number of places*: the number of cards to move is read from a variable. This is a GOTOIF as explained above, e.g. GOTOIF 1 1 V103 where V103 holds the number of places to move: positive for forward, negative for backward. We will abbreviate this as MOV V103, but note that like all operations it always needs three variable cards to consume, the first two being constants. Also comprehend that though MOV consumes a variable card, that is compensated for, so MOV k moves the cards $1 + k - 1 = k$ places. Thus MOV 1 simply goes to the next variable card.

The MOV operation is critical. By moving the variable cards by a dynamic number of places i, they can be made to rest on a card with the correct fixed storage name Vi on it. If the next operation stores a value x, it will use this fixed name, so we get $Vi \leftarrow x$. In effect, this is the dynamic addressing meant by array notation $V[i] \leftarrow x$. Once the value has been set, the variable cards need to be moved back to where they would have been before this intervention. The program may then proceed as per usual.

Recall that storing a value is accomplished using PLUS. Since PLUS V1 V2 V3 adds V2 to V3 and stores the result in V1. We abbreviate this as $V1 \leftarrow V2 + V3$. PLUS V1 V2 0

[25] M. Davis and V. Davis 2005.

will store V2 in V1. Abbreviate that as $V1 \leftarrow V2$ Note that this storage also requires and consumes exactly three variable cards.

The following method, just one of many possible implementations of the above idea, gives us a fixed *heap* of storage which can be read from and written to using a dynamic index. (A homogeneous heap is described, but the example can easily be extended to split it into blocks of arrays using prefixes.) Packs of pre-made operation and variable cards simplify this. Since Babbage contemplated the ability to manufacture cards using the engine, a program could use parameters to generate storage packs, but that takes us into speculative territory. So let's assume they are made by hand here, in duplicates. They can be reused across programs by physically removing and inserting them.

1. A pack of pre-made variable cards, of the desired size, is inserted at the front of the variable cards to form the heap. This could enable every storage variables in the machine, or a subset of them. Let's use all 1,000 variables of a 1,000 variable engine in our example to fix ideas.
2. A pack of pre-made operation cards is inserted at the head of the operation cards, to initialize the heap. When these initialization operations execute they will move over the variables past the heap, to the first ordinary program variable.
3. The programmer may then add variable and operation cards as per usual, after the above pre-made packs. The program will execute as expected.
4. At a point in the program where a dynamic memory address is used, a pre-made *reference operation pack* can be inserted, with a corresponding pre-made *reference variable pack.*
5. Just before these insertions, the dynamic reference must be initialized to set the dynamic index desired. For storing, also the value to store. These initializations use reserved variable names (the programmer must exercise due diligence to avoid collisions of these names with other possible uses of them). We reserve the following.
   (a) V1: dynamic index, 'i'. Set it for each reference, for example by V1←V999, where V999 is where you are keeping your own dynamic index. This prevents collisions, e.g. in nested loops.
   (b) V2: value to store, 'x'. Set it by V1←V998, where V998 is where your value is. Again this prevents collisions.
   (c) V3: absolute position of the next user variable card. This may be read off that card, including the heap variable cards.
   (d) V4: number of positions to move back into the heap (set internally, see below).
   (e) V5: number of positions to move forward from inside the heap (set internally, see below).
6. After these packs, if reading rather than writing a dynamic address, the value read will be placed in V2. Get it now to where you want it, say V997←V2;

The internal details of the dynamic reference packs are fixed and need not concern the programmer. They ensure that the variable cards will be backtracked to the correct fixed variable name and then restored to the next variable card in the *user* program, so that the next user specified operation executes as expected, and subsequent ones. This is a little bit tricky to implement, since the reference operations also use a variable card inside the heap, which positional gain must be backtracked, and several variables outside it, which must be skipped over when restoring the position. The programmer must just get the initialization correct.

The number of variable cards involved here is necessarily a multiple of three, since any operation must consume three cards. Internally the heap variable cards will have the following structure:

V1	V2	0	for setting the value V1 ← V2
1	1	V5	for MOV back to the next user variable
...			
Vi	V2	0	for setting the value Vi ← V2
1	1	V5	for MOV back to the next user variable
...			
V1000	V2	0	for setting the value V1000 ← V2
1	1	V5	for MOV back to the next user variable

For 1,000 variables there are $1000 \times 3 \times 2 =$ 6,000 cards in the heap. Only the variable cards need to be moved for the dynamic references to work. The operation cards execute sequentially. They may include loops and conditional statements for other reasons, but those have nothing to do with implementing dynamic references.

In outline, a user program would look as follows, using a 1,000 variable heap. Dynamic references are enabled by insertion of pre-made card packs where they are needed. It is critical here to skip over the correct number of variable cards when returning from the heap. This requires accounting for the variable cards used in the dynamic references themselves.

	Operations will execute sequentially	Variables move	
Heap pre-made packs	GOTOIF	1 1 6001 *Heap Var Cards*	Seek to the end of the heap below. 6000 cards.
User program	*Operations*	*Variables*	Not shown
Dynamic reference setup	PLUS	V1 i 0	V1 ← 1
	PLUS	V2 x 0	V2 ← x
	PLUS	V3 k 0	V3 ← $k+1$ where k is the absolute position of the current var card
Dynamic reference pre-made packs to enable V[i] ← x	MINUS	V6 V1 V3	V6 ← $i - k$
	PLUS	V7 V6 1	V7 ← $i-k+1$ add one for the first mov op var card triple in the heap init
	MULT	V4 3 V7	V4 ← $3(i-k+1)$ how far back to go
	MULT	V8 -3 V7	V8 ← $-3(i-k+1)$
	PLUS	V5 V8 18	V5 ← $-3(i-k+1)+15$ where 15=6*3 - 3 is the no. of extra variable cards inserted here (including the next line) *corrected* for the 3 cards which will be consumed in V[i] ← x below
	GOTOIF	1 1 V4	move vars to the right heap card
Operations are still here but variables are now in the heap	PLUS	Vi V2 0	the correct fixed name enables V[i] ← x
	GOTIF	1 1 V5	return vars to the next user var card
User Program continues	*Operations*	*Variables*	

Above we have shown dynamic assignment. Reading is similar, but requires the reading variants of the dynamic reference packs (left as an exercise for the reader).

A.4.6 Modern Bernoulli Numbers.

Bernoulli numbers are now defined in a slightly more general way, yielding the old-style Bernoulli numbers used here as a special case.

$$B_{2n}^{\dagger} = \frac{(-1)^{n-1}2(2n)!}{(2\pi)^{2n}} \sum_{k=1}^{\infty} k^{-2n} \tag{A.2}$$

(Note that these are not necessarily positive.) Or equivalently, a form similar to (4*) but dropping the nuisance prefix $1 - \frac{x}{2}$ and not doubling n.

$$\frac{x}{e^x - 1} = \sum_{n=0}^{\infty} \frac{B_n^{\dagger} x^n}{n!} \tag{A.3}$$

In particular $B_0^{\dagger} = 1$, $B_1^{\dagger} = -\frac{1}{2}$ and where $n \geq 1$, the odd Bernoullis, $B_{2n-1}^{\dagger} = 0$.

If we call the old form B_n^* (indexed $1, 2, 3, \ldots$, rather than Ada's idiosyncratic B_{2n-1}) then new can easily be converted to old or vice versa by

$$B_{2n}^{\dagger} = (-1)^{n-1} B_n^* \tag{A.4}$$

$$B_n^* = (-1)^{1-n} B_{2n}^{\dagger}. \tag{A.5}$$

The following list of the first eleven old-style B_k^* is useful, and was readily avalable in 1840 in the *Penny Cyclopaedia* article by Augustus De Morgan, which even goes up to the 49th case.[26]

$$\begin{aligned}
B_1^* &= \frac{1}{6} \\
B_2^* &= \frac{1}{30} \\
B_3^* &= \frac{1}{42} \\
B_4^* &= \frac{1}{30} \\
B_5^* &= \frac{5}{66} \\
B_6^* &= \frac{691}{2730} \\
B_7^* &= \frac{7}{6} \\
B_8^* &= \frac{3617}{510} \\
B_9^* &= \frac{43867}{798} \\
B_{10}^* &= \frac{174611}{330} \\
B_{11}^* &= \frac{854513}{138}.
\end{aligned}$$

[26] A. De Morgan 1840b.

A.4.7 Bogus Bernoulli Numbers.

Ada states an alternative direct definition of Bernoulli numbers that she does not use, and labels (3).[27] It is of some significance that this is bogus, though it might look superficially impressive.

$$B_{2n-1} = \frac{\pm 2n}{(2^{2n}-1)2^{n-1}} \left[\begin{array}{l} \frac{1}{2} n^{2n-1} \\ -(n-1)^{2n-1}\left\{1+\frac{1}{2}\cdot\frac{2n}{1}\right\} \\ +(n-2)^{2n-1}\left\{1+\frac{2n}{1}+\frac{1}{2}\cdot\frac{2n.(2n-1)}{1.2}\right\} \\ -(n-3)^{2n-1}\left\{ \begin{array}{l} 1+\frac{2n}{1}+\frac{2n.2n-1}{1.2}+ \\ +\frac{1}{2}\cdot\frac{2n.(2n-1).(2n-2)}{1.2.3} \end{array} \right\} \\ + \dots \quad \dots \quad \dots \quad \dots \end{array} \right] \quad (3.)$$

Figure A.1: Purported Bernoulli Formula

It is not, despite appearances, an infinite sum, nor could it be. It is somewhat ambiguous as it stands and should be written with the indexes properly specified, say as:

$$B_{2n-1} = \frac{\pm 2n}{(2^{2n}-1)(2^{n-1})} \left[\frac{1}{2} n^{2n-1} + \sum_{k=1}^{n-1} (-1)^k (n-k)^{2n-1} \left\{ \sum_{j=0}^{k-1} \binom{2n}{j} + \frac{1}{2}\binom{2n}{k} \right\} \right]$$

Evaluating this at $n = 1, 2, \ldots$ yields $(\frac{1}{3}, \frac{2}{15}, \ldots)$ so it does not produce the Bernoulli numbers, as might be inferred from the dysfunctional $\pm$ at the beginning (the old-style Bernoulli numbers of the notes are always positive, as implied by (2)). The source of this formula among the myriad of direct forms known in Ada's day could not be traced. Recall her statement 'I am doggedly attacking & sifting to the very bottom, all the ways of deducing the Bernoulli Numbers'.[28]

The formula was garbled. There was no good reason for including it. Nobody seems to have ever paid any attention to it, perhaps because it was not used, or because it was written down so incoherently. It would have taken Ada no more than a few minutes to do the calculations above showing that it is not valid.

[27] Menabrea 1843b, 725.

[28] AL to CB 1843-07-05.

Appendix B

How to Eat Your Own Tail.

Babbage noted that the following recurrence could be used to apply the method of differences to astronomical tables, where expressions of this type often appear. He referred to this as the 'engine eating its own tail'. Recall the elementary formulae:

$$\begin{aligned}\sin(A+B)+\sin(A-B) &= 2\cos B\sin A\\ \cos(A-B)-\cos(A+B) &= 2\sin A\sin B.\end{aligned}$$

Thus,

$$\begin{aligned}\Delta\sin(x) &= \sin(x+1)-\sin(x) \text{ by definition}\\ &= 2\cos[\frac{1}{2}(2x+1)]\sin[\frac{1}{2}(1)]\\ &= Z\cos(x+\frac{1}{2}) \text{ where Z is a constant.}\end{aligned}$$

Using the relevant formula for cos and combining constant terms we get second-order differences

$$\begin{aligned}\Delta^2\sin x &= \Delta\sin(x+1)-\Delta\sin(x)\\ &= Z\cos(x+1+\frac{1}{2})-Z\cos(x+\frac{1}{2})\\ &= Z(\cos(x+\frac{3}{2})-\cos(x+\frac{1}{2}))\\ &= Z\{-2\sin[\frac{1}{2}(x+\frac{3}{2}+x+\frac{1}{2})]\sin[\frac{1}{2}(x+\frac{3}{2}-x-\frac{1}{2})]\}\\ &= K\sin(x+1) \text{ after simplifying, where K is a constant.}\end{aligned}$$

Given that

$$\begin{aligned}\sin(x+1) &= \Delta\sin(x)+\sin(x) \text{ and}\\ \Delta\sin(x+1) &= \Delta^2\sin(x)+\Delta\sin(x)\\ &= K\sin(x+1)+\Delta\sin(x),\end{aligned}$$

one can use this relationship to work out the whole table of $\sin(x+n)$ values, assuming that $\sin(x)$ and $\Delta\sin(x)$ are first supplied, and provided that we can multiply by the constant K and remember the values we accumulate. For example

$$\sin(x+2) = \Delta\sin(x+1)+\sin(x+1)$$

$$
\begin{aligned}
&= \Delta^2 \sin(x) + \Delta \sin(x) + \sin(x+1) \\
&= K \sin(x+1) + \Delta \sin(x) + \sin(x+1),
\end{aligned}
$$

all of which values are already known, and so on.

Appendix C

Babbage's Difference Engine Papers.

C.1 A Note respecting the Application of Machinery to the Calculation of Astronomical Tables

By Charles Babbage, Esq., F.R.S., Sec. Ast. Soc., &c. &c.
Read 14th June, 1822.[1]

It is known to several of the members of this society that I have been engaged during the last few months in the contrivance of machinery, which by the application of a moving force may calculate any tables that may be required. I am now able to acquaint the society with the successful results at which I have arrived; and although it might at first view appear a bold undertaking to attempt the construction of an engine which should execute operations so various as those which contribute to the formation of the numerous tables that are constantly required for astronomical purposes, yet to those who are acquainted with the method of differences the difficulty will be in a considerable degree removed.

I have taken the method of differences as the principle on which my machinery is founded; and in the engine which is just finished I have limited myself to two orders of differences. With this machine I have repeatedly constructed tables of square and triangular numbers, as well as a table from the singular formula $x^2 + x + 41$, which comprises amongst its terms so many prime numbers.

These, as well as any others which the engine is competent to form, are produced almost as rapidly as an assistant can write them down. The machinery by which these calculations are effected is extremely simple in its kind, consisting of a small number of different parts frequently repeated.

In the prosecution of this plan, I have contrived methods by which type shall be set up by the machine in the order determined by the calculation; and the arrangements are of such a nature that, if executed, there shall not exist the possibility of error in any printed copy of tables computed by this engine. Of several of these latter contrivances I have made models; and, from the experiments I have already made, I feel great confidence in the complete success of the plans I have proposed.

C. Babbage.
Devonshire-Street, Portland-Place,
June 2, 1822.

[1] Babbage 1822b.

C.2 Observations on the Application of Machinery to the Computation of Mathematical Tables

By Charles Babbage, Esq., F.R.S., &c. &c.
Read 13th December, 1822[2]

Since I had the honour of communicating to the Astronomical Society a short account of an arithmetical engine for the calculation of tables, which has been examined by several of the members of this society, I have not added much to the practical part of the subject. I have however paid some attention to the improvements of which the machinery is susceptible, and which will, if another engine is made, be greatly improved.

The theoretical inquiries to which it has conducted me are however of a singular nature; and I shall take this opportunity of briefly explaining to the society some of the principles on which they depend, as far as the nature of the subject will permit me to do this without the introduction of too many algebraic operations, which are rarely intelligible when read to a large assembly.

Of the variety of tables which are required in the present state of science, by far the larger portion are intimately connected with that department of it which it is the peculiar object of this society to promote.

The importance of astronomical science, whether viewed as the proudest triumph of intellectual power, or considered as the most valuable present of abstract science to the comfort and happiness of mankind, equally claims for it the first assistance from any new method for condensing the processes of reasoning or abridging the labour of calculation. Astronomical tables were therefore the first objects on which I turned my attention, when attempting to improve the power of the engine, as they had formed the first motive for constructing it.

I have already stated to the society, in my former communication, that the first engine I had constructed was solely destined to compute tables having constant differences. From this circumstance it will be apparent that after a certain number of terms of a table are computed, unless, as rarely happens, it has a constant order of differences, we must stop the engine and place in it other numbers, in order to produce the next portion of the table. This operation must be repeated more or less frequently according to the nature of the table. The more numerous the order of differences, the less frequent will this operation become requisite. The chance of error in such computations arises from incorrect numbers being placed in the engine: it therefore becomes desirable to limit this chance as much as possible. In examining the analytical theory of the various differences of the sine of an arc, I noticed the property which it possesses of having any of its even orders of differences equal to the sine of the same arc increased by some multiple of its increment multiplied by a constant quantity. With the aid of this principle an engine might be formed which would require but little attendance, and I believe that it might in some cases compute a table of the form $A \sin \theta$ from the 1st value of $\theta = 0$ up to $\theta = 90°$ with only one set of figures being placed in it.

It is scarcely necessary to observe what an immense number of astronomical tables are comprised under this form, nor the great accuracy which must result from having reduced to so few a number the preliminary computations which are requisite.

In pursuing into its detail the principle to which I have alluded, which lends itself so happily to numerical application, I have traced its application to other species of tables, and am enabled to point out a course of analytical investigation which will in all probability afford ready methods for constructing tables, even of the most complicated transcendent, in a manner equally easy.

[2] Babbage 1822c.

I will now advert to another circumstance, which, although not immediately connected with astronomical tables, resulted from an examination of the engine by which they can be formed.

On considering the arrangement of its parts, I observed that a different mode of connecting them would produce tables of a new species altogether different from any with which I was acquainted. I therefore computed with my pen a small table such as would have been formed by the engine had it existed in this new shape, and I was much surprised at discovering that no analytical method was yet known for determining its nth term. The following is the first series I wrote down:

	Series.	Diff.		Series.	Diff.		Series.	Diff.
0 ...	2	0	10 ...	222	42	20 ...	924	86
1 ...	2	2		264	46		1010	86
	4	6		310	46		1096	92
	10	6		356	52	25 ...	1188	100
	16	12	15 ...	408	60		1288	108
5 ...	28	20		468	68		1396	114
	48	28		536	74		1510	114
	76	34		610	74		1624	118
	110	34		684	78	30 ...	1742	120
	144	38	20 ...	762	80		1862	122
10 ...	182	40		842	82		1984	

The equation of finite differences from which it is produced is

$$\Delta^2 u_x = \text{units fig. of } u_{x+1}$$

which is one of a class of equations never hitherto integrated. I succeeded in transforming this equation into a more analytical form: but still it presented great difficulties; I therefore undertook the investigation in a different manner, and succeeded in discovering a formula which represented its nth term. It is the following:

$$u_x = (\tilde{a}) + 206(106 + 2a - 1)$$

TABLE	
0	2
1	2
2	4
3	10
4	16
5	28
6	48
7	76
8	110
9	144

where $(\tilde{a})$ represents the number opposite a in the annexed subsidiary table, and a is the figure in the unit's place of x, and b is that number which arises from cutting off the last figure from x.

Example: let the 17th term be required, then $x = 17$, and $a = 7$, $b = 1$; the number opposite 7 in the table is

$$(\tilde{7}) = \qquad\qquad\qquad\qquad\qquad\qquad 76$$

$$106 + 2a - 1 = 10 + 14 - 1 = 23$$

$$206 = 20 \qquad 206(106 + 2a - 1) \qquad\qquad = 460$$

$$536 = u$$

I have formed other series of the same class, and have succeeded in expressing any term independent of all the rest by two distinct processes. Thus I have incidentally been able to integrate the equations I have mentioned: I will just state one other of a simple form; it is the equation

$$\Delta u_x = \text{units fig. } u_x$$

whose integral is

$$u_x = 20b + 2^a$$

where a is that one of the numbers 1, 2, 3, 4, which taken from x leaves the remainder divisible by 4, and b is the quotient of that division.

The table is as follows:

1	–	2
4	–	8
		16
		22
		24
		28
8	–	36
		42

One of the general questions to which these researches give rise is, supposing the law of any series to be known, to find what figure will occur in the kth place of the nth term. That the mere consideration of a mechanical engine should have suggested these inquiries, is of itself sufficiently remarkable; but it is still more singular, that amongst researches of so very abstract a nature I should have met with and overcome a difficulty which had presented itself in the form of an equation of differences, and which had impeded my progress several years since, in attempting the solution of a problem connected with the game of chess.

Appendix D

On the Mathematical Powers of the Calculating Engine.

By Charles Babbage
26. Dec. 1837.

The object of the present volume is to show the degree of assistance which mathematical science is capable of receiving from mechanism. I may possibly in a separate work describe minutely the mechanism I have contrived for that purpose. For the complete understanding of the following pages, it will be unnecessary to present to the reader more than a very general outline of the structure of the engine. If the reader feels indisposed to examine even that, they may pass it over and, taking certain mechanical data for granted, proceed at once to the mathematical investigation, in which they will find that all results depend on those mechanical data.

The calculating part of the engine may be divided into two portions:

1. The *Mill*, in which all operations are performed.
2. The *Store*, in which all the numbers are originally placed and to which the numbers computed by the engine are returned.

The plate represents a plan of the engine—those circles placed around the great central wheel constitute the Mill, while that portion which adjoins the longitudinal part or rack represents the Store.

D.1 Of the Mill

Two Axes with figure wheels, I the *Ingress Axis* and $''A$ the *Egress Axis*, connect the *Mill* with the *Store*. The Mill itself consists of:

1. Three Figure Axes
2. Three Carriage Axes
3. Ten Table Figure Axes
4. Digit Counting Apparatus
5. Selecting Apparatus
6. Barrels
7. Reducing Apparatus for Barrels
8. Operation Cards
9. Repeating Apparatus
10. Combinatorial Counting Apparatus

D.1.1 Of the Figure Axes

The Figure Axes A and $''A$ are connected with each other without the intervention of the central wheels, so that a number on the figure wheels of one axis may be transferred to those of the other.

These figure wheels are considerably larger than any others to allow sufficient space on their circumference for placing the pinions by which communications are made with other parts of the mill.

By means of some of these pinions, a process called *Stepping down* and another called *Stepping up* may be performed. These processes consist of shifting each digit of a number one cage lower or one cage higher, which are equivalent to the arithmetical operations of dividing or multiplying the number by ten.

Other pinions are fixed on *register axes* R and R_1 and convey the two highest figures of the dividend to the *Selecting apparatus.* The third figure axis $''A$ is placed near the Store and constitutes the egress axis. It is adjacent to the digit counting apparatus with which it communicates.

D.1.2 Carriage Axes

These Axes F, $'F$, $''F$, with their peculiar apparatus, are employed to execute the carriage of the tens when numbers are added to or subtracted from each other. The carriages F and $'F$ can be both connected with the Figure Axis A or one of them with the Figure Axis A and the other with $'A$, or they may, by means of the central wheels, be connected with any other part of the Mill. The third carriage $''F$ is connected both with the Mill and the Store and may be used with either.

Whenever the number subtracted is greater than that from which it is taken, the resulting carriages would, if effected and if the mechanism admitted, produce a carriage in the forty-first cage. This fact is taken advantage of for many purposes—it is one of very great importance, and when it happens, a *Running up* is said to occur. Connected with this part is a lever on which the *Running up* warning acts, and this lever governs many parts of the engine according to the circumstances.

D.1.3 Table Figure Axes

These Axes are ten in number; nine of them contain the table of the nine multiples of one factor in Multiplication and of the Divisor in Division. The tenth contains the complement of the Divisor in the latter operation. They are all connected with the central wheels, and the number on each figure wheel can be *stepped up* or *down* upon the other figure wheel of the same cage. The figure which, at each stepping, goes off from the bottom wheel is transferred to the top wheel.

D.1.4 Of the Digit Counting Apparatus

This is a mechanism by which the digits of any number brought into the mill may be counted, and certain calculations made as to the position of the decimal point in the result of multiplication and division. It is also used to limit the number of figures employed when the engine is making successive approximations either to the roots of equations or to the values of certain functions. It consists of three distinct systems nearly similar to each other.

D.1.5 Of the Selecting Apparatus

When a table of the nine multiples of a multiplicand has been made, it becomes necessary, in order to effect multiplication, to select successively those multiples indicated by the successive digits of the multiplier. This mechanically is not difficult. However, when in the process of division it becomes requisite to select that multiple which is next less than the dividend from which it is to be subtracted, the mechanical difficulty is of quite a different order, and hitherto, nothing but the most refined artifices have been found for accomplishing it. This refinement relates, however, entirely to the nature of those contrivances, not to the certainty of their action nor to any delicacy of workmanship.

The apparatus consists of a portion of the carrying apparatus for three figure wheels, which, by the addition of another contrivance, renders them available for the purpose of making the selection. This apparatus is placed immediately below the Table Axes.

D.1.6 Of the Barrels

The barrels are upright cylinders divided into about seventy rings, the circumference of each ring being divided into about eighty parts. A stud may be fixed on any one or more of these portions of each ring. Thus, each barrel presents about eighty vertical columns, every one of which contains a different combination of fixed studs.

These barrels have two movements:

1. They can advance horizontally by a parallel motion of their axis.
2. They can turn in either direction and to any extent on their axis.

When the barrels advance horizontally, these studs act on levers which cause various movements in the mill, the stud belonging to each ring giving a different order.

Amongst these movements, or rather these orders for movements, the following may be more particularly noticed. The advance of a barrel may order:

(a) A number with its sign to be received into the mill from the ingress axis.
(b) A number with its sign to be given off from the mill. This number may thus be either altogether obliterated from the mill or it may at the same time be received on the egress wheel, or the number may be given off from the mill to the egress wheel and at the same time be itself retained in the mill.
(c) A Variable card to be turned.
(d) An Operation card to be turned.
(e) The circular movement of the Barrel itself or of any other barrel to another vertical. This always occurs at every step from the beginning to the end of what are called *operations*. The barrels, when once ordered by the operation cards from their zero point to any given vertical, always direct themselves to be turned to another vertical preparatory to their next advance. This circular motion is, however, occasionally changed by an action arising from another source.

D.1.7 Of the Reducing Apparatus

Behind each barrel is placed a reducing apparatus. It consists of six or eight sectors which can be made to act upon the barrel and give it a rotatory movement so as to make it pass over 1, 2, 3, or any required number of verticals previously to its next advance.

The levers which put these sectors into action are acted upon by:

(a) The studs on their own barrel.
(b) The studs on any other barrel.
(c) The Operation cards.
(d) The Running up levers.

The first and third of these sources of action occur most frequently.

D.1.8 Of the Operation Cards

Those who are acquainted with the cards of a Jacquard loom will readily understand the functions performed by these cards. To those who are unacquainted with that beautiful contrivance, it may be necessary to state that the *Cards* consist of pieces of thick pasteboard, tin plate, or sheet zinc pierced with a number of holes; these cards being strung together by wire or tape hinges pass over a square prism.

This prism is situated in front of a number of levers placed in rows which govern the Reducing Apparatus and consequently the barrels. The faces of the prism are perforated so as to present an opening opposite *every* lever.

If the prism alone is made to advance horizontally against these levers, then the levers themselves will enter into the holes of the prism and be partly covered by it, but they will not be moved out of their places.

Again, if a card having as many holes as the prism has, or as there are levers opposite to it, is placed upon the advancing face of the prism, no effect can be produced on the levers by this advance of the prism. But if a card having one hole less than the prism is placed on its face, then when the prism advances, the lever opposite that hole will be pushed, and any order given for which that lever was appointed. Suppose after every order the levers to be replaced, and let the prism be turned one quarter round; then a new card will be presented to the levers, and if one or more holes of this second card are stopped up, a different order will be transmitted through the levers to the Reducing Apparatus and thence to the barrels.

Thus, by arranging a string of cards with properly prepared holes, any series of orders, however arbitrary and however extensive, may be given through the intervention of these levers.

The number of the levers acted upon by the operation card is small; they respectively direct the barrels to *commence* the following operations:

(a) The *Addition* of two numbers.
(b) The *Subtraction* of one number from another.
(c) The *Multiplication* of two numbers.
(d) The same *Multiplication limited* to a given number of the first figures.
(e) The *Division* of one number by another.
(f) The same *Division limited* to a given number of figures in the quotient.

The levers numbered 4 and 6 are rarely used; the extraction of roots being the only case in which they are required.

These cards are called into action by orders from the barrels. What they shall order when acting depends on the nature of each individual card. What repetitions they shall be subject to depends on the orders communicated to them from the Combinatorial Cards and their Counting Apparatus. Many calculations are much simplified by having two sets of Variable Cards.

D.1.9 Of the Combinatorial Cards

One or more peculiar cards may be inserted among the operation cards of certain formulae. They are called *Combinatorial Cards*. The object of these Cards is:

To govern the *Repeating Apparatus* of the Operation and of the Variable Cards and thus to direct, at certain intervals, the return of those Cards to given places and to direct the number and nature of the repetitions which are to be made by those cards.

Whenever Combinatorial Cards are used, other cards called index cards must occur amongst those of the formulae. The use of these cards is to compute the numbers which are to serve successively for the indices of the combinatorial cards. At what time the

Combinatorial Cards shall act depends on the number of repetitions the last of those cards appointed. What orders each Combinatorial Card shall give depends on the nature of each individual Card.

D.2 Of the Store

The Store may be considered as the place of deposit in which the numbers and quantities given by the conditions of the question are originally placed, in which all the intermediate results are provisionally preserved, and in which, at the termination, all the required results are found.

Various parts may be added to the store according to the purposes required. Some of them might perhaps with more convenience constitute distinct machines.

The store then may contain:

- Figure Axes
- Computing Apparatus
- Number Cards
- Card Punching Apparatus
- Printing Apparatus
- Copper Punching Apparatus
- Curve Drawing Apparatus
- Variable Cards

The Figure Axes and the Variable Cards alone are absolutely necessary for the mathematical enquiries in the present work. For the sake of simplicity, the others will be only occasionally referred to.

D.2.1 Of the Figure Axes

A number of axes, each having forty figure wheels placed in different cages one above another, are connected with the rack of the Store.

These figure wheels are each numbered from 0 to 9; they may be turned by hand so that any digit may stand opposite a fixed index. Thus, any number of not more than forty places of figures may be put upon the figure wheels of each axis.

Above the fortieth cage is another cage containing a wheel similar to a figure wheel and also having its circumference divided into ten parts. These parts have the signs (+) plus and (–) minus alternately engraved upon them. Above this wheel is a fixed character to distinguish each particular axis or rather the variable number which may be found upon its wheels. These fixed marks are $v_1, v_2, \ldots, v_{32}, \ldots$ as far as the number of quantities which can be contained in the store.

Below the lowest or units figure wheel, a small square frame appears in which may be inserted a card to be changed according to the nature of the calculation directed. On this card is written that particular variable or constant of the formula to be computed whose numerical coefficient and sign are expressed on the wheels above it.

The annexed representation will perhaps convey a clearer idea of this part of the engine.

The first line of quantities $v_1, v_2, \ldots$ are never altered; they merely indicate the particular sets of figure wheels to which they are attached.

The next line contains the signs of the quantities which are themselves expressed by writing them upon pieces of card and placing them in the squares at the bottom.

The intermediate forty cages contain the numerical coefficients.

The first variable or v_1 is in the present figure equal to zero, and previously to the setting of the engine to any problem, all the other variables $v_2, v_3, v_4, \ldots$ have the same symbols beneath them.

The second variable v_2 has beneath it the negative sign, the number 1758, and the algebraic quantity a.

	Variables of the Engine	$V1$	$V2$	$V3$	$V4$	$V5$	$V6$
Signs	Cage 41	+	+	+	+	+	+
Numerical.Coefficients	Cage 40	0	0	0	0	0	0
	Cage 39	0	0	0	0	0	0
	.	.	.	.	.	.	.
	.	.	.	.	.	.	.
	Cage 5 Tens of Thousands	0	0	0	2	0	0
	Cage 4 Thousands	0	1	4	3	0	0
	Cage 3 Hundreds	0	7	9	4	0	2
	Cage 2 Tens	0	5	7	1	3	4
	Cage 1 Units	8	1	0	6	7	
	Variables of the Question	a	x	$x3$	$\sin\theta$	$x+z$	$\sqrt{a^2+x^2}$

The symbols placed on the engine are in the annexed figure

$$
\begin{aligned}
v_1 &= 0 \\
v_4 &= -\ 23410\,x^3 \\
v_2 &= -\ 1758\,a \\
v_5 &= +\ 36\ \sin\theta \\
v_3 &= +\ 4971\,x \\
v_6 &= +\ 247\ \sqrt{a^2+x^2}
\end{aligned}
$$

In this manner any function however complicated it may be if it is considered as a whole may be placed in the store with its proper sign and its numerical coefficient; the function itself being merely written on a card and placed in the square below its coefficient. The number of variables which can be contained within the store will depend on the length of the rack and number of figure axes which can be placed round it and although a large number of variables might with perfect safety be employed yet there is obviously a practical limit arising from the weight of the rack to be moved.

One hundred variables would not give an inconveniently large rack but still the calculations of such an engine would be limited. This limitation can be entirely removed by another set of cards called Number Cards which will presently be described.

If any of the coefficients contain decimals or if the result is required with any number of decimals then all the coefficients must be considered as having the same number of decimals. If an imaginary line is drawn between any two cages, the third and fourth for example, then all below it may be considered as decimals. In order to convey to the Engine this information there exists a wheel with the numbers from 1 to 40 engraved on its edge; this wheel being set at any number the Engine will treat all the numbers put into the Store as having that number of decimals.

D.2.2 Of the Computing Apparatus

One of the Figure Axes K has its figure wheel connected with the rack differently from the others. It is also connected with the Carrying Axis $''F$ belonging to the mill, from which, however, it may at times be disconnected.

The object of this is to enable the figure wheel on the rack to make certain simple calculations without the necessity of sending the numbers into the mill. During the process of one of the great operations in the Mill, a series of numbers may be computed in the store by the method of differences, and thus a considerable saving of time is effected.

D.2.3 Of the Number Cards

The Number Cards have been introduced for the purpose of rendering the calculations absolutely unlimited by the too great number of variables and constants necessary for the solution of any problem.

The number cards are pierced with certain holes and stand opposite levers connected with a set of figure wheels placed on the Number Axis, which can be made at intervals to communicate with the rack. When these Number Cards are advanced, they push in those levers opposite to which there are no holes on the cards and thus transfer that number together with its sign, which the holes on the card represent, to the figure wheels of the axis opposite to which the Cards are placed. The number and its sign thus put upon these figure wheels may be immediately transferred to another part of the Store, and the string of number cards being turned, the next card conveys its number to the Store in the same way.

These number cards are for some purposes more convenient than figure wheels because the numbers upon a figure card are not obliterated by the act of giving them off. By turning the string of figure cards back to any given one, the number upon that card can be replaced in the Store as frequently as may be required and at any periods of time which the calculation may demand. On the other hand, the numbers placed upon figure wheels are always obliterated in the act of giving them off. If it is necessary to retain on the Store any number which is to be given off to the Mill, then it also must be given off through the rack to another store axis on whose figure wheels it must remain until, at a second operation, it is reconveyed by rack back to its original place.

D.2.4 Card Punching Apparatus

One mode of rendering permanent the results of any calculations made by the engine will obviously be by making it punch on cards certain holes similar to those just described as existing on the Number Cards. This plan will also enable us to use any intermediate computation which may be necessary in advancing towards the final results, for the cards so made may from time to time be removed from the punching apparatus and attached to the Number Cards. Other advantages will be observed when the subject of mathematical tables comes under consideration.

D.2.5 Printing Apparatus

It is desirable even when many copies of a calculation made by the engine are not wanted that the results should be themselves printed by the machine in order to ensure the absence of error from copying its answers.

A set of thin circular rings having metal types of the digits fixed at equal distances on their edges and themselves governed by the calculating wheels on which the result is placed are to be pressed down at intervals on a sheet of paper covered by another sheet of carbonised paper. This paper is fixed in a platform having proper motions for placing the printed results in the right order.

Thus, a single and correct copy may be produced, although from the nature of the process, the execution of the printing would not be of the highest order.

If, however, many printed copies are required, then it is intended that the type so arranged shall be made to impress their characters on a soft substance from which a mould, a stereotype plate may be cast.

D.2.6 Copper Plate Punching App[aratus].

If it should be deemed necessary to print tables or calculations upon copper-plate, an apparatus has been contrived for that purpose. This process is necessarily slower in its operation than the former modes of rendering the calculated results permanent. It has, however, the advantage of possessing greater clearness; although the additional cost of taking off impressions may in some instances be objectionable.

Since, however, the invention of the number cards, these modes of printing or engraving have ceased to become essential parts of a Calculating Engine. The absolute certainty of every printed result can now be obtained, although the printing mechanism be totally detached from the calculating portion of the Engine, an improvement which it was impossible to make until that point of the enquiry was attained.

The cards on which the results are punched may themselves be placed in a distinct machine, and from the holes formed in them, the new machine may either engrave or print them as it may have been prepared to operate.

D.2.7 Curve Drawing Apparatus

The discovery of laws from the examination of a multitude of tabulated and reduced observations is greatly assisted by the representation of such tables in the form of curves.

As one of the employments of a calculating engine would be to reduce collections of facts by some common formula, I thought that at the time it impressed the computed results, it would be desirable that it should mark the point of a corresponding curve upon paper or copper if preferred. The three or four first figures of the table will be expressed by the curve. The contrivances for this purpose are not difficult, and their employment does not lengthen the time of the calculation.

D.2.8 Variable Cards

The Variable Cards are appointed for the government of the various parts which constitute the Store. Like all the other cards, they act by pushing forward certain levers placed in front of them. These levers cause the motions of the several parts of the Store which have been described.

It is necessary in the present work to consider only—the Figure Axes, the Number Cards, and the card punching apparatus, Variable Cards, and Combinatorial Operation Cards.

With respect to these, the principle functions of the Variable Cards will be to direct:

(a) A Number and its sign to be given off from any Store Axis to the ingress Axis.

(b) A number and its sign to be received upon any Store Axis.

(c) Any number and its sign to be given off from the Number Cards to the wheels on the Number Axis.

(d) Any number and its sign to be given off to the Card Punching Apparatus and a corresponding Card to be punched.

The number of levers necessary for these purposes is not so large as might at first appear; consequently, the Cards need not approach an inconvenient magnitude. For example, fourteen levers and their equivalent fourteen holes will be all that is required in the third of the above divisions for eight thousand variables.

If the other appendages to the Store which have been already described should be thought necessary, a small number of additional levers must be added.

D.3 Of Operations

The operations which can be performed in the Mill are the following:

- Addition
- Subtraction
- Multiplication
- Division
- Extraction of Roots

These five processes are all which it will be necessary to consider in the present volume. Others might be added if required, and the additional mechanism would be small; new barrels with studs differently arranged would be nearly the whole change.

If these operations are merely viewed in an arithmetical light, the engine would greatly assist us in the numerical calculations to which analysis leads, but it possesses the power of treating the signs of the quantities on which it operates according to the rules of algebra, and thus its use is greatly extended.

D.3.1 Addition

There are three principles on which mechanical modes of adding numbers together may be constructed.

An axis passing through a wheel whose circumference is divided into ten parts may be conceived at certain periods to make a revolution independently of the wheel. This axis, therefore, will, whenever it moves, have an angular motion of one circumference, which is equal to ten divisions on the wheel.

Now, if a stud projects from the upper surface of the wheel and if a fixed arm projects from the axis in the same plane, then whenever the Axis revolves, its arm will at some period in its revolution come into contact with the stud on the wheel and will carry it forward until the axis stops.

If the wheel had been so placed previously to the movement of the axis that the stud upon the wheel stood one division in advance of the arm, then the wheel would have been driven over nine divisions; if the wheel had been so placed at the commencement that the stud had been six divisions in advance of the arm, then it would have been moved over four divisions. In the first case, the number nine would have been *added* to the wheel, and in the latter case, the number four. This addition might be communicated by a pinion to any other wheel.

Thus, by rightly placing the first wheel, any number from one to nine may be added to a second wheel. When such mechanism is employed, the axis moves uniformly round and picks up the wheel at a time dependent on the digit it is to add. Also, the motions of the wheel and of the axis terminate together.

Another method of adding might be imagined in which the axis should be bolted to the wheel at the commencement of its motion and carry it along for a certain part of its motion only, some subsidiary mechanism unbolting it when the wheel had reached a certain division.

In this case, whatever be the number added, the motion of the wheel commences with that of the axis but ends at different periods according to the nature of the number added. The combination of these two principles produces a third mode of adding in which neither the beginning nor the end of the additions occur at fixed points.

The three principles then by which addition may be executed mechanically may be thus stated:

(a) By the movement of a wheel which always ends its motion at a definite time independently of the particular number to be added but which commences its motion at a time determined by that number.

(b) By the movement of a wheel which always commences its motion at a definite time but ends it at a period dependent on the particular number added.

(c) By a wheel both commencing and terminating its movement at periods dependent on the number added.

The first of these principles is adopted in the present engine. The second exists in the Difference Machine. The third presents greater difficulties in the execution without equivalent advantages.

In all these plans, the number placed on the first wheel and which is to be added to some other number is itself obliterated by the operation. They might, therefore, with propriety be called apparatus for reducing figure wheels to zero. Another class of contrivances have been drawn in which the number added is itself retained so that the same number may be repeatedly added. The zero apparatus may generally be converted into an adding and retaining apparatus by the addition of two other wheels.

The process of addition requires for its completion that an apparatus should be attached to the wheels on which the sum is placed for the purpose of carrying any tens that may occur to the next highest digit. This mechanism may be constructed upon three different principles. It may be:

- Successive in its operation
- Postpone its operations
- Anticipate operations

Carriage may be successive in its operations: Those of the first class are the most simple, but the time necessary for the performance is an effective impediment to their employment in Engines like that we are considering. Let us suppose unity to be added to 999999, then:

Add together	1
	00999999
Time of 1st Carriage	00999990
	1
Time of 2nd Carr.	00999900
	1
Time of 3rd Carr.	00999000
	1
Time of 4th Carr.	00990000
	1

Time of 5th Carr.	00900000
	1
Time of 6th Carr.	00000000
	1
	01000000

The time necessary for the mere addition and leaving a notice that the first figure in the right has passed a ten (which is indicated by putting the unity below the tens figure) must be at least equal to that which is required for the figure wheel to pass over nine divisions. Now, the time for adding the first ten which is carried and giving notice of the second will be equal to one division, and so on, so that at the end of the sixth carriage, a time equal to fifteen divisions will have been passed over, and it is to be observed that another division must be allowed because the machine cannot know that there is not another carriage due to the seventh figure.

Thus, in an engine having forty figures, we should have for every addition and its carriage:

for addition	9
for 39 possible carriages	39
	48

So that, in fact, the time consumed in making the carriage of the tens would be more than four times as much as that required for addition. Modes have been contrived for shortening this time, but it is evident there is a limit beyond which it cannot be reduced.

Carriage may postpone its operations: When many additions are to be made to the same quantity, as in multiplication, time might be saved by reserving the notices of Carriage and executing them altogether.

Carriage may anticipate its operations: If the mechanism which carries could be made to foresee[1]that its own carriage of a ten to the digit above when that digit happens to be a nine would at the next step give notice of a new carriage, then a contrivance might be made by which, acting on that knowledge, it should effect both carriages at once. Thus, after the addition of the numbers below:

Time of Addition	123456
	346601
Time of 1st Carriage	469057
	1
Time of 2nd Carriage	460057
	1
	470057

If at the time of the first carriage, the engine knew the next number to which it [was] about to carry was a nine and that consequently it would *afterwards* become necessary

[1] In substituting mechanism for the performance of operations hitherto executed by intellectual labor it is continually necessary to speak of contrivances by which certain alterations in parts of the machine enable it to execute or refrain from executing particular functions. The analogy between these acts and the operations of mind almost forced upon me the figurative employment of the same terms. They were found at once convenient and expressive and I prefer continuing their use rather than substituting lengthened circumlocutions.

For instance, the expression "the engine *knows*, etc." means that one out of many possible results of its calculations has happened and that a certain change in its arrangement has taken place by which it is compelled to carry on the next computation in a certain appointed way.

to give notice of and to carry another ten, then it might be taught to execute both these carriages at the same time and consequently to anticipate the time of the second carriage.

This object is of such importance that it is worthy of any labor to obtain a plan at once simple and effective. I have contrived several; that which is adopted in the present engine is perfectly secure and, considering the very complicated nature of the conditions, it is not very deficient in simplicity. This extreme complication of the conditions will perhaps be more felt by considering the following cases:

Time of Addn.	006012345678
	101987655022
	007999990690
Time of carry	1 1
	008000000700

Time of Addn.	0004365780064128976438052
	0005634238990771023562048
	0009999918954899999990090
Time of carry	1 1 1 1
	0010000019054900000000100

The complexity arises from the circumstance that the result of any carriage may affect the value of every figure on its left hand and that the effect of any carriage may terminate at any digit and a new carriage may commence its effect at the succeeding one; and this, like its predecessor, may or may not be the precursor of many others.

D.3.2 Of the Algebraic Signs as Used in Addition

It has been already stated that above every set of forty figure wheels there is in the forty first cage a sign wheel and also that this sign wheel is divided like the figure wheels into a number of parts which is equal to some multiple of ten, the sign minus being engraved on the odd, the sign plus on the even divisions.

Now when these wheels all stand at zero the sign wheel being also zero will present the positive sign.

When a positive number is added its sign being fixed on some even number the sign wheel of the ingress axis will be advanced by an even number of digits and must consequently present some other even number on which will appear the sign *plus*.

The ingress wheels of the Mill which receive any number receive also its sign but as the number on the ingress wheels was zero and the sign plus the addition merely causes the ingress wheels to stand at the number entering and the sign to remain as it was before, namely at *plus*. For the 41st wheel stood at + or at an even number and the addition of + or any other even number will produce an even number on this wheel and consequently it will still present the sign plus. If on the other hand the number entering has the negative sign then its sign wheel stands at an odd division and as an odd number added to an even one produces an odd number the sign of the number in the ingress wheels will be negative.

The process of addition or subtraction is thus executed: An operation card advances and orders the Variable cards to advance, in the next turn a variable card containing the order for the sign plus or minus advances and places the sign on the sign wheels of $''F$ whilst at the same time it had ordered the number P to be given off from the store to I. The sign on the card may be called the algebraic sign. When the number comes in from the Store it will bring with it its own proper sign which will depend on the previous operations by which it had been formed and it will be convenient to call this latter its accidental sign.

This accidental sign will be added to the algebraic sign upon $''F$ and thus the two signs will be united and we shall have $\pm(\pm P)$ reduced to $\pm P$, any combination of the above signs being used.

If algebraic and accidental signs are of the same kind P is added to the figure wheels on A which have a positive sign on them. If the algebraic and accidental signs are different P is subtracted from the wheel on A and in consequence of a running up the negative sign is put upon the sign wheel of A.

The same process takes place with respect to $\pm(\pm Q)$ which after the combination of its two signs is transferred either additively or subtractively to the same wheels of A. In this second case a running up may or may not take place; if it do the previous sign is reversed.

The sum of the two numbers P and Q as modified by their respective signs will now be found on the figure wheels of A either in the form of a Number or its compliment.

If it is number it and its sign will be transferred to $''A$.

If it is complement by being subtracted from $''A$ it will appear as number and a negative sign will appear on the sign wheel.

If therefore when the first of two quantities, which it is proposed to add together, is placed upon the ingress wheels the sign wheel stands at an even number *no change* is made on the barrels; but if, when the sign of that quantity is negative or the sign wheel of the ingress axis stands at an odd division a *certain change* is made, then the Mill may be said to *know* that a negative quantity has entered and the barrels may give the orders necessary in consequence.

If two numbers P and Q are to be added together the result will be

$$+P + Q$$

where the two signs + + may be called the algebraic signs but if both P and Q are the result of certain preliminary computations made in the Engine it may not be known whether the numbers themselves denoted by P and Q are positive or negative. In this case we must distinguish between the known algebraic signs and the possible or contingent signs resulting from the computation — these latter may be called the *accidental signs*. The above formulae may therefore be written thus

$$+(\pm P) + (\pm Q)$$

This being premised the process of addition may be thus stated: At the first turn of the handle an operation card is turned; it advances and acting on the barrels through the reducing apparatus orders them to move to that vertical which commences addition. It also orders a variable card to advance. At the next turn of the handle the barrels advance and direct the Ingress figure wheels to receive from the Store any number and its sign which the Variable Cards may have ordered to be given off. The barrels in advancing order themselves to turn circularly after their advance so as to present the next vertical necessary for performing addition. If the *accidental sign* of the number on the ingress wheels is positive this arrangement continues but if that sign is negative it effects a change in the reducing apparatus and before the next turn the barrels have been moved to a different vertical which gives the orders necessary from the fact of the accidental sign of the number brought in being negative.

At the third turn the barrels advance and order the Figure wheels on A and on the two carriages F, J' to receive the number and its sign, both duly modified, from the ingress figure wheels whilst at the same time the Variable Card which was turned in the last turn of the handle orders a new quantity Q and its accidental sign out of the Store onto the other figure wheels of the Ingress axis. But this second advance of the barrels has ordered a circular motion for themselves by which they would turn to the next vertical for addition whilst the *accidental sign* of the quantity Q introduced on the ingress wheels again modifies this order if it happens to be negative.

The successive steps of this process will be more apparent in the following notations in which the words in the horizontal lines indicate the circumstances which occur during the respective turns of the handle and the vertical columns denote the events which successively happen in the places indicated by the headings of each column.

D.3.3 Subtraction

After the explanation of the process of addition with the accidental sign of the quantities added together it will not be difficult to understand that the operation of subtraction is carried on in a similar manner. As in addition the subtraction of a greater number from a less (in that case arising from the accidental signs) causes a running up which gives notice that the resulting quantity is the complement of a negative number; so in subtraction where the algebraic sign produces the same result a similar effect follows. The notations which follow will explain more clearly the several stages of the operation as well as the parts taken in it by the several portions of the Mill.

D.3.4 Of the Time Required for the Operations of Addition and Subtraction

Any two numbers P and Q whose sum does not exceed forty places of figures may be placed in the Engine and whatever be the nature of their accidental signs they may be brought out of the Store placed in the Mill and the operation of addition or subtraction performed as indicated by the formula

$$\pm(\pm P) \pm (\pm Q)$$

and the result may be returned into the Store in six turns of the handle. The first of these turns might take place as the final turn of any previous operation so that the time really required is six turns of the handle. If another quantity ($\pm R$) is to be added or subtracted it will require another turn and generally if n quantities with their accidental signs are to be added and subtracted the time required for the operations will be

n + 4 turns of the first mover.

The time which the edge of the figure wheel occupies to pass from one figure to the next figure is considered as the unit of time and the space between two adjacent figures is the unit of space. The cases in which that velocity is exceeded are very rare in any part of the machine and only occur when the moving part is carried on to a fixed stop. Wishing to under rather than overrate the speed of its calculations I have assumed that the velocity of the circumference of the figure wheel when moving will be ten feet or one hundred and twenty inches per minute. A single turn of the first mover will according to circumstances be performed either in fifteen or in twenty of such units. In the process of addition cycles of twenty units only are employed and as each unit requires .15708 seconds the addition of two numbers only will require [ ][2] seconds and the addition of n quantities will cost 21.99 or nearly 22 seconds.

D.3.5 Multiplication

The process of multiplication is thus performed by the Engine:

The two factors P and Q with their respective accidental signs being placed on the two sets of figure wheels No.1 and No.2 of the Ingress axis an operation card is turned. This

[2] Left blank.

directs the barrels through the reducing apparatus to move circularly to the vertical which commences multiplication.

At the next turn the barrels advance and order the reception of one of the factors Q. They also direct themselves to move on their axes to the second vertical belonging to Multiplication.

At the third turn the barrels advance and order the reception of the other factor P from the second set of Ingress wheels. At this turn the factor P is subtracted from the factor Q which had at the preceding turn been placed upon the figure wheels of F. The result of this subtraction determines which of the two factors is largest and consequently which factor is to be tabulated.

Much time is saved by this decision for supposing one factor to contain only four places of figures and the other to contain thirty five figures. Then if a table of the first nine multiples of the first factor were made its multiplication by the other factor would require thirty five additions whilst if the larger factor had been tabulated only four additions would have been necessary. The mechanical mode by which this knowledge is conveyed to the barrels is thus. The barrels after ordering the subtraction of P from Q in the axis F direct themselves to move on to another vertical; if P is less than Q no *Running up* takes place and the order thus given by the barrels is obeyed; but if P is larger than Q then a *Running up* takes place and the order given by the barrels to move to a certain vertical is enlarged and they are directed by the Running up lever to move on to a different vertical. This new vertical directs the tabulation of the second factor whilst the former vertical appointed that of the first factor.

At the next turn the number which the Mill has decided on Tabulating is added to each of the nine Table axes and to the Carriage F.

At the next turn the same number is added to eight of the Table axes and the carriage axis then to seven and so on until the nine multiples have all been formed in the nine Table axes.

During these turns the number of the digits in the multiplier has been ascertained and placed in the Counting Apparatus No. 1.

The Multiplier having been prepared is now stepped down and acts upon the Selecting Apparatus. Whatever digit exists in the lowest figure of the Multiplier, by means of the Selecting Apparatus the corresponding Multiple is added over to the product axis $'A$. During this addition each multiple is stepped up one cage higher in its own figure wheels; also the multiplier being stepped down its next figure is conveyed to the selecting apparatus preparatory to the next addition to the product.

The product is thus formed beginning with its lowest figure on the figure wheels of the axis A, but in the progress of stepping up the highest figure on the table axes may at length reach the top or 40th cage of their axes. At the next stepping the highest figure would of course be stepped off the machine and lost but this is prevented by a communication with the lowest figure wheel on the same axis to which the number stepped off from the highest is transferred. These figures which are so transferred to the lower cages are not added to the same axis but are by a particular mechanism added over to a set of figure wheels on the axis A and the carriage F on which the head of the product is formed.

If a number of twenty figures only is multiplied by another factor containing an equal number of digits, as the product will not exceed forty digits it will be contained on the first axis. But if both factors contain forty digits then the head or first forty digits of the product will be formed on the axis A and the tail or second forty digits of the product will be formed upon $'A$.

The signs which entered with the original factors, having been added together on the sign wheels, will give the resulting sign of the product. For if they were both positive then each sign being equivalent to an even number the sum of the two signs would give an even number or its equivalent the positive sign. If on the other hand both signs were negative then a negative sign being equivalent to an odd number the sum of two odd produces an even number which also is equivalent to the positive sign which the product ought in this case to possess. Lastly, if the sign of one factor be positive and that of the other negative, then the sum of an even and of an odd number being odd the negative sign results.

The termination of Multiplication arises from the action of the Counting apparatus which at a certain time directs the barrels to order the product thus obtained to be stepped down so that the decimal point may be in its proper place, and then to turn to another vertical, the consequence of which is the transfer of the product with its proper sign to the wheels on the egress axis from whence they may be removed by Variable cards and the barrels to any axes in the Store.

The great difficulties in constructing the Calculating Engine have in general arisen from certain conditions I had laid down as desirable for the perfect security and certainty of its action rather than from the difficulty of inventing mechanical contrivances for performing the functions of its several parts. There are few of its actions which have not been contrived and in many instances drawn according to several different principles. The only exceptions I should be inclined to make as to this relative difficulty are perhaps the mode of carrying the tens by anticipation and amongst the operations the mode of dividing one number by another.

The following principles were proposed:

No movement should depend upon the action of a spring.

This condition placed great difficulties in the way of many contrivances and the modes of obviating them are very various. It is not however proposed to exclude all springs, for those which merely retain a lever or a wheel that has been already brought to its exact place by other means are unexceptionable. In one sense they relieve the machine from the continual drag to which stiff friction if substituted would expose it, and on the other they save a multitude of moveable axes which would otherwise become necessary and they also enable wheels to remain retained until certain indefinite actions take place.

Another principle is that:

Every movement shall be of such a kind that the engine shall either break itself or stop itself or execute the intended motion.

A third principle is:

That the several operations shall be executed in the shortest possible time.

This last principle has caused greater difficulties than either of the former — in fact the whole history of the invention has been a struggle against time. As soon as any contrivance has been made which was unexceptionable as to the mechanism the question has always arisen: Can it not be executed in less time by some other contrivance? Thus every advance has but raised up a new object of rivalry, itself to be superseded by some more rapid means, nor can I hope that I have nearly reached the limit. If I have approached it, it is more than I have a right to expect as the pioneer in this difficult career. Another age must be the judge of that as well as of the other questions relating to the Engine. I will only observe that the rapidity of its performance must be judged of 1st by the velocity which the general structure of the machine enables us to employ with safety. This may in the present engine be greater or less than that I have assigned to it. 2ndly By the number of parts in the time of each cycle which are requisite beyond that necessary for the execution of addition and carriage. Nine units are absolutely necessary for addition and one for carriage and I believe that it may be stated that no engine in which all the digits are added at once can possibly require less than ten units of time for a complete addition or subtraction. Besides this each cycle or turn of the handle must have certain additional units of time allowed it in which the wheels when brought to their right position are locked or rendered immoveable and in which also certain motions are given to the axes by which these operations are executed. In the present engine when addition alone is to be executed these extra units amount to five, and when carriage also is required they are ten in number. Thus one of the cycles in which any number can be transferred from one set of figure wheels to another or in which it can be stepped up or down contains fifteen units of time. The other cycle in which addition or subtraction and consequently carriage must be made amounts to twenty units of time.

But it is not merely important that the elements of the operations should be performed in as short a time as possible. The engine ought to be so contrived that in their combination for executing the great operations of multiplication and division the result obtained should be as quickly reached as possible. Several modes have been examined for multiplication and that described has been finally adopted as the shortest.

The reader who only looks at the mechanical difficulties in contriving such an engine will doubtless be surprised when he examines the history of the contrivances, which will be described in a future work, at the lavish rejection of inventions which has taken place in order to achieve rapidity in computation. But the Analyst who is aware that the last resource in all our difficulties is the conversion of the most intractable expressions into infinite series will more readily appreciate the importance of even a small abbreviation of the time of an operation, when he is informed that it is one of the objects of this engine to arrive at the numerical value of the coefficients of such series, however complicated the laws by which they are formed, and that proceeding again from these results as established data, it is proposed to assign the values of such expressions for any magnitudes of the variables.

In the progress of the invention, it has frequently occurred that mechanical means have either been found altogether insufficient to shorten the time of certain parts of operations, or that it was only possible to effect this object by such great additions of mechanism as were themselves inexpedient.

In this difficulty, I have sometimes had recourse to an arithmetical, or to an algebraical artifice, by which in several instances it has been surmounted without any increase of mechanism. It has sometimes occurred that after the mechanism has been contrived and even drawn, some well known property of number has enabled me to supersede it. This happened with respect to the means of printing any table *true to the last figure.* It is well known that though the logarithms of

$$\left.\begin{array}{c}2\\[1ex]3\\[1ex]4\end{array}\right\}\text{are}\qquad\left\{\begin{array}{l}.301029957\\[1ex].477121517\\[1ex].602059913\end{array}\right.$$

yet if a table to only seven places of figures is to be printed, the logarithms ought in such a table to be printed:

log. of 2 is	.3010300
3	.4771213
4	.6020600

In the drawings of the first machine which I contrived for calculating tables by differences there were two arms, one of which after the logarithm had been computed went round to feel if its eighth digit were greater or less than 5. In the former case the arm caused unity to be added to the seventh figure but in the latter case it was inoperative.

After the printing of the seventh figure another arm went in an opposite direction, and subtracted the unity thus added, so that the calculation of the next logarithm might be carried on without impediment from this temporary change. I did not observe until after several contrivances for this purpose had been drawn, that the simple addition of 5 to the eighth figure of the first logarithm computed by the engine would, without any mechanism, ensure the correctness of the seventh figure throughout the whole of the succeeding logarithms of the table.

Another instance of the assistance to be derived from arithmetic occurred in the operation of multiplication. It was found that if the table itself was stepped on its own axes then every alternate turn of the actual multiplication might be saved. The mechanical difficulty which opposed the execution of this plan arose from the necessity of placing those numbers which the digits of the multiplier successively expressed upon the figure wheels of the Selecting apparatus. For as this number might be a nine, it was necessary to allow the whole time of one addition, and the movements consequent upon it, or a cycle of fifteen in order to place the number in its proper place; and after this an equal time must be allowed in order to reduce the same wheel to zero, as a preparation for the next multiple. Under these circumstances a remedy without increasing the standard velocity seemed beyond the reach of mechanism.

The following arithmetical artifice removed the obstacle. Let us suppose the multiplier to be the number 584269. If this number be placed on a Stepping axis, and stepped down, the lowest figure 9 will be stepped off and lost, but if, instead of this, it be received upon the figure wheel of the Selecting Apparatus, then the first multiple would be rightly placed. If now a second Stepping took place, without obliterating the 9 placed on the Selecting apparatus, the number 2 or the second digit of the Counting apparatus would be added to the 9, and the result would be a 1, which is not the multiple required.

If this process were continued we should have after each stepping the following corresponding number in the Selecting apparatus:

After	1st		2nd		3rd		4th		5th		6th
	5										
	8	on Selecting Wheel	5								
	4		8	on Selecting Wheel	5						
	6		4		8	on Selecting Wheel	5	on Selecting Wheel		on Selecting Wheel	
	2	9	6	1	4	7	8	1	5	9	4

So that if we do not obliterate the number at each turn, after putting it in the Selecting Apparatus, the multiplier 584629 will have the same effect as if it had been the multiplier 491719. This led me to enquire what number must be substituted for any given factor, in order that when so stepped, it may produce successively the digits of the given factor.

The question reduced itself to this: *edcba* being the digits of a given factor, what must *EDCBA*, the digits of a number to be found be, in order, that by stepping that number on itself, we may have successively the digits of *edcba* of the given factor.
The answer to this question gives the equations

$$A = a$$

$$B = b - a$$

$$C = c - b$$

$$D = d - c$$

$$E = e - d$$

and in the particular case:

$$A = 9, B = 7, C = 6, D = 2, E = 4, F = 7$$

and if we place the number 742679 on the Stepping axis, the successive results will be, after

	1st		2nd		3rd		4th		5th		6th
7											
4		7									
	on		on		on		on		on		on
2	Selecting	4	Selecting	7	Selecting		Selecting		Selecting		Selecting
	Wheel		Wheel		Wheel		Wheel		Wheel		Wheel
6		2		4		7					
7	9	6	6	2	2	4	4	7	8	0	5

Now the same equations inform us how to form the number thus required. For if we place the given multiplier upon a set of figure wheels, and subtract from it, without attending to the carriage of tens, the same number once stepped up, we shall have precisely those figures which are required, thus:

Factor	584269
Factor stepped up	584269
Subtraction without carriage	5742679

which number, being used instead of the given factor, will produce its successive digits in the figure wheel of the Selecting Apparatus.

The same artifice has been employed in another part of the engine, for the purpose of overcoming a similar difficulty in the motion of the circular plate carrying the steel types for punching the copper plates.

With respect to the time employed in the operation of multiplication: if p equal the number of digits in P, the smallest factor, d equal to the number of decimals in the engine, and if n is equal to the next whole number which is greater than $\frac{p}{12}$; then the time of multiplication without stepping the product is

$$(12 + n) \text{ Cycles of } 20 + (p + 4) \text{ Cycles of } 15.$$

But if the product is to be stepped to its proper place previously to its being carried into the Store, then the time of the operation is

$$(12 + n) \text{ Cycles of } 20 + (p + d + 4) \text{ Cycles of } 15.$$

These give respectively in turns of the handle

$$\frac{20(12 + n) + 15(p + 4)}{382}$$

for Mult without Stepping, and

$$\frac{20(12 + n) + 15(p + d + 4)}{382}$$

for Mult with Stepping.

D.3.6 Division

Of the numerous difficulties which the contrivance of an engine to perform arithmetical operations presented, none certainly offered more reasonable grounds of difficulty than that structure of it, by which it should be enabled to perform the operation of division. Contrary to the plan which might perhaps have been expected, I did not commence the attempt with the case of some simple number as 2, 3 or 5 for the divisor, but went at once to the case of large numbers both for divisor and dividend.

The first rough outline of a system by which it could be accomplished offered an instance of the most complicated mechanism I had ever imagined. But before it was submitted to that simplifying process which invariably succeeds the original conception, another principle occurred apparently so simple in its means, that little could be expected beyond it as to principle, and my whole efforts then became directed to the combinations of mechanism, a great part of which already existed in the engine for other purposes.

It has already been stated, that when any number which happens to be greater than another is subtracted from that other number, an event which has been termed a Running up takes place, and that in consequence of this event, any order which may be desired can be given to any part of the engine. Bearing this fact in mind, the division of one number by another may be conceived to take place in the following manner:

Let the number 00432101234 be divided by 2018.

In turn			
		Reg^r	
1. Barrel orders div^r to be subt^d from	00432101234		
highest figure of dividend and to	00201800000		
register that it has subtracted once	00230301234	+1	
2. Do. Do.	00201800000		
	00028501234	+1	
3. Do. Do.	00201800000		
	99826701234	+1	A Running up
4. Addition of Divisor	00201800000		happens which
Unregister one	00028501234	-1	directs the add^n of
5. Step up dividend	0028501234		the divisor and
6. Barrel orders div^r to be subtr. as	0020180000		then the stepping
before	0008321234	+1	up of dividend.
7. Do. Do.	0020180000		
	9988141234	+1	A second Running
8. Add^n of Divisor, Unregister one	0020180000		up and the same
9. Step up Dividend	0008321234	-1	orders as above.
	008321234		
10. Subt Div^r	002018000		
	006303234	+1	
11. Do.	002018000		
	004285234	+1	
12. Do.	002018000		
	002267234	+1	
13. Do.	002018000		
	000249234	+1	
14. Do.	002018000		
	998231234	−1	Running up and
15. Add Divisor	002018000		the same orders as
16. Step up Dividend	000249234	+1	above.

	00249234		
17. Subtract Divisor	00201800		
	00047434	+1	
18. Do.	00201800		
	99845634	–1	
19. Add Divisor	00201800		
20. Step up Divisor	00047434		
	0047434		
21. Subtract Divisor	00201800		
	00027254	+1	
22. Subtract Divisor	0020180		
	0007074	+1	
23. Do.	0020180	+1	
	9986894	–1	Running up the
24. Add Divisor	0020180		same orders as
25. Step up Divisor	0007074		above
	0070740		
26. Subtract Divisor	0020180	+1	
	0049560		
27. Do.	0020180		
	0029380	+1	
28. Do.	0020180		
	0009200	+1	
29. Do.	0020180		
	0009200	+1	
30. Add Divisor	0020180		
	0009200		
	009200	–1	

In the above view the first direction given by the barrels is to subtract the divisor from the dividend. This order is repeated until at the end of the third subtraction, it is found that a *Running up* has taken place, a fact which of course indicates that the divisor has been subtracted once too often. The number of times the divisor has been subtracted is marked by a register axis. Now it is in consequence of this fact of a Running up that the engine *knows* it has subtracted the divisor exactly once too often, and that knowledge is conveyed by the running up lever to the barrels, in the form of an order to move to such verticals, that at the succeeding turns they shall direct the erroneous subtraction and registration to be corrected, and the remainder of the dividend to be stepped up in order to undergo a similar process for the discovery of the next digit in the quotient.

It may perhaps be observed that in the above example another *Running* up occurs in the fourth turn at the ninth line in consequence of addition, and it might be expected that an order should consequently follow. But this is provided against by the very general nature of the Running up levers, which are capable, under different circumstances, of conveying any order that may be required to any part of the engine. Immediately upon the occurrence of a running up these levers, besides the directions already stated, convey through the barrels an order to themselves to be inefficient at the next running up, thus the runnings up in the 8th, 15th, 19th, 24th, etc., turns are inefficient, whilst in the 3rd, 7th, 14th, 18th, 23rd, and 29th, they are efficient.

The first figure in the quotient being thus obtained, a repetition of the same process will of course produce the succeeding figures.

The complete understanding of this principle is of some importance, for it not merely relates to the process of division, but it is capable of being applied to any tentative arithmetical process. I had applied it to the extraction of roots, but other improvements induced me to dispense with it in that instance. Its application is not limited to existing rules, but I believe I may venture to state, that whenever new methods are contrived for overcoming arithmetical difficulties by tentative processes, it will be found available for those new purposes, and that even in cases where the tentative processes relate to algebraic expressions, it may yet be useful.

I have given a simple case in order to explain the nature of this principle. Division however is not executed by this tedious process, which would require for its performance a number of turns more than equal to the sum of the digits in the quotient. The process really used may be thus described.

The operation card ordering division, having advanced, directs the barrels to the proper verticals.

The divisor P is first received from the ingress wheel upon several parts of the mill.

The dividend R next enters. Several processes are gone through in order to ascertain whether it will be necessary to step the dividend up or down, and how many steppings there ought to be of either kind. This depends upon the number of digits of the quantities operated upon, and is principally executed by the counting apparatus which also computes the number of figures which must occur in the quotient.

At about the eighth turn the formation of the table commences, and also the selecting apparatus is prepared.

If the stepping down is completed, the actual division may commence about the eighteenth turn.

By means of the selecting apparatus the two first figures of the dividend are compared with the two first figures of each of the nine multiples of the divisor, and that multiple which has its two first figures either nearest in magnitude or equal to the two first figures of the dividend is selected and subtracted from the dividend.

The remainder is then stepped up, and the same process repeated until the counting apparatus interferes to finish the process.

In case the two first figures of the dividend are exactly the same as the two first figures of one of the multiples of the divisor, there may arise a doubt whether that multiple or the preceding one should be subtracted; this will depend on the third figure of each quantity. But in this case, the mill always takes for granted that the multiple selected is the true one. If however the contrary happens to be the case, then the subtraction of the multiple which is really too large will cause a running up on the dividend, and in consequence of this, orders will be transmitted to the barrels to direct the addition of the first multiple, and the subtraction of unity from the multiple registered in the quotient, by which means the quotient being set right, the first process is then continued.

After the end of the actual division the quotient, and the remainder if required, are given off to the egress wheels.

As the result of a multiplication of a number containing forty digits by another of equal extent is a product containing eighty digits, and as this product can be conveyed into the store on two sets of figure wheels, a provision has been made for bringing such numbers back into the mill when they are to become dividends, so that the result of the division of a number containing eighty by one of forty digits, shall have the forty figures of its quotient true to the last figure.

In deciding on the number of the first figures of the dividend to be compared with those on the table, it was considered that the infrequency of the running up when two only are used, was such that the resulting loss of time would be more than counterbalanced by the mechanism required in case three figures should be compared. If however it should ever be thought expedient to save that portion of time, it is quite possible by adapting a carrying axis to each table axis to render division a direct process, in which no figure can ever be assumed in the quotient which requires a correction afterwards.

The process of Division occupies more time than that of Multiplication, a circumstance partly arising from the greater extent of the preliminary process, but chiefly from the necessity of occupying one turn of the hand for stepping down, which is executed in multiplication during the time of the addition.

Whenever therefore by proper arrangement of a formula or by other artifices multiplication can be substituted for division time will be gained by the change. Examples of this exchange will be found frequently in the following pages.

The time occupied by the process of division will depend on the numbers concerned, and on the position of the decimal point, and in a small degree on the accidental number of the runnings up.

The following formula will give the number of turns of the handle . . . [incomplete]

D.3.7 Extraction of the Square Root

I had at one period in the progress of the Engine determined on making it perform the operation of extracting the square root of numbers. For this purpose, the usual method was in some degree modified and adapted to the structure of the Mill. Advantage was taken of the effects of running up to alter the various directions necessary for the process, and notations were made for the purpose. The time occupied in obtaining each figure was about four turns for every unit it contained, and the process therefore rather long. When successive improvements had considerably diminished the time of the action of the engine, I gave up the process at first intended for the extraction of the square root, and the grounds of the change were:

That with the shortened cycle, the process of obtaining the square root by successive approximations would not be longer than the direct extraction in the former state of the engine.

That considerable simplification occurred in the machinery from giving it up.

That the extraction of the cube and all other roots would then be performed according to the same principles.

In the engine now described, the extraction of square and other roots will be performed by approximation. If the quantity whose square root is required has a negative sign, the arithmetical root of the number will be found, but a particular notice will be given by which the circumstance that the root so found must be multiplied by $\sqrt{-1}$, will also be indicated. The mode of executing these approximations will be explained in a future page.

D.4 Of Computing the Numerical Value of Algebraic Formulae

It will have been observed that the orders by which the engine has been directed to execute the elementary arithmetical operations already described have been of two kinds.

First, Operation Cards having certain holes by which they govern the barrels and the Variable Cards.

These operation cards are always identical with each other, as long as the same operation is to be performed, and they are totally independent of the particular numbers which they may direct to become subject to them. In the cases of multiplication and division, the operation cards will indeed be turned and advanced at the requisition of the barrels in a longer or shorter time according to the nature of the numbers, but the forms of these cards will always remain unaltered. In this sense, the operation cards partake of that generality which belongs to the algebraic signs they represent.

For the processes of addition and subtraction, two operation cards only are necessary. One of these is required for each quantity, and assigns to it the algebraic sign of the process. Thus any number of additions and subtractions are ordered by the same number of operation cards.

Besides these two operation cards, it is necessary to have three Variable Cards. The two first direct the two variables to be operated upon to move from the Store into the mill, the third variable card directs some particular Store axis to receive the computed sum of the two quantities.

Whatever be the number of quantities connected by the signs plus or minus, the same number of operation cards is required, and the same number increased by one expresses that of the Variable Cards.

In the processes of multiplication and division, one operation card is required to order the barrels to commence the process. If it should be necessary, under any circumstances, to have two different modes of performing these operations as is at present proposed, for the purpose of saving time in approximations, then another card may be attached, which is to be a blank card, and therefore ineffective in the most frequent case, but which is capable of bringing in the less usual case when it is required.

Besides these Operation cards, two variable cards will be required to order in the two factors, and three others to order the three parts of the product into their respective places on the Store.

In Division, three variable cards will be required to order the dividend out of the Store into the Mill, and one more to direct the quotient to its right place in the Store.

The following table exhibits the number of cards necessary for each operation.

	Operation Cards	Variable Cards
To add two quantities	2	3
To subtract one quantity from another	2	3
To Multiply two quantities and retain 80 figs of product	1	5
To Divide one quantity of 80 figs by another quantity	1	5

In any operation, it is only necessary that the operation card ordering its commencement should be advanced. This sets the barrels in action and provides, through the variable cards if necessary, the first quantity to be operated upon; the barrels then call in the other quantities required at the proper times, until the operation is completed, the Mill cleared of all numbers, and the computed result placed in its intended situation in the Store. The last vertical on the barrels belonging to each operation directs them to move to its zero point.

Such being the arrangement, the calculation of the numerical quantities resulting from two or more successive operations may be readily performed by placing the operation cards strung together in their proper order on their revolving prism, and the variable cards marking the numbers in the Store, on which are to be the subject of those operations, and also in the order in which they are required, on their own revolving prism, and then putting on the final verticals of one of the barrels a stud which orders the operation cards to advance.

This arrangement being made, after the first advance of the operation cards, the barrels will direct the progress of the operation, calling into the mill by the advance of the variable cards the numbers to be operated upon, and replacing in the Store the computed result by the same means. The barrel, by the return to the final verticals of this operation, orders another operation card to advance, which, giving a new order to the barrels, these latter again conduct that operation, demanding at proper times the fit numbers from the Store, and conveying back to it their new result, and finally, by returning to their final verticals, they again demand a fresh order from the operation cards.

D.4.1 Verification of the Numbers Placed in the Store

The numerical quantities intended to be operated upon may be placed by hand on a particular set of figure wheels in the Store. Or they may be punched previously by a separate machine on number cards, and thus conveyed into the Store.

In either case, the possibility of error exists, and although it might reasonably be required that no care should be spared in inserting in the engine correct data for the question whose solution is demanded, yet it is important that the fulfilment of this condition should not depend solely on the superintendent of the Engine, but that any mistake he might make should be open to detection when the calculated results are returned to those who supplied the formula and constants with which they were to be computed.

Fortunately, this condition, which really becomes of great importance when the number of the given constants amount to hundreds or thousands, can be easily fulfilled.

It is merely necessary to order by means of cards that each constant, as it enters the engine, shall pass through the printing apparatus previously to its taking its intended place in the Store. By this means, the numbers really put into the Engine by the attendants will be printed, and they may be returned with the computed results printed also by the Engine upon the same paper. If the calculation consists of the reduction of a multitude of observations, it may be convenient to print, in an adjacent column, each observation entered into the engine and its corrected result.

Thus, whatever may be the formula ordered to be computed, the employer of the engine will always possess ample means of detecting any error in the constants inserted in it by the attendant.

D.4.2 Verification of the Formulae Placed on the Cards

The methods which have been contrived for accomplishing this object are not so simple as that for insuring the accuracy of the constants, nor are they, without the use of considerable machinery, so absolutely certain in their results.

It must, however, be observed, that if care is demanded from the attendants for the insertion of numbers which are changed at every new calculation of a formula, any neglect would be absolutely unpardonable in combining the proper cards in proper order, for the much more important purpose of constructing the formula itself, the arrangement of whose cards is never changed at any after time. This verification might therefore be reasonably left to the diligence of the superintendent.

There are, however, several subsidiary modes by which he may himself check his own proceedings.

The operation cards, for example, may be made of four different colors corresponding to the four operations of arithmetic:

- A White Card indicating Addition.
- A Yellow Card indicating Subtraction.
- A Blue Card indicating Multiplication.
- A Green Card indicating Division.

When therefore the chain of operation cards is finished, their total number ought to be equal to that of the operations in the formula. The number of white cards ought to agree with the number of additions, the number of yellow cards with that of the subtractions, whilst the number of Blue cards marks the number of multiplications, and that of green cards the number of divisions.

Since the number of quantities which enter the mill to be operated upon, as well as that of the results to be returned to the Store, is known, the number of the variable cards is known. These might be colored like the operation cards, and those cards which ordered numbers out of the Mill into the Store might be colored black. If this were the case, before any two sets of formula cards were used, they might be compared as to the succession of their colors, when, if correctly placed, it would be found that the same succession of colors would occur in each, but that in the variable cards each color would be separated from the preceding by one or more black cards.

Another mode of partially verifying a formula consists in assigning such numerical values to the constant quantities as shall render its value easily computed by the pen. If either the operation or the variable cards have been wrongly put together, the formula given to the Engine will be different from that which was intended, and the numerical result will in all probability be totally different from that computed by the pen. If trials of three or four simple cases have been made, and are found to agree with the results given by the engine, it is scarcely possible that there can be any error amongst the cards.

When the formula to be computed is very complicated, it may be algebraically arranged for computation in two or more totally distinct ways, and two or more sets of cards may be made. If the same constants are now employed with each set, and if under these circumstances the results agree, we may then be quite secure of the accuracy of them all.

It should, however, never be forgotten, that whatever attention may have been bestowed on securing the accuracy of the cards, that object, when once attained, can never again become doubtful.

If, notwithstanding these methods of verification, it should still be considered desirable to have greater security, it is certainly possible to attain it by making the engine itself print the formula inserted in it, as well as the constants to be employed in that formula.

At two different stages in the progress of the invention, I had contrived mechanism by which this might be accomplished, but as it related to a much more advanced stage, I did not make any drawings of it. Since the introduction of Formula cards, these contrivances have become susceptible of great improvement. For it is now no longer necessary to add mechanism for that purpose to an engine already sufficiently complex; but a separate machine should be constructed, by placing the formula cards in which the formula might itself be printed from them.

Such a machine does not appear to me to present much difficulty, as far as regards selecting the signs and variables. The chief inconvenience would arise from the uncertain extent of the compound quantities operated upon. This might in a great measure be obviated by only requiring it to print the series of successive substitutions which are to be made in the Engine. The reader will understand the force of these remarks better when he has perused the mathematical part of this volume.

But even this assistance might fail in some of the more complicated cases, for a certain degree of doubt always remains whether we have ourselves worked out rightly the developed results which we wish to convert into number. This doubt may be almost altogether removed when we avail ourselves of the *Combinatorial Cards*, because in such instances, although the ultimate results are extremely complicated, yet the instructions communicated to the Engine by cards are very few, and exceedingly simple, and may therefore be repeatedly verified in a very short time. It appears however to me that the formula printing machine might by some improvements itself ultimately work out many of such algebraic developments.

D.4.3 Extent of the Numerical Power of the Engine

In deciding on the extent of the numbers with which the engine should compute, the first consideration was to look at the number of figures which in the present state of mathematical enquiry are required in the most extensive calculations. It then became desirable to look forward to the probable increase which improved observations might require. Finally, the mechanical structure of the engine, or its necessary arrangement, might put limits to this extent, or render the time employed in given calculations longer, at certain definite intervals.

But even if it were thus possible to satisfy all practical wants, there would still remain a desideratum to render the mechanism philosophically perfect as to its power of converting algebraic expressions into numbers, without any limit as to their magnitude or extent.

The result of my reflections has been that numbers containing more than thirty places of figures will not be required for a long time to come. I have, however, made the drawings of the Engine for forty places of figures. All additions and subtractions may be made with such numbers, and the products, amounting to eighty places, may be preserved in the Store and brought back into the mill to be divided by numbers of forty places of figures, thus retaining forty places in the quotient.

I have however considered thirty places of figures as the standard, and upon that basis have made all the estimates of the time required for the computation of formulae. In order to fulfill the condition of the conversion of algebraic formulae into number without reference to the number of digits in the quantities employed, I have availed myself of certain arithmetical principles, by the aid of which, without adding to the mechanism, calculations, if requiring any extent of digits, may be performed; these principles will be explained in a future page. At present, it may be sufficient to observe that if it is required to compute formulae having in each quantity not more than eighty figures, it will be sufficient to employ about four times the time requisite for the calculation of the same formulae, having only half that number of digits in each quantity. If three times the number of digits (or 120 digits) be required in each number, then about nine times the time will be requisite, and in general, if the number of figures required to be used in the given quantities be expressed as n times the number forty, then the time required by the engine will be:

$$40 \times n^2$$

It was important, at least for the theoretical perfection of the engine, that no additional mechanism should be required for any extent of numerical calculation, beyond that which was considered to be safely practicable in the present state of the mechanical arts.

This point has been accomplished with respect to numbers, by increasing the time nearly in proportion to the square of the number of times the sum of the digits is a multiple of forty.

As to the extent in point of successions of the algebraical operations, no practical limit seems to exist. The number of cards which indicate the operations, and those that govern the quantities, may be increased to any extent without loading the engine, and in the more complicated developments, the combinatorial cards may be made use of, and expansions, which by the usual means would require above twenty thousand cards, may be performed with little more than two hundred.

D.4.4 Time Required in Computing Formulae

The most constant difficulty in contriving the engine has arisen from the desire to reduce the time in which the calculations were executed to the shortest which is possible.

It is not to be presumed that such an attempt has succeeded. How near the approach has been made must remain for after times to determine. It is however, desirable to state the mode in which the time has been estimated and to point out sources of mistake which might perhaps be overlooked by those who have not deeply studied the subject.

The first step is to assume as a definition, the unit of space and the unit of time. I have assumed for the first of these, the space (measured on a circle), between the figures on the figure wheel. This is [...]. The unit of time is assumed to be that portion of a second required for the passage of one figure to another on the same wheels.

The time occupied by one turn of the handle of the engine, which is called a cycle, consists of the time necessary for the directive action, the time requisite for executing those directions, the time necessary for addition, the time necessary for locking the wheels moved, the time necessary for carriage and the time required for locking the wheels after carriage.

It sometimes occurs that carriage is not required, in this case the cycle consists of fifteen units of time. When carriage also is to be used, twenty units of time are necessary.

The first mode of reducing the time which presents itself is by diminishing the unit of space, or the distance between the figures. The first inconvenience which presents itself, is that the machinery becomes more delicate and requires better workmanship. The general scale on which it is constructed becomes that of clockwork, and finally is reduced to watch-work. This evil will of course be diminished as the mechanical arts advance in perfection.

Another consequence of the diminution of the unit of space is, that as the *time lost* at each gearing is not proportionally diminished, it becomes necessary to make more frequent corrections by means of additional locking apparatus. The time required for these lockings increases the length of the cycle. In all the general plans as far as the the unit of space was assumed to be .62832 and a correction was applied to destroy the loss of time at every third wheel. When the two counteracting principles of a short unit and many lockings were distinctly perceived the present unit was adopted as that most fitted for the present state of mechanical art and a locking was placed at every second wheel.

The time necessary for the action of the directive part consists in that employed in advancing the barrels, and in turning and advancing the several classes of cards. The advance of the barrels and that of the cards can be made contemporaneous, and the extent of their motion will depend on the mode by which the several parts are thrown into gear, and the time will depend upon the space, and the mass of matter to be moved.

The time necessary for the circular motion of the barrels from one vertical to another, and also that required for the turning over several cards to arrive at the one wanted, when it is at some distance from that last used, must also be considered. This operation may be executed during the addition part of the cycle, provided the distance between the vertical in the barrels and the cards on the prism is not too large. When the unit of space .62832 was employed, fourteen verticals could be passed over in a cycle and cards. But when that unit was reduced to it was found that only 5 and 7 verticals would be passed and cards in the same portion of the cycle. The consequence of this diminished time will be that under certain circumstances it may become necessary to employ an extra turn of the handle in order to bring the barrels or the cards to their proper positions. This then forms the first counteracting principle to be considered when it is proposed to shorten the unit of space.

The corrections for loss of time arising from the original structure and subsequent wear of the teeth of the wheels are made by means of the locking apparatus. Levers terminating in wedges are forced between the inclined planes of the locking teeth, and thus bring the wheels to which they are attached accurately to their places. The depth of the locking teeth depends on the extent of the correction, and the time occupied in moving them will depend partly on that depth, and partly on the clearance required when they are unlocked. If therefore the same extent of motion is carried through a short and through a long train of wheels, the time occupied in bringing the last wheel to its true place will in the first case arise from few lockings moving through certain spaces, whilst in the long train it will result from more numerous lockings, each of which moves through a smaller space.

If the space required for clearance could with safety be diminished in the same proportion as the depth of the lockings, then in point of time merely, it would be indifferent whether the lockings were few or many, and their number would then be determined by the simplicity of mechanism, and by other principles. The unlocking of the several wheels may take place in a much shorter time than that required for locking them.

The time requisite for executing the orders of the directive part consists in that during which the original source of movement is connected with the several parts which are ordered to be moved. The time of addition comes within this class, and so also does that in which certain gearings are made and broken by the lifting or descent of certain axes. As these gearings can usually be made at the same moment, their number does not augment the time they employ. But if, in order to save mechanism, those axes move to several points of rest and consequently pass over considerable space, then more time must necessarily be allowed. When the number of figure wheels in each cage was reduced from four to two, the time employed in lifting these axes was also reduced.

The time consumed in carriage varies more than that of any other portion of the cycle, according to the different mechanism employed for that purpose. The mere time of adding the units caused by carriage may occupy one unit of time or it may require forty. The time for the previous adjustment will also vary but in a less degree. It is therefore on this point that by far the most important saving can be effected, a saving which would justify even very expensive machinery.

In the earlier stages of the invention of the engine, the great extent of the mechanism for carriage, which it was necessary to attach to each set of figure wheels, became a serious obstacle to its advancement. It was the pressure of this difficulty which led me to reflect on the great assistance derived in analysis, from separating the symbols of operation from those of quantity, and by a distant analogy to attempt the introduction of a carrying apparatus, which might be attached, as occasion required, to any set of figure wheels.

When I reflected on the fact that the quantity of mechanism necessary for performing the operation of mere addition without carriage was not above the tenth part of that requisite for carriage itself, the importance of such a separation became evident, and this led me to separate the machine into two parts, one (the Mill) in which operations were performed, the other (the Store) in which the quantities used and produced were inserted.

In its present state the time by which the cycle is increased when carriage is required is five units, only one of which, however, is employed in the actual carriage. The rest of the five is filled up by the necessary motions for preparing carriage, and for the additional lockings consequent upon it.

The introduction of the principle of hoarding carriages until they amount to nine, has in the case of multiplication enabled me to reduce that operation still further by substituting *nine short* and *one long* cycle, for every *ten long* cycles previously employed.

The time necessary for locking the wheels after carriage will be nearly the same as that requisite for locking them before. According to the nature of the mechanism a gearing or two more or less will make the only difference.

Previously to deciding on the absolute time of each cycle, another question must be entered upon. Since the most important circumstance which affects the duration of a cycle by a time amounting, even in the most favourable circumstances yet contrived, to one third of that in which addition can be performed, is the time which is required for carriage, it may reasonably be enquired whether, by adopting a different radix for the arithmetical notation, we may not in some measure avoid the loss of time. I accordingly at one period considered the advantage of adopting the arithmetical notation of which the base is 100 instead of ten.

The immediate consequence of such a plan was that carriage occupied on the average about one ninth of the usual time. Some drawings were made for this view of the subject but after much consideration the plan was given up as undesirable even in an engine of differences.

Having thus examined the elements of the operations, and pointed out the chief ingredients of the time they occupy, the next consideration is their combination into operations. As far as I have been able to examine the processes of addition and subtraction, it does not appear that much is to be gained, since the lowest limit must be one turn for each process and the limit is at present $n + 4$ turns for n operations. The small saving which might possibly be contrived is not of the same importance as any economy in the time of performing the larger operations.

With respect to multiplication, two methods may be followed, either the process may be commenced by taking that multiple out of the table which corresponds to the lowest or units figure of the multiplier, and then proceeding to the multiple corresponding to the tens, and so on; or we may begin by selecting the multiple corresponding to the highest figure, then that to the second figure and so on.

The first of these modes of proceeding requires only one carriage, because at each stepping down of the product one figure is carried off to another axis, and no subsequent addition to the product can ever alter the figures so transferred. If, however, we commence at the highest figure of the multiplier, two carriages become necessary, because any multiple, down even to that which the units figure directs, may by possibility change all the others up to the highest, and since the product of forty figures by forty produces a number of eighty places of figures, this must be placed upon two sets of figure wheels and consequently require two carriages.

The operation of division does not appear to admit of the same two methods as those which are applicable to subtraction. It appears to be necessary to perform subtraction, and then step up the remainder, and any attempts I made to unite the two processes so as to perform them in one cycle have proved ineffectual.

The following are the times expressed in turns of the handle in which these operations are accomplished by the engine...

The time occupied by the engine in computing the same formulae will be different according to the mode in which it is arranged. This will be apparent from a single case, let

$$x = \frac{a^2c + b^2a + c^2b}{abc}$$

This will, in its present form, require eight multiplications, two additions, and one division.

If, however, it be put into the form

$$x = \frac{a}{b} + \frac{b}{c} + \frac{c}{a}$$

it will only require three divisions, and two additions. In the latter case the time of computing it will be but little more than half that required in the former.

The difference is still greater in other cases. If it is required to find the arithmetical value of

$$a^6 + 5a^5b + 10a^4b^2 + 20a^3b^3 + 10a^2b^4 + 5ab^5 + b^6$$

each term will require five multiplications besides that which arises from its numerical coefficient, so that even committing the time so occupied, the computation will require thirty-five multiplications and six additions whilst if it be put under the form

$$(a + b)^6$$

the same calculation may be made by five multiplications and one addition.

The consequences resulting from this circumstance are important, and deserve the attention of those who are engaged in extending the domain of analysis, as well as of those who look forward to the effects which are likely to be produced by the complete control which mechanism now gives us over number.

Whenever engines of this kind exist in the capitals and universities of the world, it is obvious that all those enquirers who wish to put their theories to the test of number, will apply their efforts so to shape the analytical results at which they have arrived, that they shall be susceptible of calculation by machinery in the shortest possible time, and the whole course of their analysis will be directed towards this object. Those who neglect the indication will find few who will avail themselves of formulae whose computation requires the expense and the error attendant on human aid.

Appendix E

Babbage Materials at Turin.

Babbage left a wealth of material at the Turin Academy of Sciences that he brought with him in September 1840. It is still preserved in the archives there.[1]

1. Cart. 22338-22346: four letters to Angelo Sismonda in English, with some translations which, according to the Archive's inventory indications, were due to Fortunato Prandi.
   (a) September 3, 1840, cart. 22338; translation in cart. 22344;
   (b) December 1, 1840, cart. 22339-22340; translation in cart. 22345;
   (c) March 24, 1841, cart. 22341-22342; translation in cart. 22346;
   (d) March 24, 1842, cart. 22343.
2. Cart. 24587: from the Sclopis archive.
3. Plana.II.13: a miscellaneous volume of Babbage writings, letters to Plana and manuscripts in English: starting from number 4, almost all of them are inscribed to Plana.
   (e) A letter to Sir Humphry Davy on the application of machinery to the purpose of calculating and printing mathematical tables;
   (f) Observations on the notation employed in the calculus of functions;
   (g) Observations on the analogy which subsists between the calculus of functions and other branches of the analysis;
   (h) On the application of analysis to the discovery of local theorems and porisms;
   (i) Examination of some questions connected with games of chance;
   (j) Abstract of a paper entitled "Observations on the temple of Serapis at Pozzuoli";
   (k) (Private) Minutes of the Council of the Royal Society relating to the report of the Committee on Mr. Babbage's Calculating machine (February 12, 1829);
   (l) (Private) Report of the Committee, appointed by the Council of the Royal Society, to consider the subject referred to in Mr. Steward's letter, relative to Mr. Babbage's calculating engine, and to report thereupon;
   (m) Eight pages manuscript in French, translation of an undated letter by Babbage, of which the original is missing; twelve pages manuscript in French, translation of the Royal Society report above;
   (n) On the advantage of a collection of numbers to be entitled "The constants of nature and arts";
   (o) Letter of November 30, 1835 from Babbage to Plana: original manuscript;
   (p) Another copy of the writing referred to in previous n. 9;
   (q) Laws of mechanical notation.

Mss. 2776 - Charles Babbage Papers About the Analytical Engine.

[1] Losano 2014.

No.	Description	Date
1	Plan of method of adding	10 May 1837.
2	Adding, axes & wheels without framing.	
3	Frame of adding wheels.	
4	Section of adding wheels and framing.	
5	Section of adding and sign, axes & wheels, frames etc. complete.	
6	Impression from original woodcut, from which No. 7 is a Stereotype plate with the letters & signs inserted.	
7	Stereotype cast from wooden block of No. 6 with the letters & signs inserted. From this Stereotype plates No. 2, 3, 4 & 5 were formed by scraping away portions of the lines.	
8	Plan of adding wheels & long & short pinions.	April 1837.
9	Long pinions in the position for adding, etc. Adding wheels; & short pinions behind long pinions.	
10	Large pinions in the position for stepping adding wheels & short pinions behind long pinions.	
11	Plan of carriage by anticipation.	4 December 1836.
12	Fixed and movable wires of the anticipating carriage 1/3 the real wire.	
13	Carriage by anticipation.	
14	The general plan of Mr. Babbage's great calculating engine (large format plate, printed).	–
15	Exp. of punching on copper for Mr. Babbage's calculating engine, 1822 (printed plates, small format, 2 copies).	1822.
16	Elevation of a position of Mr. Babbage's engine for calculating tables by the method of differences as arranged in 1834 (2 copies).	1834.
17	Notes respecting the calculating engine (manuscript of 4 unnumbered pages).	–
18	Method of carrying; intended for the great machine. Designed about 4 December 1836 (manuscript of 2 unnumbered pages and 2 blank).	4 December 1836.
19	Calculating engine (manuscript of 4 unnumbered pages).	16 September 1840.
20	The calculating engine consists of two parts ... (manuscript of 4 unnumbered pages).	–
21	Various forms of development (manuscript of 4 unnumbered pages, bearing number 2).	29 July 1840.

22	Notation of the computation of the coefficients of the powers of x in the development of: $$\frac{A_0 + A_1x + A_2x^2 + ... + A_ix^i}{a + bx + cx^2} = \lambda_0 + \lambda_1x + \lambda_2 + x^2 + ... + \lambda_m x^m.$$ (large format manuscript of 3 unnumbered pages and 1 blank, dated: 14 August 1837, bearing number 3).
23	Notation of the coefficients of the cosines of $n\theta$ in the product of: $$(A_0 + A_1 \cos\theta + A_2 \cos 2\theta + ...)(a + b\cos\theta) = \lambda_0 + \lambda\cos\theta + \lambda_2 \cos 2\theta + ... + \lambda_n \cos n\theta.$$ (large format manuscript of 3 unnumbered pages and 1 blank, dated: 7 September 1837, bearing number 14).
24	Determination of x from three equations: $$ax + by + cz + d = 0$$ $$a'x + b'y + c'z + d' = 0$$ $$a''x + b''y + c''z + d'' = 0$$ (large format manuscript of 4 unnumbered pages, dated: 23 July 1840, bearing number 18).
25	Seven unnumbered sheets with calculations.
26	Samples of punch cards.

E.1 Notes Respecting the Analytical Engine

The designs of this engine are contained in upwards of 90 Drawings each three feet wide by two feet high. The Mechanical Notations by which the various movements are represented occupy about four hundred large sheets of paper. The original sketches, descriptions and notes are contained in four small volumes. A few impressions from about twelve large wood cuts and one large engraving on stone have been taken. Besides these there are nine other volumes of Mechanical Notations and about fifty or sixty Algebraic Notations. The whole of these papers are by no means necessary for the understanding or construction of the Engine. They form the history of the progress of the mind in inventing it and an abstract of the chief points at which definite discoveries have been made, is arranged in seven large sheets of paper each divided into horizontal lines, into twelve parts of months and divided into vertical columns to separate the various distinct portions of the Engine as for ex. Adding, Carrying etc. These papers will be invaluable to whoever in future years may be destined to construct the Engine. They will enable him to see the reason for rejecting many improvements which occurred and the cause for which those which exist were approved—without the possession of these papers no just idea can be formed of the quantity of thought expended in the engine and in all probability anyone who attempted to improve it would.

The design of the calculating Engine was commenced in September 1834 and carried on continually with the exception of five months during which the drawings were discontinued whilst the author was making experiments on the Great Western Railway. The principles on which each separate portion of the engine acts were successively discovered and their mode of application improved so that there are drawings of about 15 different mechanical contrivances for addition and a still larger number for Carriage.

The general combination has undergone as many changes. The two plans brought to Turin are dated January 1837, the 13th Gen. Plan, and August 1840, the 25th Gen. Plan. The points most remarkable for their importance and contrivance are

1. The method of addition of which there is a model. This is remarkable for its simplicity.

2. The method of carrying the tens of which there are three wood cuts [with] description and notation. This is remarkable for its extreme generality and for the shortness of time in its operation.

3. The principle of division and its reduction to a direct process without having more than three carriages in the whole engine.

4. The plan of communicating all orders to the engine by means of cards with perforations. This renders the number of operations as well as of variables and of given numbers absolutely unlimited without increasing the size of the Engine.

The engine is drawn as if it were to have 40 figures which [is] so that it can multiply any two numbers each not exceeding 40 digits and divide the product containing 80 digits by any other number of 40 figures. If calculations requiring above 40 figures are wanted the same Engine can calculate the same formulae to

Places	Time Factor
80 places in	4 times the time
120 places in	9 times the time
160 places in	16 times the time

so that it is neither limited in extent of figures nor in the number of its variables or operations. ...[2]

E.2 On the Calculating Engine, 1840

Calculating Engine.
Turin 15 Sep. 1840[3]

The most general conception of this engine will scarcely be apprehended except by those who are familiar both with mechanical and mathematical Science. Those who are acquainted with the Loom are aware that by means of cards having certain holes punched in them it is possible to weave with the same loom any design however complicated. The design is translated into cards and these direct the motions of the threads. By similar means the operations of the Calculating Engine are directed. Any formula however complicated is translated into cards and their being placed in the Engine it works out that particular formula and gives its numerical value. In one respect however it advances beyond the cards of Jacquard. For in the case of the development of functions it backs the cards more or less so as to repeat the operations according to any assigned laws thus enabling a small number of cards to supply the place of a great many and rendering it unnecessary to effect the algebraic Development of a function in order to make the cards the laws of that development only being communicated to the cards.

[2] Incomplete MS.
[3] Losano 2014.

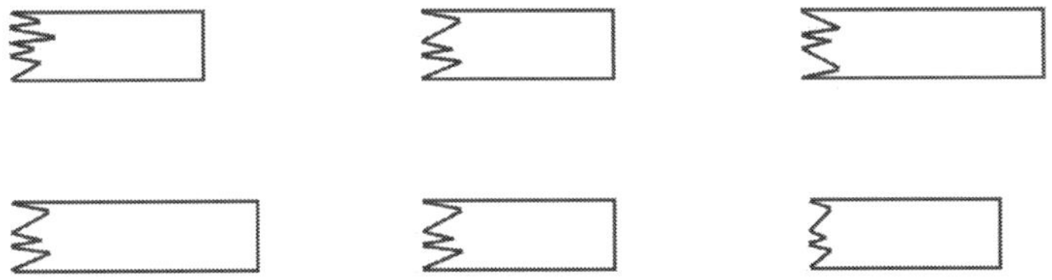

Figure E.1: Positions

E.2.1 Cards

The Cards are placed in six sided prisms which revolve pierced with holes; they are at times caused to revolve so as to make one or more cards pass forward or backward. They have also an advancing motion by which they act in levers placed opposite to them.

E.2.2 Reducing apparatus

This consists of four sectors of [...] teeth acting on a central wheel which by means of one or of two gearings can give the barrel a circular motion in either direction of 1, 2, 3, 4, 5 teeth.

E.2.3 Barrels

These consist of cylinders divided by a number of vertical lines into 31 Vertical spaces called Verticals. These are again divided into horizontal rings. Opposite each ring are two levers one of which being advanced the other moves in the opposite direction so that they may have the three following positions

Each Vertical may have opposite every ring either one or two studs which may act on these levers. Thus each Vertical may have on it a series of studs which by means of those levers give various directions to different parts of the mill by throwing certain wheels axes and levers either into or out of gear. Some of the lowest rings on these barrels govern the reducing apparatus either of their own or other barrels.

E.2.4 Card counting apparatus

This apparatus is intended to cause the prisms of the Cards to pass through a greater or less number of spaces and thus to bring any required card into that position in which it will act as soon as it is ordered to advance. It either itself computes the law of its own action (if it is simple) or it receives [that] from the Store in which it may have already been computed.

E.2.5 Decimal counting apparatus

This is a part which determines the position of the decimal point and consequently how many times the product of two factors must be stepped up or down.

E.2.6 Running up levers

This is a system of levers ordering various changes which are contingent upon certain events. They sometimes order the change of one law of calculation to a different one by acting on the Card Counting Apparatus and moving the cards either forwards or backwards. The consequences produced by these various levers will depend on the will of the constructor of the cards, but they are always called into action by the alteration of the 40 figure wheel of some carriage, which in fact corresponds to a change of sign in some quantity employed. Examples of saving time in calculations

$$\frac{1}{1}+\frac{1}{2}+\frac{1}{3}+\frac{1}{4}+\ldots \text{ to } 100000$$

$$\frac{2+1}{1\cdot 2}+\frac{4+3}{3\cdot 4}+\ldots$$

$$p\left[p\left(p+a\right)+b\right]+c \quad \text{(2 Multiplications + 3 Additions)}$$

instead of

$$p^3+ap^2+bp+c \quad \text{[(2 + 2 + 1) Multiplications + 3 Additions]}$$

In trying whether any given formula can be calculated by the Engine it must always be remembered that it was never proposed to execute by the Engine anything which cannot be executed by pen. It was only proposed to calculate in less time and with absolute certainty. It is therefore necessary to write out the formula in the same manner by pen and then to give the laws which govern it by means of proper cards to the Engine.

[Power series of the function

$$f(x)=\frac{a+bx+cx^2}{a'+b'x+c'x^2}$$

so that

$$\frac{d^n}{dx^n}\left(\frac{a+bx+cx^2}{a'+b'x+c'x}\right)=\frac{A+B_nx+C_nx^2+\cdots+x^{n+2}}{(a'+b'x+c'x^2)^{n+1}}$$

where A_{n+1}, B_{n+1} may be found in terms of A_n, B_n.]

E.3 Structure of the Calculating Engine

The Calculating Engine consists of two parts: 1st The mill which consists of a series of axes and wheels surrounding a great central column of wheels. 2nd The Store which consists of a rack communicating by means of the Ingress and Egress wheels [...] with the central wheel and having a series of columns of figure wheels marked V_1, V_2, V_3 etc., on each of which any number of not more than 40 places of figures can be expressed by the inferior wheel of each cage and any other similar number can be expressed on the superior wheel of the same cage. The mill has the power of performing the operations of addition and subtraction, multiplication and division and of combining the signs of the numbers employed according to the usual rules, each number employed having 40 places of figures.

It requires on average	**Minutes**
For Additions & Subtractions	1/8
For Multiplications with Decimals	3
For Multiplications without Decimals	2
For Divisions	5

The mill has also the power of turning backward or forward any number of cards according to any laws or even arbitrary numbers which may be prescribed for it. The Store receives on each of its columns any function whatever (this must be written on a piece of paper and cannot be changed during the calculation of the same formula) its numerical coefficient and its algebraic sign. The Store when ordered by the advance of its Variable Cards can give up any of the numerical coefficients and their signs to the mill and receive any result with their signs from the Mill and place them on any variable. These same Variable Cards can also order the reduction to zero of any of those coefficients and can transport any of them from their own to any other column or variable. It is these powers of the Store which enable it to perform analytical operations and to make an unlimited series of substitutions.

When the Calculating Engine is requested to calculate any formula the first step is to make an algebraic notation by which the number and nature of the Operations Cards as well as of those of the Variables will become known. When these cards are once made they may be replaced in the Engine at any future time and new constants being put in the machine the resultants will be calculated and furnished. The history of such calculations is thus after the cards are put into the engine, those of operations in the Mill, those of the Variables in the Store and after the constants belonging to the formula have been put into the store. Then at the first turn of the handle the first operation card advances.

This card pushes a lever which by acting on the Reducing Sectors causes the barrels to revolve on their axis through a certain angle dependent on the nature of the operation. If this operation is Multiplication the barrels are ordered to a certain vertical the 4th for example: they then advance and by acting upon the levers influenced by the studs on that vertical they give certain orders to prepare the Mill for multiplication. One of these orders will be through its own reducing apparatus to turn itself to a new vertical (say the 6th) another order will be to the Variables Cards to advance in the next turn in order that the first factor may be given to the Mill.

The 6th Vertical advances and the Variable Cards advance at the same time. This orders the first factor to be received into the mill and upon one of the Carriage F. It also orders itself to turn to a new vertical (the 8th) and also the Variable Cards to revolve one space. The 8th Vertical advances and also the next Variable Card. These order the second factor to be received into the Mill and to be given out from the Store. The second factor is at this turn, subtracted from the first upon the Axis F.

If the first factor is the largest then there is no running up but if the reverse is the case then the running up order the greatest factor to be taken and in the succeeding nine turns a table of its nine Multiples is made upon the axes $F_1, F_2, ...F_9$.

After the table is made the multiples corresponding to the several digits of the other factor are taken from the table commencing with the highest; this is stepped up or in fact multiplied by 10. The result of the next figure is then taken and being added to this the sum is again stepped up and 10.

When the multiples corresponding to the whole of the digits of the second factor are thus added together, the successive sums being each stepped up or multiplied by 10; the product must now be stepped down 10 as to place its decimal point at the proper height.

The barrel now turns to a vertical which orders the Variable Cards to advance in order that they may determine upon what variables the product must be placed. As the product of two numbers each of 40 digits may occupy two Variables it requires [two] columns.

The barrel having at its last turn reduced to zero all the multiples and any other numbers which may remain in the Mill now orders the Operation Cards to advance at the next turn and orders itself to its zero point.The next operational Cards may order multiplication again or it may be order Division then the Reducing Sectors acting on the barrels (by order of the Operational Card) are turned to the third vertical which commences Division. The 3rd Vertical prepares the Mill for division and orders the Variable Cards to advance. This vertical also orders the Reducing Apparatus to direct its own barrel to move circularly to 5th Vertical.

When the 5th Vertical advances, the Mill receives the division from the Store. It then orders the [...] and that the other succeeding vertical. These order into the Mill the Dividend which may exist upon two Variables since it may have 80 figures.

The Mill now commences by a peculiar apparatus to count the number of digits in Divisor and Dividend at the same time it makes a table of the nine Multiples of the Divisor. Having ascertained the number of digits in each it steps one up to the same level as the other. It then proceeds to compare the two first figures of the Dividend with the two first of each multiple and subtracts from it that multiple which has its two first figures either equal to or next less than it. It registers this digit of the Quotient on the bars A. This process is repeated until the quotient is formed.

But as it may sometimes occur when the two first figures of dividend are equal to the two first [figures of the multiple] this subtraction is too great; in such cases the fact is known to the Mill by the running up of one of the Carriages and the next figures of the quotient are registered on another set of figure wheels which are considered as negative digits. When a new run up occurs the quotient figures are again registered upon the positive line.

At the end of the quotient the two parts, the positive and the negative registered on separate sets of figure wheels are united together by subtracting one from the other. In the mean time, the remainder if any is transferred to the Store but if this remainder is not to be used the Variable Card which indicates its locality in the store is a blank card and gives no orders to any variable to receive it consequently it is reduced to zero in the Mill and does not appear in the Store. The Barrels now order themselves to zero and also order the Operational Card to give them a new command.

Addition and Subtraction are thus performed. The algebraic sign is ordered by Card; that is the Mill has orders to add or subtract as the sign + or minus is directed by Card. But as the numbers brought in from the Store to be operated upon may have been formed by previous processes it is possible one or more of them may have the negative sign. These signs which come in with the coefficient from the Store are called the accidental signs and are united in the Mill with the Algebraic signs and then the proper arithmetical operation is performed and the result with its right sign transmitted to the Store.

Whilst the Mill is pursuing the two operations of division and multiplication, the Carriage F is not occupied; it is connected also with the store and by means of proper Variable Cards it can (during the greater part of the time occupied by those processes) itself carry on arithmetical additions and subtractions in the Store. Thus whilst one term of an infinite series is preparing by Division the numerator and denominator the next may be made in the Store. Or any other numerical calculations (by addition and subtraction) for preparing the indices to be communicated to Card Counting Apparatus made at such times.

E.4 Turin Police Surveillance in 1840.

(State Archive of Turin, Cabinet
of Police - Cuneo, 1842, Folder I.a and II.a)

A

Royal Command of the City and Province of Turin, Police Office, N. 6549.	Most Illustrious First Secretary of State for Internal Affairs, Turin, September 21, 1840

Most Ill. Sir. Sir. Hon. Most Col.

The notorious Fortunato Prandi of Camerana arriving from Lyon, arrived here by courier on the tenth of the current month in company of a certain Mr. Babbage, English engineer, and went to lodge at the Swiss pension; the next day he rented two furnished rooms under the porticos of Po in the Hospital of the Poor building number 22 on the 2nd floor, and established himself there with said Englishman for whom he serves as interpreter. The latter has the intention of presenting soon at the Congress of Scientists a machine of his invention for enabling mathematical calculations, of which I transmit to Your Most Illustrious Lordship here attached a printed copy.

Having had the aforementioned Prandi carefully surveilled from the day of his arrival onwards, the following emerged.

In the first two days he visited several Churches and various cafés of the Capital, on the 13th he went to see Sir Cavaliere Ponte residing in Casa Falcione on the Platani avenue, but stayed there only for a few moments then went to Luciano's spice shop, and to the café of S. Carlo; at 14 he went to the Office of General Inspection of drafts, and stayed there for a good two hours, then entered the merchant Caglieri's shop, and after having purchased some handkerchiefs returned home from where he did not leave even on the 15th.

On the 16th he went for a walk under the porticos, and on the Po bridge with the aforementioned Englishman Babbage, and after lunch they did the same from the Valentino and Porta Nuova side and then went to the Carignano theater; on the 17th he went out in company of his brother Chaplain of the Royal Tax Collectors, went to see a certain Giorgio Prandi owner of the inn of Moro where he remained talking for half an hour, then went to visit, still in company of his brother, the Dora bridge, from there they both went to the Ministry of War, where they stayed about fifteen minutes, then went to lunch in the barracks of the Tax Collectors with a certain Mr. Verdi Carlo Principal Commissioner of the Tax Collectors Brigade, and afterwards returned home; on the morning of the 18th around 10 o'clock, he went to Mr. Giuseppe Pomba's printing shop from where he exported[4] various books, returned home and received a visit around two in the afternoon from 14 or 15 scientists, including Professor S. Martin Director of the Royal Botanical Establishment of S. Salvario who came expressly with others to view the Englishman Babbage's machine; on the 19th, around $11\frac{1}{2}$ he left the house with his brother priest and another person unknown to the observer, and went with them to speak to the porter of the Ministry of Foreign Affairs, and then to that of the Ubene house now Ferrero house at Borgo Nuovo; at one in the afternoon he went with Babbage to the Valentino boat and crossed the Po, but the observer not having been able to board the same boat at the same time lost sight of them on the opposite bank of the river.

Yesterday finally around $11\frac{1}{2}$ he departed in a carriage with the Englishman and his brother[5] heading towards Porta Nuova.

From the information obtained, however, I learn that he went to visit a relative of his named Vacchetta who has a country house in the vicinity of Loggia, upon his return surveillance will continue on his further movements, and I reserve the right to inform you of the results.

Meanwhile I submit to you in relation to your adjacent sheet the present report and I have the honor to renew the acts of my most distinguished respect.

The Lieutenant General Commander
D. Planargia

B

Turin, 4 October in the evening

[4] Perhaps 'took with him' is meant here.
[5] [Ed.] Prandi's brother, the priest.

To Mr. Count Lazzari
1st Officer at the Ministry of Interior

Sir Count,

The King orders me to inform you this very evening that, having been informed that the Police have put obstacles preventing Mr. *Prandi* from going to Genoa where he intended to accompany the English scholar Mr. Babege [sic], who departs tomorrow to embark in that city for England, desires that these obstacles be [sic] removed, and that Mr. Prandi may freely depart tomorrow for Genoa with his friend Mr. Babage [sic], and after his departure for England, return to his elderly father in Cameran for the time that he is permitted to remain in the King's States.

Pardon, Mr. Count, for the haste with which I have written and tell [sic] you.

Your most Ob. Servant
De Sonnaz

C

Royal Command of the City and Province
of Turin
N. 6916.
Turin, October 5, 1840

Most Illustrious First Secretary of State for Internal Affairs
Turin

Most Ill. Sir. Sir. Hon. Most Col.

Following up on my memo of September 21 past n. 6549, and according to the provision contained therein I submit to Your Most Ill. Lordship the further results obtained through the continued surveillance to which the notorious Prandi was subjected: they are the following:

On the morning of the 21st of said month he left the house around 10 o'clock, with his brother went to the letter post and then returned; at two-thirty in the afternoon he went to visit the Royal Gallery of Paintings, then to the lithographer and bookseller Giacinto *Marietti* and then visited all the book stalls existing under the porticos of Po, and having met the count of *Benevello* walked with him for a good half hour: said Count was accompanied by a priest who I was informed is a professor at the University.

At five o'clock Prandi accompanied by his brother, by the Englishman *Babbage*, and by the aforementioned priest, got into a carriage, and went to dinner at Count *Balbo*'s country house located beyond the Rubatto opposite the Valentino; then they returned home around nine o'clock.

Here I will not conceal that it has been necessary to note quite often that Count *Benevello* is in contact with persons suspected in political matters. The persons who attended Count Balbo's dinner numbered eighteen and it was noted that almost all spoke English, and this perhaps to prevent the servants from understanding their conversations.

On the 22nd, *Prandi* left the house at eleven o'clock also with his brother, and went to *Spalla* House where a certain Marquis *Federici* lives and remained there for almost an hour; upon leaving there he went to H.M.'s Palace and stayed about half an hour, then returned to his lodging; at one in the afternoon he went to Doragrossa in the courtyard of the San Simone inn, climbed the right staircase to the first floor, and there asked for Cavaliere *Pron*, but not having found him at home returned under the porticos of Po, where he joined Count *Pettiti* and five other individuals among whom was noted the aforementioned University professor priest; with this group he went to visit the *Garneri* Factory, and then a machine existing in

Vinaj House for working cotton; after taking leave from the others Prandi went to dinner at Feder's with his brother and the aforementioned priest, and they were not seen leaving again by the Secret agent, which suggests they left through the courtyard of Casa Garneri.

On the 23rd, at $9\frac{1}{2}$ Prandi after going to the letter post, got into Audisio's rental horse carriage in piazza Vittorio Emanuele and had himself driven to the vineyard of a certain Mr. Devieux or Durieu, general contractor of the Mount Cenis road, where he spent the day; said vineyard is situated in the vicinity of Moncalieri on the left going, almost at the beginning of the climb that leads to that Municipality.

On the following day the 24th he left the house around 8 in the morning and went to speak to the porter of Saluzzo House, or Massa House; having returned home he went out again with his priest brother, went to see Cavaliere Pron; shortly after leaving there he met and spoke with Mr. Pansa the shipping agent and with Mr. Vacchetta who is believed to be his relative, then returned home from where he left around noon with the priest and another gentleman heading towards the Post street; and there taking leave from them he began to run precipitously as if he wanted to catch up with someone from afar and thus disappeared from the sight of the agent who did not see him again for the rest of the day.

On the 25th he left the house alone around 9 in the morning, and spoke in Po street with a certain Mr. *Dettoma* machine manufacturer, then went to Fontana printing shop in Fieno street, and then to the booksellers Rejcend, Picco, Gianini and Fiora [sic] successors of Pomba, then by a certain Gioliti, retired carabiniere, who has a shack under the porticos of the Fair; having returned home he was visited by two gentlemen whose identity is unknown; half an hour later a carriage drawn by two horses arrived before his door with coachman and servant dressed in light blue cloth with black trim on the hat. Prandi then got in with the aforementioned two gentlemen and they went to the Royal Park; at two-thirty he returned home on foot; and at four he went out again with his brother, and Mr. Babbage and they went to H.M.'s Palace where they stopped for ten minutes at the foot of the grand staircase on the right, then they went to Dora Grossa street to the shop of Messrs. Garneri and Vivano, from there they went to the tailor Castelli's house; then around six o'clock they went to dinner at the Pastore trattoria, and stayed there until nine when they returned home.

On the 26th, he left the house around noon, bought writing paper from a bookseller near his lodging, and returned immediately. Going out again at $5\frac{1}{2}$ he went to Casa Garneri to inquire of that porter about Count Sonnaz, then went to Gianini and Fiore bookshop where he stayed very briefly and from there returned to his house without leaving again.

On the morning of the 27th Prandi and his traveling companion Babbage who are lodged together, were visited around 9 o'clock by the Intendant Cavaliere *Dionisio* and his lawyer brother both State Secretaries at the Grand Chancellery as well as by lawyer *Vacchetta*; the first two left at 10 and the last remained until $10\frac{1}{2}$ when he left with the Englishman Babbage and the Prandi brothers. These last two went to hear mass at the Church of the Most Holy Trinity then went to speak with a certain Beily horse rental and afterwards returned home; at two in the afternoon Prandi accompanied the Englishman to Casa Chanaz but he did not enter; he only stayed at the door waiting for him then they returned home; around 6 o'clock the aforementioned Prandi brothers went out again from home went for a walk and then to dinner at the barracks of the Tax Collectors, of which the priest is chaplain and has his lodging there: at $9\frac{1}{2}$ he returned.

On the 28th, he departed by carriage for Rivoli with the Englishman Babbage and a certain Mr. Gianolio merchant in riverfront goods for a commercial speculation, and returned home at nightfall.

Fortunato Prandi having left on the 29th from home at 10 o'clock, went to Contrada Nuova to the bookseller Giacomo Serra, purchased a book, and returned immediately home. Going out again at 11 with the priest his brother they went to the letter post, and after mailing quite a bundle returned again home, and did not leave except at 2 to go to the Museum where he stayed until $4\frac{1}{2}$, then having returned home was not seen leaving again.

On the 30th, he left home at 11 in the morning and went to the bookseller and printer Pomba and then to the Guards' hall of the Royal Palace, where he stayed for nearly a quarter hour, leaving there he visited all the booksellers' stalls existing under the porticos of the Royal Secretariats, and there purchased four books which he took home; leaving again at three he went to Angennes street to visit the family of a certain Mr. Rossi, an employee, who is temporarily in Genoa, and stayed there until 6 o'clock; having then returned home he was not seen leaving again.

On the first instant he left home at 9 with an Engineering officer, and went to Gianini and Fiore bookshop where he purchased a book, and then returned home alone; at 10 o'clock he went out again with his brother, and they went together to the grand Chancellery, and after having stayed there 20 minutes they went to the Red Ox inn to speak with some coachmen; afterwards the priest left his brother, who immediately transferred himself to the Royal Palace, where he stayed for a good hour and a half then returned home. Leaving again at 5, he went to dinner at the Indies, then went for a walk until $8\frac{1}{2}$ and finally returned home from where he was not seen leaving again.

On the second he left his lodging at $9\frac{1}{2}$, with his priest brother, and after having gone to Calosso café they went to Prosecutor Isnardi, from where they left with a certain Gamberoux, and went to Mr. Andreoletti's house, then returned home with the priest. Then leaving alone he went to Angennes Street to find lawyer Goroni, having returned home again he then left with the priest and another gentleman, with whom he went to Marietti bookshop, to the Passport Office, and then to a certain Tron residing in Cossato house, from there he went to dinner at the Indies, and then returned home around 8.

On the 3rd, Prandi left home at $8\frac{1}{2}$ and went to jeweler Canavero, and to trinket-seller Antonio Buri, then to cavaliere Bonelli; from there he went to the house of a certain Piovano; leaving there, he met lawyer Cerruti, and went with him to Casa Paesana to Prosecutor Gilli, then entered Vassallo café and stayed there a quarter hour; having returned afterwards to home, he left again with his priest brother around $3\frac{1}{2}$, with whom he went to Arleri spice shop, then after a brief walk went with the priest to dinner at the Pastore trattoria, and left around $7\frac{1}{2}$, heading alone to Piossasco house to visit a certain Mr. Pron, from there returned home.

On the 4th he left at eight, and went to pay a visit to Count Saluzzo, then to the Royal Palace to Count Robilant then returned home; at 11, he went out with his priest brother, and went to the General Agency of Royal Taxes, then to the most esteemed surgeon Major Ejnaudi. From there he went to Casa Garneri to find a certain Mr. Brunati, then returned home from where he left with the Englishman Babbage around three in the afternoon got into a carriage together, and took the road to Moncalieri. From the information obtained there is reason to believe they went to dinner at Count Balbo's.

It does not appear that Prandi attended either the Congress of Scientists or any other meeting.

This morning finally at 14 hours he departed in company of Babbage heading by courier to Genoa.

I have the honor to present to you the sentiments of my most profound respect.

The Lieutenant General Commander
D. Planargia

P.S. Babbage having found himself indisposed at the moment of getting into the carriage, Prandi countermanded the departure.

D

Royal Command of the City and Province of Turin N. 7020	Most Ill. Sir First Secretary of State for Internal Affairs Turin

Turin, October 9, 1840

Most Ill. Sir. Sir. Hon. Most Col.

Returning to Fortunato Prandi who repeatedly formed the subject of reports from this Command, and especially of the last one of the 5th current month n. 6916, I have the honor to add that, having left home on said day around nine in the morning, he returned to the General Agency of Royal Taxes, then to this Ministry, and then to the office of lawyer Paoletti located in Consolata street, and from there to the house of Cav. Veggi residing in Bellezia street. After having stayed there for the space of about twenty minutes he went to the confectioner Basso's shop from where he returned home. Leaving again around six in the afternoon accompanied by his priest brother and the Englishman Babbage, he went to the Feder Hotel, then to dinner at the Indies hotel: having returned afterwards from Feder he had a hand cart brought to his house, and had it loaded with two suitcases and as many trunks, directed towards the aforementioned hotel.

After completing this operation he went to visit Cavaliere[6] and professor Griffa; from there, after having stayed a short while in the San Filippo café, he went to Moschino street, number 4, to the house of lawyer Daziano Head of Division at the Ministry of War, and brother of the other lawyer who some years ago gave reason for the Police to concern themselves with him given his liberal opinions.

On the 6th of said month he left the Feder hotel at 9½ in the morning and went to the passport office, as well as to the couriers office, and then to the bookseller's shop.

Subsequently after having walked for some time under the porticos, he went to visit a certain Madame Rossi residing at Borgo Nuovo in Casa Ponzio Vaglia where he remained for a good two hours: at 5 in the afternoon he returned to the hotel, dined there, and going out again went to Cavaliere Bonafous and to Brachi's shop from where he returned to the hotel.

Yesterday he left his lodging at 9 in the morning, walked under the porticos visiting the book stalls until 11, then got into a carriage with another gentleman and transferred to Count Balbo's country house where he must have had dinner: at 3 he was back at the hotel, and at five departed by means of a private carriage towards the city of Mondovì, the Englishman Babbage departed by courier heading towards Genoa.

Before departing Prandi told the notorious Pietro Tealdi that he would return to Turin in ten days.

With this circumstance I permit myself to observe to you that in the summary of all the persons with whom the activated surveillance would show Prandi to have had relations, it appears that the majority of them profess and nurture sentiments contrary to the current system of government. That furthermore finding the same Prandi, as was supposed, very limited in pecuniary means, it seems that in this case he would not have been able to provide for the expenses of long journeys, long stays in this City, and so much indulgence in dinners, and pleasure outings without receiving aid that might perhaps be provided to him by the enemy sect of revolutionary Propaganda.

And in finally announcing to you that the individual in question received some time ago from His Majesty[7] through Cavaliere Promis a medal, given that he would be usefully

---

[6] [Ed.] A Knight.

[7] Original: S.M. (Sua Maestà).

occupied and would take interest in certain matters beneficial to the Royal States, I have the honor of reiterating my profound respects.

The Lieutenant General Commander
D. Planargia[8]

Royal Command of the City and Province
of Turin[9]
Police Office
N. 7351

To the Most Illustrious First Secretary of State
for Internal Affairs
Turin

Turin, October 25, 1840

Most Illustrious and Honored Sir,

Continuing to report on the surveillance, which I had conducted without pause on the movements of the well-known Fortunato Prandi[10], I have the honor to inform Your Most Illustrious Lordship that he arrived in this Capital on the 18th instant, taking lodging at the Swiss pension.

On said day, he went to the Feder Hotel[11] where Mr. Tealdi was lodging, with whom he went morning and evening to Count Robillante's[12], to the St. Carlo and Fiorio cafés[13]; they then walked together briefly in Carignano Square[14] before entering the theater.

On the 19th, leaving his residence at 7:30, he went to Café Fiorio, then to Feder to speak with Tealdi where he stayed for about two hours; upon leaving there, he went to Papa's shoe shop in New Street, then to St. Carlo café, then alone for a stroll under the Po porticos[15] and finally returned to his lodgings.

On the 20th, he went to the General Office of Taxes, then to the Messrs. Quartero's private residence on Platani Street, and Rolla's in Palma Street No. 2; afterward he walked with Mr. Fino, spoke with Cavaliere Gattinara, visited Marengo the shoemaker, went to Café Fiorio, passed by the post office, then to Righini's shop, and to the War Secretary's office, where he stayed for an hour, then proceeded to visit Count Balbo and Cavaliere Sauli, to Schiapati the bookseller, and at 6 o'clock to Mottura's restaurant where he remained until 7:15, after which he returned to his lodgings.

On the 21st, he left his residence at $7\frac{1}{2}$, went to Mr. Donnaz's in Meridiana Street no. 9, to Café Fiorio, to the Office of Economy, to lawyer Dallosta and to Bass the confectioner, then walked with Cavaliere Martini, went to the Royal Tax Office, to Lorenzo Musso's shop in New Street, then walked with Cavaliere Riccardi, spoke with Mr. Vigitelli, and subsequently went to dine at Pastore tavern before returning home.

On the 22nd, he visited the Leardi gentlemen at the Baldissero house on Fieno Street, Cavaliere Bricherasio in Dora Grossa at the Moriondo house, then went to Café Fiorio, and then proceeded to lawyer Mattirolo's in Printers' Street, to Ferrari's shop in New Street, then walked with Mr. Cellaris from Mondovì and another unknown person, then with

[8] Military commander of Turin.
[9] Capital city of Piedmont.
[10] Subject under surveillance.
[11] A prominent hotel in Turin.
[12] A member of the local nobility.
[13] Popular cafés in Turin.
[14] Historic square in Turin.
[15] Covered walkways of Po Street in Turin.

Abbé Ghigo, and afterward went to the Indies tavern where he dined and stayed until $6\frac{1}{2}$; subsequently, he went for a walk until $7\frac{1}{2}$ before proceeding to the Carignano Theater.

In reporting to Your Most Illustrious Lordship the results of the surveillance conducted on the subject in question, I believe it my duty not to leave you unaware that the same individual, perhaps fearing being watched by the police, wishes to evade their attention through frequent changes of dress, and I have the honor of reiterating my sincere protestations of my most profound respect.

The Lieutenant General Commander
D. Planargia

Royal Command of the City and Province
of Turin
N. 7554

To the Most Illustrious First Secretary of State
for Internal Affairs
Turin

Turin, November 3, 1840

Most Illustrious and Most Respected Sir,

In the continuation of the surveillance that was activated by this Office regarding the movements of the well-known Mr. Fortunato Prandi, subsequent to my last report of October 25, no. 7351, it emerged that on the 23rd of the same month, he left his house around eight in the morning, and after spending some time at Café Fiorio, he went to the Chamber of Commerce located on St. Carlo Street, Lascaris house, from where he then left with lawyer *Ferrero* took with him lawyer *Audifredi*, then to the Brondello house in Deposito Street. After returning home, he changed clothes and, dressed entirely in black, returned to Café Fiorio, then went to the Gioannin[sic] and Fiore booksellers, from there to the Ministry of War, where he remained for over an hour, then accompanied by a priest believed to be his brother and another individual completely unknown to the investigator, went to dine at *Mottura* tavern, and around seven o'clock returned to his residence.

On the morning of the 24th, he went to lawyer Paroletti's house, and after staying there briefly, proceeded to Café Fiorio, walked for some time with Examining Magistrate *Petiti*, from whom he then took leave, met lawyer *Muratore*, and with him went to Bianchi's shop in *Dora grossa*, from there he went alone to lawyer Trombetta's; and leaving there, he met Cavaliere and lawyer *Tesauri* from Fossano, with whom he stayed for about half an hour; then he headed to the General Office of Royal Taxes, and from there to the couriers' office, upon leaving which, he went to Count *Saluzzo*'s house, and after a brief stay there, went walking under the porticos with Cavaliere *Lisio*, former cavalry officer dismissed in 1821; then returned home where he dined; afterward he went out with a certain Mr. *Cantiglio* to St. Carlo café where he stayed for about an hour, and finally returned home again to rest.

On the 25th of said month, leaving home around eight o'clock, he went to Café Fiorio, and then to Count *Balbis*'s House, returned to Café Fiorio, visited lawyer *Andriani*, walked with Messrs. *Bensi*, merchant *Schioppo* and lawyer *Caraglio*, dined at Mottura Tavern, around $11\frac{1}{2}$ went to the Exchange café, and then to Carignano Theater.

On the 26th he left home at 9 o'clock, went to *Secondo Festa*'s shop, the couriers' office, and *Gabetti*'s shop, conferred with lawyers *Brofferio* and *Cerutti*, went to Café *Barone*, and then to the Fiorio brothers' café, proceeded to the Gioannini[16] and Fiore bookshops, to

[16] [sic] appears in original.

the *Bonafous* brothers' office, to the botanical garden at Susa gate, then to Bianchi's shop, around 5 o'clock went to dine at the Indies Tavern from where he returned to his lodging.

On the 27th he left home around $7\frac{1}{2}$, walked with Captain *Folicone* attached to the General Staff, went with him to the Exchange café where he stayed until around ten o'clock, then visited Count *Masini*, went to Bianchi's shop, then to tailor *Collo*'s workshop, walked with lawyer *Cerutti*, and subsequently, in meeting with an unknown priest, went with him to dine at Pastore tavern, and from there returned to his lodging.

On the 28th he went out at $10\frac{1}{2}$ and went to Café Fiorio, from there he went to the house of Count *St. Fermo*, subsequently to those of Cavaliere *Sauli*, and Surgeon *Einaudi*, then to the General Office of Royal Taxes, and finally to bookseller *Bona*'s shop, from where he returned to his lodging not to go out again that evening.

On the 29th he went out around $7\frac{1}{2}$, visited the St. Martino la Motta house, went to Café *Calosso*, to the War Secretary's office, to Abbé *Chiari*'s house, to *Frisetti* and Gioannin [sic] and Fiore shops, went for a walk with Messrs. *Borani*, *Bellini*, *Ricardi* and lawyer *Decaroli*, went to dine with a certain Mr. *Ansaldi* at Mottura tavern, and upon leaving went with the same person to the Exchange café and then to Carignano Theater.

On the 30th he left his lodging around 9, went to Café *Calosso*, to the post office, walked with Cavaliere *Vinaj*, and having returned to his lodging did not go out again until 5, at which time he had a voluminous paper package delivered to *Gianini* and Fiore's shop, where he actually stayed afterward until $5\frac{1}{2}$, then went to dine at the Universe tavern, where he remained until 7, to then revisit Café Fiorio, and return to his lodging to rest.

On the 31st he went early in the morning to Count *Saluzzo*, lawyer *Audifredi*, Senator *Costa*, Count *Balbis*, the *Bonafous* brothers, lawyer *Brofferio*, Count *di Robilante*, bookseller *Gabetti*, Cavaliere *Sauli*, Count *di St. Giorgio*, Cavaliere *La Marmora*, Count *d'Agliano*, booksellers *Giannini* and Fiore, went to the London café, to St. Filippo's, and then at 4 o'clock departed with the French courier heading toward London.

In presenting to Your Most Illustrious Lordship this report to complete this matter, I have the honor of reaffirming myself with most distinguished respect and highest consideration.

The Lieutenant General Commander of the City and Province of Turin
D. Planargia

Appendix F

Menabrea on the Analytical Engine.

F.1 1842. Notions Sur La Machine Analytique De M. Charles Babbage.

L. F. Menabrea (1842). "Notions sur la Machine Analytique de M. Charles Babbage". French. In: *Bibliothèque Universelle de Génève* 41 (1842), pp. 352–76

BIBLIOTHÈQUE UNIVERSELLE

DE

GENÈVE.

Nouvelle Série.

Tome Quarante-unième.

On souscrit à Genève,
AU BUREAU DE LA BIBLIOTHÈQUE UNIVERSELLE,
CHEZ B. GLASER,
Rue de la Pélisserie, 133.

PARIS,
CHEZ ANSELIN, SUCCESSEUR DE MAGIMEL,
Rue Dauphine, 36.

1842

NOTIONS SUR LA MACHINE ANALYTIQUE DE M. CHARLES BABBAGE, par Mr. L.-F. Menabrea, capitaine du génie militaire.

Les travaux qui appartiennent à plusieurs branches des sciences mathématiques, quoique paraissant, au premier abord, être uniquement du ressort de l'esprit, peuvent néanmoins se diviser en deux parties distinctes : l'une qu'on peut appeler mécanique, parce qu'elle est sujette à des lois précises et invariables, susceptibles d'être traduites physiquement, tandis que l'autre qui exige l'intervention du raisonnement, est plus spécialement du domaine de la pensée. Dès lors on pourra se proposer de faire exécuter par le moyen de machines la partie mécanique du travail, et réserver à la seule intelligence celle qui dépend de la faculté de raisonner. Ainsi la rigueur à laquelle sont soumises les règles du calcul numérique a dû, depuis longtemps, faire songer à employer des instruments matériels, soit pour exécuter entièrement ces calculs, soit pour les abréger. De là sont nées plusieurs inventions dirigées vers ce but, mais qui ne l'atteignent, en général, qu'imparfaitement. Ainsi la machine de Pascal, tant vantée, n'est maintenant qu'un simple objet de curiosité qui, tout en prouvant une grande force d'intelligence dans son inventeur, présente en elle-même peu d'utilité. Elle n'exécutait que les quatre premières opérations de l'arithmétique, et encore se bornait-elle, en réalité, aux deux premières, car la multiplication et la division étaient le résultat d'une série d'additions et de soustractions. Le grand inconvénient de la plupart de ces machines est d'exiger l'intervention continuelle de la main de l'homme pour en régler les mouvements, et c'est en cela qu'est la source des erreurs ; de sorte que, si dans l'exécution des grands calculs numériques leur usage ne s'est pas propagé, c'est qu'elles ne résolvaient point le double problème que présente la question, celui d'obtenir l'exactitude des résultats unie à l'économie du temps.

Frappé de ces réflexions, Mr. Charles Babbage a consacré plusieurs années à réaliser une pensée gigantesque. Il ne s'est proposé rien moins que de construire une machine capable d'exécuter, non-seulement les calculs arithmétiques ; mais encore les calculs analytiques, dont les lois seraient connues. L'imagination est d'abord effrayée d'une telle entreprise, mais à mesure que l'on réfléchit avec plus de calme, le succès en paraît moins impossible, et l'on sent qu'il peut dépendre de la découverte de quelque principe, assez général pour que, si on l'applique à la machine, celle-ci soit capable de traduire mécaniquement les opérations qui lui seraient indiquées par l'écriture algébrique. L'illustre inventeur, lors d'un voyage qu'il fit à Turin, ayant bien voulu me communiquer quelques-unes de ses considérations à cet égard, je viens, avec son consentement, rendre compte des impressions qu'elles ont produites dans mon esprit. On ne doit point s'attendre à trouver ici une description de la machine de Mr. Babbage : elle exigerait de longues études pour être comprise, mais je tâcherai d'en faire saisir le but et d'exposer les principes sur lesquels est fondée son exécution.

D'abord je dois prévenir qu'elle est entièrement différente de celle dont on trouve un aperçu dans le traité de l'Economie des machines du même auteur. Mais, comme cette dernière a fait naître la pensée de la machine en question, je crois qu'il est utile, avant tout, de rappeler en peu de mots quelles ont été les premières tentatives de Mr. Babbage, et quelles sont les circonstances qui y ont donné lieu.

On sait que le gouvernement français, voulant faciliter l'extension du système décimal, avait ordonné de construire des tables logarithmiques et trigonométriques d'une étendue immense. Mr. de Prony, qui avait été chargé de la direction de cet ouvrage, le répartit en trois sections, à chacune desquelles furent appliqués des hommes spéciaux. Dans la première section l'on combinait les formules de manière à les rendre propres aux calculs arithmétiques; dans la deuxième, ces mêmes formules étaient calculées pour des valeurs de la variable, prises de distance en distance; enfin dans la troisième section, composée de près de 80 individus, qui pour la plupart ne connaissaient que les deux premières règles de l'arithmétique, on interpolait, au moyen de simples additions ou soustractions, des valeurs intermédiaires à celles qui étaient calculées par la 2e sect.

Un travail analogue à celui que nous venons de citer ayant dû se répéter en Angleterre, Mr. Babbage pensa que les opérations faites par la troisième section pouvaient être exécutées par une machine, et c'est ce qu'il a réalisé au moyen d'un mécanisme dont une partie a été confectionnée, et que l'on peut appeler *machine aux différences* à cause du principe sur lequel elle est fondée. Pour en donner une idée, il suffira de considérer la série des nombres carrés entiers : 1, 4, 9, 16, 25, 36, 49, 64, etc. En faisant la différence de chacun de ces nombres avec le suivant, on aura une nouvelle série que nous appellerons série des différences premières, composée des nombres : 3, 5, 7, 9, 11, 13, 15, etc. Si l'on soustrait de chacun de ces nombres son précédent, l'on obtiendra les différences deuxièmes, qui sont toutes constantes et égales à 2. La suite de ces opérations et des résultats auxquels elles conduisent peut être représentée par le tableau suivant :

	A Colonne des nombres carrés.	B Différences premières.	C Différences deuxièmes.
	1		
	. . . .	3	
	4	. . . .	2 *b*
	. . . .	5	
a	9	. . . .	2 *d*
	. . . .	7	
c	16	. . . .	2
	. . . .	9	
	25	. . . .	2
	. . . .	11	
	36		

D'après la manière dont les deux dernières colonnes B et C ont été formées, il est facile de voir que, voulant, par exemple, passer du nombre 5 au suivant 7, il faudra ajouter au premier la différence constante 2; de même, si du carré 9 on veut passer au suivant 16, on ajoutera au premier la différence 7, ou en d'autres termes, la différence précédente 5, plus la différence constante 2; ou bien encore, ce qui revient au même, pour avoir le nombre 16, il suffira de sommer ensemble les trois nombres 2, 5 et 9, placés dans la direction oblique *a b*. D'une manière analogue, on obtiendra le nombre 25 en sommant ensemble les trois nombres placés dans la direction oblique *d c*: en commençant par la somme 2 + 7, on a la différence première 9, consécutive à 7; en y ajoutant 16, on aura le carré 25. On voit donc, qu'étant donnés les trois nombres 2, 5 et 9, toute la série des nombres carrés suivants, ainsi que celle de leurs différences premières, pourra s'obtenir par de simples additions. Maintenant, pour concevoir comment ces opérations peuvent être reproduites par une machine, supposons que celle-ci ait trois cadrans que nous désignerons par *A*, *B*, *C*, et sur chacun desquels soient tracées mille divisions, par exemple, qui seront parcourues par une aiguille. Les deux cadrans *C* et *B* auront, en outre, chacun une sonnerie qui battra un nombre de coups égal à celui des divisions indiquées par l'aiguille. Pour chaque coup du compteur du cadran *C* l'aiguille *B* devra avancer d'une division; il en sera de même de l'aiguille *A*, qui avancera d'une division pour chaque coup du cadran *B*. Telle est la disposition générale du mécanisme. Cela posé, au commencement de la série d'opérations que nous voulons exécuter, mettons l'aiguille C sur la division 2, l'aiguille B sur la division 5, et l'aiguille A sur la division 9. Faisons sonner le compteur du cadran C : il battra deux coups, et en même temps l'aiguille B marchera de deux divisions. Alors celle-ci marquera le nombre 7, qui suit le nombre 5 dans la colonne des différences premières. Faisons sonner à son tour le compteur du cadran B : il battra 7 coups, pendant lesquels l'aiguille A avancera de 7 divisions qui, ajoutées au nombre 9 qu'elle marquait déjà, donneront le nombre 16, soit le carré consécutif à 9. En recommençant l'opération à partir de l'aiguille C, qu'on laissera toujours sur la division 2, l'on voit qu'en la continuant d'une manière indéfinie, on pourra reproduire la série des carrés des nombres entiers au moyen d'un simple mécanisme.

Le théorème sur lequel est fondée la construction de la machine que nous venons de décrire, est un cas particulier d'un autre théorème plus général, et qui consiste en ce que si, dans un polynome quelconque dont la plus haute puissance de la variable soit m, l'on fait croître cette même variable par degrés égaux, que l'on calcule les valeurs correspondantes du polynome, puis, que l'on prenne successivement leurs différences première, deuxième, troisième, etc., ainsi qu'il a été fait pour la série des carrés, les différences $m^{\text{ième}}$ seront toutes égales entre elles. De sorte que, pour reproduire la série des valeurs du polynome au moyen d'une machine analogue à la précédente, il suffira qu'elle ait $(m+1)$ cadrans ayant les uns avec les autres les relations indiquées. Comme les différences peuvent être positives ou négatives, la machine contiendra un mécanisme qui fera avancer ou reculer l'aiguille, selon que le nombre à ajouter algébriquement aura le signe *plus* ou le signe *moins*.

Lorsque d'un polynome on passe à une série d'une infinité de termes ordonnés suivant les puissances ascendantes de la variable, pour appliquer la machine au calcul de la fonction représentée par cette série, il semblerait au premier abord que le mécanisme devrait contenir une infinité de cadrans, ce qui rendrait la chose impossible. Mais la difficulté disparaîtra pour bien des cas, si on observe que, pour un grand nombre de fonctions, l'on parvient à rendre convergentes les séries qui les représentent, de sorte que, selon le degré d'approximation que l'on désire, on pourra se borner à ne calculer qu'un petit nombre des termes de la série, les autres pouvant être négligés. De cette manière la question est ramenée au cas primitif d'un polynome fini. C'est ainsi que l'on peut calculer la suite des logarithmes des nombres. Mais comme, à mesure que la variable augmente de valeur, les termes que l'on avait d'abord négligés prennent un accroissement qui finirait par influer sur le degré d'approximation que l'on veut conserver, il faut, à certains intervalles calculer la valeur de la fonction par des procédés divers, et partir de ces résultats pour en déduire, par le moyen de la machine, les autres valeurs intermédiaires. Comme on le voit, cette machine remplace ici la troisième section des travailleurs dont il a été parlé au sujet des tables calculées par ordre du gouvernement français, et elle remplit ainsi le but proposé.

Telle est la première machine imaginée par Mr. Babbage. On voit que son emploi est limité aux cas où les nombres demandés sont susceptibles d'être obtenus par de simples additions ou soustractions, qu'elle n'est pour ainsi dire que l'expression d'un théorème particulier d'analyse, et qu'enfin elle ne s'étend point à la solution d'une infinité d'autres questions qui sont du ressort de l'analyse mathématique. C'est en songeant au vaste champ qui lui restait encore à parcourir, que Mr. Babbage, renonçant à ses premiers essais, conçut le plan d'un autre système de mécanisme dont l'usage devait avoir la généralité de l'écriture algébrique même, et que, pour cette raison, il nomme *machine analytique*.

Maintenant que j'ai exposé l'état de la question, il est temps de faire connaître le principe sur lequel est fondée la construction de cette machine. Lorsqu'on emploie l'analyse pour résoudre quelque problème, il y a, en général, deux espèces d'opérations à exécuter : d'abord le calcul numérique des différents coefficients, et ensuite leur distribution par rapport aux quantités qui doivent en être affectées. Si l'on a, par exemple, à former le produit de deux binomes $(a + b\,x)\,(m + n\,x)$ le résultat sera représenté par $a\,m + (a\,n + b\,m)\,x + b\,n\,x^2$, expression dans laquelle on devra d'abord calculer $a\,m$, $a\,n$, $b\,m$, $n\,b$, sommer ensuite $a\,n + b\,m$, et enfin distribuer les coefficients ainsi obtenus par rapport aux puissances de la variable. Si l'on veut reproduire ces opérations au moyen d'une machine, elle devra donc posséder deux facultés distinctes : 1° celle d'exécuter les calculs numériques, 2° celle de distribuer convenablement les valeurs ainsi obtenues. Mais si la main de l'homme était obligée d'intervenir pour diriger chacune de ces opérations partielles, il n'y aurait rien de gagné sous le rapport de l'exactitude et de l'économie de temps; il faudra donc encore que la machine ait la propriété d'exécuter elle-même toutes les opérations successives nécessaires pour la solution du problème qu'on lui propose, une fois qu'on y aura introduit les *données numériques primitives* de ce même problème. Ainsi, puisque, une fois que la nature du calcul à exécuter ou celle du problème à résoudre sont indiquées, la machine doit, en vertu de son propre pouvoir, passer d'elle-même par toutes les opérations intermédiaires qui conduisent au résultat proposé, elle exclura toutes les méthodes de tâtonnement et d'essai, et n'admettra que les procédés directs de calcul. Il doit en être ainsi, car la machine n'est point un être qui pense, mais un simple automate qui agit suivant les lois qu'on lui a tracées. Cela posé, une des premières recherches de l'auteur a dû être celle d'un moyen de faire la division d'un nombre par un autre, sans employer la méthode de tâtonnement indiquée par les règles ordinaires de l'arithmétique. Cette combinaison n'a pas été la moins difficile à obtenir; c'est d'elle que dépendait le succès de toutes les autres. Dans l'impossibilité où je suis de décrire ici le procédé par lequel on arrive à ce but, on devra se borner à admettre que les quatre premières opérations de l'arithmétique, c'est-à-dire l'addition, la soustraction, la multiplication et la division, sont susceptibles d'être exécutées d'une manière directe par l'intermédiaire de la machine. Cela étant, la machine sera par là même capable de faire toute espèce de calcul numérique, car en définitive ces calculs se réduisent tous aux quatre opérations que nous venons d'énoncer. Pour concevoir maintenant comment la machine peut fonctionner suivant les lois établies, donnons d'abord une idée de la manière dont les nombres y sont matériellement représentés.

Figurons-nous une pile ou colonne verticale composée d'un nombre indéfini de disques circulaires, tous traversés dans leur centre, par un axe commun autour duquel chacun d'eux peut prendre un mouvement de rotation indépendant. Si, sur le contour de chacun de ces disques, on écrit les dix chiffres qui composent notre alphabet numérique, en disposant, suivant une même verticale, une série de ces chiffres, on pourra exprimer de cette manière un nombre quelconque. Car il suffira de supposer que le premier disque représente les unités, le deuxième les dizaines, le troisième les centaines, ainsi de suite. Lorsque deux nombres auront été écrits de cette manière sur deux colonnes distinctes, on pourra se proposer de les combiner arithmétiquement entre eux, et d'obtenir le résultat inscrit sur une troisième colonne. En général, si l'on a une série de colonnes composées de disques et que nous désignerons par V_0, V_1, V_2, V_3, V_4, etc., on peut demander, par exemple, de diviser le nombre écrit sur la colonne V_1, par celui de la colonne V_4, et d'obtenir le résultat sur la colonne V_7. Pour faire cette opération, il faudra donner à la machine deux dispositions distinctes : la première par laquelle elle est prête à exécuter *une division*, et la seconde par laquelle on lui indique les colonnes sur lesquelles elle doit opérer, et celle où le résultat doit être écrit. Lorsqu'à cette division devra, par exemple, succéder une addition de deux nombres pris sur d'autres colonnes, les deux dispositions de la machine devront changer à la fois. S'il s'agit, au contraire, d'exécuter une suite d'opérations de la même nature, alors la première disposition demeurera la même, la seconde seule changera. Ainsi, parmi les arrangements que peuvent prendre les diverses pièces de la machine, on en distinguera deux principaux : 1° *la disposition relative aux opérations*, 2° *la disposition relative aux variables*; par cette dernière on entend celle qui indique les colonnes sur lesquelles on doit opérer. Quant aux opérations mêmes, elles s'exécutent dans un appareil spécial désigné sous le nom de *moulin*; il est lui-même composé d'un certain nombre de colonnes semblables à celles des variables. Lorsque deux nombres doivent être combinés ensemble, la machine commence par les effacer sur les colonnes où ils étaient écrits, c'est-à-dire à mettre *zéro* sur chaque disque dans les deux lignes verticales où les nombres étaient écrits; puis elle les transporte dans le moulin. Là, l'appareil étant disposé pour l'opération demandée, celle-ci s'exécute, et, une fois qu'elle est achevée, le résultat est lui-même transporté sur la colonne des variables qui aura été indiquée. Ainsi le moulin est la partie de la machine qui exécute; les colonnes des variables constituent celle où s'écrivent et s'ordonnent les résultats. D'après ce qui vient d'être dit, l'on voit que tous les résultats fractionnaires et irrationnels seront représentés par des fractions décimales. En supposant chaque colonne composée de quarante disques, cette extension serait suffisante pour tous les degrés d'approximation dont on a généralement besoin.

Maintenant l'on demandera comment la machine peut d'elle-même et sans avoir recours à la main de l'homme, prendre les dispositions successives convenables pour opérer. La solution de ce problème a été empruntée à l'appareil de Jacquard, en usage pour la confection des étoffes brochées; voici de quelle manière :

Dans les tissus on distingue généralement deux espèces de

fils : d'abord, les chaînes ou fils longitudinaux, ensuite la trame ou fil transversal qui est guidé par l'instrument que l'on nomme navette et qui croise avec les chaînes. Lorsque l'on veut faire une étoffe brochée, il faut empêcher tour à tour certains fils de croiser avec la trame, suivant un ordre déterminé par la nature du dessin qu'il s'agit de reproduire. Jadis cette opération était longue et difficile et exigeait que l'ouvrier, attentif au dessin qu'il devait copier, ordonnât lui-même les mouvements que ces fils devaient prendre. De là provenait le prix élevé de ce genre d'étoffes, surtout lorsqu'il y entrait des fils de différentes couleurs. Pour simplifier cette manufacturation, Jacquard imagina de faire communiquer chacun des groupes de fils qui devaient agir ensemble avec un levier distinct pour chaque groupe. Tous ces leviers se terminent par des tiges réunies en un faisceau qui a généralement la forme d'un parallélipipède à base rectangulaire ; ces tiges sont cylindriques et séparées entre elles par de petits intervalles. L'opération du soulèvement des fils se réduira ainsi à mouvoir convenablement ces divers bras de levier. Pour cela, on prend une feuille de carton de forme rectangulaire, un peu plus grande que la section du faisceau. Si l'on applique cette feuille contre la base du faisceau de leviers, en lui imprimant un mouvement de translation, ceux-ci se mouvront tous en même temps, ainsi que les fils qui leur correspondent. Mais si le carton, au lieu d'être *plein*, était percé dans les endroits où les extrémités des leviers viennent le rencontrer, alors, dans son mouvement le carton étant traversé de part en part par les leviers, ceux-ci resteraient en place. De cette manière on voit qu'il est aisé de déterminer la position de ces trous dans le carton, de manière que, à un instant donné, il y ait un certain nombre de leviers et par conséquent de fils soulevés, tandis que les autres restent en place. En supposant que cette opération soit successivement répétée suivant une loi indiquée par le dessin qu'on veut exécuter, l'on conçoit que ce dessin puisse être reproduit sur l'étoffe. Pour cela il n'y a qu'à composer une série de cartons suivant la loi requise, et à les disposer convenablement à la suite les uns des autres ; puis en les faisant passer sur un arbre à section polygonale, qui à chaque coup de la navette présenterait une nouvelle face, laquelle serait transportée parallèlement à elle-même contre le faisceau de leviers, l'opération du soulèvement des fils se ferait d'une manière régulière. On voit qu'ainsi l'on pourra confectionner les étoffes brochées, avec une rapidité et une précision qu'il était difficile d'obtenir auparavant.

Ce sont des dispositions analogues à celle que nous venons de décrire, qui ont été introduites dans la machine analytique. Elle contient deux espèces principales de cartons : 1° les cartons des opérations, par lesquels la machine est disposée de manière à exécuter une série déterminée d'opérations, telles que additions, soustractions, multiplications et divisions; 2° les cartons des variables, qui indiquent à la machine des colonnes sur lesquelles les résultats doivent être écrits. Les cartons étant mis en mouvement, disposent successivement les diverses pièces de la machine selon la nature des opérations à faire, et la machine les exécute en même temps au moyen des mécanismes dont elle est composée.

Pour mieux concevoir la chose, prenons pour exemple la résolution de deux équations du premier degré à deux inconnues. Soient donc les deux équations suivantes où x et y sont les inconnues :

$$\begin{cases} mx + ny = d \\ m'x + n'y = d' \end{cases}$$

on en déduit $x = \dfrac{dn' - d'n}{n'm - nm'}$, et pour y une valeur analogue. Représentons toujours par V_0, V_1, V_2, etc., les différentes colonnes qui contiennent les nombres, et supposons qu'on ait choisi les huit premières colonnes pour y écrire les nombres qui sont représentés par m, n, d, m', n', d' n et n', ce qui suppose $V_0 = m$, $V_1 = n$, $V_2 = d$, $V_3 = m'$, $V_4 = n'$, $V_5 = d'$, $V_6 = n$, $V_7 = n'$. La série des opérations commandées par les cartons, et les résultats obtenus, seront représentés par le tableau suivant :

NOMBRE des OPÉRATIONS.	CARTONS des OPÉRATIONS. — Signes indiquant la nature des opérations.	CARTONS DES VARIABLES. Colonne soumise aux opérations.	CARTONS DES VARIABLES. Colonne recevant le résultat des opérations	MARCHE des OPÉRATIONS.
1	$\times$	$V_2 \times V_4 =$	V_8	$= dn'$
2	$\times$	$V_5 \times V_1 =$	V_9	$= d'n$
3	$\times$	$V_4 \times V_0 =$	V_{10}.......	$= n'm$
4	$\times$	$V_1 \times V_3 =$	V_{11}.......	$= nm'$
5	—	$V_8 - V_9 =$	V_{12}.......	$= dn' - d'n$
6	—	$V_{10} - V_{11} =$	V_{13}.......	$= n'm - n'm'$
7	:	$\frac{V_{12}}{V_{13}} =$	V_{14}.......	$= x = \frac{dn' - d'n}{mn' - m'n}$

Comme les cartons ne font qu'indiquer comment et sur quelles colonnes la machine doit agir, il est clair qu'il faudra encore, dans chaque cas particulier, introduire les données numériques du calcul. Ainsi, dans l'exemple que nous avons choisi, on devra préalablement écrire les valeurs numériques de m, n, d, m', n', d' dans l'ordre et sur les colonnes indiquées, après quoi l'on fera agir la machine qui donnera la valeur de l'inconnue x pour ce cas particulier. Pour avoir la valeur de y, il faudra faire une autre série d'opérations analogues aux précédentes. Mais on voit qu'elles se réduiront à quatre seulement, car le dénominateur de l'expression de y, sauf le signe, est le même que celui de x, et égal à $m'n - mn'$. Dans le tableau précédent l'on remarquera que la colonne des opérations indique de suite quatre *multiplications*, deux *soustractions* et une *division*. Ainsi l'on pourra, au besoin, n'employer que trois cartons des opérations; pour cela il suffira d'introduire dans la machine un appareil qui, par exemple, après la première multiplication, retienne le carton relatif à cette opération, et ne lui permette d'avancer, pour être remplacé par un autre, que lorsque cette même opération aura été répétée quatre fois. Dans l'exemple précédent, nous avons vu que, pour trouver la valeur de x, il fallait commencer par écrire les coefficients m, n, d, m', n', d' sur huit colonnes, en répétant ainsi deux fois n et n'; d'après la même méthode, si l'on voulait également calculer y, on aurait dû écrire ces mêmes coefficients sur douze colonnes différentes. Mais il est possible de simplifier cette opération et d'éviter les chances d'erreurs, qui augmentent à mesure que la quantité des nombres à écrire, avant de faire opérer la machine, devient plus grand. Pour cela souvenons-nous que tout nombre écrit sur une colonne, afin d'être combiné arithmétiquement avec un autre nombre, doit être effacé de la colonne sur laquelle il se trouve et transporté dans le *moulin*. Or, dans l'exemple que nous avons discuté, prenons les deux coefficients m et n', qui chacun doivent entrer dans deux produits différents, c'est-à-dire m dans mn' et md', n' dans mn' et $n'd$. Ces coefficients seront écrits sur les colonnes V_0 et V_4. Si l'on commence la série des opérations par le produit de m par n', on effacera ces nombres sur les colonnes V_0 et V_4 pour les transporter dans le moulin qui les multipliera ensemble, puis ordonnera à la machine d'écrire le résultat sur la colonne V_6 par exemple. Mais comme ces nombres doivent encore servir chacun à une opération, il faudra qu'ils se trouvent de nouveau écrits quelque part; pour cela, en même temps que le moulin effectue leur produit, la machine les écrira de nouveau sur deux colonnes qui lui seront indiquées par les cartons, et comme dans le cas actuel rien ne s'oppose à ce qu'ils reprennent leurs premières places, on les supposera inscrits sur V_0 et V_4, d'où ils ne disparaîtront enfin, pour ne plus être reproduits, que lorsqu'ils auront subi les autres combinaisons auxquelles ils doivent être soumis.

On voit donc que l'ensemble des opérations nécessaires pour résoudre les deux équations du premier degré en question, pourra en définitive être représenté par le tableau suivant :

Colonnes sur lesquelles sont écrites les données primitives du problème.	Nombre des opérations.	CARTONS des opérations. Nombre des cartons des opérations	CARTONS des opérations. Signes indiquant la nature des opér.	CARTONS DES VARIABLES. COLONNES soumises aux opérations	CARTONS DES VARIABLES. Colonnes qui reçoivent le résultat des opérations.	CARTONS DES VARIABLES. Indication des nouvelles colonnes sur lesquelles sont écrites les variables.	MARCHE des OPÉRATIONS.
$V_0 = m$	1	1	×	$V_0 \times V_4 =$	$V_6 \ldots$	V_0 sur V_0 V_4 *id.* V_4	$V_6 = mn'$
$V_1 = n$	2	»	×	$V_3 \times V_1 =$	$V_7 \ldots$	V_1 *id.* V_1 V_3 *id.* V_3	$V_7 = m'n$
$V_2 = d$	3	»	×	$V_2 \times V_4 =$	$V_8 \ldots$	V_2 *id.* V_2 V_4 *id.*....	$V_8 = dn'$
$V_3 = m'$	4	»	×	$V_5 \times V_1 =$	$V_9 \ldots$	V_1 *id.*.... V_5 *id.* V_5	$V_9 = d'n$
$V_4 = n'$	5	»	×	$V_0 \times V_5 =$	$V_{10} \ldots$	V_0 *id.*.... V_5 *id.*....	$V_{10} = d'm$
$V_5 = d$	6	»	×	$V_2 \times V_3 =$	$V_{11} \ldots$	V_2 *id.*.... V_3 *id.*....	$V_{11} = dm'$
	7	2	—	$V_0 \times V_7 =$	$V_{12} \ldots$	V_6 *id.*.... V_7 *id.*....	$V_{12} = mn' - m'n$
	8	»	—	$V_8 - V_9 =$	$V_{13} \ldots$	V_8 *id.*.... V_9 *id.*....	$V_{13} = dn' \cdot d'n$
	9	»	—	$V_{10} - V_{11} =$	$V_{14} \ldots$	V_{10} *id.*.... V_{11} *id.*....	$V_{14} = d'm - dm'$
	10	3	:	$\frac{V_{13}}{V_{12}} =$	$V_{15} \ldots$	V_{13} *id.*.... V_{12} *id* V_{12}	$V_{15} = \frac{dn' - d'n}{mn' - m'n} = x$
	11	»	:	$\frac{V_{14}}{V_{12}} =$	$V_{16} \ldots$	V_{12} *id.*.... V_{14} *id.*....	$V_{16} = \frac{d'm - dm'}{mn' - m'n} = y$

Afin de diminuer autant que possible les chances d'erreur dans l'écriture des données numériques du problème, on les écrit successivement sur une des colonnes du moulin ; puis, au moyen de cartons disposés à cet effet, ces mêmes nombres vont se placer sur les colonnes convenables, sans que l'opérateur ait aucunement à s'en inquiéter ; de cette manière toute son attention se reportera sur la simple écriture de ces mêmes nombres.

D'après ce qui vient d'être exposé, on voit que l'ensemble des colonnes des variables peut être considéré comme un *magasin* de nombres qui y sont accumulés par le moulin et qui, obéissant aux ordres transmis à la machine par le moyen des cartons, passent alternativement du moulin au magasin et du magasin au moulin, pour y subir les transformations requises par la nature du calcul à exécuter.

Jusqu'ici, il n'a point encore été parlé des signes des résultats, et la machine serait loin d'être parfaite si elle n'était pas capable d'exprimer et de combiner entre elles les quantités positives et négatives. Pour arriver à ce but, au-dessus de chaque colonne, tant du moulin que du magasin, se trouve un disque semblable à ceux qui composent les colonnes. Selon que le chiffre correspondant au nombre écrit sur la pile inférieure est pair ou impair, ce dernier sera considéré comme positif ou négatif. Cela posé, voici une manière de concevoir comment les signes peuvent se combiner algébriquement dans la machine. Lorsqu'un nombre devra être transporté du magasin dans le moulin et *vice versâ*, il le sera toujours avec son signe, ce qui se fera par l'intermédiaire des cartons, ainsi qu'il a été dit précédemment. Soient donc deux nombres placés avec leurs signes respectifs dans le moulin et sur lesquels on doive opérer arithmétiquement. Supposons qu'il s'agisse d'abord de les sommer ensemble : les cartons des opérations commanderont l'addition ; si les deux nombres ont le même signe, l'un des deux s'effacera entièrement sur la colonne où il était écrit et ira s'ajouter sur la colonne qui contient l'autre nombre; la machine, au moyen d'un appareil, pourra empêcher pendant cette opération tout mouvement dans le disque des signes qui appartient à la colonne sur laquelle se fait l'addition, et alors le résultat restera avec le signe qu'avaient les deux nombres donnés. Lorsque les deux nombres ont deux signes différents, l'addition qui était ordonnée par le carton se change en soustraction par l'intermédiaire des mécanismes qui sont mis en jeu à cause de cette différence de signe. Comme la soustraction ne peut s'opérer que sur le plus grand des deux nombres, si on fait en sorte que le disque des signes du plus grand nombre n'ait aucun mouvement pendant que le plus petit de ces nombres s'efface de dessus sa colonne pour être soustrait de l'autre, il restera un résultat qui aura le signe de ce dernier, ainsi que cela doit être. Les combinaisons auxquelles donne lieu la soustraction algébrique, sont analogues aux précédentes. — Passons à la multiplication. Quand les deux nombres à multiplier ont le même signe, le résultat est positif; si les deux signes sont différents, le produit doit être négatif. Pour que la machine agisse conformément à cette loi, il suffira de concevoir que, sur la colonne qui contient le produit des deux nombres donnés, les chiffres indiquant leurs signes se soient ajoutés sur le disque des signes du produit; alors il est clair que si les chiffres des signes sont ou tous les deux pairs, ou tous les deux impairs, leur somme sera un nombre pair, et par conséquent exprimera une quantité positive; si, au contraire, les deux chiffres des signes sont l'un pair et l'autre impair, leur somme sera un nombre impair, et par conséquent exprimera une quantité négative. — Pour la division, au lieu d'ajouter les chiffres des disques, il faudra les soustraire l'un de l'autre, ce qui donnera lieu à des résultats analogues aux précédents, c'est-à-dire que, si ces chiffres sont tous les deux pairs ou impairs, le reste de cette soustraction sera pair; il sera impair dans le cas contraire. Lorsque je parle d'ajouter ou de soustraire l'un de l'autre les nombres exprimés par les chiffres des signes, je n'entends que faire avancer ou rétrograder l'un des disques d'un nombre de divisions égal à celui qui est exprimé par le chiffre de l'autre disque. On voit donc, d'après ce qui vient d'être exposé, qu'il est possible de combiner mécaniquement les signes des quantités, de manière à obtenir des résultats conformes à ceux qu'indique l'algèbre [1].

[1] N'ayant point eu le loisir de discuter avec Mr. Babbage la manière d'introduire dans sa machine la combinaison des signes algébriques, je n'ai point la prétention d'exposer ici la méthode dont il se sert à cette fin; mais j'ai cru devoir y suppléer, pensant que cet écrit eût été imparfait si je n'avais indiqué un des moyens que l'on peut employer pour résoudre ce point essentiel du problème en question.

La machine est non-seulement capable d'exécuter les calculs numériques qui dépendent d'une formule algébrique donnée, mais elle est encore propre aux calculs analytiques dans lesquels on a une ou plusieurs variables à considérer. Pour cela il faut admettre que l'expression analytique sur laquelle on doit opérer soit développée suivant les puissances de la variable, ou suivant des fonctions déterminées de cette même variable, comme seraient, par exemple, les fonctions circulaires ; il en doit être de même du résultat auquel on veut arriver. Or, si l'on suppose qu'au-dessus de chaque colonne du magasin on ait écrit les puissances ou les fonctions de la variable ordonnées selon la loi du développement que l'on considère, les coefficients de ces différents termes pourront être écrits sur les colonnes inférieures qui leur correspondent. On aura, de cette manière, représenté un développement analytique. Maintenant, en supposant que la position des différents termes qui le composent soit invariable, le problème se réduira à calculer leurs coefficients selon les lois voulues par la nature de la question. Pour mieux faire comprendre la chose, prenons l'exemple suivant, bien simple, dans lequel il s'agit de multiplier $(a + b\, x^1)$ par $(A + B \cos^1 x)$. Nous commencerons par écrire x^0, x^1, $\cos^0 x$, $\cos^1 x$ au-dessus des colonnes V_0, V_1, V_2 et V_3 ; puis comme, d'après la forme des deux fonctions à combiner, les termes qui doivent composer le produit sont de la nature suivante, $x^0 . \cos^0 x$, $x . \cos^1 x$, $x^1 . \cos^0 x$, $x^1 \cos^1 x$, on les écrira au-dessus des colonnes V_4, V_5, V_6 et V_7. Les coefficients de x^0, x^1, $\cos^0 x$, $\cos^1 x$ étant donnés, par le moyen du moulin on les fera passer sur les colonnes V_0, V_1, V_2 et V_3. Telles sont les données primitives du problème ; c'est maintenant à la machine qu'il appartient d'en opérer la solution, c'est-à-dire de trouver les coefficients qui doivent être inscrits sur V_4, V_5, V_6, V_7. Dans ce but, comme la loi de formation de ces mêmes coefficients est connue, la machine agira par l'intermédiaire des cartons, de la manière indiquée dans le tableau suivant :

Colonnes au-dessus desquelles sont écrites les fonctions de la variable.	COEFFICIENTS		CARTONS des opérations		CARTONS DES VARIABLES.			
	Donnés.	A former.	Nombre des opérations.	Nature de l'opération.	COLONNES soumises aux opérations.	COLONNES sur lesquelles sont ecrits les resultats des operations.	Indication des nouvelles colonnes sur lesquelles sont ecrits les coefficients soumis aux operations.	RÉSULTATS DES OPÉRATIONS.
$x^0 \ldots\ldots V_0$	a	»	»	»	»	»	»	» »
$x' \ldots\ldots V_0$	b	»	»	»	»	»	»	» »
$\text{Cos}\,{}^0 x \ldots V_2$	A	»	»	»	»	»	»	» »
$\text{Cos}\,{}^1 x \ldots V_3$	B	»	»	»	»	»	»	» »
$x^0 \cos{}^0 x.\,V_4$		aA	1	$\times$	$V_0 \times V_2 =$	$V_4 \ldots\ldots$	V_0 sur V_0; V_2 *id.* V_2	$V_4 = aA$ coefficients de $x^0 \cos{}^0 x$
$x^0 \cos{}^1 x..V_5$		aB	2	$\times$	$V_0 \times V_3 =$	$V_5 \ldots\ldots$	V_0...*id.*....; V_3 *id.* V_3	$V_5 = aB$ ......*id.*.....$x^0 \cos{}^1 x$
$x' \cos{}^0 x...V_6$		bA	3	$\times$	$V_1 \times V_2 =$	$V_6 \ldots\ldots$	V_1 *id.* V_1; V_2...*id.*.....	$V_6 = bA$ ......*id.*.....$x' \cos{}^0 x$
$x' \cos{}^1 x...V_7$		bB	4	$\times$	$V_1 \times V_3 =$	$V_7 \ldots\ldots$	V_1...*id.*.....; V_3 *id.*.....	$V_7 = bB$ ......*id.*.....$x' \cos{}^1 x$

24

On doit concevoir maintenant que le principe développé dans l'exemple précédent puisse s'appliquer en général à toute espèce d'opération que l'on pourrait se proposer sur les séries soumises au calcul. Il suffira de connaître la loi de formation des différents coefficients, d'inscrire cette loi dans les cartons de la machine, qui alors exécutera elle-même tous les calculs convenables pour arriver au résultat proposé. S'il s'agissait, par exemple, d'une série récurrente, la loi de formation des différents termes étant la même pour tous, les mêmes opérations qui se feront pour l'un d'eux se répéteront pour les autres; seulement il y aura changement de lieu de l'opération, c'est-à-dire qu'elle se fera sur des colonnes différentes. Comme généralement toute expression analytique est susceptible d'être exprimée par une série ordonnée selon certaines fonctions de la variable, on voit que la machine embrassera tous les calculs analytiques qui se réduisent en définitive à la formation des coefficients suivant certaines lois, et à leur distribution par rapport aux variables.

De ce qui vient d'être exposé l'on peut déduire cette conséquence importante, c'est que, puisque les cartons indiquent seulement la nature des opérations à faire et les colonnes des variables sur lesquelles elles doivent être exécutées, ces cartons auront eux-mêmes toute la généralité d'analyse, dont ils ne sont qu'une simple traduction. Examinons encore quelques-unes des difficultés que la machine doit surmonter pour que son assimilation à l'analyse soit plus complète. Il y a certaines fonctions qui doivent changer de nature lorsqu'elles passent par zéro ou l'infini, ou dont les valeurs ne sauraient être admises au delà de ces limites. Lorsque ces cas se présentent, la machine au moyen d'une sonnerie peut avertir que le passage par zéro ou l'infini a lieu, et alors elle s'arrête, jusqu'à ce que le gardien l'ait remise en action pour telle autre opération qu'on voudra lui ordonner. Si cette opération est prévue, la machine, au lieu de sonner, se disposera de manière à présenter les nouveaux cartons relatifs à l'opération qui doit succéder au passage par zéro et l'infini. Ces nouveaux cartons peuvent faire suite aux premiers et n'entrer en jeu que lorsque les deux circonstances indiquées précédemment auront lieu. Considérons un terme de la forme ab^n; puisque les cartons ne sont qu'une traduction de la formule analytique, il faudra, dans le cas actuel, que leur nombre soit le même, quelle que soit la valeur de n, c'est-à-dire quel que soit le nombre des multiplications à faire pour élever b à la puissance $n^{\text{ième}}$ (nous supposons pour le moment que n soit un nombre entier). Or, comme l'exposant n indique que b doit être multiplié n fois par lui-même, toutes ces opérations étant de la même nature, il suffira d'employer pour cela un seul carton des opérations, celui qui ordonne la multiplication.

Mais lorsque n sera donné pour le cas particulier que l'on veut calculer, il sera encore nécessaire que la machine borne ses multiplications aux nombres convenables. Or, voici comment l'opération peut être disposée. Les trois nombres a, b et n seront écrits sur autant de colonnes distinctes du magasin, que nous désignerons par V_0, V_1, V_2; le résultat ab^n viendra se transcrire sur la colonne V_3. Lorsque le nombre n aura été introduit dans la machine, un carton ordonnera à un *compteur* de marquer $(n - 1)$ et en même temps fera exécuter la multiplication de b par b. Quand celle-ci sera achevée, le compteur aura effacé une unité et ne marquera plus que $(n - 2)$, et la machine ordonnera de nouveau au nombre b écrit sur la colonne V_1 de se multiplier avec le produit b^2 écrit sur la colonne V_3, ce qui donnera b^3. Alors une nouvelle unité s'effacera sur le compteur, et les opérations indiquées se répéteront jusqu'à ce que le compteur ne marque plus que zéro. Ainsi le nombre b^n se trouvera écrit sur la colonne V_3, et la machine poursuivant le cours de ses autres opérations, ordonnera le produit de b^n par a, et le calcul demandé se trouvera achevé sans que le nombre de cartons à employer ait dû varier avec la valeur de n. Si n était négatif, les cartons, au lieu d'ordonner la multiplication de a par b^n, ordonneraient la division, ce qui peut se concevoir, puisque tout nombre peut être écrit avec son signe respectif et réagir de cette manière sur la nature des opérations à exécuter. Enfin si n était fractionnaire et de la forme $\frac{p}{q}$, on emploierait une colonne de plus pour écrire q, et la machine ferait agir deux appareils, l'un pour élever b à la puissance n, l'autre pour extraire la racine $q^{\text{ième}}$ du nombre ainsi obtenu.

On peut demander, par exemple, de multiplier une expression de la forme $ax^m + bx^n$ par une autre $Ax^p + Bx^q$, puis de réduire le produit aux moindres termes lorsque leurs indices sont égaux. Les deux facteurs étant ordonnés par rapport à x, le résultat général de la multiplication serait $Aax^{m+p} + Abx^{p+n} + aBx^{m+q} + bBx^{n+q}$. Jusqu'ici l'opération ne présente pas de difficultés; mais supposons que l'on ait $m = p$ et $n = q$, et que l'on veuille réduire les deux termes moyens à un seul $(Ab + Ba)\,x^{m+q}$. Pour cela les cartons pourront ordonner que $m + q$ et $n + p$ soient transportés dans le moulin pour y être soustraits l'un de l'autre; si le reste est nul, comme cela doit avoir lieu dans l'hypothèse admise, le moulin ordonnera à d'autres cartons de lui porter les coefficients Ab et aB pour les sommer ensemble, et les donner ainsi pour coefficient au terme unique $x^{n+p} = x^{m+q}$.

Cet exemple nous montre comment les cartons peuvent reproduire toutes les opérations qu'exécute l'intelligence pour arriver à un résultat déterminé, lorsque ces opérations sont elles-mêmes susceptibles d'être nettement précisées.

Examinons encore l'expression suivante:

$$2.\ \frac{2^2.\,4^2.\,6^2.\,8^2.\,10^2 \ldots\ldots (2n)^2}{1^2.\,3^2.\,5^2.\,7^2.\,9^2 \ldots\ldots (2n-1)^2(2n+1)^2}$$

que l'on sait devenir égale au rapport de la circonférence au diamètre, lorsque n est infini. On peut demander à la machine non-seulement d'exécuter le calcul de cette expression fractionnaire, mais encore d'indiquer immédiatement que sa valeur devient celle même du rapport de la circonférence au diamètre lorsque n est infini, cas dans lequel le calcul serait impossible. Observons que l'on exigerait ainsi que la machine interprétât un résultat non évident par lui-même, ce qui n'est point dans ses attributions, puisqu'elle n'est point un être pensant. Cependant lorsque le cos. de $n = \infty$ a été prévu, un carton peut ordonner immédiatement la substitution de la valeur de π (π étant le rapport de la circonférence au diamètre), sans passer par la série des calculs indiqués. Il suffirait pour cela qu'il y eût dans la machine un carton spécial destiné à former immédiatement le nombre π sur la colonne qui lui serait indiquée. C'est ici le cas de parler d'une troisième espèce de cartons que l'on peut appeler *cartons des nombres*. Il y a certains nombres, tels que celui qui exprime le rapport de la circonférence au diamètre, les nombres de Bernoulli, etc., qui se présentent fréquemment dans les calculs. Pour n'être pas obligé de les calculer chaque fois que l'on doit les employer, on peut combiner certains cartons destinés à les donner tout faits dans le moulin, d'où ils vont ensuite se placer dans les colonnes du magasin qui leur sont destinées. Par ce moyen, la machine sera encore susceptible des simplifications que présente l'emploi des tables numériques. Il serait également possible d'introduire, au moyen de ces cartons, les logarithmes des nombres; mais peut-être ici ne serait-ce pas la méthode la plus convenable et la plus prompte, car la machine pourrait se prêter à d'autres combinaisons plus expéditives, fondées sur la rapidité avec laquelle s'exécutent les quatre premières opérations de l'arithmétique. Pour donner une idée de cette rapidité, il suffira de dire que Mr. Babbage pense pouvoir, par le moyen de sa machine, faire en *trois minutes* le produit de deux nombres composés de vingt chiffres chacun.

Peut-être le nombre immense de cartons qu'exigerait la solution d'un problème un peu compliqué, pourrait-il paraître un obstacle. Mais il ne semble pas devoir en être ainsi : le nombre de cartons que l'on peut employer n'a pas de limites. Il y a certaines étoffes qui pour être confectionnées n'exigent pas moins de *vingt mille* cartons, et cette quantité peut certainement être de beaucoup dépassée.

En résumant ce qui vient d'être dit sur la machine analytique, on peut conclure qu'elle est fondée sur deux principes : le premier, qui consiste en ce que tout calcul arithmétique dépend en définitive de quatre opérations principales, l'addition, la soustraction, la multiplication et la division ; le second, en ce que tout calcul analytique peut se réduire à calculer les coefficients des différents termes d'une série. Si ce dernier principe est vrai, toutes les opérations de l'analyse sont du domaine de la machine. D'un autre côté, l'emploi des cartons offre une généralité égale à celle des formules algébriques, car, de même qu'une formule de ce genre indique simplement la nature et l'ordre des opérations à faire pour arriver à un résultat déterminé, les cartons ne font que commander à la machine ces mêmes opérations ; mais pour que les mécanismes puissent agir, il faut encore introduire dans chaque cas particulier les données numériques du problème. Ainsi une même série de cartons servira pour toutes les questions d'une même nature, dans lesquelles il n'y aura que les données numériques de changées. De cette manière les cartons ne sont qu'une traduction des formules algébriques, ou pour mieux dire, une autre forme d'écriture analytique.

Comme la machine a un mode d'agir qui lui est propre, dans chaque cas particulier il sera nécessaire de disposer la série des calculs conformément aux moyens qu'elle possède ; car tel procédé qui serait extrêmement facile pour un calculateur, deviendrait long et compliqué pour la machine et *vice versâ*.

Considéré sous le point de vue le plus général, l'objet essentiel de la machine étant de calculer, d'après les lois qui lui sont dictées, la valeur de coefficients numériques qu'elle doit ensuite distribuer convenablement sur les colonnes qui représentent les variables, il s'ensuit que l'interprétation des formules et des résultats est en dehors de ses attributions, à moins toutefois que cette interprétation ne soit elle-même susceptible d'être exprimée par le moyen des symboles dont elle fait usage. Ainsi elle n'est point elle-même l'être qui pense, mais on peut la considérer comme l'être qui exécute les conceptions de l'intelligence. Les cartons reçoivent l'empreinte de ces conceptions, et transmettent aux différents mécanismes dont se compose la machine les ordres nécessaires pour agir. Une fois que la machine sera construite, la difficulté se reportera donc sur la confection des cartons; mais comme ceux-ci ne sont que la traduction de formules algébriques, par le moyen de simples notations il sera facile d'en confier l'exécution à un ouvrier. Ainsi tout le travail intellectuel se bornera à la préparation des formules, qui devront être aptes à être calculées par la machine.

Maintenant, en admettant qu'une telle machine soit exécutable, on pourra demander quelle en sera l'utilité? En résumé, elle présentera les avantages suivants: 1° Exactitude; on sait que les calculs numériques sont généralement l'écueil de la solution des problèmes, car les erreurs s'y glissent aisément, et il n'est pas toujours facile de les reconnaître. Or, la machine, par la nature même de son mode d'agir, qui n'exige aucunement l'intervention de la main de l'homme durant le cours de ses opérations, présente toute espèce de garantie au sujet de l'exactitude; d'ailleurs, elle porte son contrôle avec elle-même, car à la fin de chaque opération elle donne imprimés, non-seulement le résultat, mais encore les données numériques de la question, de sorte qu'il est aisé de vérifier si cette question a été posée avec exactitude. 2° Économie de temps: pour s'en convaincre, il suffira de rappeler qu'une multiplication de deux nombres composés chacun de 20 chiffres demande tout au plus trois minutes. D'ailleurs, lorsque l'on devra faire une longue série de calculs identiques, comme ceux qu'exige la formation de tables numériques, on pourra mettre en jeu la machine de manière à donner plusieurs résultats à la fois, ce qui abrégera de beaucoup l'ensemble des opérations. 3° Economie d'intelligence : un simple calcul d'arithmétique exige le concours d'une personne ayant quelque capacité ; lorsqu'on passe à des calculs plus compliqués, et qu'on veut faire usage de formules algébriques dans des cas particuliers, il faut déjà posséder des connaissances qui supposent des études mathématiques préliminaires de quelque étendue. Or, la machine, pouvant faire elle-même toutes ces opérations purement matérielles, épargne le travail d'intelligences qui peuvent être employées plus utilement. Ainsi la machine pourra être considérée comme une vraie manufacture de chiffres, qui prêtera son secours aux sciences et aux arts utiles qui s'appuient sur les nombres. Or, qui pourrait prévoir les conséquences d'une telle invention? En effet, combien d'observations précieuses restent inutiles aux progrès des sciences, parce qu'il n'y a pas de forces suffisantes pour en calculer les résultats! Que de découragement la perspective d'un long et aride calcul ne jette-t-elle pas dans l'âme de l'homme de génie qui ne demande que du temps pour méditer et qui se le voit ravi par le matériel des opérations! Et pourtant c'est par la voie laborieuse de l'analyse qu'il doit arriver à la vérité; mais il ne peut la suivre sans être guidé par des nombres, car sans les nombres il n'est pas donné de pouvoir soulever le voile qui couvre les mystères de la nature. Ainsi la pensée de former un instrument capable d'aider la faiblesse de l'homme dans de telles recherches, est une conception qui, venant à se réaliser, marquerait une époque glorieuse dans l'histoire des sciences. Toutes les pièces, tous les rouages qui composent cet immense appareil ont été combinés, leur action a été étudiée, mais ils n'ont pu être encore assemblés. La confiance que doit inspirer le génie de Mr. Babbage rend légitime l'espoir que cette entreprise sera couronnée de succès ; en rendant hommage à l'intelligence qui la dirige, faisons des vœux pour qu'une telle œuvre s'accomplisse.

F.2 1842. Sketch of the Analytical Engine of Mr. Charles Babbage, by L. F. Menabrea, Captain of military engineering [Newly Translated].

[A new translation by GT of Menabrea 1842, in which the original text is restored without any alterations or additions by AAL. See section F.1 for the original French text. To aid discussion, section headings have been added and clearly demarcated in square brackets.]

F.2.1 [Calculators.]

The subject matter of several branches of mathematical science, although appearing, at first sight, to be solely within the purview of the mind, can nevertheless be divided into two distinct parts: one that can be deemed mechanical, because it is subject to precise and invariable laws, capable of being translated physically, while the other, which requires the intervention of reasoning, is more specifically within the domain of intellect. Consequently, we can propose to have the mechanical part of the work carried out by means of machines, and to reserve for intelligence alone that which requires the faculty of the intellect. Thus the rigor to which the rules of numerical calculation are subject must have, for a long time, led to the idea of using physical instruments, either to carry out these calculations entirely, or to shorten them. From this have been born several inventions directed towards this goal, but which generally fall short of it. Thus Pascal's much-vaunted machine is now only a mere object of curiosity which, while proving great force of intellect in its inventor, affords little utility in itself. It only performs the first four operations of arithmetic, and even then it is limited, in reality, to the first two, because multiplication and division are the result of a series of additions and subtractions. The great disadvantage of most of these machines is that they require continual manual intervention to regulate their operations, and it is in this that the source of errors lies; so that, if in the performance of large numerical calculations their use has not caught on, it is because they do not resolve the two-fold problem that the question presents, that of obtaining precise results within reasonable time.

Struck by these impressions, Mr. Charles Babbage devoted several years to the realization of a gigantic idea. He proposed nothing less than to construct a machine capable of executing not only arithmetical calculations but also any analytical calculations whose laws are known. The imagination at first recoils at such a task, but as one reflects more calmly, success appears less impossible, and one feels that it may depend on the discovery of some principle, so general that, if applied to the machine, the latter would be capable of mechanically translating those operations which could be expressed by algebraic formulae. The illustrious inventor, during a journey which he made to Turin, having been kind enough to communicate to me some of his considerations on this subject, I have come, with his consent, to give an account of the impressions which they have produced in my mind. We should not expect to find here a full description of Mr. Babbage's machine: it would require long study to be understood, but I will try to make its purpose clear and to explain the principles on which its execution is based.

F.2.2 [Log Tables.]

First I must warn that it is entirely different from that of which one finds an outline in the *Treatise on the Economy of Machinery* by the same author.[1] But, as the latter gave birth to the idea of the machine in question, I believe that it is useful, first of all, to recall briefly what were the first attempts of Mr. Babbage, and what are the circumstances which gave rise to them.

[1] Ed. Babbage 1832.

It is well-known that the French government, wishing to facilitate the adoption of the decimal system, had ordered the construction of logarithmic and trigonometric tables on an immense scale. Mr. de Prony, who had been charged with the direction of this work, divided it into three sections, to each of which were applied specialists. In the first section the formulas were combined in such a way as to make them suitable for arithmetic calculations; in the second, these same formulas were calculated for values of the variable, taken from distance to distance; Finally, in the third section, composed of nearly 80 individuals, most of whom only knew the first two rules of arithmetic, values intermediate to those calculated by the 2nd section were interpolated by means of simple additions or subtractions.

F.2.3 [The Difference Engine.]

An exercise similar to that which we have just mentioned having had to be repeated in England, Mr. Babbage thought that the operations made by the third section could be executed by a machine, and this is what he has achieved by means of a mechanism of which a part has been made, and which one can call a *difference engine* because of the principle on which it is based. To give an idea of it, it will suffice to consider the series of whole square numbers: 1, 4, 9, 16, 25, 36, 49, 64, etc. By making the difference of each of these numbers with that following, one will have a new series which we will call the series of first differences, composed of the numbers: 3, 5, 7, 9, 11, 13, 15, etc. If we subtract the previous number from each of these numbers, we obtain the second differences, which are all constant and equal to 2. The sequence of these operations and the results to which they lead can be represented by the following table:

	A *Column of square numbers*	**B** *First differences*	**C** *Second differences*
	1		
		3	
	4		2 b
		5	
	9		2 d
a		7	
	16		2
c		9	
	25		2
		11	
	36		

From the way in which the last two columns B and C have been formed, it is easy to see that, wanting, for example, to pass from the number 5 to the next 7, it will be necessary to add to the first the constant difference 2; in the same way, if from the square 9 we want to pass to the next 16, we will add to the first the difference 7, or in other words, the previous difference 5, plus the constant difference 2; or again, which amounts to the same thing, to have the number 16, it will suffice to add together the three numbers 2, 5 and 9, placed in the oblique direction a b. In an analogous manner, we will obtain the number 25 by adding together the three numbers placed in the oblique direction of: starting with the sum 2 + 7, we have the first difference 9, consecutive to 7; adding 16 to it, we will have the square 25. We see, therefore, that given the three numbers 2, 5 and 9, the whole series of the following square numbers, as well as that of their first differences, can be obtained by simple additions.

Now, to understand how these operations can be reproduced by a machine, let us suppose that it has three dials which we will designate by A, B, C, and on each of which are traced a thousand divisions, for example, which will be traversed by a needle. The two dials C and B will also each have a chime which will strike a number of strokes equal to that of the divisions indicated by the needle. For each stroke of the counter of dial C, needle B must advance by one division; the same will be true of needle A, which will advance by one division for each stroke of dial B. Such is the general arrangement of the mechanism. This being the case, at the beginning of the series of operations that we wish to perform, let us place needle C on division 2, needle B on division 5, and needle A on division 9. Let us make the counter on dial C strike: it will beat two strokes, and at the same time needle B will advance two divisions. Then this will mark the number 7, which follows the number 5 in the column of first differences. Let us in turn make the counter on dial B strike. It will beat 7 strokes, during which needle A will advance 7 divisions which, added to the number 9 that it already marked, will give the number 16, that is, the square consecutive to 9. By starting the operation again from needle C, which will always be left on division 2, we see that by continuing it indefinitely, we can reproduce the series of squares of whole numbers by means of an elementary mechanism.

The theorem on which the construction of the machine we have just described is based is a particular case of another more general theorem, which consists in the fact that if, in any polynomial whose highest power of the variable is m, we make this same variable increase by equal degrees, that we calculate the corresponding values of the polynomial, then, that we take successively their first, second, third, etc. differences, as was done for the series of squares, the m-th differences will all be equal to each other. So that, to reproduce the series of values of the polynomial by means of a machine analogous to the preceding one, it will suffice that it have $(m + 1)$ dials having with each other the relations indicated. As the differences can be positive or negative, the machine will contain a mechanism which will advance or retreat the needle, according as the number to be added algebraically has the *plus* sign or the *minus* sign.

When we pass from a polynomial to a series of an infinity of terms ordered according to the ascending powers of the variable, in order to apply the engine to the calculation of the function represented by this series, it would seem at first sight that the mechanism would have to contain an infinity of dials, which would make the thing impossible. But the difficulty will disappear for many cases, if we observe that, for a large number of functions, we may use convergent series to represent them, so that, according to the degree of approximation which we desire, we can get by with calculating only a small number of the terms of the series, the others being negligible. In this way the question is reverted to the fundamental case of a finite polynomial. It is thus that we can calculate the sequence of logarithms of numbers. But since, as the variable increases in value, the terms that were initially neglected take on an increase that would end up influencing the degree of approximation that one wishes to preserve, it is necessary, at certain intervals, to calculate the value of the function by alternative processes, and to start from these results to deduce, by means of the machine, the other intermediate values. As can be seen, this machine here replaces the third section of workers who were mentioned in connection with the tables calculated by order of the French government, and it thus fulfils the proposed goal.

This is the first machine invented by Mr. Babbage. Its use is limited to cases where the numbers required may be obtained by simple additions or subtractions, that it realizes, so to speak, only the evaluation of a particular theorem of analysis, and that finally it does not extend to the solution of an infinity of other questions which are within the scope of mathematical analysis. It was in thinking of the vast field which still remained for him to cover, that Mr. Babbage, transcending his first attempts, conceived the plan of another machine whose use was to have the generality of any algebraic formula, and which, for this reason, he calls the *analytical engine*.

F.2.4 [Analytical Expressions.]

Now that I have explained the state of the question, it is time to disclose the principle on which the construction of this machine is based. When analysis is used to solve some problem, there are, in general, two kinds of operations to be performed: first, the numerical calculation of the different coefficients, and then their distribution in relation to the quantities which must be affected by them. If, for example, we have to form the product of two binomials $(a + bx)(m + nx)$ the result will be represented by $am + anx + bmx + bnx^2$, an expression in which we must first calculate am, an, bm, nb, then add $an + bm$, and finally distribute the coefficients thus obtained in relation to the powers of the variable. If we wish to reproduce these operations by means of a machine, it must therefore possess two distinct faculties: 1° that of executing the numerical calculations, 2° that of suitably distributing the values thus obtained. But if manual intervention were required to direct each of these partial operations, there would be nothing gained in terms of accuracy and economy of time; it will therefore still be necessary for the machine to have the property of executing itself all the successive operations necessary for the solution of the problem proposed to it, once the *raw numerical data* of this same problem have been inserted. Thus, since, once the nature of the calculation to be performed or that of the problem to be solved are specified, the machine must, by virtue of its own power, pass of its own accord through all the intermediate operations which lead to the proposed result, it will exclude all methods of trial and error, and will admit only direct methods of calculation. This must be so, because the machine is not a being which thinks, but a simple automaton which acts according to the laws which have been traced for it. This being said, one of the author's first investigations must have been that of a means of dividing one number by another, without using the trial and error method indicated by the ordinary rules of arithmetic. This combination was not a little hard to obtain; the success of all the others depended on it. Since it is impossible for me to describe here the process by which this goal is achieved, we must limit ourselves to declaring that the first four operations of arithmetic, that is to say, addition, subtraction, multiplication and division, are capable of being executed in a direct manner by means of the machine. This being the case, the machine will thereby be capable of performing any kind of numerical calculation, because ultimately these calculations are all reduced to the four operations that we have just stated. To now conceive how the machine can function according to the established laws, let us first give an idea of the manner in which the numbers are materially represented there.

F.2.5 [The Magazine Store.]

Let us imagine a vertical pile or column composed of an indefinite number of circular disks, all crossed in their center by a common axis around which each of them can take an independent rotational movement. If, on the outline of each of these disks, we write the ten digits which compose our numerical alphabet, by arranging, along the same vertical, a series of these digits, we can express any number in this way. For it will suffice to suppose that the first disk represents the units, the second the tens, the third the hundreds, and so on. When two numbers have been written in this way on two distinct columns, we can propose to combine them arithmetically with each other, and to obtain the result written on a third column. In general, if we have a series of columns composed of disks and which we will designate by V_0, V_1, V_2, V_3, V_4, etc., we can ask, for example, to divide the number written on column V_1, by that of column V_4, and to obtain the result on column V_7. To do this operation, it will be necessary to give the machine two distinct configurations: the first by which it is ready to execute *a division*, and the second by which we indicate the columns on which it must operate, and that where the result must be written. When this division must, for example, be followed by an addition of two numbers taken from other columns, the two configurations of the machine must change at the same time. If, on the contrary, it is a question of executing a series of operations of the same nature, then the first configuration will remain the same, the second only will change. Thus, among the arrangements that the various parts of the machine can take, two principal ones will be distinguished: 1° *the arrangement relative to the operations*, 2° *the arrangement relative to the variables*; by the latter we mean that which specifies the columns on which one must operate. As for the operations themselves, they are executed in a special apparatus called the *mill*; it is itself composed of a certain number of columns similar to those of the variables. When two numbers must be combined together, the machine begins by erasing them on the columns where they were written, that is to say, by putting *zero* on each disk in the two vertical lines where the numbers were written; then it transports them into the mill. There, the apparatus being arranged for the requested operation, this is executed, and, once it is completed, the result is itself transported to the column of the variables which will have been indicated. Thus the mill is the part of the machine which executes; the columns of the variables constitute that where the results are stored to and retrieved from. From the above, we infer that all fractional and irrational results will be represented by decimal fractions. Assuming each column is composed of forty disks, this extension would be sufficient for all the degrees of approximation that are generally needed.

F.2.6 [Cards.]

Now we will ascertain how the machine can of itself and without having recourse to manual intervention, make the successive arrangements suitable for operating. The solution of this problem has been borrowed from the Jacquard apparatus, used for manufacturing brocaded fabrics; here is how:

In fabrics, two types of threads are generally distinguished: first, the warp or longitudinal thread, then the weft or transverse thread which is guided by the instrument called a shuttle and which crosses with the warp. When we want to make brocaded fabric, we must prevent certain threads in turn from crossing with the weft, following an order determined by the nature of the design that we are trying to reproduce. Formerly, this operation was long and difficult and required that the worker, attentive to the design that he had to copy, himself order the movements that these threads should take. From there came the high price of this type of fabric, especially when threads of different colors were included. To simplify this manufacturing, Jacquard thought of connecting a lever to each group of threads that had to act together. All these levers end with rods joined together in a bundle that generally has the shape of a parallelepiped with a rectangular base; these rods are cylindrical and separated from each other by small intervals. The operation of lifting the wires will thus be reduced to moving these various lever arms properly. To do this, we take a sheet of cardboard of rectangular shape, a little larger than the section of the bundle. If we apply this sheet against the base of the bundle of levers, giving it a translational movement, they will all move at the same time, as well as the wires which correspond to them. But if the cardboard, instead of being *solid*, were perforated in the places where the ends of the levers meet it, then, in its movement the cardboard being passed through by the levers, they would remain in place.

In this way we see that it is easy to place these holes in the cardboard, so that, at a given moment, there are a certain number of levers and consequently of threads raised, while the others remain in place. Assuming that this operation is successively repeated according to a law indicated by the design that we want to execute, we can see that this design can be reproduced on the fabric. To do this, we only have to compose a series of cards according to the required law, and to arrange them properly one after the other; then by passing them over a polygonal section shaft, which at each stroke of the shuttle would present a new face, which would be transported parallel to itself against the bundle of levers, the operation of lifting the threads would be done systematically. We see that in this way we can make brocaded fabric, with a speed and precision that it was difficult to obtain before.

F.2.7 [Program Cards.]

There are arrangements similar to those we have just described, which have been introduced into the analytical engine. It contains two principal types of cards: 1° the cards of operations, by which the machine is arranged so as to execute a determined series of operations, such as additions, subtractions, multiplications and divisions; 2° the cards of variables, which indicate to the machine columns on which the results must be written. The cards being set in motion, successively arrange the various parts of the machine according to the nature of the operations to be performed, and the machine executes them at the same time by means of the mechanisms of which it is composed.

F.2.8 [Example: Simultaneous Equations.]

To better understand the matter, let us take as an example the solution of two first-degree equations with two unknowns. Let us therefore consider the following two equations where x and y are the unknowns:

$$mx + ny = d$$
$$m'x + n'y = d'$$

we deduce

$$x = (dn' - d'n)/(n'm - nm')$$

and for y an analogous value. Let us still represent by V_0, V_1, V_2 etc., the different columns which contain the numbers, and let us suppose that we have chosen the first eight columns to write the numbers which are represented by $m, n, d, m', n', d'n$ and dn', which supposes $V_0 = m$, $V_1 = n$, $V_2 = d$, $V_3 = m'$, $V_4 = n'$, $V_5 = d'$, $V_6 = n$, $V_7 = n'$.[2] The series of operations requested by the cards, and the results obtained, will be represented by the following table:

As the cards only indicate how and on which columns the machine must act, it is clear that it will still be necessary, in each particular case, to load the numerical data of the calculation. Thus, in the example we have chosen, we must first write the numerical values of m, n, d, m', n', d' in the order and on the columns indicated, after which we will make the machine execute, which will give the value of the unknown x for this particular case. To obtain the value of y, it will be necessary to do another series of operations analogous to the preceding ones. But we see that they will be reduced to four only, because the denominator of the expression of y, except the sign, is the same as that of x, and equal to $m'n - mn'$. In the preceding table we will notice that the column of operations indicates in sequence four *multiplications*, two *subtractions* and one *division*. Thus we can, if necessary, use only three cards of operations; for this it will be sufficient to introduce into the machine a device which, for example, after the first multiplication, retains the card relating to this operation, and does not allow it to advance, to be replaced by another, until this same operation has been repeated four times. In the previous example, we saw that, to find the

[2] Ed: V_7 and V_8 duplicate V_1 and V_4, apparently because of the values being erased when they are transferred to the mill, but later we read that there are unspecified ways of preventing that.

	Op. cards	Var. Cards		
Nr of the op	Symbols ind. the nature of the op.	Cols. on which ops. are to be performed	Cols. which receive results	Progress of the operations
1	$\times$	$V_2 \times V_4 =$	V_8	$= dn'$
2	$\times$	$V_3 \times V_1 =$	V_9	$= d'n$
3	$\times$	$V_4 \times V_0 =$	V_{11}	$= n'm$
4	$\times$	$V_1 \times V_3 =$	V_{12}	$= nm'$
5	$-$	$V_8 \times V_9 =$	V_{13}	$= dn' - d'n$
6	$-$	$V_{10} \times V_{11} =$	V_{14}	$= n'm - nm'$
7	$\div$	$\frac{V_{12}}{V_{13}} =$	V_{16}	$= \frac{dn'-d'n}{n'm-nm'}$

value of x, it was necessary to begin by writing the coefficients m, n, d, m', n', d' on eight columns, thus repeating n and n' twice; according to the same method, if we also wanted to calculate y, we would have had to write these same coefficients on twelve different columns. But it is possible to simplify this operation and avoid the chances of errors, which increase as the quantity of numbers to be written, before operating the machine, becomes greater. For this let us remember that any number written on a column, in order to be combined arithmetically with another number, must be erased from the column on which it is found and transported into the mill. Now, in the example we have discussed, let us take the two coefficients m and n', which must each enter into two different products, that is, m into mn' and md, n' into mn' and $n'd$. These coefficients will be written on columns V_0 and V_4. If we begin the series of operations with the product of m by n, we will erase these numbers on columns V_0 and V_4 to transport them to the mill which will multiply them together, then order the machine to write the result on column V_6 for example. But as these numbers must each still serve an operation, they will have to be written again somewhere; for this, at the same time as the mill produces their product, the machine will write them again on two columns which will be indicated to it by the cards, and as in the current case nothing prevents them from resuming their first places, they will be assumed to be written to V_0 and V_4, from where they will finally disappear, to no longer be reproduced, after they have undergone the other combinations to which they must be subjected.[3]

We can therefore see that all the operations necessary to resolve the two first degree equations in question can ultimately be represented by the following table.[4]

In order to reduce as much as possible the chances of error in writing the numerical results of the problem, they are written successively on one of the columns of the mill; then, by means of cards arranged for this purpose, these same numbers will be placed on the appropriate columns,[5] without the operator having to worry about it in any way; in this way all his attention will be focused on the simple specification of these same numbers.

From what has just been explained, we see that the set of columns of variables can be considered as a store of numbers which are accumulated there by the mill and which, obeying the orders transmitted to the machine by means of the cards, pass alternately from the mill to the store and from the store to the mill, to undergo the transformations required

[3] Ed: the numbers are always erased from the (magazine) store when they are transferred to the mill store, unless the programmer wishes to retain them, which must be specified on the variable card, but those specifics are omitted here.

[4] Ed: Here $\ldots = 0$.

[5] Ed: of the magazine store.

Columns for raw data	No. of the operation	Number of Cards	Operation	Columns acted on by each operation	Columns that receive the result	Indication of Change		Statement of results
$V_0 = m$	1	1	$\times$	$V_0 \times V_4 = V_6$	V_6	$V_0 = V_0$	$V_4 = V_4$	$V_6 = mn'$
$V_1 = n$	2	"	$\times$	$V_3 \times V_1 = V_7$	V_7	$V_1 = V_1$	$V_3 = V_3$	$V_7 = m'n$
$V_2 = d$	3	"	$\times$	$V_2 \times V_4 = V_8$	V_8	$V_2 = V_2$	$V_4 = V_4$	$V_8 = dn'$
$V_3 - m'$	4	"	$\times$	$V_5 \times V_1 = V_9$	V_9	$V_1 = V_1$	$V_5 = V_5$	$V_9 = d'n$
$V_4 = n'$	5	"	$\times$	$V_0 \times V_5 = V_{10}$	V_{10}	$V_0 = \ldots$	$V_5 = V_5$	$V_{10} = d'm$
$V_5 = d$	6	"	$\times$	$V_2 \times V_3 = V_{11}$	V_{11}	$V_2 = \ldots$	$V_3 = \ldots$	$V_{11} = dn'$
	7	2	$-$	$V_0 \times V_7 = V_{12}$	V_{12}	$V_6 = \ldots$	$V_7 = \ldots$	$V_{12} = mn' - m'n$
	8	"	$-$	$V_8 - V_9 = V_{13}$	V_{13}	$V_8 = \ldots$	$V_9 = \ldots$	$V_{13} = dn' - d'n$
	9	"	$-$	$V_{10} - V_{11} = V_{14}$	V_{14}	$V_{10} = \ldots$	$V_{11} = \ldots$	$V_{14} = d'm - dm'$
	10	3	$\div$	$\frac{V_{13}}{V_{12}} = V_{15}$	V_{15}	$V_{13} = \ldots$	$V_{12} = V_{12}$	$V_{15} = \frac{dn'-d'n}{mn'-m'n} = x$
	11	"	$\div$	$\frac{V_{14}}{V_{12}} = V_{16}$	V_{16}	$V_{12} = \ldots$	$V_{14} = \ldots$	$V_{16} = \frac{d'm-dm'}{mn'-m'n} = y$

by the nature of the calculation to be executed.

F.2.9 [Propagating Signs.]

Up to this point, the signs of the results have not yet been spoken of, and the machine would be far from perfect if it were not capable of expressing and combining positive and negative quantities. To achieve this goal, above each column, both of the mill and of the store, there is a disk similar to those which compose the columns. According to whether the figure corresponding to the number appearing below it on the column is even or odd, the latter will be considered positive or negative. Given this, here is how the signs can be combined algebraically in the machine. When a number must be transported from the store to the mill and *vice versa*, it will always be with its sign, which will be done by means of the cards, as has been said previously. Let us therefore consider two numbers placed with their respective signs in the mill and on which one must operate arithmetically. Let us suppose that it is first a question of adding them together: the cards of the operations will control the addition; if the two numbers have the same sign, one of the two will be completely erased on the column where it was written and will be added on the column which contains the other number; the machine, by means of a device, will be able to prevent during this operation any movement in the disk of signs which belongs to the column on which the addition is made, and then the result will remain with the sign which the two given numbers had. When the two numbers have two different signs, the addition which was ordered by the card changes into subtraction by means of the mechanisms which are put into play because of this difference in sign. As the subtraction can only be carried out on the larger of the two numbers, if we make sure that the disk of signs of the larger number has no movement while the smaller of these numbers is erased from its column to be subtracted from the other, there will remain a result which will have the sign of the latter, as it should be. The combinations in which algebraic subtraction takes place are analogous to the preceding ones. Let us pass to multiplication. When the two numbers to be multiplied have the same sign, the result is positive; if the two signs are different, the product must be negative. For the machine to act in accordance with this law, it will suffice to conceive that, on the column which contains the product of the two given numbers, the figures indicating their signs are added on the disc of the signs of the product; then it is clear that if the figures of the signs are either both even, or both odd, their sum will be an even number, and consequently will express a positive quantity; if, on the contrary, the two figures of the signs are one even and the other odd, their sum will be an odd number, and consequently will express a negative quantity. For division, instead of adding the digits of the disks, it will be necessary to subtract them from each other, which will give rise to results similar to the previous ones, that is to say, if these digits are both even or odd, the remainder of this subtraction will be even; it will be odd in the opposite case. When I speak of adding or subtracting from each other the numbers expressed by the digits of the signs, I only mean to advance or retrograde one of the disks by a number of divisions equal to that which is expressed by the digit of the other disk. We see therefore, from what has just been explained, that it is possible to mechanically combine the signs of the quantities, so as to obtain results in conformity with those indicated by algebra.[6]

F.2.10 [Example: Multiplying Polynomials.]

The machine is not only capable of carrying out numerical calculations which depend on a given algebraic formula, but it is also suitable for analytical calculations in which there are one or more variables to consider. For this it must be admitted that the analytical expression on which one must operate is developed according to the powers of the variable, or according to determined functions of this same variable, as would be, for example, the circular functions; it must be the same for the result which one wishes to arrive at. Now, if one supposes that above each column of the magazine one has written the powers or the functions of the variable ordered according to the law of the development which one considers, the coefficients of these different terms can be written on the lower columns which correspond to them. One will have, in this way, represented an analytical development. Now, by supposing that the position of the different terms which compose it is invariable, the problem will be reduced to calculating their coefficients according to the laws required by the nature of the question. To make the matter better understood, let us take the following example, a very simple one, in which it is a question of multiplying $(a+bx^1)$ by $(A+Bcos^1x)$. We will begin by writing x^0, x^1, cos^0x, cos^1x above the columns V_0, V_1, V_2 and V_3; then as, according to the form of the two functions to be combined, the terms which must compose the product are of the following nature, $x^0, cos^0x, x^0cos^1x, x^1cos^0x, x^1cos^1x$, we will write them above the columns V_4, V_5, V_6 and V_7. The coefficients of x^0, x^1, cos^0x, cos^1x being given, by means of the mill we will pass them over the columns V_0, V_1, V_2 and V_3. Such are the primitive data of the problem; it is now up to the machine to solve it, that is to say, to find the coefficients which must be entered on V_4, V_5, V_6, V_7. For this purpose, as the law of formation of these same coefficients is known, the law of formation of these same coefficients is known, the machine will act through the cards, in the manner indicated in the following

6 Not having had the leisure to discuss with Mr. Babbage the manner of introducing into his machine the combination of algebraic signs, I do not claim to expose here the method which he uses for this purpose; but I believed it necessary to supplement it, thinking that this writing would have been imperfect if I had not indicated one of the means which one can employ to resolve this essential point of the problem in question.

table.[7]

		Coeffs.		Op. Cards		Var. Cards				
Cols. above which are written the functions of the var		Given	To be formed	Nr. of the op.	Nature of the op.	Cols. on which ops. are to be performed	Cols. for Results	Changed cols.		Results of the ops
x^0	V_0	a	"	"	"	"	"	"		"
x^1	V_1	b	"	"	"	"	"	"		"
$\cos^0 x$	V_2	A	"	"	"	"	"	"		"
$\cos^1 x$	V_3	B	"	"	"	"	"	"		"
$x^0 \cos^0 x$	V_4		aA	1	$\times$	$V_0 \times V_2 = V_4$	V_4	$V_0 = V_0$	$V_2 = V_2$	$aA\ x^0 \cos^0 x$
$x^0 \cos^1 x$	V_5		aB	2	$\times$	$V_0 \times V_3 = V_5$	V_5	$V_0 = 0$	$V_3 = V_3$	$aB\ x^0 \cos^1 x$
$x^1 \cos^0 x$	V_6		bA	3	$\times$	$V_1 \times V_2 = V_6$	V_6	$V_1 = V_1$	$V_2 = 0$	$bA\ x^1 \cos^0 x$
$x^1 \cos^1 x$	V_7		bB	4	$\times$	$V_1 \times V_3 = V_7$	V_7	$V_1 = 0$	$V_3 = 0$	$bB\ x^1 \cos^1 x$

We must now understand that the principle developed in the preceding example can be applied in general to any type of operation that we could propose on the series submitted to calculation. It will suffice to know the law of formation of the different coefficients, to enter this law in the cards of the machine, which will then itself execute all the appropriate calculations to arrive at the proposed result. If it were, for example, a recurring series, the law of formation of the different terms being the same for all, the same operations that will be done for one of them will be repeated for the others; only there will be a change of place of the operation, that is to say that it will be done on different columns. As generally any analytical expression is capable of being expressed by a series ordered according to certain functions of the variable, we see that the machine will embrace all the analytical calculations which are ultimately reduced to the formation of the coefficients according to certain laws, and to their distribution in relation to the variables.

From what has just been explained, we can deduce this important consequence, that since the cards only indicate the nature of the operations to be performed and the columns of the variables on which they must be executed, these cards will themselves have all the generality of analysis, of which they are only a simple translation. Let us examine again some of the difficulties that the machine must overcome so that its assimilation to analysis is more complete. There are certain functions which must change nature when they pass through zero or infinity, or whose values cannot be admitted beyond these limits. When these cases occur, the machine can warn by means of a bell that the passage through zero or infinity has taken place, and then it stops, until the operator has put it back into action for some other operation that one wishes to order it to perform. If this operation is planned, the machine, instead of ringing, will arrange itself so as to present the new cards relating to the operation which must follow the passage through zero and infinity. These new cards can follow the first ones and only come into play when the two circumstances indicated previously take place.

[7] Ed: see below.

F.2.11 [Exponentiation and Loops.]

Let us consider a term of the form ab^n; since the cards are only a translation of the analytical formula, it will be necessary, in the present case, that their number be the same, whatever the value of n, that is to say whatever the number of multiplications to be made to raise b to the n-th power (we assume for the moment that n is an integer). Now, as the exponent n indicates that b must be multiplied n times by itself,[8] all these operations being of the same nature, it will be sufficient to use for this a single card of the operations, the one which orders the multiplication.

But when n is given for the particular case that we wish to calculate, it will still be necessary for the machine to limit its multiplications to the appropriate numbers. Now, here is how the operation can be arranged. The three numbers a, b and n will be written on as many distinct columns of the magazine, which we will designate by V_0, V_1, V_2, the result ab^2 will be transcribed on column V_3. When the number n has been introduced into the machine, a card will order a counter to *mark* $(n-1)$ and at the same time will execute the multiplication of b by b.[9] When this is completed, the counter will have erased a unit and will only mark $(n-2)$, and the machine will again order the number b written on column V_1 to multiply itself with the product b^2 written on column V_3, which will give b^3. Then another unit will be subtracted from the counter, and the operations indicated will be repeated until the counter only marks zero. Thus the number b^n will be written on column V_3, and the machine, continuing the course of its other operations, will order the product of b^n by a, and the calculation requested will be accomplished without the number of cards to be used having to vary with the value of n. If n is negative, the cards, instead of ordering the multiplication of a by b^n, would order the division, which is conceivable, since any number can be written with its respective sign and react in this way on the nature of the operations to be performed. Finally, if n were fractional and of the form p/q, one more column would be used to write q, and the machine would operate two devices, one to raise b to the power p,[10] the other to extract the q-th root of the number thus obtained.[11]

We can ask, for example, to multiply an expression of the form $ax^m + bx^n$ by another $Ax^p + Bx^q$, then to reduce the product to the least terms when their indices are equal. The two factors being ordered with respect to x, the general result of the multiplication would be

$$Aax^{m+p} + Abx^{p+n} + aBx^{m+q} + bBx^{n+q}$$

Up to now the operation does not present any difficulties; but let us suppose that we have $m = p$ and $n = q$, and that we want to reduce the two middle terms to a single $(Ab + Bd)x^{m+q}$. For this the cards can order that $m + q$ and $n + p$ are transported into the mill to be subtracted from each other; if the remainder is zero, as must happen in the accepted hypothesis, the mill will order other cards to bring it the coefficients Ab and aB to add them together, and thus give them as coefficient to the single term $x^{n+p} = x^{m+q}$.[12]

This example shows us how the cards can reproduce all the operations that intelligence performs to arrive at a specific result, when these operations are themselves capable of being clearly specified.

8 Ed. really, $n-1$ times, e.g. b^2 is b multiplied by b once.

9 Ed. This is not quite accurate. First V3 would have to be initialized either to b or 1. Then the index would be set and the line to be repeated, either $n-1$ or n times, would be $V3 < -V3 * V2$.

10 Ed: the text has a typo, n.

11 Ed: how to calculate roots is left unexplained.

12 Ed: this conditional 'IF' operator is left unspecified.

F.2.12 [Constants for Convergent Series.]

Let us examine again the following expression:[13]

$$2\frac{2\,4^2 6^2 8^2 10^2 ... (2n)^2}{1^2 3^2 5^2 7^2 9^2 ... (2n-1)^2 (2n+1)^2}$$

which we know becomes equal to the ratio of the circumference to the diameter, when n is infinite.[14] We can ask the machine not only to perform the calculation of this fractional expression, but also to indicate immediately that its value becomes that of the ratio of the circumference to the diameter when n is infinite, a case in which the calculation would be impossible. Let us observe that we would thus require the machine to interpret a result that is not self-evident, which is not within its remit, since it is not a thinking being. However, when the cos of $n = \infty$[15] has been predicted, a card can immediately order the substitution of the value of π (π being the ratio of the circumference to the diameter), without going through the series of calculations indicated. For this, it would be sufficient for there to be in the machine a special card intended to immediately form the number π on the column that would be indicated to it. This is the case for speaking of a third type of card that we can call *number cards*. There are certain numbers, such as that which expresses the ratio of the circumference to the diameter, the Bernoulli numbers, etc., which frequently occur in calculations.[16] In order not to be obliged to calculate them each time they are to be used, certain cards may be combined, which are intended to give them ready-made in the mill, from whence they are then placed in the columns of the store intended for them.[17] By this means, the machine will still be susceptible of the simplifications presented by the use of numerical tables.

F.2.13 [Constant Cards.]

It would also be possible to introduce, by means of these cards, the logarithms of numbers; but perhaps here it would not be the most suitable and prompt method, for the machine could lend itself to other more expeditious combinations, based on the rapidity with which the first four operations of arithmetic are executed. To give an idea of this rapidity, it will suffice to say that Mr. Babbage thinks he can, by means of his machine, make in *three minutes* the product of two numbers composed of twenty digits each.

Perhaps the immense number of cards that would be required to solve a somewhat complicated problem might seem an obstacle. But this does not seem to be the case: the number of cards that can be used has no limits. There are certain fabrics that require no less than *twenty thousand* cards to be made, and this quantity can certainly be greatly exceeded.

Summarizing what has just been said about the analytical engine, we can conclude that it is based on two principles: the first, which consists in the fact that all arithmetic calculation ultimately depends on four principal operations, addition, subtraction, multiplication and division; the second, in that all analytical calculation can be reduced to calculating the coefficients of the different terms of a series. If this last principle is true, all the operations

[13] Ed: Wallis' very slowly converging approximation for π.

[14] Ed: that is, π.

[15] Ed: sic, 'cos' should be 'case', a printer's typo blithely translated (faithfully and absurdly) by Ada Lovelace (who does not herself introduce a typo here) in Menabrea 1843a,b and often reprinted without comment, e.g. in Babbage 1961. Silently corrected in Bowden 1953. Noticed in Hyman 1982, Stein 1985 and subsequently.

[16] Ed: Thus the example of tabulated Bernoulli numbers is already suggested here. However Babbage had already been considering them as an example for the engine since 1835.

[17] Ed: The mill processes a card for a constant by first loading the value into one of its own columns (registers) and only then transfers it to the magazine.

of analysis are within the domain of the machine. On the other hand, the use of cards offers a generality equal to that of algebraic formulas, because, just as a formula of this kind simply indicates the nature and order of the operations to be carried out to arrive at a determined result, the cards only command the machine to carry out these same operations; but in order for the mechanisms to act, it is still necessary to introduce in each particular case the numerical data of the problem. Thus the same series of cards will serve for all questions of the same nature, in which only the numerical data will be changed. In this way, the cards are only a translation of the algebraic formulas, or rather, another form of analytical composition.

As the machine has its own mode of action, in each particular case it will be necessary to arrange the series of calculations in accordance with the means it possesses; because such a process which would be extremely easy for a calculator,[18] would become long and complicated for the machine and *vice versa*.

Considered from the most general point of view, the essential object of the machine being to calculate, according to the laws dictated to it, the value of numerical coefficients which it must then distribute suitably over the columns which represent the variables, it follows that the interpretation of the formulas and the results is outside its capabilities, unless however this interpretation is itself capable of being expressed by means of the symbols which it uses. Thus it is not itself a being which thinks, but one can consider it as the entity which executes the conceptions of intelligence. The cards receive the imprint of these conceptions, and transmit to the different mechanisms of which the machine is composed the orders necessary to act. Once the machine is built, the difficulty will therefore fall on the making of the cards; but as these are only the translation of algebraic formulas, by means of simple notations it will be easy to entrust their execution to a worker. Thus all intellectual work will be limited to the preparation of formulae, which must be suitable for calculation by the machine.

F.2.14 [Usefulness.]

Now, assuming that such a machine is feasible, one may ask what will be its use? In summary, it will present the following advantages: 1° Accuracy; it is known that numerical calculations are generally the pitfall of the solution of problems, because errors easily slip in, and it is not always easy to recognize them. Now, the machine, by the very nature of its mode of action, which does not require manual intervention during the course of its operations, offers every kind of guarantee concerning accuracy; moreover, it carries its control with itself, because at the end of each operation it gives printed, not only the result, but also the numerical data of the question, so that it is easy to verify if this question has been posed with exactitude. 2° Efficiency: to be convinced of this, it will suffice to recall that a multiplication of two numbers each composed of 20 digits requires at most three minutes. Moreover, when a long series of identical calculations must be made, such as those required for the formation of numerical tables, the machine can be used in such a way as to give several results at once, which will greatly shorten the whole of the operations. 3° Economy of intelligent labor: a simple arithmetic calculation requires the assistance of a person with some capacity; when one moves on to more complicated calculations, and one wishes to use algebraic formulas in particular cases, one must already possess knowledge which presupposes preliminary mathematical studies of some extent. Now, the machine, being able to perform all these purely mechanical operations itself, saves the labor of the intelligent, who can be employed more usefully. Thus the machine can be considered as

[18] Ed: A human calculator.

a true factory of numbers,[19] which will lend its assistance to the sciences and useful arts which are based on numbers. Now, who could foresee the consequences of such an invention? Indeed, how many valuable observations remain useless to the progress of science, because there are not sufficient labourers to calculate their implications! How much discouragement does the prospect of a long and arid calculation not cast into the soul of the man of genius who asks only for time to meditate and who sees it robbed of him by the drudgery of the operations! And yet it is by the laborious path of analysis that he must arrive at the truth; but he cannot follow it without being guided by numbers, because without numbers it is not possible to lift the veil which covers the mysteries of nature. Thus the thought of creating an instrument capable of augmenting the weakness of man in such research, is an idea which, once realized, would mark a glorious epoch in the history of science. All the parts, all the cogs which make up this immense machine have been conceived, their action has been studied, but they have not yet been assembled. The confidence which Mr Babbage's genius must inspire confirms the hope that this enterprise will be crowned with success; paying homage to the intelligence which directs it, let us complete the work.

[19] This is perhaps the source of Babbage's phrase 'Enchantress of Number', most likely a reference to his engine.

F.3 1843. Sketch of the Analytical Engine Invented by Charles Babbage, esq. By L. F. Menabrea, of Turin, Officer of the Military Engineers [ed. Lovelace].

Translated, edited and annotated from Menabrea 1842 by Ada Lovelace.[20] See section F.1 for the original French text and section F.2 for a translation without emendations and alterations.

[20] Menabrea 1843a,b.

666

Article XXIX.

Sketch of the Analytical Engine invented by Charles Babbage *Esq. By* L. F. Menabrea, *of Turin, Officer of the Military Engineers.*

[From the *Bibliothèque Universelle de Génève*, No. 82. October 1842.]

[BEFORE submitting to our readers the translation of M. Menabrea's memoir 'On the Mathematical Principles of the Analytical Engine' invented by Mr. Babbage, we shall present to them a list of the printed papers connected with the subject, and also of those relating to the Difference Engine by which it was preceded.

For information on Mr. Babbage's "*Difference* Engine," which is but slightly alluded to by M. Menabrea, we refer the reader to the following sources:—

1. Letter to Sir Humphry Davy, Bart., P.R.S., on the Application of Machinery to Calculate and Print Mathematical Tables. By Charles Babbage, Esq., F.R.S. London, July 1822. Reprinted, with a Report of the Council of the Royal Society, by order of the House of Commons, May 1823.

2. On the Application of Machinery to the Calculation of Astronomical and Mathematical Tables. By Charles Babbage, Esq.—Memoirs of the Astronomical Society, vol. i. part 2. London, 1822.

3. Address to the Astronomical Society by Henry Thomas Colebrooke, Esq., F.R.S., President, on presenting the first Gold Medal of the Society to Charles Babbage, Esq., for the invention of the Calculating Engine.—Memoirs of the Astronomical Society. London, 1822.

4. On the Determination of the General Term of a New Class of Infinite Series. By Charles Babbage, Esq.—Transactions of the Cambridge Philosophical Society.

5. On Mr. Babbage's New Machine for Calculating and Printing Mathematical Tables.—Letter from Francis Baily, Esq., F.R.S., to M. Schumacher. No. 46, Astronomische Nachrichten. Reprinted in the Philosophical Magazine, May 1824.

6. On a Method of expressing by Signs the Action of Ma-

chinery. By Charles Babbage, Esq.—Philosophical Transactions. London, 1826.

7. On Errors common to many Tables of Logarithms. By Charles Babbage, Esq.—Memoirs of the Astronomical Society, London, 1827.

8. Report of the Committee appointed by the Council of the Royal Society to consider the subject referred to in a communication received by them from the Treasury respecting Mr. Babbage's Calculating Engine, and to report thereon. London, 1829.

9. Economy of Manufactures, chap. xx. 8vo. London, 1832.

10. Article on Babbage's Calculating Engine.—Edinburgh Review, July 1834. No. 120. vol. lix.

The present state of the Difference Engine, which has always been the property of Government, is as follows:—The drawings are nearly finished, and the mechanical notation of the whole, recording every motion of which it is susceptible, is completed. A part of that Engine, comprising sixteen figures, arranged in three orders of differences, has been put together, and has frequently been used during the last eight years. It performs its work with absolute precision. This portion of the Difference Engine, together with all the drawings, are at present deposited in the Museum of King's College, London.

Of the ANALYTICAL ENGINE, which forms the principal object of the present memoir, we are not aware that any notice has hitherto appeared, except a Letter from the Inventor to M. Quetelet, Secretary to the Royal Academy of Sciences at Brussels, by whom it was communicated to that body. We subjoin a translation of this Letter, which was itself a translation of the original, and was not intended for publication by its author.

Royal Academy of Sciences at Brussels. General Meeting of the 7th and 8th of May, 1835.

"A Letter from Mr. Babbage announces that he has for six months been engaged in making the drawings of a new calculating machine of far greater power than the first.

"'I am myself astonished,' says Mr. Babbage, 'at the power I have been enabled to give to this machine; a year ago I should not have believed this result possible. This machine is intended to contain a hundred variables (or numbers susceptible of changing); each of these numbers may consist of twenty-five figures, $v_1, v_2, \ldots\ldots v_n$ being any numbers whatever, n being less than a hundred; if $f(v_1, v_2, v_3, \ldots v_n)$ be any given function which can be formed by addition, subtraction, multiplication, division, extraction of roots, or elevation to powers, the machine will calculate its numerical value; it will afterwards substitute this value in the place of v, or of any other variable, and will calculate this second function with respect to v. It will reduce to tables almost all equations of finite differences. Let us suppose that we have observed a thousand values of a, b, c, d, and that we wish to calculate them by the formula $p = \sqrt{\frac{a+b}{c\,d}}$, the machine must be set to calculate the formula; the first series of the values of a, b, c, d must be adjusted to it; it will then calculate them, print them, and reduce them to zero; lastly, it will ring a bell to give notice that a new set of constants must be inserted. When there exists a relation between any number of successive coefficients of a series, provided it can be expressed as has already been said, the machine will calculate them and make their terms known in succession; and it may afterwards be disposed so as to find the value of the series for all the values of the variable.'

"Mr. Babbage announces, in conclusion, 'that the greatest difficulties of the invention have already been surmounted, and that the plans will be finished in a few months.'"

In the Ninth Bridgewater Treatise, Mr. Babbage has employed several arguments deduced from the Analytical Engine, which afford some idea of its powers. See Ninth Bridgewater Treatise, 8vo, second edition. London, 1834.

Some of the numerous drawings of the Analytical Engine have been engraved on wooden blocks, and from these (by a mode contrived by Mr. Babbage) various stereotype plates have been taken. They comprise—

1. Plan of the figure wheels for one method of adding numbers.
2. Elevation of the wheels and axis of ditto.
3. Elevation of framing only of ditto.
4. Section of adding wheels and framing together.
5. Section of the adding wheels, sign wheels and framing complete.
6. Impression from the original wood block.

7. Impressions from a stereotype cast of No. 6, with the letters and signs inserted. Nos. 2, 3, 4 and 5 were stereotypes taken from this.

8. Plan of adding wheels and of long and short pinions, by means of which *stepping* is accomplished.

N.B. This process performs the operation of multiplying or dividing a number by any power of ten.

9. Elevation of long pinions in the position for addition.

10. Elevation of long pinions in the position for stepping.

11. Plan of mechanism for carrying the tens (by anticipation), connected with long pinions.

12. Section of the chain of wires for anticipating carriage.

13. Sections of the elevation of parts of the preceding carriage.

All these were executed about five years ago. At a later period (August 1840) Mr. Babbage caused one of his general plans (No. 25) of the whole Analytical Engine to be lithographed at Paris.

Although these illustrations have not been published, on account of the time which would be required to describe them, and the rapid succession of improvements made subsequently, yet copies have been freely given to many of Mr. Babbage's friends, and were in August 1838 presented at Newcastle to the British Association for the Advancement of Science, and in August 1840 to the Institute of France through M. Arago, as well as to the Royal Academy of Turin through M. Plana.—EDITOR.]

THOSE labours which belong to the various branches of the mathematical sciences, although on first consideration they seem to be the exclusive province of intellect, may, nevertheless, be divided into two distinct sections; one of which may be called the mechanical, because it is subjected to precise and invariable laws, that are capable of being expressed by means of the operations of matter; while the other, demanding the intervention of reasoning, belongs more specially to the domain of the understanding. This admitted, we may propose to execute, by means of machinery, the mechanical branch of these labours, reserving for pure intellect that which depends on the reasoning faculties. Thus the rigid exactness of those laws which regulate numerical calculations must frequently have suggested the employment of material instruments, either for executing the whole of such calculations or for abridging them; and thence have arisen several inventions having this object in view, but which have in general but partially attained it. For instance, the much-admired machine of Pascal is now simply an object of curiosity, which, whilst it displays the powerful intellect of its inventor, is yet of little utility in itself. Its powers extended no further than the execution of the four* first operations of arithmetic, and indeed were in reality confined to that of the two first, since multiplication and division were the result of a series of additions and subtractions. The chief drawback hitherto on most of such machines is, that they require the continual intervention of a human agent to regulate their movements, and thence arises a source of errors; so that, if their use has not become general for large numerical calculations, it is because they have not in fact resolved the double problem which the question presents, that of *correctness* in the results, united with *œconomy* of time.

Struck with similar reflections, Mr. Babbage has devoted some years to the realization of a gigantic idea. He proposed to himself nothing less than the construction of a machine capable of executing not merely arithmetical calculations, but even all those of analysis, if their laws are known. The imagination is at first astounded at the idea of such an undertaking; but the more calm reflection we bestow on it, the less impossible does success appear, and it is felt that it may depend on the discovery of some principle so general, that if applied to machinery, the latter may be capable of mechanically translating the operations which may be indicated to it by algebraical notation. The illustrious inventor having been kind enough to communicate to me

* This remark seems to require further comment, since it is in some degree calculated to strike the mind as being at variance with the subsequent passage (page 675), where it is explained that *an engine which can effect these four operations* can in fact effect *every species of calculation*. The apparent discrepancy is stronger too in the translation than in the original, owing to its being impossible to render precisely into the English tongue all the niceties of distinction which the French idiom happens to admit of in the phrases used for the two passages we refer to. The explanation lies in this: that in the one case the execution of these four operations is the *fundamental starting-point*, and the object proposed for attainment by the machine is the *subsequent combination of these* in every possible variety; whereas in the other case the execution of some *one* of these four operations, selected at pleasure, is the *ultimatum*, the sole and utmost result that can be proposed for attainment by the machine referred to, and which result it cannot any further combine or work upon. The one *begins* where the other *ends*. Should this distinction not now appear perfectly clear, it will become so on perusing the rest of the Memoir, and the Notes that are appended to it.—Note by Translator.

some of his views on this subject during a visit he made at Turin, I have, with his approbation, thrown together the impressions they have left on my mind. But the reader must not expect to find a description of Mr. Babbage's engine; the comprehension of this would entail studies of much length; and I shall endeavour merely to give an insight into the end proposed, and to develope the principles on which its attainment depends.

I must first premise that this engine is entirely different from that of which there is a notice in the 'Treatise on the Œconomy of Machinery,' by the same author. But as the latter gave rise* to the idea of the engine in question, I consider it will be a useful preliminary briefly to recall what were Mr. Babbage's first essays, and also the circumstances in which they originated.

It is well known that the French government, wishing to promote the extension of the decimal system, had ordered the construction of logarithmical and trigonometrical tables of enormous extent. M. de Prony, who had been entrusted with the direction of this undertaking, divided it into three sections, to each of which were appointed a special class of persons. In the first section the formulæ were so combined as to render them subservient to the purposes of numerical calculation; in the second, these same formulæ were calculated for values of the variable, selected at certain successive distances; and under the third section, comprising about eighty individuals, who were most of them only acquainted with the two first rules of arithmetic, the values which were intermediate to those calculated by the second section were interpolated by means of simple additions and subtractions.

An undertaking similar to that just mentioned having been entered upon in England, Mr. Babbage conceived that the operations performed under the third section might be executed by a machine; and this idea he realized by means of mechanism,

* The idea that the one engine is the offspring and has grown out of the other, is an exceedingly natural and plausible supposition, until reflection reminds us that no *necessary* sequence and connexion need exist between two such inventions, and that they *may* be wholly independent. M. Menabrea has shared this idea in common with persons who have not his profound and accurate insight into the nature of either engine. In Note A. (see the Notes at the end of the Memoir) it will be found sufficiently explained, however, that this supposition is unfounded. M. Menabrea's opportunities were by no means such as could be adequate to afford him information on a point like this, which would be naturally and almost unconsciously *assumed*, and would scarcely suggest any inquiry with reference to it.—Note by Translator.

which has been in part put together, and to which the name Difference Engine is applicable, on account of the principle upon which its construction is founded. To give some notion of this, it will suffice to consider the series of whole square numbers, 1, 4, 9, 16, 25, 36, 49, 64, &c. By subtracting each of these from the succeeding one, we obtain a new series, which we will name the Series of First Differences, consisting of the numbers 3, 5, 7, 9, 11, 13, 15, &c. On subtracting from each of these the preceding one, we obtain the Second Differences, which are all constant and equal to 2. We may represent this succession of operations, and their results, in the following table:—

	A. Column of Square Numbers.	B. First Differences.	C. Second Differences.
	1		
		3	
	4		2 *b*
		5	
a	9		2 *d*
		7	
c	16		2
		9	
	25		2
		11	
	36		

From the mode in which the two last columns B and C have been formed, it is easy to see that if, for instance, we desire to pass from the number 5 to the succeeding one 7, we must add to the former the constant difference 2; similarly, if from the square number 9 we would pass to the following one 16, we must add to the former the difference 7, which difference is in other words the preceding difference 5, plus the constant difference 2; or again, which comes to the same thing, to obtain 16 we have only to add together the three numbers 2, 5, 9, placed obliquely in the direction *a b*. Similarly, we obtain the number 25 by summing up the three numbers placed in the oblique direction *d c*: commencing by the addition 2 + 7, we have the first difference 9 consecutively to 7; adding 16 to the 9 we have the square 25. We see then that the three numbers 2, 5, 9 being given, the whole series of successive square numbers, and that of their first differences likewise, may be obtained by means of simple additions.

Now, to conceive how these operations may be reproduced by a machine, suppose the latter to have three dials, designated as A, B, C, on each of which are traced, say a thousand divisions, by way of example, over which a needle shall pass. The two dials, C, B, shall have in addition a registering hammer, which is to give a number of strokes equal to that of the divisions indicated by the needle. For each stroke of the registering hammer of the dial C, the needle B shall advance one division;

similarly, the needle A shall advance one division for every stroke of the registering hammer of the dial B. Such is the general disposition of the mechanism.

This being understood, let us at the beginning of the series of operations we wish to execute, place the needle C on the division 2, the needle B on the division 5, and the needle A on the division 9. Let us allow the hammer of the dial C to strike; it will strike twice, and at the same time the needle B will pass over two divisions. The latter will then indicate the number 7, which succeeds the number 5 in the column of first differences. If we now permit the hammer of the dial B to strike in its turn, it will strike seven times, during which the needle A will advance seven divisions; these added to the nine already marked by it, will give the number 16, which is the square number consecutive to 9. If we now recommence these operations, beginning with the needle C, which is always to be left on the division 2, we shall perceive that by repeating them indefinitely, we may successively reproduce the series of whole square numbers by means of a very simple mechanism.

The theorem on which is based the construction of the machine we have just been describing, is a particular case of the following more general theorem: that if in any polynomial whatever, the highest power of whose variable is m, this same variable be increased by equal degrees; the corresponding values of the polynomial then calculated, and the first, second, third, &c. differences of these be taken (as for the preceding series ot squares); the mth differences will all be equal to each other. So that, in order to reproduce the series of values of the polynomial by means of a machine analogous to the one above described, it is sufficient that there be $(m + 1)$ dials, having the mutual relations we have indicated. As the differences may be either positive or negative, the machine will have a contrivance for either advancing or retrograding each needle, according as the number to be algebraically added may have the sign *plus* or *minus*.

If from a polynomial we pass to a series having an infinite number of terms, arranged according to the ascending powers of the variable, it would at first appear, that in order to apply the machine to the calculation of the function represented by such a series, the mechanism must include an infinite number of dials, which would in fact render the thing impossible. But in many cases the difficulty will disappear, if we observe that for a great number of functions the series which represent them may be rendered convergent; so that, according to the degree of approximation desired, we may limit ourselves to the calculation of a certain number of terms of the series, neglecting the rest. By this method the question is reduced to the primitive case of a finite polynomial. It is thus that we can calculate the succession of the logarithms of numbers. But since, in this particular instance, the terms which had been originally neglected receive increments in a ratio so continually increasing for equal increments of the variable, that the degree of approximation required would ultimately be affected, it is necessary, at certain intervals, to calculate the value of the function by different methods, and then respectively to use the results thus obtained, as data whence to deduce, by means of the machine, the other intermediate values. We see that the machine here performs the office of the third section of calculators mentioned in describing the tables computed by order of the French government, and that the end originally proposed is thus fulfilled by it.

Such is the nature of the first machine which Mr. Babbage conceived. We see that its use is confined to cases where the numbers required are such as can be obtained by means of simple additions or subtractions; that the machine is, so to speak, merely the expression of one* particular theorem of analysis; and that, in short, its operations cannot be extended so as to embrace the solution of an infinity of other questions included within the domain of mathematical analysis. It was while contemplating the vast field which yet remained to be traversed, that Mr. Babbage, renouncing his original essays, conceived the plan of another system of mechanism whose operations should themselves possess all the generality of algebraical notation, and which, on this account, he denominates the *Analytical Engine.*

Having now explained the state of the question, it is time for me to develope the principle on which is based the construction of this latter machine. When analysis is employed for the solution of any problem, there are usually two classes of operations to execute: firstly, the numerical calculation of the various coefficients; and secondly, their distribution in relation to the quantities affected by them. If, for example, we have to obtain

* See Note A.

the product of two binomials $(a + b x)$ $(m + n x)$, the result will be represented by $a m + (a n + b m)\ x + b n x^2$, in which expression we must first calculate $a m$, $a n$, $b m$, $b n$; then take the sum of $a n + b m$; and lastly, respectively distribute the coefficients thus obtained, amongst the powers of the variable. In order to reproduce these operations by means of a machine, the latter must therefore possess two distinct sets of powers: first, that of executing numerical calculations; secondly, that of rightly distributing the values so obtained.

But if human intervention were necessary for directing each of these partial operations, nothing would be gained under the heads of correctness and œconomy of time; the machine must therefore have the additional requisite of executing by itself all the successive operations required for the solution of a problem proposed to it, when once the *primitive numerical data* for this same problem have been introduced. Therefore, since from the moment that the nature of the calculation to be executed or of the problem to be resolved have been indicated to it, the machine is, by its own intrinsic power, of itself to go through all the intermediate operations which lead to the proposed result, it must exclude all methods of trial and guess-work, and can only admit the direct processes of calculation*.

It is necessarily thus; for the machine is not a thinking being, but simply an automaton which acts according to the laws imposed upon it. This being fundamental, one of the earliest researches its author had to undertake, was that of finding means for effecting the division of one number by another without using the method of guessing indicated by the usual rules of arithmetic. The difficulties of effecting this combination were far from being among the least; but upon it depended the success of every other. Under the impossibility of my here explaining the process through which this end is attained, we must limit ourselves to admitting that the four first operations of arithmetic, that is addition, subtraction, multiplication and division, can be performed in a direct manner through the intervention of the machine. This granted, the machine is thence capable of performing every species of numerical calculation, for all such calculations ultimately resolve themselves into the four operations

* This must not be understood in too unqualified a manner. The engine is capable, under certain circumstances, of feeling about to discover which of two or more possible contingencies has occurred, and of then shaping its future course accordingly.—NOTE BY TRANSLATOR.

we have just named. To conceive how the machine can now go through its functions according to the laws laid down, we will begin by giving an idea of the manner in which it materially represents numbers.

Let us conceive a pile or vertical column consisting of an indefinite number of circular discs, all pierced through their centres by a common axis, around which each of them can take an independent rotatory movement. If round the edge of each of these discs are written the ten figures which constitute our numerical alphabet, we may then, by arranging a series of these figures in the same vertical line, express in this manner any number whatever. It is sufficient for this purpose that the first disc represent units, the second tens, the third hundreds, and so on. When two numbers have been thus written on two distinct columns, we may propose to combine them arithmetically with each other, and to obtain the result on a third column. In general, if we have a series of columns* consisting of discs, which columns we will designate as V_0, V_1, V_2, V_3, V_4, &c., we may require, for instance, to divide the number written on the column V_1 by that on the column V_4, and to obtain the result on the column V_7. To effect this operation, we must impart to the machine two distinct arrangements; through the first it is prepared for executing *a division*, and through the second the columns it is to operate on are indicated to it, and also the column on which the result is to be represented. If this division is to be followed, for example, by the addition of two numbers taken on other columns, the two original arrangements of the machine must be simultaneously altered. If, on the contrary, a series of operations of the same nature is to be gone through, then the first of the original arrangements will remain, and the second alone must be altered. Therefore, the arrangements that may be communicated to the various parts of the machine, may be distinguished into two principal classes:

First, that relative to the *Operations*.

Secondly, that relative to the *Variables*.

By this latter we mean that which indicates the columns to be operated on. As for the operations themselves, they are executed by a special apparatus, which is designated by the name of *mill*, and which itself contains a certain number of columns, similar to those of the Variables. When two numbers are to be

* See Note B.

combined together, the machine commences by effacing them from the columns where they are written, that is it places *zero* * on every disc of the two vertical lines on which the numbers were represented; and it transfers the numbers to the mill. There, the apparatus having been disposed suitably for the required operation, this latter is effected, and, when completed, the result itself is transferred to the column of Variables which shall have been indicated. Thus the mill is that portion of the machine which works, and the columns of Variables constitute that where the results are represented and arranged. After the preceding explanations, we may perceive that all fractional and irrational results will be represented in decimal fractions. Supposing each column to have forty discs, this extension will be sufficient for all degrees of approximation generally required.

It will now be inquired how the machine can of itself, and without having recourse to the hand of man, assume the successive dispositions suited to the operations. The solution of this problem has been taken from Jacquard's apparatus †, used for the manufacture of brocaded stuffs, in the following manner:—

Two species of threads are usually distinguished in woven stuffs; one is the *warp* or longitudinal thread, the other the *woof* or transverse thread, which is conveyed by the instrument called the shuttle, and which crosses the longitudinal thread or warp. When a brocaded stuff is required, it is necessary in turn to prevent certain threads from crossing the woof, and this according to a succession which is determined by the nature of the design that is to be reproduced. Formerly this process was lengthy and difficult, and it was requisite that the workman, by attending to the design which he was to copy, should himself regulate the movements the threads were to take. Thence arose the high price of this description of stuffs, especially if threads of various colours entered into the fabric. To simplify this manufacture, Jacquard devised the plan of connecting each group of threads that were to act together, with a distinct lever belonging exclusively to that group. All these levers terminate in rods, which are united together in one bundle, having usually the form of a parallelopiped with a rectangular base. The rods are cylindrical, and are separated from each other by small in-

* Zero is not *always* substituted when a number is transferred to the mill. This is explained further on in the memoir, and still more fully in Note D.—NOTE BY TRANSLATOR.

† See Note C.

tervals. The process of raising the threads is thus resolved into that of moving these various lever-arms in the requisite order. To effect this, a rectangular sheet of pasteboard is taken, somewhat larger in size than a section of the bundle of lever-arms. If this sheet be applied to the base of the bundle, and an advancing motion be then communicated to the pasteboard, this latter will move with it all the rods of the bundle, and consequently the threads that are connected with each of them. But if the pasteboard, instead of being plain, were pierced with holes corresponding to the extremities of the levers which meet it, then, since each of the levers would pass through the pasteboard during the motion of the latter, they would all remain in their places. We thus see that it is easy so to determine the position of the holes in the pasteboard, that, at any given moment, there shall be a certain number of levers, and consequently of parcels of threads, raised, while the rest remain where they were. Supposing this process is successively repeated according to a law indicated by the pattern to be executed, we perceive that this pattern may be reproduced on the stuff. For this purpose we need merely compose a series of cards according to the law required, and arrange them in suitable order one after the other; then, by causing them to pass over a polygonal beam which is so connected as to turn a new face for every stroke of the shuttle, which face shall then be impelled parallelly to itself against the bundle of lever-arms, the operation of raising the threads will be regularly performed. Thus we see that brocaded tissues may be manufactured with a precision and rapidity formerly difficult to obtain.

Arrangements analogous to those just described have been introduced into the Analytical Engine. It contains two principal species of cards: first, Operation cards, by means of which the parts of the machine are so disposed as to execute any determinate series of operations, such as additions, subtractions, multiplications, and divisions; secondly, cards of the Variables, which indicate to the machine the columns on which the results are to be represented. The cards, when put in motion, successively arrange the various portions of the machine according to the nature of the processes that are to be effected, and the machine at the same time executes these processes by means of the various pieces of mechanism of which it is constituted.

In order more perfectly to conceive the thing, let us select as an example the resolution of two equations of the first degree with two unknown quantities. Let the following be the two equations, in which x and y are the unknown quantities:—

$$\begin{cases} m x + n y = d \\ m' x + n' y = d'. \end{cases}$$

We deduce $x = \dfrac{d n' - d' n}{n' m - n m'}$, and for y an analogous expression. Let us continue to represent by V_0, V_1, V_2, &c. the different columns which contain the numbers, and let us suppose that the first eight columns have been chosen for expressing on them the numbers represented by m, n, d, m', n', d', n and n', which implies that $V_0 = m$, $V_1 = n$, $V_2 = d$, $V_3 = m'$, $V_4 = n'$, $V_5 = d'$, $V_6 = n$, $V_7 = n'$.

The series of operations commanded by the cards, and the results obtained, may be represented in the following table:—

Number of the operations.	Operation-cards. Symbols indicating the nature of the operations.	Cards of the variables. Columns on which operations are to be performed.	Cards of the variables. Columns which receive results of operations.	Progress of the operations.
1	$\times$	$V_2 \times V_4 =$	V_8	$= d n'$
2	$\times$	$V_5 \times V_1 =$	V_9	$= d' n$
3	$\times$	$V_4 \times V_0 =$	V_{10}	$= n' m$
4	$\times$	$V_1 \times V_3 =$	V_{11}	$= n m'$
5	$-$	$V_8 - V_9 =$	V_{12}	$= d n' - d' n$
6	$-$	$V_{10} - V_{11} =$	V_{13}	$= n' m - n m'$
7	$\div$	$\dfrac{V_{12}}{V_{13}} =$	V_{14}	$= x = \dfrac{d n' - d' n}{n' m - n m'}$

Since the cards do nothing but indicate in what manner and on what columns the machine shall act, it is clear that we must still, in every particular case, introduce the numerical data for the calculation. Thus, in the example we have selected, we must previously inscribe the numerical values of m, n, d, m', n', d', in the order and on the columns indicated, after which the machine when put in action will give the value of the unknown quantity x for this particular case. To obtain the value of y, another series of operations analogous to the preceding must be performed. But we see that they will be only four in number, since the denominator of the expression for y, excepting the sign, is the same as that for x, and equal to $n' m - n m'$. In the preceding table it will be remarked that the column for operations indicates four successive *multiplications*, two *subtractions*, and one *division*. Therefore, if desired, we need only use three operation cards; to manage which, it is sufficient to introduce into the machine an apparatus which shall, after the first multiplication, for instance, retain the card which relates to this operation, and not allow it to advance so as to be replaced by another one, until after this same operation shall have been four times repeated. In the preceding example we have seen, that to find the value of x we must begin by writing the coefficients m, n, d, m', n', d', upon eight columns, thus repeating n and n' twice. According to the same method, if it were required to calculate y likewise, these coefficients must be written on twelve different columns. But it is possible to simplify this process, and thus to diminish the chances of errors, which chances are greater, the larger the number of the quantities that have to be inscribed previous to setting the machine in action. To understand this simplification, we must remember that every number written on a column must, in order to be arithmetically combined with another number, be effaced from the column on which it is, and transferred to the *mill*. Thus, in the example we have discussed, we will take the two coefficients m and n', which are each of them to enter into *two* different products, that is m into $m n'$ and $m d'$, n' into $m n'$ and $n' d$. These coefficients will be inscribed on the columns V_0 and V_4. If we commence the series of operations by the product of m into n', these numbers will be effaced from the columns V_0 and V_4, that they may be transferred to the mill, which will multiply them into each other, and will then command the machine to represent the result, say on the column V_6. But as these numbers are each to be used again in another operation, they must again be inscribed somewhere; therefore, while the mill is working out their product, the machine will inscribe them anew on any two columns that may be indicated to it through the cards; and, as in the actual case, there is no reason why they should not resume their former places, we will suppose them again inscribed on V_0 and V_4, whence in short they would not finally disappear, to be reproduced no more, until they should have gone through all the combinations in which they might have to be used.

We see, then, that the whole assemblage of operations requisite for resolving the two* above equations of the first degree, may be definitively represented in the following table:—

* See Note D.

Columns on which are inscribed the primitive data.	Number of the operations.	Cards of the operations.		Variable cards.			Statement of results.
		Number of the Operation cards.	Nature of each operation.	Columns acted on by each operation.	Columns that receive the result of each operation.	Indication of change of value on any column.	
$^1V_0 = m$	1	1	×	$^1V_0 \times {}^1V_4 =$	$^1V_6$	$\left\{\begin{matrix} ^1V_0 = {}^1V_0 \\ ^1V_4 = {}^1V_4 \end{matrix}\right\}$	$^1V_6 = m\,n'$
$^1V_1 = n$	2	„	×	$^1V_3 \times {}^1V_1 =$	$^1V_7$	$\left\{\begin{matrix} ^1V_3 = {}^1V_3 \\ ^1V_1 = {}^1V_1 \end{matrix}\right\}$	$^1V_7 = m'\,n$
$^1V_2 = d$	3	„	×	$^1V_2 \times {}^1V_4 =$	$^1V_8$	$\left\{\begin{matrix} ^1V_2 = {}^1V_2 \\ ^1V_4 = {}^0V_4 \end{matrix}\right\}$	$^1V_8 = d\,n'$
$^1V_3 = m'$	4	„	×	$^1V_5 \times {}^1V_1 =$	$^1V_9$	$\left\{\begin{matrix} ^1V_5 = {}^1V_5 \\ ^1V_1 = {}^0V_1 \end{matrix}\right\}$	$^1V_9 = d'\,n$
$^1V_4 = n'$	5	„	×	$^1V_0 \times {}^1V_5 =$	$^1V_{10}$	$\left\{\begin{matrix} ^1V_0 = {}^0V_0 \\ ^1V_5 = {}^0V_5 \end{matrix}\right\}$	$^1V_{10} = d'\,m$
$^1V_5 = d'$	6	„	×	$^1V_2 \times {}^1V_3 =$	$^1V_{11}$	$\left\{\begin{matrix} ^1V_2 = {}^0V_2 \\ ^1V_3 = {}^0V_3 \end{matrix}\right\}$	$^1V_{11} = d\,m'$
	7	2	−	$^1V_6 - {}^1V_7 =$	$^1V_{12}$	$\left\{\begin{matrix} ^1V_6 = {}^0V_6 \\ ^1V_7 = {}^0V_7 \end{matrix}\right\}$	$^1V_{12} = m\,n' - m'\,n$
	8	„	−	$^1V_8 - {}^1V_9 =$	$^1V_{13}$	$\left\{\begin{matrix} ^1V_8 = {}^0V_8 \\ ^1V_9 = {}^0V_9 \end{matrix}\right\}$	$^1V_{13} = d\,n' - d'\,n$
	9	„	−	$^1V_{10} - {}^1V_{11} =$	$^1V_{14}$	$\left\{\begin{matrix} ^1V_{10} = {}^0V_{10} \\ ^1V_{11} = {}^0V_{11} \end{matrix}\right\}$	$^1V_{14} = d'\,m - d\,m'$
	10	3	÷	$^1V_{13} \div {}^1V_{12} =$	$^1V_{15}$	$\left\{\begin{matrix} ^1V_{13} = {}^0V_{13} \\ ^1V_{12} = {}^1V_{12} \end{matrix}\right\}$	$^1V_{15} = \dfrac{d\,n' - d'\,n}{m\,n' - m'\,n} = x$
	11	„	÷	$^1V_{14} \div {}^1V_{12} =$	$^1V_{16}$	$\left\{\begin{matrix} ^1V_{14} = {}^0V_{14} \\ ^1V_{12} = {}^0V_{12} \end{matrix}\right\}$	$^1V_{16} = \dfrac{d'\,m - d\,m'}{m\,n' - m'\,n} = y$
1	2	3	4	5	6	7	8

In order to diminish to the utmost the chances of error in inscribing the numerical data of the problem, they are successively placed on one of the columns of the mill; then, by means of cards arranged for this purpose, these same numbers are caused to arrange themselves on the requisite columns, without the operator having to give his attention to it; so that his undivided mind may be applied to the simple inscription of these same numbers.

According to what has now been explained, we see that the collection of columns of Variables may be regarded as a *store* of numbers, accumulated there by the mill, and which, obeying the orders transmitted to the machine by means of the cards, pass alternately from the mill to the store, and from the store to the mill, that they may undergo the transformations demanded by the nature of the calculation to be performed.

Hitherto no mention has been made of the *signs* in the results, and the machine would be far from perfect were it incapable of expressing and combining amongst each other positive and negative quantities. To accomplish this end, there is, above every column, both of the mill and of the store, a disc, similar to the discs of which the columns themselves consist. According as the digit on this disc is even or uneven, the number inscribed on the corresponding column below it will be considered as positive or negative. This granted, we may, in the following manner, conceive how the signs can be algebraically combined in the machine. When a number is to be transferred from the store to the mill, and *vice versâ*, it will always be transferred with its sign, which will be effected by means of the cards, as has been explained in what precedes. Let any two numbers then, on which we are to operate arithmetically, be placed in the mill with their respective signs. Suppose that we are first to add them together; the operation-cards will command the addition: if the two numbers be of the same sign, one of the two will be entirely effaced from where it was inscribed, and will go to add itself on the column which contains the other number; the machine will, during this operation, be able, by means of a certain apparatus, to prevent any movement in the disc of signs which belongs to the column on which the addition is made, and thus the result will remain with the sign which the two given numbers originally had. When two numbers have two different signs, the addition commanded by the card will be changed into a subtraction through the intervention of mechanisms which are brought into play by this very difference of sign. Since the subtraction can only be effected on the larger of the two numbers, it must be arranged that the disc of signs of the larger number shall not move while the smaller of the two numbers is being effaced from its column and subtracted from the other, whence the result will have the sign of this latter, just as in fact it ought to be. The combinations to which algebraical subtraction give rise, are analogous to the preceding. Let us pass on to multiplication. When two numbers to be multiplied are of the same sign, the result is positive; if the signs are different, the product must be negative. In order that the machine may act conformably to this law, we have but to conceive that on the column containing the product of the two given numbers, the digit which indicates the sign of that product, has been formed by the mutual addition of the two digits that respectively indicated the signs of the two given numbers; it is then obvious that if the digits of the signs are both even, or both odd, their sum will be an even number, and consequently will express a positive number; but that if, on the contrary, the two digits of the signs are one even and the other odd, their sum will be an odd number, and will consequently express a negative number. In the case of division, instead of adding the digits of the discs, they must be subtracted one from the other, which will produce results analogous to the preceding; that is to say, that if these figures are both even or both uneven, the remainder of this subtraction will be even; and it will be uneven in the contrary case. When I speak of mutually adding or subtracting the numbers expressed by the digits of the signs, I merely mean that one of the sign-discs is made to advance or retrograde a number of divisions equal to that which is expressed by the digit on the other sign-disc. We see, then, from the preceding explanation, that it is possible mechanically to combine the signs of quantities so as to obtain results conformable to those indicated by algebra*.

The machine is not only capable of executing those numerical calculations which depend on a given algebraical formula, but it is also fitted for analytical calculations in which there are one or several variables to be considered. It must be assumed that the analytical expression to be operated on can be developed according to powers of the variable, or according to determinate functions of this same variable, such as circular functions, for instance; and similarly for the result that is to be attained. If we then suppose that above the columns of the store, we have inscribed the powers or the functions of the variable, arranged according to whatever is the prescribed law of development, the coefficients of these several terms may be respectively placed on the corresponding column below each. In this manner we shall have a representation of an analytical development; and, supposing the position of the several terms composing it to be invariable, the problem will be reduced to that of calculating their coefficients according to the laws demanded by the nature of the question. In order to make this more clear, we shall take the

* Not having had leisure to discuss with Mr. Babbage the manner of introducing into his machine the combination of algebraical signs, I do not pretend here to expose the method he uses for this purpose; but I considered that I ought myself to supply the deficiency, conceiving that this paper would have been imperfect if I had omitted to point out one means that might be employed for resolving this essential part of the problem in question.

following* very simple example, in which we are to multiply $(a + b\,x^1)$ by $(\mathrm{A} + \mathrm{B}\cos^1 x)$. We shall begin by writing $x^0, x^1, \cos^0 x, \cos^1 x$, above the columns $\mathrm{V}_0, \mathrm{V}_1, \mathrm{V}_2, \mathrm{V}_3$; then, since from the form of the two functions to be combined, the terms which are to compose the products will be of the following nature, $x^0.\cos^0 x,\ x^0.\cos^1 x,\ x^1.\cos^0 x,\ x^1.\cos^1 x$; these will be inscribed above the columns $\mathrm{V}_4, \mathrm{V}_5, \mathrm{V}_6, \mathrm{V}_7$. The coefficients of $x^0, x^1, \cos^0 x, \cos^1 x$ being given, they will, by means of the mill, be passed to the columns $\mathrm{V}_0, \mathrm{V}_1, \mathrm{V}_2$ and V_3. Such are the primitive data of the problem. It is now the business of the machine to work out its solution, that is to find the coefficients which are to be inscribed on $\mathrm{V}_4, \mathrm{V}_5, \mathrm{V}_6, \mathrm{V}_7$. To attain this object, the law of formation of these same coefficients being known, the machine will act through the intervention of the cards, in the manner indicated by the following table:—

† Columns above which are written the functions of the variable.	Coefficients.		Cards of the operations.		Cards of the variables.			
	Given.	To be formed.	Number of the operations.	Nature of the operation.	Columns on which operations are to be performed.	Columns on which are to be inscribed the results of the operations.	Indication of change of value on any column submitted to an operation.	Results of the operations.
$x^0 \ldots\ldots\ldots {}^1\mathrm{V}_0$	a	"	"	"	"	"	"	" "
$x^1 \ldots\ldots\ldots {}^1\mathrm{V}_1$	b	"	"	"	"	"	"	" "
$\mathrm{Cos}^0 x \ldots {}^1\mathrm{V}_2$	A	"	"	"	"	"	"	" "
$\mathrm{Cos}^1 x \ldots {}^1\mathrm{V}_3$	B	"	"	"	"	"	"	" "
$x^0 \cos^0 x \ldots {}^0\mathrm{V}_4$	...	$a\,\mathrm{A}$	1	×	${}^1\mathrm{V}_0 \times {}^1\mathrm{V}_2 =$	${}^1\mathrm{V}_4$	$\left\{\begin{matrix} {}^1\mathrm{V}_0 = {}^1\mathrm{V}_0 \\ {}^1\mathrm{V}_2 = {}^1\mathrm{V}_2 \end{matrix}\right\}$	${}^1\mathrm{V}_4 = a\,\mathrm{A}$ coefficients of $x^0 \cos^0 x$
$x^0 \cos^1 x \ldots {}^0\mathrm{V}_5$	...	$a\,\mathrm{B}$	2	×	${}^1\mathrm{V}_0 \times {}^1\mathrm{V}_3 =$	${}^1\mathrm{V}_5$	$\left\{\begin{matrix} {}^1\mathrm{V}_0 = {}^0\mathrm{V}_0 \\ {}^1\mathrm{V}_3 = {}^1\mathrm{V}_3 \end{matrix}\right\}$	${}^1\mathrm{V}_5 = a\,\mathrm{B}$ ... ... $x^0 \cos^1 x$
$x^1 \cos^0 x \ldots {}^0\mathrm{V}_6$	...	$b\,\mathrm{A}$	3	×	${}^1\mathrm{V}_1 \times {}^1\mathrm{V}_2 =$	${}^1\mathrm{V}_6$	$\left\{\begin{matrix} {}^1\mathrm{V}_1 = {}^1\mathrm{V}_1 \\ {}^1\mathrm{V}_2 = {}^0\mathrm{V}_2 \end{matrix}\right\}$	${}^1\mathrm{V}_6 = b\,\mathrm{A}$ ... ... $x^1 \cos^0 x$
$x^1 \cos^1 x \ldots {}^0\mathrm{V}_7$	...	$b\,\mathrm{B}$	4	×	${}^1\mathrm{V}_1 \times {}^1\mathrm{V}_3 =$	${}^1\mathrm{V}_7$	$\left\{\begin{matrix} {}^1\mathrm{V}_1 = {}^0\mathrm{V}_1 \\ {}^1\mathrm{V}_3 = {}^0\mathrm{V}_3 \end{matrix}\right\}$	${}^1\mathrm{V}_7 = b\,\mathrm{B}$ ... ... $x^1 \cos^1 x$

It will now be perceived that a general application may be made of the principle developed in the preceding example, to every species of process which it may be proposed to effect on series submitted to calculation. It is sufficient that the law of formation of the coefficients be known, and that this law be inscribed on the cards of the machine, which will then of itself execute all

* See Note E.

† For an explanation of the upper left-hand indices attached to the V's in this and in the preceding Table, we must refer the reader to Note D, amongst those appended to the memoir.—NOTE BY TRANSLATOR.

the calculations requisite for arriving at the proposed result. If, for instance, a recurring series were proposed, the law of formation of the coefficients being here uniform, the same operations which must be performed for one of them will be repeated for all the others; there will merely be a change in the locality of the operation, that is it will be performed with different columns. Generally, since every analytical expression is susceptible of being expressed in a series ordered according to certain functions of the variable, we perceive that the machine will include all analytical calculations which can be definitively reduced to the formation of coefficients according to certain laws, and to the distribution of these with respect to the variables.

We may deduce the following important consequence from these explanations, viz. that since the cards only indicate the nature of the operations to be performed, and the columns of Variables with which they are to be executed, these cards will themselves possess all the generality of analysis, of which they are in fact merely a translation. We shall now further examine some of the difficulties which the machine must surmount, if its assimilation to analysis is to be complete. There are certain functions which necessarily change in nature when they pass through zero or infinity, or whose values cannot be admitted when they pass these limits. When such cases present themselves, the machine is able, by means of a bell, to give notice that the passage through zero or infinity is taking place, and it then stops until the attendant has again set it in action for whatever process it may next be desired that it shall perform. If this process has been foreseen, then the machine, instead of ringing, will so dispose itself as to present the new cards which have relation to the operation that is to succeed the passage through zero and infinity. These new cards may follow the first, but may only come into play contingently upon one or other of the two circumstances just mentioned taking place.

Let us consider a term of the form $a b^n$; since the cards are but a translation of the analytical formula, their number in this particular case must be the same, whatever be the value of n; that is to say, whatever be the number of multiplications required for elevating b to the nth power (we are supposing for the moment that n is a whole number). Now, since the exponent n indicates that b is to be multiplied n times by itself, and all these operations are of the same nature, it will be sufficient to employ one single operation-card, viz. that which orders the multiplication.

But when n is given for the particular case to be calculated, it will be further requisite that the machine limit the number of its multiplications according to the given values. The process may be thus arranged. The three numbers a, b and n will be written on as many distinct columns of the store; we shall designate them V_0, V_1, V_2; the result $a\,b^n$ will place itself on the column V_3. When the number n has been introduced into the machine, a card will order a certain registering-apparatus to mark $(n-1)$, and will at the same time execute the multiplication of b by b. When this is completed, it will be found that the registering-apparatus has effaced a unit, and that it only marks $(n-2)$; while the machine will now again order the number b written on the column V_1 to multiply itself with the product b^2 written on the column V_3, which will give b^3. Another unit is then effaced from the registering-apparatus, and the same processes are continually repeated until it only marks zero. Thus the number b^n will be found inscribed on V_3, when the machine, pursuing its course of operations, will order the product of b^n by a; and the required calculation will have been completed without there being any necessity that the number of operation-cards used should vary with the value of n. If n were negative, the cards, instead of ordering the multiplication of a by b^n, would order its division; this we can easily conceive, since every number, being inscribed with its respective sign, is consequently capable of reacting on the nature of the operations to be executed. Finally, if n were fractional, of the form $\frac{p}{q}$, an additional column would be used for the inscription of q, and the machine would bring into action two sets of processes, one for raising b to the power p, the other for extracting the qth root of the number so obtained.

Again, it may be required, for example, to multiply an expression of the form $a\,x^m + b\,x^n$ by another $A\,x^p + B\,x^q$, and then to reduce the product to the least number of terms, if any of the indices are equal. The two factors being ordered with respect to x, the general result of the multiplication would be $A\,a\,x^{m+p} + A\,b\,x^{n+p} + B\,a\,x^{m+q} + B\,b\,x^{n+q}$. Up to this point the process presents no difficulties; but suppose that we have $m=p$ and $n=q$, and that we wish to reduce the two middle terms to a single one $(A\,b + B\,a)\,x^{m+q}$. · For this purpose, the cards may order $m + q$ and $n + p$ to be transferred into the mill, and there subtracted one from the other; if the remainder is nothing, as would be the case on the present hypothesis, the mill will order other cards to bring to it the coefficients $A\,b$ and $B\,a$, that it may add them together and give them in this state as a coefficient for the single term $x^{n+p} = x^{m+q}$.

This example illustrates how the cards are able to reproduce all the operations which intellect performs in order to attain a determinate result, if these operations are themselves capable of being precisely defined.

Let us now examine the following expression:—

$$2 \cdot \frac{2^2 . 4^2 . 6^2 . 8^2 . 10^2 \ldots\ldots (2\,n)^2}{1^2 . 3^2 . 5^2 . 7^2 . 9^2 \ldots\ldots (2\,n-1)^2 . (2\,n+1)^2},$$

which we know becomes equal to the ratio of the circumference to the diameter, when n is infinite. We may require the machine not only to perform the calculaiton of this fractional expression, but further to give indication as soon as the value becomes identical with that of the ratio of the circumference to the diameter when n is infinite, a case in which the computation would be impossible. Observe that we should thus require of the machine to interpret a result not of itself evident, and that this is not amongst its attributes, since it is no thinking being. Nevertheless, when the cos of $n = \infty$ has been foreseen, a card may immediately order the substitution of the value of π (π being the ratio of the circumference to the diameter), without going through the series of calculations indicated. This would merely require that the machine contain a special card, whose office it should be to place the number π in a direct and independent manner on the column indicated to it. And here we should introduce the mention of a third species of cards, which may be called *cards of numbers*. There are certain numbers, such as those expressing the ratio of the circumference to the diameter, the Numbers of Bernoulli, &c., which frequently present themselves in calculations. To avoid the necessity for computing them every time they have to be used, certain cards may be combined specially in order to give these numbers ready made into the mill, whence they afterwards go and place themselves on those columns of the store that are destined for them. Through this means the machine will be susceptible of those simplifications afforded by the use of numerical tables. It would be equally possible to introduce, by means of these cards, the logarithms of numbers; but perhaps it might not be in this case either the shortest or the most appropriate method; for the machine might be able to perform the same calculations by other more expeditious combinations, founded on the rapidity with which it executes the four first operations of arithmetic. To give an idea of this rapidity, we need only mention that Mr. Babbage believes he can, by his engine, form the product of two numbers, each containing twenty figures, in *three minutes*.

Perhaps the immense number of cards required for the solution of any rather complicated problem may appear to be an obstacle; but this does not seem to be the case. There is no limit to the number of cards that can be used. Certain stuffs require for their fabrication not less than *twenty thousand* cards, and we may unquestionably far exceed even this quantity *.

Resuming what we have explained concerning the Analytical Engine, we may conclude that it is based on two principles: the first, consisting in the fact that every arithmetical calculation ultimately depends on four principal operations—addition, subtraction, multiplication, and division; the second, in the possibility of reducing every analytical calculation to that of the coefficients for the several terms of a series. If this last principle be true, all the operations of analysis come within the domain of the engine. To take another point of view: the use of the cards offers a generality equal to that of algebraical formulæ, since such a formula simply indicates the nature and order of the operations requisite for arriving at a certain definite result, and similarly the cards merely command the engine to perform these same operations; but in order that the mechanisms may be able to act to any purpose, the numerical data of the problem must in every particular case be introduced. Thus the same series of cards will serve for all questions whose sameness of nature is such as to require nothing altered excepting the numerical data. In this light the cards are merely a translation of algebraical formulæ, or, to express it better, another form of analytical notation.

Since the engine has a mode of acting peculiar to itself, it will in every particular case be necessary to arrange the series of calculations conformably to the means which the machine pos-

* See Note F.

sesses; for such or such a process which might be very easy for a calculator, may be long and complicated for the engine, and *vice versâ*.

Considered under the most general point of view, the essential object of the machine being to calculate, according to the laws dictated to it, the values of numerical coefficients which it is then to distribute appropriately on the columns which represent the variables, it follows that the interpretation of formulæ and of results is beyond its province, unless indeed this very interpretation be itself susceptible of expression by means of the symbols which the machine employs. Thus, although it is not itself the being that reflects, it may yet be considered as the being which executes the conceptions of intelligence*. The cards receive the impress of these conceptions, and transmit to the various trains of mechanism composing the engine the orders necessary for their action. When once the engine shall have been constructed, the difficulty will be reduced to the making out of the cards; but as these are merely the translation of algebraical formulæ, it will, by means of some simple notations, be easy to consign the execution of them to a workman. Thus the whole intellectual labour will be limited to the preparation of the formulæ, which must be adapted for calculation by the engine.

Now, admitting that such an engine can be constructed, it may be inquired: what will be its utility? To recapitulate; it will afford the following advantages:—First, rigid accuracy. We know that numerical calculations are generally the stumbling-block to the solution of problems, since errors easily creep into them, and it is by no means always easy to detect these errors. Now the engine, by the very nature of its mode of acting, which requires no human intervention during the course of its operations, presents every species of security under the head of correctness; besides, it carries with it its own check; for at the end of every operation it prints off, not only the results, but likewise the numerical data of the question; so that it is easy to verify whether the question has been correctly proposed. Secondly, economy of time: to convince ourselves of this, we need only recollect that the multiplication of two numbers, consisting each of twenty figures, requires at the very utmost three minutes. Likewise, when a long series of identical computations is to be performed, such as those required for the formation of

* See Note G.

numerical tables, the machine can be brought into play so as to give several results at the same time, which will greatly abridge the whole amount of the processes. Thirdly, economy of intelligence: a simple arithmetical computation requires to be performed by a person possessing some capacity; and when we pass to more complicated calculations, and wish to use algebraical formulæ in particular cases, knowledge must be possessed which pre-supposes preliminary mathematical studies of some extent. Now the engine, from its capability of performing by itself all these purely material operations, spares intellectual labour, which may be more profitably employed. Thus the engine may be considered as a real manufactory of figures, which will lend its aid to those many useful sciences and arts that depend on numbers. Again, who can foresee the consequences of such an invention? In truth, how many precious observations remain practically barren for the progress of the sciences, because there are not powers sufficient for computing the results! And what discouragement does the perspective of a long and arid computation cast into the mind of a man of genius, who demands time exclusively for meditation, and who beholds it snatched from him by the material routine of operations! Yet it is by the laborious route of analysis that he must reach truth; but he cannot pursue this unless guided by numbers; for without numbers it is not given us to raise the veil which envelopes the mysteries of nature. Thus the idea of constructing an apparatus capable of aiding human weakness in such researches, is a conception which, being realized, would mark a glorious epoch in the history of the sciences. The plans have been arranged for all the various parts, and for all the wheel-work, which compose this immense apparatus, and their action studied; but these have not yet been fully combined together in the drawings* and mechanical notation †. The confidence which the genius of Mr. Babbage must inspire, affords legitimate ground for hope that this enterprise will be crowned with success; and while we render homage to the intelligence which directs it, let us breathe aspirations for the accomplishment of such an undertaking.

* This sentence has been slightly altered in the translation in order to express more exactly the present state of the engine.—NOTE BY TRANSLATOR.

† The notation here alluded to is a most interesting and important subject, and would have well deserved a separate and detailed Note upon it, amongst those appended to the Memoir. It has, however, been impossible, within the space allotted, even to touch upon so wide a field.—NOTE BY TRANSLATOR.

NOTES BY THE TRANSLATOR.

Note A.—Page 674.

The particular function whose integral the Difference Engine was constructed to tabulate, is

$$\Delta^7 u_x = 0.$$

The purpose which that engine has been specially intended and adapted to fulfil, is the computation of nautical and astronomical tables. The integral of

$$\Delta^7 u_x = 0$$

being $$u_x = a + b\,x + c\,x^2 + d\,x^3 + e\,x^4 + f\,x^5 + g\,x^6,$$

the constants a, b, c, &c. are represented on the seven columns of discs, of which the engine consists. It can therefore tabulate *accurately* and to an *unlimited extent*, all series whose general term is comprised in the above formula; and it can also tabulate *approximatively* between *intervals of greater or less extent*, all other series which are capable of tabulation by the Method of Differences.

The Analytical Engine, on the contrary, is not merely adapted for *tabulating* the results of one particular function and of no other, but for *developing and tabulating* any function whatever. In fact the engine may be described as being the material expression of any indefinite function of any degree of generality and complexity, such as for instance,

$$F\,(x, y, z, \log x, \sin y, x^p, \&c.),$$

which is, it will be observed, a function of all other possible functions of any number of quantities.

In this, which we may call the *neutral* or *zero* state of the engine, it is ready to receive at any moment, by means of cards constituting a portion of its mechanism (and applied on the principle of those used in the Jacquard-loom), the impress of whatever *special* function we may desire to develope or to tabulate. These cards contain within themselves (in a manner explained in the Memoir itself, pages 677 and 678) the law of development of the particular function that may be under consideration, and they compel the mechanism to act accordingly in a certain corresponding order. One of the simplest cases would be, for example, to suppose that

$$F\,(x, y, z, \&c.\ \&c.)$$

is the particular function

$$\Delta^n u_x = 0$$

which the Difference Engine tabulates for values of n only up to 7. In this case the cards would order the mechanism to go through that succession of operations which would tabulate

$$u_x = a + b\,x + c\,x^2 + \ldots .\, m\,x^{n-1},$$

where n might be any number whatever.

These cards, however, have nothing to do with the regulation of the particular *numerical* data. They merely determine the *operations** to be effected, which operations may of course be performed on an infinite variety of particular numerical values, and do not bring out any definite numerical results unless the numerical data of the problem have been impressed on the requisite portions of the train of mechanism. In the above example, the first essential step towards an arithmetical result, would be the substitution of specific numbers for n, and for the other primitive quantities which enter into the function.

Again, let us suppose that for F we put two complete equations of the fourth degree between x and y. We must then express on the cards the law of elimination for such equations. The engine would follow out those laws, and would ultimately give the equation of one variable which results from such elimination. Various *modes* of elimination might be selected; and of course the cards must be made out accordingly. The following is one mode that might be adopted. The engine is able to multiply together any two functions of the form

$$a + bx + cx^2 + \ldots . px^n.$$

This granted, the two equations may be arranged according to the powers of y, and the coefficients of the powers of y may be arranged according to powers of x. The elimination of y will result from the successive multiplications and subtractions of several such functions. In this, and in all other instances, as was explained above, the particular *numerical* data and the *numerical* results are determined by means and by portions of the mechanism which act quite independently of those that regulate the *operations*.

In studying the action of the Analytical Engine, we find that the peculiar and independent nature of the considerations which in all mathematical analysis belong to *operations*, as distinguished from *the objects operated upon* and from the *results* of the operations performed upon those objects, is very strikingly defined and separated.

It is well to draw attention to this point, not only because its full appreciation is essential to the attainment of any very just and adequate general comprehension of the powers and mode of action of the Analytical Engine, but also because it is one which is perhaps too little kept in view in the study of mathematical science in general. It is, however, impossible to confound it with other considerations, either when we trace the manner in which that engine attains its results, or when we prepare the data for its attainment of those results. It were much to be desired, that when mathematical processes pass through the human brain instead of through the medium of inanimate mechanism, it were equally a necessity of things that the reasonings connected with *operations* should hold the same just place as a clear and well-defined branch of the subject of analysis, a fundamental but yet independent

* We do not mean to imply that the *only* use made of the Jacquard cards is that of regulating the algebraical *operations*. But we mean to explain that *those* cards and portions of mechanism which regulate these *operations*, are wholly independent of those which are used for other purposes. M. Menabrea explains that there are *three* classes of cards used in the engine for three distinct sets of objects, viz. *Cards of the Operations*, *Cards of the Variables*, and certain *Cards of Numbers*. (See pages 678 and 687.)

ingredient in the science, which they must do in studying the engine. The confusion, the difficulties, the contradictions which, in consequence of a want of accurate distinctions in this particular, have up to even a recent period encumbered mathematics in all those branches involving the consideration of negative and impossible quantities, will at once occur to the reader who is at all versed in this science, and would alone suffice to justify dwelling somewhat on the point, in connexion with any subject so peculiarly fitted to give forcible illustration of it, as the Analytical Engine. It may be desirable to explain, that by the word *operation*, we mean *any process which alters the mutual relation of two or more things*, be this relation of what kind it may. This is the most general definition, and would include all subjects in the universe. In abstract mathematics, of course operations alter those particular relations which are involved in the considerations of number and space, and the *results* of operations are those peculiar results which correspond to the nature of the subjects of operation. But the science of operations, as derived from mathematics more especially, is a science of itself, and has its own abstract truth and value; just as logic has its own peculiar truth and value, independently of the subjects to which we may apply its reasonings and processes. Those who are accustomed to some of the more modern views of the above subject, will know that a few fundamental relations being true, certain other combinations of relations must of necessity follow; combinations unlimited in variety and extent if the deductions from the primary relations be carried on far enough. They will also be aware that one main reason why the separate nature of the science of operations has been little felt, and in general little dwelt on, is the *shifting* meaning of many of the symbols used in mathematical notation. First, the symbols of *operation* are frequently *also* the symbols of the *results* of operations. We may say that these symbols are apt to have both a *retrospective* and a *prospective* signification. They may signify either relations that are the consequence of a series of processes already performed, or relations that are yet to be effected through certain processes. Secondly, figures, the symbols of *numerical magnitude*, are frequently *also* the symbols of *operations*, as when they are the indices of powers. Wherever terms have a shifting meaning, independent sets of considerations are liable to become complicated together, and reasonings and results are frequently falsified. Now in the Analytical Engine the operations which come under the first of the above heads, are ordered and combined by means of a notation and of a train of mechanism which belong exclusively to themselves; and with respect to the second head, whenever numbers meaning *operations* and not *quantities* (such as the indices of powers), are inscribed on any column or set of columns, those columns immediately act in a wholly separate and independent manner, becoming connected with the *operating mechanism* exclusively, and re-acting upon this. They never come into combination with numbers upon any other columns meaning *quantities*; though, of course, if there are numbers meaning *operations* upon n columns, these may *combine amongst each other*, and will often be required to do so, just as numbers meaning *quantities* combine with each other in any variety. It might have been arranged that all numbers meaning *operations* should have appeared on some separate portion of the engine from that which presents numerical *quantities*; but the present mode is in some cases more simple, and offers in reality quite as much distinctness when understood.

The operating mechanism can even be thrown into action independently of any object to operate upon (although of course no *result* could then be developed). Again, it might act upon other things besides *number*, were objects found whose mutual fundamental relations could be expressed by those of the abstract science of operations, and which should be also susceptible of adaptations to the action of the operating notation and mechanism of the engine. Supposing, for instance, that the fundamental relations of pitched sounds in the science of harmony and of musical composition were susceptible of such expression and adaptations, the engine might compose elaborate and scientific pieces of music of any degree of complexity or extent.

The Analytical Engine is an *embodying of the science of operations*, constructed with peculiar reference to abstract number as the subject of those operations. The Difference Engine is the embodying of *one particular and very limited set of operations*, which (see the notation used in note B) may be expressed thus, $(+, +, +, +, +, +)$, or thus, $6(+)$. Six repetitions of the one operation, $+$, is, in fact, the whole sum and object of that engine. It has seven columns, and a number on any column can add itself to a number on the next column to its *right-hand*. So that, beginning with the column furthest to the left, six additions can be effected, and the result appears on the seventh column, which is the last on the right-hand. The *operating* mechanism of this engine acts in as separate and independent a manner as that of the Analytical Engine; but being susceptible of only one unvarying and restricted combination, it has little force or interest in illustration of the distinct nature of the *science of operations*. The importance of regarding the Analytical Engine under this point of view will, we think, become more and more obvious, as the reader proceeds with M. Menabrea's clear and masterly article. The calculus of operations is likewise in itself a topic of so much interest, and has of late years been so much more written on and thought on than formerly, that any bearing which that engine, from its mode of constitution, may possess upon the illustration of this branch of mathematical science, should not be overlooked. Whether the inventor of this engine had any such views in his mind while working out the invention, or whether he may subsequently ever have regarded it under this phase, we do not know; but it is one that forcibly occurred to ourselves on becoming acquainted with the means through which analytical combinations are actually attained by the mechanism. We cannot forbear suggesting one practical result which it appears to us must be greatly facilitated by the independent manner in which the engine orders and combines its *operations*: we allude to the attainment of those combinations into which *imaginary quantities* enter. This is a branch of its processes into which we have not had the opportunity of inquiring, and our conjecture therefore as to the principle on which we conceive the accomplishment of such results may have been made to depend, is very probably not in accordance with the fact, and less subservient for the purpose than some other principles, or at least requiring the cooperation of others. It seems to us obvious, however, that where operations are so independent in their mode of acting, it must be easy by means of a few simple provisions and additions in arranging the mechanism, to bring out a *double* set of *results*, viz.—1st, the *numerical magnitudes* which are the results of operations performed on *numerical data*. (These results are the *primary* object of the engine). 2ndly, the *symbolical results* to be attached to those numerical results, which symbolical results are not less the necessary and logical consequences of operations performed upon *symbolical data*, than are numerical results when the data are numerical*.

If we compare together the powers and the principles of construction of the Difference and of the Analytical Engines, we shall perceive that the capabilities of the latter are immeasurably more extensive than those of the former, and that they in fact hold to each other the same relationship as that of analysis to arithmetic. The Difference Engine can effect but one particular series of operations, viz. that required for tabulating the integral of the special function

$$\Delta^n u_z = 0;$$

and as it can only do this for values of n up to 7†, it cannot be considered as being the most *general* expression even of *one particular* function, much less as being the expression of any and all possible functions of all degrees of generality. The Difference Engine can in reality (as has been already partly explained) do nothing but *add*; and any other processes, not excepting those of simple subtraction, multiplication and division, can be performed by it only just to that extent in which it is possible, by judicious mathematical arrangement and artifices, to reduce them to a *series of additions*. The method of differences is, in fact, a method of additions; and as it includes within its means a larger number of results attainable by *addition* simply, than any other mathematical principle, it was very appropriately selected as the basis on which to construct *an Adding Machine*, so as to give to the powers of such a machine the widest possible range. The Analytical Engine, on the contrary, can either add, subtract, multiply or divide with equal facility; and performs each of these four operations in a direct manner, without the aid of any of the other three. This one fact implies everything; and it is scarcely necessary to point out, for instance, that while the Difference Engine can merely *tabulate*,

* In fact such an extension as we allude to, would merely constitute a further and more perfected development of any system introduced for making the proper combinations of the signs *plus* and *minus*. How ably M. Menabrea has touched on this restricted case is pointed out in Note B.

† The machine might have been constructed so as to tabulate for a higher value of n than seven. Since, however, every unit added to the value of n increases the extent of the mechanism requisite, there would on this account be a limit beyond which it could not be practically carried. Seven is sufficiently high for the calculation of all ordinary tables.

The fact that, in the Analytical Engine, the same extent of mechanism suffices for the solution of $\Delta^n u_z = 0$, whether $n = 7$, $n = 100{,}000$, or $n =$ any number whatever, at once suggests how entirely distinct must be the *nature of the principles* through whose application matter has been enabled to become the working agent of abstract mental operations in each of these engines respectively; and it affords an equally obvious presumption, that in the case of the Analytical Engine, not only are those principles in themselves of a higher and more comprehensive description, but also such as must vastly extend the *practical* value of the engine whose basis they constitute.

and is incapable of *developing*, the Analytical Engine can *either tabulate or develope*.

The former engine is in its nature strictly *arithmetical*, and the results it can arrive at lie within a very clearly defined and restricted range, while there is no finite line of demarcation which limits the powers of the Analytical Engine. These powers are co-extensive with our knowledge of the laws of analysis itself, and need be bounded only by our acquaintance with the latter. Indeed we may consider the engine as the *material and mechanical representative* of analysis, and that our actual working powers in this department of human study will be enabled more effectually than heretofore to keep pace with our theoretical knowledge of its principles and laws, through the complete control which the engine gives us over the *executive manipulation* of algebraical and numerical symbols.

Those who view mathematical science not merely as a vast body of abstract and immutable truths, whose intrinsic beauty, symmetry and logical completeness, when regarded in their connexion together as a whole, entitle them to a prominent place in the interest of all profound and logical minds, but as possessing a yet deeper interest for the human race, when it is remembered that this science constitutes the language through which alone we can adequately express the great facts of the natural world, and those unceasing changes of mutual relationship which, visibly or invisibly, consciously or unconsciously to our immediate physical perceptions, are interminably going on in the agencies of the creation we live amidst: those who thus think on mathematical truth as the instrument through which the weak mind of man can most effectually read his Creator's works, will regard with especial interest all that can tend to facilitate the translation of its principles into explicit practical forms.

The distinctive characteristic of the Analytical Engine, and that which has rendered it possible to endow mechanism with such extensive faculties as bid fair to make this engine the executive right-hand of abstract algebra, is the introduction into it of the principle which Jacquard devised for regulating, by means of punched cards, the most complicated patterns in the fabrication of brocaded stuffs. It is in this that the distinction between the two engines lies. Nothing of the sort exists in the Difference Engine. We may say most aptly that the Analytical Engine *weaves algebraical patterns* just as the Jacquard-loom weaves flowers and leaves. Here, it seems to us, resides much more of originality than the Difference Engine can be fairly entitled to claim. We do not wish to deny to this latter all such claims. We believe that it is the only proposal or attempt ever made to construct a calculating machine *founded on the principle of successive orders of differences*, and capable of *printing off its own results*; and that this engine surpasses its predecessors, both in the extent of the calculations which it can perform, in the facility, certainty and accuracy with which it can effect them, and in the absence of all necessity for the intervention of human intelligence *during the performance of its calculations*. Its nature is, however, limited to the strictly arithmetical, and it is far from being the first or only scheme for constructing *arithmetical* calculating machines with more or less of success.

The bounds of *arithmetic* were however outstepped the moment the idea of applying the cards had occurred; and the Analytical Engine does not occupy common ground with mere "calculating machines." It holds a position wholly its own; and the considerations it suggests are most interesting in their nature. In enabling mechanism to combine together *general* symbols, in successions of unlimited variety and extent, a uniting link is established between the operations of matter and the abstract mental processes of the *most abstract* branch of mathematical science. A new, a vast, and a powerful language is developed for the future use of analysis, in which to wield its truths so that these may become of more speedy and accurate practical application for the purposes of mankind than the means hitherto in our possession have rendered possible. Thus not only the mental and the material, but the theoretical and the practical in the mathematical world, are brought into more intimate and effective connexion with each other. We are not aware of its being on record that anything partaking in the nature of what is so well designated the *Analytical* Engine has been hitherto proposed, or even thought of, as a practical possibility, any more than the idea of a thinking or of a reasoning machine.

We will touch on another point which constitutes an important distinction in the modes of operating of the Difference and Analytical Engines. In order to enable the former to do its business, it is necessary to put into its columns the series of numbers constituting the first terms of the several orders of differences for whatever is the particular table under consideration. The machine then works *upon* these as its data. But these data must themselves have been already computed through a series of calculations by a human head. Therefore that engine can only produce results depending on data which have been arrived at by the explicit and actual working out of processes that are in their nature different from any that come within the sphere of its own powers. In other words, an *analysing* process must have been gone through by a human mind in order to obtain the data upon which the engine then *synthetically* builds its results. The Difference Engine is in its character exclusively *synthetical*, while the Analytical Engine is equally capable of analysis or of synthesis.

It is true that the Difference Engine can calculate to a much greater extent with these few preliminary data, than the data themselves required for their own determination. The table of squares, for instance, can be calculated to any extent whatever, when the numbers *one* and *two* are furnished; and a very few differences computed at any part of a table of logarithms would enable the engine to calculate many hundreds or even thousands of logarithms. Still the circumstance of its requiring, as a previous condition, that any function whatever shall have been numerically worked out, makes it very inferior in its nature and advantages to an engine which, like the Analytical Engine, requires merely that we should know the *succession and distribution of the operations* to be performed; without there being any occasion *, in order to obtain data on which it can work, for our ever having gone through either the same particular operations which it is itself to effect, or any others. Numerical data must of course be given it, but they are mere arbitrary ones; not data that could only be arrived at through a systematic and necessary series of previous numerical calculations, which is quite a different thing.

* This subject is further noticed in Note F.

To this it may be replied that an analysing process must equally have been performed in order to furnish the Analytical Engine with the necessary *operative* data; and that herein may also lie a possible source of error. Granted that the actual mechanism is unerring in its processes, the *cards* may give it wrong orders. This is unquestionably the case; but there is much less chance of error, and likewise far less expenditure of time and labour, where operations only, and the distribution of these operations, have to be made out, than where explicit numerical results are to be attained. In the case of the Analytical Engine we have undoubtedly to lay out a certain capital of analytical labour in one particular line; but this is in order that the engine may bring us in a much larger return in another line. It should be remembered also that the cards when once made out for any formula, have all the generality of algebra, and include an infinite number of particular cases.

We have dwelt considerably on the distinctive peculiarities of each of these engines, because we think it essential to place their respective attributes in strong relief before the apprehension of the public; and to define with clearness and accuracy the wholly different nature of the principles on which each is based, so as to make it self-evident to the reader (the mathematical reader at least) in what manner and degree the powers of the Analytical Engine transcend those of an engine, which, like the Difference Engine, can only work out such results as may be derived from *one restricted and particular series of processes*, such as those included in $\Delta^n u_x = 0$. We think this of importance, because we know that there exists considerable vagueness and inaccuracy in the mind of persons in general on the subject. There is a misty notion amongst most of those who have attended at all to it, that *two* "calculating machines" have been successively invented by the same person within the last few years; while others again have never heard but of the one original "calculating machine," and are not aware of there being any extension upon this. For either of these two classes of persons the above considerations are appropriate. While the latter require a knowledge of the fact that there *are two* such inventions, the former are not less in want of accurate and well-defined information on the subject. No very clear or correct ideas prevail as to the characteristics of each engine, or their respective advantages or disadvantages; and, in meeting with those incidental allusions, of a more or less direct kind, which occur in so many publications of the day, to these machines, it must frequently be matter of doubt *which* "calculating machine" is referred to, or whether *both* are included in the general allusion.

We are desirous likewise of removing two misapprehensions which we know obtain, to some extent, respecting these engines. In the first place it is very generally supposed that the Difference Engine, after it had been completed up to a certain point, *suggested* the idea of the Analytical Engine; and that the second is in fact the improved offspring of the first, and *grew out* of the existence of its predecessor, through some natural or else accidental combination of ideas suggested by this one. Such a supposition is in this instance contrary to the facts; although it seems to be almost an obvious inference, wherever two inventions, similar in their nature and objects, succeed each other closely in order of *time*, and strikingly in order of *value*; more especially when the same individual is the author of both. Nevertheless the ideas which led to the Analytical Engine occurred in a manner wholly independent of any that were connected with the Difference Engine. These ideas are indeed in their own intrinsic nature independent of the latter engine, and might equally have occurred had it never existed nor been even thought of at all.

The second of the misapprehensions above alluded to, relates to the well-known suspension, during some years past, of all progress in the construction of the Difference Engine. Respecting the circumstances which have interfered with the actual completion of either invention, we offer no opinion; and in fact are not possessed of the data for doing so, had we the inclination. But we know that some persons suppose these obstacles (be they what they may) to have arisen *in consequence* of the subsequent invention of the Analytical Engine while the former was in progress. We have ourselves heard it even *lamented* that an idea should ever have occurred at all, which had turned out to be merely the means of arresting what was already in a course of successful execution, without substituting the superior invention in its stead. This notion we can contradict in the most unqualified manner. The progress of the Difference Engine had long been suspended, before there were even the least crude glimmerings of any invention superior to it. Such glimmerings, therefore, and their subsequent development, were in no way the original *cause* of that suspension; although, where difficulties of some kind or other evidently already existed, it was not perhaps calculated to remove or lessen them that an invention should have been meanwhile thought of, which, while including all that the first was capable of, possesses powers so extended as to eclipse it altogether.

We leave it for the decision of each individual (*after he has possessed himself* of competent information as to the characteristics of each engine), to determine how far it ought to be matter of regret that such an accession has been made to the powers of human science, even if it *has* (which we greatly doubt) increased to a certain limited extent some already existing difficulties that had arisen in the way of completing a valuable but lesser work. We leave it for each to satisfy himself as to the wisdom of desiring the obliteration (were that now possible) of all records of the more perfect invention, in order that the comparatively limited one might be finished. The Difference Engine would doubtless fulfil all those practical objects which it was originally destined for. It would certainly calculate all the tables that are more directly necessary for the physical purposes of life, such as nautical and other computations. Those who incline to very strictly utilitarian views, may perhaps feel that the peculiar powers of the Analytical Engine bear upon questions of abstract and speculative science, rather than upon those involving every-day and ordinary human interests. These persons being likely to possess but little sympathy, or possibly acquaintance, with any branches of science which they do not find to be *useful* (according to *their* definition of that word), may conceive that the undertaking of that engine, now that the other one is already in progress, would be a barren and unproductive laying out of yet more money and labour; in fact, a work of supererogation. Even in the utilitarian aspect, however, we do not doubt that very valuable practical results would be developed by the extended faculties of the Analytical Engine: some of which results we think we could now hint at, had we the space; and others, which it may not yet be possible to foresee, but which would be brought forth by the daily increasing requirements of science, and by a more intimate practical acquaintance with the powers of the engine, were it in actual existence.

On general grounds, both of an *à priori* description as well as those founded on the scientific history and experience of mankind, we see strong presumptions that such would be the case. Nevertheless all will probably concur in feeling that the completion of the Difference Engine would be far preferable to the non-completion of any calculating engine at all. With whomsoever or wheresoever may rest the present causes of difficulty that apparently exist towards either the completion of the old engine, or the commencement of the new one, we trust they will not ultimately result in this generation's being acquainted with these inventions through the medium of pen, ink and paper merely; and still more do we hope, that for the honour of our country's reputation in the future pages of history, these causes will not lead to the completion of the undertaking by some *other* nation or government. This could not but be matter of just regret; and equally so, whether the obstacles may have originated in private interests and feelings, in considerations of a more public description, or in causes combining the nature of both such solutions.

We refer the reader to the 'Edinburgh Review' of July 1834, for a very able account of the Difference Engine. The writer of the article we allude to, has selected as his prominent matter for exposition, a wholly different view of the subject from that which M. Menabrea has chosen. The former chiefly treats it under its mechanical aspect, entering but slightly into the mathematical principles of which that engine is the representative, but giving, in considerable length, many details of the mechanism and contrivances by means of which it tabulates the various orders of differences. M. Menabrea, on the contrary, exclusively developes the analytical view; taking it for granted that mechanism is able to perform certain processes, but without attempting to explain *how*; and devoting his whole attention to explanations and illustrations of the manner in which analytical laws can be so arranged and combined as to bring every branch of that vast subject within the grasp of the assumed powers of mechanism. It is obvious that, in the invention of a calculating engine, these two branches of the subject are equally essential fields of investigation, and that on their mutual adjustment, one to the other, must depend all success. They must be made to meet each other, so that the weak points in the powers of either department may be compensated by the strong points in those of the other. They are indissolubly connected, though so different in their intrinsic nature that perhaps the same mind might not be likely to prove equally profound or successful in both. We know those who doubt whether the powers of mechanism will in practice prove adequate in all respects to the demands made upon them in the working of such complicated trains of machinery as those of the above engines, and who apprehend that unforeseen practical difficulties and disturbances will arise in the way of accuracy and of facility of operation. The Difference Engine, however, appears to us to be in a great measure an answer to these doubts. It is complete as far as it goes, and it does work with all the anticipated success. The Analytical Engine, far from being more complicated, will in many respects be of simpler construction; and it is a remarkable circumstance attending it, that with very *simplified* means it is so much more powerful.

The article in the 'Edinburgh Review' was written some time previous to the occurrence of any ideas such as afterwards led to the invention of the Analytical Engine; and in the nature of the Difference Engine there is much less that would invite a writer to take exclusively, or even prominently, the mathematical view of it, than in that of the Analytical Engine; although mechanism has undoubtedly gone much further to meet mathematics, in the case of this engine, than of the former one. Some publication embracing the *mechanical* view of the Analytical Engine is a desideratum which we trust will be supplied before long.

Those who may have the patience to study a moderate quantity of rather dry details, will find ample compensation, after perusing the article of 1834, in the clearness with which a succinct view will have been attained of the various practical steps through which mechanism can accomplish certain processes; and they will also find themselves still further capable of appreciating M. Menabrea's more comprehensive and generalized memoir. The very difference in the style and object of these two articles, makes them peculiarly valuable to each other; at least for the purposes of those who really desire something more than a merely superficial and popular comprehension of the subject of calculating engines. A. A. L.

Note B.—Page 676.

That portion of the Analytical Engine here alluded to is called the storehouse. It contains an indefinite number of the columns of discs described by M. Menabrea. The reader may picture to himself a pile of rather large draughtsmen heaped perpendicularly one above another to a considerable height, each counter having the digits from 0 to 9 inscribed on its *edge* at equal intervals; and if he then conceives that the counters do not actually lie one upon another so as to be in contact, but are fixed at small intervals of vertical distance on a common axis which passes perpendicularly through their centres, and around which each disc can *revolve horizontally* so that any required digit amongst those inscribed on its margin can be brought into view, he will have a good idea of one of these columns. The *lowest* of the discs on any column belongs to the units, the next above to the tens, the next above this to the hundreds, and so on. Thus, if we wished to inscribe 1845 on a column of the engine, it would stand thus:—

1
3
4
5

In the Difference Engine there are seven of these columns placed side by side in a row, and the working mechanism extends behind them; the general form of the whole mass of machinery is that of a quadrangular prism (more or less approaching to the cube); the results always appearing on that perpendicular face of the engine which contains the columns of discs, opposite to which face a spectator may place himself. In the Analytical Engine there would be many more of these columns, probably at least two hundred. The precise form and arrangement which the whole mass of its mechanism will assume is not yet finally determined.

We may conveniently represent the columns of discs on paper in a diagram like the following:—

V_1	V_2	V_3	V_4	&c.
○	○	○	○	&c.
0	0	0	0	
0	0	0	0	
0	0	0	0	&c.
0	0	0	0	
□	□	□	□	&c.

The V's are for the purpose of convenient reference to any column, either in writing or speaking, and are consequently numbered. The reason why the letter V is chosen for this purpose in preference to any other letter, is because these columns are designated (as the reader will find in proceeding with the Memoir) the *Variables*, and sometimes the *Variable columns*, or the *columns of Variables*. The origin of this appellation is, that the values on the columns are destined to change, that is to *vary*, in every conceivable manner. But it is necessary to guard against the natural misapprehension that the columns are only intended to receive the values of the *variables* in an analytical formula, and not of the *constants*. The columns are called Variables on a ground wholly unconnected with the *analytical* distinction between constants and variables. In order to prevent the possibility of confusion, we have, both in the translation and in the notes, written Variable with a capital letter when we use the word to signify a *column of the engine*, and variable with a small letter when we mean the *variable of a formula*. Similarly, *Variable-cards* signify any cards that belong to a column of the engine.

To return to the explanation of the diagram: each circle at the top is intended to contain the algebraic sign + or −, either of which can be substituted* for the other, according as the number represented on the column below is positive or negative. In a similar manner any other purely *symbolical* results of algebraical processes might be made to appear in these circles. In Note A. the practicability of developing *symbolical* with no less ease than *numerical* results has been touched on.

The zeros beneath the *symbolic* circles represent each of them a disc, supposed to have the digit 0 presented in front. Only four tiers of zeros have been figured in the diagram, but these may be considered as representing thirty or forty, or any number of tiers of discs that may be required. Since each disc can present any digit, and each circle any sign, the discs of every column may be so adjusted† as to express any positive or negative number whatever within the limits of the machine; which limits depend on the *perpendicular* extent of the mechanism, that is, on the number of discs to a column.

* A fuller account of the manner in which the *signs* are regulated, is given in Mons. Menabrea's Memoir, pages 682, 683. He himself expresses doubts (in a note of his own at the bottom of the latter page) as to his having been likely to hit on the precise methods really adopted; his explanation being merely a conjectural one. That it *does* accord precisely with the fact is a remarkable circumstance, and affords a convincing proof how completely Mons. Menabrea has been imbued with the true spirit of the invention. Indeed the whole of the above Memoir is a striking production, when we consider that Mons. Menabrea had had but very slight means for obtaining any adequate ideas respecting the Analytical Engine. It requires however a considerable acquaintance with the abstruse and complicated nature of such a subject, in order fully to appreciate the penetration of the writer who could take so just and comprehensive a view of it upon such limited opportunity.

† This adjustment is done by hand merely.

Each of the squares below the zeros is intended for the inscription of any *general* symbol or combination of symbols we please; it being understood that the number represented on the column immediately above, is the numerical value of that symbol, or combination of symbols. Let us, for instance, represent the three quantities a, n, x, and let us further suppose that $a = 5$, $n = 7$, $x = 98$. We should have—

V_1	V_2	V_3	V_4 &c.
+ *	+	+	+
0	0	0	0
0	0	0	0 &c.
0	0	9	0
5	7	8	0 &c.
[a]	[n]	[x]	[ ]

We may now combine these symbols in a variety of ways, so as to form any required function or functions of them, and we may then inscribe each such function below brackets, every bracket uniting together those quantities (and those only) which enter into the function inscribed below it. We must also, when we have decided on the particular function whose numerical value we desire to calculate, assign another column to the right-hand for receiving the *results*, and must inscribe the function in the square below this column. In the above instance we might have any one of the following functions:—

$$a x^n, \quad x^{a^n}, \quad a \,.\, n \,.\, x, \quad \frac{a}{n} x, \quad a + n + x, \text{ \&c. \&c.}$$

Let us select the first. It would stand as follows, previous to calculation:—

V_1	V_2	V_3	V_4 &c.
+	+	+	+
0	0	0	0 &c.
0	0	0	0
0	0	9	0
5	7	8	0 &c.
[a]	[n]	[x]	[ax^n] &c.

$a x_n$

The data being given, we must now put into the engine the cards proper for directing the operations in the case of the particular function chosen. These operations would in this instance be,—

Firstly, six multiplications in order to get x^n ($= 98^7$ for the above particular data).

Secondly, one multiplication in order then to get $a \,.\, x^n$ ($= 5.98^7$).

In all, seven multiplications to complete the whole process. We may thus represent them:—

$$(\times, \times, \times, \times, \times, \times, \times), \text{ or } 7\,(\times).$$

The multiplications would, however, at successive stages in the solution of the problem, operate on pairs of numbers, derived from *different* columns. In other words, the *same operation* would be performed

* It is convenient to omit the circles whenever the signs + or − can be actually represented.

on different *subjects of operation*. And here again is an illustration of the remarks made in the preceding Note on the independent manner in which the engine directs its *operations*. In determining the value of $a x^n$, the *operations* are *homogeneous*, but are distributed amongst different *subjects of operation*, at successive stages of the computation. It is by means of certain punched cards, belonging to the Variables themselves, that the action of the operations is so *distributed* as to suit each particular function. The *Operation-cards* merely determine the succession of operations in a general manner. They in fact throw all that portion of the mechanism included in the *mill*, into a series of different *states*, which we may call the *adding state*, or the *multiplying state*, &c. respectively. In each of these states the mechanism is ready to act in the way peculiar to that state, on any pair of numbers which may be permitted to come within its sphere of action. Only *one* of these operating states of the mill can exist at a time; and the nature of the mechanism is also such that only *one pair of numbers* can be received and acted on at a time. Now, in order to secure that the mill shall receive a constant supply of the proper pairs of numbers in succession, and that it shall also rightly locate the result of an operation performed upon any pair, each Variable has cards of its own belonging to it. It has, first, a class of cards whose business it is to *allow* the number on the Variable to pass into the mill, there to be operated upon. These cards may be called the *Supplying-cards*. *They* furnish the mill with its proper food. Each Variable has, secondly, another class of cards, whose office it is to allow the Variable to *receive* a number *from* the mill. These cards may be called the *Receiving-cards*. *They* regulate the location of results, whether temporary or ultimate results. The Variable-cards in general (including both the preceding classes) might, it appears to us, be even more appropriately designated the Distributive-cards, since it is through their means that the action of the operations, and the results of this action, are rightly *distributed*.

There are *two varieties* of the *Supplying* Variable-cards, respectively adapted for fulfilling two distinct subsidiary purposes: but as these modifications do not bear upon the present subject, we shall notice them in another place.

In the above case of $a x^n$, the Operation-cards merely order seven multiplications, that is, they order the mill to be in the *multiplying state* seven successive times (without any reference to the particular columns whose numbers are to be acted upon). The proper Distributive Variable-cards step in at each successive multiplication, and cause the distributions requisite for the particular case.

For x^{an}	the operations would be	34 (×)
... $a.n.x$	... ...	(×, ×), or 2 (×)
... $\frac{a}{n}.x$	... ...	(÷, ×)
... $a+n+x$	... ...	(+, +), or 2 (+)

The engine might be made to calculate all these in succession. Having completed $a x^n$, the function x^{an} might be written under the brackets instead of $a x^n$, and a new calculation commenced (the appropriate Operation and Variable-cards for the new function of course coming into play). The results would then appear on V_5. So on for any number of different functions of the quantities a, n, x. Each *result* might either permanently remain on its column during the succeeding calculations, so that when all the functions had been computed, their values would simultaneously exist on V_4, V_5, V_6, &c.; or each result might (after being printed off, or used in any specified manner) be effaced, to make way for its successor. The square under V_4 ought, for the latter arrangement, to have the functions $a x^n$, x^{an}, $a n x$, &c. successively inscribed in it.

Let us now suppose that we have *two* expressions whose values have been computed by the engine independently of each other (each having its own group of columns for data and results). Let them be $a x^n$, $b \cdot p \cdot y$. They would then stand as follows on the columns:—

V_1	V_2	V_3	V_4	V_5	V_6	V_7	V_8	V_9
+	+	+	+	+	+	+	+	+
0	0	0	0	0	0	0	0	0
0	0	0	0	0	0	0	0	0
0	0	0	0	0	0	0	0	0
0	0	0	0	0	0	0	0	0
a	n	x	$a x^n$	b	p	y	$b p y$	$\frac{a x^n}{b p y}$

We may now desire to combine together these two *results*, in any manner we please; in which case it would only be necessary to have an additional card or cards, which should order the requisite operations to be performed with the numbers on the two result-columns, V_4 and V_8, and the *result of these further operations* to appear on a new column, V_9. Say that we wish to divide $a x^n$ by $b \cdot p \cdot y$. The numerical value of this division would then appear on the column V_9, beneath which we have inscribed $\frac{a x^n}{b p y}$. The whole series of operations from the beginning would be as follows (n being $= 7$):—

$$\left\{7(\times), 2(\times), \div\right\}, \text{or} \left\{9(\times), \div\right\}.$$

This example is introduced merely to show that we may, if we please, retain separately and permanently any *intermediate* results (like $a x^n$, $b.p.y.$), which occur in the course of processes having an ulterior and more complicated result as their chief and final object $\left(\text{like } \frac{a x^n}{b p y}\right)$.

Any group of columns may be considered as representing a *general* function, until a *special* one has been implicitly impressed upon them through the introduction into the engine of the Operation and Variable-cards made out for a *particular* function. Thus, in the preceding example, V_1, V_2, V_3, V_5, V_6, V_7 represent the *general* function $\phi(a, n, b, p, x, y)$ until the function $\frac{a x^n}{b \cdot p \cdot y}$ has been determined on, and *implicitly* expressed by the placing of the right cards in the engine. The actual working of the mechanism, as regulated by these cards, then *explicitly* developes the value of the function. The inscription of a function under the brackets, and in the square under the result-column, in no way influences the processes or the results, and is merely a memorandum for the observer, to remind him of what is going on. It is the Operation and the Variable-cards only, which in reality determine the function. Indeed it should be distinctly kept in mind that the inscriptions within *any* of the squares, are quite independent of the mechanism or workings of the engine, and are nothing but arbitrary memorandums placed there at pleasure to assist the spectator.

The further we analyse the manner in which such an engine performs its processes and attains its results, the more we perceive how distinctly it places in a true and just light the mutual relations and connexion of the various steps of mathematical analysis, how clearly it separates those things which are in reality distinct and independent, and unites those which are mutually dependent. A. A. L.

Note C.—Page 677.

Those who may desire to study the principles of the Jacquard-loom in the most effectual manner, viz. that of practical observation, have only to step into the Adelaide Gallery or the Polytechnic Institution. In each of these valuable repositories of scientific *illustration*, a weaver is constantly working at a Jacquard-loom, and is ready to give any information that may be desired as to the construction and modes of acting of his apparatus. The volume on the manufacture of silk, in Lardner's Cyclopædia, contains a chapter on the Jacquard-loom, which may also be consulted with advantage.

The mode of application of the cards, as hitherto used in the art of weaving, was not found, however, to be sufficiently powerful for all the simplifications which it was desirable to attain in such varied and complicated processes as those required in order to fulfil the purposes of an Analytical Engine. A method was devised of what was technically designated *backing* the cards in certain groups according to certain laws. The object of this extension is to secure the possibility of bringing any particular card or set of cards into use *any number of times successively* in the solution of one problem. Whether this power shall be taken advantage of or not, in each particular instance, will depend on the nature of the operations which the problem under consideration may require. The process is alluded to by M. Menabrea in page 680, and it is a very important simplification. It has been proposed to use it for the reciprocal benefit of that art, which, while it has itself no apparent connexion with the domains of abstract science, has yet proved so valuable to the latter, in suggesting the principles which, in their new and singular field of application, seem likely to place *algebraical* combinations not less completely within the province of mechanism, than are all those varied intricacies of which *intersecting threads* are susceptible. By the introduction of the system of *backing* into the Jacquard-loom itself, patterns which should possess symmetry, and follow regular laws of any extent, might be woven by means of comparatively few cards.

Those who understand the mechanism of this loom will perceive that the above improvement is easily effected in practice, by causing the prism over which the train of pattern-cards is suspended, to revolve *backwards* instead of *forwards*, at pleasure, under the requisite circumstances; until, by so doing, any particular card, or set of cards, that has done duty once, and passed on in the ordinary regular succession, is brought back to the position it occupied just before it was used the preceding time. The prism then resumes its *forward* rotation, and thus brings the card or set of cards in question into play a second time. This process may obviously be repeated any number of times. A. A. L.

NOTE D.—Page 680.

We have represented the solution of these two equations, with every detail, in a diagram* similar to those used in Note B.; but additional explanations are requisite, partly in order to make this more complicated case perfectly clear, and partly for the comprehension of certain indications and notations not used in the preceding diagrams. Those who may wish to understand Note G. completely, are recommended to pay particular attention to the contents of the present Note, or they will not otherwise comprehend the similar notation and indications when applied to a much more complicated case.

In all calculations, the columns of Variables used may be divided into three classes:—

1st. Those on which the data are inscribed:

2ndly. Those intended to receive the final results:

3dly. Those intended to receive such intermediate and temporary combinations of the primitive data as are not to be permanently retained, but are merely needed for *working with*, in order to attain the ultimate results. Combinations of this kind might properly be called *secondary data*. They are in fact so many *successive stages* towards the final result. The columns which receive them are rightly named *Working-Variables*, for their office is in its nature purely *subsidiary* to other purposes. They develope an intermediate and transient class of results, which unite the original data with the final results.

The Result-Variables sometimes partake of the nature of Working-Variables. It frequently happens that a Variable destined to receive a final result is the recipient of one or more intermediate values successively, in the course of the processes. Similarly, the Variables for data often become Working-Variables, or Result-Variables, or even both in succession. It so happens, however, that in the case of the present equations the three sets of offices remain throughout perfectly separate and independent.

It will be observed, that in the squares below the *Working*-Variables nothing is inscribed. Any one of these Variables is in many cases destined to pass through various values successively during the performance of a calculation (although in these particular equations no instance of this occurs). Consequently no *one fixed* symbol, or combination of symbols, should be considered as properly belonging to a merely *Working*-Variable; and as a general rule their squares are left blank. Of course in this, as in all other cases where we mention a *general* rule, it is understood that many particular exceptions may be expedient.

In order that all the indications contained in the diagram may be completely understood, we shall now explain two or three points, not hitherto touched on. When the value on any Variable is called into use, one of two consequences may be made to result. Either the value may *return* to the Variable after it has been used, in which case it is ready for a second use if needed; or the Variable may be made zero. (We are of course not considering a third case, of not unfrequent occurrence, in which the same Variable is destined to receive the *result* of the very operation which it has just supplied with a number.) Now the ordinary rule is, that the value *returns* to the Variable; unless it has been foreseen that no use for that value can recur, in which case

* See the diagram of page 711.

zero is substituted. At the *end* of a calculation, therefore, every column ought as a general rule to be zero, excepting those for results. Thus it will be seen by the diagram, that when m, the value on V_0, is used for the second time by Operation 5, V_0 becomes 0, since m is not again needed; that similarly, when $(m\,n' - m'\,n)$, on V_{12}, is used for the third time by Operation 11, V_{12} becomes zero, since $(m\,n' - m'n)$ is not again needed. In order to provide for the one or the other of the courses above indicated, there are *two* varieties of the *Supplying* Variable-cards. One of these varieties has provisions which cause the number given off from any Variable to *return* to that Variable after doing its duty in the mill. The other variety has provisions which cause *zero* to be substituted on the Variable, for the number given off. These two varieties are distinguished, when needful, by the respective appellations of the *Retaining* Supply-cards and the *Zero* Supply-cards. We see that the *primary* office (see Note B.) of both these varieties of cards is the same; they only differ in their *secondary* office.

Every Variable thus has belonging to it *one* class of *Receiving* Variable-cards and *two* classes of *Supplying* Variable-cards. It is plain however that only the *one* or the *other* of these two latter classes can be used by any one Variable for *one* operation; never *both* simultaneously; their respective functions being mutually incompatible.

It should be understood that the Variable-cards are not placed in *immediate contiguity* with the columns. Each card is connected by means of wires with the column it is intended to act upon.

Our diagram ought in reality to be placed side by side with M. Menabrea's corresponding table, so as to be compared with it, line for line belonging to each operation. But it was unfortunately inconvenient to print them in this desirable form. The diagram is, in the main, merely another manner of indicating the various relations denoted in M. Menabrea's table. Each mode has some advantages and some disadvantages. Combined, they form a complete and accurate method of registering every step and sequence in all calculations performed by the engine.

No notice has yet been taken of the *upper* indices which are added to the left of each V in the diagram; an addition which we have also taken the liberty of making to the V's in M. Menabrea's tables of pages 681, 684, since it does not *alter* anything therein represented by him, but merely *adds* something to the previous indications of those tables. The *lower* indices are obviously indices of *locality* only, and are wholly independent of the operations performed or of the results obtained, their value continuing unchanged during the performance of calculations. The *upper* indices, however, are of a different nature. Their office is to indicate any *alteration* in the value which a Variable represents; and they are of course liable to changes during the processes of a calculation. Whenever a Variable has only zeros upon it, it is called ${}^0V$; the moment a value appears on it (whether that value be placed there arbitrarily, or appears in the natural course of a calculation), it becomes ${}^1V$. If this value gives place to another value, the Variable becomes ${}^2V$, and so forth. Whenever a *value* again gives place to *zero*, the Variable again becomes ${}^0V$, even if it have been ${}^nV$ the moment before. If a *value* then again be substituted, the Variable becomes ${}^{n+1}V$ (as it would have done if it had not passed through the intermediate ${}^0V$); &c. &c. Just before any calculation is commenced, and after the data have been given, and everything adjusted and prepared for setting the mechanism in action, the upper indices of the Variables for data are all unity, and those for the Working and Result-variables are all zero. In this state the diagram represents them*.

There are several advantages in having a set of indices of this nature; but these advantages are perhaps hardly of a kind to be immediately perceived, unless by a mind somewhat accustomed to trace the successive steps by means of which the engine accomplishes its purposes. We have only space to mention in a general way, that the whole notation of the tables is made more consistent by these indices, for they are able to mark a *difference* in certain cases, where there would otherwise be an apparent *identity* confusing in its tendency. In such a case as $V_n = V_p + V_n$ there is more clearness and more consistency with the usual laws of algebraical notation, in being able to write $^{m+1}V_n = {}^qV_p + {}^mV_n$. It is also obvious that the indices furnish a powerful means of tracing back the derivation of any result; and of registering various circumstances concerning that *series of successive substitutions*, of which every *result* is in fact merely the final consequence; circumstances that may in certain cases involve relations which it is important to observe, either for purely analytical reasons, or for practically adapting the workings of the engine to their occurrence. The series of substitutions which lead to the equations of the diagram are as follow:—

$$\overset{(1.)}{{}^1V^*_{16}} = \overset{(2.)}{\frac{{}^1V_{14}}{{}^1V_{12}}} = \overset{(3.)}{\frac{{}^1V_{10} - {}^1V_{11}}{{}^1V_6 - {}^1V_7}} = \overset{(4.)}{\frac{{}^1V_0 \cdot {}^1V_5 - {}^1V_2 \cdot {}^1V_3}{{}^1V_0 \cdot {}^1V_4 - {}^1V_3 \cdot {}^1V_1}} = \frac{d'm - dm'}{mn' - m'n}$$

$$\overset{(1.)}{{}^1V_{15}} = \overset{(2.)}{\frac{{}^1V_{13}}{{}^1V_{12}}} = \overset{(3.)}{\frac{{}^1V_8 - {}^1V_9}{{}^1V_6 - {}^1V_7}} = \overset{(4.)}{\frac{{}^1V_2 \cdot {}^1V_4 - {}^1V_5 \cdot {}^1V_1}{{}^1V_0 \cdot {}^1V_4 - {}^1V_3 \cdot {}^1V_1}} = \frac{dn' - d'n}{mn' - m'n}.$$

There are *three* successive substitutions for each of these equations. The formulæ (2.), (3.), and (4.) are *implicitly* contained in (1.), which latter we may consider as being in fact the *condensed* expression of any of the former. It will be observed that every succeeding substitution must contain *twice* as many V's as its predecessor. So that if a problem require n substitutions, the successive series of numbers for the V's in the whole of them will be $2, 4, 8, 16 \ldots 2^n$.

The substitutions in the preceding equations happen to be of little value towards illustrating the power and uses of the upper indices; for owing to the nature of these particular equations the indices are all unity throughout. We wish we had space to enter more fully into the relations which these indices would in many cases enable us to trace.

M. Menabrea incloses the three centre columns of his table under the general title *Variable-cards*. The V's however in reality all represent the actual *Variable-columns* of the engine, and not the cards that belong to them. Still the title is a very just one, since it is through the special action of certain Variable-cards (when *combined* with the more generalised agency of the Operation-cards) that every one of the particular relations he has indicated under that title is brought about.

Suppose we wish to ascertain how often any *one* quantity, or combi-

* We recommend the reader to trace the successive substitutions backwards from (1.) to (4.), in Mons. Menabrea's Table. This he will easily do by means of the upper and lower indices, and it is interesting to observe how each V successively ramifies (so to speak) into two other V's in some other column of the Table; until at length the V's of the original data are arrived at.

nation of quantities, is brought into use during a calculation. We easily ascertain *this*, from the inspection of any vertical column or columns of the diagram in which that quantity may appear. Thus, in the present case, we see that all the data, and all the intermediate results likewise, are used twice, excepting $(m n' - m' n)$, which is used three times.

The *order* in which it is possible to perform the operations for the present example, enables us to effect all the eleven operations of which it consists, with only *three Operation-cards*; because the problem is of such a nature that it admits of each *class* of operations being performed in a group together; all the multiplications one after another, all the subtractions one after another, &c. The operations are $\{6(\times), 3(-), 2(+)\}$.

Since the very definition of an operation implies that there must be *two* numbers to act upon, there are of course *two Supplying* Variable-cards necessarily brought into action for every operation, in order to furnish the two proper numbers. (See Note B.) Also, since every operation must produce a *result*, which must be placed *somewhere*, each operation entails the action of a *Receiving* Variable-card, to indicate the proper locality for the result. Therefore, at least three times as many Variable-cards as there are *operations* (not *Operation-cards*, for these, as we have just seen, are by no means always as numerous as the *operations*) are brought into use in every calculation. Indeed, under certain contingencies, a still larger proportion is requisite; such, for example, would probably be the case when the same result has to appear on more than one Variable simultaneously (which is not unfrequently a provision necessary for subsequent purposes in a calculation), and in some other cases which we shall not here specify. We see therefore that a great disproportion exists between the amount of *Variable* and of *Operation*-cards requisite for the working of even the simplest calculation.

All calculations do not admit, like this one, of the operations of the same nature being performed in groups together. Probably very few do so without exceptions occurring in one or other stage of the progress; and some would not admit it at all. The *order* in which the operations shall be performed in every particular case, is a very interesting and curious question, on which our space does not permit us fully to enter. In almost every computation a great *variety* of arrangements for the succession of the processes is possible, and various considerations must influence the selection amongst them for the purposes of a Calculating Engine. One essential object is to choose that arrangement which shall tend to reduce to a minimum the *time* necessary for completing the calculation.

It must be evident how multifarious and how mutually complicated are the considerations which the workings of such an engine involve. There are frequently several distinct *sets of effects* going on simultaneously; all in a manner independent of each other, and yet to a greater or less degree exercising a mutual influence. To adjust each to every other, and indeed even to perceive and trace them out with perfect correctness and success, entails difficulties whose nature partakes to a certain extent of those involved in every question where *conditions* are very numerous and inter-complicated; such as for instance the estimation of the mutual relations amongst *statistical* phænomena, and of those involved in many other classes of facts. A. A. L.

Diagram belonging to Note D.

Number of Operations.	Nature of Operations.	Variables for Data.						Working Variables.									Variables for Results.	
		$^1V_0$	$^1V_1$	$^1V_2$	$^1V_3$	$^1V_4$	$^1V_5$	$^0V_6$	$^0V_7$	$^0V_8$	$^0V_9$	$^0V_{10}$	$^0V_{11}$	$^0V_{12}$	$^0V_{13}$	$^0V_{14}$	$^0V_{15}$	$^0V_{16}$
		+	+	+	+	+	+	+	+	+	+	+	+	+	+	+	+	+
		0	0	0	0	0	0	0	0	0	0	0	0	0	0	0	0	0
		0	0	0	0	0	0	0	0	0	0	0	0	0	0	0	0	0
		0	0	0	0	0	0	0	0	0	0	0	0	0	0	0	0	0
		0	0	0	0	0	0	0	0	0	0	0	0	0	0	0	0	0
		m	n	d	m'	n'	d'										$\frac{dn' - d'n}{mn' - m'n} = x$	$\frac{d'm - dm'}{mn' - m'n} = y$
1	×	m	......	......	......	n'	......	mn'										
2	×	......	n	......	m'	......	......	......	$m'n$									
3	×	......	......	d	......	......	......	......	......	dn'								
4	×	......	0	......	......	......	d'	......	......	......	$d'n$							
5	×	0	......	......	......	......	0	......	......	......	......	$d'm$						
6	×	......	......	0	0	......	......	......	......	......	......	......	dm'					
7	−	......	......	......	......	......	......	0	0	......	......	......	......	$(mn' - m'n)$				
8	−	......	......	......	......	......	......	......	......	0	0	......	......	......	$(dn' - d'n)$			
9	−	......	......	......	......	......	......	......	......	......	......	0	0	......	......	$(d'm - dm')$		
10	÷	......	......	......	......	......	......	......	......	......	......	......	......	$(mn' - m'n)$	0	......	$\frac{dn' - d'n}{mn' - m'n} = x$	
11	+	......	......	......	......	......	......	......	......	......	......	......	......	0	......	0	......	$\frac{d'm - dm'}{mn' - m'n} = y$

Note E.—Page 684.

This example has evidently been chosen on account of its brevity and simplicity, with a view merely to explain the *manner* in which the engine would proceed in the case of an *analytical calculation containing variables*, rather than to illustrate the *extent of its powers* to solve cases of a difficult and complex nature. The equations of page 679 are in fact a more complicated problem than the present one.

We have not subjoined any diagram of its development for this new example, as we did for the former one, because this is unnecessary after the full application already made of those diagrams to the illustration of M. Menabrea's excellent tables.

It may be remarked that a slight discrepancy exists between the formulæ

$$(a + b\,x')$$
$$(\mathrm{A} + \mathrm{B}\cos' x)$$

given in the Memoir as the *data* for calculation, and the *results* of the calculation as developed in the last division of the table which accompanies it. To agree perfectly with this latter, the data should have been given as

$$(a\,x^0 + b\,x')$$
$$(\mathrm{A}\cos^0 x + \mathrm{B}\cos' x).$$

The following is a more complicated example of the manner in which the engine would compute a trigonometrical function containing variables. To multiply

$$\mathrm{A} + \mathrm{A}_1\cos\theta + \mathrm{A}_2\cos 2\theta + \mathrm{A}_3\cos 3\theta + \ldots$$

by $\mathrm{B} + \mathrm{B}_1\cos\theta$

Let the resulting products be represented under the general form

$$\mathrm{C}_0 + \mathrm{C}_1\cos\theta + \mathrm{C}_2\cos 2\theta + \mathrm{C}_3\cos 3\theta + \ \ldots\ldots \quad (1.)$$

This trigonometrical series is not only in itself very appropriate for illustrating the processes of the engine, but is likewise of much practical interest from its frequent use in astronomical computations. Before proceeding further with it, we shall point out that there are three very distinct classes of ways in which it may be desired to deduce numerical values from any analytical formula.

First. We may wish to find the collective numerical value of the *whole formula*, without any reference to the quantities of which that formula is a function, or to the particular mode of their combination and distribution, of which the formula is the result and representative. Values of this kind are of a strictly arithmetical nature in the most limited sense of the term, and retain no trace whatever of the processes through which they have been deduced. In fact, any one such numerical value may have been attained from an *infinite variety* of data, or of problems. The values for x and y in the two equations (see Note D.), come under this class of numerical results.

Secondly. We may propose to compute the collective numerical value of *each term* of a formula, or of a series, and to keep these results separate. The engine must in such a case appropriate as many columns to *results* as there are terms to compute.

Thirdly. It may be desired to compute the numerical value of various *subdivisions of each term*, and to keep all these results separate. It may be required, for instance, to compute each coefficient separately from its variable, in which particular case the engine must appropriate *two* result-columns to *every term that contains both a variable and coefficient.*

There are many ways in which it may be desired in special cases to distribute and keep separate the numerical values of different parts of an algebraical formula; and the power of effecting such distributions to any extent is essential to the *algebraical* character of the Analytical Engine. Many persons who are not conversant with mathematical studies, imagine that because the business of the engine is to give its results in *numerical notation*, the *nature of its processes* must consequently be *arithmetical* and *numerical*, rather than *algebraical* and *analytical*. This is an error. The engine can arrange and combine its numerical quantities exactly as if they were *letters* or any other *general* symbols; and in fact it might bring out its results in algebraical *notation*, were provisions made accordingly. It might develope three sets of results simultaneously, viz. *symbolic* results (as already alluded to in Notes A. and B.); *numerical* results (its chief and primary object); and *algebraical* results in *literal* notation. This latter however has not been deemed a necessary or desirable addition to its powers, partly because the necessary arrangements for effecting it would increase the complexity and extent of the mechanism to a degree that would not be commensurate with the advantages, where the main object of the invention is to translate into *numerical* language general formulæ of analysis already known to us, or whose laws of formation are known to us. But it would be a mistake to suppose that because its *results* are given in the *notation* of a more restricted science, its *processes* are therefore restricted to those of that science. The object of the engine is in fact to give the *utmost practical efficiency* to the resources of *numerical interpretations* of the higher science of analysis, while it uses the processes and combinations of this latter.

To return to the trigonometrical series. We shall only consider the four first terms of the factor $(A + A_1 \cos\theta + \&c.)$, since this will be sufficient to show the method. We propose to obtain separately the numerical value of *each coefficient* C_0, C_1, &c. of (1.). The direct multiplication of the two factors gives

$$\left.\begin{array}{l} BA + BA_1\cos\theta + BA_2\cos 2\theta + BA_3\cos 3\theta + \ldots\ldots\ldots\ldots \\ + B_1A\cos\theta + B_1A_1\cos\theta.\cos\theta + B_1A_2\cos 2\theta.\cos\theta + B_1A_3\cos 3\theta.\cos\theta \end{array}\right\}(2.)$$

a result which would stand thus on the engine:—

Variables for Data.

V_0	V_1	V_2	V_3	V_{10}	V_{11}
A	A_1	A_2	A_3	B	B_1
	$\cos\theta$	$\cos 2\theta$	$\cos 3\theta$		$\cos\theta$

Variables for Results.

V_{20}	V_{21}	V_{22}	V_{23}	V_{31}	V_{32}	V_{33}	V_{34}
B A	B A_1	B A_2	B A_3	B_1 A	$B_1 A_1$	$B_1 A_2$	$B_1 A_3$
	$\cos\theta$	$\cos 2\theta$	$\cos 3\theta$	$\cos\theta$	$(\cos\theta . \cos\theta)$	:	$(\cos 3\theta . \cos\theta)$
						$(\cos 2\theta . \cos\theta)$	

The variable belonging to each coefficient is written below it, as we have done in the diagram, by way of memorandum. The only further reduction which is at first apparently possible in the preceding result, would be the addition of V_{21} to V_{31} (in which case B_1 A should be effaced from V_{31}). The whole operations from the beginning would then be—

First Series of Operations.	Second Series of Operations.	Third Series, which contains only one (final) operation.
${}^1V_{10} \times {}^1V_0 = {}^1V_{20}$	${}^1V_{11} \times {}^1V_0 = {}^1V_{31}$	${}^1V_{21} + {}^1V_{31} = {}^2V_{21}$, and V_{31} becomes $=0$.
${}^1V_{10} \times {}^1V_1 = {}^1V_{21}$	${}^1V_{11} \times {}^1V_1 = {}^1V_{32}$	
${}^1V_{10} \times {}^1V_2 = {}^1V_{22}$	${}^1V_{11} \times {}^1V_2 = {}^1V_{33}$	
${}^1V_{10} \times {}^1V_3 = {}^1V_{23}$	${}^1V_{11} \times {}^1V_3 = {}^1V_{34}$	

We do not enter into the same detail of *every* step of the processes as in the examples of Notes D. and G., thinking it unnecessary and tedious to do so. The reader will remember the meaning and use of the upper and lower indices, &c., as before explained.

To proceed: we know that

$$\cos n\theta . \cos\theta = \tfrac{1}{2}\cos \overline{n+1}\,\theta + \tfrac{1}{2}\overline{n-1} . \theta \quad . \quad . \quad .. \quad (3.)$$

Consequently, a slight examination of the second line of (2.) will show that by making the proper substitutions, (2.) will become

C_0	C_1	C_2	C_3	C_4
B A	$+$B A_1 . $\cos\theta$	$+$B A_2 . $\cos 2\theta$	$+$B A_3 . $\cos 3\theta$	
	$+B_1$ A . $\cos\theta$			
$+\frac{1}{2} B_1 A_1$		$+\frac{1}{2} B_1 A_1 . \cos 2\theta$		
	$+\frac{1}{2} B_1 A_2 . \cos\theta$		$+\frac{1}{2} B_1 A_2 . \cos 3\theta$	
		$+\frac{1}{2} B_1 A_3 . \cos 2\theta$		$+\frac{1}{2} B_1 A_3 . \cos 4\theta$

These coefficients should respectively appear on

V_{20} V_{21} V_{22} V_{23} V_{24}.

We shall perceive, if we inspect the particular arrangement of the results in (2.) on the Result-columns as represented in the diagram, that, in order to effect this transformation, each successive coefficient upon V_{32}, V_{33}, &c. (beginning with V_{32}), must through means of proper cards be divided by *two**; and that one of the halves thus

* This division would be managed by ordering the number two to appear on any separate new column which should be conveniently situated for the purpose, and then directing this column (which is in the strictest sense a *Working-Variable*) to divide itself successively with V_{32}, V_{33}, &c.

obtained must be added to the coefficient on the Variable which precedes it by ten columns, and the other half to the coefficient on the Variable which precedes it by twelve columns; V_{32}, V_{33}, &c. themselves becoming zeros during the process.

This series of operations may be thus expressed:—

Fourth Series.

$$\begin{cases} {}^1V_{32} \div 2 + {}^1V_{22} = {}^2V_{22} = B\,A_2 + \tfrac{1}{2}\,B_1\,A_1 \\ {}^1V_{32} \div 2 + {}^1V_{20} = {}^2V_{20} = B\,A + \tfrac{1}{2}\,B_1\,A_1 \,\ldots\ldots\ldots = C_0 \end{cases}$$

$$\begin{cases} {}^1V_{33} \div 2 + {}^1V_{23} = {}^2V_{23} = B\,A_3 + \tfrac{1}{2}\,B_1\,A_2 \,\ldots\ldots\ldots = C_3{}^{*} \\ {}^1V_{33} \div 2 + {}^2V_{21} = {}^3V_{21} = B\,A_1 + B_1\,A + \tfrac{1}{2}\,B_1\,A_2 = C_1 \end{cases}$$

$$\begin{cases} {}^1V_{34} \div 2 + {}^0V_{24} = {}^1V_{24} = \tfrac{1}{2}\,B_1\,A_3 \,\ldots\ldots\ldots\ldots\ldots = C_4 \\ {}^1V_{34} \div 2 + {}^2V_{22} = {}^3V_{22} = B\,A_2 + \tfrac{1}{2}\,B_1\,A_1 + \tfrac{1}{2}\,B_1\,A_3 = C_2. \end{cases}$$

The calculation of the coefficients C_0, C_1, &c. of (1.), would now be completed, and they would stand ranged in order on V_{20}, V_{21}, &c. It will be remarked, that from the moment the fourth series of operations is ordered, the Variables V_{31}, V_{32}, &c. cease to be *Result*-Variables, and become mere *Working*-Variables.

The substitution made by the engine of the processes in the second side of (3.) for those in the first side, is an excellent illustration of the manner in which we may arbitrarily order it to substitute any function, number, or process, at pleasure, for any other function, number or process, on the occurrence of a specified contingency.

We will now suppose that we desire to go a step further, and to obtain the numerical value of each *complete* term of the product (1.), that is of each *coefficient and variable united*, which for the $(n+1)$th term would be $C_n . \cos n\,\theta$.

We must for this purpose place the variables themselves on another set of columns, V_{41}, V_{42}, &c., and then order their successive multiplication by V_{21}, V_{22}, &c., each for each. There would thus be a final series of operations as follows:—

Fifth and Final Series of Operations.

$$\begin{aligned} {}^2V_{20} \times {}^0V_{40} &= {}^1V_{40} \\ {}^3V_{21} \times {}^0V_{41} &= {}^1V_{41} \\ {}^3V_{22} \times {}^0V_{42} &= {}^1V_{42} \\ {}^2V_{23} \times {}^0V_{43} &= {}^1V_{43} \\ {}^1V_{24} \times {}^0V_{44} &= {}^1V_{44} \end{aligned}$$

(N.B. that V_{40} being intended to receive the coefficient on V_{20} which has *no* variable, will only have $\cos 0\,\theta\,(=1)$ inscribed on it, preparatory to commencing the fifth series of operations.)

From the moment that the fifth and final series of operations is ordered, the Variables V_{20}, V_{21}, &c. then in their turn cease to be *Result*-

* It should be observed, that were the rest of the factor $(A + A\cos\theta + \&c.)$ taken into account, instead of *four* terms only, C_3 would have the additional term $\tfrac{1}{2}\,B_1\,A_4$; and C_4 the two additional terms, $B\,A_4$, $\tfrac{1}{2}\,B_1\,A_5$. This would indeed have been the case had even *six* terms been multiplied.

Variables and become mere *Working*-Variables; V_{40}, V_{41}, &c. being now the recipients of the ultimate results.

We should observe, that if the variables cos θ, cos $2\,\theta$, cos $3\,\theta$, &c. are furnished, they would be placed directly upon V_{41}, V_{42}, &c., like any other data. If not, a separate computation might be entered upon in a separate part of the engine, in order to calculate them, and place them on V_{41}, &c.

We have now explained how the engine might compute (1.) in the most direct manner, supposing we knew nothing about the *general* term of the resulting series. But the engine would in reality set to work very differently, whenever (as in this case) we *do* know the law for the general term.

The two first terms of (1.) are

$$(B\,A + \tfrac{1}{2}\,B_1\,A_1) + (\overline{B\,A_1 + B_1\,A + \tfrac{1}{2}\,B_1\,A_2}\,.\cos\theta) \ldots\ldots (4.)$$

and the general term for all after these is

$$(B\,A_n + \tfrac{1}{2}\,B_1\,.\,\overline{A_{n-1} + A_{n+1}})\cos n\,\theta \ldots\ldots\ldots\ldots\ldots (5.)$$

which is the coefficient of the $(n+1)^{\text{th}}$ term. The engine would calculate the two first terms by means of a separate set of suitable Operation-cards, and would then need another set for the third term; which last set of Operation-cards would calculate all the succeeding terms *ad infinitum*; merely requiring certain new Variable-cards for each term to direct the operations to act on the proper columns. The following would be the successive sets of operations for computing the coefficients of $n + 2$ terms:—

$$(\times, \times, \div, +), (\times, \times, \times, \div, +, +), n\,(\times, +, \times, \div, +).$$

Or we might represent them as follows, according to the numerical order of the operations:—

$$(1, 2 \ldots 4), (5, 6 \ldots 10), n\,(11, 12 \ldots 15).$$

The brackets, it should be understood, point out the relation in which the operations may be *grouped*, while the comma marks *succession*. The symbol + might be used for this latter purpose, but this would be liable to produce confusion, as + is also necessarily used to represent one class of the actual operations which are the subject of that succession. In accordance with this meaning attached to the comma, care must be taken when any one group of operations recurs more than once, as is represented above by $n\,(11 \ldots 15)$, not to insert a comma after the number or letter prefixed to that group. $n, (11 \ldots 15)$ would stand for *an operation n, followed by the group of operations* $(11 \ldots 15)$; instead of denoting *the number of groups which are to follow each other*.

Wherever a *general term* exists, there will be a *recurring group* of operations, as in the above example. Both for brevity and for distinctness, a *recurring group* is called a *cycle*. A *cycle* of operations, then, must be understood to signify any *set of operations* which is repeated *more than once*. It is equally a *cycle*, whether it be repeated *twice* only, or an indefinite number of times; for it is the fact *of a repetition occurring at all* that constitutes it such. In many cases of analysis there is a *recurring group* of one or more *cycles*; that is, a *cycle of a cycle*, or a *cycle of cycles*. For instance: suppose we wish to divide a series by a series,

$$(1.)\qquad \frac{a + b\,x + c\,x^2 + \dots\dots}{a' + b'\,x + c'\,x^2 + \dots\dots},$$

it being required that the result shall be developed, like the dividend and the divisor, in successive powers of x. A little consideration of (1.), and of the steps through which algebraical division is effected, will show that (if the denominator be supposed to consist of p terms) the first partial quotient will be completed by the following operations:—

$$(2.)\qquad \Big\{(+), p\,(\times, -)\Big\} \quad \text{or} \; \Big\{(1), p\,(2, 3)\Big\},$$

that the second partial quotient will be completed by an exactly similar set of operations, which acts on the remainder obtained by the first set, instead of on the original dividend. The whole of the processes therefore that have been gone through, by the time the *second* partial quotient has been obtained, will be,—

$$(3.)\qquad 2\Big\{(\div), p\,(\times, -)\Big\} \quad \text{or}\; 2\Big\{(1), p\,(2, 3)\Big\},$$

which is a cycle that includes a cycle, or a cycle of the second order. The operations for the *complete* division, supposing we propose to obtain n terms of the series constituting the quotient, will be,—

$$(4.)\qquad n\Big\{(\div), p\,(\times, -)\Big\} \quad \text{or}\; n\Big\{(1), p\,(2, 3)\Big\}.$$

It is of course to be remembered that the process of algebraical division in reality continues *ad infinitum*, except in the few exceptional cases which admit of an exact quotient being obtained. The number n in the formula (4.), is always that of the number of terms we propose to ourselves to obtain; and the nth partial quotient is the coefficient of the $(n-1)$th power of x.

There are some cases which entail *cycles of cycles of cycles*, to an indefinite extent. Such cases are usually very complicated, and they are of extreme interest when considered with reference to the engine. The algebraical development in a series, of the nth function of any given function, is of this nature. Let it be proposed to obtain the nth function of

$$(5.)\qquad \phi\,(a, b, c \,\dots\dots\, x),\; x \text{ being the variable.}$$

We should premise that we suppose the reader to understand what is meant by an nth function. We suppose him likewise to comprehend distinctly the difference between developing *an* nth *function algebraically*, and merely *calculating an* nth *function arithmetically*. If he does not, the following will be by no means very intelligible; but we have not space to give any preliminary explanations. To proceed: the law, according to which the successive functions of (5.) are to be developed, must of course first be fixed on. This law may be of very various kinds. We may propose to obtain our results in successive *powers* of x, in which case the general form would be

$$C + C_1\,x + C_2\,x^2 + \&c.,$$

or in successive powers of n itself, the index of the function we are ultimately to obtain, in which case the general form would be

$$C + C_1 n + C_2 n^2 + \&c.,$$

and x would only enter in the coefficients. Again, other functions of x or of n instead of *powers*, might be selected. It might be in addition proposed, that the coefficients themselves should be arranged according to given functions of a certain quantity. Another mode would be to make equations arbitrarily amongst the coefficients only, in which case the several functions, according to either of which it might be possible to develop the nth function of (5.), would have to be determined from the combined consideration of these equations and of (5.) itself.

The *algebraical* nature of the engine (so strongly insisted on in a previous part of this Note) would enable it to follow out any of these various modes indifferently; just as we recently showed that it can distribute and separate the numerical results of any one prescribed series of processes, in a perfectly arbitrary manner. Were it otherwise, the engine could merely *compute the arithmetical* nth *function*, a result which, like any other purely arithmetical results, would be simply a collective number, bearing no traces of the data or the processes which had led to it.

Secondly, the *law* of development for the nth function being selected, the next step would obviously be to develope (5.) itself, according to this law. This result would be the first function, and would be obtained by a determinate series of processes. These in most cases would include amongst them one or more *cycles* of operations.

The third step (which would consist of the various processes necessary for effecting the actual substitution of the series constituting the *first function*, for the *variable* itself) might proceed in either of two ways. It might make the substitution either wherever x occurs in the original (5.), or it might similarly make it wherever x occurs in the first function itself which is the equivalent of (5.). In some cases the former mode might be best, and in others the latter.

Whichever is adopted, it must be understood that the result is to appear arranged in a series following the law originally prescribed for the development of the nth function. This result constitutes the second function; with which we are to proceed exactly as we did with the first function, in order to obtain the third function; and so on, $n - 1$ times, to obtain the nth function. We easily perceive that since every successive function is arranged in a series *following the same law*, there would (after the *first* function is obtained) be a *cycle, of a cycle, of a cycle*, &c. of operations *, one, two, three, up to $n - 1$ times, in order to get the nth function. We say, *after the first function is obtained*, because (for reasons on which we cannot here enter) the *first*

* A cycle that includes n other cycles, successively *contained one within another*, is called a cycle of the $n + 1$th order. A cycle may simply *include* many other cycles, and yet only be of the second order. If a series follows a certain law for a certain number of terms, and then another law for another number of terms, there will be a cycle of operations for every new law; but these cycles will not be *contained one within another*,—they merely *follow each other*. Therefore their number may be infinite without influencing the *order* of a cycle that includes a repetition of such a series.

function might in many cases be developed through a set of processes peculiar to itself, and not recurring for the remaining functions.

We have given but a very slight sketch, of the principal *general* steps which would be requisite for obtaining an nth function of such a formula as (5.). The question is so exceedingly complicated, that perhaps few persons can be expected to follow, to their own satisfaction, so brief and general a statement as we are here restricted to on this subject. Still it is a very important case as regards the engine, and suggests ideas peculiar to itself, which we should regret to pass wholly without allusion. Nothing could be more interesting than to follow out, in every detail, the solution by the engine of such a case as the above; but the time, space and labour this would necessitate, could only suit a very extensive work.

To return to the subject of *cycles* of operations: some of the notation of the integral calculus lends itself very aptly to express them: (2.) might be thus written:—

$$(6.)\qquad (\div), \Sigma(+1)^p(\times, -) \text{ or } (1), \Sigma(+1)^p(2, 3),$$

where p stands for the variable; $(+1)^p$ for the function of the variable, that is, for ϕp; and the limits are from 1 to p, or from 0 to $p-1$, each increment being equal to unity. Similarly, (4.) would be,—

$$(7.)\qquad \Sigma(+1)^n \left\{(\div), \Sigma(+1)^p(\times, -)\right\}$$

the limits of n being from 1 to n, or from 0 to $n-1$,

$$(8.)\qquad \text{or } \Sigma(+1)^n \left\{(1), \Sigma(+1)^p(2, 3)\right\}.$$

Perhaps it may be thought that this notation is merely a circuitous way of expressing what was more simply and as effectually expressed before; and, in the above example, there may be some truth in this. But there is another description of cycles which *can* only effectually be expressed, in a condensed form, by the preceding notation. We shall call them *varying cycles*. They are of frequent occurrence, and include successive cycles of operations of the following nature:—

$$(9.)\ p(1, 2, ..m), \overline{p-1}(1, 2...m), \overline{p-2}(1, 2..m) ... \overline{p-n}(1, 2..m),$$

where each cycle contains the same group of operations, but in which the number of repetitions of the group varies according to a fixed rate, with every cycle. (9.) can be well expressed as follows:—

$$(10.)\qquad \Sigma p(1, 2, ..m), \text{ the limits of } p \text{ being from } p-n \text{ to } p.$$

Independent of the intrinsic advantages which we thus perceive to result in certain cases from this use of the notation of the integral calculus, there are likewise considerations which make it interesting, from the connections and relations involved in this new application. It has been observed in some of the former Notes, that the processes used in analysis form a logical system of much higher generality than the applications to number merely. Thus, when we read over any algebraical formula, considering it exclusively with reference to the processes of the engine, and putting aside for the moment its abstract signification as to the relations of quantity, the symbols $+$, $\times$, &c., in reality represent (as their immediate and proximate effect, when the formula is applied to the engine) that a certain prism which is a part of the mechanism (see Note C.), turns a new face, and thus presents a new card to act on the bundles of levers of the engine; the new card being perforated with holes, which are arranged according to the peculiarities of the operation of addition, or of multiplication, &c. Again, the *numbers* in the preceding formula (8.), each of them really represents one of these very pieces of card that are hung over the prism.

Now in the use made in the formulæ (7.), (8.) and (10.), of the notation of the integral calculus, we have glimpses of a similar new application of the language of the *higher* mathematics. Σ, in reality, here indicates that when a certain number of cards have acted in succession, the prism over which they revolve must *rotate backwards*, so as to bring those cards into their former position; and the limits 1 to n, 1 to p, &c., regulate how often this backward rotation is to be repeated.

A. A. L.

Note F.—Page 688.

There is in existence a beautiful woven portrait of Jacquard, in the fabrication of which 24,000 cards were required.

The power of *repeating* the cards, alluded to by M. Menabrea in page 680, and more fully explained in Note C., reduces to an immense extent the number of cards required. It is obvious that this mechanical improvement is especially applicable wherever *cycles* occur in the mathematical operations, and that, in preparing data for calculations by the engine, it is desirable to arrange the order and combination of the processes with a view to obtain them as much as possible *symmetrically* and in cycles, in order that the mechanical advantages of the *backing* system may be applied to the utmost. It is here interesting to observe the manner in which the value of an *analytical* resource is *met* and *enhanced* by an ingenious *mechanical* contrivance. We see in it an instance of one of those mutual *adjustments* between the purely mathematical and the mechanical departments, mentioned in Note A. as being a main and essential condition of success in the invention of a calculating engine. The nature of the resources afforded by such adjustments would be of two principal kinds. In some cases, a difficulty (perhaps in itself insurmountable) in the one department, would be overcome by facilities in the other; and sometimes (as in the present case) a strong point in the one, would be rendered still stronger and more available, by combination with a corresponding strong point in the other.

As a mere example of the degree to which the combined systems of cycles and of backing can diminish the *number* of cards requisite, we shall choose a case which places it in strong evidence, and which has likewise the advantage of being a perfectly different *kind* of problem from those that are mentioned in any of the other Notes. Suppose it be required to eliminate nine variables from ten simple equations of the form—

$$a x_0 + b x_1 + c x_2 + d x_3 + \ldots\ldots = p \quad (1.)$$

$$a^1 x_0 + b^1 x_1 + c^1 x_2 + d^1 x_3 + \ldots\ldots = p^1 \quad (2.)$$

&c. &c. &c. &c.

We should explain, before proceeding, that it is not our object to consider this problem with reference to the actual arrangement of the data on the Variables of the engine, but simply as an abstract question of the *nature* and *number* of the *operations* required to be performed during its complete solution.

The first step would be the elimination of the first unknown quantity x_0 between the two first equations. This would be obtained by the form—

$$(a^1 a - a a^1) x_0 + (a^1 b - a b^1) x_1 + (a^1 c - a c^1) x_2 + \\ + (a^1 d - a d^1) x_3 + \ldots\ldots\ldots\ldots = a^1 p - a p^1,$$

for which the operations 10 (×, ×, —) would be needed. The second step would be the elimination of x_0 between the second and third equations, for which the operations would be precisely the same. We should then have had altogether the following operations :—

$$10\,(\times, \times, -), 10\,(\times, \times, -), = 20\,(\times, \times, -).$$

Continuing in the same manner, the total number of operations for the complete elimination of x_0 between all the successive pairs of equations, would be—

$$9 . 10\,(\times, \times, -) = 90\,(\times, \times, -).$$

We should then be left with nine simple equations of nine variables from which to eliminate the next variable x_1; for which the total of the processes would be—

$$8 . 9\,(\times, \times, -) = 72\,(\times, \times, -).$$

We should then be left with eight simple equations of eight variables from which to eliminate x_2, for which the processes would be—

$$7 . 8\,(\times, \times, -) = 56\,(\times, \times, -),$$

and so on. The total operations for the elimination of all the variables would thus be—

$$9 . 10 + 8 . 9 + 7 . 8 + 6 . 7 . + 5 . 6 + 4 . 5 + 3 . 4 + 2 . 3 + 1 . 2 = 330.$$

So that *three* Operation-cards would perform the office of 330 such cards.

If we take n simple equations containing $n - 1$ variables, n being a number unlimited in magnitude, the case becomes still more obvious, as the same three cards might then take the place of thousands or millions of cards.

We shall now draw further attention to the fact, already noticed, of its being by no means necessary that a formula proposed for solution should ever have been actually worked out, as a condition for enabling the engine to solve it. Provided we know the *series of operations* to be gone through, that is sufficient. In the foregoing instance this will be obvious enough on a slight consideration. And it is a circumstance which deserves particular notice, since herein may reside a latent value of such an engine almost incalculable in its possible ultimate results. We already know that there are functions whose numerical value it is of importance for the purposes both of abstract and of practical science to ascertain, but whose determination requires processes so lengthy and so complicated, that, although it is possible to arrive at them through great expenditure of time, labour and money, it is yet on these accounts practically almost unattainable; and we can conceive there being some results which it may be *absolutely impossible* in practice to attain with any accuracy, and whose precise determination it may prove highly important for some of the future wants of science in its manifold, complicated and rapidly-developing fields of inquiry, to arrive at.

Without, however, stepping into the region of conjecture, we will mention a particular problem which occurs to us at this moment as being an apt illustration of the use to which such an engine may be turned for determining that which human brains find it difficult or impossible to work out unerringly. In the solution of the famous problem of the Three Bodies, there are, out of about 295 coefficients of lunar perturbations given by M. Clausen (Astro^e. Nachrichten, No. 406) as the result of the calculations by Burg, of two by Damoiseau, and of one by Burckhardt, fourteen coefficients that differ in the nature of their algebraic sign; and out of the remainder there are only 101 (or about one-third) that agree precisely both in signs and in amount. These discordances, which are generally small in individual magnitude, may arise either from an erroneous determination of the abstract coefficients in the development of the problem, or from discrepancies in the data deduced from observation, or from both causes combined. The former is the most ordinary source of error in astronomical computations, and this the engine would entirely obviate.

We might even invent laws for series or formulæ in an arbitrary manner, and set the engine to work upon them, and thus deduce numerical results which we might not otherwise have thought of obtaining. But this would hardly perhaps in any instance be productive of any great practical utility, or calculated to rank higher than as a kind of philosophical amusement. A. A. L.

Note G.—Page 689.

It is desirable to guard against the possibility of exaggerated ideas that might arise as to the powers of the Analytical Engine. In considering any new subject, there is frequently a tendency, first, to *overrate* what we find to be already interesting or remarkable; and, secondly, by a sort of natural reaction, to *undervalue* the true state of the case, when we do discover that our notions have surpassed those that were really tenable.

The Analytical Engine has no pretensions whatever to *originate* any thing. It can do whatever we *know how to order it* to perform. It can *follow* analysis; but it has no power of *anticipating* any analytical relations or truths. Its province is to assist us in making *available* what we are already acquainted with. This it is calculated to effect primarily and chiefly of course, through its executive faculties; but it is likely to exert an *indirect* and reciprocal influence on science itself in another manner. For, in so distributing and combining the truths and the formulæ of analysis, that they may become most easily and rapidly amenable to the mechanical combinations of the engine, the relations and the nature of many subjects in that science are necessarily thrown into new lights, and more profoundly investigated. This is a decidedly indirect, and a somewhat *speculative*, consequence of such an invention. It is however pretty evident, on general principles, that in devising for mathematical truths a new form in which to record and throw themselves out for actual use, views are likely to be induced, which should again react on the more theoretical phase of the subject. There are in all extensions of human power, or additions to human knowledge, various *collateral* influences, besides the main and primary object attained.

To return to the executive faculties of this engine: the question must arise in every mind, are they *really* even able to *follow* analysis in its whole extent? No reply, entirely satisfactory to all minds, can be given to this query, excepting the actual existence of the engine, and actual experience of its practical results. We will however sum up for each reader's consideration the chief elements with which the engine works:—

1. It performs the four operations of simple arithmetic upon any numbers whatever.

2. By means of certain artifices and arrangements (upon which we cannot enter within the restricted space which such a publication as the present may admit of), there is no limit either to the *magnitude* of the *numbers* used, or to the *number* of *quantities* (either variables or constants) that may be employed.

3. It can combine these numbers and these quantities either algebraically or arithmetically, in relations unlimited as to variety, extent, or complexity.

4. It uses algebraic *signs* according to their proper laws, and developes the logical consequences of these laws.

5. It can arbitrarily substitute any formula for any other; effacing the first from the columns on which it is represented, and making the second appear in its stead.

6. It can provide for singular values. Its power of doing this is referred to in M. Menabrea's memoir, page 685, where he mentions the passage of values through zero and infinity. The practicability of causing it arbitrarily to change its processes at any moment, on the occurrence of any specified contingency (of which its substitution of $(\frac{1}{2}\cos.\overline{n+1}\,\theta + \frac{1}{2}\cos.\overline{n-1}\,\theta)$ for $(\cos.n\theta.\cos.\theta)$ explained in Note E., is in some degree an illustration), at once secures this point.

The subject of integration and of differentiation demands some notice. The engine can effect these processes in either of two ways:—

First. We may order it, by means of the Operation and of the Variable-cards, to go through the various steps by which the required *limit* can be worked out for whatever function is under consideration.

Secondly. It may (if we know the form of the limit for the function in question) effect the integration or differentiation by direct* substitu-

* The engine cannot of course compute limits for perfectly *simple* and *uncompounded* functions, except in this manner. It is obvious that it has no power of representing or of manipulating with any but *finite* increments or decrements; and consequently that wherever the computation of limits (or of any other functions) depends upon the *direct* introduction of quantities which either increase or decrease *indefinitely*, we are absolutely beyond the sphere of its powers. Its nature and arrangements are remarkably adapted for taking into account all *finite* increments or decrements (however small or large), and for developing the true and logical modifications of form or value dependent upon differences of this nature. The engine may indeed be considered as including the whole Calculus of Finite Differences; many of whose theorems would be especially and beautifully fitted for development by its processes, and would offer peculiarly interesting considerations. We may mention, as an example, the calculation of the Numbers of Bernoulli by means of the *Differences of Nothing.*

tion. We remarked in Note B., that any *set* of columns on which numbers are inscribed, represents merely a *general* function of the several quantities, until the special function have been impressed by means of the Operation and Variable-cards. Consequently, if instead of requiring the value of the function, we require that of its integral, or of its differential coefficient, we have merely to order whatever particular combination of the ingredient quantities may constitute that integral or that coefficient. In $a\,x^n$, for instance, instead of the quantities

V_0	V_1	V_2	V_3
a	n	x	ax^n

$$\underbrace{\qquad\qquad\qquad}_{ax^n}$$

being ordered to appear on V_3 in the combination $a\,x^n$, they would be ordered to appear in that of

$$a\,n\,x^{n-1}.$$

They would then stand thus:—

V_0	V_1	V_2	V_3
a	n	x	anx^{n-1}

$$\underbrace{\qquad\qquad\qquad}_{a\,n\,x^{n-1}}$$

Similarly, we might have $\frac{a}{n}\,x^{(n+1)}$, the integral of $a\,x_n$.

An interesting example for following out the processes of the engine would be such a form as

$$\int \frac{x^n\,d\,x}{\sqrt{a^2-x^2}},$$

or any other cases of integration by successive reductions, where an integral which contains an operation repeated n times can be made to depend upon another which contains the same $n-1$ or $n-2$ times, and so on until by continued reduction we arrive at a certain *ultimate* form, whose value has then to be determined.

The methods in Arbogast's *Calcul des Dérivations* are peculiarly fitted for the notation and the processes of the engine. Likewise the whole of the Combinatorial Analysis, which consists first in a purely numerical calculation of indices, and secondly in the distribution and combination of the quantities according to laws prescribed by these indices.

We will terminate these Notes by following up in detail the steps through which the engine could compute the Numbers of Bernoulli, this being (in the form in which we shall deduce it) a rather complicated example of its powers. The simplest manner of computing these numbers would be from the direct expansion of

$$\frac{x}{\varepsilon^x-1}=\frac{1}{1+\frac{x}{2}+\frac{x^2}{2.3}+\frac{x^3}{2.3.4}+\&c.} \qquad (1.)$$

which is in fact a particular case of the development of

$$\frac{a + bx + cx^2 + \&c.}{a' + b'x + c'x^2 + \&c.}$$

mentioned in Note E. Or again, we might compute them from the well-known form

$$B_{2n-1} = 2 \cdot \frac{1.2.3 \ldots 2n}{(2\pi)^{2n}} \cdot \left\{1 + \frac{1}{2^{2n}} + \frac{1}{3^{2n}} + \ldots\right\} \qquad (2.)$$

or from the form

$$B_{2n-1} = \frac{\pm 2n}{(2^{2n}-1)2^{2n-1}} \left\{ \begin{array}{l} \frac{1}{2} n^{2n-1} \\ -(n-1)^{2n-1}\left\{1 + \frac{1}{2} \cdot \frac{2n}{1}\right\} \\ +(n-2)^{2n-1}\left\{1 + \frac{2n}{1} + \frac{1}{2} \cdot \frac{2n.(2n-1)}{1.2}\right\} \\ -(n-3)^{2n-1}\left\{ \begin{array}{l} 1 + \frac{2n}{1} + \frac{2n.2n-1}{1.2} + \\ + \frac{1}{2} \cdot \frac{2n.(2n-1).(2n-2)}{1.2.3} \end{array} \right\} \\ + \ldots \qquad \ldots \qquad \ldots \qquad \ldots \end{array} \right\} \qquad (3.)$$

or from many others. As however our object is not simplicity or facility of computation, but the illustration of the powers of the engine, we prefer selecting the formula below, marked (8.). This is derived in the following manner:—

If in the equation

$$\frac{x}{e^x - 1} = 1 - \frac{x}{2} + B_1 \frac{x^2}{2} + B_3 \frac{x^4}{2.3.4} + B_5 \frac{x^6}{2.3.4.5.6} + \ldots . \qquad (4.)$$

(in which B_1, B_3 . . . . , &c. are the Numbers of Bernoulli), we expand the denominator of the first side in powers of x, and then divide both numerator and denominator by x, we shall derive

$$1 = \left(1 - \frac{x}{2} + B_1 \frac{x^2}{2} + B_3 \frac{x^4}{2.3.4} + \ldots\right)\left(1 + \frac{x}{2} + \frac{x^2}{2.3} + \frac{x^3}{2.3.4} \ldots\right) (5.)$$

If this latter multiplication be actually performed, we shall have a series of the general form

$$1 + D_1 x + D_2 x^2 + D_3 x^3 + \ldots\ldots\ldots\ldots\ldots\ldots (6.)$$

in which we see, first, that all the coefficients of the powers of x are severally equal to zero; and secondly, that the general form for D_{2n} the co-efficient of the $2n+1$th *term*, (that is of x^{2n} any *even* power of x), is the following:—

$$\left. \begin{array}{l} \frac{1}{2.3 \ldots 2n+1} - \frac{1}{2} \cdot \frac{1}{2.3 \ldots 2n} + \frac{B_1}{2} \cdot \frac{1}{2.3 \ldots 2n-1} + \frac{B_3}{2.3.4} \cdot \frac{1}{2.3 \ldots 2n-3} + \\ \quad + \frac{B_5}{2.3.4.5.6} \cdot \frac{1}{2.3 \ldots 2n-5} + \ldots + \frac{B_{2n-1}}{2.3 \ldots 2n} \cdot 1 = 0 \end{array} \right\} (7.)$$

Multiplying every term by $(2.3\ldots 2n)$, we have

$$\left.\begin{array}{l} 0=-\frac{1}{2}\cdot\frac{2n-1}{2n+1}+B_1\left(\frac{2n}{2}\right)+B_3\left(\frac{2n.2n-1.2n-2}{2.3.4}\right)+ \\ \quad +B_5\left(\frac{2n.2n-1\ldots\ldots\ldots 2n-4}{2.3.4.5.6}\right)+\ldots.+B_{2n-1} \end{array}\right\} (8.)$$

which it may be convenient to write under the general form:—

$$0=A_0+A_1B_1+A_3B_3+A_5B_5+\ldots+B_{2n-1}\ldots\ldots. (9.)$$

A_1, A_3, &c. being those functions of n which respectively belong to B_1, B_3, &c.

We might have derived a form nearly similar to (8.), from D_{2n-1} the coefficient of any *odd* power of x in (6.); but the general form is a little different for the coefficients of the *odd* powers, and not quite so convenient.

On examining (7.) and (8.), we perceive that, when these formulæ are isolated from (6.) whence they are derived, and considered in themselves separately and independently, n may be any whole number whatever; although when (7.) occurs *as one of the* D's in (6.), it is obvious that n is then not arbitrary, but is always a certain function of the *distance of that* D *from the beginning*. If that distance be $= d$, then

$$2n+1=d, \text{ and } n=\frac{d-1}{2} \text{ (for any } even \text{ power of } x.)$$

$$2n=d, \text{ and } n=\frac{d}{2} \text{ (for any } odd \text{ power of } x.)$$

It is with the *independent* formula (8.) that we have to do. Therefore it must be remembered that the conditions for the value of n are now modified, and that n is a perfectly *arbitrary* whole number. This circumstance, combined with the fact (which we may easily perceive) that whatever n is, every term of (8.) after the $(n+1)$th is $=0$, and that the $(n+1)$th term itself is always $= B_{2n-1}\cdot\frac{1}{1} = B_{2n-1}$, enables us to find the value (either numerical or algebraical) of any nth Number of Bernoulli B_{2n-1}, *in terms of all the preceding ones*, if we but know the values of $B_1, B_3 \ldots. B_{2n-3}$. We append to this Note a Diagram and Table, containing the details of the computation for B_7, (B_1, B_3, B_5 being supposed given).

On attentively considering (8.), we shall likewise perceive that we may derive from it the numerical value of *every* Number of Bernoulli in succession, from the very beginning, *ad infinitum*, by the following series of computations:—

1st Series.—Let $n=1$, and calculate (8.) for this value of n. The result is B_1.

2nd Series.—Let $n=2$. Calculate (8.) for this value of n, substituting the value of B_1 just obtained. The result is B_3.

3rd Series.—Let $n=3$. Calculate (8.) for this value of n, substituting the values of B_1, B_3 before obtained. The result is B_5. And so on, to any extent.

The diagram* represents the columns of the engine when just prepared

* See the diagram at the end of these Notes.

for computing B_{2n-1}, (in the case of $n = 4$); while the table beneath them presents a complete simultaneous view of all the successive changes which these columns then severally pass through in order to perform the computation. (The reader is referred to Note D, for explanations respecting the nature and notation of such tables.)

Six numerical *data* are in this case necessary for making the requisite combinations. These data are 1, 2, n ($=4$), B_1, B_3, B_5. Were $n=5$, the additional datum B_7 would be needed. Were $n=6$, the datum B_9 would be needed; and so on. Thus the actual *number of data* needed will always be $n+2$, for $n=n$; and out of these $n+2$ data, $(\overline{n+2}-3)$ of them are successive Numbers of Bernoulli. The reason why the Bernoulli Numbers used as data, are nevertheless placed on *Result*-columns in the diagram, is because they may properly be supposed to have been previously computed in succession by the *engine* itself; under which circumstances each B will appear as a *result*, previous to being used as a *datum* for computing the succeeding B. Here then is an instance (of the kind alluded to in Note D.) of the same Variables filling more than one office in turn. It is true that if we consider our computation of B_7 as a perfectly *isolated* calculation, we may conclude B_1, B_3, B_5 to have been arbitrarily placed on the columns; and it would then perhaps be more consistent to put them on V_4, V_5, V_6 as data and not results. But we are not taking this view. On the contrary, we suppose the engine to be *in the course of* computing the Numbers to an indefinite extent, from the very beginning; and that we merely single out, by way of example, *one amongst* the successive but distinct series' of computations it is thus performing. Where the B's are fractional, it must be understood that they are computed and appear in the notation of *decimal* fractions. Indeed this is a circumstance that should be noticed with reference to all calculations. In any of the examples already given in the translation and in the Notes, some of the *data*, or of the temporary or permanent results, might be fractional, quite as probably as whole numbers. But the arrangements are so made, that the nature of the processes would be the same as for whole numbers.

In the above table and diagram we are not considering the *signs* of any of the B's, merely their numerical magnitude. The engine would bring out the sign for each of them correctly of course, but we cannot enter on *every* additional detail of this kind, as we might wish to do. The circles for the signs are therefore intentionally left blank in the diagram.

Operation-cards 1, 2, 3, 4, 5, 6 prepare $-\frac{1}{2}\cdot\frac{2n-1}{2n+1}$. Thus, Card 1 multiplies *two* into n, and the three *Receiving* Variable-cards belonging respectively to V_4, V_5, V_6, allow the result $2n$ to be placed on each of these latter columns (this being a case in which a triple receipt of the result is needed for subsequent purposes); we see that the upper indices of the two Variables used, during Operation 1, remain unaltered.

We shall not go through the details of every operation singly, since the table and diagram sufficiently indicate them; we shall merely notice some few peculiar cases.

By Operation 6, a *positive* quantity is turned into a *negative* quantity, by simply subtracting the quantity from a column which has only zero upon it. (The sign at the top of V_8 would become — during this process.)

Operation 7 will be unintelligible, unless it be remembered that if we were calculating for $n = 1$ instead of $n = 4$, Operation 6 would have completed the computation of B_1 itself; in which case the engine, instead of continuing its processes, would have to put B_1 on V_{21}; and then either to stop altogether, or to begin Operations 1, 2....7 all over again for value of n ($= 2$), in order to enter on the computation of B_3; (having however taken care, previous to this recommencement, to make the number on V_3 equal to *two*, by the addition of unity to the former $n = 1$ on that column). Now Operation 7 must either bring out a result equal to zero (if $n = 1$); or a result *greater* than *zero*, as in the present case; and the engine follows the one or the other of the two courses just explained, contingently on the one or the other result of Operation 7. In order fully to perceive the necessity of this *experimental* operation, it is important to keep in mind what was pointed out, that we are not treating a perfectly isolated and independent computation, but one out of a series of antecedent and prospective computations.

Cards 8, 9, 10 produce $-\frac{1}{2}\cdot\frac{2n-1}{2n+1} + B_1\frac{2n}{2}$. In Operation 9 we see an example of an upper index which again becomes a value after having passed from preceding values to zero. V_{11} has successively been ${}^0V_{11}$, ${}^1V_{11}$, ${}^2V_{11}$, ${}^0V_{11}$, ${}^3V_{11}$; and, from the nature of the office which V_{11} performs in the calculation, its index will continue to go through further changes of the same description, which, if examined, will be found to be regular and periodic.

Card 12 has to perform the same office as Card 7 did in the preceding section; since, if n had been $= 2$, the 11th operation would have completed the computation of B_3.

Cards 13 to 20 make A_3. Since A_{2n-1} always consists of $2n - 1$ factors, A_3 has three factors; and it will be seen that Cards 13, 14, 15, 16 make the second of these factors, and then multiply it with the first; and that 17, 18, 19, 20 make the third factor, and then multiply this with the product of the two former factors.

Card 23 has the office of Cards 11 and 7 to perform, since if n were $= 3$, the 21st and 22nd operations would complete the computation of B_5. As our case is B_7, the computation will continue one more stage; and we must now direct attention to the fact, that in order to compute A_7 it is merely necessary precisely to repeat the group of Operations 13 to 20; and then, in order to complete the computation of B_7, to repeat Operations 21, 22.

It will be perceived that every unit added to n in B_{2n-1}, entails an additional repetition of operations (13...23) for the computation of B_{2n-1}. Not only are all the *operations* precisely the same however for every such repetition, but they require to be respectively supplied with numbers from the very *same pairs of columns*; with only the one exception of Operation 21, which will of course need B_5 (from V_{23}) instead of B_3 (from V_{22}). This identity in the *columns* which supply the requisite numbers, must not be confounded with identity in the *values* those columns have upon them and give out to the mill. Most of those values undergo alterations during a performance of the operations (13....23), and consequently the columns present a new set of values for the *next* performance of (13....23) to work on.

At the termination of the *repetition* of operations (13...23) in computing B_7, the alterations in the values on the Variables are, that

$V_6 = 2n - 4$ instead of $2n - 2$.
$V_7 = 6$ . . . . 4.
$V_{10} = 0$ . . . . 1.
$V_{13} = A_0 + A_1 B_1 + A_3 B_3 + A_5 B_5$ instead of $A_0 + A_1 B_1 + A_3 B_3$

In this state the only remaining processes are first: to transfer the value which is on V_{13}, to V_{24}; and secondly to reduce V_6, V_7, V_{13} to zero, and to add* *one* to V_3, in order that the engine may be ready to commence computing B_9. Operations 24 and 25 accomplish these purposes. It may be thought anomalous that Operation 25 is represented as leaving the upper index of V_3 still = unity. But it must be remembered that these indices always begin anew for a separate calculation, and that Operation 25 places upon V_3 the *first* value *for the new calculation*.

It should be remarked, that when the group (13...23) is *repeated*, changes occur in some of the *upper* indices during the course of the repetition: for example, ${}^3V_6$ would become ${}^4V_6$ and ${}^5V_6$.

We thus see that when $n = 1$, nine Operation-cards are used; that when $n = 2$, fourteen Operation-cards are used; and that when $n > 2$, twenty-five Operation-cards are used; but that no *more* are needed, however great n may be; and not only this, but that these same twenty-five cards suffice for the successive computation of all the Numbers from B_1 to B_{2n-1} inclusive. With respect to the number of *Variable*-cards, it will be remembered, from the explanations in previous Notes, that an average of three such cards to each *operation* (not however to each Operation-*card*) is the estimate. According to this the computation of B_1 will require twenty-seven Variable-cards; B_3 forty-two such cards; B_5 seventy-five; and for every succeeding B after B_5, there would be thirty-three additional Variable-cards (since each repetition of the group (13...23) adds eleven to the number of operations required for computing the previous B). But we must now explain, that whenever there is a *cycle of operations*, and if these merely require to be supplied with numbers from the *same pairs of columns* and likewise each operation to place its *result* on the *same* column for every repetition of the whole group, the process then admits of a *cycle of Variable-cards* for effecting its purposes. There is obviously much more symmetry and simplicity in the arrangements, when cases do admit of repeating the Variable as well as the Operation-cards. Our present example is of this nature. The only exception to a *perfect identity* in *all* the processes and columns used, for every repetition of Operations (13...23) is, that Operation 21 always requires one of its factors from a new column, and Operation 24 always puts its result

* It is interesting to observe, that so complicated a case as this calculation of the Bernoullian Numbers, nevertheless, presents a remarkable simplicity in one respect; viz., that during the processes for the computation of *millions* of these Numbers, no other arbitrary modification would be requisite in the arrangements, excepting the above simple and uniform provision for causing one of the data periodically to receive the finite increment unity.

on a new column. But as these variations follow the same law at each repetition, (Operation 21 always requiring its factor from a column *one* in advance of that which it used the previous time, and Operation 24 always putting its result on the column *one* in advance of that which received the previous result), they are easily provided for in arranging the recurring group (or cycle) of Variable-cards.

We may here remark that the average estimate of three Variable-cards coming into use to each operation, is not to be taken as an absolutely and literally correct amount for all cases and circumstances. Many special circumstances, either in the nature of a problem, or in the arrangements of the engine under certain contingencies, influence and modify this average to a greater or less extent. But it is a very safe and correct *general* rule to go upon. In the preceding case it will give us seventy-five Variable-cards as the total number which will be necessary for computing any B after B_5. This is very nearly the precise amount really used, but we cannot here enter into the minutiæ of the few particular circumstances which occur in this example (as indeed at some one stage or other of probably most computations) to modify slightly this number.

It will be obvious that the very *same* seventy-five Variable-cards may be repeated for the computation of every succeeding Number, just on the same principle as admits of the repetition of the thirty-three Variable-cards of Operations (13...23) in the computation of any *one* Number. Thus there will be a *cycle of a cycle* of Variable-cards.

If we now apply the notation for cycles, as explained in Note E, we may express the operations for computing the Numbers of Bernoulli in the following manner:—

$$
\begin{array}{llll}
(1...7), (24, 25) \ldots\ldots\ldots\ldots\ldots\ldots \text{ gives } B_1 & = \text{1st number}; & (n \text{ being} & = 1). \\
(1...7), (8...12), (24, 25) \ldots\ldots\ldots\ldots\ldots\ldots B_3 & = \text{2nd} \ldots\ldots; & (n \ldots\ldots & = 2). \\
(1...7), (8...12), (13...23), (24, 25) \ldots\ldots\ldots B_5 & = \text{3rd} \ldots\ldots; & (n \ldots\ldots & = 3). \\
(1...7), (8...12), 2(13...23), (24, 25) \ldots\ldots\ldots B_7 & = \text{4th} \ldots\ldots; & (n \ldots\ldots & = 4). \\
\ldots\ldots\ldots\ldots\ldots\ldots\ldots\ldots\ldots\ldots\ldots\ldots & & & \\
\ldots\ldots\ldots\ldots\ldots\ldots\ldots\ldots\ldots\ldots\ldots\ldots & & & \\
(1...7), (8...12), \sum(+1)^{n-2}(13...23), (24, 25) \ldots B_{2n-1} & = n\text{th} \ldots\ldots; & (n \ldots\ldots & = n).
\end{array}
$$

Again,

$$
(1...7), (24, 25), \underset{\text{limits 1 to } n}{\sum(+1)^n} \left\{ (1...7), (8...12), \underset{\text{limits 0 to } (n+2)}{\sum(n+2)} (13...23), (24, 25) \right\}
$$

represents the total operations for computing every number in succession, from B_1 to B_{2n-1} inclusive.

In this formula we see a *varying cycle* of the *first* order, and an ordinary cycle of the *second* order. The latter cycle in this case includes in it the varying cycle.

On inspecting the ten Working-Variables of the diagram, it will be perceived, that although the *value* on any one of them (excepting V_4 and V_5) goes through a series of changes, the *office* which each performs is in this calculation *fixed* and *invariable*. Thus V_6 always prepares the *numerators* of the factors of any A; V_7 the *denominators*. V_8 always receives the $(2n-3)$th factor of A_{2n-1}, and V_9 the $(2n-1)$th. V_{10} always decides which of two courses the succeeding processes are to follow, by feeling for the value of n through means of a subtraction; and so on; but we shall not enumerate further. It is desirable in all calculations, so to arrange the processes, that the *offices* performed by the Variables may be as uniform and fixed as possible.

Supposing that it was desired not only to tabulate B_1, B_3, &c., but A_0, A_1, &c.; we have only then to appoint another series of Variables, V_{41}, V_{42}, &c., for receiving these latter results as they are successively produced upon V_{11}. Or again, we may, instead of this, or in addition to this second series of results, wish to tabulate the value of each successive *total* term of the series (8), viz: A_0, $A_1 B_1$, $A_3 B_3$, &c. We have then merely to multiply each B with each corresponding A, as produced; and to place these successive products on Result-columns appointed for the purpose.

The formula (8.) is interesting in another point of view. It is one particular case of the general Integral of the following Equation of Mixed Differences:—

$$\frac{d^2}{dx^2}\left(z_{n+1}\, x^{2n+2}\right) = (2n+1)(2n+2)\, z^n x^{2n}$$

for certain special suppositions respecting z, x and n.

The *general* integral itself is of the form,

$$z_n = f(n).x + f_1(n) + f_2(n).x^{-1} + f_3(n).x^{-3} + \ldots$$

and it is worthy of remark, that the engine might (in a manner more or less similar to the preceding) calculate the value of this formula upon most *other* hypotheses for the functions in the integral, with as much, or (in many cases) with more, ease than it can formula (8.).

A. L. L.

Diagram for the computation by the Engine of the Numbers of Bernoulli. See Note G. (page 722 *et seq.*)

Number of Operation.	Nature of Operation.	Variables acted upon.	Variables receiving results.	Indication of change in the value on any Variable.	Statement of Results.	Data			Working Variables.										Result Variables.			
						$^1V_1$ ○ 0 0 0 1 [1]	$^1V_2$ ○ 0 0 0 2 [2]	$^1V_3$ ○ 0 0 0 4 [n]	$^0V_4$ ○ 0 0 0 0	$^0V_5$ ○ 0 0 0 0	$^0V_6$ ○ 0 0 0 0	$^0V_7$ ○ 0 0 0 0	$^0V_8$ ○ 0 0 0 0	$^0V_9$ ○ 0 0 0 0	$^0V_{10}$ ○ 0 0 0 0	$^0V_{11}$ ○ 0 0 0 0	$^0V_{12}$ ○ 0 0 0 0	$^0V_{13}$ ○ 0 0 0 0	$^1V_{21}$ ○ B_1 in a decimal fraction. [B_1]	$^1V_{22}$ ○ B_3 in a decimal fraction. [B_3]	$^1V_{23}$ ○ B_5 in a decimal fraction. [B_5]	$^0V_{24}$ ○ 0 0 0 0 [B_7]
1	×	$^1V_2 \times {}^1V_3$	$^1V_4, {}^1V_5, {}^1V_6$	$\left\{\begin{matrix} ^1V_2 = {}^1V_2 \\ ^1V_3 = {}^1V_3 \end{matrix}\right\}$	$= 2n$	...	2	n	$2n$	$2n$	$2n$											
2	−	$^1V_4 - {}^1V_1$	$^2V_4$	$\left\{\begin{matrix} ^1V_4 = {}^2V_4 \\ ^1V_1 = {}^1V_1 \end{matrix}\right\}$	$= 2n - 1$	1	...	...	$2n - 1$													
3	+	$^1V_5 + {}^1V_1$	$^2V_5$	$\left\{\begin{matrix} ^1V_5 = {}^2V_5 \\ ^1V_1 = {}^1V_1 \end{matrix}\right\}$	$= 2n + 1$	1	...	...	...	$2n + 1$												
4	÷	$^2V_5 \div {}^2V_4$	$^1V_{11}$	$\left\{\begin{matrix} ^2V_5 = {}^0V_5 \\ ^2V_4 = {}^0V_4 \end{matrix}\right\}$	$= \frac{2n-1}{2n+1}$	...	...	...	0	0	...	...	...	...	...	$\frac{2n-1}{2n+1}$						
5	÷	$^1V_{11} \div {}^1V_2$	$^2V_{11}$	$\left\{\begin{matrix} ^1V_{11} = {}^2V_{11} \\ ^1V_2 = {}^1V_2 \end{matrix}\right\}$	$= \frac{1}{2} \cdot \frac{2n-1}{2n+1}$	...	2	...	...	...	...	...	...	...	...	$\frac{1}{2} \cdot \frac{2n-1}{2n+1}$						
6	−	$^0V_{13} - {}^2V_{11}$	$^1V_{13}$	$\left\{\begin{matrix} ^2V_{11} = {}^0V_{11} \\ ^0V_{13} = {}^1V_{13} \end{matrix}\right\}$	$= -\frac{1}{2} \cdot \frac{2n-1}{2n+1} = A_0$	...	...	...	...	...	...	...	...	...	...	0	....	$-\frac{1}{2} \cdot \frac{2n-1}{2n+1} = A_0$				
7	−	$^1V_3 - {}^1V_1$	$^1V_{10}$	$\left\{\begin{matrix} ^1V_3 = {}^1V_3 \\ ^1V_1 = {}^1V_1 \end{matrix}\right\}$	$= n - 1 \ (= 3)$	1	...	n	...	...	...	...	...	...	$n - 1$							
8	+	$^1V_2 + {}^0V_7$	$^1V_7$	$\left\{\begin{matrix} ^1V_2 = {}^1V_2 \\ ^0V_7 = {}^1V_7 \end{matrix}\right\}$	$= 2 + 0 = 2$	...	2	...	...	...	...	2										
9	÷	$^1V_6 \div {}^1V_7$	$^3V_{11}$	$\left\{\begin{matrix} ^1V_6 = {}^1V_6 \\ ^0V_{11} = {}^3V_{11} \end{matrix}\right\}$	$= \frac{2n}{2} = A_1$	...	...	...	...	...	$2n$	2	...	...	...	$\frac{2n}{2} = A_1$						
10	×	$^1V_{21} \times {}^3V_{11}$	$^1V_{12}$	$\left\{\begin{matrix} ^1V_{21} = {}^1V_{21} \\ ^3V_{11} = {}^3V_{11} \end{matrix}\right\}$	$= B_1 \cdot \frac{2n}{2} = B_1 A_1$	...	...	...	...	...	...	...	...	...	...	$\frac{2n}{2} = A_1$	$B_1 \cdot \frac{2n}{2} = B_1 A_1$	....	B_1			
11	+	$^1V_{12} + {}^1V_{13}$	$^2V_{13}$	$\left\{\begin{matrix} ^1V_{12} = {}^0V_{12} \\ ^1V_{13} = {}^2V_{13} \end{matrix}\right\}$	$= -\frac{1}{2} \cdot \frac{2n-1}{2n+1} + B_1 \cdot \frac{2n}{2}$	...	...	...	...	...	...	...	...	...	...	....	0	$\left\{-\frac{1}{2} \cdot \frac{2n-1}{2n+1} + B_1 \cdot \frac{2n}{2}\right\}$				
12	−	$^1V_{10} - {}^1V_1$	$^2V_{10}$	$\left\{\begin{matrix} ^1V_{10} = {}^2V_{10} \\ ^1V_1 = {}^1V_1 \end{matrix}\right\}$	$= n - 2 \ (= 2)$	1	...	...	...	...	...	...	...	...	$n - 2$							
13	−	$^1V_6 - {}^1V_1$	$^2V_6$	$\left\{\begin{matrix} ^1V_6 = {}^2V_6 \\ ^1V_1 = {}^1V_1 \end{matrix}\right\}$	$= 2n - 1$	1	...	...	...	...	$2n - 1$											
14	+	$^1V_1 + {}^1V_7$	$^2V_7$	$\left\{\begin{matrix} ^1V_1 = {}^1V_1 \\ ^1V_7 = {}^2V_7 \end{matrix}\right\}$	$= 2 + 1 = 3$	1	...	...	...	...	...	3										
15	÷	$^2V_6 \div {}^2V_7$	$^1V_8$	$\left\{\begin{matrix} ^2V_6 = {}^2V_6 \\ ^2V_7 = {}^2V_7 \end{matrix}\right\}$	$= \frac{2n-1}{3}$	...	...	...	...	...	$2n - 1$	3	$\frac{2n-1}{3}$									
16	×	$^1V_8 \times {}^3V_{11}$	$^4V_{11}$	$\left\{\begin{matrix} ^1V_8 = {}^0V_8 \\ ^3V_{11} = {}^4V_{11} \end{matrix}\right\}$	$= \frac{2n}{2} \cdot \frac{2n-1}{3}$	...	...	...	...	...	...	...	0	...	...	$\frac{2n}{2} \cdot \frac{2n-1}{3}$						
17	−	$^2V_6 - {}^1V_1$	$^3V_6$	$\left\{\begin{matrix} ^2V_6 = {}^3V_6 \\ ^1V_1 = {}^1V_1 \end{matrix}\right\}$	$= 2n - 2$	1	...	...	...	...	$2n - 2$											
18	+	$^1V_1 + {}^2V_7$	$^3V_7$	$\left\{\begin{matrix} ^2V_7 = {}^3V_7 \\ ^1V_1 = {}^1V_1 \end{matrix}\right\}$	$= 3 + 1 = 4$	1	...	...	...	...	...	4										
19	÷	$^3V_6 \div {}^3V_7$	$^1V_9$	$\left\{\begin{matrix} ^3V_6 = {}^3V_6 \\ ^3V_7 = {}^3V_7 \end{matrix}\right\}$	$= \frac{2n-2}{4}$	...	...	...	...	...	$2n - 2$	4	...	$\frac{2n-2}{4}$	...	$\left\{\frac{2n}{2} \cdot \frac{2n-1}{3} \cdot \frac{2n-2}{3} = A_3\right\}$						
20	×	$^1V_9 \times {}^4V_{11}$	$^5V_{11}$	$\left\{\begin{matrix} ^1V_9 = {}^0V_9 \\ ^4V_{11} = {}^5V_{11} \end{matrix}\right\}$	$= \frac{2n}{2} \cdot \frac{2n-1}{3} \cdot \frac{2n-2}{4} = A_3$	...	...	...	...	...	...	...	...	0								
21	×	$^1V_{22} \times {}^5V_{11}$	$^0V_{12}$	$\left\{\begin{matrix} ^1V_{22} = {}^1V_{22} \\ ^0V_{12} = {}^2V_{12} \end{matrix}\right\}$	$= B_3 \cdot \frac{2n}{2} \cdot \frac{2n-1}{3} \cdot \frac{2n-2}{3} = B_3 A_3$	...	...	...	...	...	...	...	...	...	...	0	$B_3 A_3$	....	......	B_3		
22	+	$^2V_{12} + {}^2V_{13}$	$^3V_{13}$	$\left\{\begin{matrix} ^2V_{12} = {}^0V_{12} \\ ^2V_{13} = {}^3V_{13} \end{matrix}\right\}$	$= A_0 + B_1 A_1 + B_3 A_3$	...	...	...	...	...	...	...	...	...	...	....	0	$\left\{A_3 + B_1 A_1 + B_3 A_3\right\}$				
23	−	$^2V_{10} - {}^1V_1$	$^3V_{10}$	$\left\{\begin{matrix} ^2V_{10} = {}^3V_{10} \\ ^1V_1 = {}^1V_1 \end{matrix}\right\}$	$= n - 3 \ (= 1)$	1	...	...	...	...	...	...	...	...	$n - 3$							
					Here follows a repetition of Operations thirteen to twenty-three.																	
24	+	$^4V_{13} + {}^0V_{24}$	$^1V_{24}$	$\left\{\begin{matrix} ^4V_{13} = {}^0V_{13} \\ ^0V_{24} = {}^1V_{24} \end{matrix}\right\}$	$= B_7$	...	...	...	...	...	...	...	...	...	...	....	....	....	......	......	......	B_7
25	+	$^1V_1 + {}^1V_3$	$^1V_3$	$\left\{\begin{matrix} ^1V_1 = {}^1V_1 \\ ^1V_3 = {}^1V_3 \\ ^5V_6 = {}^0V_6 \\ ^5V_7 = {}^0V_7 \end{matrix}\right\}$	$= n + 1 = 4 + 1 = 5$; by a Variable-card. by a Variable card.	1	...	$n + 1$	...	...	0	0										

F.4 1855-04-20. Menabrea to the Editor of *Cosmos les Mondes*.

Miscellaneous[21]

The appearance in Cosmos of the speech by Count Rosse, where the noble president of the Royal Society passionately argued for Mr. Babbage's calculating machine, has earned us the honor of a letter from Mr. Menabrea, a highly distinguished Italian scholar who has also dealt with this question. Here is the letter from the Turin mathematician:

"While browsing the April 6th issue of your interesting journal Cosmos, I was pleased to learn that there is serious thought in England of implementing Mr. Charles Babbage's Analytical Engine. In the speech where Count Rosse reports on the steps that have already been taken in this matter, I particularly noted the following passage:

'Before making a decision and launching a campaign, I wrote to several eminent men renowned for their knowledge, asking them if, in their opinion, one could make a significant advance in the realm of theoretical and practical science if the ideas of Mr. Babbage, as presented in the small essay published by him under the title Menabrea, were completely realized. Their responses were unanimously and strongly affirmative.'

Since I am not entirely unfamiliar with this issue, allow me, Sir, to share with you some details that might be of interest to those who are engaged in science. I am indeed the true author of the small piece that Count Rosse attributes to Mr. Babbage under the pseudonym Menabrea. Many years ago, during a trip to Italy, Mr. Babbage stopped for a while in Turin, where he kindly explained to me the main features of his Analytical Engine, which differs fundamentally from the already well-known Difference Engine. To be precise, the problem proposed by Mr. Babbage was so singular that my first feeling was one of doubt. But upon reflection, I convinced myself that the idea of this illustrious scientist was perfectly reasonable and, with the author's consent, I decided to explain the fundamental principles upon which the remarkable instrument in question is based, in an article published in the *Bibliothèque universelle de Genève*, No. 82, October 1842, page 352 and following.

Some time later, an English translation of this same work appeared, titled: *Sketch of the Analytical Engine, invented by Charles Babbage Esq. by L. F. Menabrea of Turin, officer of the military engineers, with notes by the translator.* (Extracted from the *Scientific Memoirs*, vol. iii. — London, printed by Richard and John E. Taylor — Read Liancourt[22] Fleet-street, 1843.) The notes that accompanied the translation of my little memoir were extremely remarkable and revealed, in their author, an uncommon sagacity. I was unaware of the name of this author when, to my great astonishment, I learned from Mr. Babbage himself that the translation and the notes were the work of Lady Ada Lovelace, daughter of Lord Byron, a lady equally distinguished by the loftiness of her mind as by her beauty, and whom death took a few years ago in the prime of her youth.[23]

I believed this work to be forgotten; but since it has just served as a starting point for the efforts now underway with the English government to construct Mr. Babbage's invention, I felt it necessary to take this opportunity to pay homage to the memory of the daughter of the great poet and to draw the attention of scholars, not to my modest work, but to the notes and commentaries that accompany the translation and which serve to make known the purpose and power of the Analytical Engine. This invention resembles nothing conceived to this day, and I could not, within the limits of a single letter, explain its principles; but suffice it to say that, through its means, one could execute a whole series of analytical and numerical operations required for solving a problem as specified, in the same way as in the Jacquard loom, by using punched cards, the designs of a brocaded fabric are produced.

---

21 Menabrea 1855. See also the alternative translation in Babbage 1889.

22 Sic. Red Lion Court.

23 Menabrea would later write in his memoirs, posthumously published in 1971, that he thought the notes were actually supplied by Babbage, Menabrea 1971, 36-7.

Accept, Sir, the expression of the distinguished consideration with which I have the honor to be.

L. F. Menabrea

Member of the Royal Academy of Sciences and of the Sardinian Parliament."

F.5 1884-07-28. Comptes Rendus.

APPLIED MECHANICS. – On the analytical machine of Charles Babbage.
Note by General L. F. MENABREA.[24]

I request the Academy's permission to recall here to its memory a machine design which, many years ago, when announced, produced a great sensation in the scientific world: it is Charles Babbage's Analytical Engine, to which he devoted a large part of his life and almost all his fortune, and died before being able to complete its construction. It should not be confused with the Difference Engine by the same author, nor with other more or less perfected calculating machines. The one in question has a greater scope; its purpose is to perform analytical and arithmetic operations whose laws are defined for it and to print the results, thus saving humans the purely mechanical work required by these operations. The moment to commemorate Charles Babbage's invention seems all the more opportune as the use of machines to record and calculate the results of observations relating to Physics and Astronomy tends to spread. Thus the London meteorological office is equipped with a series of ingenious devices, some of which serve to integrate, others to calculate even trigonometric functions, and provide exact results that could not be obtained with such speed and precision through direct human work. The Analytical Engine encompasses all such operations. It is given the formula relating to the question to be solved; it develops it, applies the coefficients to the variables, executes the arithmetic calculations and displays the printed final numerical result that refers to it. The machine not being a thinking being, but a simple automaton, one cannot ask anything of it that requires the use of intelligence; thus trial-and-error methods are not within its scope.

It consists of different parts, each corresponding to a determined order of operations. To fix ideas, let us take a very simple example, that of the product of two binomials

$$(a + bx)(m + nx) = am + (an + bm)x + bnx^2.$$

The first part of the machine will receive the numbers designated by a, b, m, n and will transport them to a second part, which is the operator and which Babbage calls the mill, where the indicated operations are executed; from there the numerical results expressing the coefficients of the powers of x pass to a third part, which is that of the variables, where each power of x receives its respective coefficient. All these operations are executed by the machine, which acts under the direction of a mechanical controller whose nature I will soon indicate. The numbers are arranged in vertical columns composed of circular disks traversed in their center by a vertical axis, but all able to rotate independently of each other. On the contour of each are traced the ten digits of the arithmetic alphabet, and they successively represent units, tens, hundreds; so that, by arranging one of the digits of each disk on the same vertical generatrix of the column, one thus writes any number. In Babbage's machine, each column being composed of twenty disks,[25] one can, consequently, have numbers of twenty digits. The various parts of the machine are composed of analogous columns that can receive the pre-calculated numbers that are of usual employment.

I will now explain what consists the apparatus directing the operations, or the controller.

[24] Menabrea 1884. Translated from French.
[25] Ed. Extended to forty by Babbage.

The idea was taken from a well-known apparatus, that of Jacquard, which serves in the manufacture of brocaded fabrics. Since, by means of perforated cards corresponding to levers that raise the various threads of a fabric's chain, one can create varied patterns, the author of the analytical machine thought one could apply this same idea to the execution of a calculation. The formulas and their developments, appropriately expressed on cards, are presented to the machine which, being set in motion, like in a Jacquard loom, performs the analytical and mathematical operations requested of it. The solutions of problems by means of the analytical machine must be able to be expressed in algebraic form or in convergent series: the rule of signs, passage through zero or infinity, are within the machine's scope.

Arbogast's method of derivations is considered by Babbage as very appropriate for these various operations.

The speed of execution obtained is such that Babbage claimed to be able to perform with it, in three minutes, a multiplication of two numbers of twenty digits each. The properties of this analytical machine seem a bit fantastic. Well no, Gentlemen: it is partly constructed, and what exists of it could be utilized today. It belongs to General Babbage, son of the author. Several attempts have been made to complete it, but have failed due to the expense and the difficulties inherent to the enterprise, as it would require total dedication, and its study is not easy.

At the beginning of my modest scientific career, I was in personal contact with Babbage, who explained his system to me. I believed I understood it and I made it, many years ago, the subject of an article in the *Bibliothèque universelle de Genève*, n° 82, October 1842. My description was approved by Babbage himself and was translated into English in the *Scientific Memoirs*, vol. III, without the translator's name; but this translation was accompanied by notes of the highest interest, which develop in a luminous manner what I had only been able to indicate incompletely. At first, I had believed that Babbage himself was the author; but, in a letter of August 28, 1843 which follows, he undeceived me by giving me the name of my mysterious translator, who was none other than a very noble and very beautiful English lady, whose name will be transmitted to posterity on the wings of one of the greatest poets of our century: it was Lady Ada Lovelace, the only daughter of Lord Byron.

In the aforementioned Notes, one can see how many questions relating to transcendental analysis can be treated; one finds, among others, the Table of operations, executed by the machine, for the calculation of Bernoulli numbers. It is the above-mentioned translation which, even recently, has served as a basis for discussions which, more than once, have opened on the Analytical Engine.

May these memories, which I exhume at the end of my career, provoke the completion of work which would be precious for Science and a triumph for mechanical art, while at the same time a tribute paid to the memory of a man of genius, as well as to that of the noble lady who, by her example, demonstrated that the fairer half of humankind can have, for the higher Sciences, aptitudes equal to those of the other half which, modestly, chooses to call itself the stronger sex.

[Letter from Babbage, 1843-08-28, see section I.148, p. 425.]

Appendix G

Gilt by Association: Ada's Calculus Homework.

A more technical treatment of the details of Ada's elementary calculus correspondence with Augustus de Morgan is presented below for those who care to follow at this level, along with a reconstruction of the calendar of correspondence, the dates of which are often uncertain.[1]

The correspondence contains a basic confusion early on circa August 1840, at which stage Ada was working through the preliminary chapters of De Morgan's *Differential and Integral Calculus*. Earlier they had discussed fractions, roots, and logarithms. She writes to insist that she believes a function should not be represented by a continuous curve in the Cartesian plane, but rather by something we would recognize as a 'bar chart' at discrete points: 'this curve (according to my interpretation) is alluded to as "the representation of a function, or functions". This does not appear to me to hold good, I should have said that the perpendicular straight lines are the representation of the function, & I do not see any precise relation that the existing curve holds to them'.[2] This shows that she was starting with very little recollection of whatever mathematical instruction she had previously received and was at this stage a rank beginner. Confidently advancing her idiosyncratic interpretation of things, a position to be argued down from, is a persistent pattern throughout.

After some fragmented notes exchanged about the definition of 'nearly equal' using fractions and equations fitted to known points, we next see Ada on November 10 (62r-67v) working on the first few pages of George Peacock's *Collection of Examples of the Applications of the Differential and Integral Calculus* (1820). She was having some difficulty understanding elementary derivatives and (very) basic theorems to simplify their calculation.[3] She was still struggling with this material on November 18 (146r-148v). In the meantime De Morgan had directed her to his *Elements of Algebra* to work up some more basics (see Figure G.1, p. 298). Although her letter shows her wrestling with the definition of functions and simple functional equations—'I feel I only half understand that little Chapter X, and it has already cost me more trouble with less effect than most things have'—it is not clear how much of the earlier material she had mastered—equations, algebraic symbols, exponents, numerical solutions of equations, limits, expressions, series, equality in general. In parallel, she tried limits and derivatives, infinite and divergent series, and complex numbers.

Functional equations proved to be more than a passing problem. She wrote, 'That I do not comprehend at all the means of deducing from a Functional Equation the form which will satisfy it, is I think clear from my being quite unable to solve the example at the end of the Chapter [X]'.[4] This is a mechanical 'substitute, rearrange and simplify' problem: show

---

[1] Transcripts and images of originals are available at `https://www.claymath.org/online-resources/ada-lovelaces-mathematical-papers/`.

[2] n.d. f.11. ADM's reply is on the same undated sheet.

[3] Peacock, Herschel, and Babbage 1820.

[4] Of A. De Morgan 1835. [1840]-11-18? AL to ADM. Wednesday. 146r-148v.

CONTENTS.

Figure G.1: De Morgan's Elements: Contents 1

8 CONTENTS.

CHAPTER VIII.

CHAPTER IX.

CHAPTER X.

CHAPTER XI.

CHAPTER XII.

CHAPTER XIII.

Figure G.2: De Morgan's Elements: Contents 2

that the equation

$$\phi(x+y)+\phi(x-y)=2\phi x\times\phi y$$

is satisfied by

$$\phi(x)=\frac{1}{2}(a^x+a^{-x}).$$

After ten further days of futile work on this problem she wrote again. 'I do not know when I have been so tantalized by anything, & should be ashamed to say how much time I have spent upon it, in vain. These Functional Equations are complete Will-o'-the-Wisps to me. The moment I fancy I have really at last got hold of something tangible & substantial, it all recedes further & further & vanishes again into thin air'.[5] By now De Morgan, who must have gone through several rounds of trying to explain the problem, decided after a final attempt that enough was enough. 'I think you have got all you were meant to get from the chapter on functions [X]. The functional equations which can be fully solved are few in number'. This particular example was emphasized by Dorothy Stein, and also noted by Karen Hunger Parshall.[6]

By mid-January of 1841, Ada was struggling with the definition of derivatives, running aground on the most elementary things. 'It had not struck me that, calling $(x+\theta)=v$, the form $[(x+\theta)^n-x^n]/\theta$ becomes $(v^n-x^n)/(v-x)$'.[7] A month later De Morgan had to laboriously explain to her that in the Mean Value Theorem, as we now call it, the θ solution depends in general on the particular interval $(a, a+h)$:[8]

$$\frac{\varphi(a+h)-\varphi a}{h}=\varphi'(a+\theta h)$$

. There are many such failures in these letters to understand material that is straightforward, and usually already explained in the textbook. None of these are challenging exercises.

A trip to France in Spring of 1841 appears to have interrupted Ada's studies, but when she returned in Summer matters went on much as before, as she laboured her way through the first part of De Morgan's *Differential Calculus*. The arbitrary constants added to indefinite integrals gave her much trouble—we will spare the reader the details, which can be pursued readily in the transcripts along with many other examples. By November, after wrestling with inter alia differential equations, she was close to the end, confessing to De Morgan in one exchange that 'I cannot think how I could be so negligent as to forget that ϵ^{-x} is a function of $(-x)$ which is itself a function of x'.[9] This after a year and a half of study. She may have abandoned work at page 165 of the text (that is where the correspondence ceases).

Finally, it is necessary to consider an attempt to reevaluate Ada's calculus homework by Hollings et al, who believe that their background in the history of mathematics, and (supposedly, but see below) in the work of De Morgan himself, is relevant.[10] It is impossible to mistake the attempt in this to 'pull rank' on Stein and those who have followed her. Their work follows the pattern of an artfully-phrased letter De Morgan wrote to Ada's mother, who had asked him for an opinion about her daughter's mathematical prospects.[11]

The request had placed the mathematician in a delicate position. Lady Byron and the Lovelaces were people of substance and social status few would choose to annoy, even if they could get away with it, quite aside from parental relations. De Morgan resolved

[5] [1840]-11-27. AL to ADM. No day. 149r-151v.
[6] Stein 1985, 89-90, Parshall 1989.
[7] [1841-01]. AL to ADM. No day. 91r-95r
[8] [1841-02-22]. ADM to AL. Monday Evening. 42r-45.
[9] [1841]-11-08. AL to ADM. Monday. 134r-135v.
[10] Hollings, Martin, and Rice 2017a,b, 2018.
[11] 1844-01-21. De Morgan to Lady Byron. See section I.157, p. 429.

this by gently pointing out that Ada was not material for a Cambridge Wrangler (which in his day was precisely what a mathematician was in England), but exactly because she possessed *other* virtues, 'utterly out of the common way for any beginner'. Given her 'desire of distinction', she might instead *become* an 'original mathematical investigator, perhaps of first-rate eminence'. It is an ingenious turn by De Morgan and must be read (in the cold light of the uncomprehending letters we have seen here) as the logical result of a thousand 'why' questions and an irresolvable inability to master algebra. Perhaps she might invent something else?

The Hollings et al version of this reads like a bespoke résumé for Ada, organized into a succession of virtues: 'A keen eye for detail', 'Independent thinking', 'A prescient speculation', 'Research-awareness'. Summoning these in Ada's output takes Hollings et al far and wide to topics the reader will find scant sight of in what is after all *merely elementary homework*. Their technique is to flank her with the company of advanced ideas, a *gilt by association* recasting of *guilt-by-association*. Multiple flanks are contrived, stretching to William Hamilton and his quaternions, only invented in 1843, and Peacock's 'principle of the permanence of equivalent forms'. They also expend much ink on the supposed implications of correctly dating her inherently-uncertain correspondence. Several mathematical errors, even unto the 'History of Mathematics', and logical fallacies are introduced in the process.

Ada had been reading De Morgan's article on 'Negative and Impossible Quantities' in the *Penny Cyclopaedia*. In a letter, she mentioned further extensions to complex numbers, which may be considered (two-dimensional) tuples of real numbers, to higher dimensions. "It cannot help striking me that this extension of Algebra ought to lead to a further extension similar in nature, to Geometry in Three-Dimensions; & that again perhaps to a further extension into some unknown region, & so on ad-infinitum possibly".[12]

This is less interesting than it may appear. There are countless students of mathematics who have had similar ideas when they first encounter the extension of real to complex numbers. Doing something about it is what counts: understanding that the laws of multiplication need to be worked out (pairwise doesn't work, just as it doesn't for complex numbers) to preserve useful properties of that operation. Ada did not even attempt to investigate that let alone solve it, leaving her sentence as a whimsical speculation. This vague mention is all she ever says about it anywhere.

As it happened, De Morgan had worked on this problem himself for the case of three dimensions and failed. He had even explicitly mentioned this fact in his Penny Cyclopaedia article. This is where Ada had seen the idea, for she added a few sentences later 'You do hint in parts of page 136 at the possibility of something of this sort'. It is rather more than a *hint*.

> The subject-matter of the preceding algebra is geometry of only two dimensions; whereas it might be supposed that the application would never be complete until it embraced geometry of three dimensions. No such extension has however yet been made; though it is not unreasonable to suppose that it may be made at some future time.[13]

Hollings et al, in their technical paper, refer to Ada's 'strikingly accurate prediction' (in truth no more than an undeveloped speculation) but concede paragraphs later that she was echoing De Morgan.[14] However, in their book for a more popular audience they completely omit this important qualification and proceed to riff on *quaternions*. William Rowan Hamilton was soon able to prove that a *four* dimensional algebra can be constructed, and Graves

12 [1841]-09-19. AL to ADM. Sunday. 127r-129r. This was noticed by Stein 1985, 79-80.

13 A. De Morgan 1840a, 136. Stein 1985, 79-80 did not call attention to this explicit statement by De Morgan.

14 Hollings, Martin, and Rice 2017b.

would follow later with an eight dimensional one. Hollings et al. make much of the added speculation about dimensions beyond three by Ada but credit is not due where it is not merited: that speculative extension is obvious and had occurred to many others. Indeed it can be shown, as Ada did not attempt to, and had no conception of, that no other dimensions preserve the properties we desire than four and eight. That fact means that this speculation echoed from De Morgan is inaccurate for the three dimensions it actually mentions, and wrong by chance about every other dimension except four and eight, neither of which is mentioned. Regardless, Hamilton's solemn portrait is introduced by Hollings et al in full colour. None of it has anything to do with Ada, who did not work on the problem or show any sign of comprehending what was needed to solve it. This gives the impression that she kept company with advanced research in her era, when she certainly did not. Crowning this, the authors describe *qua*ternions as 'a three-dimensional version of complex numbers'.[15]

Riffing on quaternions is rhythmically accompanied by another theme introduced by Hollings et al, known to posterity as the 'Principle of Permanence of Equivalent Forms'. Again, there is vastly less here than they try to claim. Ada had been puzzled by the fact that $(1+x)^n \times (1+x)^m = (1+x)^{n+m}$ is true for real exponents in general, and not just for whole numbers.[16] De Morgan had justified this in his text with the statement that 'When an algebraical multiplication, or other operation, such as has hitherto been defined, can be proved to produce a certain result in cases where the letters stand for whole numbers, then the same result must be true when the letters stand for fractions, or incommensurable [irrational] numbers, and also when they are negative'.[17] Ada was nonplussed, stating 'I am not at all sure that I like the assumption,' 'It seems to me somewhat a large one, & much more wanting of proof than many things which in Mathematics are rigorously & scrupulously demonstrated'.

The reader will note that the principle De Morgan introduces above—which Ada does no more than confess that she does not understand—is limited by him there to *real* numbers and their sub-classes, though he could have extended it to complex numbers too with suitable conditions. It is this limited claim—which does not invoke *all conceivable numbers*, including those which might be invented (or discovered, depending on the politics of the reader) in the future—that she actually expresses surprise at. Moreover it is undeniably true for the numbers De Morgan had specified when applied to $(1+x)^n \times (1+x)^m = (1+x)^{n+m}$. Today we would simply *define* the operation $a^p = e^{p \log a}$ for any real numbers $a > 0$ and p and then the result immediately follows.

Breezing past these facts, Hollings et al proceed to elaborate a much stronger form of De Morgan's simple principle, which he had *not* invoked in his text, Peacock's 'Principle of Permanence of Equivalent Forms' for *all* numbers (though De Morgan used looser language in his later *replies* to Ada, which she did *not* comment on). This overly strong claim proved to be false once you consider quaternions, those four-dimensional numbers not yet invented (or discovered) at the time these letters were written. Quaternions unlike all the reals do not in general commute. If they did then $xy = yx$ would always be true for any x and y. We don't need to go into that since none of it has anything to do with Ada—except to note that by this technique she is again transported into far more advanced realms than she belongs in, there to be associated with results she did not understand let alone obtain, with pages of material and reams of ideas spuriously introduced.

The authors would even have us believe that Ada was able to get the better of her mentor. Considering the (intuitively obvious, previously mentioned) dependence of the θ in

[15] Hollings, Martin, and Rice 2018, 65-6 though they at least get the dimensions right in Hollings, Martin, and Rice 2017a.

[16] [1840]-11-18? AL to ADM. Wednesday. 146r-148v.

[17] A. De Morgan 1835, 212.

the Mean Value Theorem on the interval in question $(a, a+h)$. Ada had written that 'I neither see the truth of this assertion, nor do I perceive the importance of it (supposing it is true) to the rest of the argument'. De Morgan had patiently explained how one can find that relationship analytically.[18]

Let ψ be the inverse function of ϕ' so that $\psi\phi'x = x$. Then

$$\frac{\phi(a+h)-\phi a}{h} = \phi'(a+\theta h)$$

$$\psi\left(\frac{\phi(a+h)-\phi a}{h}\right) = \psi\phi'(a+\theta h) = a+\theta h$$

$$\theta = \frac{\psi\left(\frac{\phi(a+h)-\phi a}{h}\right)-a}{h}$$

{Say that this is not a function of a and h, if you dare!}

Hollings et al triumphantly point out that the function ϕ' might not be invertible. 'In this case it was actually the professor, and not the student, who was in error'. This is a straightforward logical fallacy. Even if De Morgan, a major figure in 19th century mathematics, had been wrong (he was not) the failure of his explanation would not mean that Ada was correct (she was not). Worse, their 'history of mathematics' has let them down, even though what they miss is contained in the very same volume of *Differential Calculus*. De Morgan had explicitly defined what he meant by 'function', using an expansive definition which comprehends *multiple values*.[19] His function above gives a set of values. This is not the modern usage, but that's why one needs to bear the 'History of Mathematics' in mind.

> we see then that a function may have more values than one for the same value of the variable, and we know from algebra that such functions will arise from the inversion of any direct operation, except only addition.

The authors add insult to injury by attributing this non-sequitur to Ada! 'Consequently, Lovelace's doubts about the validity of his assertion and its use in proving Taylor's Theorem turned out to be well founded'. Ada made no comments about De Morgan's rejoinder. She had merely stated that she did not 'see the truth' of the original statement.

The authors even attempt to situate Ada in a *research* context by late 1841, noting that De Morgan had sent her a paper of his from 1839 on the 'Foundation of Algebra'. The mere fact that he had sent her a paper establishes nothing one way or another. She may simply have asked him for it after seeing a reference somewhere, or it may have come up in one of their conversations. Any idea that he could have expected her to make progress with it is refuted by her reply, in which she confesses that she was unable to understand it. This is no surprise since it concerns her old enemy functional equations.

> In the treatise you sent me on the "Foundation of Algebra", I cannot make out in the least (page 5) ... about the general solution of $\phi^2 x = ax$. I suspect I do not understand the notation $f^{-1}x$. ... I conceive $f^{-1}x$ or $\phi^{-1}x$ may mean "the quantity which having had an operation f or ϕ performed with & upon it, is = x." ... I have considered much over the last half of this page 5, & I can't understand it.[20]

De Morgan defined $f^{-1}x$ in his text, and in the same letter Ada had already used $\cos^{-1} x$.[21]

[18] [1841]-02-19. AL to ADM. Friday. 100r-102v. [1841-02-22]. ADM to AL. Monday Evening. 42r-45.

[19] A. De Morgan 1842b, 52.

[20] [1841]-11-08. AL to ADM. Monday. 134r-135v. She means A. De Morgan 1842a, 177.

[21] The 'last half' is about those will-o'-the wisp functional equations: 'A general solution of $\phi^2 x = ax$ can be expressed when any particular solution ϖx is known. For if $f\varpi f^{-1}x$ be the general solution, we have

$$\phi^2 x = f\varpi^2 f^{-1}x = faf^{-1}x = ax, \text{ or } fax = afx :$$

The authors interpret the mere sharing of this paper—in which Ada had not understood even the fundamentals on page 5 (really 177), involving simple algebraic substitute-and-simplify ideas—very ambitiously. 'Lovelace seemed to be on the cusp of blossoming into real mathematical maturity'. They claim that this is reinforced by the fact that De Morgan also repeated some of his (in retrospect) speculative ideas about divergent series in a letter to her a year earlier.[22] But he had already 'explained' this material in chapter 19 of his *Elements of Algebra* text, which she had been reading with great difficulty at the time.[23] This topic from *a year earlier* could hardly place Ada on any 'cusp' in October 1841, but it does allow the authors to expand on divergent series for quite some time, without shedding any light on Ada's thoughts about them.

To see how absurd this 'cusp' of 'real mathematical maturity' idea really is, in the very same letter Ada had confessed that 'I cannot think how I could be so negligent as to forget that ϵ^{-x} is a function of $(-x)$ which is itself a function of x'.[24] Another remark shows that she still could not understand arbitrary constants of integration. In a worked example De Morgan had found the solution of an indefinite arc-length integral to be $\sqrt{2kx - x^2} + 2k \text{ vers}^{-1}\frac{x}{k}$ + constant.[25] The text proceeds 'Any value of this constant may be used. In fact, if the constant be made = p ...' Ada was puzzled, 'I do not see how it can be said ... that the constant may have any value p'. These arbitrary constants were a remarkably persistent problem for her.[26] De Morgan surely had the patience of a saint to put up with this after a year and half of study. However that may be, if Ada was on the cusp of anything it was at the very beginning of even understanding what algebra is. Many more examples from this date and later—any date really—could be given.

Finally we must turn to what the authors present as the centrepiece of their analysis, a reordering of the chaotically undated letters to De Morgan, which they believe refutes Dorothy Stein's argument that they contain strong evidence that Ada could not have mastered the more complicated mathematics in Menabrea 1843a, specifically the example involving Bernoulli numbers. Stein had called out a letter dated late 1842, 'shortly before she [Ada] translated the Menabrea memoir'. This we have already seen above: it contains Ada's stubborn confusion about elementary functional equations which had led De Morgan to give up on further explanation. But Stein's dating of the letter is probably wrong, as Hollings et al triumphantly point out, but if so that's because it is incorrectly dated by Ada herself. Based on the content it may date from late 1840.[27] Stein's dating was likely inferred from the stated (partial) date Wednesday 11-16, which following the calendar would put it in 1842. In this case Ada perhaps inked the *wrong date* on the letter: it may well have been Wednesday 11-18 instead, which was in 1840. Hollings et al argue at great length that this redating to 1840, stretching the gap between the letter and the 1843 notes on Menabrea, invalidates Stein's conclusions about Ada's mathematical weaknesses.

This is another logical fallacy. Even if the argument for a conclusion is shown to be invalid (hold that thought), the conclusion may still be true. One may reach the right conclusion for the wrong reasons. It may be possible to perform a simple repair of the argument if necessary. Other arguments entirely might be found. Examining Stein's argument, we see that though she mentions the (dubious) proximity of the letter to the date of the notes, she

so that it is only necessary that f and a should be convertible. Since then $(-1)^{\frac{1}{2}}x$ is a particular solution of $\phi^2 x = -x$, a general solution is $f\{-1\}^{\frac{1}{2}} f^{-1}x$, where $f(-x) = -fx$.' Here 'convertible' means commutative.

22 1840-10-15. ADM to AL. Thursday Evening. 18r-19v. For divergent series with mention of De Morgan's by-no-means-clear ideas about them see Hardy 1949.

23 A. De Morgan 1835, 201-2.

24 [1841]-11-08. AL to ADM. Monday. 134r-135v.

25 A. De Morgan 1842b, 141. Recall that $\text{vers}^{-1}(u) = \cos^{-1}(1-u)$.

26 See also [1841?]. AL to ADM. No day. 152r-153r.

27 [1840]-11-18? AL to ADM. Wednesday. 146r-148v.

does not formally depend on that, and explicitly warned her reader that the dating of the letters is chaotic. She does not argue that the proximity is *necessary* to conclude that Ada probably did not have the skills to write the note on the Bernoulli numbers. If we remove the proximity by redating, the letter may nevertheless be *prima facie* evidence of a lack of skill if we have no reason to believe that her fundamental mathematical comprehension improved markedly between 1840 and 1843.

The particular example that Stein had chosen may easily be replaced with a *long list* of examples of mathematical incomprehension, some of which we have noticed above. Many more can be added. They stretch right through the correspondence from beginning to end. Recall that in October 1841, close to the very end of the extant correspondence, she had not understood arbitrary constants of integration, had confessed to forgetting that e^{-x} is a function of x, and was still completely at sea with simple functional equations, exactly the class of problem concerned in the letter that Stein had cited.[28] After a year and half of work.

More directly, if we wish to know whether or not Ada wrote the note about the Bernoulli numbers, or rather how much of it, and which parts, we can simply examine the note itself. Perplexingly, Hollings et al do not. Instead, they refer the reader to Misa 2016. The reader who follows that reference may be astonished to discover that Misa does not address that question directly either. Instead he presents wildly inaccurate assertions about the 'substantial depth to Ada's studies beginning in the 1830s'.[29] This is followed by a vague history of Bernoulli numbers in general, and an inadequate analysis of the contents of Note G which blithely repeats the incorrect recurrence relation for the old-style Bernoulli numbers used by Ada, which fails to comprehend that they cannot, by that definition, be negative, and fails to notice the mathematical nonsense she introduced in her adumbration of Babbage's explanation.

Ultimately Misa concedes—after some token resistance, and contrary to Hollings et al—that Babbage did the essential mathematics. He retreats to the position that 'the transformation of the general formula [of the Bernoullis] into a step-by-step algorithm remains Ada's achievement, as the letters clearly indicate'. This secondary position, no longer a mathematical claim, is also false, as recorded in considerable detail elsewhere here: Babbage provided the substance of the program.[30] Misa has understood neither the program and its origins, nor the nature of its limitations and how they may be overcome.[31]

Given this reevaluation of Ada's chronologically reordered calculus homework and the direct analysis of the written mathematics of Note G, attempts to resurrect her mathematical reputation on the basis of that homework come to nothing and cannot be rescued by special pleading and 'gilt by association'. Nor will it do to promote Ada by way of compensation to the office of the world's first programmer, since that was Babbage himself.

---

28 [1841]-11-08. AL to ADM. Monday. 134r-135v.

29 A statement like 'Her letters to De Morgan are filled with specific questions about differential calculus, limits, Leibnitz's notation, three-dimensional geometry, notation of functions, and standards of reasoning and proof' is actively misleading. Those are rank, persistent, uncorrectable confusions rather than merely 'specific questions'. Indeed, if Misa read the letters, he has not comprehended them, since he informs the reader that 'Ada's calculus studies with De Morgan likely drew on the calculus textbook he was writing during these years'. As we have seen, De Morgan's texts are mentioned *constantly* in this correspondence, being the *basis* for it, including his *Differential Calculus*.

30 See section 276, p. 71 for an overview of Note G and section A.4, p. 112 for full mathematical details.

31 See subsection A.4.3, p. 121.

G.1 Calendar of the Ada Lovelace and Augustus De Morgan Correspondence.

Ada tried to work through various mathematical texts, almost all by Augustus de Morgan, and corresponded with him for advice, when she had questions, confusions, ideas of her own (mostly mistaken) or simply got stuck. Sometimes she just shared news and made social plans. The notes must have been exchanged between mid-1840 and late-1842, but many completely lack dates, or only have incomplete ones, and the last date we can take with assurance is from late 1841. Moreover, where they do have dates, errors by the author sometimes occur. Years may be worked out from day of week, where that is given, assuming no correspondent mistakes and a universal calendar (Babbage's friend Plana devised exactly such a universal calendar, which still works today). Often years must be assigned using the sequence of content instead, by the balance of probabilities. Some uncertainty must remain.

Letters are identified uniquely below by their assigned 'folion' physical page numbers, *nr* or *nv*, where *r* is recto (front-side), *v* is verso (reverse-side). This allows them to be reordered by assigned date, not their erroneous sequence in the archives.

Despite the trivial content, the Clay Institute paid for the correspondence to be transcribed.[32] The timeline here is based on those transcripts, but has been reordered approximately, using the deduced and imputed dates, to make it more comprehensible. The per-item summaries given here were created from scratch and are not based on any of the summaries given there.

G.1.1 [1840]. ADM to AL. Thursday Morning. 1r-2r.

Music. Zero, fractions and 0/0.

G.1.2 1840-08-01. ADM to AL. 3r-4r.

Roots and variable coefficients.

G.1.3 [1840?]. AL to ADM. Friday Morning. 54r-57v.

Types of Logarithms.

G.1.4 [1840?]. ADM to AL. Wednesday Morning. 5r-5v.

Logarithms.

G.1.5 [1840?]. ADM to AL. No day. 9r-11.

Study of mathematics. Graphing curves. [Part of f.11 is by AL.]

G.1.6 1840-08-17. ADM to AL. Monday. 7r-8v.

p. 16 of [Differential and Integral Calculus], definition of 'nearly equal' using fractions. Origins of Differential and integral calculus. Archimedes already had a primitive version sans algebra. Refers to his 'Negative and impossible quantities' in the *Penny Cyclopaedia*.

G.1.7 [1840?]. ADM to AL. No day. 12r-12v.

Formula for an equation which passes through given points.

[32] Transcripts and images of originals are available at `https://www.claymath.org/online-resources/ada-lovelaces-mathematical-papers/`.

G.1.8 [1840]-09-13. AL to ADM. Sunday. 48r-49v, 164r-163v.

[1840 inferred from day of week and month-day.] Equation to a curve.

G.1.9 1840-09-[14]. ADM to AL. Monday. 14r-15v.

[De Morgan dated this as Monday 15th but that was a Tuesday in 1840. The year is confirmed by mention of William Frend's stroke. Corrected to the 14th.] Functional equations.

G.1.10 1840-09-27. ADM to AL. Sunday Morning. 16r-16v.

Limits and derivatives.

G.1.11 [1840?]-11-10 AL to ADM. 62r-67v.

[Year inferred to be 1840 from content.] Peacock's Examples.[33] Derivatives. Algebra. Erratum. 'I am not at all sure that I like the assumption in the last paragraph of page 212 [of Elements of Algebra]. It seems to me somewhat a large one, & much more wanting of proof than many things which in Mathematics are rigorously & scrupulously demonstrated' [Concerns $(1+x)^n \times (1+x)^m = (1+x)^{n+m}$. 'When an algebraical multiplication, or other operation, such as has hitherto been defined, can be proved to produce a certain result in cases where the letters stand for whole numbers, then the same result must be true when the letters stand for fractions, or incommensurable numbers, and also when they are negative'.] Mentions "The note in page 2" with theorems from Maclaurin's Fluxions. "I feel I only half understand that little Chapter X [on functions], and it has already cost me more trouble with less effect than most things have."

G.1.12 1840-10-15. ADM to AL. Thursday Evening. 18r-19v.

Infinite and divergent series. Primes.

G.1.13 1840-11-14. ADM to AL. 20r-22r.

Statements true for all (complex) numbers. Functional equations.

G.1.14 [1840-11?]. ADM to AL. Friday Evening. 24r-25v.

Statements true for all (complex) numbers again. Limit of $\log z$. Functional equation.

G.1.15 [1840]-11-18? AL to ADM. Wednesday. 146r-148v.

[Incorrectly dated 11-16. Likely 11-18, if this really was a Wednesday, and in 1840. Otherwise it would have to be 1842 and the content makes this unlikely. So the year 1840 is inferred from the content.] Definition of derivatives. "had I thought a little more upon what I read some weeks ago, before I wrote my last letter to you, I should not have sent the question about $du = \phi(x)dx$" Does not see why $(1+x)^n \times (1+x)^m = (1+x)^{n+m}$ is true for all real numbers merely because it is true for whole numbers. Euler's proof of the Binomial Theorem. 'That I do not comprehend at all the means of deducing from a Functional Equation the form which will satisfy it, is I think clear from my being quite unable to solve the example at the end of the Chapter [X of Elementary Algebra] "Shew that the equation $\phi(x+y) + \phi(x-y) = 2\phi x \times \phi y$ is satisfied by $\phi x = \frac{1}{2}(a^x + a - x)$".' Brings up "The note in page 2" (of Peacock) with elementary theorems from Maclaurin's Fluxions again. 'I go on well with the Trigonometry, & have nearly finished the Number & Magnitude'.

[33] Peacock, Herschel, and Babbage 1820.

G.1.16 [1840]-11-27. AL to ADM. No day. 149r-151v.

[Year inferred from content by tracing a sequence back to the letter from 1840-11-14.] 'These Functional Equations are complete Will-o'-the-Wisps to me. The moment I fancy I have really at last got hold of something tangible & substantial, it all recedes further & further & vanishes again into thin air'.

G.1.17 [1840-11?]. ADM to AL. Monday Morning. 27r-27v.

Functional equation on p. 206 of *Elements*. 'I think you have got all you were meant to get from the chapter on functions. The functional equations which can be fully solved are few in number'.

G.1.18 [1840-12?]. ADM to AL. No day. 29r-29v.

[Date hazarded from content.] Limits of ratios.

G.1.19 [1841]-12-17. AL to ADM. Friday. 68r-69v.

[Date inferred from content. Appears to be incorrectly dated 12-17 since that was a Friday in 1841 not 1840, when it was a Thursday, but the content seems to belong to late 1840.] [Elements of] Algebra. Where convergence of series begins, page 231.

G.1.20 [1840]-12-22. AL to ADM. Tuesday. 70r-73v.

[Year inferred from day of week.] Incommensurables (irrationals).

G.1.21 [1840-12?]. ADM to AL. Thursday. 31r-32r.

[Date hazarded from content.] Limits of geometrical series.

G.1.22 [1841]-01-04. AL to ADM. Monday. 74r-76v.

[Year inferred from day of week.] Power series.

G.1.23 [1841]-01-10. AL to ADM. Sunday. 77r-83v.

[Year inferred from day of week.] Series. Has been working on Differential Calculus, pp. 46, 47, 48. Demonstration of the Logarithmic & Exponential Series in your Algebra.

G.1.24 [1841?]-01-14. AL to ADM. No day. 154r-155r.

[Year inferred from content and month.] 'amusing myself with Murphy's Theory of Equations'.

G.1.25 [1841]-01-17. AL to ADM. Sunday. 84r-87r.

[Year inferred from day of week.] Limits.

G.1.26 [1841-01]. AL to ADM. No day. 91r-95r.

[First part is missing. Date inferred from content, see below.] Binomial theorem and derivative of x^n. Alternate forms of the definition of derivative as a limit. 'It had not struck me that, calling $(x+\theta) = v$, the form $\frac{(x+\theta)^n - x^n}{\theta}$ becomes $\frac{v^n - x^n}{v-x}$'. Monday 25th coming up. [Must be mid January then.] Past XMas (regrets not inviting ADM)

G.1.27 [1841-01?]. ADM to AL. No day. 35r-36r.

[Year inferred from content.] Equating polynomial coefficients. Will invite Lord L if he is not afraid of Algebra. [Mentions Monday evening, so followup to 91r-95r.]

G.1.28 [1841]-01-22. AL to ADM. Friday. 88r-88v.

[Year inferred from day of week.] Lord L is not afraid of Algebra. [follow-up to 35r-36r.]

G.1.29 [1841-01?]. ADM to AL. Wednesday. 33r-34r.

[Date inferred from content.] Derivative of x^n and the Binomial Theorem.

G.1.30 [1841]-02-03. AL to ADM. Wednesday. 89r-90v.

[Year inferred from day of week.] page 69 of the Differential Calculus. monotonicity, limits.

G.1.31 [1841]-02-06. AL to ADM. Saturday. 96r-98v.

[Year inferred from day of week.] Signs of actual numbers do not alter algebraic expressions. pages 70, 71, 72, 73 [of Differential Calculus]

G.1.32 1841-02-06. ADM to AL. No day. 37r-37v.

Taylor's Theorem.

G.1.33 1841-02-11. ADM to AL. No day. 38r-39r.

Continuity and discontinuity.

G.1.34 [1841]-02-19. AL to ADM. Friday. 100r-102v.

[Year inferred from day of week.] Confused about 'θ itself is a function of x and ω' in the Mean Value Theorem. "Calculus of Finite Differences" up to page 82. Chapter on Algebraical Development. Logarithms in Number & Magnitude.

G.1.35 [1841-02-22]. ADM to AL. Monday Evening. 42r-45.

[Dated from the death of William Frend the previous day, mentioned in content.] θ is a function of a and h in $\varphi(a+h) = \varphi a + h\varphi'(a+\theta h)$.

G.1.36 [1841]-02-22. AL to ADM. Monday. 104r-105v.

[Year inferred from day of week.] Law for the Coefficients of Δu_n.

G.1.37 [1841-02-24?]. ADM to AL. Wednesday Evening. 40r-42v.

[Date inferred from content.] Misprint (mistake) in her work. Mentions 'Sunday Evening (the day after Mr. Frend's funeral)'. [1841-02-27?]

G.1.38 [1841?] AL to ADM. Sunday Evening. 58r-60v.

[Year inferred from content.] Completed a chapter [V, On Implicit Differentiation, pp.85–96] of Differential Calculus.

G.1.39 [1841-05?]. AL to ADM. Saturday Morning. 50r-52r.

[After Paris trip in April 1841, based on content.] Resuming Differential Calculus. pp. 100-103.

G.1.40 [1841?]. AL to ADM. No day. 152r-153r.

[Year tentatively inferred from content] page 100 [of Differential Calculus]. Confused about constants of integration.

G.1.41 [1841?]-07-04. AL to ADM. No day. 106r-107v.

[Year tentatively inferred from content.] page 106 [of Differential Calculus]. Integral of 1/x and 1/(1-x).

G.1.42 [1841?]-07-04? AL to ADM. Sunday. 108r-109v.

[Year inferred from content. Impossible date 07-06 written, given weekday. Assuming it really is Sunday, it must be 07- 04.] page 108 [of Differential Calculus]. Simple integrals.

G.1.43 [1841?]-07-05? AL to ADM. Monday. 110r-111r.

[Year inferred from content. Incorrectly dated as Monday 07-06. If it really is Monday then it is 07-05.] Differential calculus [follow-up to 1841?-07-04?]

G.1.44 [1841]-07-11. AL to ADM. Sunday. 112r-114r.

[Year inferred from day of week.] Integration blunders. 'I am still occupied on pages 108, 109, 110, & hope to complete to page 112' of Differential Calculus.

G.1.45 [1841]-08-15. AL to ADM. Sunday. 115r-118r.

[Year inferred from day of week.] Integrals Logarithms of negative quantities. Application of the Differential & Integral Calculus to the subject of Accelerating Force, & Accelerated Motion

G.1.46 [1841]-08-16. AL to ADM. Monday morning. 119r-119v.

[Year inferred from day of week.] Integration and chain rule.

G.1.47 [1841]-08-21. AL to ADM. [Saturday]. 121r-122v.

[Year inferred from content. Wrongly dated Friday, which was really the 20th, by Ada.] Derivatives and integrals.

G.1.48 [1841?]-08-29? AL to ADM. Sunday Morning. 144r-145v.

[08-28 Sunday must be incorrect, as this would make it 1843. 1841 tentatively inferred from content. Perhaps 08-29, which was a Sunday in 1841.] Basic definition of integrals. Indefinite versus definite integrals.

G.1.49 [1841?]-09-06? AL to ADM. Monday. 46r-47v.

[Year inferred from content. Date written incorrectly as Monday 09-04, which must be 09-06 if Monday is correct, otherwise this is 1843 and that's too late.] Differentials.

G.1.50 [1841]-09-09. AL to ADM. Thursday Morning. 123r-125v.

[Year inferred from day of week.] 'I am plagued over page 135 of the Calculus'.

G.1.51 [1841]-09-19. AL to ADM. Sunday. 127r-129r.

[Year inferred from day of week.] Negative & Impossible Quantities. Trigonometrical Chapters of the Differential Calculus. Extensions of complex numbers. 'It cannot help striking me that this extension of Algebra ought to lead to a further extension similar in nature, to Geometry in Three-Dimensions; & that again perhaps to a further extension into some unknown region, & so on ad-infinitum possibly'. [However De Morgan had said this himself in the text.]

G.1.52 [1841]-10-27. AL to ADM. Wednesday. 130r-131r.

[Year inferred from day of week.] Regrets no mathematics for a fortnight.

G.1.53 [1841]-11-04. AL to ADM. Thursday. 132r-133v.

[Year inferred from day of week.] Chapter 9 of Differential Calculus, p. 156.

G.1.54 [1841]-11-08. AL to ADM. Monday. 134r-135v.

[Year inferred from day of week.] Solving differential equation. 'I cannot think how I could be so negligent as to forget that ϵ^{-x} is a function of $(-x)$ which is itself a function of x'.

G.1.55 [1841]-11-10. AL to ADM. Wednesday. 136r-136v.

[Year inferred from day of week.] Very anxious to see ADM.

G.1.56 [1841]-11-11. AL to ADM. Thursday. 138r-139r.

[Year inferred from day of week.] Must reschedule.

G.1.57 [1841-11?] AL to ADM. Wednesday. 140r-141r.

[Date inferred from content.] Busy with chapter on Operation [Article by De Morgan in Penny Cyclopaedia].

G.1.58 [1841]-11-21. AL to ADM. Sunday. 142r-143r.

[Year inferred from day of week.] Exercises page 158 and 159. 'I left off at page 165 of the Calculus'.

G.1.59 ????. AL to ADM. Tuesday. 156r-156v.

Must postpone meeting. No mathematical.

G.1.60 ????. AL to ADM. No day. 157r-158r.

Appears non-mathematical. Appears to have written a paper mentioning 'Irish Priests & Douai'

G.1.61 ????. AL to ADM. Tuesday morning. 159r-160v.

Missed meeting, tries to arrange another. Non mathematical.

G.1.62 ????. AL to ADM. Tuesday night. 161r-162v.

More about meeting, blames servants. Non mathematical.

G.1.63 ????. AL to ADM. Thursday. 166r-166v.

Sends 'papers' on Integral Developments and Accelerating Force. Non mathematical.

G.1.64 Misc.

[loose undated notes on basic calculus by ADM then by AL]

Appendix H

Ada as Mathematician: Precis.

Ada Lovelace published no original mathematical research of any kind. She left no manuscript material, no notes, no diaries, nothing with any interesting mathematical content. She was not a childhood prodigy either. In mid-1840, at age 25, she was taking basic lessons in trigonometry, algebra and calculus from Augustus de Morgan. As far as we know, that only lasted a year and few months, until the end of 1841 (some dates are uncertain). The letters show her struggling with basic concepts: defining and graphing functions, derivatives, the mean value theorem, functional equations.

She struggled fruitlessly throughout with anything involving algebraic manipulation, including straightforward substitute-rearrange-simplify problems. De Morgan advised her to move on to other things. At the end of 1841 she was still struggling. "I cannot think how I could be so negligent as to forget that ϵ^{-x} is a function of $(-x)$ which is itself a function of x'.[1] She showed no sign of ever understanding constants of integration. Her work is well below average even for a beginner. There have been recent attempts by Hollings et al (2017/8) to seize on casual remarks and puzzlements in these letters, in order to walk her in the company of some of the great figures of mathematics—a kind of 'gilt by association'. Taking a few examples, these would have her

1. Anticipating quaternions, merely because she repeated, without any elaboration, one of De Morgan's own remarks in his textbook about the then-open question of extending complex numbers to higher dimensions.

2. Correcting a 'fallacious' proof by De Morgan related to the mean value theorem, which she had only expressed puzzlement at the conclusion of. This is a simple non-sequitur since a fallacious proof does not invalidate a conclusion, and the proof is not a fallacy at all once his definition of a function as something potentially taking multiple values is understood, by means of reading his text.

3. Inducted into 'research-level mathematics' because De Morgan sent her a paper he had written mentioning potential uses of divergent series. This was a topic he had already mentioned in his text book for all students and signifies nothing whatsoever about Ada, who replied to him that she could not make head or tail of it.

Many more examples can be given. However, the best way to assess Ada's mathematical ability is the only work published under her name, in 1843: the annotated translation of Menabrea's 1842 paper on the Analytical Engine. The mathematical manipulation in it is routine and unoriginal but still very revealing. Menabrea's printer had misprinted 'cas n=∞', as 'cos n=∞', an absurdity. Ada carefully and accurately translated this *misprint*

[1] [1841]-11-08. AL to ADM. Monday. 134r-135v.

as 'cos n=∞' rather than correcting it to 'the case n=∞'. This was *not* a typo on her part. Nor was it an isolated example. While attempting to introduce notation for counting finitely repeated operations in programs, she deployed Σ to add their occurrences up, but confused the *index* of the term with its *bounds*, three times in a row—a tyro's error. She grandly referred to this as a 'technique from the integral calculus', again three times in succession, whereas the Σ is just shorthand for a finite sum. Needing a $\phi(i)$ as a generalizable multiplier function in the sum, she introduced a bizarre $(+1)^i$ substitution for $\phi(i)$ in the simplest case, without any explanation of that intent, apparently not realizing that 1 would do since it is just a constant function of i. This appears all the more mysterious given her confusion, noted before, of the index variable with its bounds.

The most revealing material of all is the note on the program for calculating the Bernoulli numbers, which has a lot of elementary mathematics purporting to explain the numbers themselves, and the derivation of a recurrence relation used in the program to calculate the next Bernoulli number from previously calculated ones. Many have argued (Huskey and Huskey, Hyman, Hollings, Toole, Eisenger, Baum, Fuegi etc.) that this shows her mathematical *sophistication.* It shows the exact opposite, as anyone who actually reads it closely enough to understand it will discover. In the first place she had to ask Babbage to help her with all of the mathematics, from the definition of the numbers through to the derivation of the recurrence relation, as their correspondence conclusively and unambiguously shows. He also provided the program. Years later Babbage dissimulated that he had worked all this out for her to 'save her the trouble. On the contrary, it was she who had pleaded with him. She was hopelessly stuck. What he gave her was an outline, which she then attempted to explicate. In the process she made a succession of mistakes and descended to gibberish when she got to the recurrence relation itself.

The first definition given by Ada requires the Bernoulli numbers to be positive (we recognize this as old-fashioned now).

$$B_{2n-1} = \frac{2(2n)!}{(2\pi)^{2n}} \sum_{k=1}^{\infty} k^{-2n} > 0 \tag{2}$$

Since this make their calculation difficult in practice, an alternative definition must be used. For unexplained reasons she exhibits one which takes up a lot of space in badly mangled notation. It is completely bogus, as some simple hand calculations show, and she does *not* use it. Instead the definition below will be manipulated to solve for the B_{2k-1} as a recurrence:

$$\frac{x}{e^x - 1} = 1 - \frac{x}{2} + B_1 \frac{x^2}{2!} + B_3 \frac{x^4}{4!} + B_5 \frac{x^6}{6!} + \cdots \tag{4}$$

This is actually the wrong formula. Incredibly, nobody has remarked on this before! It permits every other B_{2k-1} to be negative. Each term involving B_{2k-1} needs a multiplier $(-1)^{n-1}$. Ada noticed she was getting every other 'Bernoulli number' negative when she checked a few by hand, and dissimulated that she only cared for the 'magnitudes', and the machine would get it right! Of course it would not.

Much worse was to come. Equation (4) can be manipulated by some simple procedures in which coefficients are gathered and named to get a series of equations to zero (which we will pass over), culminating in

$$0 = (A_0 + A_1 B_1 + \cdots + A_{2n-3} B_{2n-3}) + B_{2n-1} \tag{9}$$

where the B_{2k-1} are the incorrectly signed 'Bernoulli numbers', and the A_{2k-1} are factors which also depend on n (they should be written accordingly $A_{2k-1,n}$ though Ada does not do so and gets herself into trouble later as a result). Here Ada comes unstuck. It is a simple matter of arithmetic to deduce an expression for B_{2n-1} from the above, but she does not see this. She must have heard in conversation with Babbage about 'linear dependence'. Not knowing what that means, she persuades herself that it is a statement about the index n of the terms in an equation. Perhaps because lines (*line*ar) can be used to measure distance, she thinks that this is the statement that 'n is then not arbitrary' but rather 'a certain function of the distance of that [term] from the beginning'. She believes that this obtains in some of the intermediate equations prior to (9) and proceeds to derive some completely useless expressions stating n in terms of that supposed distance d.

$$2n + 1 = d, \text{ and } n = \frac{d-1}{2} \text{ (for any \textit{even} power of } x\text{)}$$

$$2n = d, \text{ and } n = \frac{d}{2} \text{ (for any \textit{odd} power of } x\text{)}$$

These are left hanging in the breeze, as they must be, along with the boggled 'linear dependence'. Ada proceeds to locate *linear independence* in equations like (9), since there she detects that the 'conditions for the value of n are now modified, and that n is a perfectly arbitrary whole number' (oddly, the 'distance' of n from the initial term is silently dropped). She feels able to deduce (simplifying a little) from this supposed linear independence 'combined with the fact (which we may easily perceive) that whatever n is, every term ... after the $(n+1)$th is $= 0$, and that the $(n+1)$th term itself is always $= B_{2n-1} \cdot \frac{1}{1} = B_{2n-1}$, enables us to find the value (either numerical or algebraical) of any nth Number of Bernoulli B_{2n-1}, *in terms of all the preceding ones*, if we but know the values of $B_1, B_3 \ldots B_{2n-3}$'. Interpreting this sympathetically, she seems to mean no more than that one can imagine phantom terms $0 = A_{n+1,n}B_{2n+1,n} = A_{n+2,n}B_{2n+3} = \ldots$ and $A_{2n-1,n} = 1$ (where the coefficient dependency on n has been clarified), which were omitted from what is really a linear dependency. Those terms and the coefficient 1 can be inserted but they are phantoms because they are quite useless, and it is a non-sequitur that restoring them 'enables us to find the value (either numerical or algebraical) of any nth Number of Bernoulli B_{2n-1}'. Finding B_{2n-1} only requires (9) and arithmetic. It is anyone's guess why Ada detected an advantage in writing 1/1 instead of 1.

As it happens, she fails at the end of this frightful gibberish to correctly perform the subtraction required and in practice, without clearly stating so, uses the *wrong* expression

$$B_{2n-1} = A_{0,n} + A_{1,n}B_1 + \cdots + A_{2n-3,n}B_{2n-3}$$

This should of course be

$$B_{2n-1} = -(A_{0,n} + A_{1,n}B_1 + \cdots + A_{2n-3,n}B_{2n-3})$$

It would have been trivial to check by hand that she had bungled repeatedly. Calculating the first four or so 'Bernoulli numbers', a few minutes work, shows that they do not match the published enumeration in her former tutor Augustus de Morgan's *Penny Cyclopaedia* article of 1840. For these reasons and other reasons related to the lack of dynamic addressing in the Analytical Engine at that stage, the Bernoulli number program exhibited in the notes simply doesn't work. Fixing the mathematics is the easy part, altering the engine is much more difficult.

It is remarkable that not a single previous commentator has noticed this nonsense, whistling through it rather than understanding the details. The Bernoulli example, so often referenced as proof of Ada's abilities, actually shows that she had not mastered the elementary mathematics required for it. She had asked Babbage to provide her with the definitions and derivation of the recurrence, because she could not do that herself (consulting De Morgan's textbooks, which she had already used, should have been sufficient). Since he could not have perpetrated the nonsense noticed here, she must have inserted her own failed attempts to understand how the argument that he outlined worked.

Appendix I

Letters, Diaries and Notes.

I.1 1812.08 JFW Herschel to JW Whittaker.

August. 1812

Dear Whittaker,[1]

Behold thou art destined to be greatly honoured, I have chosen thee as a vessel of wrath, wherein I mean to discharge the overflowings of my anger — be prepared for the worst, for I tell thee beforehand I am in a most horribly ill humour, not indeed with thee, for that, I confess to thee, I do not think practicable, but with myself, & all the world (saving & excepting thy own sweet person). Why the deuce did I ever leave Cambridge? Nine times a day does my melancholy ghost wander along the sedgy banks of the Cam, (if you meet it knock it down boldly) and bewail and groan and squeal in a most lamentable style, & repeat in ten thousand plaintive tones "Ye Gods annihilate both space and time!" In fact I am throwing away time with a prodigality truly abominable. I hope you are employing yours better. Where are you? what are you doing? Quae circumvolitas agilis thyma? (Which interpreted) from what bunch of thyme are you sucking muddy inspiration, a tumbler head over heels into the water?

I have been for my part asleep almost ever since I left you. I make it a point of conscience to get drunk every day, and live in other respects, a very irregular life, although travelling. I was this evening at Stonehenge, but having a head none of the clearest, saw not a single stone. May his Satanic Majesty fetch me quickly if I know whether it is made of stone or cheese. I am now writing to you with "a dreadful noise of waters in my ears" which I suppose must mean a swimming in my head, occasioned however by anything but water. Mathematics I have utterly discarded; – a peaceful repose to their manes! You see I am in a hopeful condition. I do not expect to return to Cambridge before Xmas. I have a commission for you. Go round the College & on each of the fellows bestow in my name a Malediction of 3 pounds weight at the least, for a heavy Malediction I find is always most adviseable. I had the happiness of an introduction to a dulcinea who reads the [...] Cel— and [...] Analytique. What say you to that? To use Playfair's expression she "possesses a degree of Math. Knowledge rarely found in the other sex and is at once an ornament & example to her own—to her country & to the human race!!" No mortal however would have been able to read a page of La Grange in her eyes, or to discover the Analyst in her manner, her conversation or her appearance crushing every human beastly indulgence of every liberal and social feeling.

[1] John William Whittaker (1790–1854). A St. John's friend of both Herschel and Babbage, party to gossipy letters. See section J.7, p. 443.

[...] She is confoundedly not beautiful. Of course one ought never positively to affirm the reverse of beauty to belong to any woman. Now my dearly beloved I have written myself sober, so let me give you 1 3/4 words of advice. Read. Write. Imprison yourself. Speak to no one. Cram. Repeat. Do all this and you will then be fitted for a fellow of a College, to spend (or rather lose) an unhappy existence in solitary idleness. To die forgotten and unlamented, to leave your fame to the keeping of your Bedmaker who may perhaps when drunk retail an anecdote of some trivial nature of you, and to be succeeded by [...] one to whom you were in life an obstacle [...] death not a cypher. These are thy trophies Cambridge. Sweet, protecting Alma, whose nourishment is from the vitals of thy sons.* Think of my words Whittaker when you first begin to curse the hour of your birth – they are prophetic.

Farewell. If you have a mind to let me know what the devil is busy about at Cambridge, & that within 5 or 6 days, direct to me at Sir Wm Watkins Dawlish near Chudleigh Devon. If not; so.

Yours Sincerely
J.F.W. Herschel

* What do you say to the idea of Alma Mater overlaying the mind of her children (as an old sow rendered careless by some excessive and free gluttony, overlays her pigs.)

Whittaker Esq.
St. Johns Coll.
Cambridge

I.2 1813-02-08. JFW Herschel to C. Babbage.

Dear Babbage,

After leaving Cambridge, I fell into a sort of indifference with regard to "Analysis" which lasted me till yesterday, when I drew up a sketch of my memoir about the series of tangents, & today I filled up the skeleton with as much flesh as I could scrape together, & very sorry I am to say that after all it will be but a sort of meagre anatomy, fit only for a scarecrow. Yet [much] the best part of it consists in some beautiful infinite products, which I fear will interfere with your memoir [mathematics omitted][...] [...] I have an host of questions to ask about the A. S. – you cannot well imagine how anxious I am on this subject. I trumpet its fame as much as I can, when I meet with any one who can at all enter into the subject. It is true, my opportunities are not very frequent, but what there are are precious. For Heaven's sake keep up a solemn serious appearance about it at Cambridge, and be for some time very cautious whom you admit. We really now must begin to be somewhat in earnest & avoid "every thing common or unclean." What say you to determining an annual subscription, that each member may know what he has to pay. For instance 3 guineas (or more, or less) per An. or else a donation of £50 at first entrance in lieu of it. Do you not think 1 meeting in a term enough? I think that no member should be elected without a written notice from himself that he wishes it; and a written certificate from 1 Member that he is a fit person. Perhaps it would be as well (as soon as it can be done) to contrive to have another secretary resident in London—at all events to transfer the "seat of empire" thither as soon as possible. For I repeat it again & again, we must not be a "Cambridge" Analytical Society. [...]

J.F.W. Herschel
Feb. 8, 1813

write your answer soon, & how the printing proceeds — soon.

I.3 1813-02-19. C. Babbage to JFW Herschel.

[Cambridge] 19th
Feb. 1813

Dear Herschel,

I have received your letter and am rather ashamed of not having answered it before but in fact I have been very idle. I shall proceed to answer your questions in order.

Not above one of the equations relative to B which you sent me was in my memoir & several might I believe be deduced from some theorems which I have given in a very general form.

The first sheet of the memoir printed it looks very splendid but I am afraid we shall go on but slowly. The preface was left in London with Maule Senr. he has promised to make his observations freely on it and send it to me when they are finished. I expect a letter from him daily it will contain a new method of solving the problem of the Knight which he mentioned to me some time since as communicated to him by Brandreth.

I want you here much & as we must give some attention to the preface. I think it may be made a good one. Maule seems to think it would be advisable not to affix our names to memoirs. His brother is of the same opinion and several members coincide with him. You know my opinion on this subject but I am very ready to give it up, however we will consult on the subject when we meet. We have held no meeting yet nor does any appear necessary. The Society did not appear thunderstruck as I expected at the appearance of the first sheet.

I made a thesis for my act on equations at which Jephson muttered something which no mortal could hear. Higman says it contained something like gravitas. Now this in one sense may signify a stinking smell. I had but finished until past 6 o'clock and performed my part of the farce very well. My first opp. did not understand his arguments and puzzled the Mod.; my second understood his arguments but could not find words to express them; and Jephson who had words could not understand them. My first opp. [honestly] was very ludicrous. The Mod. did not ask me one question, which was lucky, as I could not have answered one; in the moral argument I quoted St Luke in English and in Greek; and spoke half-Latin half-English. This together with the singularity of the arguments made Macfarlan and every one in the schools laugh most heartily. Grant Greenwood and myself have projected an expedition to Whittlesea mere in four boats in about a fortnight or three weeks. I hope you will be here to join us as we intend to have much sport. The astronomer Sir Joseph Banks is to be invited.

I have been rather idle with my act &c. but shall read a good deal I hope in the next fortnight. I admire Brinkley's proof of Lagrange's theorems.[2] I should suppose he means by a theorem for taking fluxions per saltum nothing more than finding the nth fluxions without finding all the intermediate ones, this you know La Croix gives a theorem for performing. I have been making some speculations on the powers of different numbers such as this placing over each other the successive powers of 5 in the first column the number 5 constantly recurs; in the 2d column the number 2 recurs; in the 3rd the two figures 1,6 constantly recur; in the 4th column 4 figures (3,5,8,0) recur and in general in the (n+1)th column 2^n figures recur in order.

C Babbage

I.4 1813-02-20. JFW Herschel to JW Whittaker.

Feby 20, 1813.

[2] Brinkley 1807.

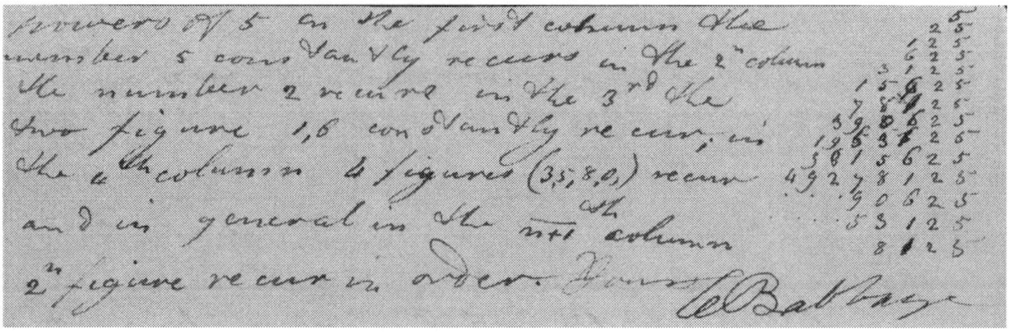
powers of 5 in the first column the number 5 constantly recurs in the 2nd column the number 2 recurs in the 3rd the two figures 1,6 constantly recur, in the 4th column 4 figures (3,5,8,0) recur and in general in the nth column 2n figures recur in order. Yours C Babbage

Figure I.1: [Powers of numbers.]

Dear Whittaker, I had resolved to defer writing to you till the cloud of engagements, dinners, parties, &c &c in which I have been involved since my arrival here was in some degree dissipated and I could sit down with a cool head and a tolerably good conscience, untormented by the reproaches of neglected analysis, to devote a connected hour to your correspondence. But as I see no end of the present state of affairs, I must be contented to send you a patchwork of abrupt & miscellaneous nonsense, just by way of shewing that I still remember with pleasure the friend I have left at Cambridge & do not from hence conclude me idle. Every moment that I have nothing else to do I devote (at the utmost not 3 hours in the day) to the only study which affords me full and entire satisfaction — but Analysis is not the object of human existence, nor the bond of human society. To mix in a superficial world, a polished surface is the one thing needful, — at all events, that degree of acquaintance with the ordinary affairs of life and capacity of entering with relish into them which may obliterate every trace of the midnight lamp and well-conned book from a man's appearance and address.

œque pauperibus prodest, locupletibus œque
œque neglectum pueris senibusque nocebit[3]

Sunday. Feb. 22. As we advance in life or in acquirements, we are continually taking more extensive views of the objects which surround us. This is a truism, but experience only shews us that it is such. When I look back on my college existence, short as is the period which has separated me from it, it seems but as the vivid & evanescent dream of a child — just so I used to look back on the time I spent at school — and just so shall we all, one time or other, look back on the period of our present existence, with a mixture of pleasure & pity. We work our way onwards through a fog which magnifies trifles, which dilates the circle of objects immediately around us into a wide amphitheatre in which we stand like the hero of a fairy tale, surrounded with giants and dwarfs and disproportionate existences. A truce to rhapsodizing. I intended to come up time enough to sit for a fellowship, & you will oblige me if you would make enquiries about the time when I should return to College, as I have forgotten what Hornbuckle told me. I want much to know how you were satisfied with your act. Has anything more absurd than usual (Babbage's act excluded, of course) taken place in the schools? Did the moderator make his first appearance drunk or has any one called him fool yet? There are two momentous questions, which I beg you will duly answer. In short I must know all what is going on at Cambridge or (as one should say with propriety) "in the flesh?"

Farewell. Believe me ever,

[3] Horace *Ars Poetica*, 25: "It profits poor and rich alike, and neglected, it will harm young and old alike."

Yours most sincerely
J.F.W. Herschel

PS. Would you have the goodness to enquire if there are any letters for me & desire them to be forwarded to me here. Remember me to all friends, Greenwood — Babbage, McArthur &c. When you see Jeremy Wood, or Hornbuckle, kiss them.

I.5 1813-02-25. JFW Herschel to C. Babbage.

Feb. 25, 1813

Dear Babbage,

When men with common pursuits in which they are deeply interested, correspond on the subject of those pursuits, the trifling ceremonials of an ordinary correspondence may in great measure be waived. I do not therefore think it necessary to allow two or three weeks or any stated time to elapse after receiving an answer to one letter before I receive another. In this point I fancy we agree, and I shall ∴ waste no more paper in unnecessary explanations &c. ——. It is singular that I have been employing myself about the recurrence of figures in the powers of numbers at the same time with you – Here are all the numbers under 20 places of figures whose powers end in the same figures as the original numbers 1, 5, 25, 625, 9625, 890625, 2890625, ... and so on to 92256259918212890626. Again 6, 76, 376, 9376, 109376, 7109376, ..., 60774374008178710937 6. These form 3 scales of numbers – the first contains only 1; the figures of the others are all determined according to a certain law, not (so far as I can perceive) capable of an algebraic expression.

I have been endeavouring to find some rapidly converging series for obtaining π and in doing so have fallen upon some strange theorems which relate to the summation of certain series of nth functions. I will give you a specimen [mathematics omitted here]

J.F.W. Herschel

I.6 1813-02-26. JWW Whittaker to JFW Herschel.

Given in our Garret at St. John's Coll. Camb.
26th Feb. 1813

My dear Fellow,

Your letter pleased ye cockles of my heart & puzzled me as to ye inner man: indeed I feared you would forget our treaty of correspondence, as we had not renewed it at your late departure. I am glad you are mixing in ye gaieties of ye flesh a little, it can do you no harm — perhaps good: It strikes me however from your letter that you will find few places so convenient for reading of any kind as Cambridge & gives me hopes that ultimately you may be induced to fix your chief residence in this place. 'Tis a weak notion that every person who does this must at last assimilate himself to ye people amongst whom he is thrown: though true of weak minds, which are incapable of philos. attainments & exalted views, you will not suppose that J.F.W.H. will be either captivated with a swinish life, or be likely to imitate y^e [Blicks] & y^e [Blands], & ye KC's at Camb.

McArthur rec^d a letter from Roscoe a day or two since, who was hindered from coming up to Division of Term by a sudden cold. He intends however as soon as he is free from it, w^h he expects to be in a very short time to come amongst us, & I hope he will appear in a few days. I have just been paying Wood's bill; he was uncommonly gracious & facetious. I enquired when you ought to be up for ye Fellowship. He answered, by Mid-lent Sunday, w^h is March 28th. So you must be forth coming a few days before that time.

I have kept my act, w^h went off ([...] no small quant'y of bad Latin) much better than I expected. Yesterday I had a second opp^y against 9th & 3rd [...]. Jephson had y^e enormity to ask me 2 things w^h no mortal could have expected; viz: to find y^e flux^n x^y & to exterminate y^e 3^d term of equation x^n in $x^{n-1}+$ &c $= 0$.

I got through them both one way or other. Babbage had a very crowded audience. He kept y^e freshmen in a roar of Laughter, you can judge what at. Jephson asked him if he recollected some formula — & Babbage coolly answered 'Non recolligi illud expressio'. Abominable Quantities — cosinus & methodus.

I have just been amusing myself with a parody on Scott. It is supposed to be a conversation between ye spirit of ye Cam, & ye Observatory at St John's. It will be as amusing as any prove, so you shall have it.

(Cam) Sleepest thou? Brother.

(Obs) Brother, nay —
In my leads the moon-beams play.
From Spinning House to Castle End,
Where streets out-stretch, or allies bend.
Bulldogs fierce, their night-rounds pacing,
Oer the Proctors ministry,
W—s, & gownsmen swiftly chacing,
Scour it deft & merrily;
Up & mark their nimble feet,
Up & list their voices Sweet!

(Cam) Tears of a divine defeated
Mix with my polluted stream,
The Margaret Professor, bible-cheated
Mourns beneath ye moons pale beam.
Tell me thou who view'st ye Stars,
When shall cease these bible-wars?
What shall be ye Doctor's fate?
Whence shall he a mitre get?

(Obs.) Arthur's slow main his course doth roll
In utter darkness round ye Pole,
The planet-stars I cannot see;
For Catton[4] never comes near me:
Long time my hinges have been rusty,
Speculums broken, lenses dusty
Ill can I read their high decree!
But no kind influence deign they shower
On [Carnus'] tide, or Granta's tower,
Till pride be quell'd & conscience free.

My only apology for this, if you ask "cui bono?" is Dulce est desipere in loco.

Have you heard from Grahame? He has owed me a letter for a long time. I suppose he must be very [...] By ye bye I must not forget a joke of [Mc...]. One day being in ye box he saw a no. of people crossing ye court, & entering ye divinity schools; so he spake thus in a whisper "[...] Opp. Who is keeping an act in ye Divinity schools?" "Can't tell, Sir, I think

4 Likely the Rev. Thomas Catton (1758–1838) the astronomer.

a Queens man—" "Ah yes, it must be so, there's that D–d swagger bellied fellow Barnes walking across ye court" This was so loud that every body near ye Mods box could hear it, & I think is an indubitable proof that Mac. is a real blackguard.

I have not seen Babbage lately so I cannot inform you how ye Book is getting on. I am very busy with Woodhouse's Sph. Trig. — can make neither head nor tail of Horney's [art] MSS. I have invented a beautiful proof that in Sph. Δ's, ye sines of ye L s are :: to ye sines of opp. sides — it is so simple too that I think it cannot be new. Macte virtute[5]

With neatest love & esteem, y^rs &c:
J. W. Whittaker

I.7 1813-04. C. Babbage to Herschel.

[Before May 1, 1813]

Dear Herschel,

The devil, no doubt, to punish my Analytical propensities, has been describing on my body certain small circles, and has extended a vast circumference on the seat of honor. In fact, I am covered with boils and have one of large magnitude applied a posteriori which prevents me from riding, walking, sitting and almost anything else but laying flat on my face. In this pleasant attitude I now address you.

In the first place then I like your remarks on the preface very much; though I differ in some respects for which I will give you my reasons.

The lone sheet which I intend for the beginning is metaphysical, and the conclusion will be a little so likewise.

Now I am inclined to think that you have not applied much to that subject and from some of your opinions I think you have rather a prejudice against it. For these reasons I propose leaving the decision whether they shall be included to Ryan and Higman, both of whom have read very widely and thought much on that subject. Ryan is here and Higman will be during some part of the summer. I will put my ideas into [mathematical] language and they will polish and correct it by their observations. Higman was first man in the first class at Trinity. You very justly exclaim "Explode variation." And seem likewise to desire to explode circles. Now though from my present bodily sufferings I would willingly explode all circles of whatever magnitude, yet I think it scarcely possible to find a general term denoting sines tan. and sec. without using the word circular functions. With respect to the use of Parseval's Theorem you will find its use as applied to partial differential equations in the 3d vol of La Croix. By the bye, have you the 2nd and 3rd vols of a new edition and have you Arbogast or Plaff. Your observations on the Theorie Analytique des Probabilités shall certainly be inserted. I am of your opinion as far as I have read. I have a curious question in the doctrine of Chances which I do not ever remember seeing treated on. It is this. At hazard the whole game depends on chance. At chess the whole depends on skill. Now between these are a vast number of games as Whist, backgammon &c. which depend partly on skill and partly on chance. It is required to assign the relative portion due to each. At chess $\frac{skill}{chance} = \frac{1}{0}$. At hazard $\frac{skill}{chance} = \frac{0}{1}$. If you have met with any solutions of such questions I wish you would send me a hint as to the methods employed. If they have not been considered before it would be well to mention them in the preface.

I have just received your letter. The theorems you have lately discovered are very curious. They do not (to me at least) appear to rest on perfectly legitimate principles. From a method

[5] Bravo.

of reasoning in some respects similar, I have deduced theorems some false, some true, and many doubtful. [mathematics omitted]

Greenwood was here some days since, he dined with me. I saw likewise Northmore but could not prevail on him to stay. I remain yours very sincerely

C. Babbage.

Peacock is at Hastings with 6 pupils if you write to him direct the letter to be left at the Post office. I am rather of your opinion with respect to saying something more relative to the society in the Preface but I will think about it. The Physician who attends me has ordered me to take nitric acid diluted of course. I hope this will dissolve the circles which are very painful. When you write let me know how the printing goes on. I am glad of your solution of [...]

J. F. W. Herschel Esqre
Slough
near Windsor

I.8 1813-05-01. C. Babbage to JFW Herschel.

May 1, 1813

Dear Herschel,

Your Memoir is in the hands of two compositors one just released from mine. I propose that they shall print 8 pages and I will send it for your correction when it is returned again from the printer. I will inspect the proofs to see that your corrections are attended to. My chief cause of writing is that the compositor has stated the very great difficulty which occurs in your mem. on account of $\sqrt{-1}$ being printed in the manner annexed ($\sqrt{-1}$). This might be remedied by printing it in either of these forms $\surd(-1)$ or $\surd{-1}$ the first I think more elegant even than yours, and I think no mistake possibly arise from the second. Send me your decision by return of post that the compositors may not be delayed.

Where shall you be in June (4th of 5th)? Can I meet you as I should like to talk with you for a few hours before I go into Devonshire. On my return here I found you had departed.

I have had much conversation with old Maule on some very strange theorems which contradict some conclusions of Lorgna's and lead to absurdities.[6] The reasoning is simple and neither he Peacock nor myself can find any sort of objection to it.

I will only send you a few of the results equally strange, elegant, and doubtful.

[mathematics omitted]

Yours truly,
C Babbage
May 1st 1813

Get if possible the 3rd Vol of Trans. of Royal Acad. at Turin for years 1786–1787 and look at a paper on series of sines by Lorgna.[7]

---

6 Antonio Maria Lorgna (1735–1796).

7 Antonio Maria Lorgna 'Méthode pour sommer les séries réciproques de sinus ou co-sinus d'arcs en progression arithmétique', *Mémoires de l'Académie Royale des Sciences de Turin*, Vol. 3 (1786–7), 215.

I.9 1813-05-04. JW Whittaker to JFW Herschel.

St. John's Coll.
4th May 1813

Dear Herschel,

I was rather surprised on returning to this place to find that you had quitted it for good. I hope this is an exaggeration of y^r Cambridge friends, who say that you were frightened away by ye Johnians fellows, after you had dined a single day with them. You had intentions of horrifying them; have they been before-hand with you? I must not forget to congratulate – I mean, to condole with you on y^r mishap w^h has befallen you on being elected fellow; but being afraid you would find me alike Eliphaz, ye comforter of Job I will dispense with this special exercise of my friendship. With respect to [Jones] — I really suspect it will turn out a mishap. I am informed that he is quite altered.

I arrived here on Saturday night just in time to sleep in my own warm rooms, after a most uncomfortable & cold journey. On rising in ye morning I found sufficient to amuse me & all Cambridge. Some evil designing persons, who are yet unknown, had taken ye liberty during ye darkness of ye night to pull down with iron crows & other implements 3 out of 4 of ye new stone pillars at ye gate of ye fellows gardens at Trinity, & to wrench up ye iron rails out of their sockets. They then went to ye gate leading into Clare Hall walks, from ye exterior pillars of w^h they rolled ye mighty stone spheres w^h had been balanced there for many years. To crown all with impiety & horrible sacrilege, they ascended to ye Clock of St Mary's, & forcibly twisted ye fingers thereof at right angles to their former position. You will easily conceive how foolish ye Clock looked on Sunday Morning. But I will defy you to form any idea of ye comicalness of ye scene when ye Big-wigs, walked into Church. Each looked up to see ye Hour, & each was struck with wonder, surprize, & unutterable horror on seeing ye fingers of ye Clock pointing at King's Chapel. I hope it did not spoil their Devotions or hinder them from listening to the Sermon. Trinity has offered a reward of £100 for those who discover ye perpetrators.

So you saw Grahame at Huntingdon. When you write, I hope to hear an account of him. How he looks & what account he gives of himself. I received a letter from him which I was at Bradford dated Edinburgh. He mentioned nothing of his visit to ye South, nor of his interview with you. I wonder why he did not — perhaps he thought I had not heard of his being in London. A whole page of his letter is taken up with an humorous exhortation to wedlock — a subject on w^h he is most particularly eloquent.

I want to have your opinion as to ye justice of Jephson's behaviour to Babbage. Having publickly reproved him in ye schools, he refuses to hear every thing said in his favour, or in extenuation of his offence. I very strongly suspect that Babbage deserved ye treatment he has received.

Roscoe is here — he came ye day after I did — he looks very thin, & in a delicate state of health. No alteration in me, except that I am turned Methodist, & y^e spirit has moved me to take no more snuff.

With y^e sincerest friendship I am &c.
J.W.W.
excuse haste & inaccuracy

I.10 1813-05-22. JFW Herschel to JW Whittaker.

Friday night $\frac{1}{2}$ after 2 o'clock

Dearly beloved Whittaker, I am gloriously drunk, d–n me, and hope you are the same. Seriously speaking however, I am a little warm in the head, and a good deal in the heart. These are the moments a man should chuse for communicating with his friends. I remember a passage in a German publication which struck me forcibly when I read it "Whatever you read, read when your heart is warm; — with a kiss, a squeeze of the hand, — or, if you have nothing better, — with a quarrel". True it is I have not yet arrived at the good fortune of either a kiss, or a squeeze of the hand, but simply the society, & that too for the first time, of three of the most charming girls the world possesses. I shall tumble now straightways into my bed, which stands within a yard of me — but I could not sleep in peace till I had told you I was devilish happy, & given you a word of advice. Videlicet, leave Cambridge and also Mathematics, to shift for themselves; and mix more in the society of females than you seem to do.

What signifies the life of man
If t'were na' for the lasses o[8]

You are now in your garret poring over some infernal old cram book, or getting some MS. by heart for examination. Fling them to the devil, my beloved. I wish for your sake you may be last in the last class. But good night. You are sleepy & alas! so am I, I shall finish this to morrow.

Yours affectionately. H.

Saturday,

I do not know whether I ought not to commit the stuff I wrote last night to the flames. However you shall have it for on serious consideration I approve of the advice therein contained, however oddly expressed. Now I will answer your letter as I may. Grahame is looking very well, though I saw him but for ten minutes, he was not in the best of spirits however. A certain acquaintance of ours, who got roughly handled by Jephson in the schools, you well know was very wroth about it.[9] He richly deserved his punishment however in my opinion, if I have heard the story right. Upon my soul it is a pity the man is so intemperate. He is like a hot poker, which you may play with till you burn your fingers unless you take a devilish deal of care — but this entre nous. However if you should happen to meet with him, I would thank you just to mention that I am in hopes of hearing something of him before he leaves Cambridge. If he does not write to me, I must to him, which I do not want to do.

I am reading harder than ever, and just in my own way. It is inconceivable what a difference there is in my feelings when I now sit down to my books and when I used to be cramming and cursing old Wood or Newton, or any of those dull and obsolete writers. I have written this to set your back up, & I have no doubt of my success. Do you stay up the long Vacation? I believe I shall come among you again before the end of the year, but I have not determined.

Yours sincerely
J F. W. H.

Slough. Sat^{y}. May 22 1813.

8 Robert Burns, 'Green Grow the Rashes, O.', 1784.
9 Charles Babbage.

I.11 1813-07-02. JFW Herschel to JW Whittaker.

Slough
Friday. July 2. 1813

My sweet Whittaker, Thy letter though unintelligible to my understanding, nevertheless hath found favor in my sight. I am glad to see the mist clearing from thine eyes, & to find that thou in some degree beholdest things in their proper forms & colours. What the devil have university distinctions to do with thee or thou with them. Truly the Mammon of unrighteousness dwelleth there. The devil walketh abroad in the shape of a Big-wig seeking what he may devour. But do thou not fear, tho' the ungodly triumph. Thou & I are of the number of the elect, the familiar friends of the Lord, who have no occasion to stand on the ceremonies of a common acquaintance — full of honey — fellow labourers in the beehive of the church. For verily the church is the Lord's beehive, and many are the drones which do infest the same. There is a sarcastic savour in the former part of thy letter which I like not. I will not be quizzed. Irony stinketh in my intellectual nose, & with reason. For is it not the excrement of a splenetic disposition? I thank thee for the copy of Lord Byron's epigram. There is more of bitterness in it than wit it is cutting but not keen. Nevertheless I like it much, & one may see that it expresses the uppermost feelings in the writer's mind — feelings for which I admire him. I am truly some-what terrified at the state of the public sentiment with regard to our heaven born ruler. All with one voice & with one accord, wherever I have been, even here, under the sacred nose of Royalty, cry out that he des-[missing] in a right line to heaven. I do [torn] but when (I should have said if) [torn] raree show for kingdoms. But these are secondary considerations. Religious rancour is a monster of more heads than civil discontent, and the chain which confines it is worn to the last link. Tell me sincerely, do you not perceive in the present situation of our domestic affairs every symptom which has been held forward as the characteristic of decaying greatness. Indecision in our measures abroad, pusillanimity in our conduct at home. With famine and discord gnawing at our vitals, we fancy ourselves arbiters of the destiny of Europe, & are surprised if any one whispers that we hold but a secondary rank in the scale of power. Would you estimate the state of public taste, go to our theatres, & return humiliated. When shall we have to welcome the appearance of a volume of common sense from the press? But Look at our overgrown metropolis and the hundred tumours of the second magnitude which have arisen in every limb of this diseased & corrupted body each tainting the little circle around it, each daily increasing in extent & foulness. The nation is apoplectic. Choked with its own population. Overloaded with its useless manufactures & decayed commerce, it stands in need of no simple or common remedy. The sin has been crying, and the expiation must be vast, & sweeping and satisfactory.

Enough. We cannot set these deranged parts of the great machine to rights, but the united voice of many may make it known that something is wrong, which perhaps our dreamers never dreamt of — which never entered into the inspiration of our enthusiasts. Let them go on, poor Devils! Let them convert the Jews, & pour forth the bitter vials of religious controversy [torn] draught of religious fanaticism [torn] is it to us, so long as we are not obliged to take a part in the game. But it is well if the game ends here, if we are not at last obliged to stake our own happiness & lives against their blindness. For where did you ever see a religionist whose only object was to climb up into Heaven. He never will be contented without a tug at the rest, to drag them neck & heels after him, or clamber on their shoulders to earthly preferment. Join with me, Whittaker in the wish & prayer, that religion, as established by law may never entirely usurp the superiority & controul over religion, established by nature.

Thou complainest of ennui. Art thou not ashamed of thy self, at thy time of life, even to think of such a thing? Employ thyself. As for me, I execrate this mortal life daily &

hourly but, while it lasts I mean to use it, as if (mind only as if) I should never get another. I remember an idea of Sallust's where after remarking on the shortness of our existence he says that this should be an incitement to us "memoriam quam possumus longam facere".[10] To live the lives of others, & to do something which shall prolong the memory of our own. (Mind Mr Critic I do not mean this for a translation). Were it not for some such thought as this I believe I might be easily tempted to close my eyes at once on the dark scene of misery & despair which stares us in the face in every quarter.

Between Mathematics & music I divide my whole time. When wearied with the former, I fly to my pianoforte, & drown all thought in horrible discord, for I know little yet of the matter. Sometimes (as now) for variety I tieze a Correspondent with my nonsense, or my melancholy, or whatever happens to be floating on the surface of my chaotic ideas. Meanwhile my habits are becoming more sedentary & insulated, and my reflections daily of a more sombre cast. If this lasts (as I do not think it will) I know not, nor care what will become of it.

I hope soon to hear from you, & for God's sake write me something to make me laugh. A hearty fit of laughter (not arising from spleen) would do me a world of good. If you fail to attempt it, or fail in your attempt, my curse be upon you, & I trust you will never prosper in anything you undertake. Yours, for all that, most sincerely

J. F. W. Herschel

P.S. Tell me Bromhead's direction. I had a letter from him about the Math^c boy wch I answered, & directed to Southampton buildings near Lincoln — should I have said near Lincoln's Inn? London

J. W. Whittaker Esqr
St John's Coll.
Cambridge

I.12 1813-08-02. JFW Herschel to JW Whittaker.

Mon^y Aug. 2 1813

Dear Whittaker

I scribble this in a desperate hurry, just to request that as soon as you receive it you would repair (or send) to Babbage's Gyp, Boning, and having extorted from him the key of his master's penetralia take from thence the 3^d vol of Lacroix, and as many of the Numbers of the Ecole Polytechnique as you can lay hands on. For the Lacroix I have his permission. For the rest, my own fiat must suffice. I am in perfect despair for want of them. Having then obtained them, pack up & send your prize by the very first conveyance, — writing on the outside "To be delivered instantly". For God's sake lose no time. I have sent this by the coach as by that means I gain a day on the post. Had I time I would say more. As it is I can only request you to remember me affectionately to Greenwood, if he is with you, — and to subscribe myself

Yours till death
shall stop my breath.
J. F. W. Herschel

Turn over

Too late for the coach to-night, I have now time to open this, and add a few lines, "levi calamo" (with a goose quill?) Grahame's Father has been staying with us for a few

[10] We can make the memory as long as we can.

days. Grahame is at Edinburgh, making the most rapid and expanding progress at the bar. He whom we used to consider as volatile and inconstant, seems now to have become fixed and earnest. There is no part of his profession — even the driest and most laborious — apparently the most disgusting to a man of his disposition, which he refuses. Congratulate me, my dear Whittaker on this. I would to God you could become acquainted with his Father. I never am in his company, even for half an hour, but I feel myself the wiser and better for it. Do not laugh, my good fellow, at my feeling myself wiser and better. I am not the same as when last I saw you. During these last two or three months I have thought more than in all my life before, and I venture to hope, to good purpose. Do not however suppose from this that I am going to turn Methodist or high churchman, or even religionist in the received sense of the word. Heaven forbid! But I will not trouble you with my creed at present. I am sleepy, and must wish you good night.

Tuesday Aug 3.

The printing of the A. S. [Analytical Society] goes on so slowly that I am afraid Smith is gone to sleep in the world. He wants to have his memory fly-flapped. Write me, will it be disagreeable to you, occasionally to step in, and see how they get on, and at the same time impress on Smith the urgent necessity of getting the thing done before the middle of October.

J. W. Whittaker Esqr
St John's Coll.
Cambridge

I.13 1813-08-06. JFW Herschel to JW Whittaker.

Friday Aug 6. 1813

My sweetest, prettiest, loveliest and most adorable Whittaker, art thou insane. - Didst thou write thy letter to torment me? In the beginning to raise confident hopes of at last getting the object of my search, and at last in a sort of P.S. to overwhelm me in despair by telling me that another had been beforehand with me! However let me prevent mischief as far as I can. I did not dream of Babbage having all the 15 N^{os} of the Ecole: Pol. But for God's sake do not send me such a load of lumber. Just take the trouble to look at the indices, and if you see any Papers of Laplace's, particularly one "Sur diverses points d'Analyse" send them as soon as possible. What are fortifications to me? At all events do not send more than 3 or 4 Vols.

I like your account of yourself much, & I assure you that your letter afforded me more than one hearty laugh. I shall not send you my creed till I can take a large sheet and a long time to fill it, and till I happen to be just in the humour for it. Believe me, it requires a good deal of pretty deep thinking to hatch a reasonable creed. Religious eggs are made of damned refractory stuff, and it is not half an hours incubation, urged by the heat of fanaticism or temporal ambition, that has ever brought them to perfection. Such work produces (I appeal to experience) nothing but monsters and anamorphoses, – or turns all outright into stench and rottenness. I cannot conceive how the devil you could gather from any thing I said in my last letter that I am in love, - I never was more free from any possibility of such a thing. Beware beware Whittaker! What mean all these visits to the Thackerays. The intoxicating vapour once inhaled how slight an Effort does it require to fancy all others involved in its influence. I will add a word or two to Greenwood at the tail of this, so farewell, and write oftener.

Herschel.

PS. for "Totus in hoc sum", read "omnis in hoc sum", You pretend to quote Horace indeed! By the bye, you never told me by what strange concatenation, the bible Society introduced you to the Thackerays. Be candid, & tell me fairly by what rascally trick did you manage this?

PPS. I heard not long ago from Babbage of Higman's victory at Trinity, & should have penned him a congratulatory address, had I known where to write to him

Dear Greenwood, - One of us two owes the other a letter, I believe I am the debtor, and as nothing is more easy than to pay in words I shall break the ice upon the chance of it. Whittaker writes me word that you are turned composer. I will give you a problem. Compose a piece of music in the time 5/4. I have often attempted it, & never yet could make anything of it. My musical mania, which had forsaken me for some time is now returned upon me with double virulence and I am playing from morning to night. I gave great scandal to a very religious personage who has been staying with us, by fluting a whole Sunday morning instead of going to church. Can you inform me of any decent work on the principles of harmony & Thorough Bass. I do not mean one of your elementary codes of nonsense which profess much & teach nothing, but something from which a man with a superficial knowledge, might be enabled to derive real improvement. If you know of any thing of the sort you would oblige me by the communication.

It is now near 3 months since I have heard of you. If you mean to cut – well & good. If not I shall hope not to be long without a few words from your own proper pen.

J. F. W. Herschel

J. W. Whittaker Esqr.
St John's Coll.
Cambridge

I.14 1813? JFW Herschel to JW Whittaker.

Dear Whittaker. As I conclude from not hearing anything from you (although I know by Greenwood that you got my last letter) that you have sent the books I have tormented you about, I have waited with the greatest impatience for this good while in hopes of seeing them, and at length I think it right to apprise you that they are not arrived. If your patience is not exhausted, do have the goodness to see whether they have not got them at the Inn, mislaid in some corner and forgotten. If so, put in the Landlord's face, and call him Horse, in my name. I have not time to add anything, but only to subscribe myself

Yours till death
Shall stop my breath
J. F. W. H.

PS. What do you think of the plague in London?

J. W. Whittaker Esqr.
St John's Coll.
Cambridge

I.15 1814-01-10. JFW Herschel to JW Whittaker.

Jan. 16. 1814

Dear Whittaker,

I am mentally enjoying the idea of the funk you are in now examination is coming on, and I cannot deny myself the satisfaction of telling you how heartily I laugh whenever I think of it. I desire that you will not forget to write to me as soon as it is over and tell me how affairs stand and give me an epitome of all the scandal in circulation. By the way, the principle on which we enjoy scandal, and delight in hearing our neighbours ill spoken of is one of the noblest which actuates the human mind, a position which however strange it may seem I am well prepared to defend, and would do, but that I have much talk to throw off. In about a week I am going up to enter at Lincoln's Inn. I do not think you ever imagined me serious in my threatening to take this step, but so it is - I never was more in earnest. God send quarrels among the good people of this nation, and pour forth the bitter vials of litigation like water on every side–and above all may the spirit of scandal and indiscriminate [...] stalk abroad, so that we lawyers may never want work. Amen.

I have a few words of advice for you respecting your conduct in the Senate House. Keep all tight below, that nature give not the better of you. Get drunk at both the Moderator's Rooms, but yet not so bad as to roll about, or overtop the modesty of nature. When you get a problem you cannot do, grin in the examiner's face & tell him you know a trick worth two of that. For "my dear boy" this is a wicked world we live in and we have always need to keep in mind Shakespeare's speech "Come what come may, Time & the hour get through the blackest day." For my part, I now lay it down as a rule, never to care for any thing.

Tho' peers may laugh & Papists swagger
I care not, I, one single D–n.

You remember the work we had to expand $\tan\theta$ in a series of powers of θ? I have at last succeeded in getting a general expression for the coefficients of θ^{2n-1} [...].

But poor devil you have had enough of mathematics by this time, Well, reflect that [...] or 2 short weeks, you may say with Moore[11]

Science, to you
will I now bid a last & a careless adieu
Still flying from Nature to study her laws
And dulling delight by exploring its cause &c.

Remember me to all the people who may or mayn't enquire about me. Let me know whether Babbage is pluckt. I have written two tragedies since I saw you. In about 3 weeks or a month it is my expectation once more to bask in the sunshine of your kingly eyes. Yours so long as this machine is to me.

Herschel

PS. I have just read the Heroine, a ludicrous novel lately published.[12] Oh W. let me recommend to drive away spleen during the Senate house week.

J. W. Whittaker Esqr.
St. John's Coll.
Cambridge

[11] Thomas Moore, 'To the Invisible Girl'.
[12] Barrett 1813.

I.16 1814-01-21. C. Babbage to JFW Herschel.

Pet. Coll. Evening before Execution.[13]

Dear Herschel,

I confess I do not deserve a letter from you but it was really impossible to see you before I left Cambridge, as I resolved on the expedition in the evening and set off the next morning. I have had a copy of Smith's bill and have been once or twice to enquire about the paper being charged as much but have not found him at home. I believe most of the Society are acquainted with its amount and I had proposed the following method of raising it:

- 3 per head after degree – £30..0..0
- 2 at Midsummer – £20..0..0
- 3 at next Oct. – £30..0..0
- 4 at Jan. offer – £10..0..0
- for our Copies – £25..0..0
- £115..0..0

By this means Smith would have his money early at the end of a twelvemonth which is the credit given by the trade and part of it might be had before. I confess I don't agree with your plans, here are two reasons

1. 1st I have not the cash.
2. 2ndly had they been at first proposed I should not have agreed to them, besides I do not think it would be fair to call on Mill for a share of £34 when he at the time of undertaking it only expected to share 9 that this has actually become *£*25 was equally unexpected by all and should therefore be equally borne, besides it will be a long time before we derive any advantage from them if ever we do.

However now you mention equations in my mem. [...]
I wish when you are in town you would buy for me at Allen's in Plough Court Lombard Street (at no other shop) two ounces of Hyper. Oxymuriate of Potash and likewise a small quantity according to the price of triple Prussiate of Potash and iron in chrystals. Also some Gallic Acid in Chrystals. I have been Chemicising a good Deal and have met with much success in cutting burning and defacing my own hands. Such things must be endured. I hope you will soon come and assist me. I have been endeavouring to put a stop to my idleness, and to the continual evolution of mad schemes from my brain. I have had some tough work in defending you against some (I don't [smear]) Simeonites who raised a hue and cry against your paper because they could not do anything in it. I hope you will soon revisit these denizens of knavery, pomposity and bigotry. Ten good men would have saved Sodom and Gomorrah. Quere: how many would this place require? I remain yours very sincerely

C Babbage

[Zeus] I don't know what but let us eat and drink for to-morrow we may be plucked. Quere the odds?

[13] Annotated by Herschel: in answer to one of mine of Jan. 12, 1814.

- 1st Class
  - Adams Sid[ney].
  - Leigh Pemb[roke]. — Gone mad.
  - Gwatkin [St.] John's.
  - Shelford Benet's [Corpus Christi] in a devil of a funk reported to have degraded NB (report not true)
  - Walker Trin. too high clasped[14]
  - Williams John's half dead cannot go in[15]
- Clasped not giving much satisfaction, 75 men clasped, 160 going out excluding a few to be plucked.

I.17 1814-01-28. JW Whittaker to JFW Herschel.

Saturday 28th Jan. 1814
St. John's Coll.

Dear Herschel,

Our manifold troubles are at last over, & we breathe again in peace & quietness.

Your advice about the Astronomy failed of being any good to me, though I followed it, & crammed up several things out of your manuscript of the Tides. The astronomy was of no use at all, very little indeed was asked, & I did every thing except one question which was to determine the incl of 6s angle to the Ecliptic by observn on solar spots. Woodhouse's Trig. did me more good than all my Newton, though 2 or 3 things in the 2nd vol. were asked w^h wh I knew. We flatter ourselves we failed in not having read De Moivre's Chances or Emerson's Increments w^h told most amazingly. I do not send the Tripos because you will have seen it in the Papers, wherein you would see me thirteenth, & beaten by some most awful asses. But you want to hear some Academical scandal, & truly were I to give you all, I should exhaust more paper than Deighton or Nicholson can supply me with.

On the bracketing day Barnes came into the Senate house & kicked up a dust; There followed a row royal — between him & the Examiners & proctors. Old Cresswell[16] trembled with fury, & McFarlan[17] was savage in his ire. It was the intention of Ashbridge[18] & the [...] examiners to pluck a score or so, but the violence of Barnes caused them to stop when they had plucked 6, of whom St John's has two of them.

What do you think of the nous of a man in the S[enate] House, who worked this problem "to find the time of the sails of a windmill going round given the force of the wind!"? He began the square of the time $\times$(rad of sail)^3 & got an eq^n involving the length of a year!

McFarlan's evening paper was very difficult, & he was thought somewhat <u>ursine</u> in his behaviour & in the exiguous supply of cake & wine, which precluded the possibility of my following your second artl. of advice. In the course of the examination two things were asked out of the Analytical Memoirs. Babbage, who is in the gulph, but would without doubt have been plucked if he had not been clasped, worked one of them, & report says it was the only thing he did. Gwatkin & Wilkinson are Smith's prize-men. The last wrangler sat but was disappointed.

---

14 False, he was a wrangler.
15 False, he was a wrangler.
16 Possibly Daniel Creswell, admitted to Trinity in 1798.
17 George Macfarlan MA Trinity, admitted in 1798, 9th Wrangler in 1802.
18 Possibly John Ashbridge MA 1813 who was made a fellow in 1812.

I suppose the weather has been rough the same all over the country. We had most severe frost during the examination. Surely nothing is more abominable than those morning problems, kneeling down in the gallery, & writing with frozen fingers & hearing the distant hum of pounds, shillings & pence problems below. I never hated an occupation so much in my life. McFarlan made a most excellent speech, in w^{h} he praised old Woodhouse most extravagantly, & something about Calculus [...]. He then fell foul of the private tutors, the Bakerian race; "qui problemata concinna & difficiles nugas fingunt, juventutem Academice ineptilis refarcire studentes."[19] He was hissed when it was concluded. They talk of subscribing for a piece of plate to be presented to him. I would subscribe for an halter most willingly.

I would have written to Grahame on this occasion but I did not think I could do it with decency. I have reasons for thinking he meant to cut me before he rec^{d} my gratulatory epistle. However I have sent him the Camb. Paper.

I have scribbled enough & my fingers wax weak so I will subscribe myself yours till death.

JWW

On Tuesday I go to Babylon

J. F. William Herschel Esq.
at Mr. Barker's
No. 5 Henrietta Street, Covent Garden
London

I.18 1814-07-07. JFW Herschel to JW Whittaker.

Satdy July 7$^{?}$

Dear Whittaker – here is a devil of a piece of work. Over-take him—stop him—catch him—g–d d–n him. Is it not monstrous, my beloved, that an honest, well-meaning Christian cannot lay himself down, with a belly-full of good resolutions, & a cool cup of blessed forgetfulness to prevent indigestion, (just to take a quiet nap – with his left hand on his cod-piece) – but that cursed scoundrel must forthwith take to his heels, and run–scamper–fly–with long strides of a week at a time – as if the devil was behind him? Such such hebdomadal strides has he taken, since I got your letter, & here I sit just waking from my dream to answer it. Well, if I live to get another, and happen to be in a better humour for scribbling — it shall not be so again — that is all I can promise.

I have been these three weeks in London, and this fortnight at Slough (make that out how you can). I have lived long enough I find. What should such a poor snivelling democratic dog do in this aristocratic world? I found all the population in Town frantic & in high monarchy humour – & arriving here, I found just the same, & so it is every where.

I like your description of the murder of Hamlet, but being engaged in profound investigations respecting the distillation of goose-livers, I have not yet had leisure to consider fully the nature of the case. Upon my life it is an oppressive thing to be so deeply involved in Scientific pursuits. I have not even had time to read the newspapers, or to celebrate in one single ode the exploits of the everlasting Wellington (which resound through all the habitable globe, & undiscovered islands), but when these goose-livers are dispatched, I hope — to get married – and then I shall have nothing farther to do in this world of wickedness than to write odes, and —.

You see how urgent my affairs are at present, & how impossible it will be for me to execute my plan of visiting the lakes, but by all that is hirsute and [gashly] I will have

[19] Who invent clever problems and difficult trifles, seeking to fill the youth with academically inept things.

my revenge some other time (when I have disburdened my shoulders of the load of these goose-livers)

PS. Should be glad to know where to direct to Greenwood

Sunday Morning, Date unknown. Latitude and Longitude not determined. Wind NNE. Current of ideas running but 2 knots an hour.

By what I have written you will be enabled to form a pretty correct estimate of the State of my head, which is at present full of the most incongruous phantasms. I am this moment reading the 2nd vol of Lacroix's new edition — analysing a series of metallic alloys calculating the orbit of the Satellites of Uranus from about 500 observations of my father — making a course of experiments on the solubility of salts — reading Newton's Optics, & learning Italian for the purpose of cramming up some MSS of Huygens on the resolution of algebraic [equations] which have fallen in-to my hands. Add to this that I am composing 3 treatises in Greek against Arianism, & correcting the press of a novel entitled

The
Victim of Villainy
or
Travelling Tetrahedrons.

the production of a particular friend of mine. So you see God hath not left himself without a witness. But I forgot, Reverend Sir, I crave your forgiveness, and entreat you to forget that the writer of these words was & is

Yours ever
J. F. W. H.

Query: Did any mortal so subscribe himself?

You say you are in a part of the world full of mines. Now if you have a mind to go on a wild goose chase, I will send you on one. I know that somewhere in the neighbourhood of your place, is a mine in [which] is contained a curious sort of native carbonate of Baryta. If you know any one who is conversant in mineralogy in that part, you could possibly procure me a specimen – JFWH.

J. W. Whittaker Esqr.
St. John's Coll.
near Cambridge

I.19 1814-08-07. JFW Herschel to C. Babbage.

7 Aug 1814

Dear Babbage,

Almost immediately after I had dispatched my last I received yours. The degree of intimacy which has hitherto subsisted between us calls for somewhat farther than the common forms of cold congratulation, which flutter about on these occasions, wafted by every wind of acquaintance – and drop in with every post. I did not expect – careless, and unthinking as we have walked along the same path together, to the entrance of life, that our courses would so soon divide. But sooner or later the time must come to all of us when the duties of men shall take place of desultory pursuits and boyish amusements. The sooner the decided step is taken — the sooner the plunge is made — the sooner do we get rid of the weariness of idle expectation & the anxious load of choice. Your step has been decided — your plunge deep indeed.

"I am married, & have quarreled with my Father". Good God Babbage, how is it possible for a man calmly to sit down & write two such sentences, add a few more which look like justification, and pass off to functional equations. But stop — I am not going to write you a sermon. Three months hence will be time enough for that. I am happy to see however that you seriously intend to set about being useful in this world. To one who like ourselves has existed more in theory than in practice — who has made the beautiful & the abstract his cynosure in mockery of base utility — the resolution must ever be difficult to form, & more so to follow. I can only wish you may have energy & determination enough to go through with it. For my own part I feel that I never shall.

I would to God that I could propose anything to you in the line you speak of. You are aware that a practical knowledge of Mechanics is there of far more importance than any (except a very profound one) of Chemistry. Whatever information I can obtain on this point, which can be I fear but little, you may depend on my exerting myself to collect. Let me however hear from you soon.

On running over the above, I perceive that I have been sermonising and that I have not yet thrown off my Metaphysical chrysalis. When this takes place I know not in what variety of colours I shall glitter. Till then however believe me ever

Yours most Sincerely,
J.F.W. Herschel

Aug 7, 1814. Shaw: near Newbury

PS. You have not informed me who the lady is? I presume it can be no secret.

I.20 1816-08-14. JFW Herschel to Henry Wilkinson.

Bath Aug 14. 1816.

Dear Wilkinson[20]

Whittaker writes me word that you were kind enough to offer to undertake the tiresome task of superintending the printing of the latter part of our Lacroix, in his absence. I do assure you that I am more obliged to you than I know how to express for this as I have no one else to whom I could trust it, & it would be excessively inconvenient to me in the state of locomotion I am in at present, to have all the proofs sent after me about the country. There are a few points however which I did not explain to Whittaker and which it will be necessary to attend to, respecting the references to notes. I have given orders to the printers Devil to send with every proof sheet a copy of the last struck off, so as to complete a copy of the whole work as it proceeds. By this means you will be able to see how near the end of the alphabet the references to Notes are got, and to continue them accordingly. Those to be inserted in the latter part are
See Note ( ) after Art - 255
?" ( ) after Art - 270
?" ( ) d[itt]o - 287
?" ( ) d[itt]o - 320
?" ( ) d[itt]o - 339.

[20] WILKINSON, HENRY. Adm. pens. (age 18) at ST JOHN'S, June 7, 1810. S. of Capt. Richard. B. at Chatham. Matric. Michs. 1810; Scholar, 1811; B.A. (2nd Wrangler and 2nd Smith's prize) 1814; M.A. 1817. Fellow, 1814-21. Adm. at Lincoln's Inn, Nov. 18, 1814. Called to the Bar, Feb. 8, 1821. Assistant Master at Sedbergh School, 1819-38. Married Agnes, dau. of Arthur Shepherd, of Shaw End, Westmorland. Died Mar. 31, 1838. M.I. at Sedbergh. (F. P. White; Inns of Court.)

and it will be as well to mark them in the compositor's copy of the French Edition to obviate the possibility of mistakes.

Will you write me word as soon as convenient whether the printing of my part of the MS is begun, & whether any more copy is likely to be soon wanted and I will send the Appendix or at least part of it as soon as there is occasion for it.

How do affairs in general go on at Cambridge. Were you at the B[isho]p of Landaff's dinner? and is he going to leave Cambridge? When you write, tell me all the news you can hear, from all quarters relating to Cambridge. I have been passing a month in the Isle of Wight — Jacob, Whitcombe, Romilly &c with a whole tribe of Cantabs were at Ryde. I should suppose College must be quite empty.

I am on my way at present to Dawlish with my father. I expect to see Babbage. He is I hear intent upon nothing in the world but sailing, and functions. He has just sent me a copy of his second paper in the Phil. Transactions, which is abstract enough to choke any reasonable man and to turn the brains of any analytical one who has contrived to escape madness hitherto. Remember me to all friends. I expect to be back in October unless something very unlooked for takes place. Till then believe me

Yours very sincerely
J.F.W.Herschel

Address to me "Sir Wm Watson's, Dawlish Devon"

PS. Is Hornbuckle, or Jackson in College or does any one supply their place?

I.21 1821-12-02. JFW Herschel to C. Babbage.

Dec 2, 1821, Slough

Dear Babbage

I have once more to thank you for an instance of active friendship in giving me the earliest intelligence of Vince's death,[21] and for your warmly expressed wishes relative to myself. It would I allow be a spectacle edifying to the world & rather novel in the history of science to see our triumvirate in a state of amicable contention for a Professorship of Mathematics — like 3 puppy-dogs of a litter pulling at a bone, half in play & half in earnest, and old Towser popping out from behind a corner & running away with it from all of them — & I almost grieve that I cannot bring myself to make one in a trio which would afford so much amusement both to the parties & the lookers on. You are not a man to be satisfied without reasons, so I shall tell you mine situation. And first then my own views of life & tastes for the good things it affords are very materially altered within no very long period. I have no longer the keen relish for abstract mathematical studies I formerly felt, & except excited to their application by something wh. interests me more, am perfectly content to relinquish the distant fame & equivocal enjoyment they afford to others. Moreover I am equally well satisfied to leave the decision of the question between the ancients & moderns to combatants who have better stomach for the fight, contenting myself with the knowledge that if the moderns cannot silence the ancients, they are sure to outlive them, & that in the final adjustment of all things, this too will come right. But as it is very possible that to be a Lucasian Professor without knowing or caring one halfpenny about mathematics, I must add too that a regular genial residence at Cambridge (especially under the necessity circumstances would then impose upon me of filling occasionally the office of Moderator

21 Samuel Vince (1749–1821), Plumian Professor of Astronomy and Experimental Philosophy and Copley medallist for 1780.

&c.) would be in the highest degree irksome to me. I shall always remember Cambridge with pleasure on account of the friendships I have formed there, but if I once so far mistook the nature of this feeling as to suppose I could contentedly pitch my tent there, be assured I am no longer under the influence of any such illusion. A larger world & a more varied scene are necessary for my happiness. & as far as mere science is concerned I had rather pass my days among those who are advancing eagerly & rapidly & running a race with ardour, than in goading up hill the sluggish paces of any established institution under the sun. And first then you must have forgotten that I have already twice offered myself as a candidate for a professorship at Cambridge and twice been obliged to retire without even a promise of support among the heads and can you recommend to me to woo for the third time the favour of these coy Seniors. I will not believe you serious but I will confess to you that my repugnance to it is so great that if I had positive promises of 7 votes out of the 15 which I believe there are, secured for me by my friends, I question whether I could bring myself personally to ask for the 8th. It is in vain to say that I now stand on different grounds & have perhaps stronger claims than heretofore. It is possible I may, but I may fairly calculate on the permanence of their ignorance and must even shrink from any endeavour on the part of friends to explain my pretensions or point out my merits (such as they are) to them.

But I don't want you to suppose that I have no better reason for declining to offer than being nettled at a former rejection. There are much more rational ones.

(My own views of life . . . . under the sun,)[22]

I.22 1821-12-20. C. Babbage to JFW Herschel.

Dec 20/21[23]

Dear Herschel

Does the R.S. meet tonight? I think not and unless there should be something of extraordinary interest I should not go.

Can you come to me in the evening as early as you like. I want to explain my Arithmetical engine and to open to you sundry vast schemes which promise to reach the third and fourth generation. You have witnessed the wave-like succession of my plans and I can laugh with you at the rapidity and evanescence of the past, but there are reasons which incline me to reject the strong probability of failure derived from ~~past~~ experience and I am confident that when you hear them you will allow their weight. Do let me see you for I cannot rest until I have [and] have communicated to you a world of new thought. Most truly yours

Mrs B is better Monday.[24]

C Babbage

I.23 1822-03? JFW Herschel to C. Babbage.

Dear Babbage,[25]

Bad news — I am sorry to have to tell you that some rascally Frenchman has been presenting to the Institute 'A Machine for performing readily all the most difficult operations of Arithmetics and forming all sorts of tables'? — D[amn] his soul! The Inst. have, I am

[22] Ecclesiastes 1:14.

[23] Later Annotation, with further note 'from postmark'.

[24] Dec 20 of 1821 was a Thursday, the usual day of the week that the Royal Society met, matching 'Does the R.S. meet tonight?' Perhaps Babbage meant that his wife Georgiana had recovered by Monday 17th.

[25] Undated, Royal Society: HS.2.363.

told, approved it. It does not however follow that it is the same as yours and everybody will at least swear to your independent invention of it. [...]

I.24 1822-04-09. C. Babbage to JFW Herschel.

April 9th 1822[26]

Dear Herschel

I was much more glad to find from your letter that your interest in the success of the machine has rekindled than I was sorry at the news you sent me. It is scarce probable that another person should have contrived the same means for the same purpose nor is it probable that anyone should have copied from mine, however when they both appear if such a thing has taken place I think it will not be very difficult to find it out.[27] If any one has found out other principles of applying machinery to arithmetic and has produced such results as my engines will procure he deserves to be amply rewarded and I wish him success and what is much more to the purpose if he has contrived such and if they put the type together and if he is a Frenchman I am quite sure he will be well rewarded. Supposing the worst for me this is a case of two independent inventors and each must be content with that portion of fame due to the engines he produces. It will be quite otherwise with any future contrivers however ingenious; the egg has once stood upright and my future imitators may if they choose set it erect by breaking a different end. As for the fame to be derived from the contrivance of an engine to make tables by differences and even to ~~print~~ produce them ready printed I think it will be much greater than it deserves and I care little about it. The same cause (namely want of sufficient information on the subject) which will overrate its difficulty in the opinion of the multitude will in our scientific world underrate the value of those other machines I have contrived for reducing different parts of mathematics to the dominion of gravity: every moment I devote to this enquiry convinces me that although at present they are few and in an infant state yet that there is a vast field of discovery open and that the reputation to be derived from these ought deservedly to be greater and more permanent.

I have given up a profitable situation in life that I might devote myself to abstract speculation for this: the world if they know it would call me a fool. I have met with such little success in my attempts to render this task even moderately profitable that I shall scarcely be induced to repeat them. But having by good luck hit upon a plan which is actually useful in itself I think it quite fair to claim from this cui bono aqe that reward which they ought according to their own principles to bestow. This I will endeavor to procure as far as I reasonably can particularly as it is not very probable I shall ever have another opportunity. To return to you your information, whence came it? I cannot well write to Paris to enquire into its truth but I wish you would consider a little whether you could not. It must be so done as not to give any color to any romance which might hereafter be invented of my fishing their inventions: all I want to know is what the machine does; whether it goes by jack-up or req[uires] superintendence whether it gives written numbers or requires some mode of counting [...][28] like the Chinese thing. All this if it has been publicly reported at the institute can be fairly asked but I would not yet mention my own.

I hope to have it home in about ten days. The printing part in its new form gets on and I have no doubt about it suceeding, it is much simpler than the first contrivance and

[26] Later annotation.

[27] Though Lindgren 1990, 69, claims 'this piece of bad news is mentioned by a worried Babbage', the text shows that he was positively *insouciant* rather than worried!

[28] Water stain.

in some experiment I made at home promised very well. I have a machine for the roots of equations possible, and I will almost venture to say impossible, but it is not yet perfect and on this I want to talk with you.

Another idea concerning a table of sines to make by once setting an engine is not quite ripe but you shall have the embrio [embryo]: you know that[29]

$$\Delta^2 \sin\theta = -(2\sin\frac{h}{2})^2 \sin(\theta + h)$$

Suppose for a moment that in the difference of any two arcs in the table is such that

$$(2\sin\frac{h}{2})^2 = .0001$$

then it would be [a] way to make a machine in which the second difference should be may [be obtained] by transferring the preceding tabular number cutting off the four figures at the end and such an one would make without interruption a table of $\sin h$ $\sin 2h$ &c. [...]

Kater was much pleased with his visit.[30]

Very sincerely Yours
CB

I.25 1822-07-31. JFW Herschel to C. Babbage.

31st July 1822[31]

Dear Babbage,

M, Le Gros came to me this morning wishing to have my opinion on the propriety of executing in Gun-metal rather than in Cast Iron, two cross pieces aa,bb destined to receive a brass Arbor in a machine for cutting wheels, of which he is making a model for you. His reasons are 1st that he is unprovided with the proper tools chisels &c. for cutting Cast Iron & he and his workmen are not used to it, it would therefore take him more time a great deal than Gun Metal; 2dly the chance of finding the iron unsound & 3dly that if so the whole expense of time & work would be to be incurred again ∴ he says even in his own affair he would choose Gun metal. I told him, that if he with the authority have given him, had other scruples to use his judgement. I could not rely on mine but would write to you—so now the case is before you.

I found your letter to Sir H. Davy on my table 3 days ago. I question whether the mention of those Engines of which you have only sketches & general conceptions, especially those destined to the solution of equations of differences and those which work series reducible to no law but that which the mechanism itself determines, will tend so much to advance your speculations in the eye of the world as the more tangible & practical results to which you have already come. But as I know you think me a thingy fellow, this is only by the bye. [...] Laplace has sent me his last researches on Gases which he desires me to communicate to you [...]. Le Gros will wait for yours.[32] [...]

[29] Babbage has a stray bracket which has been corrected. He uses the standard result

$$\Delta^2 \sin k\theta = -4\left\{\sin^2\frac{kh}{2}\right\}\sin(\theta + kh).$$

[30] Captain Henry Kater (1777–1835).

[31] Addressed from No.9 Downing Street to CB at No.3 The Den, Teignmouth. Royal Society Herschel Collection 2.175. Based on Roberts 2014c.

[32] Babbage's decision.

I.26 1822-08-03. C. Babbage to JFW Herschel.

My dear Herschel,[33]

Your letter reached me this day (3rd August). I hope the answer will find you before you commit yourself to the waves.

Will you inform Legros that he may manufacture the cross pieces in gun-metal and if he should again apply to you I would rather trust to your opinion than my own as you have the advantage of seeing what has been done.

I rather agree with you about the Letter to Sir H. Davy but I wrote it at his suggestion. If you wish for other copies of it the publisher will on application send you as many as you like. I begin to be sick of the machine and have just got a chemical scheme into my head which I am very desirous of trying; [...]

I.27 1823-06-22. John Millington to C. Babbage.

[Resolutions of a meeting on 13th June 1823 of the Astronomical Society.[34]]

At a meeting of the Council of the Astronomical Society of London, held at their apartments in Lincoln's Inn Fields on Friday the 13th June 1823. The Council having taken into its consideration the advantages likely to arise to Astronomy from the important invention of Mr Babbage as communicated in his papers to the Society.

Resolved unanimously

That the Gold Medal of the Society be presented to Charles Babbage Esqre as a token of the high estimation in which it holds his valuable Invention of an engine for calculating and printing Mathematical and Astronomical Tables.

It was further Resolved

That all Medals awarded by the Council be presented at the Anniversary Meeting of the Society next ensuing such award; and that they be given to the Parties personally or to such proxies as they may appoint to receive them.

To Charles Babbage Esqre FRS &c. &c.

Sir,

I have the honor to transmit you the above copies of resolutions passed by the Council of the Astronomical Society of London and am Sir with the highest respect your very obedt servt

John Millington
Secty to the Astl Society

I.28 1823-06-27. C. Babbage to JFW Herschel.

27th June 1823,
Devonshire Street[35]

Dear Herschel,

[33] Based on Roberts 2014c.
[34] British Library Add MS 37183 f39. Based on Roberts 2014c.
[35] Royal Society Herschel Collection 2.184. Based on Roberts 2014c.

I had some conversations this morning with the Chancellor of the Exr, who treated me in a most liberal and gentlemanlike manner. He seems quite convinced of the utility of the machine and that it ought to be encouraged. At present he is to procure for me from the Civil Contingencies £1500 and next session if I want more to complete it he is willing that more should be granted or that I should have a committee of the House if a larger sum were wanted than that fund could be charged with.

He acquired the knowledge of the Astronomical Medal two days since from Gilbert so that you got it for me just at the fortunate moment. Mr Brougham had given him very just views on the subject of the machine and the manner in which he spoke of it was more gratifying than the grant itself. As I have liberal people to deal with I shall not be annoyed about pence and the particular mode in which I may think it right to distribute them and I shall I hope be able to bring the thing to perfection or at least to a good practicable working state and that in a few years we shall have new (but not patent) stereotype logarithmic tables as cheap as potatoes. Mr Ryan expects you on Monday night when I hope to tell you more of my interview with the nation's purse-bearer.

It was quite delightful to see Gilbert's joy when I told him of my success. He proposed that we should have a frolic in consequence of the good news which is to consist in an excursion to see the great telescope.

Give our kind regards to your Mother and believe me dear Herschel

most sincerely yours
C. Babbage

I.29 1823-07-05. JFW Herschel to C. Babbage.

5th July 1823,
Slough[36]

Dear Babbage,

I enclose the paper,[37] which I have not the least hesitation in signing, as however strongly it is worded, (I mean when the Degree of perfection to which the Engine may be brought is spoken of) you have done so much that I can easily believe you have it in your power to go as far as is said and I have no scruple to declare the faith that is in me to all the world.

Sir S. Young[38] manifested a great wish to see the Engine & I gave him your address for that purpose. He had been [busying?] himself in spreading a knowledge of it from the account I put into him, among his parliamentary friends, and according to his own statement had convinced so far 14 or 15 of its ability as to induce them to promise to vote for it if it came before the House.

Can you send me 3 or 4 more copies of your letter to Davy[39] as I have exhausted my stock.

Yours Truly
JFW Herschel

[36] Royal Society Herschel Collection 2.186. Based on Roberts 2014c.
[37] The report of the Royal Society ad-hoc committee on the proposed Difference Engine.
[38] Likely Sir Samuel Young Bart. F.R.S. (1766–1826), a friend of Herschel and a member of the Astronomical Society, having joined in 1821.
[39] Babbage 1822a.

I.30 1828-03-06. W. Whewell to Herschel.

TRIN. COLL. March 6, 1828.[40]

Dear Herschel,

I hope you will be as glad to hear as I am to tell you that Babbage was to-day elected our Lucasian professor. It was entirely in consequence of your letter, which I got yesterday, that I bestirred myself in the matter, for before that I did not consider myself sufficiently authorized to press his claims upon the electors. Peacock, Higman, and I, wrote a letter to each of them which was sent last night about eleven, and to-day at 2 the election took place. Nothing could be more distinguished than the mode of giving him the chair, for though at such a period one might have naturally supposed a large proportion of the votes to be engaged, out of 11 electors only 2 voted for one other candidate, and 1 for a third. I hope Babbage will take the matter in the same spirit in which I am sure it was received here: for I am quite clear that the Heads elected him in a very cordial veneration for his talents and confidence in his goodwill to the office and its duties. I suppose you or some of his friends will write to him immediately and inform him of what has been done. It was at first suggested that a difficulty would arise in consequence of the impossibility of his being admitted immediately, but it appears that the words of the deed require him to be admitted "*proximo opportuno tempore*," which allow any reasonable latitude of interpretation. In what I had to say to the electors about the matter I referred them to their own interviews with him on the former occasion[41] for information as to his intentions with regard to the manner of discharging his Lucasian functions.

I rejoice very much at the event of this business, for it is very honourable to him, and, as appears to me, to the electors. Nothing but the wish to obtain the best professor for the University and to do justice to Babbage can be supposed to have influenced them; and having been obliged to act in some haste, and to take a responsibility which might have been disagreeable, if matters had turned out otherwise, I am very glad to find that, judging favourably of our Heads, and not despairing of the result, we have been led to that which will I hope be a source of satisfaction both to our new professor and to us.

Yours very truly, W. Whewell.

I.31 1828-03-09. Herschel to C. Babbage.

Slough March 9, 1828[42]

Dear Babbage

You are elected Lucasian Professor at Cambridge. Thus writes Whewell (Mar 6) (in consequence of some previous correspondence) "Peacock Higman and I, wrote a letter to each of the electors which was sent last night and today at 2 the Election took place. Nothing could be more distinguished than the mode of giving him the chair for though at such a period one might have admittedly supposed a large proportion of the votes to have been engaged — and of 11 Electors only 2 voted for one other candidate and 1 for a third. I hope B. will take the matter in the same spirit in which it was viewed here for I am quite clear that the heads elected him with a very cordial veneration for his talents, a confidence in his

---

40 Todhunter 1876, 90-1.

41 Babbage had actively campaigned for the Lucasian professorship in late 1826, even preparing for such a bid earlier than that, losing out to Airy.

42 Royal Society Herschel Collection, HS/2/222.

good will to the office and its duties". "The words of the deed require him to be admitted professor 'opportuno tempore' which allows any reasonable latitude of interpretation. In the words I had to say to the electors I referred them to their own interviews with him on the former occasion for information as to his intentions with regard to the manner of discharging his Lucasian functions. I rejoice very much at the event of this business for it is very honorable to him and as appears to me to the electors. Nothing but the wish to obtain the best professor for the university and to do justice to Babbage can be supposed to have influenced them."

Read also what the Vice-chancellor writeth (Mar. 6) "I lose not a moment in imparting to you the agreeable intelligence that our friend Babbage has just been elected L. P. by a great majority. There were 11 members present of whose votes Mr.B. had 8, Mr. Maddy[43] 2 (the Masters of St John's & Catherine Hall) and Mr Bird[44] 1 — the Master of Sidney. The Provost of Kings, the Master of Peterhouse, the President of Queens and the Masters of Trinity and Downing were absent. You will be enabled therefore to ascertain how all the rest voted." ... "I shall hope to hear from you or Mr Babbage or both very soon." The V.C. is Dr. Davy our Caius friend to whom I suppose you will write "proxima opportuno tempore"

Read further what comfortable words [George] Peacock writeth (Mar. 6) "My dear H. I cannot express to you the joy I feel at the election of our dear friend Babbage to the Lucasian Professorship which took place today. When Airy showed me your letter" (I mentioned the subject in a letter I had occasion to write to him) "last week I was on the point of leaving Cambridge for 3 days & I desired him to tell Higman to call upon the Heads and declare B. a candidate. He called upon one or two but was discouraged in consequence of some objections being started to a point of form, as it was considered necessary that he should be upon the spot to be admitted. Upon my return on Tuesday I determined to renew the examination of this question & to make the attempt. On the receipt of your letter yesterday we consulted together, Higman, Whewell and myself and determined to propose B." (I had written to W. finding from your Mother that you could not have got my line mentioning the vacancy caused by Woodhouse's death and Airy's accession to the Plumian — and Fitton[45] also wrote to Sedgwick, suggesting the propriety of assuming that you would accept the office if conferred) — "as I considered that a private proposition of his friends without his knowledge could not in any way compromise his character or his delicacy — we called on the VC who was very favorable as was also the Master of C.C.C. We tried some others without success, but as the day was far advanced we addressed letters to all the Heads signed with our 3 names, and this morning made all the letter interest we could produce in his favor & you may judge of my delight when informed by the Registrar he was elected by a large majority" (numbers as before) "this announcement was received with universal satisfaction, except among the friends of Maddy who is a very respectable man and for whose disappointment I cannot help feeling a little pity & sorrow & even Dr. Fitton himself must confess that the election could not have taken place under circumstances more honorable to Babbage"

"The statutes direct that The Professor should be admitted 'tempore quovis proxime opportuno' B. ought therefore to return to this country as soon as he conveniently can and when he receives the news he should write to the VC and return his thanks to the electors and pledge himself to the performance of the duties of the Professorship at least to the same extent that he did in Nov. 1826. You may tell him that we will get him rooms in College

43 Likely Watkin Maddy (1798–1857) the astronomer, second wrangler in 1820, author of *The Elements of the Theory of Plane Astronomy* (1826).

44 Possibly Charles Smith Bird, for whom see section J.2, p. 441.

45 William Henry Fitton (1780–1861), MD turned geologist, a long-time associate of both Babbage and Herschel.

and that nothing will be wanting on our part to render his occasional residence during his lectures & examinations as agreeable as we can."

I have no paper to add comments, but shall only offer my congratulations and hope soon to hear from you. I wrote you at Rome & Florence. Your mother is in Town & well, also your children. [Joseph] Clement's health continues good — Machine gets on slowly and more is done on paper than in brass. Remember me to [Monichini & Scarpellini] and all the [-inis] and [-anis] in Rome who think me worth inquiring about. I have entirely quitted London and am living at Slough. My mother has taken a house at [Kensington][46]

I.32 1828-04-02. C. Babbage to JFW Herschel.

Dear Herschel,[47]

Nothing ever surprised me so completely as my Mother's letter announcing my election to the Lucasian professorship. The unceasing exertions of your friendship have become so habitual to me through life that I could not for a moment doubt the prime mover in an event which I look upon with feelings of the highest pride and gratification.

I do not yet feel strong enough in mind or body yet to return to England nor do I think such a course immediately necessary. I shall however shorten my tour and shall return with far different feelings from those which I anticipated: as far as Cambridge is concerned at least. I shall forget the past and shall try to do credit to the recommendation of my friends.

I shall go there with most cordial feelings and I have not the least doubt in meeting with the kindest reception.

I have not yet answered your letter about another plague which I consigned to your care and which will I fear have given you much trouble — I mean the machine. The plan for communicating the motion by degrees from the prime mover by means of steps instead of by a long arm may do but Clement must remember that the time be carefully attended to. I think the time of rest is 1/8 rev (he has probably made the change to avoid interference with some axis). The other plan for printing a given number of figures by means of 3 sector wh[eel]s is in principle the same as that which I left him and will do; he has probably developed it a little. The drawings appear to advance but he seems still making tools. I have not thought about the machine except [when] some one has enquired and ~~I~~ whenever I have thought about it, it has been to wish I had never undertaken it.

Now however a new and most unexpected difficulty has arisen on finding that Lord Goderich ceased to be at the head of the government. I wrote to W.W. Whitmore[48] to get some communication from Lord Goderich to his successors. It seems (I hope it arises from my imperfect instructions to my brother in-law) that Lord G. contends that there never was any understanding that I should have more money. Now if after the interview I had with Lord Goderich some years since, he could have had any such understanding or indeed any ~~the~~ remotest understanding that I was to be out-of-pocket by carrying it on — then will I utterly give up the little faith that I have in public men and in future be as cautious in my explicit dealings with them as I am with an Italian Inn-keeper. I have not yet answered W.W.W[hitmore]'s letter — this fact has not vexed me as it might have done. The result will be that I shall finish the machine and be more than half ruined by it. But I can scarcely bring myself to believe that Lord Goderich will not recall to his memory the bearing of that conversation. If he do not I will repay the money the government advanced. However I must think of these things when I return.

---

46 Perhaps 26 Lower Phillimore Place, Kensington, London.

47 Annotated: Rcvd, April 17 1828. Royal Society Herschel Collection, HS/2/224.

48 Babbage's brother-in-law.

Vesuvius has been giving indications of an eruption sending up stones 800 feet high and all but running over with lava — 160 passports were signed by the Neapolitan ambassador and 50 carriages started, fortunately I was at Tivoli and returned two days after the first rush and waiting for the post of the third day found on consideration the chances not worth running. Everest is here.[49] I shall probably go in 3 weeks.[50] You can send letters there if necessary. I suppose I must return here but am quite unsettled yet. I find it a great effort to write even a letter and am quite oppressed by the necessity of writing six or seven.

Remember me most kindly to all our friends they have been very attentive to my mother who constantly speaks of them.

Tell Fitton I can imagine the new professor escapes a tremendous lecture by being abroad, but that ~~he~~ I can fancy every word of it and that it is, or will be, all excellent — and further that I will try to act upon his advice — tell him also I shall be glad to hear from him although I do not deserve it.

Believe me dear Herschel ever most sincerely yours

C.B.
Rome 2 April 1828

I.33 1828-05-09. C. Babbage to JFW Herschel.

May 9 / 28,
Naples Santa Lucia
opposite Vesuvius gently smoking[51]

Dear Herschel,

I have just received yours of the 10-17 April. The accounts you give of the progress of the machine are by no means discouraging. As to the work done on the drawings it is a species of work which makes but little show. As to the actual steel and brass cut up for my use I fear it is little and that Clement is spending much time in making tools. Now this is to a certain extent necessary ~~but~~ and requires considerable supplies of money but I should wish you incidentally if possible to find out whether it is not Clement's intention to make me pay for the construction of these tools and then to keep them as his own property — from the multitude he is making it looks so. On reflecting upon my first interview with Lord Goderich I cannot believe he will not represent to the present government that it was distinctly understood thet I was not to be a loser by this machine. I trusted to his being an exception and thought him a sincerely honest man. Perhaps I was a fool for [so doing] and I will profit by my experience which in this instance will be rather costly.

[On] the whole I must look at my present position and [what is] to be done I will give you my own view and rely upon [you] to undertake the troublesome part a little [longer]:

1st I most heartily wish I had never thought of the infernal machine.

2nd I must finish it at any expense and as I am [in] an isolated way I feel [a][52] little uneasiness at the prospect of sacrificing a large portion of my fortune.

3 Of the machine itself, I think the drawings advance more rapidly then they did before, and the work on the whole does make considerable progress, more indeed in your hands then it did in mine. The money expended is large with regard to the labor part but this I cannot help and as I must complete it I would not have its progress checked on any account —

---

49 Sir George Everest (1790–1866) the geographer, for whom the mountain is named.

50 Perhaps to Naples?

51 Royal Society Herschel Collection HS 2/226. Based on Roberts 2014c.

52 This insertion seems the most accurate reading of Babbage's intentions, otherwise one must believe he was *happy* to pay for it all himself—he was not.

indeed if any sacrifice could terminate it instantly I would gladly make it. It is the wear and tear of moderate annoyances which I most dread. It would be impossible to pay for it out of income. I have therefore decided upon consuming capital and have made arrangements at Messrs Praeds so that you may continue to draw for money. This has all along been the case — I thought I was advancing instead of consuming capital.

I left Clement so well acquainted with all the mechanical actions of the machine that I have not the least fear of his making such changes as he may think necessary. The change in the spring attached to the screw carrying the copper plate I do not understand but have no fear of its being rightly rejected.

The change in the gradual form of motion to the horl calc axis I have already said will do. The plans of the sectors 6/24, 3/24, 2/24 is the development of a scheme I left and will do.

The change proposed for the calc. wh[eel]s by making them revolve on a sheath is not amiss but the hanging wh. which is attached to them will also [...] in the same.

The change which I most doubt is that of rejecting the mode of communication of motion from the horl colc. axis to the vert. calc axis [...]

I think it possible I may be in England about the end of [November] but have decided nothing. The Duke of Buckingham with whom I dined the day before yesterday has kindly offered to give me a cruise in his yacht but I have not yet decided—he returns in a week. I should certainly not have returned had not the Lucasian given me duties which I hope to execute well. I have it frequently in my mind and am collecting books and thoughts for an oddish set of lectures — but I cannot seriously work for two hours together, it is the most insufferable exertion and generally when I attempt it leaves me almost knocked up in body [...] when you write to Smedley press on him the impossibility of my doing for the [...] what he expected. I will finish the Binomial part of the principles of mechanics but it must take time and he must in the mean while get a Lardner or some other person to help him with the rest. I have tried to begin it but it will not do. I am so void of energy that in six months I have not required one word either of German or Italian. I regret every day that we are not travelling together. I think [I would] be energized if you or any friend with similar pursuits [were] with me but as it is I am losing time and learning little. [However] time passes and we must see out the dream of existence. [...] Fitton has done little for the Geol. Soc. in getting rooms. What will [become] of the Royal, will it dissolve into thin air for its character [abroad] now is not at present [...] much higher than it was at home those attacks in the Newspapers have spread far wider than you could have imagined. I remember Wollaston and some of our friends ridiculing (in the council) the effect of <u>public opinion</u> on the RS. It is a lever of far greater power than they imagine and I am inclined to think nothing but great wealth or perfect integrity can enable any man to resist it. I am heartily glad I have been absent during the late campaign of the R.S. and wish I could follow your example of living out if its vortex. I am now writing in view of Vesuvius which smokes very little. I was at it summit a few days since and saw about an acre of liquid lava in its crater. I fear we are not to have an eruption. [Would] that you were here for alone I have hardly energy to geologize. I left Everest at Rome but just starting for Venice we shall probably meet again in [the] North of Italy. I do not know [where] to desire you to write, perhaps I may get another letter here — perhaps [I will] again pass through Florence. Remember me most kindly [to Sir John] Franklin, South, Baily and all our friends. Ever most [...]

9th May 1828 Naples.

I.34 1828-07-24. JFW Herschel to J. Clement.

Mr Clement (copy) with an order for £180 dated London July 24 1828

July 24, 1828[53]

Sir,

I recd yours of the 21st on my return to Slough after a short absence and in the course of the week hope to be able to see you. With respect to the money 300£ you require, I transmit a draft for ~~150~~ 180£ which is the outside of what I consider myself authorised to do by Mr Babbage's instructions ~~which were on this point decidedly distinct~~. Having no longer ago than the 12th June given an order for 250 making in all paid to you up to that date since Mr Babbage's departure 800£, being at the average rate of about ~~100~~ 120£ per month, which certainly appears to me a large sum for the work done. That however may be, and probably is, owing to my inexperience in the detail of work of that description, and I am very far from intending that I consider it as beyond its just value. I mean only that ~~in such~~ under the circumstances I am no way desirous of taking on myself the responsibility of making ~~larger~~ [disbursements] beyond my instructions ~~authorise, till Mr B[abbage]s return [...].~~ Were I more perfectly acquainted with the complicated mechanism of the engine, I should feel no scruple in pushing forward the work with its utmost speed, but though I have studied it much, I find it impossible to master all its details, so as to feel full confidence in the absence of the inventor that nothing we do shall require alteration, and as he will now (as I expect from his last letter) not be more than four months longer absent, I should recommend that you would endeavour during the remainder of that time to ~~guard~~ keep ~~as near~~ within the average of expenditure ~~I have~~ above mentioned ~~as conveniently may be~~.

I remain Sir
Your obedt. Servt
JFW Herschel

I.35 1829? Richard Penn to C. Babbage.

Great George Street[54]

My Dear Sir,

I am truly sorry to hear of the annoyances to which you are exposed by the conduct of Mr Clement. These annoyances, great as they may be, would have been much increased if the settlement now proposed had been deferred to a later period.

I feel scarcely qualified to advise on such a subject but it appears to me at first sight that (as Mr Clement has chosen by his offensive conduct to forfeit all claim to that liberality on your part which disposed you at first not to question the amount of his charges but merely to require the particulars of them) your best course would now be to force the matter to a more hostile arbitration. I cannot but think that if you should do so, Mr Clement's low cunning will be defeated because I conceive that under such an arbitration he would not be allowed to charge both for his own time, and also a Trademan's profit on the articles, and I am sure that an investigation made in this spirit would greatly reduce the amount of his charges; and shew that you purchased paid for everything which is now done. A conclusive settlement made now seems to be absolutely necessary for your future comfort in this business, and I would advise you to speak to Mr Warburton on the subject.[55]

[53] Royal Society Herschel Collection HS/27/48.

[54] British Library Add MS 37184 f281. Based on Roberts 2014a. Richard Penn FRS (1784–1863), who resided at 5 Great George Street, was employed in the Colonial Office. He was the grandson of the founder of Pennsylvania, William Penn(1644–1718). Not dated but likely circa April 1829.

[55] Possibly Henry Warburton (1784–1858), who was the radical MP for Bridport, 1826-1841 and a follower of the utilitarian Jeremy Bentham. Later he was President of the Geological Society, 1843-1845

Yours Faithfully
Richd Penn

I suffer so much from cramp in my fingers that I have been obliged to dictate this to an amanuensis.

I.36 1829-04-22. Bryan Donkin to C. Babbage.

22nd April 1829,
No. 6 Paragon, New Kent Road[56]

Dear Sir,

Mr Rennie and myself have had a conference with Mr Clement this morning upon the subject of his Bill for the Calculating Machine, and the questions in your note respecting the same.

With regard to the Bill delivered, Mr Clement says, he only meant it as information to you, of what the amount of a more detailed Bill would be, and he promised us that he would immediately set about preparing an account from his Books containing his charge for materials and Time, or Labour employed upon the machine for every week since the Commencement to the present time.

Question 1st To whom do the Tools belong.
Answer—To Mr Clement. Mr C says that all the Lathes have been made at his own expense.

Question 2nd To whom do the patterns belong.
Answer—To Mr Babbage.

Question 3rd To whom do the Drawings belong.
Answer—To Mr Babbage.

Question 4th At whose risk is the Machine Tools and Drawings.
Answer—The Machine, Drawings, and Patterns are at Mr Babbage's risk, the Tools at Mr Clement's.

Question 5th Ought they to be insured, and by whom?
Answer—They ought, and by the respective Owners.

Question 6th How secured against Mr Clement's creditors?
Answer—This is a legal question, which we cannot answer.

Question 7th How is the amount of work now done to be ascertained, in order that it may be known what has been added at the next settlement?
Answer—To describe the present state of the machine and parts in their unfinished state would be very difficult. Would it not be better for Mr Clement in future to deliver Mr Babbage weekly, or Monthly Bills? Mr Babbage may procure a description of the present state of the machine.

Question 8th Ought not Mr Clement to engage, not to make another Machine without Mr Babbage's written permission?
Answer—Mr Clement declines doing so.

[56] British Library Add MS 37184 f266. Copy of the report of the arbitrators in the dispute between Babbage and Clement, Messrs Rennie and Donkin. Based on Roberts 2014a.

Mr Clement says the large drawing board cost him £50, and that Mr Babbage may have it at that sum, or that he, Mr Clement, will keep it.

We learned from Mr Clement that neither his House, Shop, Tools or Stock are insured, we would therefore recommend that you lose no time in insuring your own Property.

With regard to the Security against Mr Clement's Creditors, altho' it is desirable, I do not think you need be under much apprehension, as I have always understood that Mr Clement was a very careful and prudent man. Mr Rennie and myself are of Opinion that Mr Clement's Bill for the Slide Rest etc. should be allowed.

I am
Dear Sir
Yours most respectfully
Bryan Donkin

I.37 1829-04-11. C. Babbage to Bryan Donkin.

[To B. Donkin of 6 Paragon, New Kent Road.]

My Dear Sir,[57]

I enclose my notes as to the Calc. Machine. I have written to Mr Rennie and have sent him a copy. I hope he will be able to appoint an early day.

I have hitherto in conversation with Mr Clement always treated the work as my own private property for two reasons: first I thought his charges might be more moderate than if they were paid by government, and secondly I did not feel myself quite authorised to state the contrary. Since I have been acquainted with the result of my letter to the Duke of Wellington and the report of the Council of the R.S. I can scarcely continue to consider this to be the case.

I send another bill relating to the machine except two articles which I wish you would look over. The rest is charged at £77-4s. It was agreed for at £50 but Mr C assures me that he has only charged the cost price to himself. I believe you are perfectly aware of my feelings towards Mr Clement he is a most excellent workman and draftsman and ought to be well paid.

Believe me
My Dear Sir
Yours V. faithfully
CB

I.38 1829-04-28. Treasury Minute.

6th May 1829,
Treasury Chambers[58]

Sir,

The Lords Commissioners of His Majesty's Treasury having had under consideration your letter dated 16 Feby last enclosing copy of a resolution of the Royal Society, with a report of a Committee relative to Mr Babbage's Calculating Engine I have it in command to transmit herewith copy of their Lordship's minutes of the 28 Ultimo on the subject.

[57] British Library Add MS 37184 f252. Based on Roberts 2014a.

[58] British Library Add MS 37184 f283. Copy of a letter from J. Stewart to Secretary of Royal Society. Based on Roberts 2014a.

I am sir
Your obedient Servant
J.Stewart

[Extract from Treasury Minute Dated 28th April 1829]

Read a letter from the Secretary of the Royal Society dated 10th February last, transmitting a copy of a resolution by the President, and Council of the Royal Society, on a report of the Committee on Mr Babbage's Calculating Machine, who were desired to consider the question referred by this Board to the Royal Society, whether the Progress made by Mr Babbage in the construction of his machine confirms their former opinion that it will prove adequate to the important object which it was intended to attain.

My Lords read again the former Papers on this subject, and refer to the Minute of 11th July 1823 directing, upon grounds stated therein that the sum of £1500 should be paid to Mr Babbage to enable him to bring this machine to perfection.

My Lords observe by the report of the Committee composed of persons most capable of judging that on the best estimate they can give, three fifths of the work necessary to complete this machine has been accomplished, and on the assurance of Mr Babbage, which they as Practical Mechanics see no reason to doubt, they state that £6000 has been already expended by Mr Babbage on this undertaking.

On these grounds My Lords are of opinion that they are fully justified in directing a further payment of £1500 to Mr Babbage to enable him to complete the machine by which such important benefit to Science may be expected to result, and they are pleased to direct Mr Sargent to pay to Mr Babbage fifteen hundred pounds accordingly out of any money in his hands on account of Civil Contingencies.

I.39 1831? C. G. Jarvis to C. Babbage.

16 Lambeth Terrace

Sir,[59]

I have thought of the spiral barrel you were speaking of the other day and no difficulty of any consequence has presented itself. I am anxious to name this because you might mention the barrel to others who might see some trifling objection and condemn it at once. I have also been thinking on the subject of patents which was new to me, but the more I think of it the more I find myself convinced of its complexity, and the danger there is of coming to a wrong conclusion respecting it.

I understand that it is proposed by Messrs Donkin and Field that the drawing of each article belonging to the Calc. Mache. should be numbered and the time expended upon that particular drawing noted down in a book. I do not see any useful purpose this can answer and I do not think I can submit to it. If they know that the quantity of paper covered is not by any means a measure of my assiduity or efficiency, and they must be convinced that no one can judge of my exertions but the individual under whose immediate observation I am employed. My keeping an account of my time in this manner, therefore, seems to me like writing my own condemnation. I could not explain to Messrs Donkin and Field that the greater part of the time charged was expended in considering the propriety of adding a 'hair breadth' to the head of that screw, or taking a 'thirty-second' off that boss; they would know that these things were of no consequence whatever to the efficiency of the thing itself, and not being acquainted with me would most probably doubt the truth of my statement.

[59] British Library Add MS 37185 f419. Based on Roberts 2014a.

The goodness of a machine depends upon the soundness of the principles upon which it is constructed; and although a man who does not know enough of the Science of Mechanics to ascertain the magnitude and direction of a force resulting from two other forces whose directions are different may be a very good tool-maker, we call upon him for too much when we expect him qualified to arrange and construct a new and complicated machine. It should be borne in mind that the inventor of a machine and the maker of it have two distinct ends to attain. The object of the first is to make the machine as complete as possible. The object of the second—and we have no right to expect he will be influenced by any other feeling—is to gain as much as possible by making the machine: and it is his interest to make it as complicated as he is permitted to make it. I am fully aware how far these observations may do me injury. But they are made, sir,—whether well or ill-judged—for your good: and if you choose to destroy me with my own weapons I leave you to seek your reward from your own feelings.

These considerations and others which I do not think proper to commit to paper have almost determined me to leave Mr Clement. In the expectation of doing so I will take this opportunity of suggesting what appears to me to be improvements

1. A dead stop for the barrels; without which they will be very liable to overrun when the number of springs urging them forward is much greater than those retarding them.

2. That the inclined planes for the barrels be cut so that the barrels may be dressed in any way whatever which may be done easily.

I must observe that I can see no advantage in allowing 6 turns of 1st axis for every calculation as Mr C proposes doing; all he wishes may be done by the barrels.

I am Sir,
Very Respectfully
Your Obedient Servant
C.G. Jarvis

I.40 1832-09-14. J. Stewart to C. Babbage.

14th September 1832
Treasury Chambers[60]

Sir,

I am commanded by the Lords Commissioners of His Majesty's Treasury to acquaint you that My Lords have had under consideration the memorandum drawn up by Mr Clement the Engineer whom you employ in constructing the Calculating Engine which was transmitted in your letter of 14th July last in which paper Mr Clement gives an Estimate of the expense of removing the Engine to the premises which have been erected for its reception, and also presses a personal claim for his own benefit on the ground of the loss and expense which a change of residence will entail upon him.

It appears by this Memorandum that Mr Clement desires that a Lease of the Premises may be given to him for three or four years, and estimates the expense of removal at the sum of £350 in which sum is included the expense of a second removal and fixing of tools etc. at the expiration of the time specified in the lease. He presses also a personal claim for £130 for interest on Money laid out in his furniture, and for wear and tear of his furniture and requires various alterations to be made in the House in East Street for his accommodation.

[60] British Library Add MS 37187 f134. Based on Roberts 2014a. Reformatted for clarity.

He also claims to himself an annuity from the Public of £660 per Annum with additions for rent, rates and taxes in East Street.

My Lords cannot but express their suprize that Mr Clements should have advanced so unreasonable a claim, which no doubt must have been quite unexpected by you, or you would have brought it under the notice of this Board when you recommended the removal of the Engine from Prospect Place to East Street, where a Building has been erected at a large Public expense. My Lords desire therefore that you will acquaint Mr Clement that his demand is considered by their Lordships unreasonable and inadmissable.

My Lords are ready to defray the actual expense of the removal of the Engine and all the tools, not only the tools for which Government have paid, and which consequently are the property of the Public, but also all the tools and property which belongs to Mr Clement which he may wish to remove to East Street. If any trifling alterations in the Apartments in East Street are judged by the Board of Works to be reasonable for Mr Clement's accommodation, my Lords will sanction the expenses of making them. They will also be ready to defray the expense of removing Mr Clement's property from East Street, when the Engine is completed, and they will take into consideration the propriety of indemnifying Mr Clements for any actual loss which it may be clearly shewn he shall have sustained by the injury of his Property in the removal, if it should be necessary for him to reside on the Premises in East Street.

But Mr Clement must recollect that £12,340 of Public Money has been expended on the Engine, the greater part, if not the whole, of which sum has passed through his hands, from which a very large amount of profit has accrued to him, and that his employment by you has constituted, and still continues to constitute, as My Lords are informed, the principal part of his business, and that he has not been restricted from using for his general business any tools for which the public have paid, and that the further expense of completing the engine will yield to him a considerable profit.

Under these circumstances, My Lords cannot entertain Mr Clement's Claim for interest on the money laid out in furniture, and for wear and tear of it, or for any Annuity or addition to the profit he derives upon his accounts, My Lords would, however, suggest for your consideration whether it would not be a more economical and at the same time not a materially inconvenient arrangement, if Mr Clements were to continue to reside at his own house, in which case My Lords would not object to allow him two pounds per week to defray the expense of conveyance to and from East Street.

I am Sir,
Your obedient Servt
J. Stewart

I.41 1832. C. Babbage to Lord Somerset.

My dear Lord Duke,[61]

I need scarcely assure you that the interest you take in the calculating Eng. is most gratifying to me. it is ~~however but~~ part of that unabated kindness which I have experienced for years and which I trust you know me well enough to ~~know/feel that~~ perceive is not the less deeply felt because it is but little expressed. You have pointed out clearly the great source of the difficulty in my position between the Government on the one hand and the engineer on the other hand. I quite agree with you about the probable feeling of the Secy of State or of the Treasury.

[61] British Library Add MS 37187 f302. Draft only. Based on Roberts 2014a.

It is singular that I should not have perceived the full difficulty of my position until the great mechanical difficulty was overcome. But I have since the execution of the part already completed been compelled to perceive that the discovery is too much in advance of the age to be appreciated and that amongst all civilised countries England is that [in] which there exist the greatest number of practical engineers who can appreciate the mechanical part, whilst at the same time it is of all others that country in which the governing powers are most incompetent to understand the merit either of the mechanical or mathematical.

The suggestion you throw out relative to making the interest of the engineer to finish the work as soon as possible is one deserving of the greatest consideration. I had already thought of it in case I should employ a new one but the practical mode of effecting this is difficult and requires an examination of the detail (perhaps by engineers).

There exists one other difficulty and I am sure it could not have escaped you although you have not alluded to it — I mean my own feelings. Some years since I was forced to make up my mind to give up the Engine entirely before any thing had been executed by which the public could be convinced of its ~~utility~~ practicability at the same time I had resolved to repay the whole sum which the government had spent upon it then amounted to about £5000. I have not forgotten nor can I ever forget that circumstance and I confess that I am deeply indignant at the treatment I have experienced from the country.

Perhaps it is appointed by nature that those who ~~are supposed to~~ possess the envied power of invention are doomed to expiate it by being the victims of intense feelings. If so my expiation is complete ~~for I have little of feeling after both those of indifference to the point of my unjust country~~. It has however ~~and perhaps the public neglect~~ had the effect of making me cherish more highly those marks of private esteem which I meet with from my personal friends.

Believe me
My dear Lord Duke
Very sincerely yours
C.B.

I.42 1833-06-21. Lady Byron to Dr King.

[...] We [herself and Ada] both went to see the thinking machine (for such it seems) last Monday.[62] It raised several Nos. to the 2nd & 3rd powers, and extracted the root of a Quadratic Equation. I had but faint glimpses of the principles by which it worked. Babbage said it had given him notions with respect to general laws which were never before presented to his mind. For instance, the machine would go on counting regularly, 1, 2, 3, 4, &c— to 10,000—and then pursue its calculation according to a new ratio, which was, I think, 10,002, 10,005, 10,009—but I am only certain that the numbers were no longer successive ones, and that their differences were neither in arithmetical nor Geometrical ratio, as far as I could apprehend. If this occult principle of change existed in the law according to which the machine was constructed (for Babbage discovered it to be latent in the mathematical formula originally applied by him) it may be consistent with the general laws of our solar system that the sun shall not rise tomorrow.[63] He said, indeed, that the exceptions which took place in the operation of his Machine, & which were not to be accounted for by any errors or derangement of structure, would follow a greater number of uniform experiences than the world has known of days & nights. There was sublimity in the views thus opened

[62] Lovelace Papers, LP 336.

[63] Lady Byron misunderstands what Babbage was doing. He had arbitrarily added a feedback mechanism to computationally change the differences based on the most significant digit carried from the previous calculation, see Babbage 1824a.

of the ultimate results of intellectual power. [...]

I.43 1834. C. G. Jarvis to C. Babbage.

I cannot help enforcing the necessity of some alteration in the method pursued for the construction of the Calculating Engine, that is, if it is wished or intended to be finished within any reasonable time. The plan I wish to recommend is, that the designs and drawings be all made on your premises and under your immediate inspection, working drawings made from them and directions given in writing to various persons, to construct various parts of the machine, which parts might all be going on at the same time and the entire machine be speedily completed. In this case whenever any difficulty occurred you might be at once appealed to, whenever it was found very difficult to produce nearly that effect—which is a very common case in machinery—your decision would at once put an end to all perplexity and hesitation and the work would proceed; instead of having to wait days, weeks or months for an arrangement which half an hour would complete [...]

To a man who although inactive and unenterprising loves money, it must be agreeable to construct a newly invented machine the cost of the parts of which cannot be taxed; and still more agreeable to be able to charge for time expended upon arranging the parts of that machine without the possibility of the useful employment of that time being disputed, and to doze over the construction year after year [...] It may be suspected I enforce my opinion from interested motives. It is perhaps difficult to divest oneself of selfish feelings and even Messrs Donkin and Field were somewhat under their influence when they prohibited the paying of draftsmen more than a certain sum whatever their worth [they were afraid their own draftsmen might demand higher salaries], and therefore all I shall say on the subject is, that my plan may be followed without my being in any way a gainer by its adoption. In fact the probability is that I shall be out of Britain before the machine is proceeded with in any way.[64]

I.44 1834-03-09. Ada Byron to Dr William King.

Sunday, 9 March 1834
Fordhook

Dear Dr King[65]

I must thank you for your kind & affectionate letter of encouragement & advice. How far I am really & permanently awaking to a sense of religious duty & religious obligations, time alone can prove. I cannot but feel very distrustful of myself. To you, as well as to Mrs King, do I consider myself to be deeply indebted — more so than I can express. You may remember perhaps that in more than one of our "pleasant walks," you said much on the subject of controlling our imaginations and our thoughts. I often think of this now, as I cannot but perceive that it is a paramount duty for one in my circumstances to exercise this sort of self-government, and indeed upon its performance depends my future welfare. I must cease to think of living for pleasure or self gratification; and there is but one sort of excitement, if indeed it can be called by that name, which I think allowable for me at present, viz: that of study & intellectual improvement.

[64] British Library, Add. Ms. 37,188, f 437. Cited in Hyman 1982, 131-2.

[65] Lovelace Papers, 172. Ada Lovelace's letters are full of girlish underlining, repeated, sometimes doubled or more, without noticeably altering the meaning of the contents. This redundancy has not been reproduced in her letters given here.

I find that nothing but very close & intense application to subjects of a scientific nature now seems at all to keep my imagination from running wild, or to stop up the void which seems to be left in my mind from a want of excitement. I am most thankful that this strong source of interest does seem to be supplied to me now almost providentially, & think it is a duty vigorously to use the resources thus as it were pointed out to me. If you will do me so great a favour as to give me the benefit of your advice and suggestions as to the plan of study most advisable for me to follow, I shall be most grateful.

I may say that I have time at my command, & that I am willing to take any trouble. It appears to me that the first thing is to go through a course of Mathematics, that is to say Euclid, and Arithmetic & Algebra; and as I am not entirely a beginner in these subjects, I do not anticipate any serious difficulties, particularly if I may be allowed to apply to you in any extreme case. My wish is to make myself well acquainted with Astronomy, Optics &c; but I find that I cannot study these satisfactorily for want of a thorough acquaintance with the elementary parts of Mathematics [...]. In short, here I am, ready to be directed! I really want some hard work for a certain number of hours every day [...].

Yours most gratefully & affectionately
A.A. Byron

I.45 1834-07-08. Ada Byron to Mary Somerville.

Tuesday Evening, 8 July[66]

My dear Mrs Somerville

I am only afraid I must have seemed to you very presuming for venturing to suggest such an arrangement as I proposed, and I assure you I had little difficulty in making up my mind to be so impudent. Nothing but my great anxiety about the machine could have made me so I believe. But I think you must be fond enough of those things, to sympathize with my eagerness about them. I am afraid that when a machine, or a lecture, or anything of the kind, comes in my way, I have no regard for time, space, or any ordinary obstacles. This is the only excuse I can offer, and you must always tell me whenever I encroach on your very great kindness. [...]

Believe me ever,
Your sincerely obliged
A. Ada Byron

I.46 1834-09-01. A. DeMorgan to William Frend.

[...] I was very sorry to find when I came home that Mr. Woolgar[67] had been very uncourteously received by B[abbage] with my note. That unfortunate man will never rest until he succeeds in getting nobody's good word. He calculates very wrong (for a calculating machine maker) if he thinks such a thrower of stones as himself can stand alone in the world. It takes all his analysis and his machine to boot to induce me to say I will ever have any communication with him again, which nothing should induce me to do except the consideration that men of real knowledge should have more allowance made for them than some charlatans I know. I make no doubt Mr. Woolgar will detail to you the reception he

66 Somerville collection, Bodleian, 17.

67 From S. E. De Morgan 1882, 78–81. John Webb Woolgar (1795–1851) of Lewes, was a solicitor and fellow of the Royal Astronomical Society.

received. . . . Apropos of logarithms, give my kind regards to all your circle, and believe me, dear sir,

Yours sincerely,
A. De Morgan.
5 Upper Gower Street

I.47 1834-11-08. Ada Byron to Mary Somerville.

Monday Morning, 8 November
Fordhook[68].

My dear Mrs Somerville

[...] May I trouble you the first time you see either Mr Babbage, or his son, to say how exceedingly obliged I am to the latter for his unexpected kindness in sending me the plates & account of the Machine, which is exactly what I was in want of & is a very great help to me. I am very busy copying the Steam Engine paper I carried off from Mr Babbage's the other day. I have finished two & expect to complete the 3rd in a few days. I hope he is not in want of them, & would have no scruples in claiming them immediately! Believe me with kind regards to your daughters.

Yours most sincerely
A. Ada Byron

I.48 1834-11-25. Somerville to Ada Byron.

My dear Miss Byron[69]

If you are not otherwise engaged will you meet me at Mr Babbage's on Friday at 12? We are to see the drawings of his new machine and to have a few hours conversation with him a pleasure which I know you appreciate as highly as I do. I promise to inform Mr Babbage of our arrangement, and therefore should be glad to have a line in answer to this. If Friday is not convenient for you we shall fix another day. I beg you will offer my kind regards to Lady Byron and believe me, ever yours

Mary Somerville.

R.H. Chelsea,
Tuesday 25th Novr.

I.49 1835-05. C. Babbage to A. Quetelet.

A LETTER TO M. QUETELET FROM M. CH. BABBAGE CONCERNING THE CALCULATING MACHINE[70]

General Session of May 7 and 8.

---

68 Somerville collection, Bodleian, 17.

69 Beinecke, Yale University, Box: 26, Folder: Group 477, F-6. The MS is reproduced in Baum 1986, Ch. 4 plates.

70 Babbage 1835. However this letter was excerpted there and additional material is in Babbage 1989a, 12-14.

M. Quetelet read a letter which he had just received from M. Ch. Babbage, a correspondent of the Academy, concerning the calculating machine, of which various circumstances, regardless of the author's wishes, continue to delay its completion.

"During the last six months," says M. Babbage, "I have invented a new machine of much greater power: I have abandoned all other research and am presently making the drawings. I am progressing rapidly, but it is probable that I will not have it built here. I am myself astonished at the power which I have managed to give it, and which I would not have believed possible to achieve a year ago.

This machine is intended to handle *one hundred* variables (or numbers susceptible to change); and each of these numbers may be composed of 25 figures. It will then execute each of the following operations, and the combinations which may result:

$v_1, v_2, \dots v_{100}$ being any numbers, the machine will be able to:

- Add v_i and v_k;
- Subtract v_i from v_k;
- Multiply v_i by v_k;
- Divide v_i by v_k;
- Extract the root v_i of v_k;
- Reduce v_k to zero.

For example, if $f(v_1, v_2, \dots v_n)$, $n < 100$, is any given function that can be formed by addition, multiplication, subtraction, division, root extraction, the machine will calculate its numerical value. It will then substitute this value in place of v_1, or any other variable, and calculate this second function with respect to v_1.

It appears that it can reduce to tables almost all equations of finite differences.

Suppose that you have, through observation, a thousand values of a, b, c, d, and that you wish to calculate them using the formula:

$$p = \sqrt{\frac{a+b}{cd}};$$

the machine will be set up for the calculation of the formula, and one will input the first series of values of a, b, c, d; then, it will calculate them, print them, and reduce them to zero; finally, it will ring a bell to indicate that a new series of constants must be input.

When there exists a relation between any number of successive coefficients of a series (provided it can be expressed by addition, subtraction, multiplication, division, powers or root extraction), the machine will calculate and display the terms successively; and it may afterward be arranged to find the value of the series for all values of the variable.

The greatest difficulties of the invention have already been overcome; but I will need several more months to complete all the details and construct the drawings. However, the work is already well advanced."

Mr. Babbage then informs us that he has just received a letter from Sir J. Herschel, written from the Cape of Good Hope, which contains some details relating to meteorology. From comparisons, already made between observations gathered in India and at the Cape, Sir J. Herschel arrived at some curious results concerning a general motion of the atmosphere; Mr. Babbage is waiting for a report from him on this subject. Sir J. Herschel has arranged for the Philosophical Society of South Africa to take four days a year (December 21, March 21, June 21, and September 21) during which they will take hourly readings of the barometer, thermometer, etc. The readings are to start 6 hours before noon on the 21st, taking the whole day, and continuing until 6 hours after noon of the 22nd. If the 21st is a Sunday, they will not begin until the 22nd. It is important that the same days are chosen at different parts of the earth.

The persons who will take part in these observations can send their results to Sir J. Herschel or Mr. Babbage, using the Academy as intermediary.

I.50 1835-06-10. C. Babbage to Ada Lovelace.

My Dear Madam[71].

I enclose cards for my two last parties this season. I hope you intend to patronise the "Silver Lady". She is to appear in new dresses and decorations.[72]

I have been studying Turbans and have pinned up one which one of my fair friends has pronounced "not so bad as might have been expected." I suppose perseverance will do much for decorating the out- as well as the inside of heads, and perhaps next year I may succeed better.

I am My Dear Madam
Very faithfully Yours
C Babbage
Dorset St.
10 June 1835

I.51 1835-08-02. Charles Babbage to Nathaniel Bowditch.

Dorset St
Manchester Sq.
London
2 August 1835[73]

My Dear Sir,

I can scarcely fancy that I am addressing this letter, I will not say to a stranger, but to a friend whose person I have not yet seen. The unaffected simplicity of your son's character whilst it won my esteem also impressed me with a high respect for those who had formed it and I occasionally indulge myself in imagining your domestic circle and fancying the remarks of my young friend on the persons and the things of our other world.[74] I know well from your son's anxiety to study the Calculating Engine that I can scarcely write to you on a topic of greater interest nor will I attempt to conceal my own opinion of the importance it now possesses ~~in my own estimation~~.

But I must commence by informing you that I speak of an entirely different Engine from that which your son is acquainted with. Shortly after he left England in September last I got possession of the drawings of the old Engine which I had not seen for two years. In explaining these to my son many new views struck me and a totally new and more extensive arrangement occurred to me which appeared of such importance that I thought it worthwhile before I proceeded to reexamine every part of the old Engine. I soon found many new modes of adding, of carrying, and of performing other processes which were mechanically much more simple than the ancient methods. These new simplifications enable me to add new

71 Lovelace Papers, 168
72 A 12 inch high *danseuse* automaton, with a fluttering bird on its finger, constructed by John Joseph Merlin (1735–1803). It had been recently purchased at auction by Babbage. He had originally seen it as a child.
73 Dibner Library, National Museum of American History, Smithsonian Institution, MSS 32 A. Published in Williams 1988.
74 Henry Ingersoll Bowditch (1808-1892).

powers to the Engine and yet to make it upon the whole less complicated than the former one. It is somewhat curious that in the Engine I am now drawing there scarcely occurs a single contrivance which has been used in the former one. For example, in the old Engine when addition takes place a memorandum is made by the transit of a lever from one side to the other of a wheel in order to indicate that a ten is to be carried. The proper arms then pick up these memoranda in succession beginning with the carriage to the tens then to the hundreds etc. In the new Engine after addition all the carriages are affected at once (at the same instant) the engine foreseeing if necessary when a carriage will itself cause carriage to several nines above and acting upon this foresight it makes any number of them at once.

It is impossible to explain without figures this mechanism which is not so complicated as the former and which is in my opinion one of the happiest of the minor contrivances in either of the engines. I have had a draftsman constantly at work since September and I have given up all other pursuits for the sake of this and the progress I have already made surpasses my expectations. It is indeed more than I ever before made in several years. This arises partly from a more enlarged experience, partly by having a better tempered draftsman, partly by having all my drawings in my own house instead of at a distance of four miles. My intention is to finish a complete set of mechanical drawings of the new engine and a variety of the Mechanical Notations to explain its operations. I have already got over all the difficulties of the first order and most of those of the second and feel very confident that in twelve months more if I can carry it [...] I have [...], that I shall have completed the drawings so that the invention cannot ever be lost. I will now give you an outline of its power presuming that except in very peculiar cases the old engine (although it worked some complicated equations of finite differences), only added one number to another and that it effected its calculations only by the artful use of addition.

- It will add any number to any other
- subtract ... from ...
- ... multiply ... by ...
- ... divide ... by ...
- It will at any time reduce any number in any part of the engine to zero
- It will find any power of any number
- It will extract the square root
- It will extract any root by successive approximations.

In the engine I am drawing each number or variable may have thirty places of figures and above 140 variables may be put into it at one time. I have chosen so large a number of variables because it enables me to put ~~into it~~ a table of lives wholly into the Engine and also the rate per cent and to calculate the values of any Annuity on one, two, or three lives for all differences of age.

The engine will find the numerical value of any algebraic function of any number of quantities, for ex. let a, b, c, d, ... be given quantities it will calculate and print the value of a'

$$a' = f(a, b, c, d, ...)$$

it will also find that of b'

$$b' = f_1(a, b, c, ...)$$
$$c' = f_2(a, b, c, ...)$$

etc. etc. and having printed these it will without any new setting reduce a, b, c, etc. to zero and substitute in their places a', b', c', etc and proceed to find the values of a'', b'', c'' etc. and substituting these proceed to the values of a''' b''' c''' etc. and so on writing each set.

Of course it can calculate the values of π, e, the numbers of Bernoulli etc. Also it can calculate any logarithm or any difference of any logarithm or sum any converging series provided an algebraic relation exists between any number of its successive coefficients.

If the general term of a series be

$$\frac{(-1)^n(a+bn+cn^2+...pn^i)}{(a'+b'n+c'n^2+...rn^{i+2})}$$

it will find the sum. In developing functions if

$$\phi(x) = a + (a' + bx)y + (a'' + b'x + cx^2)y^2 + ...$$

whenever any algebraic relation can be found between given numbers of successive coefficients the engine will calculate and print in succession as many as may be required. If the development be

$$\phi(x) = A + B\Psi fx + C\Psi f^2x + D\Psi f^3x + ...$$

the numerical value of A, B, C, etc. may be found under this form.

If $f(x)$ be any algebraic function of x then the Engine will give the numerical values of $f(x)$, $f^2(x)$, $f^3(x)$, $f^4(x)$, ... etc and it will if required make tables of the values of these for any values of x.

The engine will print the value of every constant or variable put into it at the commencement of a calculation on the same plate on which it prints the calculated results. Thus if the superintendent has put in wrong numbers his error will be detected by comparing those which the Engine printed with those given to him to put into it. This is a most important power for if any calculations are so complicated as to require it, or if tables previously printed by the engine are to be employed, the engine will print every number first in and every calculation made in one piece of copper. Some of the calculations then made may again be employed in the Engine for new calculations. But they will be again printed on a second plate of copper with the new results and these new results may be used again and so on. Thus any error committed by the superintendent in putting in the figures will inevitably be detected by comparing the successive plates.

When means of large series of observations are required to be taken according to any law they may be done thus: The superintendent sits before the engine and puts in the first number as soon as it disappears he puts in the 2nd then 3rd etc. etc.

Now the engine may be set so that it shall print several columns: in the 1st the number actually first in will be printed, in the second the mean of every two, in the 3rd the mean of every three then of every 5 or every ten as may be required. And at the same time in another part of the engine the curves formed by the observations themselves as well as by all the means shall be engraved in a plate of copper.

You will perceive by the first statement that the whole system of equations of finite differences is now reduced to wheel works and that it is only a question of time to arrive at the value of any term.

The Calculating part of the engine consists of a directive part, in which the formula is set, a driving part and an arithmetical part in which about 140 numbers containing above 4000 figures are placed. It has scarcely a single spring, in which respect it differs entirely from the former one and it is much simpler and less liable to wear or get out of order.

You will be able to appreciate the influence of such an engine on the future progress of science. I live in a country which is incapable of estimating it and I am much displeased with the treatment I have experienced. You who have yourself made great pecuniary sacrifices for the production of a great and important work will understand my feelings. I have made large sacrifices and am prepared to make still larger. There is however a limit at which my duty to my children demands any forbearance and I place that at the completion of the drawings which is in fact the completion [of] the invention. These will remain when I am gone and when some more enlightened country than my own shall at some future period become sensible of the value of the engine or ambitious of the fame of calling it into real existence, those drawings will undoubtedly be sought and from them a machine will be made which shall give to all nations and to all after time tables at once perfect and enduring and which shall execute calculations from which all human agents would shrink in hopeless despair. I have thus given you an outline of the work I am employed in, the views that it has opened and which it continually suggests are amongst the most interesting and important I have met with.

With my kindest regards to your Son
I am
My dear Sir
With greatest esteem
Sincerely Yours
Charles Babbage

I.52 1835-11-01. Ada Lovelace to Mary Somerville.

1 November
Ockham Park[75]

My Dear Mrs Somerville

You must not think that I have forgotten you, although it is so long since I have written to you. - I am ashamed to think how long. We only came home last week. [...]

I now write, partly to ask news, partly to give news, principally &. to remind you of your promises to visit us. Will you name your own time, & I can only say the sooner the better, & if your stay is long it will be best of all [...] .

I now read Mathematics every day & am occupied in Trigonometry & in preliminaries to Cubic & Biquadratic Equations. So you see that matrimony has by no means lessened my taste for those pursuits, nor my determination to carry them on, although it has necessarily diminished the time I have at command. But I suspect it is no bad thing to be limited in that respect.

By the bye I have a mathematical question to ask you, which I hope is not too trifling to be beneath your notice...

$$\sin(a+b) = \frac{\sin a \cos b + \sin b \cos a}{R}$$

$$\cos(a+b) = \frac{\cos a \cos b - \sin a \sin b}{R}$$

The proposition these formulas relate to is, "The sines & cosines of two arcs a & b being given, the sine and cosine of the sum or difference of these arcs may be found by the

[75] Somerville Papers, Bodleian, Dep. c, 367.

following formulas &c.—["]. The expressions for the sine & cosine of the sum, as above, are then demonstrated by a figure; after which occurs the following passage: "The values for the sine & cosine of *the differences* $(a - b)$ may easily be *deduced from these two formulas.*" Now however easily deduced, I have not succeeded in doing it. The required values are given in the book, though not the *method* of deducing them from the above. They are, $\sin(a - b) = \ldots$

I suppose the deduction depends on some principle which I have overlooked, but I do not perceive how from consideration of the first two formulas, the last two can be deduced.

I am sorry to find that your copy of Legendre's Geometry is still in my possession. I thought I had returned it & my only excuse is—my marriage. [...]

I have troubled you with a very long letter, & though I could add much more, yet I think it would not be justifiable to occupy another minute of your time.

Believe me,
Yours Affectionately
A A King

I.53 1835-11-18. Ada Lovelace to Mary Somerville.

[...] I have another trigonometrical question to ask you, & am encouraged by your kindness to trouble you with these things, which is almost a shame too, when you are so busy.[76] I believe this question is as simple to answer as the last, & am afraid you must think me a great dunce not to make out such easy things.

"Since the arc a is formed from the sum of the two arcs $(a - b)$ and b, by the preceding formulas we shall have:

$$R \sin a = \sin(a - b) \cos b + \cos(a - b) \sin b$$
$$R \cos a = \cos(a - b) \cos b - \sin(a - b) \sin b$$

and from these we find

$$R \sin(a - b) = \sin a \cos b - \sin b \cos a$$
$$R \cos(a - b) = \cos a \cos b + \sin a \sin b["]*$$

My query is, how the latter two formulas are deduced from the former? [...]

I.54 1835? C. Babbage to Plana.

[This exists in a French translation only, made for Plana from the original English. It has been translated back to English here.]

My dear Sir,

I am much obliged to you for the kindness and consideration you have shown my friend Mr. Martin, and I sincerely regret not seeing you in England to be able to express my gratitude to you and at the same time explain to you the nature of the calculating machine, of which you have shown me so much desire to be informed. The first machine I had begun about 14 years ago by order of our government. I had completed part of it long ago; but circumstances having prevented me from completing it for several years, I have not

[76] Somerville Papers, Bodleian, Dep. c, 367.

occupied myself with it since. I send you herewith copies of a report made in 1829 by the Royal Society.

Since last September I have imagined an entirely new machine, and I have continually occupied myself in making the drawings, which are now well advanced. By the old machine I had proposed to calculate tables by means of differences, and in some cases by equations of differences; but in this last respect it was not of absolutely general application. I will now give you an idea of the new machine. Each number, or variable put into the machine, can have 30 figure places. More than 100 variables or constants can be put into the machine at once. The machine has the property

1. of adding one variable to another,
2. of subtracting idem,
3. of multiplying idem,
4. of dividing idem,
5. of reducing any number that can be encountered to zero,
6. of extracting the square root of any number.

The machine can therefore receive at once any number of constants, less than 100. And it can be arranged so that by simply turning the handle (without making the least effort of mind) it will give the numerical value of all the algebraic functions of these constants as:

$$f(a, b, c, d, ...)$$

Currently after calculating and printing the value of this function, it can reduce the number a to zero and put in its place the numerical value of the function, and then calculate the new value: so if

$$\begin{aligned} a' &= f(a, b, c, ...) \\ a'' &= f(a', b, c) \\ a''' &= f(a'', b, c, ...) \end{aligned}$$

all these values of a, a', a'', a''' ... etc. will be printed and calculated. But if you had wanted at each new calculation to replace b, c, d, etc. by other values taken from the numbers a, b, c, etc. the machine would have had to be arranged so as to produce these other values b', c', d', etc. and b'', c'', d'', etc. and it would have substituted them, and would have used the result.

You see by this that the whole theory of finite difference equations is reduced to a simple mechanism, and that several developments which could not be undertaken are now possible. For example if

$$\Psi(x) = A_0 + A_1x + A_2x^2 + A_3x^3 + ...$$

then if we can find some algebraic relation between the coefficients of the k terms, such as

$$F(A_{v+1}, A_{v+2}, etc., A_{v+k}) = 0$$

however complicated this equation may be, the machine will proceed to its calculation, and will print successively all its numerical coefficients, without the slightest mistake. If it had been necessary to keep the development in a more algebraic form, as if

$$\Psi(x) = a + (a_1 + by)x$$

$$+ (a_2 + b_1 y + cy^2)x^2$$
$$+ (a_3 + b^2 y + c_1 y^2 + dy^3)x^3 ...$$

then if some similar relation can exist between a, a_1, a_2 and b, b_1, b_2, etc. the machine will calculate them all and print them successively.

We are not limited to developments according to the powers of x, but the application is general because if

$$\psi(x) = A + B$$

any relations being given between successive coefficients, the machine will calculate them up to any extent. The most difficult parts of the machine are already imagined and drawn; perhaps the drawings of the entire machine will be finished in a year.

I have spent a great deal of money in my researches, and have not yet received any encouragement from my country. I am not in a position to have a machine made according to the designs, and I do not care to do so; for when the advantage of such a machine shall be appreciated, whether in my lifetime or not, recourse will always be had to my designs. The subject is indeed too far beyond the understanding of the persons who first occupied themselves with it; and the men at the head of our affairs, who would know its value, are too busy keeping their places, to pay the least attention to subjects of high science.

I will give you a few more details concerning the machine. It will calculate logarithms, or their differences of any order whatsoever, without passing through any of the antecedents. It will calculate the values of π, of e and of Bernoulli numbers, up to 30 places, and perhaps up to 50, by giving it more time. It will calculate according to a given date, tables of annuities (life annuities) on the life or any sort of combination between the life of one, two or three individuals of any age whatsoever. If you have a number of observations for example on the tides, on the barometer, or of any other kind whatsoever, you can employ a secretary, who will put into the machine at times determined by the machine itself, all these numerical observations. The machine will print on different columns. The first column will include all the numbers that the secretary has put into the machine, so that if he has made any mistake, you will be able to notice it. The second column will include the averages of the observations taken two by two. The third, the averages of these same observations three by three. The fourth, the averages taken 4 by 4 etc. and so on.

In another part of the machine, the curves of these same observations, and their averages, will have been engraved on copper. Observe that it is not necessary that these averages be arithmetic; but they can be expressed by any other algebraic means. The machine will sum any series whose general term is

$$\frac{a + bn + cn^2 + dn^3 + ... + rn^i}{a' + b'n + c'n^2 + d'n^3 + ... + r'n^{i+2}} \cdot (-1)^n$$

in which n varies, and $a, a', b, b', ...$ are any numbers. By using this machine as a difference machine, it can add 12 differences, each of 30 figure places and make a calculation with the first eight figures engraved on copper, in about a minute.

I have not told you of all the uses to which my machine can be applied; but from the general idea that I have just given you, you will understand that it has nothing more to do with the old one.

I am, my dear Sir, yours
Charles Babbage

I.55 1835-10-03. Plana to Babbage.

[...] If it were possible to grasp the constitution of a machine by a single description; certainly the details set out in your letter, in the report of the Committee of the Royal Society, and in periodical works, should suffice.[77] But in fact, the machine requires seeing and examination to imprint a faithful picture on the imagination. I understand very well that, deprived of these effective means, I am not able to form correct ideas about the artifices imagined by you, Sir, to achieve a goal that could be unattainable, even after having reflected on it for a long time. Provided that it is, I feel that the machine is invented, largely designed, and ready to be delivered to posterity. If obstacles prevent its execution in the course of your life, by the report, made in 1829, on your first machine, I see with pleasure, that our illustrious colleagues of the Royal Society have spoken of your discovery from the feelings which come from the contemplation of a great difficulty overcome. However, I was seized by a deep feeling of regret on reading in your letter, that you have not yet received any encouragement! It is astonishing, that the men of England, who have the power to offer genius the rewards it deserves, have not yet seized the honorable opportunity which you offer them, to point out and recommend you to posterity. I dare, however, to hope that such a comparable apathy will be overcome by your perseverance, and that you will have the satisfaction of seeing your efforts crowned by success, and at least in part, rewarded. How can one remain indifferent to the sight of a machine capable of calculating the ratio of circumference to diameter, the base of Napier logarithms, and Bernoulli numbers, with 30 figures? Doubtless you can also calculate the tables on elliptic transcendents that you see in Legendre's work, and the numerical coefficients that one obtains by the return of a given sequence. Now all this is astonishing, for anyone who is aware of the strength of the legislative power and the weakness of the executive power of algebraic analysis. Doubtless, the tables for solving the problems of ballistics are, by your machine, facilitated beyond what one could hope for. Indeed, I sincerely congratulate you. I deeply regret being detained by family considerations; otherwise, I would come to London first to procure the pleasure of making your personal acquaintance and then to confer with you on your machine and on a whole host of other subjects which require [...]. I like to have the illusion and hope that I will be able to make this journey another year. Now I console myself for this loss by thinking that I am fulfilling this duty of observing Halley's Comet here: I am often annoyed by the state of the sky, which [...] the sight of this star. But that is a perpetual phenomenon: the whole of the astronomers of Europe will leave to their posterity a series of observations capable of perfecting the theory of this comet. [...]

I.56 1835-11-18. AL to Mary Somerville.

18 November 1835[78]

My Dear Mrs Somerville

I have another trigonometrical question to ask you, & am encouraged by your kindness to trouble you with these things, which is almost a shame too, when you are so busy. I believe this question is as simple to answer as the last, & am afraid you must think me a great dunce not to make out such easy things.

> Since the arc a is formed from the sum of the two arcs $(a-b)$ and b, by the preceding formulas we shall have:

[77] British Library Add MS 37189 f173. Based on Roberts 2014a.

[78] Somerville Papers, Bodleian, Dep. c. 367.

$$R \sin a = \sin(a-b)\cos b + \cos(a-b)\sin b$$
$$R \cos a = \cos(a-b)\cos b - \sin(a-b)\sin b$$

and from these we find

$$R \sin(a-b) = \sin a \cos b - \sin b \cos a$$
$$R \cos(a-b) = \cos a \cos b + \sin a \sin b$$

My query is, how the latter two formulas are deduced from the former?

Mr Greig comes on Saturday next, I am happy to say; and I think it very probable that Mr Babbage will be here next week. I wish you were at present a plus instead of a minus quantity. The head of my picture is now completed, but I would rather not give my opinion of it, that your judgment may be unbiased.[79] Mrs Carpenter would be very glad of your opinion & so should we. The picture is going to Town this week. Would it be asking too great a favour of you, to call at Mrs Carpenter's [...]

I.57 1835-11-30. C. Babbage to Plana.

My dear Sir,[80]

I received your two parcels from the Sardinian Ambassador delivered at my house free from any expense whatever. I only mention this at your own wish, as I should always be most happy to receive any Communication from you by post or by any ordinary channel and should consider myself well rewarded for the trifling expense attendant in such conveyances.

I have distributed almost all the copies you sent of your valuable paper and together with my own thanks I am commissioned by many of your friends in England to return their best acknowledgements. You seem to have seized on the mathematical principles which the Calculating Engine embraces most completely. It will as you observe revert any given series and give the numerical coefficients and it will also prepare the Numbers and differences of any particular function at such intervals as will enable the former engine to fill up the whole table by ordinary differences. It is impossible by description to convey to you any idea of the mechanical contrivances but I hope you will yourself be able to examine them next summer and I assure you nothing will give me more delight than to give you every opportunity of judging of it for yourself and of discussing with you the varied uses to which it can be applied.

I am at present most busily employed in continuing the drawings of the new Engine and I expect it will cost my draftsman and myself about twelve months of hard and continuous labors to complete them. I shall then rest satisfied that the invention cannot perish and that the wants of Analysts will ultimately become sufficiently great to demand its completion.

The next meeting of the British Association will be at Bristol which is twelve hours distance from London and the time fixed on is about the end of August. I hope you may be able to be present and that you will find many things worthy of your visit to England.

Dorset St.
Manchester Square
London

[79] Presumably the portrait of Ada painted that year by Alfred Edward Chalon.

[80] Turin Academy of Sciences, Plana.II.13, Losano 2014, 122. The British Library contains the following letters from Plana to Babbage: 12.5.1838; 02.13.1839; 02.28.1839; 03.1 7.1841; 05.25.1842; 06.29.1844, Bulferetti 1968a.

I am My dear Sir
With sincere esteem
Yours most truly
C. Babbage
30 Novber 1835

I.58 1836-01-14. Spring Rice to C. Babbage.

14th January 1836,
Downing Street[81]

My dear Sir,

I was out of town and confined to the house by illness when you addressed a letter to the Chancellor of the Exchequer[82] under cover to me. I did not therefore receive it till after the period at which you stated an answer was necessary to effect its object. The Chancellor of the Exchequer is now anxious to come to some definite result with you on the subject of the Calculating Machine but in the first place allow me to recall to your attention what has occurred in reference to the statement addressed by you to His [Grace the] Duke of Wellington in Dec 1834.

You will recollect that on receiving the above paper he communicated with you requesting you to inform him whether you wished to make it official or whether you wished to substitute any other for it, as he is unwilling that the paper which was placed in his hands by Sir R. Peel should be registered and made official without your entire connivance. The conclusion to be drawn from the statement alluded to is that you have invented a machine of much more extensive powers and capable of calculations of a nature far more complicated than those to be performed by the machine at present in progress: and you request to be informed whether Govt would undertake to defray the expense of this new machine.[83]

The Chancellor of the Exchequer is most sensible of the authority of any recommendation of yours. But considering that the first machine has been now nearly 13 years in progress, that during that time a very considerable amount of public money has been spent on it, that the attention of Parlt. has been drawn to the subject and public expectation and justly excited and as the second machine, so far from being calculated to supersede the first, promises to extend its range and utility, the Chancellor of the Exchequer would feel himself bound to look to the completion of the first machine before he could propose to Parliament or the Treasury to enter on the consideration of a second one.

During the last session of Parliament it was asked in the House of Commons what progress has been made in the construction of the machine, and when the completion of the work might be expected. This enquiry will probably be expected in the approaching Session and in order to give him the means of ~~asking~~ answering it, he proposes to follow the same course which has been taken throughout the conduct of this transaction and to obtain a further Report from the Royal Society with respect to the present state of the machine and the period at which its completion may be looked for. The Royal Society will be authorized to employ any practical mechanist or engineer, if they think fit so to do, to assist their enquiries and shew their report before him; the Chancellor of the Exchequer will be most ready to communicate with you respecting the best mode of attaining this result.

The Chancellor of the Exchequer desires me not to conclude without expressing to you his great anxiety that this matter may be brought to a final settlement at once satisfactory

[81] British Library Add MS 37189 f273. Based on Roberts 2014a.
[82] Then Thomas Spring Rice (1790–1866).
[83] Pencil annotation: I never did make this request nor ever thought of doing so. CB.

to yourself and to the public. No one is more alive than he is to the importance of the object sought and to the genius displayed in seeking it but these considerations make it most necessary to take care that the just hopes and expectations of the public may not be disappointed by inadvertent steps on his part.

Believe me, my Dear Sir
very faithfully yours
(signed) Spring Rice

I.59 1836-02-01. C. Babbage to Spring Rice.

1st February 1836,
Trinity College Cambridge[84]

My dear Sir,

The conversation we had at Cambridge will have explained my feelings relative to the Engine. Although neither the public character of the R.S nor its conduct towards myself afford me the least ground of confidence in its decision yet I am so perfectly confident in the merits of the Calc. Engine that I shall make no objection to the reference to that body. ~~However highly skilled the individuals may be who examine it.~~

With respect to the Statement addressed to the Duke of Wellington in Dec 1834 I always considered it official and I think it would have been unfair in me not to have informed the govt of the new Engine. I do not on ~~looking through~~ examining this paper see that it ~~leads to / can be considered as~~ contains any application ~~to the Govt~~ to take up the new engine and I certainly never intended to apply for what might be considered a favor when I have been treated ~~which~~ with what I ~~consider~~ look upon as injustice.

It will be desirable to state distinctly the points on which the opinion of the R.S. ~~are~~ is required and perhaps to point out the necessity of consulting machine makers on the practical part of the subject.

[Insert from another page] The enclosed report containing information which I ought to communicate may be of some use but I ~~I shall be in town on Thursday night and~~ will call in Downing St on Friday morning and shall be happy to give you any information you may require in order to enable the Chanr. of the Ex to ~~put~~ shape the reference as he may think best.

I am My dear Sir
Yours very truly
C. Babbage
Trinity College
Cambridge

I.60 1836-02-01. C. Babbage to Spring Rice.

[Report to the Rt Honble the Chancellor of the Exchequer on the Calculating Machine]

2 February 1836[85]

In a Statement addressed to the Duke of Wellington in Dec'ber ~~1835~~ 1834 I thought it my duty to inform the government of the invention of a new Calculating Engine altogether differing from that whose construction ~~I have~~ I had for many years been superintending.

[84] British Library Add MS 37189 f291. Draft. Based on Roberts 2014a.
[85] British Library Add MS 37189 f292. Based on Roberts 2014a.

The general conception of that engine having been imagined, ~~many~~ most of the greatest difficulties overcome and many sketches and drawings having been made, I at the time stated that the new Engine did not supersede the old but on the contrary that it would add to its value.

Since that period I have given unremitting attention to the plans and designs of that new engine and have had numerous drawings both of its elementary parts and of their modes of combination executed. ~~The result is that of~~ This more advanced state of the drawings of the new Engine has now proved that it is capable not only of accomplishing all those other complicated calculations which I had intended but that it performs all those calculations which were peculiar to the old Engine both in less time and to a greater extent—in ~~short~~ fact that it completely supersedes the old Engine. Nor is this the only conclusion to which the present state of the drawings leads. I had in my former statement mentioned the fact that it differs essentially in all its elements from the first engine: the mechanical simplicity ~~of~~ to which those elements have been reduced is such that I am led to this ~~conclusion~~ opinion.

If it were desired to construct an engine merely possessing the powers ~~of~~ intended for that which is already partly made I believe any practical makers of machinery who could bestow sufficient time on the enquiry would ~~to~~ arrive at the conclusion that it would be more economical to construct such an engine on the new principles than to finish the one already partly executed and I am quite sure that one so constructed would be a much better instrument.

The fact of a new ~~engine~~ machine superseding an old one in a very few years is one of constant occurrence in our manufacturing and instances might be pointed out in which the advance of invention has become so rapid and the demand for machinery so great that half finished machines have been thrown aside as useless before their completion.

It is now nearly fourteen years since I undertook for the government to superintend the making of the first Calculating Engine. During nearly four years its ~~progress~~ construction ~~was~~ has been absolutely stopped and instead of being employed in overcoming the physical I have been harassed by the impediments of what might be called the moral difficulties of the question. It is painful to reflect that in the time so employed the first Engine might, under more favorable circumstances, have been completed.

In making this report I wish distinctly to state that I do not entertain the slightest doubt of the success of the first Engine nor do I intend it as any application to finish the one or ~~make~~ to construct the other: but I make it from a conviction that the information it contains ought to communicated to those who must decide the question relative to the Calculating Machine.

C. Babbage
Dorset Pl.
Manchr Sq.

I.61 1836-01-18. Ada Lovelace to C. Babbage.

18 January 1836
Ockham[86]

Dear Mr Babbage

I will not any longer delay thanking you for the Minerals, which I received a few days ago from St James's Square. I hope I shall also find courage enough to take to pieces one

[86] British Library Add MS 37189 f281.

of the wooden crosses, though even with the other before me, I doubt if I have ingenuity to put it together again.

I have got on delightfully with Lardner's Trigonometry, which as far as I have yet gone I think excellent.[87] His book seems at all events to be the one most suited to my mind. I have had quantities of formulae to work out myself, & have destroyed a great deal of paper.

Lord King desires to be remembered to you, & we both hope to see you here again before long. Believe me,

Yours truly
A.A. King

I.62 1838. Ada Lovelace to C. Babbage.

2 March
Ockham[88]

Dear Mr Babbage

I have just received amongst a number of packages from Town, the new edition of your work which you have so kindly sent me; & I will not delay another day thanking you for it, as I am much gratified by the kind recollection it shows.[89]

You may possibly have heard of the very tedious & suffering illness which has occupied so many months, since a Miss King has been added to our family. Though I am now to all appearance perfectly well again & am in fact most wonderfully improved, yet I am still quite far from being really strong. But for these untoward circumstances, Lord King or I myself should probably have written to you long ago, & have strongly urged a visit or visits to Ockham.

With Lord K's very kind remembrances, & hoping soon to renew my acquaintance with you. Believe me,

Yours sincerely
Augusta Ada King

I.63 1838-07-26. C Babbage to Viscount Melbourne.

26th July 1838[90]

My Lord,

Seventeen years have elapsed since I first undertook to superintend for the government the construction of an engine which I had invented for the calculation and printing of ~~mathematical~~ astronomical and analytical tables. In return for the sacrifices of time and fortune which that invention has cost me I have as yet received ~~from my country~~ nothing but ~~unmerited reproaches and unmerited~~ neglect.

During the last five years the construction of that machine has been suspended and I have repeatedly urged upon the government both by ~~formal~~ direct application and through my friends the necessity of some final decision on the subject. If the question has now

87 Lardner 1828. Ada's correspondence with A. De Morgan in 1840 and after shows that she did not get far with the subject.
88 British Library Add MS 37190 f386.
89 Presumably Babbage 1838.
90 Annotated in red ink. British Library Add MS 37190 f496. Based on Roberts 2014a.

become more difficult because I have invented [a] superior mechanism which supersedes that already partly executed this consequence has arisen from that very delay against which I have repeatedly remonstrated.

I now appeal to your Lordship for the last time to ask for no favor but to ask for that which it is an injustice to withhold from me—a decision.

I am your Lds
Obd Sert
CB

I.64 1839-11. Ada Lovelace to C. Babbage.

[...] quite made up my mind to have some instruction next year in Town, but the difficulty is to find the man. I have a peculiar way of learning, & I think it must be a peculiar man to teach me successfully.[91]

Do not reckon me conceited, for I am sure I am the very last person to think over-highly of myself; but I believe I have the power of going just as far as I like in such pursuits, & where there is so very decided a taste, I should almost say a passion, as I have for them, I question if there is not always some portion of natural genius even. At any rate the taste is such that it must be gratified. I mention all this to you because I think you are or may be in the way of meeting with the right sort of person, & I am sure you have at any rate the will to give me any assistance in your power.

Lord L[92] desires all sorts of reminiscences, & that I am to take care & remind you about coming to Ockham.

Yours most sincerely
Augusta Ada Lovelace

I.65 1839-11-29. C. Babbage to Ada Lovelace.

Dear Lady Lovelace[93]

I make no most ungrateful returns for your kind letter from London. I have lately been ever more than usually occupied by the Engine.

I allowed myself ten days in Cheshire and finding this did not do I was obliged to go to Brighton for five days which restored me to the calculating state and have been working very hard ever since.

I have just arrived at an improvement which will throw back all my drawings full six months unless I succeed in carrying out some new views which may shorten the labour.

I have now commenced the description of the Engine so that I am fully occupied.

I think your taste for mathematics is so decided that it ought not to be checked. I have been making enquiry but cannot find at present any one at all to recommend to assist you. I will however not forget the search. The London World is very quiet at present. Mrs De Morgan has just added a new philosopher to its population and Mr Sheridan Knowles has written a most popular play called 'Love' to which I have been a frequent attendant. I met the author yesterday at a dinner at Mr Rogers.[94]

[91] British Library Add MS 37191 f871.
[92] William, Lord Lovelace.
[93] Lovelace Papers, 168.
[94] The banker-poet Samuel Rogers (1763–1855).

I could not by possibility have visited you this year in the West, but I cherish the hope of getting a few days at Ockham when I can indulge in a little recreation.

Pray forgive my epistolary negligence and believe me with best regards to Lord Lovelace.

Ever very sincerely yours.
C Babbage

I.66 1840-02-16. Ada Lovelace to Babbage.

16 February 1840[95]

My Dear Mr Babbage

Have you ever seen a game, or rather puzzle, called Solitaire? There is an Octagonal Board, like the enclosed drawing, with 37 little holes upon it in the position I have drawn them, & 37 little pegs to fill the little holes. One peg is abstracted to begin with, and then the remaining ones hop over & take each other. For instance if peg No 19, the center one is taken out to begin, then peg 6 may hop over peg 12 into the empty hole 19, & peg 12 is taken off the board; or, peg 21 might hop over peg 20 into 19, & then peg 20 goes off the board. The pegs are only allowed to hop over each other at right angles, not diagonally. The puzzle is to leave only one on the board. People may try thousands of times, and not succeed in this, leaving three, four, five, or many more even which have no neighbours to give them a lift off the board. I have done it by trying & observation & can now do it at any time, but I want to know if the problem admits of being put into a mathematical formula, & solved in this manner. I am convinced myself that it does, though I cannot do it. There must be a definite principle, a compound I imagine of numerical & geometrical properties, on which the solution depends, & which can be put into symbolic language. I believe that much depends, to begin with, on the particular peg first abstracted, & am inclined to think there is but one which will admit of subsequent success. I will not name which — I believe these boards are to be had at every toy-shop. I have numbered the holes in my drawing for the sake of convenience of reference. The real boards are not numbered.

Do not forget that I am to have the "Stratagems of Chess." I hope you are bearing me in mind, I mean my mathematical interests. You know this is the greatest favour any one can do me. Perhaps, none of us can estimate how great. Who can calculate to what it might lead; if we look on beyond this present condition especially?

You know I am by nature a bit of a philosopher, & a very great speculator, so that I look on through a very immeasurable vista, and though I see nothing but vague & cloudy uncertainty in the foreground of our being, yet I fancy I discern a very bright light a good way further on, and this makes me care much less about the cloudiness & indistinctness which is near. Am I too imaginative for you? I think not. I do not think you possess half my forethoughts and powers of foreseeing. Though I see nothing but cloudy and vague uncertainty in the foreground of our being, yet I discern a very bright light a good way further on. Immutable truths, intrinsic beauty, symmetry and logical completeness. A new and powerful language. That brain of mine is something more than mortal, as time will show. The devil's in it if I've not sucked out some of the life-blood from the mysteries of the universe. Elaborate and scientific music.

[95] Add. MSS 37191 f.331.

I.67 1840-03-06. F. Prandi to C. Babbage.

March 6th [n.y.], 46 Berness Street[96]

Dear Sir,

If you are likely before next Monday to show your Calculating Machine to any friends, I should consider it as a great favour if you would allow me to join the party together with some gentlemen of the suite of the Duke of Lucca, and perhaps the Duke himself, who will leave for Italy at the beginning of next week. As I have scarcely the honor of being known to you, I should not venture to make this request, were I not induced to hope that from regard to Lyell and many other common friends, you will perhaps excuse the liberty which I thus take.

I have the honor to be
Your very Obedient Servt
F. Prandi

I.68 1840-03-14. Ada Lovelace to Babbage.

Saturday, 14 March
Ockham[97]

My Dear Mr Babbage

Can you be so kind as to tell us your friend Prandi's address?[98] We want to ask him here for Satdy next, & cannot you add yourself? Or if not that day, is the following Tuesday, the 17th, quite out of the question?

I fear not withstanding the Eglintoun Tournament, that the real days of Chivalry are gone bye, or you would never resist a lady's entreaties as you do. We shall not I think be in Town quite in time for your party the 21st; probably a day or two later.

I am reading a book which is really well & lucidly explained, & to me very interesting; Mosely's Mechanics applied to the Arts, &c. Do you know anything of him? If I can judge from a man's writing, he is no common person.[99]

Should there seem no chance after I go to Town, of the much desired great Unknown[100] being found for me, I have some idea of having instead for this season some German lessons. I know a little of it already, & have always intended to know more. Indirectly I think it would bear on some of my objects [...]

Yours most sincerely
A. A. Lovelace

I.69 1840? C. Babbage to Lord Northampton.

My dear Lord,[101]

---

[96] British Library Add MS 37201 f308. Based on Roberts 2014a.
[97] British Library, Babbage Papers.
[98] Fortunato Prandi.
[99] H. Moseley 1834. Henry Moseley (1801-1872) was a mathematician. At this time Ada certainly lacked the required background needed to read Moseley's book with understanding.
[100] That is, a mathematics tutor.
[101] British Library Add MS 37191 f456. Based on Roberts 2014a.

In your situation as PRS [President of the Royal Society] renders it natural that I should apply to you as the representative of the interests of Science in any Communication with the Government.

It is now about 18 years[102] since ~~I commenced~~ at the desire of the government I began to superintend the construction of an engine I had invented for the Caln. and printing of Astronomical ~~and nautical~~ tables ~~that~~. During 12 years I ~~worked at that~~ gave my whole time and undivided attention to that work. ~~That~~ I have refused during that ~~time~~ period several situations of considerable pecuniary profit one of which would have produced an income of at least £2000 a year. ~~That~~ I declined these ~~pos~~ offers because I was unwilling to delay for even a few months the completion of that engine. ~~I have carried~~ The enquiries which enabled me to make these mechanical inventions could only be carried on at great expense ~~and~~. I have made and am making great personal sacrifices ~~for that complete them~~ to meet that expenditure. The construction of the ~~machine~~ engine for the government has now discontinued ~~but not from any fault of mine~~. I am unable to obtain any decision ~~as to its ultimate destination~~ I have been frequently and publickly reproached with having myself recd. for my own use that money which had been paid by government to the operatives who constructed the engine and I have recd. no assistance in refuting that ~~representation~~ statement ~~and~~ but on the contrary the circumstances of my being totally omitted in the list of those who have received the few rewards ~~pensions and honors bestowed~~ which have been bestowed on Science ~~confirmed~~ added credit to the misstatement and it received ~~still further~~ full confirmation from a printed report of ~~a statement which appears in~~ a report of a Com. of the H of C.

I do feel deeply the injustice I have suffered and I now apply to the R.S. to assist in refuting them.

The Mastership of the M[int] is ~~now~~ vacant; it is a situation ~~for~~ which ~~my pursuits~~ it will be admitted that my pursuits ~~have~~ qualified me to ~~fill~~ occupy. It was formerly filled by a predecessor in the chair which for ten years I held. It was filled by Sir Isaac Newton and although he was in Parliament during a small portion of the time he occupied it yet he did not hold it as a political appointment. Neither do I wish you to apply for it to Lord M. as a political appointment but as one for which my knowledge qualifies me and which my services deserve.

Should it be objected to [your application] that it is an office whose occupant is changed with every administration I answer that if it were distinctly stated that the appointment was made for the reason assigned I do not believe that any succeeding administration would withdraw one given to Science.

I.70 1840-03-25. Alexander Dallas Bache to C. Babbage.

25th March 1840,
Philadelphia[103]

[...] Why have you not published on the subject of the new Calculating Machine? Henry and I have been looking with impatience for the appearance of the sketch which was begun and partly in shape when you were at Newcastle. [...]

I.71 1840-09-30. C. Babbage to Angelo Sismonda.

Turin

[102] This dates the letter to roughly 1840.

[103] British Library Add MS 37191 f354. Based on Roberts 2014a. Bache (1806–1867) was a great-grandson of Ben Franklin and a surveyor. He had attended the British Association at Newcastle in 1838.

30 Sep. 1840[104]

Dear Sismonda,[105]

I cannot leave Turin without expressing to you my thanks for the kind attentions, which in the midst of very pressing demands on your time, you have still found opportunities for bestowing on your friends. Allow me to leave with you the accompanying plates[106] both as a memorial of a meeting which has been so gratifying to myself and also of a wish that it may frequently be renewed. I am Dear Sismonda,

Very sincerely yours
Charles Babbage

I.72 1840?-11-27. Ada Lovelace to Augustus De Morgan.

Ashley-Combe
Nov 27th[107]

Dear Mr. De Morgan,

I have I believe made some little progress towards the comprehension of the chapter on Notation of Functions, & I enclose you my demonstration of one of the Exercises at the end of it: "Show that the equation $\varphi(x + y) = \varphi x + \varphi y$ can be satisfied by no other solution than $\varphi x = ax$".[108] At the same time I am by no means satisfied that I do understand these Functional Equations perfectly well, because I am completely baffled by the other exercise: "Show that the equation $\varphi(x + y) + \varphi(x - y) = 2\varphi x \times \varphi y$ is satisfied by $\varphi x = \frac{1}{2}(a^x + a^{-x})$ for every value of a".

I do not know when I have been so tantalized by anything, & should be ashamed to say how much time I have spent upon it, in vain. These Functional Equations are complete Will-o'-the-Wisps to me. The moment I fancy I have really at last got hold of something tangible & substantial, it all recedes further & further & vanishes again into thin air.

But now for this perplexing $\varphi x = \frac{1}{2}(a^x + a^{-x})$. I believe I have left no method untried; but I cannot get further than as below, with any certainty:[109]

$$\begin{aligned} \varphi(x + y) + \varphi(x - y) &= 2\varphi x \times \varphi y \\ \therefore 2\varphi x &= \frac{\varphi(x + y) + \varphi(x - y)}{\varphi y} \end{aligned}$$

Since x and y may have any values whatever, (at least such I conclude is of course intended), let $y = 0$. We have then

[104] Turin Academy of Sciences, Babbage Collection, cart. 22338.

[105] Prof. Angelo Sismonda (1807–1878), a geologist and mineralogist. President of the Geological Section of the 1840 Turin conference. The conference catalog lists him as 'Professor of Mineralogy, Director of the Natural History Museum of the Royal University, Member of the Royal Academy of Sciences and the Mining Council, in Turin'.

[106] Newly-invented photographic plates, in which his friend JFW Herschel had a hand.

[107] Lovelace De Morgan Correspondence f.149–151. In a letter from Nov. 16 Ada mentions that she is working on this problem: prove that $f(x) = (a^x + a^{-x})/2$ is a solution of $f(x+y)+f(x-y) = 2f(x)f(y)$. One may attempt to date this letter by the sequence to which it appears to belong, one of which contains an explicit year, though this involves supposing that the letter previous to this was *incorrectly* dated, see section G.1, p. 306. Formatting and punctuation have been mildly adjusted for clarity.

[108] A. De Morgan 1835, Chap. X.

[109] Note that by φx Ada means $\varphi(x)$.

$$2\varphi x = \frac{\varphi(x) + \varphi(x)}{\varphi(0)}$$
$$\therefore 2\varphi x \times \varphi(0) = \varphi(x) + \varphi(x) \text{ or}$$
$$2\varphi x \times \varphi(0) = 2\varphi x$$

[Thus] $\varphi(0)$ must $= 1$, or $= a^0$, since a^0 is the only function of 0 which can $= 1$. I think so far is correct in itself, but whether it be the right road to the rest is another question. At any rate, I have not succeeded in proving it such.

To assume that since $\varphi(0) = a^0$, $\varphi(x+y) = a^{x+y}$, $\varphi(x-y) = a^{x-y}$, $\varphi y = a^y$ appears to me scarcely warrantable; and besides in that case it must be equally assumed that $\varphi x = a^x$, (there being the same ground for the one assumption as for the others), and we should then have,

$$a^x = \frac{a^{x+y} + a^{x-y}}{a^y}$$
$$\therefore a^x = \frac{a^{x+y}}{a^y} + \frac{a^{x-y}}{a^y}$$
$$\therefore a^x = a^x + a^{x-2y}$$

most clearly absurd, independent of it being discordant with the book.

Once I thought I had hit on something very clever indeed, and wrote as follows:

$$\varphi(x+y) = \varphi((x+y)\cdot 1)$$
$$= \{\varphi(1)\}^{x+y} \text{ by equation } (\varphi a)^n = \varphi(na) \text{ page 205,}$$

entirely forgetting that the φ of that equation had nothing whatever to do with the φ of any other equations; (a disagreeable truth which did not occur to me until 24 hours later). I then had

$$\varphi(x-y) = \varphi((x-y)\cdot 1) = \{\varphi(1)\}^{x-y}$$
$$\varphi(y) = \varphi(y\cdot 1) = \{\varphi(1)\}^{y} \text{ and}$$
$$2\varphi x = \frac{\{\varphi(1)\}^{x+y} + \{\varphi(1)\}^{x-y}}{\{\varphi(1)\}^{y}}$$
$$= \frac{\{\varphi(1)\}^{x+y}}{\{\varphi(1)\}^{y}} + \frac{\{\varphi(1)\}^{x-y}}{\{\varphi(1)\}^{y}}$$
$$= \{\varphi(1)\}^{x} + \{\varphi(1)\}^{x-2y} \text{ and supposing x = y, then}$$
$$2\varphi x = \{\varphi(1)\}^{x} + \frac{1}{2}\{\varphi(1)\}^{-x} \text{ or calling } \varphi(1) = a,$$
$$\varphi x = \frac{1}{2}(a^x + a^{-x})$$

But besides my unwarrantable assumption of $\varphi((x+y).1) = \{\varphi(1)\}^{x+y}$, there was this in the result which was unsatisfactory, that it was necessary to assume $x = y$, and the result seemed to hold good in that case alone. Also, when to verify, I tried $x = 1, \therefore \varphi(1) = \frac{1}{2}(a^1 + a^{-1})$, which ought to have come out $a = a$, I could make neither head or tail of it. Well, I abandoned this, & tried all sorts of other resources. I understand to work out something by means similar to those in page 205 and in the problem I send; but equally unsuccessfully.

I also in equation (A) page 204, changed $\varphi(x+y)$ into $\varphi(x-y)$ and investigated this, thinking I might derive a hint possibly from it. In short, many & various are the experiments I have made, but I will not detail any more. Indeed I think you may be possibly heartily sick of what I have detailed. But I wished to show you that I have not failed from want of trying, at least; & also to give you the chance of smiling at my expense.

I shall have to trouble you with another letter shortly, on other knotty points. Really I do not give you a sinecure. Your letters are however well bestowed, in as far as the use they are of to me, can make them so, and the great encouragement that such assistance is to me to continue my studies with zeal & spirit. We are to return to Surrey very soon. I expect to have occasion to trouble you again before we go, & after that I shall hope to see you & M^{rs} De Morgan in town, where I intend to be for two or three days in about a fortnight.

Yours very truly
A. A. L.

I.73 1840-12-01. C. Babbage to Sismonda.

1 Dorset St.
Manchester Square
London 1 Dec'ber 1840[110]

Dear Sismonda,

I should have been glad to have availed myself of any circumstance connected with science to have recalled myself to the memory of a fellow laborer whose unwearied zeal in the cause I have personally witnessed and whose friendship I am most happy to have acquired. The origin however of this letter although of a personal nature is not in my opinion altogether foreign to the great interests of Science. Our friend Prandi has communicated to me a paragraph of a letter of yours from which, although it is expressed in such prudent terms that I might perhaps be accused of vanity in putting upon it that interpretation which a knowledge of mankind has taught me it may mean, yet on the other hand if I neglected to state to you my own opinions on the subject I should reproach myself with a want of that openness which is equally due to truth and to friendship. Whatever our friend may have said on the subject has arisen entirely from the independent suggestions of his own mind; I have no right either to blame him for their expression or to doubt that it has been done with his usual delicacy and discretion. I ought to state in the first place that however highly I should esteem any mark of distinction conferred on me as a successful cultivator of Science by a Sovereign[111] whose enlightened support of it has commanded my sincere respect, yet that such a favor becomes the more acceptable from its being entirely unsought and unexpected on my part. The pause I took to arrange the contract and to see its faithful execution in the construction of the steam-engines for one of your vessels of war were justly due to a Sovereign in alliance with my own—to a King whose appreciation of science had already excited my admiration long before I was honored by those delicate and refined attentions which his Majesty condescended to bestow on me during my short residence in his capital. Having formed in your hospitable city friendships which only life can terminate—having received from all classes of your countrymen attentions far greater than any personal merits of my own could claim—I feel that the respect with which I was treated was yielded to the objects I pursued rather than to the individual who cultivated them, and that the higher in rank as well as the most elevated in understanding but expressed in their kindness towards myself their own deep convictions of the importance of science to the happiness of

[110] Turin Academy of Sciences, Babbage Collection, cart. 22339-22340.
[111] King Carlo Alberto of Sardinia.

mankind. I feel therefore that any circumstance of the kind to which you allude would not only be an additional link connecting me with persons whom I admire but that it would also have a much wider influence as a mark of approbation of those pursuits which lead to the advancement of the arts of civilization and of peace.

You perhaps, being yourself a parent, will both understand and pardon me if I mention another motive. I have a mother far advanced in years, whose unimpaired faculties are acutely sensitive to any public tribute to the character of her only son. Could your King have been an invisible witness of the conversation with my venerable parent on my return to England his heart would I am sure have been gratified by seeing that the kindness with which he had treated me when at Turin called forth tears of joy and gratitude when I related the particulars to my aged mother. I am confident a distinction so honorably conferred, so entirely free from any of those political or family motives which too frequently destroy the value of such distinctions in my own country would throw a gleam of sunshine over her remaining years.

I am at present actively employed in my great work, but I propose revising the paper on the Temple of Serapis for the printer and hope before the summer to offer you a copy. Let me request you to recall me with kindness to the recollection of many of our common friends and especially do not fail to express my respect and personal regard for two of the members of the illustrious house of Salluce. I am with the greatest regard and esteem

Very sincerely yours
Dear Sismonda
Charles Babbage

I.74 1841? C. Babbage to Unknown [Antinori?]

[Had hopes for an honour which] would have undoubtedly made much effect in my country, and it is even probable that it would have opened the path to the construction of this machine which, as Plana wrote, encompasses all the executive of analysis [...].[112] It would have also helped to elevate science in England; and I hope to be able to add that H.M. himself [Carlo Alberto] would have in some way added to his glory, being the first to appreciate and worthily reward a discovery that is destined to change the aspect of modern ~~society~~ analysis. But it was perhaps much more than Sardinian forms and customs allowed to do for a simple private individual like me [...]. Thus, being unable to ask our Government permission to wear a decoration that is inferior to the one I refused to accept from itself, I am left to regret, not only being unable to draw any advantage from this distinguished mark of favor, but also being forced to appear indifferent to the kindness of a Sovereign to whom I profess a very particular gratitude. [...] I have always maintained, through my writings and my public speeches, that a nation like England, whose power and greatness are essentially based on the excellence of its commerce and its manufactures should have high-ranking personal honors to give as reward to individuals who have notably contributed to the improvement of Sciences and Arts[113] [...]. In accordance with this principle, when, some years ago, the English Government offered me a decoration of inferior rank, I did not want to accept it, so as not to appear to approve the maxim that military art and politics have right to the highest honors of the State, while the Sciences, whatever their degree of perfection, do not deserve to be admitted to them. [...]

[112] Draft. British Library Add MS 37191.
[113] Meaning engineering.

I.75 1841-01-05. Ada Lovelace to C. Babbage.

Tuesday, 5 January
Ockham[114]

My Dear Mr Babbage

You have put me into a Dilemma, because I should naturally say —Come on Friday, & come again rather later.

If you come this week, you will I believe find us quite alone; if later, suppose about the 15th, there will probably be company. Now we like so much to have you in either circumstances; when alone the pleasure of monopolizing you is so great; when in company, you make the company so tenfold agreeable, that I cannot choose between them. I can only say we wish to see you at all times, & as much as possible [...]

I much wish to have you here, & talk with you over some of my own doings etc. Today, I have been working much at Mathematics. [...]

I must show you a certain book called my Mathematical Scrap-Book. But pray do not think of coming for so very short a time as only 3 nights. It would be shameful!

Some day or other, you will have to put me in possession of the main points relating to your engine. I have more reasons than one for desiring this.

Yours most sincerely
A. A. Lovelace

I.76 1841-01-07. Ada Lovelace to Lady Byron.

Ockham
Monday night, 11th Jany[115]

Dearest Mama

[...] I have had a mathematical week since I last wrote. On Friday we again expect company. This alteration of company & mathematics suits me astonishingly. It is wonderful how I enjoy company (which I never did in my life before) in consequence of the Mathematics. The latter is now become in good earnest a very serious pursuit. I work well, & with a facility strange to myself. I have gained a strength of head & attention which I never expected; and I have made some curious observations as to the effects of the study. [...] I cannot but reflect on the vast change 6 months have made in me; a change perhaps not the least evident to any one else, but most deeply felt by myself. I see ground for any degree of hope for the future. Those sort of changes & improvements continue (the same Causes operating) in an increasing Ratio.

I go on most delightfully with Mr. De Morgan. What can I ever do to repay him? I sometimes think with scruples on the subject. He is certainly most kind, & does what no one else I believe could do so well for me. No two people ever suited better. I am not likely either to trouble him less, at present. I have some hard & dry work & much of it, before me; but the greater the difficulty, & the less immediately inviting the subject, the more doggedly does that large Concentrativeness of mine, backed by Firmness, set about the matter. [...]

[114] British Library Add MS 37191 f532.
[115] Lovelace Papers, 42.

I.77 1841-01-12. Ada Lovelace to C. Babbage.

Ockham
Tuesday 12th Jany[116]

My Dear Mr. Babbage

If you will come by the Railway on Friday, we will send the carriage to meet you at Weybridge, for the Train that leaves Town about 4 o'clock & arrives at Weybridge a few minutes before 5 o'clock.

Bring warm coats or cloaks, as the carriage will be probably an open one. If you are a Skater, pray bring Skates to Ockham; that being the fashionable occupation here now, & one I have much taken to.

I am very anxious to talk to you. I will give you a hint on what. It strikes me that at some future time (it might be even within 3 or 4 years, or it might be many years hence), my head may be made by you subservient to some of your purposes & plans. If so, if ever I could be worthy or capable of being used by you, my head will be yours. And it is on this that I wish to speak most seriously to you. You have always been a kind and real & most invaluable friend to me, & I would that I could in any way repay it, though I scarcely dare so exalt myself as to hope however humbly, that I can be intellectually worthy to attempt serving you.

Yours most sincerely
A. A. Lovelace

You must stay some days with us. Now don't contradict me.

I.78 1841-02-22. Ada Lovelace to C. Babbage.

Ockham Park
Monday, 22nd Feby[117]

My Dear Mr. Babbage

We are to move to Town on Thursday; & I hope to see you as soon afterwards as you like, — the sooner the better. Remember that one o'clock is the best hour for a call.

I believe I shall perhaps pass Sunday Evening with Mr. & Mrs. De Morgan, but this is not yet quite fixed, & if it should not take place, will you come & spend it in St. James' Sqre. You see I am determined to celebrate the Sabbath Mathematically in one way or other. I have been at work very strenuously since I saw you, & quite as successfully as heretofore.

I am now studying attentively the Finite Differences. [...] And in this I have more particular interest, because I know it bears directly on some of your business. Altogether I am going on well, & just as we might have anticipated

I think I am more determined than ever in my future plans; and I have quite made up my mind that nothing must be suffered to interfere with them. I intend to make such arrangements in Town as will secure me a couple of hours daily (with very few exceptions), for my studies.

I think much of the possible (I believe I may say the probable) future commission between us; and it is an anticipation I increasingly like to dwell on. I think great good may be the

[116] British Library Add MS 37191 f543.
[117] British Library Add MS 37191 f566.

result to both of us; and I suspect that the idea, (which by the bye is one that I believe I have long entertained, in a vague & crude form), was one of those happy instincts which do occur to one sometimes so unaccountably & fortunately. At least, in my opinion, the results may ultimately prove it such.

Believe me
Yours most sincerely
A. Ada Lovelace

I.79 1841-02-24. Ada Lovelace to Lady Byron.

[...] The Mathematics & Mr. De Morgan going on very well indeed.[118] You would be much pleased to see the heap of papers of my writing, which have now accumulated into honorable & substantial evidence of my steady industry for some months past. And by the bye, it is a very strange thing, that so disorderly as I generally am about papers & such things, there is yet never any disorder in any part of my mathematical papers or proceedings. This is a curious phenomenon! [...]

I.80 1841-03-17. G. Plana to Babbage.

17 March 1841,
Turin[119]

My dear and very honorable friend,

Since your departure from Turin I have often thought of you and on the system of your ideas which you outlined for me. Above all I am sincerely touched by the signs of friendship which you have shown to me and I am incapable of forgetting what I owe to you in this matter. I admire in you the qualities of the heart as well as those of the spirit. After the report which I made to you about my pangs at home caused by the unhappy marriage of my daughter, you must understand that time alone can bring me back to a normal state where there is more or less equilibrium between the bad and the good. I am approaching a point without looking for it where I am neither happy nor unhappy. I have taken the path of devoting myself to research which through perseverance can render useful the sustained efforts. But the immensity of Science and the short duration of the life of an individual increases the difficulty of directing the use of our intellectual faculties. I must tell you that I must limit my investigations to those special researches which are suggested to me by my own studies, and I must ignore a host of other truths compacted inside a huge sphere of which I am confined as a punishment to its surface. I regret that I am not in London close to you or your friends able there to taste the sorrow of the friendship for a chance to be able to enjoy the spectacle of a crowd of discoveries which increase the power and well-being of a mass of men. But such is my position, and I must accept it.

It has been several months since I deposited in the Archives of the Academy of Science the model of your calculating machine with the drawings and papers which you had entrusted me with for such a destination. Since then I have nothing further of any relevance to report in this letter. You alone have an intimate knowledge of this invention, and of its details. Every one else draws back in the face of difficulties of carrying it out. Let us hope that a happy combination will put you yourself into such a state as to be able to overcome all the obstacles. Moreover I hope that it will render to you a complete justice, and that you

[118] Lovelace Papers, 42.
[119] British Library Add MS 37191. f583. Based on Roberts 2014a.

will have no need to publish declarations of acts of ingratitude comparable to those which I read on page 188 of your work 'The Ninth Bridgewater Treatise'. My wife and daughter command me to send their regards and urge me to recall on their behalf the pleasure of the moments spent in company with you. Farewell for now, my dear Babbage; send me your news, and believe in the inalterable sentiments of friendship and of the high esteem with which I have the honour of being

Your very devoted colleague and friend
J. Plana

Greetings from me to M. Prandi.

I.81 1841-03-24. C. Babbage to Sismonda.

London, 24 March 1841[120]

My dear Sir,

I felt much gratified by your congratulations on the honor which the King of Sardinia has been pleased to confer upon me, and still more for the warm interest you have taken in all that relates to me. The great object of my visit to Turin was to convey to Plana and to some of the analysts of Italy the principles on which I had contrived an engine to perform as he has beautifully expressed it "The whole Executive of Analysis". The discovery is so much in advance of my own country and I fear even of the age, that it is very important for its success that the fact should not rest upon my own unsupported authority. I therefore selected the meeting at Turin as the time of publication partly from the celebrity of its Academy and partly from my high estimation of Plana, and I had hoped that a report on the principles on which the Calculating Engine is founded could have been already made to the Royal Academy. I am aware of the difficulty of the subject and of Plana's ill health; but you have amongst you several others who favored me with their attention who are quite able and I hope not unwilling to assist him. If you could in any way accelerate this report it would be of great importance. I have already troubled you too much with my affairs but I must rely once more on your kindness. When the collection of portraits is published will you do me the favor of sending with my copy twelve impressions of my own portrait which I had ordered when at Turin. The cost of these as well as of the porcelain colors you so kindly procured for me and of any fees which may be owed by me to the Chancellery of the Order of St. Maurice and St. Lazare you will very much oblige me by defraying and I will remit the amount through a banker. I beg you to present my best respects to my valued friend the Count De Salluce and his brother: assure them and my many other friends that I look back to the period which made me personally acquainted with them as one of the happiest of my life.

Accept my warmest thanks for all the good offices you have done me and believe me with the greatest regard.

Dear Sismonda
Yours very Sincerely
Charles Babbage

P.S. I have just received a letter from Plana but I fear no Report has been made to the Academy.

[120] Turin Academy of Sciences, cart. 22341-22342.

I.82 1841-06-08. Sismonda to C. Babbage.

8th June 1841,
Turin[121]

My dear Mr Babbage,

Your letter of 24th March has filled my heart and soul very greatly and with much pleasure, I have hurriedly given acknowledgements to our Plana, and also to those mathematicians who had the honour of knowing you durin g your short stay here in our city. The small model of your very ingenious and very surprising calculating machine has been deposited at the Academy since when it has been shown at the meetings.[122] Plana who made the presentation spoke in very flattering terms about its inventor. In relation to this then he has arranged for a report to be published. He has told me what you have written in this respect but I have not insisted in knowing what that was, realising that M. Menabrea is busy preparing a report which when it is finished he will submit for your evaluation and thus obtain your approval, [following which] he wants to have it published in the Library of Geneva series. [...]

[Angelo] Sismonda

I.83 1841-07-31 C. Babbage to A. von Humboldt.

31 July 1841[123]

My dear Sir,

It is now a very long time since we met and it has been a frequent subject of regret. The drawings ~~and experiments~~ of the great analytical engine have so occupied me during the last eight years that I have been unable to leave England until last autumn/year. On talking over that engine with the Prussian Minister at Berne, M. Bunsen, he much wished me to meet you at Paris but I found afterwards that your visit was postponed.

I am very desirous to make you acquainted with the objects which I have proposed to accomplish and ~~without entering into details further that you may like/wish~~ to convince you that the means exist of constructing an engine which as Plana has ~~written to me~~ described it gives us as complete a power over the executive of analysis as we have hitherto possessed over the legislative portion. The greater part of the drawings exist by which an engine might be made which shall reduce to numbers any explicit function whose law of formation can be assigned. For instance the eqn. of finite differences expressing the law of formation of the numbers of Bernouilli [sic] being given to it, it will work out and print the successive numbers.

This engine is unfortunately far too much in advance of my own country to meet with the least support. I have at an expense of many thousands of pounds caused the drawings to be executed and I have carried on [experiments] for its perfection ~~the former engine for constructing tables by the method of differences they will not finish~~ and unless however some country more enlightened ~~government~~ than my own should take up the subject, there is no chance of that machine ever being executed during my own life and I am doubtful how to dispose of those drawings after its termination. The respect in which Science is held in

[121] British Library Add MS 37191 f625. Based on Roberts 2014a.
[122] This was probably a wooden model of the anticipatory carry mechanism.
[123] British Library Add MS 37191 f638. Based on Roberts 2014a. See also Collier 1970, 1990, 174-5 for an extract from this letter.

your country together with the ~~personal regard~~ support afforded to it by your King to whom I had the honor of being introduced by yourself in 1828, induce me to wish strongly to talk over with you the whole question. I would willingly meet you at Paris but as it is impossible to transport all the drawings and apparatus this would not put you fully in possession of the subject.

I will therefore propose to you a plan which even without any reference to the Calcg Engine would give me the greatest pleasure. If you could conveniently spare a week or two ~~in London~~ for England, I shall be most happy to meet you on the arrival of the steamboat and convey you to my own house where we can in the mornings discuss the whole question with all the advantage which the machinery and drawings will supply. The society of London will I know interest you in the evenings and perhaps the excitement of our elections may amuse you; but whether you may prefer the repose of philosophy or the society of London or a mixture of both I shall be equally happy to receive you as a valued guest and as far as your time admits to ~~partake~~ share them with you. My present intentions are to remain in London until about the 12 of Augt.

I.84 1841-08-26. F. Prandi to C. Babbage.

26th August 1841,
46 Berness Street London[124]

My Dear Babbage,

Thank you for your letter of last Sunday. I am delighted to hear that you have had a good journey, and have found Humboldt well disposed to aid you in your important object. A few hours before the arrival of your welcome letter, your mother had sent me a parcel, by which I had, with great pleasure, perceived that, in spite of Plana's inactivity, Menabrea had not been idle. As I feel convinced that you will not regret the expense, having at present no private opportunity, I send you that parcel by post herewith. You will see that it contains the Sketch of an account of the Calculating Engine, by Menabrea, dated May 22nd. How Sir Augustus Foster has contrived to detain the paper so long, I am quite at a loss to conceive; but I am vexed at the delay, because I think that if [you] had received it in due time, you might have turned it to account.[125] I hope you w[ill] meet Menabrea either at Florence or Turin. In the mean time, you might write to him, in order to explain the cause of your long silence and ask him to inform you of his plans, addressing his answer either to Toulon or Nice. ..

I.85 1841?-09-06. Ada Lovelace to De Morgan.

Monday, 4 September
Ockham[126]

Dear Mr De Morgan

[124] British Library Add MS 37191 f641. Based on Roberts 2014a. Addressed to Hotel de l'Europe, Rue Rivoli, Paris.

[125] Foster (1780–1848) served in Turin as British minister plenipotentiary until 1840. From the subsequent letter by Menabrea it seems that Foster had promised to deliver the manuscript of the 'sketch' to Babbage, but had delayed his departure to England until 1841.

[126] Here Monday 4 is probably a mistake for Monday 6. It was only in 1843, far too late for this letter, that the Monday would be the 4th. See the incorrectly dated full transcription by Hollings `https://www.claymath.org/wp-content/uploads/2023/08/46-47.pdf`.

Will you send the enclosed to Mrs De Morgan. It explains the arrangements I have made in case she comes here, & also that Lord L— & myself have delayed our departure until Satdy next. Now to the mathematical business: I think you have hit the nail on the head, altho' my confused notion of the differential was not the only piece of puzzle & mistiness which constituted the impediments towards my comprehension [...]

I cannot help here remarking a circumstance which I think is almost invariably true respecting all my difficulties & confusions in studying [...]. I therefore have lately begun to ask myself whenever I am stopped, whether I clearly understood what the subjects of the reasoning are; & to go over carefully the verbal & symbolic representative of a thing or an idea, with the question respecting each "now what does it mean, & how was it got? Am I sure of this, is each instance involved in the subject?" This may save me much future trouble [...]

I.86 1841-09-18. Menabrea to Babbage.

September 18, 1841[127]

Sir,

Your letter dated Nice the 8th instant reached me in Chambery where a grave illness of my father obliged me to go. The delay experienced by mine, which contained my musings on your astonishing machine, did not surprise me at all, since Mr. Foster having had to postpone his departure for England, it could not have been delivered to you at the time I would have desired. I am happy to see, sir, that I have managed to grasp the principle of your admirable invention.

Certainly, I am far from claiming to have understood all its secrets, and since I read your letter, it appears even more significant to me than before. Indeed, you have mechanically realized the conception that could pass as one of the most metaphysical of all analysis, I mean the idea that one must be content with the passage of one quantity by zero to infinity. This is due to the simplicity and exactitude of the principle on which your machine is founded and which, as you think, seems to be the most general of mathematical language and even that which, perhaps, will bring analysis out of the impasse in which it currently finds itself. I approve of your intention to call your machine the analytical engine. The name *calculator* was not capable of expressing all the extent you have given it. This name will serve in the eyes of lay people to mark a difference between it and those that may have already existed under the name of calculating machine.

My intention was to go this year to the Congress in Florence where I hoped to meet you. But circumstances have determined otherwise. I would still essentially need to confer with you for the drafting of the article that I wish to dedicate to you. What I had written in my letter was only a first sketch that I proposed to modify according to your observations. If you [...] return to England I would be in that capital around the 25th of next month. In any case, I could see you passing through Chambery. If these arrangements are not possible, I would ask you to kindly write to me with some detail concerning the points of view under which you wish that your invention should be presented and the remarks that you consider most essential. On that information, I will draft the article in question and before delivering it for publication I will have it reach you in England through the Embassy. Please, Sir, believe in the sincere devotion of your very obedient servant,

L. F. Menabrea

[127] British Library Add MS 37191 f642. Translated from the French. See Bulferetti 1968a and Roberts 2014a for transcripts of the original texts.

To Mr. Charles Babbage of the Royal Society of London
Poste Restante - Florence (Tuscany)

I.87 1841-10-03. C. Babbage to Plana.

3rd October 1841,
Florence[128]

My dear Plana,

I have much regretted ~~not~~ that we could not meet at Florence, ~~and~~ as I had hoped we should have had more leisure than at Turin to discuss the principles of the Analytical Engine. ~~Had I been enabled to~~ If you had made a report on the subject to the Academy of Turin during last year it might have been of essential service to me in the discussion of the question with my own Government. As it is I must be content with the ~~rep~~ description drawn up by M. Menabrea with which I am well satisfied ~~very content~~ because he seems to have penetrated completely the principles on which it rests. With my best compliments to Madame Plana and your daughter.

I am with great regard
Ever very sincerely yours
CB

I.88 1842-01-04. F. Prandi to C. Babbage.

4th January 1842,
Hotel Bonne Femme, Turin[129]

[...] Menabrea is absent on account of his father's death which happened about a fortnight ago at Chambéry. Plana will not write anything about the Engine. He seems to think that you illude yourself, that the engine, if ever executed, will be a great curiosity, but perfectly useless. He told me that you are wrong in complaining of the Report of the Committee of the Royal Society, it being as favourable as it could be expected, and that he is quite [sure?] that neither Gauss, nor Herschel, nor Airy, nor Hamilton could ever consent in your views. He expressed great friendship and regard for you, but this is in substance what he told me concerning the engine. I suspect that the report made of it by Menabrea has disgusted him. He says he will write to you if you write to him. Pray say nothing of this. I tell you all I hear without restraint, but you must take care that you do not compromise me. [...]

I.89 1842-01-27. Menabrea to Babbage.

27 January 1842[130]

Sir,

Mr. Piolti, an architect of the Government, is going to London, and I take advantage of his courtesy to send you my reply to your letter of 24 November last.

Since that time, I have given much thought to the analytical machine and to the clarifications you were kind enough to give me regarding it, which allowed me to expand considerably

[128] British Library Add MS 37192 f645. Based on Roberts 2014a.
[129] British Library Add MS 37192 f4. Based on Roberts 2014a.
[130] British Library Add MS 37192 f22. Based on Roberts 2014a.

the field of ideas on this matter. I therefore believe I must entirely rewrite the article that I had written and of which I had already given you a communication. This is the main reason why it has not yet been published.

Other circumstances have also delayed me: suffering the misfortune of losing my father, from which I am slowly recovering, and the need to manage a number of urgent matters have disrupted my activities. Now that I find myself a bit more settled, I will be sending the last draft to the office of the *Bibliothèque Universelle*, and I hope within a few days to be able to send it to Geneva. When it is printed, I will have some copies forwarded to you through Mr. Prandi, who provided us with recent updates or via another route to your ministry near your Government. I am studying a question that consists of explaining in a clear manner, for less informed minds, the solution to numerical equations. I believe this is an important point to persuade of the utility of this invention and to demonstrate the rationality of the principle upon which it is founded.

I would still like to know, if you have reached a precise idea, if there is any means of introducing the calculation of logarithms into your machine or at least of substituting something equivalent to speed up the calculations; because I understand that logarithms are nothing more than a gear in current calculation that simplifies operations without being fundamental to them, and that, in another method, it would be possible to find a different way of calculating directly. However, my main concern is to understand what your machine is capable of; similarly, if we adapt our usual formulas to your methods, which seem to me the simplest for calculations, we will need to transform analytical expressions according to the way your machine works.

There is another observation to which I call your attention, as your machine is mainly intended for numerical calculations, such as constructing tables, the speed of these operations will necessarily require that your devices be able to perform several identical operations at the same time; let me explain: in constructing logarithmic tables, for example, when a series of intervals of logarithms is calculated, the intermediate logarithms for corresponding numbers can be calculated by groups or arithmetic sequences: so that the task being distributed among a large number of operations, the work will be completed faster. Your machine should be able to work this way, for example, calculating 50 identical values simultaneously. When you speak of Economy of Machinery, this condition requires a certain number of devices that would work simultaneously. Without a doubt, your machine has this capability, but I have not had the opportunity to consider this circumstance for myself.

You will probably receive my letter within 11 days; so you should decide if you will need to add anything to what I have already written, and you will be ready to act. Space is lacking for me to fully express my gratitude to you, and I am assured that you will believe in the sincere admiration of

Your humble servant,
L. F. Menabrea
Turin, January 27, 1842.

Mr. Sismondi and Madame Bonevello specially ask me to recall them to your memory.

To Mr. Charles Babbage, Member of the Royal Society of London
Dorset Street
Manchester Square
London

Entrusted to Mr. Piolti, architect of the Sardinian Government Service.

I.90 1842-02? Ada Lovelace to Lord Lovelace.

Saturday, 4 o'clock.[131]

I have posted your letter to the Hen. You have blundered about her stay in Town. [...] I did not forget the Medallion. Wyon[132] is the man to take my likeness; but he has of late declined all private commissions. Wheatstone thinks however that when he has been introduced to me, & knows that I am the lady in question, he will not object. Wheatstone only mentioned a lady. There is no one, Wheatstone says, at all to compare with him for medallion likeness. [...] But really I don't think I am pretty enough to be medallioned at all. [...] Wedding-cake has arrived from Mr & Mrs Kay Shuttleworth. [...]

I.91 1842-02-08. James MacCullagh to C. Babbage.

Trin: Coll: Dublin
Feb 8/1842[133]

My dear Babbage,

Allow me to thank you for your very kind letter. I should feel greatly obliged by anything you may do to get me admission into the Athenaeum Club. My name was put up there by Wheatstone a year ago or more but one time that I was in London happening to be in low spirits I rather damped his ardour, and left him, I believe, in doubt whether I wished to be a member of the club or not. I shall write to him immediately upon that point and set him right.

If I should happen to be elected (and I think I am now in a fair way) I shall probably be induced to go over a little oftener to London?

Ever truly yours
J. MacCullagh

I.92 1842-02-25. James MacCullagh to C. Babbage.

Trin: Coll: Dublin
Feb 25, 1842[134]

My dear Babbage,

You have probably heard from Sir James South[135] that I have been elected a member of the Athenæum. He tells me that I have reason to be proud of the letters which he was able to produce in my favour. One of them was from you. I suppose I ought to thank you in due form for what you were pleased to say of me—at the risk too of drawing upon yourself an additional visit every year—for no doubt I shall be disposed—for a time at least—to spend more of my vacations in London.

I suppose you are going on pretty much as usual—devoting your mornings to hard work (I wish it were not thankless)—and your evenings to Society. It is a good mixture, I am sure, and the one gives a relish to the other. But when shall we have something *in print*

---

[131] Lovelace Papers.
[132] William Wyon (1795–1851), from a family well-known for production of medallions.
[133] Scaife 1990.
[134] Scaife 1990.
[135] Astronomer, 1785–1867.

about the Machine. For you know, *that* has always been the burden of my song. Has Menabrea published anything yet? I live quite out of the world here, and never hear a word of news—scientific or political.

To talk of myself—I have grown very stupid of late, and regularly fail in everything I attempt. What the reason may be I cannot tell. But I begin to be of Newton's opinion, that after a *certain age*, a man may as well give up mathematics. Perhaps it would be better—at least for one's own happiness—to have some occupation or profession which should connect one immediately with his *fellow men*, and to make the pursuit of science a collateral object, rather than a direct one.

Believe me always, my dear Babbage,
Yours faithfully,
J. MacCullagh

I.93 1842-05-02. Ada Lovelace to C. Babbage.

Monday, 2 May
St James' Square[136]

My Dear Babbage

Will you do me the favor of showing your Calculating Machine to an old friend of mine, who is on leave of absence from India, & is to return there immediately. It is Mr Henry Siddons, Grandson of the Mrs Siddons, & first cousin to Mrs Butler, & Adelaide Kemble

He is very desirous of just seeing the small portion that was completed, of your old machine. His wishes arise from the perusal of your 9th Bridgewater Treatise: & he would ask leave to call on you for that purpose any morning this week that you can fix.

I believe that the favour would be well bestowed by you. He is not one of the idle or merely inquisitive; altho', he himself is fully aware of, he cannot pretend or expect to comprehend much of the principles on which the machine is planned & executed.

Yours ever,
A.A. Lovelace

I.94 1842-05-25. G. Plana to C. Babbage.

[...] the sphere of exact knowledge[137] grows to the point where it seems impossible to embrace it through thought [...] to be able to enjoy the spectacle of a multitude of discoveries that increase the power and happiness of a mass of people [...]. I presume that you are ceaselessly occupied with your calculating machine; I am assured that you are composing a work that will contain a complete explanation of what concerns it as well as details of its construction. You alone can properly create such a work. By thus delivering your ideas to others' meditation, one will better appreciate the goal and the means: then, many doubts will cease, and the consideration of the expense to construct this machine will be balanced by the advantages it presents. And if difficulties arise against your project, they will need to be clearly stated; and you will know how to put them to profit.

[136] British Library Add MS 37192 f75. Based on Roberts 2014a.
[137] British Library Add MS 37192 f88. Based on Roberts 2014a.

I.95 1842-08-11. Ada Lovelace to C. Babbage.

11 August
Ockham Park,
Ripley, Surrey[138]

My Dear Babbage

Will you come on Satdy the 20th,[139] to visit us for 3 or 4 days. We expect Fonblanque[140] & others. In haste,

Yours most sincerely
A. Ada Lovelace

I.96 1842-08-14. Ada Lovelace to C. Babbage.

Sunday, 14 August
Ockham Park[141]

My Dear Babbage

Will you come Sunday the 21st, instead of Satdy. I find it will suit us better. And pray stay till [Wednesday] Morning. We have friends coming whom I am very desirous you should remain for. In haste,

Yours ever
A.A. Lovelace

I.97 1842-08-11? C. Babbage to Ada Lovelace.

[circa 11th August 1842[142]]

I will endeavour to join you at Ockham on Saturday the 20th ~~if~~. But the state of drawings and a possible discussion with Sir R. Peel may ~~prevent~~ either of them deprive me of that pleasure.

I.98 1842-09. Herschel to H. Goulburn.

September 1842[143]

Sir,

The Astronomer Royal [Airy] has communicated to me your desire to know my opinion as to the probable utility of continuing to expend upon Mr Babbage's calculating machine the sums necessary for its perfection; informing me at the same time as ~~the~~ a ground for the formation of such opinion that a sum of about £16000 has been already expended on it

[138] British Library Add MS 37192.
[139] In 1842, the 20th was indeed on a Saturday.
[140] Presumably the journalist Albany William Fonblanque (1793–1872).
[141] British Library Add MS 37192. The day places this in 1842.
[142] British Library Add MS 37192 f127. Based on Roberts 2014a.
[143] Royal Society Herschel Collection Box 27 Item 51. Draft. Based on Roberts 2014c. See also Turvey 1991.

and that you understand from Mr Babbage that a sum not less than that will be necessary for its completion. He further states ~~as a reason for your wishing to ascertain my present views on the subject~~ that you understand me to have been consulted in an early stage of the business and ~~would~~ are therefore unwilling to proceed without endeavouring to ascertain my present views.

In reference to this latter point it is proper for me to observe that I was never individually consulted by Government on the subject in any stage of the business or called on to express an opinion on it otherwise than as a member of the Council of the Royal Society and of its committees to whom its consideration was referred. Under these circumstances and as a private friend of Mr Babbage it must be supposed that I feel the utmost delicacy in forming ~~and divulging~~ an opinion on a point so nearly concerning his interests and that only a permanent sense of public duty when disjointly called upon by the Government of my country which has a right to the best exercise of my judgement and faculties when seeking to be informed by them for the public advantage could decide me to enter upon such a task. To that delicacy is added a serious difficulty. During the last 10 or 12 years I have in great measure lost sight of the progress of the Engine, having resided several years out of England and since my return having been by no means in close or frequent communication with him ~~up~~on its subject. Nor does my present situation permit me to enter ~~upon~~ fully into ~~all~~ the many and great developments of which would demand the requirements of a practical engineer for which I feel some months of severe ~~and arduous~~ application would but inadequately prepare me. It is therefore at best a birds-eye view of the subject that I can take and one in which it is very possible Mr Babbage may feel that he has a right to complain of some of his most cherished conceptions ~~not being~~ being left unconsidered.

~~On the subject of~~ The utility of continuing to spend money on this machine must depend

1st On the absolute utility of the machine supposing it was completed ~~or as it~~

3dly On the probability or certainty that its completion will be secured by the expenditure of such sums that the Government may consider itself justified in devoting to it

2dly On the total amount ~~ultimately~~ henceforward to be expended ~~henceforward~~ on it as ~~weighed~~ considered in relation to the absolute utility of the machine to the claims of other branches of science on the national purse and ~~as consideration of~~ to the sum already laid out. ~~The part of the question dependent on this point may be considered as the question of its relative utility.~~

On the subject of the absolute utility of the machine when completed, my own opinion coincided fully with that of the body of whose judgement it constituted an element nor have I since found the least reason to retract or alter it. I might therefore content myself with referring to the official record of that judgement as delivered in reply to references made by the Government to the Council, as the expression of my present sentiments on that head. Nevertheless I shall particularize a few cases in which its utility will be readily perceived out of many which might be selected.

But first it is necessary for me to observe that I understand by the "Calculating ~~Engine~~ Machine" referred to in the Astronomer Royal's communication to be intended a machine destined to accomplish the objects of the Engine brought under the notice of the Royal Society, whether by the same identical mechanical combinations, or by other simpler and more efficacious suggested by subsequent experience or which ~~may bring~~ have brought under its dominion tables not within its power at that epoch. An Engine in short of which the essential character is the calculating and printing series of numbers following one another in tabular order, the principle of such calculation being the successive addition of numbers one to another by mechanism which includes as a particular cases the process of multiplication ~~and~~ as well as the calculation of series by differences.

I make this observation because I am perfectly aware that Mr Babbage's views on the applicability of mechanism to calculatory combination in general have extended very far in advance of a simply numerical tabulation, to which however if the Engine in question be not restricted, I beg to understand as delivering no opinion whatever as to any parts of its structure which may surpass this limit. And this, not as in any way undervaluing the possible utility of such combinatory mechanism, but as being no way prepared to enter ~~into so vast a~~ considerations this subject suggests. This premised, I proceed to the particular instances of expected utility alluded to above.

1. There are constantly going forward reprints, in various forms, of logarithmic, trigonometrical, and a multitude of other useful mathematical and practical tables, in all which reprints, ~~(especially if attended with recomputation)~~ there is not only danger, but almost a moral certainty of errata creeping in, whose consequences might become serious where (as in navigation) large interests, or life ~~is~~ are at stake. An undetected error in a logarithmic table is like a sunken rock at sea yet undiscovered, upon which it is impossible to say what wrecks ~~may have taken place (in the literal sense of the word)~~ may have taken place. Now were an Engine capable of doing such work accessible at reasonable cost ~~might inte~~ it is hardly to be supposed that it would not at least in a great many cases supersede the ordinary processes of ~~letter~~press printing and ~~rending~~ revision.

2. I consider it by no means improbable that the present generation will not pass away without witnessing the reconstruction of the whole map of Solar Lunar and Planetary tables, based on a more complete ~~and refined~~ development of Newtonian theory than that which has served for the construction of those now in use, on the one hand; and on the other, on the reduction (according to a uniform system and with the best elements) of all the existing observations—at least of all the British ones—such reduction being actually in progress, if not completed, under the direction and at the instance of the present Astronomer Royal. ~~seems all or nearly all Astronomical tables~~ Tables of this nature ~~which are not~~ are for the most part either arithmetical progressions or tables of natural cosines to a given radius and therefore within the power of the Engine in its improved form. And its application to their calculation and printing would be the most satisfactory proof which could be afforded of its practical success. ~~The great mass of such tables therefore might be written of almost currente calamis with hardly any computation, by the aid of a general table of such cosines from radius I up to 10,000 which would be the easiest task of such an Engine, or indeed of one infinitely less elaborate in its arithmetical part provided the printing part were perfect. [Footnote: Such a table for integer degrees would comprise 450000 numbers and two volumes of this extent one for cosines and one for tangents would greatly facilitate be a vast assistance to the general computist saving logarithmic computation in every case when tabulation of such functions to integer degrees in requisite.]~~ Now would the computation and printing of the tables in question by the direct operation of an engine such as Mr. Babbage's present any materially greater difficulty the process of multiplication of one number by a fixed succession of given numbers being the whole work of such a table. The facility with which an Arithmetical Engine could accommodate itself to all the requisites of a work of this kind so as to execute ~~the~~ it with convenience and cheapness would be a fair practical test of its utility.

~~3. The mathematical quantities known by the name of Elliptic Transcendents are continually acquiring a greater importance as a means of calculating the numerical results of geometrical investigations. Tables exist for them, but should their use become more common, these would require to be revised and extended. There can be no doubt that means might be found of computing such tables by differences as in the case of logarithms and thus subjecting them to mechanical calculation and printing, This perhaps is looking too far forward into the future,~~

3. ~~In say nothing of the great number of~~ In the numerous tables (chiefly arithmetical progressions) which the Public Service requires, ~~for the~~ the guarantee of unimpeachable correctness ~~in such tables~~ is not without its dsitinct value. ~~a very simple case will illustrate~~

~~this~~ This may be instanced by a ~~simple~~ case of recent occurrence. In the reports ~~and made by~~ of the Commissioners ~~for considering~~ appointed to consider the steps to be taken for restoring the lost standards of weights and measures, it is recommended that an edition of certain tables ~~of a peculiar~~ should be prepared and printed by Government for the use of dealers in Bullion as a probably influential means of procuring ~~publicity~~ the disuse of the Troy Pound. As vast sums would come to be computed ~~by such tables, a perfect reliance of their freedom from error indispensable becomes the most indispensable condition for their introduction.~~ by such tables it is obvious that nothing but a general impression of their absolute correctness could secure their introduction ~~and adoption~~ and so enable them to ~~accomplish~~ exert their designed ~~purpose~~ intent.

~~5. By the use of Automaton wheels various tables founded on observed laws of mortality or other statistical or observational data not reducible to any known mathematical expression might be calculated and printed, I presume Mr Babbage cannot have overlooked this application of his mechanism,~~

These instances, ~~and a great many more which it would be tedious to recapitulate~~ will suffice to shew that the prospect of utility to accrue from the completion of the Engine in question is neither illusionary nor trifling. The ultimate practical utility of the Engine must however also greatly depend on the convenience and cost of working and maintaining it, on the regulations under which its accessibility to the public may be secured and on the distinctness of the direction for its use and for the repair of such parts as may in possibility get out of order as its inventor shall furnish to accompany it ~~and~~ which ~~direction I should observe~~ must be regarded as an essential part of it, ~~These however are points on which I have no information and indeed which are [...] posterior to its construction.~~

The second point ~~or that of relative utility~~ on which as already observed the utility of continuing to spend money on the machine appears to me to depend, ~~is that in which the the question of relative utility, is one on which the fact placed before me afford no data but hypothetical ones~~ is the total proposed expenditure, as weighed against the absolute utility of the machine, and as considered in relation both to the sums already expended and to the claims which other branches of science theoretical and practical may be held to have on that portion of the national income ~~which~~ be it greater or less [than] which can be devoted to such objects I should cripple my prospect of scientific utility irrecoverably in my present and future possible advocacy of any such claims if then consulted I could ~~leave out~~ omit so very important an element.

Now the terms ~~of any problem~~ in which the question is put viz that a sum not less than 16000 will be required, leave me here without any ground to go upon except ~~up~~on one assumption which therefore I am compelled to make. Viz: that the minimum named is also the ~~ultimatum~~ ultimum, that the least grant which will accomplish the object is the last which will be asked for. ~~Even with this assumption~~ And in effect it is possible that such may be the case for, I consider that by this long and costly experience, Mr Babbage must have acquired a full practical knowledge not only of the expenses attending this peculiar style of workmanship in general, but a clear view of particular items, so that it would be not unreasonable to expect that he ~~should~~ might be ~~enabled~~ able by this time to name a sum which would certainly cover the cost and with which, if granted he would pledge himself to the completion of the Engine, life being spared and that he has only been held back from doing so by the natural reluctance which every man feels to pledge himself against contingencies. At all events I must suppose such a pledge given with respect to the sum actually in question, or declare myself unable to advance towards the formation of any opinion at all on this ~~point~~ head ~~and must be understood to decline doing so~~.

Making this assumption then the work, pecuniarily speaking, must be regarded as half done. As regards invention and experience I presume it must be considered much further advanced since it seems to me inconceivable that a man of Mr Babbage's talents bending his powers almost exclusively for 20 years to this object and ~~Expend?~~ devoting to it so vast an amount of intellectual labour as I know him to have done should not at least have fully perfected the/his idea ~~of the Engine he has undertaken~~ of the Engine ~~as have for those only I contemplate~~ he originally proposed to construct ~~whatever development his ulterior views~~

~~And the question is as to the utility of expending the latter half of its total cost in resuscitating the former which will otherwise have been sunk.~~

~~Now~~ It is clear that if, in the outset of the project, it had been decided that £32000 was not too much to expend on it, Mr Babbage is only so far advanced on the high road to his conclusion, and all is right.

If on the other hand it had been decided to expend only as far as 16000 the money is expended and the work not completed. ~~Therefore the attempt on that supposition, and in that sense of the word has failed.~~ The attempt then on that supposition and in that sense of the word has proved unsuccessful.

And it ~~has~~ now comes to be enquired whether it be for the public benefit to make a ~~first~~ second attempt with the advantage of starting from the advance gained and the experience of 20 years acquired.

~~Now though I will not presume to state~~ I think it would not be becoming in me formally to deliver an opinion on this point because it is rather a moral than an abstract one, and one which it rests solely with the executive to estimate—for, (in the case which I am now arguing) ~~they have~~ it has ~~their~~ its part, and it must be purely a matter of liberality if ~~they~~ it volunteer a second time. Yet it is my duty to observe

1st that the advantages to the Public of such an Engine remain just what they were before its construction was attempted.

2d That I have full confidence in Mr Babbage's ability to complete the Engine if the pecuniary means be afforded him.

3d That although at the outset of this matter as a member of the Council of the Royal Society and bearing in mind the claim of other branches of Science to ~~the~~ public aid I should certainly have viewed with dismay the idea of recommending ~~as a member of the Council of the Royal Society~~ an expenditure of £32000 of the public money on this object—yet judging from my present impressions I am disposed to think I should not have ~~regarded~~ considered the ~~half~~ moiety of that sum wholly inadmissible, if regarded as an absolutely as final and under proper understanding as to the accordance to the public of the privilege [2 under ~~reasonable~~ proper regulations][1 of using the Engine when completed].

4th That, as 20 years have elapsed in proceeding so far, and ~~that~~ as doubtless a considerable time must still elapse before the completion of the work and the Total Expenditure being therefore spread over a very long period, a corresponding enlargement of the period within which the equivalent benefits are to accrue, ought to be allowed. A tree of slow growth must repay its culture by a long maturity.

5th ~~That I have hitherto forborne to advert to the immitiga~~ That although ultimate success may be damped in the minds of some by an impression of excessive cost, yet there can be no dissentient opinion as to the immitigated evil~~s~~ of an outlay of £16000 terminating in an abandonment of the undertaking.[144]

~~So far in reference to the Expenditure already incurred, in weighing a proposed public expenditure for useful purposes against the expected it is necessary in so far as those benefits can be estimated as a matter of pecuniary value in order to turn the scales to shew either 1st a certainty of a saving accruing to Government itself of a sum at least equal in a short time, or 2d of a much longer saving or profit in a long one or 3ly a highly probable gain or saving to the Community to an amount so great as to render the Expenditure comparatively insignificant. In none of these points of view however do I feel prepared~~

[144] The 'sunk cost' fallacy!

~~to give any opinion much less to with such a demonstration in the actual state of the data, and to hazard an unsupported a merely conjectural opinion and expect it to be received as valid would be idle.~~

Of benefits not ~~so~~ distinctly capable of pecuniary estimation I have already stated my opinion that they are neither illusionary nor trifling.

~~As respects the claims of other branches of Science I can have only one remark to make that Science consists of parts in harmony and not in jealous conflict or competition with each other and that the most desirable and advantageous way in which it can be advanced by public aid is by an application of such aid to all its branches, liberal in proportion to their relative importance until a comparatively void period.~~

The Third ~~point~~ element into which I have resolved the question submitted to my consideration is the probability of success, the means being granted. I have already stated my opinion of Mr. Babbage's full ability to complete the machine satisfactorily, ~~life being spared~~. But in an event to be deplored equally by his friends and by the world, should this great work not be finished or very nearly so ~~I know not when the task of completing it could the case being one altogether peculiar [...] its completion by a successor could hardly be looked for considering the rare combination of inventive power, zeal and devotion to this particular object which would be required. A son indeed might complete what the Father began, but the public would have no right to expect it at his hands~~ it would be fortunate indeed if a successor should arise possessing the same combination of inventive powers and ~~absolute~~ devotion to the task which ~~so eminently~~ distinguish Mr Babbage himself.

I ought further to add that I have felt it my duty, being so applied to ~~form and deliver~~ give the subject the best consideration I am able without reference to ~~those~~ the opinions whatever they may be of my scientific colleagues. This, though necessary to secure the singleness and independence ~~of~~ which I think an opinion so demanded ought to have, yet tends to render me diffident of its value[145] and disposed to wish that as the principle of individual has been preferred to that of public reference, that principle should be extended further than to myself alone supposing it not already done.[146]

Neither have I communicated on the subject with Mr Babbage himself for a similar reason though as ~~my personal~~ a friend ~~whose~~ from whom I am desirous to have no concealment on a matter so nearly ~~relating to his interests~~ affecting him I can have no objection but rather the contrary ~~that a copy of this letter should be transmitted to him~~ to his being made acquainted with the contents of this letter,

I have the honor ~~to be~~ most respectfully ~~Sir~~ to remain
~~with the utmost respect Sir~~

Your most obedt & humble Servt
JFW Herschel

I.99 1842-11-03. Henry Goulburn to C. Babbage.

3rd November 1842,
Downing Street[147]

My dear Sir

The Solicitor General has informed me that you are most anxious to have an early and decided answer as to the determination of the Government with respect to the Completion

145 Meaning the value of Herschel's personal opinion about the engine.
146 i.e. ask other people too.
147 British Library Add MS 37192 f172. Based on Roberts 2014a. Goulburn was Chancellor of the Exchequer

of your Calculating Engine. I accordingly took the earliest opportunity of communicating with Sir R. Peel on the subject. We both regret the necessity of abandoning the completion of a machine on which so much scientific ingenuity and labour has been bestowed. But on the other hand the expense which would be necessary in order to render it either satisfactory to yourself or generally useful appears on the lowest calculation so far to exceed what we should be justified in incurring that we consider ourselves as having no other alternative. We trust by withdrawing all claim on the part of the Government to the machine as at present constructed and by placing it at your entire disposal we may so assist afresh your future exertions in the cause of science.

I have the honor to be
Dear Sir
Yours ever most faithfully
Henry Goulburn

Sir Robert Peel begs me to add that as I have undertaken to express to you our joint opinion on this matter he trusts you will excuse his not separately replying to the letter which you addressed to him on the subject a short time since.

I.100 1842-11-11. CB notes on meeting Peel

11th November 1842[148]

I first asked Sir Peel if he had seen Sir Wm Follett within the last two days assigning as my reason that if he had it might perhaps save him ~~some~~ time as in that case Sir Wm Follett would have anticipated for me some part of what I had to communicate.

I then informed Sir R.P. that many circumstances had at last forced upon me the conviction that there existed amongst men of science great jealousy of me ~~which~~ probably acted to my disadvantage. I said that I had been reluctantly forced to this conclusion of which I now had ample evidence which however I should not state unless he asked me. In reply to [an observation] of Sir R.P. in a subsequent part of the conversation I mentioned one circumstance that within a few day[s] the Secretary of one of the foreign ~~ambass~~ embassies in London had incidentally remarked to me that he had long observed a great jealousy of me in certain classes of English Society.

I then said that as he Peel must of course obtain his views of both the Difference Engine and Analytical Engine from others I thought it right without alluding to any individuals ~~and~~ not ~~asking~~ wishing to know the name of any of his scientific advisers yet for his P's sake as well as my own to ~~give~~ mention this conviction.

I then turned to the next subject the importance of the Anal. Engine. I stated my own opinion that in the future scientific history of the present day it would probably form a marked epoch, and that much depended upon the result of this interview. I added that the Difference Engine was only capable of applications to one limited part of the science (although that part was certainly of great importance and capable of more immediate practical applications than any other) but that the Analytical Engine embraced the whole science. I stated that it was in fact already invented and that it exceeded any anticipations ~~of~~ I had ever entertained respecting the powers of ~~deriving~~ assistance in applying machinery to science. I then mentioned to Sir R. Peel that one of the first mathematicians in Europe, Plana, had written to me his view respecting it — that ~~Plana~~ "We hitherto (said Plana)

[148] Recollections by CB of an Interview with Sir R. Peel on Friday 1842-11-11 at 11 am. British Library Add MS 37192 f189. Based on Roberts 2014a.

have possessed full power over the legislative department of ~~Science~~ Analysis in the Executive we have been all feeble. Your invention appears to have given us the same control over the/its executive as we have hitherto had over its legislative."

I now observed that I came to the point on which I had asked for an interview. I stated that at an early period of life I gave up the prospect of succeeding my father in a lucrative profession (that of banker in the City) in order that I might devote myself to Science

—That in ~~consequence~~ of following out that pursuit I had invented the Difference Engine.

—That the in 1822 government wished ~~me~~ to make that such a machine ~~for~~ them should be constructed.

—That at their desire I had undertaken the construction of it for them.

—That during twelve years I had devoted amidst many difficulties my whole time unceasingly to that object.

—That circumstances over which I had no control caused what was thought a temporary cessation.

—That during this interval I had been examining other combinations of machinery and had opened out views which ~~might~~ seemed likely to have the most important bearings on the machine ~~now~~ then constructing.

—That when new arrangements had been made and the government wished the work to be resumed I thought it would ~~not~~ be right improper to ~~conceal~~ withold from them the knowledge I had thus acquired.

—That I did communicate ~~these~~ that knowledge but that no decision was arrived at by the government.

—That as time advanced those views became gradually more clear and distinct so that ultimately it appeared that it would be both a shorter and a more economical course to throw aside all that had been done and to make a new Difference engine using for it some of the more simple contrivances which I had invented for the Analytical Engine. Finding still that the government after repeated applications came to no decision I confined my subsequent applications to the simple question whether they intended to call on me to complete the old difference engine or ~~intend~~ to abandon it altogether. Year after year passed on in which I was kept ~~keeping~~ me in the most harassing state of uncertainity ~~and~~ that now after ~~eight~~ nine years I had just received notice of the intention of the government altogether to abandon the engine.

— That during this time I had myself expended a large sum of money and that the public had constantly accused me of having myself received that money which government had paid to the workmen.

— That on the grounds of the great pecuniary and personal sacrifices which I had made and on that of the expectations I might reasonably have entertained upon the completion of such a machine which had now been by their decision abandoned I thought ~~I~~ had some that the services I had rendered ought not to be utterly unrequited and unrewarded.

I then adverted to another circumstance that whilst I had been thus exerting myself in advancing science many other pursuing the same career had been rewarded ~~for~~ by the government for their labors whilst I ~~who~~ had worked for them was made the marked exception. Sir R. Peel himself immediately adverted to the pensions given to Science and literature which amount only to £1200 each year. I remarked that although I partly alluded to them yet that there were other occasions on which Science was rewarded and said I would mention half a dozen names not unknown to him having him to put the amount of income derived by them which he probably knew better than I did.

Airy	Astronomer Royal	800 + Pension 300 + house &c.	1500
Buckland	Canon of Christchurch		1000
Herschel	A Baronet		—
Peacock	Dean of Ely		1800
Sedgwick	Dean of Norwich		1000
Whewell	Master of Trinity		2000

I then concluded with stating that ~~both~~ on these ~~first~~ grounds ~~of~~ time spent in the act I thought ~~that~~ I had some claim to the consideration of the government.

~~Both~~ these grounds Sir R. P. denied altogether that either of these claims entitled me to any thing. He observed that I had rendered the Difference Engine useless by inventing a better. I remarked that it would if finished do more than I had promised and that although it was undoubtedly superseded by better machinery yet that I had never stated that it was useless. I said that the general fact of machinery being superseded in several of our great branches of manufactures after a few years was perfectly well known.

In reply to the statement of the anxiety I had suffered and the vexation of finding that the public believed I had profited by the money expended I mentioned that the belief was so prevalent that ~~I~~ several of my intimate friends had asked if it were not true; and that I had even met with it on the hustings at Finsbury. Sir R.P. took up the latter remark and said that I was too sensitive to such attacks, that men of sense never cared for them. I looked at him and said you must Sir R. Peel in your own experience of public life have frequently observed that the best heads and highest minds are often the most susceptible ~~to~~ of annoyance from the injustice or the ingratitude of the public. (He seemed to admit it but I am not sure that the name[s] of Romilly,[149] Whitbread and others occurred to him.)

With respect to the other ground of claim that as one of the class of men of science I had been neglected entirely, he tried rather artfully to interpret the statement I had made of the emolument of the half dozen men I had quoted as meaning to put myself in competition ~~with~~ for the places held by the individuals Airy and Herschel (Peel seeming to think that Herschel held some office). I had commenced my statement by denying this expressly and had again repeated at the conclusion that they were all men eminent in science and that not wishing to disparage them in the slightest degree I had yet a right to be considered as belonging to that class.

Sir R. Peel's first remark however was that most of them were professional rewards. This I at once denied and said that it was perfectly well known that they were not given for professional services for that though they were eminent in science they had not any of them ever done anything to distinguish them professionally (this of course alluded to the clerical portion of the list).

Sir R P seemed excessively angry and annoyed during the whole interview but more particularly when I knocked over with some vivacity his argument about professional service. He thus proceeded to [...] attempt humbug saying that the Institutions of the country &c. admitted of certain places being given to certain professions for services not exactly prof'l and so on.

I listened to all his statements looking him steadfastly in the face. When he got aground, I still retained my view upon him as if expecting at least some argument would be produced. This position of course was not very agreeable and certainly not very dignified for a prime minister.

I then said Sir Peel if those are your views I wish you good morning.

Finding Sir R P utterly denying that I had any claim I merely remarked that I considered myself as having been treated with great injustice but that as he was of a different opinion I ~~had~~ could ~~do~~ not help myself on which I got up and wished him good morning.

[149] Perhaps Sir Samuel Romilly (1757–1818), of Huguenot descent.

I.101 1842-12-24. J. Maclaren to C. Babbage.

24th December 1842[150]

Dear Babbage,

I have looked over your papers and should have returned them sooner had I not waited for an opportunity to send them to you; you cannot do better than complete and publish them; I am competent to advise in this matter only on one point; I advise you to be magnanimous and publish without any allusion at all as to the treatment you have received from Government as to the misconception which prevails as to your undertaking of the former engine, [...]. I am afraid that a repetition of the statement here would expose you to a reputation of neglect and misconstruction, for some book tho' it would contain an account of the wonderful powers of the new Engine would shew, at least to the common reader, that the making of it was feasible. In short I think this is not the time. I am much obliged for the Phil. Mags. 2 bundles of which I return to you as those numbers I have already; the rest I shall find very useful.

Truly Yours
J. Maclaren

I.102 1842-12-31. Ada Lovelace to Lord Lovelace.

Thursday 4 o'clock
St James' Square[151]

Dearest W.

I have nothing very particular to tell you [...].

Wheatstone has been with me a long while today, & has taken my translation away with him, after reading it over with me. I hope to receive the proofs of it for correction, by & bye as I trust Taylor will not reject it. I am now translating a beautiful Italian scientific paper.

Tonight I take Hester to see Miss Kemble in *Norma*.

I seem likely to lead a very quiet life I think altogether. And I certainly care little about anything but the scientific & the musical. [...]

I.103 1843-02-07. Babbage Note.

CB to Lady Lovelace[152]
Tuesday 7th Feb 43
accept invitation
long since we met,
begin under new circumstances

I.104 1843? Ada Lovelace to C. Babbage.

Sunday,
Ockham[153]

---

[150] British Library Add MS 37192 f205. Based on Roberts 2014a. Possibly an iron founder, c.f. Hyman 1982, 199.

[151] Lovelace Papers.

[152] British Library Add MS 37192 f237. Based on Roberts 2014a.

[153] British Library Add MS 37192.

The lamentably small quantity of margin has placed my ingenuity in full play how to make all my corrections distinct & legible. Why do they not give one more room?

I am much better today. My intellect will keep me alive; & active & merry into the bargain.

I verily believe I should be in a bad way but for my great objects. If I were in circumstances that permitted of my dwelling much or long on myself & my sensations, it would be a very unpromising business. But I have happily got fairly entraînée now, into a course which leaves me little opportunity for mere self study & speculation; & which moreover gives me pressing motives to desire not only a mere continuance, but a highly active & efficient continuance, of life in this world for many more years to come. Science has thrown its net over me, & has fairly ensnared the fairy, or whatever she is.

Depend upon it that I here express a very remarkable & a very true truth. And I am quite certain that none is perhaps so capable of feeling the justice & reality of my own idea on the subject, as yourself.

A.L.
At three tomorrow.

I.105 1843-03-02. Menabrea to Babbage.

March 2, 1843[154]

Sir,

One of my friends, Chevalier Della Bocca, Captain of the General Staff and first equerry to H.R.H. the Duke of Savoy, wishing to take charge of my letter, I take advantage of his kindness to recall myself to your memory. I think, Sir, that you have received the copies of a note that I published in the *Bibliothèque Universelle of Geneva* (October issue 1842) about your analytical engine and which I sent to you from Chambery a short time ago.

I hope to have been a faithful interpreter of your thoughts in the exposition of the underlying principles of your invention, which I consider as the most beautiful effort of human genius. My friend who visits England for the first time will be very happy to find himself in contact with a man such as yourself. Please grant him your benevolence and accept the expression of respectful sentiments from your very humble and devoted servant.

L. F. Menabrea
Captain of Military Engineering

To
Mr. Charles Babbage member of the Royal Society of London
Dorset Street
Manchester Square

I.106 1843-03-16. Babbage Note.

CB to Lady L.[155]
Thursday 16 March 43
Have kept Tuesday
Talk over notes to Menabrea
Woven picture Jacquard

[154] British Library Add MS 37192 f269. Based on Roberts 2014a.
[155] British Library Add MS 37192 f278. Based on Roberts 2014a.

I.107 1843. Ada Lovelace to C. Babbage.

Thursday Morning
Ockham[156]

My Dear Babbage

I have read your papers over with great attention: but I want you to answer me the following question by return of post. The day I called on you, you wrote off on a scrap of paper (which I have unluckily lost), that the Difference Engine would do [...] (something or other), but that the Analytical Engine would do [...] (something else that is absolutely general). Be kind enough to write this out properly for me; & then I think I can make some very good Notes. I have been considering about Prince Albert; but I much doubt the expediency of it. However there is time enough to consider of this. I am anxious to hear how you are.

Yours ever Ockham
A. A. L.
Thursday morng

I.108 1843. Ada Lovelace to C. Babbage.

Ockham
Monday[157]

My Dear Babbage

I am working very hard for you; like the Devil in fact; (which perhaps I am).

I think you will be pleased. I have made what appears to me some very important extensions & improvements. Why I now write is to beg you will send down to the Square before tomorrow evening, Brooke's Formulae,[158] & also the Report of the Royal Society on your machine. I suppose you can get it easily, & I particularly want to see it, before I see you on Weddy Morg.

It appears to me that I am working up the Notes with much success; & that even if the book be delayed in its publication, a week or two, in consequence, it would be worth Mr. Taylor's while to wait. I will have it well & fully done; or not at all.

I want to put in something about Bernoulli's Numbers, in one of my Notes, as an example of how an implicit function may be worked out by the engine, without having been worked out by human head & hands first. Give me the necessary data & formulae.

Yours, ever
A. A. L.

I.109 1843-05-13. C. Babbage to Ada Lovelace.

My Dear Lady Lovelace[159]

---

[156] British Library Add MS 37192 f357.
[157] British Library Add MS 37192 f362.
[158] Brooke 1829.
[159] Lovelace Papers, 168.

What are your commands for today? I am going into the City and return to dinner at home unless you wish otherwise. [...]. Let me know that you have passed a good night. Don't bore yourself with writing—a verbal answer by my messenger or a visit from your own Lady-bird will be sufficient.

Ever truly yours
C. Babbage

I.110 1843-05-30. C. Babbage to Ada Lovelace.

My Dear Lady Lovelace[160]

I am delighted with Note D it is in your usual clear style and requires only one trifling alteration which I will make. This arises from the circumstance of our not having yet had time to examine the outline of the mechanical part. Only three kinds of Variable cards are used:

1st Those which give off a Variable from the Store to the Mill and leave zeros on the variable itself.

2nd Those which give off a Variable from the Store to the Mill and at the same time (or in the same turn of the hand) retain the same variable in the same place.

3rd Those which order any Variable on which only zeros exist to receive a result from the Mill.

I propose to omit one paragraph which I have marked in pencil and the rest is quite correct, as however I suppose you will wish to see the change I return you the sheet.

I have not yet received a proof and shall enquire about it tomorrow.

I enclose a copy of the integration. I am still working at some most entangled notations of Division but see my way through them at the expense of heavy labor, from which I shall not shrink as long as my head can bear it—I have been somewhat impeded however for the last few days. Your latest information was the most agreeable.

Ever My Dear Lady Lovelace
Sincerely Yours
C Babbage
Dorset Street
30 June 1843

I.111 1843. Ada Lovelace to C. Babbage.

Satdy
6 o'clock[161]

My dear Babbage

I have been hard at work all day intending to send you the Diagram & all quite complete.

Think of my horror then at just discovering that the Table & Diagram, (over which I have been spending infinite patience & pains) are seriously wrong, in one or two points. I have done them however in a beautiful manner, much improved upon our first edition of a Table & Diagram. But unluckily I have made some errors.

I send you all this final note H, excepting the said Table & Diagram. I also return you Note C, in which (for a wonder), I can discover nothing to alter or mend.

160 Lovelace Papers 168.
161 British Library Add MS 37192 f401.

Note E. I still retain; since I find I have (as I suspected) put into it one or two things that are inconsistent with what I have subsequently written. The sheet of Note D I return, having pinned over the effaced part, the alteration I wish to make in it in lieu of altogether omitting the passage.

I pray you to send me by my servant, (who will call in the morning for your orders), the other sheets of this Note D, as I wish to alter one or two passages in consequence of your information of this morning.

I also beg for Note A, in which I remember a wrong passage about Variable-cards. Now pray attend strictly to my requests; or you will cause me very serious annoyance.

I shall be up betimes tomorrow morning, & finish off the Table & Diagram; so as to send it you by post; together with the amendments in what my servant shall bring me down tomorrow morning from you.

You will have therefore a budget on Monday morg; and I intend to go to Town myself on Monday, for a few hours, in order to run over a few things with you finally. Be so kind as to be in the Square at two o'clock. I fear you will find me detestably perservering.

Yours ever,
A. L.
Let me know how you like my finishing up of H.[162] Mind you scrutinise all the ns very carefully. I mean those of Sheets 4 and 5.

I.112 1843-07-02. Ada Lovelace to C. Babbage.

Sunday Morning,
2 July 1843
Ockham[163]

My Dear Babbage

Unless you really have necessary business to transact with me tomorrow, pray do not come to see me; for I am full of pressing & unavoidable engagements; & moreover, I shall obliged to be in Town on Friday next, & could then see you at 12 o'clock [...].

I am reflecting much on the work & duties for you & the engine, which are to occupy me during the next two or three years I suppose; & I have some excellent ideas on the subject.

I intend to incorporate with one department of my labours a complete reduction to a system, of the principles & methods of discovery—elucidating the same with examples. I am already noting down a list of discoveries hitherto made, in order myself to examine into their history, origin, & progress. One first & main point, whenever & wherever I introduce the subject, will be first to define & to classify all that is to be legitimately included under the term discovery. Here will be a fine field for my clear, logical & accurate, mind, to work its powers upon; & to develop its metaphysical genius, which is not least amongst its qualifications & characteristics.

Let me find a note tomorrow when I arrive, which I expect to do between twelve & one. I should like to know whether to expect you or not.

Yours A.L.

---

[162] Actually Note G.
[163] British Library Add MS 37192 f337.

I.113 1843-07-02. C. Babbage to Ada Lovelace.

My Dear Lady Lovelace[164]

If you are as fastidious about the acts of your friends as you are about those of your pen, I much fear I shall equally lose your friendship and your Notes. I like much the improved form of the Bernoulli Note but can judge of it better when I have the diagram and Notation.

I am very reluctant to return the admirable and philosophic view of the Anal. Engine contained in Note A. Pray do not alter it and do let me have it returned on Monday. I send also the rest of Note D. There is still one trifling misapprehension about the Variable cards. A Variable card may order any number of Variables to receive the same number upon them at the same instant of time. But a Variable card never can be directed to order more than one Variable to be given off at once because the mill could not receive it and the mechanism would not permit it. All this it was impossible for you to know by intuition and the more I read your notes the more surprised I am at them and regret not having earlier explored so rich a vein of the noblest metal. The account of them stands thus

A Sent to Lady L.
B With C.B.
C Ditto
D Sent to Lady L.
E With C. B.
F Retained by Lady L.
G Where is it gone??
H With C.B.

I have not seen Mr. Wheatstone and am ashamed to write until I can positively put the whole of the notes in his hands. I will attend your commands tomorrow—and am

Ever Most Truly Yours
Dorset St.
C Babbage
2 July 1843

I.114 1843-07-02. Ada Lovelace to C. Babbage.

Ockham
Sundy 6 o'clock[165]

I have worked incessantly & most successfully, all day. You will admire the Table & Diagram extremely. They have been made out with extreme care & all the indices most minutely & scrupulously attended to. Lord L is at this moment kindly inking it all over for me. I had to do it in pencil.

You must bring all the Notes with you tomorrow, as I have observations to make on each one; & especially on this final one H.

There never was a Note G. I do not know why I chose H instead of G, & thus insulted the latter worthy letter.

I cannot imagine what you mean about the Variable-Cards; since I never either supposed in my own mind, that one Variable-card could give off more than one Variable at a time; nor have (as far as I can make out) expressed such an idea in any passage whatever.

---

164 Lovelace Papers, 168.
165 British Library Add MS 37192 f337.

I cannot find what I fancied I had put in Note A; so I return it whole & sound, for your speedy relief.

I send back Note D. You will find the only alteration I wished to make pinned over; in the upper part of Sheet 2.

So I now retain nothing but Note F, which I shall give you tomorrow. Lord L— has put up, I find, in a separate cover, all that belongs to Note H. (He is quite enchanted with the beauty & symmetry of the Table & diagram). No—I find I can put in Note. D with H.

I.115 1843-07-04. Ada Lovelace to C. Babbage.

Ockham
Tuesday morng[166]

My Dear Babbage

I now write to you expressly on three points; which I have very fully & leisurely considered during the last 18 hours; & think of sufficient importance to induce me to send a servant up so that you may have this letter by half after six this evening. The servant will leave Town tomorrow morning early, but will call for anything you may have for me, at eight o'clock in the morning before he goes.

Firstly; the few lines I enclosed you last night about the connexion of (8) with the famous Integral, I by no means intend you to insert, unless you fully approve the doing so. It is perhaps very dubious whether there is any sufficient pertinence in noticing at all that (8) is an Integral [...].

Secondly; Lord L suggests my signing the translation & the Notes, by which he means, simply putting at the end of the former: "translated by A. A. L.' & adding to each note the initials A. A. L.

It is not my wish to proclaim who has written it; at the same time that I rather wish to append anything that may tend hereafter to individualise & identify it, with other productions of the said A. A. L.

My third topic, tho' my last, is my most anxious & important. I have yesterday evening & this morning very amply analysed the question of the numbers of Variable-Cards, as mentioned in the final Note H (or G?). And I find that you & I between us have made a mess of it; (for which I can perfectly account, in a very natural manner). I enclose what I wish to insert instead of that which is now there. I think the present wrong passage is only about eight or ten lines, & is I believe on the second of the three great sheets which are to follow the diagram.

The fact is that if my own exposition about the Variable-Cards in Note D, had been strictly followed by myself, in Note H; this error would not have occurred. The confusion has arisen simply from the circumstance of applying to the Variable-Cards, facts which relate to the Operation-Cards. In Note D. it is very well & lucidly demonstrated that every single Operation, demands the use of at least three Variable-Cards. It does not signify whether the operations be in cycles or not. A million successive additions +, +, +, etc., etc., etc. would each demand the use of three new Variable-Cards, under ordinary circumstances. In Note H, the erroneous lines are founded on the hasty supposition that the cycle, or recurring group, of Operation-Cards (13 . . . 23), will be fed by a cycle, or recurring group, of Variable-Cards.

I enclose what I believe it ought to be. If already gone to the printer, we must alter that passage in the proofs, unless you could call at the printers & there paste over the amendment.

[166] British Library Add MS 37192 f342.

I can scarcely describe to you how very ill & harassed I felt yesterday. Pray excuse any abruptness or other unpleasantness of manner, if there were any. I am breathing well again today, & am much better in all respects; owing to Dr. L's remedies. He certainly does seem to understand the case, I mean the treatment of it, which is the main thing.

As for the theory of it, he says truly that time & Providence alone can develop that. It is so anomalous an affair altogether. A Singular Function, in very deed!

Think of my having to walk (on occasion run), to the Station, in half an hour last evening; while I suppose you were feasting & flirting in luxury & ease at your dinner. It must be a very pleasant merry sort of thing to have a Fairy in one's service, mind & limbs! I envy you! I, poor little Fairy, can only get dull heavy mortal, to wait on me!

Ever Yours
A. L.

I.116 1843. Babbage Note.

C B to Lady L[167]
Wednesy July 43
Return sheet with 2 corrections
Right about card 21
requiring new Variable

I.117 1843. Ada Lovelace to C. Babbage.

My Dear Babbage[168]

I hope you will approve of what I send. I have taken much pains with it. I have explained that there would be, in this instance & in many others, a recurring group or cycle of Variable as well as of Operation cards, & I have (I think very judiciously & easily) touched on the only departure from perfect identity which could exist during the repetitions of (13 ... 23), & yet have not committed myself by saying if these departures would require to be met by the introduction of one or more new cards, or not; but have simply indicated that as the variations follow a regular rule, they would be easily provided for. I think I have done it admirably & diplomatically. Here comes in the intrigante & the politician! Ever yours

Ockham A. L.
Tuesdy

I.118 1843-07-05. Ada Lovelace to C. Babbage.

Ockham Park
Weddy 5th July[169]

My Dear Babbage

I am much obliged by the contents of your letter, in all respects. Should you find it expedient to substitute the amended passage about the Variable-Cards, there is also one other short sentence which must be altered similarly. This sentence precedes the passage I

[167] British Library Add MS 37192 f354.
[168] British Library Add MS 37192 f348.
[169] British Library Add MS 37192 f349.

sent yesterday by perhaps half a page or more. It is where I explain that for every B after B5, operations (13 ... 23) have to be repeated; & I believe it runs as follows:

"Not only are the Operation-Cards precisely the same for this repetition, but the Variable-Cards as well with the exception of one new one necessary to introduce B5 instead of B3 for Operation 21 to act upon." I should wish to substitute what I enclose.

"Why does my friend prefer imaginary roots for our friendship?" Just, because she happens to have some of that very imagination which you would deny her to possess; & therefore she enjoys a little play & scope for it now & then. Besides this, I deny the Fairyism to be entirely imaginary; (& it is to the fairy similies that I suppose you allude).

That brain of mine is something more than merely mortal; as time will show; (if only my breathing & some other etceteras do not make too rapid a progress towards instead of from mortality).

Before ten years are over, the Devil's in it if I haven't sucked out some of the life-blood from the mysteries of this universe, in a way that no purely mortal lips or brains could do.

No one knows what almost awful energy & power lie yet undeveloped in that wiry little system of mine. I say awful, because you may imagine what it might be under certain circumstances.

Lord L sometimes says "What a General you would make!" Fancy me in times of social & political trouble, (had worldly power, rule, & ambition been my line, which it now never could be).

A desperate spirit truly; & with a degree of deep & fathomless prudence, which is strangely at variance with the daring & the enterprise of the character, a union that would give me unlimited sway & success, in all probability.

My kingdom however is not to be a temporal one, thank Heaven!

I am doggedly attacking & sifting to the very bottom, all the ways of deducing the Bernoulli Numbers. In the manner I am grappling with this subject, & connecting it with others, I shall be some days upon it. I shall then take in succession the other subjects that have been suggested to me during my late labours, & treat them similarly.

Labor ipse voluptas is in very deed my motto! And, (as I hinted just now), it is perhaps well for the world that my line & ambition is over the spiritual: & that I have not taken it into my head, or lived in times & circumstances calculated to put it into my head, to deal with the sword, poison, & intrigue, in the place of x, y, & z [...].

Yours Fairy for ever
A. A. L.
This being the Rent-day, I cannot see Lord L— to speak to him about the initials. Time enough I suppose.

I.119 1843-07-06. Ada Lovelace to C. Babbage.

Thursday
Ockham[170]

My Dear Babbage

I send you the first sheet all corrected. I have taken much pains with it, & I think it very much improved.

The printer has made one or two paragraphs where none ought to exist; & has also not put some words in Italics that ought to be so expressed. I have endeavoured to indicate all this, as I best might.

[170] British Library Add MS 37192 f355.

I send you four Foot-Notes, which are referred to in the proper places. I think you will like them; especially the first one about Pascal's Machine. I hope to send you the remainder of the translation tomorrow. My plans are again all altered. I go on Monday for the day only. Will you come at three o'clock.

I do not suppose that the Notes will take half the corrections which the translation does. I took so much more pains with them. I hope not, for it is damnably troublesome work, & plagues me. Pray let me know if my corrections are intelligible. [...]

F Knight reappeared from Holland on Tuesday, so he spent Wednesday Morg with me, & was very delightful. He is anxious about my publication, & only fears my writing anything inexperienced. I see I am more his ladye-love than ever. He is an excellent creature, & deserves to have a nice ladye-love.

Yours A.A.L.

P.S. I mean to put A.A.L. to all the Notes, but to leave the translation without.

I.120 1843-07-11. Ada Lovelace to C. Babbage.

Tuesday,
5 o'clock
St James' Square[171]

I find that I do not come up on Friday next. But if you will give me Monday Evening (I shall stay the night), it will be desirable.

Meanwhile I shall get on with the corrections, & send the papers back to you as early as possible; only hoping that nothing in them will oblige me to have your assistance. I do not think this likely.

I hope you will attend carefully to my criticisms about the Preface. I think them of consequence. If Lord L.— suggests any further ones, you shall hear. Perhaps he may. Ever yours

A.L.

I.121 1843-07-11. Babbage Note.

C. B. to Lady L[172]
Tuesday 11 July 43
Bernouillis numbers
Diagram
Proof of translation
Post answer

I.122 1843-07-13. Ada Lovelace to C. Babbage.

Thursday, 6 o'clock[173]

My Dear Babbage

171 British Library Add MS 37192 f364.
172 British Library Add MS 37192 f366.
173 British Library Add MS 37192.

Will you come at nine on Satdy morning, & stay as long as we find requisite. I name so early an hour, because we shall have much to do I think. And it certainly must not be later than ten o'clock.

I have been very unwell indeed; partly with distressing indigestion & partly with this most miserable & changed weather. This is why you have heard nothing more from me [...]

I send you a note from F K.[174] It will show you the tone of writing between us; & the sort of footing we are on. Are you not amused at "your Ladyship" which just means, by the way that I am anything but My Ladyship to him! [...]

Am I very naughty to send you such a caller-up of many & dubious speculations and associations? No.

Ever yours A.L.

I.123 1843-07-21. Ada Lovelace to C. Babbage.

My Dear Babbage[175]

I am in much dismay at having got into so amazing a quagmire & botheration with these Numbers, that I cannot possibly get the thing done today. I have no doubt it will all come out clear enough tomorrow; & I shall send you a parcel up, as early in the day as I can. So do not be uneasy. (Tho' at this moment I am in a charming state of confusion, but it is that sort of confusion which is of a very bubble nature.)

I am now going out on horseback. Tant mieux

Yours puzzle-pate

I.124 1843-07-22. Ada Lovelace to C. Babbage.

Saturday, 3 o'clock
Ockham[176]

My Dear Babbage

I send you all excepting four pages which I cannot get done today, as they necessitate one or two troublesome alterations. You will perceive therefore that the two parts of what I send do not follow. I shall send the rest by post tomorrow.

You ought to get this tonight & an additional shilling is promised upon immediate delivery. I must now explain one or two things. I am much annoyed at your having altered my Note. You know I am always willing to make any required alterations myself, but that I cannot endure another person to meddle with my sentences. If I disapprove therefore, I hope I may be able to alter in the revise, supposing you have sent away the proofs & notes.

Then I cannot agree to your not having effaced the paragraphs. In one instance, at any rate, if not in all, it is very necessary that the paragraph should be effaced; as it makes a division in the sense where there should be a perfect continuity.

In short I am somewhat disturbed about the matter altogether.

I want to make a slight alteration too in one of the smaller Notes which I sent; the one relating to the substituting zero for a number that has been packed off to the mill.

[174] Frederick Knight.
[175] British Library Add MS 37192 f382.
[176] British Library Add MS 37192 f360.

And I must then beg you to deliver to me tout suit plus vite,[177] all the documents I am at present to have in order to commence work upon. I think much about them & all other scientific matters, and I find that my plans & ideas keep gaining in clearness; & assuming more of the crystalline & less & less of the nebulous form.

Yours A.A.L.
P.S. I have been very suffering, & more ill than ever yet. But today I have rallied considerably. Hang the whole affair.

I.125 1843. Ada Lovelace to C. Babbage.

We'dy night[178]

My Dear Babbage

It is quite evident to me that you have been looking over the superseded sheet 4, instead of the corrected one. All your remarks seem to apply to the former; & the latter is passed over without notice. Pray be with me at half past nine o'clock tomorrow morning, if you can; as I am exceedingly disturbed about it. Lord L— is so vexed too at everything not being done, that I am half beside myself.

He is pressing me in several ways just now, most unfortunately. And amongst it all, I really shall be a long time, & shall lose my head for everything. Pray look carefully over the real papers. I send them together (& the superseded one I have put separately).

I really cannot believe it to be incorrect, for nothing can exceed the care with which I have gone over it. But the fact is I am plagued out of my life, just now.

P. S. I have made out a list of the operations which calculate each co-efficient for each variable, & have subjoined it to your own memorandum. Look over it carefully.

I.126 1843-07-25. Ada Lovelace to C. Babbage.

Ockham
Tuesdy Morng[179]

My Dear Babbage

The bearer will I hope deliver this by nine this evening. He is to return here as early tomorrow as you can let him have an answer. What I want to know is this: Can you be with me in Town at four on Thursday, in order that I may read over aloud, with you, all the Notes. For I cannot feel satisfied (on re-perusing them), I find: without going over them with you. before they are finally printed off.

Notes 5 & 6, I have not sent you; for I cannot quite please myself about them. I must either send them to you by post tomorrow: or take them on Thursdy. Can you afterwards give me Thursday evening, to go to the Opera. If my servant can have a reply from you tonight, he will be off by daylight tomorrow morning. If not, he must come as early as he can tomorrow. I keep back your note; wishing to consider it a little more. I think it unobjectionable, as far as I have yet considered it. Pray take care that the printing is so managed as to separate distinctly the translator's notes, from either your note, or one there is of Menabrea's own (see page 36).

177 Expeditiously.
178 British Library Add MS 37192 f379.
179 British Library Add MS 37192 f386.

I shall not put my initials to my notes but I wish them to be Translator's Notes. There are some blanks left in two or three places, for pages referred to, that I cannot put the numbers in for until I have the proofs.

Yours ever A. L.

I.127 1843-07-26. Notes by Babbage.

Wedy. 26th July[180]
Home late
Parcel & note B.
Send revise of Preface. B.[181]

I.128 1843-07-27. Ada Lovelace to C. Babbage.

Ockham
Thursdy Morng
27th July[182]

My Dear Babbage

I cannot get on satisfactorily with either the proofs or the revise, unless I have my own manuscripts for the former, & the corrected proofs for the latter.

Pray therefore get both forthwith, & let them be in St. James' Sqre before four tomorrow.

I am happy to find that the Notes will require very little correction indeed. To say the truth, I am rather amazed at them, & cannot help being struck quite malgre moi,[183] with the really masterly nature of the style, & its superiority to that of the Memoir itself.

I have made Lord L— laugh much by the dryness with which I remarked, "Well, I am very much satisfied with this first child of mine. He is an uncommonly fine baby, & will grow to be a man of the first magnitude & power."

I approve your alteration in the preface, excepting that I think the word "So" comes in both awkwardly & superfluously. Pray efface it, & let it stand, "of the money to be expended." That little word spoils it.

You will be amused, & somewhat triumphant perhaps, when I own that I entirely approve your alteration of my Foot-note to page 622, & only find the insertion of one single word (& that not in the part you muddled with), to be a desideratum.

Altogether I think things are doing very well. I expect to have all the Notes done by Monday; & the subsequent Revise of them, will probably be but a very trifling matter.

Lord L— seems pleased beyond measure, with the very learned & knowing aspect of my baby's physiognomy which he has glanced at.

Yours A. L.

I.129 1843-07-28. Ada Lovelace to C. Babbage.

St. James' Sqre
Friday, 5 o'clock[184]

[180] British Library Add MS 37192 f391.
[181] British Library Add MS 37192 f392.
[182] British Library Add MS 37192 f393.
[183] If I don't say so myself.
[184] British Library Add MS 37192 f399.

The beginning of Note G (by which I mean the Table & all that precedes it) never has been returned into my hands; a small part of the remainder was, but that I speedily gave you back, & thus it is, now printed.

The missing part must be either at your house or at the printers; & it seems to me very unlikely that you should have retained it. So altogether I would wager almost anything that it is at the office; or that if lost, it has been lost there.

At the same time, I have always fancied you were a little harum-scarum & inaccurate now & then about the exact order & arrangement of sheets, pages, & paragraphs etc.; (witness that paragraph which you so carelessly pasted over).

I suppose that I must set to work to write something better, if I can, as a substitute. The same precisely I could not recall. I think I should be able in a couple of days to do something. However I should be decidedly inclined to swear at you, I will allow.

I desire my messenger to wait; as it is possible you may have something to communicate more agreeable.

I go soon after seven. I believe I shall not be in Town myself on Monday as I expected.

Yours
A. L.

I.130 1843-07-28. Notes by Babbage.

Friday 28 July[185]
Part of Ms lost
Note G with its beautiful Table is missing.
Notes on Menabrea's Paper
Will obey summons
here for earlier hour.
Sat 29 July. Missing Mss found.[186]

I.131 1843-07-30. Ada Lovelace to C. Babbage.

[Sunday, July 30][187]

I am beyond measure vexed to find that instead of inserting my corrected Table in the Revise [...] they have left it exactly as it was before. Pray see about it immediately. It is exceedingly careless & annoying.

Out of several corrections made, not one is inserted, neither are the Upper Indices added; nor the little Foot-Note. I send you back all the latter part of the Revise, & the corresponding proofs, that you may look to the matter forthwith. I cannot account for such negligence.

I find myself in some distress for the original Memoir; which you should have sent me back. I do not think you possess half my forethoughts, & powers of foreseeing all possible contingencies (probable & improbable, just alike).

I am glad to see the sheets I return so clean on the whole. Tomorrow I expect to send you up the rest of the Revise, & Note A by my governess, in the middle of the day, & more by post.

I will work most diligently, but I wish to revise the Notes myself. You might send some one down here the moment you get them; & I would attend immediately & send them back by the same or some other special messenger.

185 British Library Add MS 37192 f398.
186 British Library Add MS 37192 f404.
187 British Library Add MS 37192 f407.

I have begun your examples of the Calculus of Functions, appended to Herschel's examples of Finite Differences. This will familiarize me a little with the subject; but I must get various other papers, referred to by you at the end of the examples.

The paper of Murphy's I want is entitled, "Elementary Principles of the Theory of Electricity."

How very careless of you to forget that Note; & how much waiting on & service you owe me, to compensate.

I am in good spirits; for I hope another year will make me really something of an Analyst. The more I study, the more insatiable do I feel my genius for it to be.

I do not believe that my father was (or ever could have been) such a Poet as I shall be an Analyst, (& Metaphysician), for with me the two go together indissolubly.

Yours A. L.

I.132 1843-08-01. Charles Lyell to C. Babbage.

I have spent two hours in negotiating with Mr. Francis & in seeking in vain after twelve o'clock for Wheatstone at K.C.[188] I began by explaining to Mr. Francis the proposed alterations of which he highly approved but having heard me out he said finally that he would not print it [...] on his responsibility [...] On Francis proposing that you should wait fourteen days for T[aylor]'s return I remarked that before that he must be out [published] to which he assented but said that they could promise in six weeks after to come out with another part of the translation.

I then asked, "Do you think that Mr. Taylor will then consent to insert Mr. B's article?"—"I do not—but we could exert ourselves to get out the translation with the notes in the present No. if Mr. G.[189] would omit the preface." We then entered upon the pros & cons of the Phil. Mag. & ended as yesterday with objecting to the translatress making her debut there. On my saying that the notes by Lady L. must go to the Phil. Trans. as a separate paper, F. expressed great regret at the loss of it to them & their work & on my dwelling on the great interest the public would take in the Preface & on my asking him whether he could not devise some way of their profiting by both, he suggested that the Preface should make a separate notice signed by you, addressed to the Public in which you say that as a translation was about to appear you wished to explain &c. He could then stitch this in to every no. of the translations & although it could not form part of the work everybody seeing your name could bind it up with the rest. He remarked that a friend of his seeing the Preface on his table, said that he could swear to it's being all Mr. Babbage's therefore why should he not put his name?

I left him, asking if he could promise to get out the whole preface [...]. before the Cork meeting of the British Association & without waiting for Taylor, provided you assented that the Preface should appear as your own or as Wheatstone's. He said he would promise but no time must be lost. As these are better terms than I expected after the beginning of the conference I hope you will agree & put your name manfully.

I.133 1843-08. Ada Lovelace to Lady Byron.

Thursday Morning
Ockham[190]

[188] British Library Add MSS 37192 f412.
[189] The printer, Gyde.
[190] Lovelace Papers.

[...] I certainly feel some relief & satisfaction in turning to the packet of clear, forcible & logical documents now before me for my final revision; after the tortuous & nefarious documents & affairs which have recently so painfully engaged much of our energy & attention. My first born is characterized chiefly by a strong sense & union of the most minute & laborious accuracy, in particulars, with a colouring & undercurrent (rather hinted at & suggested than definitely expressed) of large, general. & metaphysical views. He is not eloquent or brilliant, but he shows indication of much latent power, & of a most indomitable industry which despises no even apparently trivial minutia. He will make an excellent head of (I hope) a large family of brothers & sisters; to whom he will impart a certain staid & solid character, for which they will be much indebted to him, (tho' probably no one may ever perceive or acknowledge his quiet & insensible influence) [...]

I.134 1843. Ada Lovelace to C. Babbage.

Monday afternoon[191]

There is not much correction for Note A; (I send you the whole of it), and of these the chief part I think are merely paragraphs to be effaced, & stops to be inserted or altered. I cannot refrain from expressing my amazement at my own child. The pithy & vigorous nature of the style seem to me to be most striking; and there is at times a half-satirical & humorous dryness, which would I suspect make me a most formidable reviewer. I am quite thunder-struck at the power of the writing. It is especially unlike a woman's style surely; but neither can I compare it with any man's exactly.

It is my intention at some future time, when I have knowledge enough, to review Whewell's works. Do you not pity him? And yet the beauty of it would be that I should never say one single ostensibly cutting, or harsh, or condemnatory thing.

By the way, I want to ask your advice about a more immediately practical matter. I wish to review Ohm's little work. I have much to say upon it. My review however would not be a very long one. Could it be put into the Philosophical Magazine, do you suppose? I do not know if they review books at all, in that publication.

I shall not probably begin the review just yet. But I am likely to have it done sometime this autumn; or within the year. Of course I shall submit it to you, & also perhaps to some others. I don't yet know quite enough to be sure always of the solidity or oppositeness of all my views.

By the way, there is a thing which must be seen to in the Notes & Translation; owing to the former not having been yet properly paged I am unable to give the correct numerical references to the pages in various places, as they ought ultimately to be. Likewise the Notes should have A. A. L. appended to them. I do not mean the Footnotes, only Notes A, B, C, D, E, F, G, the illustrious seven!

You have now the whole of the Revise. I trust that all of Note G was found. There is but a very little bit of it indeed, yet sent to me. I can correct the proof of it in an hour or two, when I get it. Expect a parcel from me tomorrow, with the remaining proofs of the Notes.

A. A. L.

I.135 1843. Ada Lovelace to C. Babbage.

Ockham

[191] British Library Add MS 37192 f339.

Tuesday afternoon[192]

I am half beside myself with hurry & work. I could not get anything done in time to send by coach this morning, & now I am obliged still to retain one sheet; which however I hope to send you by an occasion tomorrow afternoon.

Note B has plagued me to death; altho' I have made but little alteration in it. Such alterations as there are however, happen to have been very tiresome & to have demanded minute consideration & very nice adjustments. It is a very excellent Note.

I wish you were as accurate, & as much to be relied on, as I am myself. You might often save me much trouble, if you were; whereas you in reality add to my trouble not infrequently, and there is at any rate always the anxiety of doubting if you will not get me into a scrape; even when you don't.

By the way, I hope you do not take upon yourself to alter any of my corrections. I must beg you not. They all have some very sufficient reason. And you have made a pretty mess & confusion in one or two places (which I will show you sometime), where you have ventured in my M.S's, to insert or alter a phrase or word; & have utterly muddled the sense.

I could not conceive at first in one or two places what had happened to my sentences; tho' I soon saw they were patch-work & not my own; and found it so on referring to the M.S.

I fear you will think this a very cross letter. Never mind. I am a good little thing after all.

Yours ever
A. A.L.

Later P.S. It is impossible to send you anything but Notes B and C; (& this partly owing to some wrong references & blunderations of your own). Do not be afraid, for I will work like the Devil early tomorrow morning.

I.136 1843-08-02. Ada Lovelace to C. Babbage.

Ockham
Weddy, 4 o'clock[193]

After working almost incessantly since 7 o'clock this morning, until I am forced to give in from sheer inability to apply longer, I find only the sheet I enclose quite completed. I shall however send a servant up tomorrow morning by a ten o'clock train, to take you all the rest, so that you will have it almost as soon as this letter. You cannot conceive the trouble I have had with the trigonometrical Note E! In fact no one but me, I really believe, would have doggedly stuck to it, as I have been doing, in all its [wearying] minutiae.

I am very uneasy at not hearing from you as I have expected to do both yesterday & today, & fear some disaster or other. I hope all of Note G is forthcoming; & I also hope you have received all my communications safely. I think you had better do the second revise of the translation for me. If you will compare it carefully with my first revise, it can hardly be necessary I think for me to go over it again. I suppose I ought to take it for granted that no news is good news, but I am in a sad fidget.

Yours ever
A. L.

[192] British Library Add MS 37192 f414.
[193] British Library Add MS 37192 f388.

Turnover

I have sent you, since Satdy, the whole of the revised translation, and the corrected proofs of Notes A, B, C. I send with this a portion of Notes D and E. Tomorrow morning, I shall send you the remainder of these two, & all of Note F.

The very little bit I have of Note G I shall retain till I get the remainder. I hope they are getting on with all that is already sent back.

I.137 1843. Ada Lovelace to C. Babbage.

Wednesday Evening
10 o'clock[194]

I send you what I have done of Note E, which is not nearly all. So you must not judge of it, as it is. I am becoming sadly overworked, & have scarcely brains left for anything.

I wonder how you will like my further addition to the upper indices. I half fear not. But I can cancel it, if you disapprove. No more tonight, for I can neither talk, write, nor think common sense. And yet I feel more like a fairy than ever. (But I suppose that idea is uncommon & not common sense).

Yours Addlepate

I.138 1843-08-05. C. Babbage to Ada Lovelace.

My Dear Lady Lovelace[195]

I send you back the revise and its alterations. You had better return this by post to

Mr. Taylors Printing Office
Red Lion Court
Fleet St.

I shall send the final revise of the Trans. tonight so that they will then commence putting it into pages—I have suggested two amendments— I send in another cover the last of the notes, and the great Notations. I waited with some anxiety for a communication from you yesterday.

Today I saw Wheatstone and proposed to him a plan which will fulfill all your conditions and some of mine. He approves of it and thinks it will be adopted. If it is I shall write by Monday's post.

Sat. Evg.
Ever truly yours
C B
5 Aug

I.139 1843-08-06. Ada Lovelace to C. Babbage.

Sunday, 6 August
Ockham[196]

[194] British Library Add MS 37192 f390.
[195] Lovelace Papers, 168.
[196] British Library Add MS 37192.

My Dear Babbage

I did not make you any communication on Friday, because I did not feel I could do so satisfactorily to myself or to you, without another day of leisure to reflect. On the one point of not withdrawing the translation & Notes from the Memoir, or consenting to its separate publication, I was entirely & finally decided, as I think neither for your advantage nor my own, to do so; added to my opinion that it would under the circumstances be dishonorable & unjustifiable.

But in other respects, I have felt very anxious to obtain your co-operation & cordial assent & concurrence.

It was my intention therefore to write to you very strongly on some points in which I consider you mistake your own interests. I will not however, now do so; at any rate until I hear what the present plan is; for my representations may be unnecessary & irrelevant.

Having decided as an axiom that the Memoirs are to contain the paper, I intended to leave it to you (only subject to Lord L's approbation) whether it should come out in this number, or be delayed until the next to give you more time for thinking over matters. This I still do.

Be assured that I am your best friend; but that I never can or will support you in acting on principles which I conceive to be not only wrong in themselves, but suicidal.

This sentence will excite your deprecation. I am well aware; & you will inwardly exclaim: "how she misunderstands me!"

No more of this just at this moment, for I have perhaps already said too much.

But, before I undertake any further labors in conjunction with you, I must have an explicit understanding on one or two points; & do not anticipate that you will hesitate much to agree to my conditions.

Yours A.L.

I.140 1843-08-08. C. Babbage to Ada Lovelace.

My Dear Lady Lovelace[197]

I leave the Mss. and also the proofs of the Notes I recd last night and promised to send this evening.

I will write to Printer to say you will send them up by post direct to them. This direct communication will save time and there is very little to spare for the Number ought to be out in the course of a very few days.

I have nothing to add at present except that you did me an injustice in supposing I wished you to break any engagement with the Editor. I wished you to ask him to allow you to withdraw from it. Had the Editor been in England I believe he would at my request have inserted my defense or forborn to have printed the paper. As it stands I have done all I can at present do to defend myself and having failed in the most important part shall make the best I can of the rest.

Ever Truly Yours
C Babbage
2 1/2 P.M. Tuesday
Dorset St.
8 Aug 1843

[197] Lovelace Papers, 168.

I.141 1843-08-08. Ada Lovelace to Lady Byron.

St. James' Sqre
Tuesday 8th Augst[198]

Dear Mama

[...] Lady Mahon being now out of Town, & I believe taking no present part in this affair, I considered that further communication from me to her, would come better in a week or two weeks from the time I received her reply just at the moment. [...]

I must tell you that another reason why I have not been writing to you (as I did not perceive immediate call for it), is that I have been harassed & pressed in a most perplexing manner by the conduct of Mr. Babbage. We are in fact at issue, & I am sorry to have come to the conclusion that he is one of the most impracticable, selfish, & intemperate persons one can have to do with. I do not anticipate an absolute alienation between us; but there must ever be a degree of coolness & reserve I fancy in future. I have had in W's[199] absence, to act quite unadvised in the matter; but I am happy to find that W— & Wheatstone entirely approves my conduct & views. I declared at once to Babbage, that no power should induce me to lend myself to any of his quarrels, or to become in any way his organ; & that I should myself communicate in a direct manner with the editors on the subject, as I did not choose to commit a dishonorable breach of engagement, even to promote his advantage (if it were to his advantage which I doubted).

He was furious, I imperturbable & unmoved. He will never forgive me. I had tried to conciliate & quietly to advise & suggest, until I found that it was necessary to be very determined & explicit. Pray forgive my forcing on you at this moment, these painful circumstances in another matter.

I only want you to understand that all my time & my energy have been miserably absorbed the last few days; for what between Babbage & the editors, both pressing hard in different directions, I have been torn to pieces; & as I heard nothing more from any one, of the E. L.[200] affair I really endeavoured to forget for a few days all its tortuous & villainous implications.

I must hope that I may develop God's truth in ere fully for the use of mortals. With this single object, if I have but the determination to adhere to it in purity & steadfastness. I shall be able to weather all the storms; & to leave a name for the race & the family which may screen & cloak over even its manifold iniquities & tortuousities, when all these (if they ever do) come fully to light.

I.142 1843. Ada Lovelace to Robert Noel.

Wednesday Afternoon,
9 August 1843
Ockham Park[201]

My Dear Robert

[...] Since Lord Lovelace wrote his note (herewith enclosed) this morning, he & I have conversed earnestly on the subject of Herr Kraemer, & have in fact come to stronger conclusions respecting him, than we contained in Lord L's note. These Lord L— wishes me

198 Lovelace Papers, 42.
199 William, Lord Lovelace.
200 Medora.
201 Lovelace Papers.

therefore to state. We think [...] we had better decline entering on an engagement with him, at such a distance, & under such circumstances. My own impression of his letter, is by no means agreeable; and I should expect he would prove to be somewhat impracticable. & proud.

Should you decide against him, we must still beg of you the great favor to be on the look-out for us. Our case is becoming rather pressing [...]. The real ground of the very independent way of conducting our family arrangements, ought to be understood. It is neither the result of pride; nor of indifference to our own children & their habits. But I am now myself a completely professional person, to speak plainly; & am engaged in studies & in literary & scientific avocations, which render me quite unable (were I even fitted by nature, which I am not), to associate much personally with my children, or to exercise a favorable influence over them by attempting to do so. [...]

I shall have the pleasure of sending you, (if you will tell me how), a copy of my translation of Monr. Menabrea's description of Babbage's Calculating Engines, to which I have appended very copious Notes of my own, that are in fact longer & more abstruse than the original Memoir itself. This engagement has been in some respects arduous & troublesome; and it will probably bring me in but little return of reputation or fame, (which were indeed no part of my motive when I undertook it): for there is more of a quiet patient labour & industry in it, than of brilliancy or attractiveness. I intentionally keep myself way back. I had better feel my way gradually. I have plenty of imagination and eloquence, when the right time shall come. Meanwhile I wish to build upon strong foundations of logic, industry, & real study & training. But I am glad to have got launched, in however humble & dry a form.

I wish now to know how far you may think yourself able & willing, to organize a plan I have for you; which I fancy you exactly fitted for fulfilling, & is a great desideratum in this country. I believe that many researches are going on in Germany, as to the microscopical structure & changes in the brain, nervous matter, & also in the blood. Now, we in this country, need data, on this subject more than almost any [...].

My kindest remembrances to Mrs R. N.; & believe me ever, with many thanks. Most sincerely yours

A.A. Lovelace

I.143 1843-08-14. Ada Lovelace to C. Babbage.

Ockham Park
Monday 14th August 1843[202]

My Dear Babbage

You would have heard from me several days ago, but for the hot work that has been going on between me and the printers. This is now all happily concluded. I have endeavoured to work up everything to the utmost perfection, as far as it goes; & I am now well satisfied on the whole, since I think that within the sphere of aims I set out with, & in accordance with which the whole contents & arrangement of the Notes are shaped, they are very complete, & even admirable. I could now do the thing far better, but this would be setting out upon a wholly different basis.

I say you would have heard from me before. Your note (enclosed on Monday with my papers etc), is such as demands a very full reply from me, the writer being so old & so

[202] British Library Add MS 37192 f422.

esteemed a friend, & one whose genius I not only so highly appreciate myself but wish to see fairly appreciated by others.

Were it not for this desire (which both Lord L— & myself have more warmly at heart than you are as yet at all aware of), coupled with our long-established regard & intercourse, I should say that the less notice taken by me of that note the better: & it was only worthy to be thrown aside with a smile of contempt. The tone of it, it is impossible to misunderstand; & as I am myself always a very "explicit function of x" I shall not pretend to do so; & shall leave to you (if you please it) to continue the "implicit" style which is exceedingly marked in the said note.

As I know you will not be explicit enough to state the real state of your feelings respecting me at this time, I shall do so for you. You feel, my dear Babbage, that I have (tho' in a negative manner) added to the list of injuries & of disappointments & miscomprehensions that you have already experienced in a life by no means smooth or fortunate. You know this is your feeling; & that you are deeply hurt about it; & you endeavour to derive a poor & sorry consolation from such sentiments as "Well, she don't know or intend the injury & mischief if she has done" &c.

You say you did not wish me to "break my engagement, but merely to ask to be released from it." My dear friend, if the engagement was such that I had no right to break it without leave. I had still less right to appeal to the courtesy of parties, in order to obtain an apparent sanction & excuse for doing that which their justice & sense of their own rights could not have conceded. There is no greater sin, or deeper double-dealing in this world, than that of endeavouring thro' the influence of secondary motives to get the apparent support & consent of others towards that which on a higher & more general motive would be inadmissable. Your reply to this will be, that my principle is just, good & great: but that it did not apply to the particular case; in as much as the editors would themselves have been glad of an excuse to be released, under the circumstances. You must allow me to state that I took measures for ascertaining that point beyond all possibility of doubt, & that I found it to be very much otherwise.

You will deny & dispute this; or more probably you will immediately perceive grounds why, (from various unworthy *arriere-pensees* [ulterior motives]) the persons concerned might still wish to retain the article. Remember however, that when you do this, you have shifted your original ground: & that your question then becomes not whether Lady L— ought to oblige two parties, you & the editors, who both tho' on different grounds wish to dispose of my publication thro' another channel than that originally proposed, but whether Lady L— ought tacitly to lend herself to certain possible or probable unworthy motives entertained by the editors. Now to this the reply is perfectly plain in the opinion of all parties accustomed to fair & honorable dealings, uninfluenced by secondary motives. My engagement was unconditional, & had no reference to the motives of the parties with whom I contracted it. I have therefore no right to withdraw it on grounds subsequently thought of. If I had undertaken to do the thing specially for you, in addition to its being for them, the case would have been wholly different. But with the circumstance of your happening to be a private friend of my own, & of my therefore being too happy & delighted to make prior engagement especially pleasing & useful to you, they had nothing to do. Consequently, because my private friend wished it, (however justly), this could form no real & equitable ground for withdrawing the article.

I have now touched on all the grounds which can be taken on the supposition of its really being pernicious to your interests that I have thus allowed the article to appear. This however I cannot agree to or believe; & were you not influenced by a set of feelings which are very different from those that I myself, & the minds whom I most esteem, can consider wise, justifiable, or in harmonious accordance with man's moral nature, you would not think so. Mind, I do not say that your views may not be in reality higher, juster, &

wiser, than my own. But my moral standard, such as it is, I must stick to; as long as it is my moral standard. It would not be of any use for me to endeavour making you see thro' my glasses; for, (besides the fact that they may be as far or farther than yours, from refracting & reflecting quite truly), no one can [instantly] alter the views & modes of feeling of a life. But I do wish you to understand the fact, that I believe myself (however erroneous that belief may be), to have forwarded your interests far more by allowing the article to appear than I should have done by any of the courses you suggested. I have a right to expect from you the belief that I do sincerely & honestly take this view. For if your knowledge of me does not furnish sufficient grounds for doing so, then I can only say that no mutual knowledge of any two human beings in this life, can give stable & fixed grounds for faith & confidence. Then Adieu to all trust, & to everything most generous, in this world!

I must now come to a practical question respecting the future. Your affairs have been, & are, deeply occupying both myself & Lord Lovelace. Our thoughts as well as our conversation have been earnest upon them. And the result is that I have plans for you, which I do not think fit at present to communicate to you; but which I shall either develop, or else throw my energies, my time & pen into the service of some other department of truth & science, according to the reply I receive from you to what I am now going to state. I do beseech you therefore deeply & seriously to ponder over the question how far you can subscribe to my conditions or not. I give to you the first choice & offer of my services & my intellect. Do not lightly reject them. I say this entirely for your own sake, believe me.

My channels for developping [sic] & training my scientific & literary powers, are various, & some of them very attractive. But I wish my old friend to have the refusal.

Firstly: I want to know whether if I continue to work on & about your own great subject, you will undertake to abide wholly by the judgement of myself (or of any persons whom you may now please to name as referees, whenever we may differ), on all practical matters relating to whatever can involve relations with any fellow-creature or fellow-creatures.

Secondly: can you undertake to give your mind wholly & undividedly, as a primary object that no engagement is to interfere with, to the consideration of all those matters in which I shall at times require your intellectual assistance & supervision: & can you promise not to slur & hurry things over; or to mislay, & allow confusion & mistakes to enter into documents, &c?

Thirdly: if I am able to lay before you in the course of a year or two, explicit & honorable propositions for executing your engine, (such as are approved by persons whom you may now name to be referred to for their approbation), would there be any chance of your allowing myself & such parties to conduct the business for you; your own undivided energies being devoted to the execution of the work; & all other matters being arranged for you on terms which your own friends should approve?

You will wonder over this last query. But, I strongly advise you not to reject it as chimerical. You do not know the grounds I have for believing that such a contingency may come within my power, & I wish to know before I allow my mind to employ its energies any further on the subject, that I shall not be wasting thought & power for no purpose or result.

At the same time, I must place the whole of your relations with me, in a fair & just light. Our motives, & ways of viewing things, are very widely apart; & it may be an anxious question for you to decide how far the advantages & expediency of enlisting a mind of my particular class, in your service, can over balance the annoyance to you of that divergency on perhaps many occasions. My own uncompromising principle is to endeavour to love truth & God before fame & glory or even just appreciation: & to believe generously & unwaveringly in the good of human nature, (however dormant & latent it may often seem).

Yours is to love truth & God (yes, deeply & constantly); but to love fame, glory, honours, yet more. You will deny this; but in all your intercourse with every human being (as far as

I know & see of it), it is a practically paramount sentiment. Mind, I am not blaming it. I simply state my belief in the fact. The fact may be a very noble & beautiful fact. That is another question.

Far be it from me to disclaim the influence of ambition & fame. No living soul ever was more imbued with it than myself. And my own view of duty is, that it behooves me to place this great & useful quality in its proper relations & subordination: but I certainly would not deceive myself or others by pretending that it is other than a very important motive & ingredient in my character & nature.

I wish to add my mite towards expounding & interpreting the Almighty, & his laws & works, for the most effective use of mankind; and certainly, I should feel it no small glory if I were enabled to be one of his most noted prophets (using this word in my own peculiar sense) in this world. And I should undoubtedly prefer being known as a benefactor of this description, to being equally great in fact, but promulgating truths from obscurity & oblivion.

At the same time, I am not sure that 30 years hence, I may put even so much value as this, upon human fame. Every year adds to the unlimited nature of my trust & hope in the Creator, & decreases my value for my relations with mankind excepting as His minister: & in this point of view those relations become yearly more interesting to me. Thro' my present relations with man,, I am doubtless to become fit for relations of another order hereafter; perhaps directly with the great Power Himself. Of course my view respecting every even casual social contact & intercourse, takes a corresponding colour; & will do so increasingly, if that view should become more confirmed.

My dear friend, if you knew what sad & direful experience I have had, in ways of which you cannot be aware, you would feel that some weight is due to my feelings about God & man. As it is, you will only smile & say, "poor little thing; she knows nothing of life or wickedness!" Such as my principles are, & the conditions (founded on them), on which alone you may command my services, I have now stated them; to just such extent as I think is absolutely necessary for any comfortable understanding & cooperation between us, in a course of a systematized & continued intellectual labour. It is now for you to decide. Do not attempt to make out to yourself or me that our principles entirely accord. They do not, nor cannot at present, (for people's views as I said are not to be altered in a moment).

Will you come here for some days on Monday. I hope so. Lord L— is very anxious to see & converse with you; & was vexed that the Rail called him away on Tuesdy before he had heard from yourself your own views about the recent affair.

I sadly want your Calculus of Functions. So Pray get it for me. I cannot understand the Examples.

I have ventured inserting to one passage of Note G a small Foot-Note, which I am not sure is quite tenable. I say in it that the engine is remarkedly well adapted to include the whole Calculus of Finite Differences. & I allude to the computation of the Bernoullian Numbers by means of the Difference of Nothing, as a beautiful example for its processes.[203] I hope it is correctly the case.

This letter is sadly blotted & corrected. Never mind that however. I wonder if you will choose to retain the lady-fairy in your service or not.

Yours ever most sincerely,
A.A.L.

[203] See A. De Morgan 1842b, 308.

I.144 1843-08-15? Note by Babbage.

Morning 15[204]
Saw AAL this morning and refused all the conditions.

I.145 1843-08-15. Ada Lovelace to Lady Byron.

St. James' Sqre.
5 o'clock, Tuesdy, 15th Augst[205]

[...] W— said that if I had time, I might as well write you a line from Town. I think he said he had not mentioned to you in his this morning, that I was unexpectedly summoned by the printers, who needed a further supervision; & as it is actually to be out I understand tomorrow, there was no time for post communications.

No one can estimate the trouble & interminable labour of having to revise the printing of mathematical formulae. This is a pleasant prospect for the future; as I suppose many hundreds & thousands of such formulae will come forth from my pen, in one way or another.

You will receive a few copies (amongst a hundred that are printed separately for me). But where & how shall I send them [...].

I am uncertain as yet how the Babbage business will end. He has written unkindly to me. For many reasons however, I still desire to work upon his subjects & affairs, if I can do so with any reasonable prospect of peace. I have written to him therefore, very explicitly; stating my own conditions, without which I positively refuse to take any further part in conjunction with him, upon any subject whatever. He has so strong an idea, I suspect, of the advantage of having my pen as his servant, that he will probably yield; tho' I demand very strong concessions.

If he does consent to what I propose, I shall probably be enabled to keep him out of much hot water; & to bring his engine to a consummation, (which all I have seen of him & his habits the last 3 months, makes me scarcely anticipate it ever will be, unless some one really exercises a strong coercive influence over him). He is beyond measure careless & desultory at times. I shall be willing to be his Whipper-in during the next 3 years, if I see fair prospect of success. Much of this is W's suggestion; (altho' W— thinks B's conduct to me has recently been very blameable) [...]

I.146 1843. C. Babbage to Ada Lovelace.

My Dear Lady Lovelace[206]

I much fear the drawings will not be very intelligible. They were never published and only a few proofs were taken. I will endeavour to find a complete set and bring them with me on Monday.

The three papers you mention were all published in the Phil. Transactions. I have been to see if Maynard the Math. Bookseller can procure them for you. I have no copy except those of the two first papers interleaved, with which I will trust you with, hoping it may be returned. Ever most truly yours

C. Babbage
18 Aug. 1843
Dorset St.

[204] British Library Add MS 37192 f422. Annotation.
[205] Lovelace Papers, 42.
[206] Lovelace Papers, 168.

I.147 1843-08-25. C. Wheatstone to Ada Lovelace.

Dear Lady Lovelace,[207]

I called yesterday at the printers and was informed that a separate copy of your paper had been forwarded by post to Ockham and the new number of the Scientific Memoirs sent to St. James' Square.

I made no stipulations whatever with Mr. Taylor when he sent Menabrea's paper for translation, and the only arrangement made since has been that you should be allowed to have for the mere expense of paper and print 100 copies and Mr. Babbage 50. He is desirous of knowing what to do with the remaining 99 copies.

If you have made up your mind not to undertake the translation of Seebeck's memoir in the Transactions of the Berlin Academy, the volume which I sent for the purpose had better be returned to Mr. Taylor.

I have so much on my hands at present that I am unable to leave home, but should you remain at Ockham over next week it will give me great pleasure to pay you a visit on the ensuing Saturday (Sept 2d). I fear however, from what I hear, that you will have left before then for the Coast of England.

I.148 1843-08-28. Babbage to Menabrea.

My dear Sir,[208]

I avail myself of the kindness of Lady Murray to convey to you a translation with notes of your admirable explanation of the analytical engine. You will probably have received the rough proofs I sent to Turin by my son and I am now at liberty to give you the name of your fair commentator: she is the countess of Lovelace, the only daughter of your great poet Lord Byron.

Should you be in Turin during the short visit of lady Murray, you will do me a great favor by pointing out to her the scenery most deserving attention in your beautiful country.

Accept, my dear Sir, the sincere expression of my esteem and regard. I am your, etc.

CHARLES BABBAGE,
1, Dorset street, Manchester square, London.
28 august 1843.

I.149 1843-09-09. C. Babbage to AL.

My Dear Lady Lovelace[209]

I find it quite in vain to wait until I have leisure so I have resolved that I will leave all other things undone and set out for Ashley taking with me papers enough to enable me to forget this world and all its troubles and if possible its multitudinous Charlatans—everything in short but the Enchantress of Number.

My only impediment would be my Mother's health which is not at this moment quite so good as I could wish.

Are you at Ashley? and is it still convenient with all your other arrangements that I should join you there? and will next Wednesday or next Thursday or any other day

[207] Lovelace Papers, 175.
[208] Menabrea 1884, 181.
[209] Lovelace Papers, 168.

suit you; and shall I leave the [...] road at Taunton or at Bridgewater and have you got Arbogast Du Calcul Des Derivations with you there at Ashley. I shall bring some books about that horrible problem—the three bodies which is almost as obscure as the existence of the celebrated book "De Tribus Impostoribus".[210] So if you have Arbogast I will bring something else.

Farewell my dear and much admired Interpreter. Ever Most Truly Yours,

C. Babbage

I.150 1843. Mary Somerville to Ada Lovelace.

A thousand thanks my dear Lady Lovelace for your kindness in sending me a copy of your publication on Babbage's machine[211] the whole of which but especially the notes I have read with the greatest interest and most sincerely congratulate you on the proficiency you have made in the highest branches of mathematics & the clearness with which you have illustrated a very difficult subject and I am sure Mr. Babbage must be truly gratified in having such a commentator. I only fear you have been working too hard as I hear from all our mutual friends that you are thin and not in very strong health. I hope you will take care of yourself & remember that to be able to work long you must take it in moderation [...]

I.151 1843-09-01. Faraday to C. Babbage.

R Institution
11 Sept. 1843[212]

My dear Babbage

I think I may thank you for the translation & though I cannot understand your great work yet I can well comprehend by its effect upon those who do understand it how great a work it is. Wishing you life & health to see it in full action

I am
Yours Most Truly
M. Faraday
C. Babbage Esq
&c &c

I.152 1843-09-09. C. Babbage to Faraday.

My dear Faraday,[213].

I am not quite sure whether I thanked you for a kind note imputing to me unmeritedly the merit of a present you received I conjecture from Lady Lovelace.

I now send you what ought to have accompanied that Translation.

So you will now have to write another note to that Enchantress who has thrown her magical spell around the most abstract of Sciences and has grasped it with a force which few masculine intellects (in our own country at least) could have exerted over it. I remember

[210] *The Three Impostors*, an apocryphal anti-religious treatise.
[211] Lovelace Papers, 174.
[212] British Library Add MS 37192 f445.
[213] Faraday 1996, 164-5

well your first interview with the youthfull fairy which she herself has not forgotten and I am gratefull to you both for making my drawings rooms the Chateau D'Eu[214] of Science.

I am going for a short time to Lord Lovelace's place in Somersetshire. It is a romantic spot on the rocky coast called Ashley about 2 miles from the Post town Porlock.

I am
My dear Faraday
Ever truly yours
C. Babbage.
Dorset St.
Manch. Sq.[215]
9 Sep. 1843

I.153 1843-09-10. Ada Lovelace to C. Babbage.

Sunday, 10 September 1843[216]

My Dear Babbage

Your letter is charming, and Lord L— & I have smiled over it most approbatively. You must forgive me for showing it to him. It contains such simple, honest, unfeigned admiration for myself, that I could not resist giving him the pleasure of seeing it. I send you De Morgan's kind & approving letter about my article. I never expected that he would view my crude young composition so favourably.

You understand that I send you his letter in strictest confidence. He might perhaps not like you to see his remarks about the relative times of the invention of the two engines. I am going to inform him of my grounds of feeling satisfied of the literal correctness of my statement on that point. I cannot say how much his letter has pleased me.

You are a brave man to give yourself wholly up to Fairy-Guidance! I advise you to allow yourself to be unresistingly bewitched, neck, & crop, out & out, whole seas over, &c, &c, &c, by that curious little being! [...]

I.154 1843-09-12. C. Babbage to Ada Lovelace.

My Dear Lady Lovelace[217]

I enclose the letter of De Morgan. I hope to be able to bring with me a final proof of the Statement in which the dates of every fact are clearly set out. It is gratifying to me also; for you know I had arrived at the same conclusion. You should have written an original paper.

The postponement of that will however only render it more perfect. . . . I have been making vast resolutions of unbounded work during my visit; but in preparing the materials I find indications of the necessity of repose after a long period of excitement.

Ever my fair Interpretress,
Your faithful slave
C Babbage
Dorset St.
Tuesday 12 Sep. 1843

[214] The summer residence of King Louis-Philippe in France.
[215] Manchester Square.
[216] British Library Add MS 37192.
[217] Lovelace Papers, 168.

I.155 1843-09-15. Ada Lovelace to Lady Byron.

Friday, 15 September
Ashley[218]

[...] I was so amazed when I read the beginning of your letter this morning, at the absolute identity of the principle expressed in it, to that which I had expressed (tho' in a different form) to you in my last, that I could not help exclaiming to Babbage "this is most extraordinary!" & relating to him the circumstance.

There are however so many very remarkable circumstances that have occurred during the latter weeks and months, & which I cannot pretend to unravel, at present, but which something tells me I shall find the real clue to all in good time, that this merely adds one to their number. I am very much afraid as yet of exerting the powers I know I have over others, & the evidence of which I have certainly been most unwilling to admit, & in fact for a long time considered to be quite fanciful & absurd. I am an utter novice at present, on the threshold of a new world; & I feel my best plan is to wait modestly & humbly for God to teach me, & not to anticipate that teaching, but rather to keep behind [...]. Those powers are not ripe, & my habits & principles are still far too young, now, & green [...]

I had better continue to be simply the High-Priestess of Babbage's Engine, & serve my apprenticeship faithfully therein, before I fancy myself worthy to approach a step higher towards being the High Priestess of God Almighty Himself. And one has a fearful universe of habits to acquire. Many get habits first, & great principles afterwards. But here am I (who always did everything topsy-turvy, & certainly ought to have come into the world feet downwards) with principles highly & largely developped [sic], but habits alas that are either non-existent or positively bad, & (hardly a single good one).

So one must drag one's habits (however unwillingly), after & up to one's principles, before one thinks oneself worth anything whatever.

By the way I am particularly ill at present [...]. But the time will come, when all this will be very different [...] .

I.156 1844-01-02. Lady Byron to De Morgan.

I find both from your letter to Lady Lovelace & from Mrs. De M's to-day, that you are most kindly anxious to prevent Lady L from injuring herself by mathematical study but I feel apprehensive that this caution may be carried too far.[219] I have at all times observed that she was the better for pursuits of that description & if she would but attend to her stomach, her brain would be capable even of more than she has ever imposed on it—notwithstanding the very annoying circumstances which attended her Babbage pubn (I mean the quarrelling, & the pressure as to time) I thought the effects on the whole beneficial to her. I certainly have never known her in so satisfactory a state of mind as since that time, though I should not wish so equal an effort to be frequently demanded of her. After the weeks which I passed with her at Bristol, it is impossible for me to share in the extent of apprehension entertained by some friends who judge from report instead of observation. The consciousness of making progress in science seems to me an essential element in her happiness, & appears not less desirable to Lord Lovelace than to myself [...] .

[218] Lovelace Papers.
[219] Lovelace Papers, 67.

I.157 1844-01-21. De Morgan to Lady Byron.

My Dear Lady Byron[220]

I have received your note and should have answered no further than that I was very glad to find my apprehension (of being a party to doing mischief if I assisted Lady Lovelace's studies without any caution) is unfounded in the opinion of yourself and Lord Lovelace, who must be better judges than I am, on every point of the case but one, and may be on that one. But at the same time it is very necessary that the one point should be properly stated. I have never expressed to Lady Lovelace my opinion of her as a student of these matters: I always feared that it might promote an application to them which might be injurious to a person whose bodily health is not strong. I have therefore contented myself with very good, quite right, and so on. But I feel bound to tell you that the power of thinking on these matters which Lady L. has always shewn from the beginning of my correspondence with her, has been something so utterly out of the common way for any beginner, man or woman, that this power must be duly considered by her friends, with reference to the question whether they should urge or check her obvious determination to try not only to reach but to get beyond, the present bounds of knowledge. If you or Lord L. only think that it is a fancy for that particular kind of knowledge, which, though unusual in its object, may compare in intensity with the usual tastes of a young lady, you do not know the whole. And the same if you think that desire of distinction is the motive, science one of many paths which might be chosen to obtain it. There is easily seen to be the desire of distinction in Lady L's character: but the mathematical turn is one which opportunity must have made her take independently of that.

Had any young beginner, about to go to Cambridge, shewn the same power, I should have prophesied first that his aptitude at grasping the strong points and the real difficulties of first principles would have very much lowered his chance of being senior wrangler, secondly, that they would have certainly made him an original mathematical investigator, perhaps of first-rate eminence.

The tract about Babbage's machine is a pretty thing enough, but I could I think produce a series of extracts, out of Lady Lovelace's first queries upon new subjects, which would make a mathematician see that it was no criterion of what might be expected from her.

All women who have published mathematics hitherto have shewn knowledge, and power of getting it, but no one, except perhaps (I speak doubtfully) Maria Agnesi, has wrestled with difficulties and shewn a man's strength in getting over them. The reason is obvious: the very great tension of mind which they require is beyond the strength of a woman's physical power of application. Lady L. has unquestionably as much power as would require all the strength of a man's constitution to bear the fatigue of thought to which it will unquestionably lead her. It is very well now, when the subject has not entirely engrossed her attention: by and bye when, as always happens, the whole of the thoughts are continually and entirely concentrated upon them, the struggle between the mind and body will begin.

Perhaps you think that Lady L. will, like Mrs. Somerville, go on in a course of regulated study, duly mixed with the enjoyment of society, the ordinary cares of life &c &c. But Mrs. Somerville's mind never led her into other than the details of mathematical work. Lady L. will take quite a different route.

It makes me smile to think of Mrs. Somerville's quiet acquiescence in ignorance of the nature of force, saying "it is dv/dt" (a math. formula for it) "and that is all we know about the matter"—and to imagine Lady L. reading this, much less writing it.

Having now I think quite explained that you must consider Lady L's case as a peculiar one I will leave it to your better judgment, supplied with facts, only begging that this note

[220] Lovelace Papers, 339.

may be confidential.

All here pretty well: I hope your house is free from illness and remain Dear Lady Bryon

Yours very truly
A De Morgan
Janry. 21/44

I.158 1844-03-27. Menabrea to Babbage.

March 27, 1844[221]

Sir,

Your son has given me the English translation of the memoir that I had published in the Bibliothèque Universelle of Geneva about your admirable analytical machine. I was not expecting the very flattering reception that this work has received in England. For the rest, I attribute all the success to the importance of the subject that I treated and if I have hope that this work will not be useless, I base it on the learned notes with which they were kind enough to accompany this translation; I will not describe to you the pleasant surprise I experienced when I learned that the author of these commentaries was Lady Lovelace. The illustrious name of this amiable scholar was well known to me; but I did not think that the most rebellious questions of analysis would become so docile under her graceful hand. I beg you, Sir, to kindly offer to this noble Lady the expression of my respectful homage, of my gratitude and my admiration.

According to what your son has told me, your machine would be close to being completed; I wish this as much for the glory of human intelligence as for your particular satisfaction. Besides the immense advantages that science will derive from it, I believe it will give rise to a good industrial speculation.

I have the pleasure of seeing from time to time, Sir, your son who is very occupied with the Genoa railway.[222] I believe there is not yet a decision in this regard. The financial question is moreover the only obstacle that the project encounters.

Milan brings together, this year, the scholars of Italy and I propose to receive them with the hospitality that befits such a beautiful city. Will we have the honor of seeing you?

Please, Sir, accept the expression of sincere and respectful attachment of your very humble and very obedient servant

L. F. MENABREA
Turin, March 27, 1844

To Mr. Charles Babbage member of the Royal Society of London
Dorset Street
Manchester Square
London

I.159 1844-07-16. Menabrea to Babbage.

July 16, 1844[223]

Sir,

---

[221] From the French translations in Bulferetti 1968a and Menabrea 1884.
[222] Herschel Babbage.
[223] Bulferetti 1968a.

Baron Chiodo, General Commander in Chief of Military Engineering of the Sardinian States, is going to London to examine public works.

I take the liberty of recommending him to your kindness so that you might have the goodness to give him the necessary directions in his research. What particularly occupies him at present in our country is the serious matter of railways since there is talk of building them between Genoa, Milan and Turin. General Chiodo, whose enlightened counsel the government has called upon in this important affair, knows of the fine works of your son on this subject. We hope they will be executed, which gives us also the hope of possessing you once again in Piedmont.

I think that my letter from last April will have reached you and that since then your admirable analytical engine will have made new steps toward its realization. I had the pleasure of seeing, about a month and a half ago, sir, your son in Genoa. He was preparing to make a trip to Livorno for some maritime construction planned in that city.

Please remind your younger son, whom I met during his passage through Turin, of my remembrance. Also deign to offer my humble homage to the illustrious Lady Lovelace and accept the expression of respectful sentiments from your devoted servant and friend

L. F. MENABREA
Turin, July 16, 1844

To Mr. Charles Babbage member of the Royal Society of London
Dorset Street
Manchester Square n. 1
London

I.160 1844-10-16. Ada Lovelace to Faraday.

Ashley Combe
Porlock
Somerset
16th Octr. 1844[224]

Dear Mr Faraday,

I have never yet thanked you for the little paper you sent me this spring. I read it with the deepest attention & interest, & it has suggested to me some very curious (& perhaps important) considerations for my own future use as an Analyst, considerations which fell in with some previous trains of ideas I had been long gradually forming, but which you have called into more tangible existence in my mind.

Perhaps no one has read your paper with such full appreciation as myself of its practical bearings; or has valued it so justly, both for its contents, & as presented to me by its Author, for whom I entertain an esteem little short of reverence.

I must now however proceed to the immediate object I have in view in writing to you. I am going to address you with a frankness & an intimacy, which the mere number of hours we have ever passed together in actual personal presence, could not warrant in ordinary cases.

There are instances (very few however) where the peculiar mutual suitability of character, & of objects in life, justify mind speaking at once to mind, & leaving out of consideration the ordinarily very necessary & useful barriers of external convenances.

[224] Faraday 1996, 253-5.

You will be kind enough to think of me simply as one of God's children. The mere accidents of my being an inhabitant of this particular planet, of this particular corner of it England, & of my wearing the female form, (with a human coronet to boot stuck at the apex), these constitute only one particular case of the general formula in which God has chosen to create moral beings fitted to hold relations with each other, & with Him.

I have long been vowed to the Temple—the Temple of Truth, Nature, Science! And every year I take vows more strict, till now I am just entering those portals & those mysteries which cut of [sic] all retreat, & bind my very life & soul to unwearied & undivided science at its altars henceforward. I hope to die the High-Priestess of God's works as manifested on this earth, & to earn a right to bequeath to my posterity the following motto, "Dei Naturaeque Interpres".

The initiation however for this, is of the severest & longest description; and may be beyond my powers.

All this is highly metaphysical; but there is some literal truth in it also, in my case. This, if we hold in future more of intercourse together, you will by degrees yourself perceive.

For many years I have desired to be admitted to intercourse & friendship with you; & to become in some respects your disciple. But I determined to with-hold myself from any advances towards such an object, until I should feel I was worthy of the privilege; & until the progress of my own mind should be such that the hours & thoughts bestowed by you on me should be in reality an advancement of your objects as well as mine. I think that time has now arrived. I think that any assistance & intercourse you may consent to honour me with, would pay full interest to yourself &. for your own scientific & moral purposes in this world.

Were I not persuaded of this, I assure you I could not be so presumptuous as to expect you would direct your mind one instant on my account from its main & noble purposes; and I can only be justified in suggesting it, if the attention given me is in fact not a diversion from these.

To be again metaphysical: I have no business to deflect the wave of your existence by the smallest curve even, from it's course. But if my wave can in certain of its points follow & touch yours; so that in fact they mutually add to each other's force at those points, then all is welL

To come to the practical matter, to which all this is but the prelude. I am anxious to go thro' all your Researches, with the advantage of your showing me the examples & experiments practically, as I study each paper. It would be to me an inestimable advantage, for some of my objects.

Should you decide on bestowing on me so great a favour, I would arrange to be in Town (after my return to Surrey) on certain appointed days, say once a week or a fortnight, or something of the sort. During the intervals I should study the subjects accordingly, & should perhaps sometimes write to you on them. I need not say that my hours would be wholly at your orders, of course. My own great scientific object (but this is strictly confidential) is the study of the Nervous System, & it's relations with the more occult influences of nature.

By a masterly union of the highest abstract analysis, with most skilful & varied courses of experimental & practical science, such as I have already sketched out in my mind, I expect to bring the actions of the nervous & vital system within the domain of mathematical science, & possibly to discover some great vital law of molecular action, similar for the universe of life, to gravitation for the sidereal universe. I need not say that this (if it can be accomplished at all), is the work of a life-time. I have a very world of heterogeneous materials & considerations to bring by degrees, thro' years of labour, into one great focus;—all to bear on this (as yet) nebulous beacon of mine, but which I hope to crystallise before my death into a radiant & burning star for the light of mankind.

One reason why I desire to become more intimate with you, is my opinion of your moral

& religious feelings.

Do you agree with me, as I rather expect you do, in the impression that the highest & most penetrating degree of intellect, that species of it which is alone fitted to deal with the more subtle & occult agents of nature, is unattainable excepting thro' a high spiritual & moral development; far higher than it is usual even to aim at. I believe there is a connexion between the two that is not understood or suspected by mankind.

You are the only philosopher I have ever seen, who gave me the impression of feeling this in its full force.

Am I mistaken? All I can say is that if you & I do think alike on this point, I suspect we are very nearly a singular couple in the scientific world. I do not know to what particular sect of Christians you belong, or whether to any, nor do I think that much matters.

I am myself a Unitarian Christian; as far as regards some of their views of Christ that is.

But in truth, I cannot be said to be anything but myself. In some points I am Swedenborgian in feelings. Again in others I am slightly Roman Catholic; & I have also my alliance with the older Rosecrucians. Pray don't exclaim, "What patch-work!" None are wholly right, or wholly wrong; so in fact from all I call a somewhat on others.

I make you no apology for this long & free letter. I feel that none is needed.

I remain
Yours sincerely
Augusta Ada Lovelace

I.161 1844-10-24. Faraday to Ada Lovelace.

Royal Institution
24 Octr. 1844.[225]

Your letter ought to have been answered before but there are two circumstances which have caused delay, its high character and my want of health, for since I returned from a very forced journey to Durham I have been under the Doctor's hands: I am quickly recovering and now have the difficult pleasure of writing to you. I need not say how much I value your letter, you can feel that and even if it were possible that you did not, no words of mine would convey the consciousness to you:—the thanks which I owe you can only properly be acknowledged by an open & sincere reply and the absence of all conventional phrase. I wonder that with your high object, and with views, determinations and hopes consistent with it, all of which are justified by the mind and powers which you possess; which latter are not known to yourself only but, as I say in perfect simplicity, are now made fully manifest to others;-I wonder, that you should think as I believe you do of me. But whilst I wonder, and at the same time feel fully conscious of my true position amongst them that think, and know how unworthy I am of such estimation, I still receive it with gratitude from you, as much for the deep kindness as for that proportion of the praise which I may perhaps think myself entitled to; and which is the more valuable because of the worthiness of the giver.

That with your deep devotion to your object you will attain it I do not doubt. Not that I think your aspirations will not grow with your increasing state of knowledge and even faster than it:—but you must be continually passing from the known to the unknown, and the brightness of that which will become known as compared to the dullness or rather obscurity which now surrounds it; will be, and is worthy to be, your expected reward:—and though I may not live to see you attain even what your mind now desires, yet it will be a continually recurring thought in my imaginings, that, if you have life given you you will do so.

[225] Faraday 1996, 264-6. See also Lovelace Papers, 171, and IEE MS SC 3.

That I should rejoice to aid you in your purpose you cannot doubt; but nature is against you. You have all the confidence of unbaulked health & youth both in body & mind; I am a labourer of many years' standing made daily to feel my wearing out. You, with increasing acquisitions of knowledge, enlarge your views and intentions; I, though I may gain from day to day some little maturity of thought, feel the decay of powers, and am curtailing to a continual process of lessening my intentions and contracting my pursuits. Many a fair discovery stands before me in thought which I once intended, and even now desire, to work out; but I lose all hope respecting them, when I turn my thoughts to that one which is in hand and see how slowly, for want of time and physical powers, it advances, and how likely it is to be, not only a barrier between me & the many beyond in intellectual view, but even the last upon the list of those practically wrought out. Understand me in this:—I am not saying that my mind is wearing out; but those physico-mental faculties by which the mind and body are kept in conjunction and work together, and especially the memory, fail me; and hence a limitation of all that I was once able to perform into a much smaller extent than heretofore. It is this which has had a great effect in moulding portions of my later life, has tended to withdraw me from the communion & pursuits of men of science my contemporaries, has lessened the number of points of investigation, (that might at some time have become discoveries,) which I now pursue; and which, in conjunction with its effects, makes me say, most unwillingly, that I dare not undertake what you propose, to go with you through even my own experiments. You do not know, and should not now but that I have no concealment on this point from you, how often I have to go to my medical friend to speak of giddiness and reeling of the head &c, and how often he has to bid me cease from restless thoughts and mental occupation and retire to the seaside and to inaction.

If I were with you I could talk for hours of your letter and its contents, though it would do my head no good, for it is a most fertile source of thoughts to my mind; and whether we might differ upon this or that point or not I am sure we should not disagree. I should be glad to think that high mental powers insured something like a high moral sense, but have often been grieved to see the contrary as also on the other hand my spirit has been cheered by observing in some lowly & uninstructed creature such a healthful & honourable & dignified mind as made one in love with human nature. When that which is good mentally & morally meet in one being that that being is more fitted to work out & manifest the glory of God in the creation I fully admit.

You speak of religion & here you will be sadly disappointed in me. You will perhaps remember that I guessed & not very far aside your tendency in this respect. Your confidence in me claims in return mine to you which indeed I have no hesitation to give on fitting occasions but these I think are very few for in my mind religious conversation is generally in vain. There is no philosophy in my religions I am of a very small & despised sect of Christians known, if known at all, as Sandemanians and our hope is founded on the faith that is in Christ. But though the natural works of God can never by any possibility come in contradiction with the higher things that belong to our future existence, and must with every thing concerning Him ever glorify him still I do not think it at all necessary to tie the study of the natural sciences & religion together and in my intercourse with my fellow creatures that which is religious & that which is philosophical have ever been two distinct things.

And now my dear Lady I must conclude until I see you in town being indeed

Your true and faithful Servant
M. Faraday

I.162 1844-10-24. Ada Lovelace to Faraday.

Ashley Combe
Porlock
Somerset
Thursday 24th Octr[226]

Dear Mr Faraday,

Since I despatched my long letter to you a few days ago, it has occurred to me that I omitted the mention of two very essential points.

One is that I hope you will on no account press yourself to answer me. I ought to have specified this very particularly. My communication to you is of a nature to suggest many & various considerations. You have no time to waste in needless letter-writing. I do hope therefore that you will not answer it at all until you can do so after sufficient time has elapsed to make a reply quite satisfactory to yourself. And in the nature of that reply, consider yourself & not me. Do what is natural & agreeable to you to do; & think of me as a mere instrument. If you knew more of me, I think you would be inclined to believe (as I do) that a certain degree of scientific cooperation between us, would materially subserve the interests & objects of both parties. And for you to know truly my intellectual characteristics, I believe I ought to begin by becoming in a manner your pupil, (independant of the great advantage that would be to me in itself).

But you may feel you had rather not go out of your way for the chance of a perhaps dubious result.

My second point is that I have an ulterior object respecting your "Researches". I want to make a review & abstract of them, (for the Quarterly perhaps, or some such vehicle). I have therefore another motive for wishing go thro' them under your immediate guidance, besides the mere advantages to myself. I should wish to take those points of view in my review, which should be most consonant with your mind & views. I think in this work I could be useful to you & to science.

Now I will detain you no longer. I really am half afraid I have been unwarrantably intrusive in writing to you as I did. I mean rather, that I have exposed myself to the appearance of it. For as to the motives & incentives, they have been anything but what could displease you, if you can truly perceive them. But am I not expecting a great deal too much, in supposing that your very slight personal intercourse hitherto, could enable you to do so?

I remain
Yours truly
Augusta Ada Lovelace

I.163 1844-11-08. Ada Lovelace to Faraday.

Ashley Combe
Porlock
Somerset
Friday 8th Novr[227]

Dear Mr Faraday,

[226] Faraday 1996, 267. See also IEE MS SC 2.
[227] Faraday 1996, 276. See also IEE MS SC 2.

How much have you said to me by simply not answering my last letter! (I mean the one I wrote in reply to your two).

I feel, merely from this circumstance, how well you understand me, & how much & deeply you have felt on the subject of our communications. I seem to know all this, as well as if you had written volumes.

My object in this letter is not to give you the further remarks which I promised, & which I still defer for a little; but to explain to you some of my arrangements as they now stand.

I am going to Mr Crosse at Bromfield on the 20th, to stay a few days. I enclose you a letter from him, received today.

Now it has struck me that perhaps I might be useful to you while there. So, if there is any subject or subjects you would like me especially to observe, enquire, or consider, for you, during my stay, give me your orders, & I will show myself a faithful & intelligent agent.

Again, I have (I know) great influence with Mr C-; & hints or suggestions from me he would be apt to consider & follow. So perhaps you might like to suggest to me, how to suggest usefully to him!

Now, let none of this trouble you. I would not miss a possible opportunity of being useful to you, or useful to Science, (Science whose bride I am)! But none of this may fall in with your views, your feelings, or your powers, at this moment. If so, just put up Mr C-'s letter to return me, & say on the envelope merely 3 words to let me know that you have nothing to charge me with, but that I have not plagued you.

Remember that I am to be your assistant, if I am any thing at all to you. And you are never to let yourself be plagued by me, or in connection with me. I expect to be in Town somewhere about the 25th. I hope that you will then give me an evening, & will also let me come one morning to your philosopher's cell, just to look about me there, (but not to interrupt you). In this way, I shall soon find out how best to serve you. Trust me for that. Your mind is one so peculiarly interesting to me, that I earnestly desire to brighten & to lighten its earthly labours & anxieties.

I am again preparing some publications; & have engagements of that kind until July next. Then, I mean (unless you discourage me) to undertake your Researches for review, or at any rate as my hinge & centre for an Electrical Article (probably for the Quarterly). Meanwhile, all of this winter I shall make you & your views & occupations my study more or less; that is, in as far as you will let me.

When once I have the key to your prevailing line of interests, I know I can be a minister in your service, in a thousand ways; & that without interfering with my own views & avocations, but so as to make all my services to you co-operate with my pursuits & progress.

One thing I am now writing is again in the service of my friend Babbage; but of a very different description from the last. How I wish by the bye that his mind were (in some points) more akin to that which in you I so admire. There is in him too much self, & too little of what I would term divine love. He is capable of great devotion to individuals now & then, but there is not in him the Scripture the Christian love which I so adore. He is a great man, & has a great intellect; but he would have been a yet greater, & his intellect would have been yet higher & wider, had his moral feelings corresponded.

This, by the way, is an illustration of some of the feelings & principles I slightly sketched in my first letter to you. Babbage has I think in him the capacity however, for a high moral & religious development, tho' I doubt if it will be in this world. I am much attached to & interested in him, up to a certain point; but then comes something which creates a bar to that degree & kind of affection I feel for several others.

You see I cannot help writing to you as if you were indeed "the friend of many years" I named you in my last.

Ever most sincerely yours

A.A. Lovelace

I.164 1844-11-10. Ada Lovelace to Faraday.

Ashley Combe
Porlock
Somerset
Sunday 10th Novr[228]

Dear Mr Faraday,

I am exceedingly tickled with your comparison of yourself to a tortoise.[229] It has excited all my fun (& I assure you I have no little of that in me). I am also struck with the forcible truth of your designation of my character of mind: "elasticity of intellect". It is indeed the very truth, most happily put into language.

You have excited in my mind a ridiculous, but not ungraceful, allegorical picture, viz that of a quiet demure plodding tortoise, with a beautiful fairy gambolling round it in a thousand radiant & varying hues; the tortoise crying out, "Fairy, fairy, I am not like you. I cannot at pleasure assume a thousand aerial shapes & expand myself over the face of the universe. Fairy, fairy, have mercy on me, & remember I am but a tortoise".

What says the kind gentle fairy in reply? Somewhat as follows: "Good tortoise, then I will be to you of plain & sober hue. I can assume what form I like. I will be the beautiful phantom, glowing in colour & eloquence, when so you order me. But I will now be a little quiet brown bird at your side, & gently let you teach me how to know & aid you. But my wand is yours at pleasure, & into your hands I deliver it for your use".

So speaks the ladye-fairy. Well! forgive my fun & metaphor. I am as basinfull of sportive mirth as of science, you know. One friend of mine always calls me "The fairy". This is what suggested to me a comparison so very complimentary to myself, that I ought scarcely to have originated it, & did not.

My Lord, my husband, always calls me The Bird. So strong is the feeling in his mind of the similarity, that I really think he almost sees me under the form of that lovely species of God's creatures. An eminent physician said I was as like an Arab mare as could be, both in physical constitution & in disposition.

Oh dear! I have no business to entertain you with all this egotism. But really, your note has set the vein a-going; the allegorical species of my mind; & I can hardly stop.

Another friend always calls me "the sprite". I don't know how it is, but nobody calls me or thinks me a simple mortall But is not a mathematical fairy a very curious compound? So it is however; & there is the secret. I have in me both north & south polarities! & dame Nature has played a freak, & for once united in a beautiful harmony in one individual, faculties that usually do not co-exist.

The deepest Analysis, with brilliant Imagination for ever playing on the surface of those grave & fathomless depths. Analysis is for ever calling the gambols of my Imagination to order & confining them to a subjection to the true & the logical.

To try & be quite grave & matter of fact again: You are right. We must meet, & talk. So, will the Tortoise be good enough to appoint an evening for coming to this mischievous sprite of a thousand forms; (for venturing within her magic circle)? Thursday 28th, at 6

[228] Faraday 1996, 279-81. See also IEE MS SC 2.

[229] In a letter Faraday wrote to Christian Friedrich Schoenbein on 25 October 1844. 'I am working but I cannot get on. Work is now slow with me & one thing or another is continually occurring to prevent progress. I think I must at last entirely shut out this world for now my progress is slow & like that of the tortoise—a trifle to others stops me altogether'. It is not known if Faraday had tried the lettuce cure.

o'clock. Will that do? And I will give you some human earthly tea or coffee, or any other mortal food you may choose; & not insist on your having nothing but the dew & ambrosia I am obliged to feed my fairy nature on.

I hope this letter won't make you very angry. I mean it to make you laugh. At any rate you must, I am sure, perceive that you have a most good-natured creature to deal with in

Yours most sincerely
A.A.L.

P.S. It has struck me that as I have not specified St James' Sqre, for our meeting on the 28th, you may perhaps suppose that I have a habit of magically transporting myself over space, & that I expect you & others to do the same, & that you are to be here at six o'clock on the 28th. So I just add a P.S to say I do not demand quite so much; & that we are to meet in St James' Sqre; as becomes other mortals, & tortoises especially.

I.165 1844-11-13. Ada Lovelace to Faraday.

Wed. day
13th Novr.[230]

Agreed then!

I say Amen to every line. You must know that the fairy always takes to "hard, rough, & Straightforward" mortals.

You have yet to learn how well she can understand those who are yet wholly different from herself. That power is part of her fairy-gift. Meanwhile you would think she paid dear for her gifts (such as they are), if you indeed knew all she goes thro', (& has gone thro' in her life). I have been very ill the last 3 weeks even;—one of my too frequent attacks. I am subject to two very delightful ailments. Gastritis (alias Gastric Fever); and Asthma. I am getting better again now. I said nothing about this to you, till I was mending.

I am able to work on, more or less, under even dreadful states of suffering. Had I not this ballast, this hard food for my ideas, I should indeed be a tempest-tossed bark under my most extraordinary physical constitution. When quite unfit for speaking even, I can reflect, & sometimes write, & this keeps me all compos and happy.

My strength never fails. The vis vitae seems intense in my case. It is this curious fact which made the distinguished M.D. I mentioned, compare me to the pure Arab Blood-Horse.

I have no weak or bad constitution, tho' I have one prone to sufferings—quite a different thing. There is the wing power & endurance of the true blood far more than in those who perhaps never suffer at all. Intense vitality engenders intensely acute susceptibilities sometimes. So that in fact it is my strength that is my weakness.

I wanted you to know a little of all this, before we meet. I gain experience every month of my life; & thro' treatment expect to ward of[f] illnesses more & more effectually. You can have no idea of my fearful susceptibility to certain agents put into the stomach. And almost all the ordinary medical rules are reversed in my case. Excepting my mother, & you, I have not mentioned to any correspondent, that I have been ill lately. To say the truth, it is so very familiar an occurrence that I say as little about it even as possible. I have learned how to manage so as to keep it chiefly behind the scenes; a great comfort! I know no one so strong, so vitalized, as myself; notwithstanding all this.

There is nothing for you to answer in this letter.

How glad I shall be when we meet! Meanwhile think of me as a pleasant thought, if you can; & always remembering my mixed fairy & mortal nature. I believe that of the latter

[230] Faraday 1996, 282-3. See also IEE MS SC 2.

there is very little however. Pain & suffering have sharply & sternly schooled me, & made me dead to all merely earthly things.

Ever yours then,
The Ladye Fairy.

I.166 1844-12-01. Ada Lovelace to Faraday.

Dear Mr Faraday,[231]

My engagements & also the rain prevented my looking in on you either yesterday or Friday. I do not know when you will return to Town, but I shall beg sometime for another hour or two from you in the evening. We must talk business & science next time.

Pray do not fancy you are too plain & rough for me. You do not know how I have (for years past) dispensed with all the ordinary appurtenances & ceremonials of my situation in life. I have in fact roughed it thoroughly, as they say; & I am unused to adulation & attention; so much so that such are irksome & intolerable to me.

I have been highly trained, so to speak, to habits of independance or rather of self-dependance, in all respects. Often I go about incog to all sorts of places, and travelling &c, quite unattended. Of course I have thus to deal with various characters, to fight my own way, and to take care of myself in many ways.

All this is a very necessary training for one who dares to grasp at ambitious intellectual enterprises. There is an important connexion, tho' not a direct one, between these handy, these anti-luxurious, practical habits, & intellectual success. These practical daily habits, engender a certain state of mind & feeling, very promotive of all that is a sine qua non to success in high intellectual training.

I write in a hurry, & perhaps I am not over clear in expressing to you all I mean. If so, we can talk about it.

Ever yours
A.A.L.
St. James' Sqre
Sunday

[231] Faraday 1996, 291-2.

Appendix J

Alumni Cantabrigiensis.

Entries from J. Venn and J. A. Venn 1922–1954.

J.1 Babbage, Charles.

Adm. pens. at TRINITY, Apr. 21, 1810.
S. of Benjamin, banker, of Teignmouth, Devon.
B. Dec. 26, [1791].
Matric. Michs. 1810.
Migrated to Peterhouse, Apr. 7, 1812; B.A. 1814; M.A. 1817.
Lucasian Professor of Mathematics, 1828–39, but delivered no lectures.
[Re-adm. at Trinity, Apr. 7, 1824.]
F.R.S., 1816.

A skilled mechanician, who conceived the idea of calculating numerical tables by machinery; devoted 37 years (and a large share of his fortune) to the invention and perfection of a calculating machine based on 'the method of differences'; obtained a government grant but was refused further aid for an improved design on the ground of expense.[1]

One of the founders of the Astronomical Society, 1820; Secretary, 1820–4; later, successively Vice-President, foreign Secretary, and a Member of the Council. Associated with the starting of the British Association, 1831, one of the trustees, 1832–8; originator of the Statistical Section, at the Cambridge Meeting, 1833. Principal founder of the [Royal] Statistical Society of London, 1834.

Attended the meeting of Naturforscher at Berlin, 1828; and the Scientific Congress at Turin, 1840. A member of numerous scientific bodies all over the world. Author, scientific.

Died Oct. 18, 1871. *(T. A. Walker, 387; D.N.B.)*

J.2 Bird, Charles Smith.

Adm. pens. (age 20) at TRINITY, Oct. 21, 1815. S. of William [West Indian merchant]. B. May. 28, 1795, at Union Street, Liverpool. School, Macclesfield, Cheshire (Dr Davies). Matric. Lent, 1816; Scholar, 1818; 2nd Smith's Prizeman; B.A. (3rd Wrangler) 1820; M.A. 1827. Fellow, 1820. Ord. deacon, 1822; priest, 1823; C. of Burghfield, Berks., 1823-44; took private pupils – one of them being Lord Macaulay.

---

[1] Ed: this is false.

J.3 Greenwood, Thomas.

Adm. pens. at ST JOHN'S, Mar. 31, 1810. Of Middlesex. [2nd s. of Thomas, merchant, of London. B. 1790.] Matric. Michs. 1810; B.A. 1815; M.A. 1831. Adm. at Gray's Inn, Mar. 14, 1809. Called to the Bar, 1817. Bencher, 1837-71; Treasurer, 1841-2. Fellow of University of Durham and reader in history and polite literature there, 1833-71. Author, Cathedra Petri, a survey of Papal Supremacy; Position and Prospects of the Churches of Great Britain and Ireland with reference to the establishment of a Roman Catholic hierarchy; History of Great Latin Patriarchate, etc. Died Nov. 1, 1871, at 14, Westbourne Terrace, Hyde Park, London. (Boase, I. 1230; D.N.B.)

J.4 Herschel, John Frederick William.

Adm. pens. at St John's, Aug. 2, 1808. Of Slough, Bucks. [Only s. of Sir William, the astronomer (for whom see *D.N.B.*). B. Mar. 7, 1792, at Slough. School, Eton.] Matric. Michs. 1809; Scholar, 1809; B.A. (*Senior Wrangler* and 1st Smith's prize) 1813; M.A. 1816. Fellow, 1813–29. Hon. Fellow, 1867. D.C.L., Oxford, 1839. Adm. at Lincoln's Inn, Jan. 21, 1814. Co-founder of The Analytical Society at Cambridge, 1813. Candidate for the chair of chemistry. Author of many mathematical works, some of which were responsible for 'the restoration of mathematical science in England, by introducing the differential notation and Continental methods of analysis.' Completed his father's work on double stars and published a memoir on their orbits. His *Outlines of Astronomy* became a standard introduction to the subject. F.R.S., 1813; Copley medal, 1821; Secretary, 1824–7; Royal medal, 1833, 1836, 1840. Took an active part in formation of the R. Astronomical Society and was its first foreign Secretary and also its President; his *Catalogue of Double Stars* was honoured by its gold medal and by the Lalande Prize, 1825. One of the eight foreign members of the Institute of France. Married, Mar. 3, 1829, Margaret Brodie, dau. of the Rev. Dr Alexander Stewart, of Ross-shire. In 1833 took his family to S. Africa in order to survey the heavens from the southern hemisphere. Lived at Feldhausen, near Cape Town. Returned to England, 1838. Published *Results of Astronomical Observations made during the years 1834–8 at the Cape of Good Hope* in 1847. Created Baronet, July 17, 1838. Retired to Collingwood, Hawkhurst, Kent, 1840. 'His position as celestial explorer is unique.' An unsurpassed observer whose breadth of knowledge and power of vividly describing what he saw added incalculably to the value of his observations. Died May 11, 1871, at Collingwood. Buried May 19, in Westminster Abbey. Father of Alexander S. (1854). (Gunther, *Early Science in Cambridge*, 173; Boase, I. 1447; *D.N.B.*; *Inns of Court*; Burke, *P. and B.*, 1940; *Eagle*, VII. 359.)

J.5 Higman, John Philips.

Adm. pens. (age 19) at TRINITY, June 2, 1812. S. of Henry. B. [1793], at Kingsbridge, Devon. School, Kingsbridge (Mr Wilcocks). Matric. Michs. 1812; Scholar, 1814; B.A. (3rd Wrangler) 1816; M.A. 1819. Fellow, 1818; Tutor, 1822-34. F.R.S., May 23, 1820. F.R.A.S. Ord. priest (Ely) Oct. 28, 1821. R. of Fakenham, Norfolk, 1834-55. Author, A Syllabus of the Differential and Integral Calculus, 1826. Died Aug. 7, 1855, at Cambridge Terrace, Hyde Park, London. (Boase, I. 1466; Clergy List; G. Mag., 1855, II. 329.)

J.6 Robinson, Thomas.

Adm. pens. at TRINITY, July 9, 1808. [Youngest s. of the Rev. Thomas (1767). B. Oct. 20, 1790. School, Rugby.] Matric. Michs. 1808; Scholar, 1810; Bell Scholar, 1810; B.A. (13th

Wrangler) 1813; M.A. 1816; D.D. 1844. Lord Almoner's Professor of Arabic, 1837-54. Adm. ad eundem at Oxford, 1839. Ord. deacon (Ely) 1815; priest (Lincoln) 1816; Chaplain to the Archbishop of York. Chaplain at Seroor and Poona, where he studied Persian, and began to translate the Bible into that language. Chaplain to Bishop Heber, being transferred to Bengal, and remaining his constant companion. Archdeacon of Madras, 1828-35.[2] Bencher of the Middle Temple, 1845. Master of the Temple, 1845-69. R. of Therfield, Herts., 1853-60. Preb. of St Paul's, 1847-54. Canon of Rochester, 1854-73. Author of a translation of the Old Testament into Persian and of many books and pamphlets. Died May 13, 1873, at Rochester. Father of Charles E. R. (1847) and of the next. (Rugby Sch. Reg.; Boase, III. 231; Crockford; Middle Temple Bench Book; F. Penny; D.N.B.; Al. Oxon., which erroneously calls him Fellow of Trinity.)

J.7 Whittaker, John William.

Adm. pens. (age 19) at ST JOHN'S, Mar. 31, 1810. S. of William, gent., of Bradford, Yorkshire (and Sarah Buck). [B. c. 1790, in Manchester.] School, Bradford. Matric. Michs. 1810; Scholar, 1810; B.A. (13 th Wrangler) 1814; M.A. 1817; B.D. 1824; D.D. 1830. Fellow, 1814-26. Ord. deacon (Ely), Mar. 26, 1820; priest, Nov. 12, 1820. V. of Blackburn, Lancashire, 1822-54. Hon. Canon of Manchester, 1852-4 One of the founders of the Royal Astronomical Society, 1820. Married, June 20, 1825, Mary Haughton, eldest dau. of William Feilden (afterwards Bart.), of Feniscowles, Lancashire. Author, An Historical and Critical Inquiry into the Interpretation of the Hebrew Scriptures; Justification by Faith; The Catholic Church; etc. Died Aug. 23, 1854. Father of Philip H. (1850). (Boase, III. 1327; Foster, Index Eccles.; Clergy List; D.N.B.)

J.8 Wollaston, William Hyde.

Adm. pens. (age 15) at CAIUS, July 6, 1782. [3rd] s. of the Rev. Francis (1748), [R.] of Chislehurst, Kent (and Althea Hyde). B. [Aug. 6, 1766] at East Dereham, Norfolk. Schools, Lewisham, Kent and Charterhouse. Matric. Easter, 1784; Scholar; M.B. 1788; M.D. 1793. Fellow, 1787-1828. F.R.C.P. 1795; Censor, 1798, 1824. F.R.S. 1793; Royal medal, 1828; Copley medal, 1804; Secretary of the Society, 1804-16. F.G.S., 1812. In medical practice at Huntingdon, 1789, and later at Bury St Edmunds; finally settled in London, 1797, relinquishing medical practice for experimental science. Invented the goniometer for measuring the angles of crystals. Patented the camera lucida, 1807, which led to the invention of photography. Discovered the malleability of platinum, which produced him £30,000. A great friend of Sir Humphry Davy. Published 56 papers on pathology, physiology, chemistry, optics, astronomy, mineralogy, crystallography, astronomy, electricity, mechanics, and botany, and almost every paper marks a distinct advance in the particular science involved.' Died Dec. 22, 1828; buried at Chislehurst. Left money to the Royal Society and endowed the 'Wollaston Fund' of the Geological Society. Brother of Francis J. H. (1779) and Charles H. (1789). (Venn, II. 106; Carthusian Worthies; G. Mag., 1828, II. 550; D.N.B.; Munk, II. 438-42; Boase, Biblio. Cornub., 900; Gunther, Early Science in Cambridge; Leonard Blomefield, 41-5.)

[2] The DNB states he was appointed in 1826, Lee 1897.

Bibliography.

Adams, Beverley (2023). *The World's First Computer Programmer. The Extraordinary Life of Ada Lovelace*. Pen & Sword History. Yorkshire–Philadelphia: Pen & Sword Books Ltd, 2023.

Airy, George Biddell (1896). *Autobiography of Sir George Biddell Airy, K.C.B., M.A., LL.D., D.C.L., F.R.S., F.R.A.S., Honorary Fellow of Trinity College, Cambridge, Astronomer Royal from 1836 to 1881*. Ed. by Wilfrid Airy. Cambridge: Cambridge University Press, 1896.

Andryane, Alexander (1838). *Memoirs of a Prisoner of State in the Fortress of Spielberg*. English. Trans. by Fortunato Prandi. Translated from Mémoires d'un prisonnier d'état au Spielberg. 13, Pall-Mall East, London: H. Hooper, 1838.

Andryane, Alexandre (1837). *Mémoires d'un prisonnier d'état au Spielberg. Mêlés d'aventures, d'anecdotes, d'épisodes sur les prisons d'Allemagne et sur l'Italie moderne*. French. 4 vols. Published in four volumes from 1837-1838. Chronicles the author's imprisonment (1823-1832) in the Spielberg fortress as a political prisoner. Paris: Ladvocat, 1837.

Anonymous (1840). "On Mechanical Notation, or the Method of Expressing by Signs the Action of Machinery". In: *Encyclopaedia Britannica*. 7th ed. Part III—Description of Machines. Edinburgh: Adam and Charles Black, 1840. Chap. I, p. 445.

Appleyard, Rollo (1926). "Pioneers of Electrical Communication: Hans Christian Oersted—VI". In: *Electrical Communication* 5 (1926). Part VI of a series on pioneers of electrical communication.

Attar, Karen, Adrian Rice, and Christopher Stray, eds. (2024). *Augustus De Morgan, Polymath: New Perspectives on his Life and Legacy*. Cambridge, UK: Open Book Publishers, 2024. DOI: 10.11647/OBP.0408.

Babbage, Charles (1815). "An Essay towards the Calculus of Functions". In: *Philosophical Transactions of the Royal Society of London* 105 (1815), pp. 389–423.

— (1816). "An Essay towards the Calculus of Functions. Part II". In: *Philosophical Transactions of the Royal Society of London* 106 (1816), pp. 179–256.

— (1820). *Observations on the Notation Employed in the Calculus of Functions*. From the Transactions of the Cambridge Philosophical Society. Cambridge: J. Smith, Printer to the University, 1820.

— (1822a). *A Letter to Sir Humphry Davy, Bart. PRS on the Application of Machinery to the Purpose of Calculating and Printing Mathematical Tables*. Content is dated 1822-07-03. London, UK, 1822.

— (1822b). "A Note Respecting the Application of Machinery to the Calculation of Astronomical Tables". In: *Memoirs of the Royal Astronomical Society* 1.2 (June 1822), pp. 309–309.

— (1822c). "Observations on the Application of Machinery to the Computation of Mathematical Tables". In: *Memoirs of the Royal Astronomical Society* 1 (1822), pp. 311–314.

Babbage, Charles (1824a). *On the Determination of the General Term of a New Class of Infinite Series.* Offprint from Transactions of the Cambridge Philosophical Society, printed by J. Smith, Printer to the University. Cambridge, 1824.

— (1824b). "On the Determination of the General Term of a New Class of Infinite Series". In: *Transactions of the Cambridge Philosophical Society* 2.1 (May 3, 1824). (Incorrectly dated 1826 in Hyman 1982), pp. 218–225.

— (1826). "On a Method of Expressing by Signs the Action of Machinery". In: *Philosophical Transactions of the Royal Society* 116.1/3 (1826), pp. 250–265.

— (1827a). "Notice respecting some Errors Common to many Tables of Logarithms". In: *Memoirs of the Astronomical Society* 3 (1827), pp. 65–67.

— (1827b). *Table of Logarithms of the Natural Numbers from* 1 *to* 108000. London, UK, 1827.

— (1830). *Reflections on the Decline of Science in England, and some of its Causes.* London, UK, 1830.

— (1831). *Specimen of Logarithmic Tables, printed with different coloured inks and on variously-coloured papers.* One single copy printed. Experiment to ascertain the tints of paper and colours of inks least fatiguing to the eye. Contains 151 variously-coloured papers with stereotyped logarithm tables printed in 10 different ink colours across 21 volumes. London: n.p., 1831.

— (1832). *On the economy of machinery and manufactures.* Pall Mall East, London: Charles Knight, 1832, pp. xvi + 320 + 2.

— (1833a). *Traité sur l'économie des machines et des manufactures. [On the economy of machinery and manufactures].* French. Trans. by Éd. Biot. Paris, France: Bachelier, 1833, pp. xvi + 515.

— (1833b). *Ueber Maschinen- und Fabrikenwesen [On the economy of machinery and manufactures].* German. Trans. by G. Friedenberg. Berlin, Germany: Im Verlage der Stuhrschen Buchhandlung, 1833, pp. iii + 462.

— (1835). "Une lettre à M. Quetelet de M. Ch. Babbage relativement à la machine à calculer. [A letter to Mr. Quetelet from Mr. Ch. Babbage relative to the calculating machine]". French. In: *Acad. Roy. Bruxelles, Bulletins* 2 (1835), pp. 123–126.

— (1837). *The Ninth Bridgewater Treatise: a Fragment.* 1st ed. London, UK: John Murray, 1837.

— (1838). *The Ninth Bridgewater Treatise: a Fragment.* 2nd ed. Revised and extended. London, UK: John Murray, 1838.

— (1843). "Addition to the Memoir of M. Menabrea on the Analytical Engine". In: *Philosophical Magazine* 23 (Sept. 1843), pp. 235–9.

— (1851). *The Exposition of 1851; or, Views of the Industry, the Science and the Government of England.* London: John Murray, 1851.

— (1864). *Passages from the Life of a Philosopher.* (Reprinted in 1968 by Dawsons of Pall Mall). London: Longman, Green, 1864.

— (1889). *Babbage's Calculating Engines. Being a Collection of Papers Relating to them; their History and Construction.* Ed. by Henry Babbage. London: E. and F.N. Spon, 1889.

— (1961). *Charles Babbage and His Calculating Engine. Selected writings of Charles Babbage and others.* Ed. by Philip Morrison and Emily Morrison. New York: Dover, 1961.

— (1973). "On the Mathematical Powers of the Calculating Engine". In: *Origins of Digital Computers. Selected Papers.* Ed. by B. Randell. From an MS written in 1837. Berlin: Springer, 1973.

— (1989a). *The analytical engine and mechanical notation.* Ed. by Martin Campbell-Kelly. Vol. 3. The Works of Charles Babbage. London: William Pickering, 1989.

— (1989b). *The Difference Engine and Table Making*. Ed. by Martin Campbell-Kelly. Vol. 2. The Works of Charles Babbage. London: William Pickering, 1989.
— (1989c). "The Science of Number Reduced to Mechanism". In: ed. by Martin Campbell-Kelly. Vol. 2. The Works of Charles Babbage. Unpublished manuscript from 1822. London: William Pickering, 1989, pp. 15–32.
— (1822–1825). "Journal 1820-5". Archived 2025-03-21. Waseda University, Tokyo, Japan, 1822–1825. URL: `https://babbagedifferenceengine.blogspot.com/2014/09/extracts-from-babbages-journal.html`. Transcribed by CJ Roberts.
Babbage, Charles, J. F. W. Herschel, and George Peacock (1813). *Memoirs of the Analytical Society*. Cambridge: J. Smith, Printer to the University, 1813.
Baily, Francis (1823). "On Mr. Babbage's New Machine for Calculating and Printing Mathematical and Astronomical Tables". In: *Astronomische Nachrichten* 2.46 (Nov. 28, 1823). Also reprinted in the Philosophical Magazine, 1824-05, 409-422, pp. 409–422.
Barrett, Eaton Stannard (1813). *The Heroine, or, Adventures of Cherubina*. 3 volumes. London: Printed for H. Colburn, 1813.
Baruffi, Giuseppe Filippo (1841). *Pellegrinazioni Autunnali ed Opuscoli*. [Autumn Pilgrimages and Minor Works]. Italian. Vol. 1. Turin: Tipografia Cassone e Marzorati, 1841.
Baum, Joan (1986). *The Calculating Passion of Ada Byron*. New York: Archon Books, 1986.
Becher, Harvey W. (1995). "Radicals, Whigs and conservatives: the middle and lower classes in the analytical revolution at Cambridge in the age of aristocracy". In: *British Journal for the History of Science* 28 (1995), pp. 405–426.
Belanger, Jay and Dorothy Stein (2005). "Shadowy vision: spanners in the mechanization of mathematics". In: *Historia Mathematica* 32.1 (Feb. 2005), pp. 76–93.
Binet, Alfred and Théophile Simon (1916). *The Development of Intelligence in Children*. Translated by Elizabeth Kite. Vineland, NJ: Publications of the training school at Vineland, 1916.
Birle, Ann Lucas and Lisa A. Francavilla, eds. (2011). *Thomas Jefferson's Granddaughter in Queen Victoria's England: The Travel Diary of Ellen Wayles Coolidge, 1838–1839*. Published in association with the Massachusetts Historical Society, Boston and the Thomas Jefferson Foundation, Charlottesville. Charlottesville: University of Virginia Press, 2011.
Bowden, B. V., ed. (1953). *Faster Than Thought: A Symposium on Digital Computing Machines*. First Edition. New York, Toronto, and London: Pitman Publishing Corporation, 1953.
— (1960). "He invented the computer — before its time". In: *Think* (July 1960). An account of the author's search for information about Lady Lovelace and a biographical sketch of Charles Babbage, pp. 28–32.
Bowers, Brian (2001). *Sir Charles Wheatstone FRS 1802-1875*. Published in association with the Science Museum, London. London: The Institution of Electrical Engineers, 2001.
Brinkley, John (1807). "An investigation of the general term of an important series in the inverse method of finite differences". In: *Philosophical Transactions of the Royal Society of London* 97 (1807), pp. 114–132. DOI: `10.1098/rstl.1807.0007`. URL: `https://doi.org/10.1098/rstl.1807.0007`.
British Association for the Advancement of Science (1839). *Report of the Eighth Meeting of the British Association for the Advancement of Science; Held at Newcastle in August 1838*. London: John Murray, 1839, p. xxv.

Bromley, Allan G. (1980a). *Charles Babbage's Analytical Engine, 1838.* Technical Report 165. University of Sydney Basser Department of Computer Science, 1980.

— (1980b). *The Functional Algorithms and Control of Charles Babbage's Analytical Engine Circa 1838.* Technical Report 167. University of Sydney Basser Department of Computer Science, 1980.

— (1980c). *The Mechanism of Charles Babbage's Analytical Engine Circa 1838.* Technical Report 166. University of Sydney Basser Department of Computer Science, 1980.

— (1982a). "Charles Babbage's Analytical Engine, 1838". In: *IEEE Annals of the History of Computing* 4.3 (1982), pp. 196–217.

— (1982b). "Introduction". In: *Babbage's Calculating Engines.* Ed. by H. P. Babbage. Vol. 2. Charles Babbage Institute Reprint Series for the History of Computing. Los Angeles: Tomash Publishers, 1982.

— (1987). "The Evolution of Babbage's Calculating Engines". In: *IEEE Annals of the History of Computing* 2.9 (1987), pp. 113–136.

— (1990). "Difference and Analytical Engines". In: *Computing Before Computers.* Ed. by W. Asprey. Ames, Iowa: Iowa State University Press, 1990, pp. 59–98.

— (1991). *The Babbage Papers in the Science Museum Library.* London: Science Museum, 1991.

— (2000). "Babbage's Analytical Engine Plans 28 and 28a - The Programmer's Interface". In: *IEEE Annals of the History of Computing* 22.4 (2000), pp. 5–19.

Brooke, Charles (1829). *A Synopsis of the Principal Formulae and Results of Pure Mathematics.* Cambridge: Cambridge, 1829.

Bulferetti, Luigi (1968a). "Les correspondants italiens de Charles Babbage [The Italian correspondents of Charles Babbage]". French. In: *Actes du XIIe Congrès International d'Histoire des Sciences* XI (1968), pp. 29–38.

— (1968b). "Un Amico di Charles Babbage: Fortunato Prandi [A Friend of Charles Babbage: Fortunato Prandi]". Italian. In: *Memorie dell'Istituto Lombardo - Accademia di Scienze e Lettere. Classe di lettere, scienze morali e storiche* 30.2 (1968), pp. 83–116.

— (1970). "G. Plana e Ch. Babbage nella costruzione dell'apparato motore del piroscafo Tripoli [Giovanni Plana and Charles Babbage in the construction of the engine apparatus of the steamship Tripoli]". In: *Ricerche Storiche ed Economiche in Memoria di Corrado Barbagallo.* Ed. by Luigi De Rosa. Napoli: Edizioni scientifiche italiane, 1970.

Buxton, Henry Wilmot (1988). *Memoir of the Life and Labours of the Late Charles Babbage, Esq. F.R.S.* Ed. by Anthony Hyman. Cambridge, MA: MIT Press, 1988.

Chiaverini, Jennifer (2017). *Enchantress of Numbers: A Novel of Ada Lovelace.* Dutton, 2017.

Clausen, Thomas (1840). "Zusammenstellung der periodischen Gleichungen von Bürgs, Burckhardts und Damoiseau's Mondtafeln". German. In: *Astronomische Nachrichten* 17.406 (1840), p. 337.

Cobbett, William (1823). *The Parson and the Boy; or, The Fire-Shovel Hats in a Bustle: Being an Account of the Trial of the Reverend Thomas Jephson, Doctor of Divinity, and Tutor at Cambridge, on a Charge of Attempting to Commit an Unnatural Crime on James Welsh, a Labourer.* Anonymous publication, authorship attributed to William Cobbett. London: Printed and published by William Benbow, 1823.

Colebrooke, Henry Thomas (1824). "On presenting the Gold Medal to CHARLES BABBAGE, Esq. F.R.S." In: *Memoirs of the Astronomical Society of London* 1.2 (Feb. 1824). Published in 1825, pp. 509–512.

Collier, Bruce (1970). "The Little Engines that Could've. The Calculating Machines of Charles Babbage". PhD. Harvard, Aug. 1970. URL: `http://robroy.dyndns.info/collier/index.html` (visited on 03/09/2024).

— (1990). *The Little Engines that Could've. The Calculating Machines of Charles Babbage.* Reissue with a new foreword. Greenwood Press, 1990.
Comrie, L. J. (1946). "A Manual of Operation for the Automatic Sequence Controlled Calculator". In: *Nature* 158 (1946), pp. 567–568.
Corina, John Grenville (1994). "William King (1786–1865): Physician and Father of the Co-Operative Movement". In: *Journal of Medical Biography* 2.3 (1994). DOI: 10.1177/096777209400200309.
Crosse, Mrs. Andrew (1890). "John Kenyon and His Friends". In: *Temple Bar* 88 (1890), pp. 490–491.
Darwin, Francis, ed. (1887). *The Life and Letters of Charles Darwin, Including an Autobiographical Chapter.* Vol. 1. London: John Murray, 1887.
Davis, Martin and Virginia Davis (2005). "Mistaken Ancestry: The Jacquard and the Computer". In: *TEXTILE* 3.1 (2005), pp. 76–87. DOI: 10.2752/147597505778052594. URL: https://doi.org/10.2752/147597505778052594.
De Morgan, Augustus (1835). *The Elements of Algebra.* London: John Taylor, 1835.
— (1840a). "Negative and Impossible Quantities". In: *Penny Cyclopaedia.* Ed. by Society for the Diffusion of Useful Knowledge. Vol. 16. London: Charles Knight, 1840, pp. 130–7.
— (1840b). "Numbers of Bernoulli". In: *Penny Cyclopaedia.* Ed. by Society for the Diffusion of Useful Knowledge. Vol. 16. London: Charles Knight, 1840, pp. 365–6.
— (1842a). "On the Foundation of Algebra". In: *Transactions of the Cambridge Philosophical Society* 7 (1842).
— (1842b). *The Differential and Integral Calculus.* Library of Useful Knowledge. Previously issued in installments from 1836 onward. London: Robert Baldwin, 1842.
De Morgan, Sophia Elizabeth (1882). *Memoir of Augustus De Morgan.* London: Longmans, Green and Co., 1882, p. 428.
Diffusion of Useful Knowledge, Society for the, ed. (1840). *Penny Cyclopaedia.* London: Charles Knight, 1840.
Dubbey, J. M. (1978). *The Mathematical Work of Charles Babbage.* Cambridge: Cambridge University Press, 1978.
Elwin, Malcolm (1975). *Lord Byron's Family. Annabella, Ada and Augusta 1816-1824.* Ed. by Peter Thomson. Posthumous. John Murray, 1975.
Enros, Philip C. (1983). "The Analytical Society (1812–1813): Precursor of the Renewal of Cambridge Mathematics". In: *Historia Mathematica* 10 (1983), pp. 24–47.
Enros, Philip Charles (1979). "The Analytical Society: Mathematics at Cambridge University in the Early Nineteenth Century". PhD Thesis. Toronto, Canada: Institute for the History, Philosophy of Science, and Technology, 1979.
Essinger, J. (2014). *A Female Genius: How Ada Lovelace Lord Byron's daughter started the computer age.* London: Gibson Square, 2014.
Faraday, Michael (1996). *The Correspondence of Michael Faraday: 1841–8.* Ed. by Frank A. J. L. James. Vol. 3. The Institution of Engineering and Technology, 1996.
Farr, William (1871). "The President's Inaugural Address". In: *Journal of the Statistical Society of London* 34 (1871). Obituary of Babbage, pp. 411–417.
Fisch, Menachem (2014). "Babbage's two lives". In: *British Journal for the History of Science* 47.1 (Mar. 2014). First published online 4 July 2013, pp. 95–118. DOI: 10.1017/S000708741300037X.
Franksen, Ole Immanuel (1981). "Mr. Babbage, the Difference Engine, and the Problem of Notation: An Account of the Origin of Recursiveness and Conditionals in Computer Programming". In: *International Journal of Engineering Science* 19.12 (1981), pp. 1657–1694.

Füegi, John and Jo Francis (2003). "Lovelace & Babbage and the Creation of the 1843 'Notes'". In: *IEEE Annals of the History of Computing* 25.4 (2003), pp. 16–26.
— (2004). "Lovelace, Ada". In: *Dictionary of Nineteenth-Century British Scientists.* Ed. by Alan Lightman. 4 vols. London: Thoemmes, 2004, pp. 1268–72.
Galton, Francis (1849). *The Telotype. A Printing Electric Telegraph.* London: J. Ridgway, 1849.
— (1850). *The Telotype. A Printing Electric Telegraph.* Post-dated. London: J. Weale, June 1850.
— (1855). *The Art of Travel; or, Shifts and Contrivances Available in Wild Countries.* First edition. London: John Murray, 1855.
— (1856). *The Art of Travel; or, Shifts and Contrivances Available in Wild Countries.* Second edition. London: John Murray, 1856.
— (1908). *Memories of My Life.* London: Methuen, 1908.
Gandy, Robin (1995). "The Confluence of Ideas in 1936". In: *The Universal Turing Machine: A Half-Century Survey.* Ed. by Rolf Herken. 2nd ed. Springer-Verlag Wien, 1995.
Glaschick, Rainer (2016). *Ada Lovelace's Calculation of Bernoulli's Numbers.* Sept. 29, 2016. URL: `https://rclab.de/_media/analyticalengine/aal_noteg_glaschick_v1.2.pdf` (visited on 12/03/2024).
Grattan-Guinness, Ivor (1992). "Charles Babbage as an Algorithmic Thinker". In: *IEEE Annals of the History of Computing* 14.3 (1992), pp. 34–48.
— (2000). *The Rainbow of Mathematics: A History of the Mathematical Sciences.* Originally published as: The Norton History of the Mathematical Sciences. New York: W. W. Norton & Company, May 2000.
Gridgeman, Norman T. (1970). "Babbage, Charles". In: *Dictionary of Scientific Biography.* Ed. by Charles Coulston Gillispie. Vol. 1. New York: Scribner, 1970.
Grier, D. A. (2004). "Appreciating Charles Babbage: Emails between Allan Bromley and Maurice Wilkes". In: *IEEE Annals of the History of Computing* 26.4 (2004), pp. 62–70.
Hardy, G. H. (1949). *Divergent Series.* Oxford: Oxford University Press, 1949.
"Earliest Stereoscopic Portraits" (1854). In: *The Journal of the Photographic Society of London: Containing the Transactions of the Society and a General Record of Photographic Art and Science* 1 (1854). Ed. by Arthur Henfrey, p. 200.
Henry, Mary A. (1899). "Henry's Place in the Telegraph of England". In: *The Conservative Review* 2 (Aug. 1899).
Herken, Rolf, ed. (1995). *The Universal Turing Machine: A Half-Century Survey.* 2nd ed. Springer-Verlag Wien, 1995.
Herschel, J. F. W. (2007). *Cape Landscapes: Sir John Herschel's Sketches 1834–1838.* Ed. by Brian Warner. Cape Town: University of Cape Town Press, 2007.
Herschel, J. F. W., David S. Evans, et al. (1969). *Herschel at the Cape: Diaries and Correspondence of Sir John Herschel, 1834–1838.* Vol. 3. South African Biographical and Historical Studies. Cape Town: A. A. Balkema, 1969, pp. xxxviii + 398.
Herschel, J. F. W., Mira F. Hardcastle, et al. (1932). *Herschel Letters, Manuscripts, etc., 1838-1932.* A photo-reproduced collection of materials apparently gathered for a planned biography and catalog of documents relating to Sir John F. W. Herschel and his family. Covers primarily 1838–1871, detailing Herschel's life at Slough, England, his travels in Europe, and his scientific experiments, instruments, and observations. Largely anecdotal. n.p., 1932.
Hollings, Christopher, Ursula Martin, and Adrian Rice (2017a). "The early mathematical education of Ada Lovelace". In: *BSHM Bulletin: Journal of the British Society for the History of Mathematics* (June 1, 2017). Online only.

— (2017b). "The Lovelace–De Morgan mathematical correspondence: A critical re-appraisal". In: *Historia Mathematica* 44.3 (Aug. 2017), pp. 202–231.
— (2018). *Ada Lovelace. The Making of a Computer Scientist.* Oxford: Bodleian Library Press, 2018.
Huskey, Velma R. and Harry D. Huskey (1980). "Lady Lovelace and Charles Babbage". In: *Annals of the History of Computing* 2.4 (Oct. 1980), pp. 299–329.
Hyman, Anthony (1982). *Charles Babbage. Pioneer of the Computer.* Oxford: Oxford University Press, 1982.
— ed. (1988). *Science and Reform. Selected Works of Charles Babbage.* Cambridge: Cambridge University Press, 1988.
— (1997). *The Babbage Pages.* Archived in full. 1997. URL: `https://web.archive.org/web/20240204234410/http://projects.exeter.ac.uk/babbage/` (visited on 03/16/2024).
Laplace, Pierre-Simon (1831). *The Mechanism of the Heavens.* Ed. by Mary Somerville. Translated from the French Traité de mécanique céleste. London: John Murray, 1831.
Lardner, Dionysius (1828). *An Analytical Treatise on Plane and Spherical Trigonometry, and the Analysis of Angular Sections: Designed for the Use of Students in the University of London.* London: John Taylor, 1828.
— (1834). "Babbage's Calculating Engine". In: *Edinburgh Review* 59.120 (July 1834), pp. 263–327.
Lee, Sidney (1897). "Robinson, Thomas (1790–1873)". In: *Dictionary of National Biography, 1885–1900.* Ed. by Leslie Stephen and Sidney Lee. Vol. 49. London: Smith, Elder & Co., 1897, pp. 123–124.
Lightman, Alan, ed. (2004). *Dictionary of Nineteenth-Century British Scientists.* 4 vols. London: Thoemmes, 2004.
Lindgren, Michael (1990). *Glory and Failure: The Difference Engines of Johann Müller, Charles Babbage and Georg and Edward Scheutz.* Trans. by Craig G. McKay. MIT Press, 1990.
— (2023). *Glory and Failure: The Difference Engines of Johann Müller, Charles Babbage and Georg and Edward Scheutz.* Trans. by Craig G. McKay. Linköping Studies in Arts and Science. Linköping University, 2023.
Lloyd, Mary (1880). *Sunny Memories: Containing Personal Recollections of Some Celebrated Characters.* London: Women's Printing Society, 1880.
Losano, Mario G. (2014). *La macchina da calcolo di Babbage a Torino.* Italian. Firenze: Leo S. Olschki, 2014.
Ludgate, P. E. (1909). "On a proposed analytical machine". In: *Scientific Proceedings of the Royal Society of Dublin* 12.9 (1909), pp. 77–91.
— (1914). "Automatic calculating engines". In: *Napier tercentenary celebration: Handbook of the exhibition.* Ed. by E. M. Horsburgh. Brief description of the Babbage Analytical Engine and Ludgate's own work on difference and analytical engines. Edinburgh: Royal Society of Edinburgh, 1914, pp. 124–127.
Lyell, Charles (1881). *Life, Letters and Journals of Sir Charles Lyell, Bart.* Ed. by Katharine Murray Lyell. Vol. 2. 2 vols. Albemarle Street, London: John Murray, 1881.
Macready, William Charles (1875). *Macready's Reminiscences and Selections from His Diaries and Letters.* Ed. by Frederick Pollock. Vol. 2. 2 vols. Editor credit: Sir Frederick Pollock, Bart., One of his Executors. London: Macmillan and Co., 1875.
Maquet, Albert (1965). *L'astronome royal de Turin Giovanni Plana (1781-1864): Un homme, une carrière, un destin.* [The Royal Astronomer of Turin Giovanni Plana (1781-1864): A Man, a Career, a Destiny]. French. Vol. XXXVI. Mémoires de l'Académie royale de Belgique, Classe des Sciences 1792. Published bilingually by the

Royal Academy of Belgium (Académie royale de Belgique/Koninklijke Academie van België). Rue Ducale, 1: Palais des Académies, 1965.

Martineau, Harriet (1877). *Harriet Martineau's Autobiography.* Ed. by Maria Weston Chapman. Vol. 1. New York: 85 Fifth Avenue, The Riverside Press, Cambridge. Boston: Houghton, Mifflin and Company, 1877.

Menabrea, L. F. (1841). "On Cavendish's experiment". In: *The London, Edinburgh and Dublin Philosophical Magazine and Journal of Science* 19.121 (1841), pp. 62–63. DOI: 10.1080/14786444108650367.

— (1842). "Notions sur la Machine Analytique de M. Charles Babbage". French. In: *Bibliothèque Universelle de Génève* 41 (1842), pp. 352–76.

— (1843a). *Sketch of the Analytical Engine Invented by Charles Babbage, esq.* Trans. by Ada Lovelace. Pamphlet issue of Menabrea 1843b. London: Richard and John E. Taylor, 1843.

— (1843b). "Sketch of the Analytical Engine Invented by Charles Babbage, esq. By L. F. Menabrea, of Turin, Officer of the Military Engineers". Trans. by Ada Lovelace. In: *Scientific Memoirs* 3 (1843). With notes by Charles Babbage, not credited, and Ada Lovelace, pp. 666–731.

— (1855). "Variétés". French. In: *Cosmos, Les Mondes; revue hebdomadaire des sciences et de leurs applications* 6 (Apr. 20, 1855), pp. 421–2.

— (1884). "Sur la machine analytique de Charles Babbage". French. In: *Comptes Rendus, Acad'emie des Sciences Paris* (June 1884). 28 July 1884, pp. 179–182.

— (1971). *Memorie.* Italian. Ed. by L. Briguglio and L. Bulferetti. Florence: Giunti, 1971, p. 291.

Metropolis, N. and J. Worlton (1980). "A Trilogy of Errors in the History of Computing". In: *Annals of the History of Computing* 2.1 (1980), pp. 49–59.

Misa, Thomas J. (2016). "Charles Babbage, Ada Lovelace and the Bernoulli Numbers". In: *Ada's Legacy: Cultures of Computing from the Victorian to the Digital Age.* Ed. by R. Hammerman and A. L. Russell. Association for Computing Machinery, 2016. Chap. 1, pp. 11–31.

Moore, Dorothy Langley (1977). *Ada, Countess of Lovelace. Byron's legitimate daughter.* London: John Murray, 1977.

Morley, John (1870). "Carlyle". In: *Fortnightly Review* 8.43 (July 1870). New series, pp. 1–28.

— (1871). *Critical Miscellanies.* First Series. London: Chapman and Hall, 1871.

Morrison, Philip and Emily Morrison (1952). "The Strange Life of Charles Babbage". In: *Scientific American* 186.4 (Apr. 1952), pp. 66–73.

Moseley, Henry (1834). *A Treatise on Mechanics, Applied to the Arts: Including Statics and Hydrostatics.* 1st ed. London: John W. Parker, 1834.

Moseley, Maboth (1964). *Irascible Genius. The Life of Charles Babbage.* (Reprinted, Henry Regnery, 1970). London: Hutchinson & Co, 1964.

Müller, Johann Helfrich (1786). *Beschreibung seiner neu erfundenen Rechenmaschine, nach ihrer Gestalt, ihrem Gebrauch und Nutzen [J. H. Müller's Description of his newly-invented Calculating Machine according to its form, its use and utility].* German. Ed. by Philipp Engel von Klipstein. Frankfurt und Mainz: Varrentrapp Sohn und Wenner, 1786.

Mullett, Charles F. (1948). "Charles Babbage: A Scientific Gadfly". In: *The Scientific Monthly* 67.5 (Nov. 1948), pp. 361–371. URL: http://www.jstor.org/stable/22300.

Norton, Rictor (2015). *Newspaper Reports, 1823: Homosexuality in Nineteenth-Century England: A Sourcebook.* Last updated 6 February 2015. Feb. 2015. URL: http://rictornorton.co.uk/eighteen/1823news.htm (visited on 12/07/2025).

Parshall, Karen Hunger (1989). "Review of: Ada: A Life and a Legacy. A Life and a Legacy". In: *Historia Mathematica* 16 (1989), pp. 94–97.

Peacock, George, J. F. W. Herschel, and Charles Babbage (1820). *Collection of Examples of the Applications of the Differential and Integral Calculus.* Babbage contributed Part III, Examples of the Solution of Functional Equations. Cambridge: J. Smith, Printer to the University, 1820.

Plana, Giovanni (1832). *Théorie du mouvement de la lune.* French. 3 vols. Turin: Impérie Royale, 1832.

Playfair, Lyon (1899). *Memoirs and Correspondence of Lyon Playfair: First Lord Playfair of St. Andrews.* Ed. by Wemyss Reid. New York and London: Harper & Brothers Publishers, 1899.

Randell, B., ed. (1973). *Origins of Digital Computers. Selected Papers.* Berlin: Springer, 1973.

Roberts, CJD (2014a). *[Babbage] Correspondence in the British Library Manuscripts Department.* Archived. Sept. 2014. URL: `https://babbagedifferenceengine.blogspot.com/2014/09/correspondence-in-british-libary_21.html` (visited on 03/21/2025).

— (2014b). *Analysis of the Costs of and Government Expenditure on Babbage's 1st Difference Engine.* Archived. Sept. 21, 2014. URL: `https://babbagedifferenceengine.blogspot.com/2014/09/analysis-of-costs-of-and-government.html` (visited on 03/21/2025).

— (2014c). *Royal Society Herschel Collection.* Archived. Nov. 2014. URL: `https://babbagedifferenceengine.blogspot.com/2014/11/royal-society-herschel-collection.html` (visited on 03/21/2025).

Rosenberg, Jerry M. (1969). *The Computer Prophets.* New York: Macmillan, 1969.

Rouse Ball, W. W. (1889). *A History of the Study of Mathematics at Cambridge.* Cambridge: Cambridge University Press, 1889.

Sabine, Edward (1842). "The translation of foreign scientific memoirs". In: *Report of the Eleventh Meeting of the British Association for the Advancement of Science.* London: John Murray, 1842, p. 329.

Scaife, B. K. P. (1990). "James MacCullagh, M.R.I.A., F.R.S., 1809-47". In: *Proceedings of the Royal Irish Academy. Section C: Archaeology, Celtic Studies, History, Linguistics, Literature* 90C (1990), pp. 67–106. (Visited on 12/20/2024).

Smiles, Samuel (1881). *Brief Biographies.* New York: American Book Exchange, 1881.

Snyder, Laura J. (2011). *The Philosophical Breakfast Club: Four Remarkable Friends Who Transformed Science and Changed the World.* New York: Broadway Books, 2011.

Sotheby, Wilkinson, and Hodge (1872). *Mathematical and Scientific Library of the Late Charles Babbage, of No. 1, Dorset Street, Manchester Square.* Auction catalog. London: C. F. Hodgson & Son, 1872.

Spearman, Charles (1904). "'General Intelligence,' Objectively Determined and Measured". In: *The American Journal of Psychology* 15.2 (Apr. 1904), pp. 201–292.

Stein, Dorothy (1985). *Ada. A Life and a Legacy.* Cambridge, MA: MIT Press, 1985.

Swade, D. D. (2000). *The Cogwheel Brain.* London: Little Brown and Company, 2000.

Swade, Doron (1991). *Charles Babbage and His Calculating Engines.* Reprinted, 1994. London: Science Museum, 1991.

— (2002). *The Difference Engine. Charles Babbage and the Quest to Build the First Computer.* First published in 2000 as The Cogwheel Brain by Little, Brown. London: Penguin, 2002.

— (2003). "Calculation and Tabulation in the Nineteenth Century: Airy Versus Babbage". PhD. University College London, 2003.

Todhunter, I. (1876). *William Whewell, D.D.: An Account of His Writings with Selections from His Literary and Scientific Correspondence.* London: Macmillan and Co., 1876.

Toole, Betty Alexandra (1992). *Ada, the Enchantress of Numbers. A Selection from the Letters of Lord Byron's Daughter and Her Description of the First Computer.* 1st ed. San Francisco: Strawberry Press, 1992.

— (1998). *Ada, the Enchantress of Numbers. Prophet of the Computer Age.* 2nd ed. Revised and abridged. San Francisco: Strawberry Press, 1998.

— (2010). *Ada: The Enchantress of Numbers: Poetical Science.* 3rd ed. Sausalito, California: Critical Connection, 2010.

Tredoux, Gavan (2023a). *Francis Galton's Genius. 1865–1911.* galton.org, 2023.

— (2023b). *Francis Galton's Nature and Nurture. 1822–1865.* galton.org, 2023.

— (2023c). *Sir Francis Sacheverel Darwin.* galton.org, 2023.

Trivers, Robert (2011). *The Folly of Fools: The Logic of Deceit and Self-Deception in Human Life.* Also published in the UK as *Deceit and Self-Deception: Fooling Yourself the Better to Fool Others* (Allen Lane/Penguin). New York: Basic Books, 2011. ISBN: 978-0465027552.

Turvey, Peter J. (1991). "Sir John Herschel and the Abandonment of Charles Babbage's Difference Engine No. 1". In: *Notes and Records of the Royal Society of London* 45.2 (July 1991), pp. 165–176. URL: `http://www.jstor.org/stable/531696`.

Van Sinderen, Alfred W. (1980). "The Printed Papers of Charles Babbage". In: *Annals of the History of Computing* 2.2 (Apr. 1980), pp. 169–185.

Venn, John and J. A. Venn (1922–1954). *Alumni Cantabrigienses: A Biographical List of All Known Students, Graduates and Holders of Office at the University of Cambridge, from the Earliest Times to 1900.* 10 vols. Part I (4 volumes, to 1751) published 1922–1927; Part II (6 volumes, 1752–1900) published 1940–1954. Cambridge: Cambridge University Press, 1922–1954.

Weld, Charles Richard (1848). *History of the Royal Society, with Memoirs of the Presidents.* Vol. 1. Compiled from authentic documents. Author was Barrister-at-Law, Assistant-Secretary and Librarian to the Royal Society. London: John W. Parker, 1848.

Wilkes, Maurice V. (2002). "Charles Babbage and his world". In: *Notes and Records of the Royal Society of London* 56.3 (Sept. 22, 2002), pp. 353–365. DOI: `10.1098/rsnr.2002.0188`.

Williams, Michael R. (1988). "Babbage and Bowditch: A Transatlantic Connection". In: *IEEE Annals of the History of Computing* 9 (1988), pp. 283–290.

— (1992). "Joseph Clement: The First Computer Pioneer". In: *IEEE Annals of the History of Computing* 14.3 (1992), pp. 69–76.

Woodhouse, Robert (1803). *The Principles of Analytical Calculation.* Cambridge: Cambridge University Press, 1803.

Woolley, Benjamin (1999). *The Bride of Science: Romance, Reason and Byron's Daughter.* First published 1999 by Macmillan, an imprint of Macmillan Publishers Ltd. 25 Excelsior Place, London SW1W 5NT: Macmillan, 1999. ISBN: 0-333-72436-4.

Wright, J. M. F. (1825). *Solutions of the Cambridge Problems, from 1800 to 1820.* 2 vols. London: Black, Young, and Young, 1825.

Wright, John Martin Frederick (1827). *Alma Mater; or, Seven Years at the University of Cambridge.* 2 vols. London: Black, Young, and Young, 1827.

Yanes, Javier (2015). "Ada Lovelace: Original and Visionary, but No Programmer". In: *BBVA Open Mind* (Dec. 9, 2015). Archived. URL: `https://web.archive.org/web/20240120001719/https://www.bbvaopenmind.com/en/technology/visionaries/ada-lovelace-original-and-visionary-but-no-programmer/` (visited on 01/20/2024).

Made in the USA
Monee, IL
12 November 2026

b6dc45e6-e0f7-4521-82f5-694b57decc66R01

高寶書版集團
gobooks.com.tw

BK 077
邪教事典
史上最惡名昭彰的九大邪教團體，揭露惡魔領袖與狂熱信徒的致命共生之謎
Cults: Inside the World's Most Notorious Groups and Understanding the People Who Joined Them

作　　者　麥克斯・卡特勒（Max Cutler）、凱文・康利（Kevin Conley）
譯　　者　傅文心
責任編輯　陳柔含
封面設計　黃馨儀
內頁排版　賴姵均
企　　劃　陳玟璇、何嘉雯

發 行 人　朱凱蕾
出　　版　英屬維京群島商高寶國際有限公司台灣分公司
　　　　　Global Group Holdings, Ltd.
地　　址　台北市內湖區洲子街88號3樓
網　　址　gobooks.com.tw
電　　話　(02) 27992788
電　　郵　readers@gobooks.com.tw（讀者服務部）
傳　　真　出版部(02) 27990909　行銷部(02) 27993088
郵政劃撥　19394552
戶　　名　英屬維京群島商高寶國際有限公司台灣分公司
發　　行　英屬維京群島商高寶國際有限公司台灣分公司
法律顧問　永然聯合法律事務所
初版日期　2024年09月

國家圖書館出版品預行編目(CIP)資料

邪教事典：史上最惡名昭彰的九大邪教團體,揭露惡魔領袖與狂熱信徒的致命共生之謎/麥克斯.卡特勒(Max Cutler), 凱文.康利(Kevin Conley)著；傅文心譯. -- 初版. -- 臺北市：英屬維京群島商高寶國際有限公司臺灣分公司, 2024.09
面；　公分. --（Break；BK077）

譯自：Cults：inside the world's most notorious groups and understanding the people who joined them

ISBN 978-626-402-063-3(平裝)

1.CST: 宗教團體　2.CST: 犯罪

213.4　　113011940

culture-74343.

邪教事典

史上最惡名昭彰的九大邪教團體，
揭露惡魔領袖與狂熱信徒的致命共生之謎

Cults:
Inside the World's Most Notorious Groups
and Understanding the People Who Joined Them

麥克斯．卡特勒（Max Cutler）、
凱文．康利（Kevin Conley）　著
傅文心　譯

目錄
Contents

引言

大家都想要相信某事或某人——更崇高的理想、在世神明、天上來的聲音、超越人類的智慧。當人們渴望信仰，需要歸屬，厲害的事就發生了：國家誕生、寺廟教堂樹立、太空人登陸月球。需要歸屬是人類本能，就刻寫在DNA上。人類是群居動物，要生存下去，得仰賴彼此，也依賴中心原則，這些原則讓宗教和政治得以運作。集體行動，激發的情緒透過社群放大，信仰與歸屬可以醉人，令人忘我。

雖然少見，但人性黑暗面顯露時，又該怎麼辦呢？

自我們的Podcast節目成立起，宗旨就是關注虔誠信徒封閉社會的不同小群體（又稱邪教），以及極具魅力卻又駭人的領袖。二〇一七年九月第一集〈邪教崇拜〉（Cults）Podcast上架，立刻榮登Apple Podcast排行榜第一名。自那集首播之後，我們很訝異每一週的每段歷史都受到熱烈迴響，節目四年來下載次數超過五千五百萬次，絲毫沒有停歇之勢。聽眾興趣如潮不止，促使我們搜尋邪教極端信仰的歷史紀實，來滿足聽眾胃口，一週接著一週。這並不會不尋常，某份報告的結論寫道，如果認為自己終於發掘邪教最糟糕的一面，到頭來只會發現更為糟糕的事件，心神為之震懾不已。

多虧每週的Podcast，我們手邊隨時有多份個案研究可用。從一開始，我們就不只是觀察邪教領袖與信徒生活的原始資料，也會檢視他們的心理狀態和動機（雖然我們不是心理學家）。我們花時間拉遠距離，看他們使用哪幾種操縱手法，仔細觀察潛意識趨力，許多人因此走向極端行為，連續殺人、性偏差，甚至集體自殺。

讀者閱讀書中章節，會看到邪教領袖的相似之處，他們在不同年代吸收、引誘信眾，近乎全權掌控信徒，這也驅使領袖測試控制的極限，接著他們會越過極限——通常出於百無聊賴、嗜虐興趣，或因為領袖也開始相信自己的幻想，他們宣稱自己具有神力，能掌控生死。

本書詳述的領袖，每位都獨一無二。他們各有特色，不易混淆——冷酷無情、童年恥辱、抑制的性慾、過度相信自身才能，或是使親近之人心生恐懼能帶來歡愉。但幾乎所有領袖都具備三項明顯特質，也就是所謂的惡毒自戀暗黑三角：缺乏同理心、操控他人、過於愛自己。不管是哪位領袖，我們都無法指出他們是天生如此，還是後天養成，但我們幾乎都能輕易點出影響每位領袖喪失同理心的證據。

即使我們知道每位領袖的生命故事，幾乎挖掘出每件罪行的細節，謎團依然存在。讀者讀得越仔細，就越難回答這一核心問題：這些人具備特殊才能，可以迷倒眾生、滿口謊言、操控他人、魅惑人心、編造真實，究竟是天生惡質，還是生活迫使他們成為怪物？他們有得選嗎？要是他們對自己最糟糕的選擇也從未感到後悔，他們究竟感受到了什麼？

先從查爾斯．曼森（Charles Manson）談起。一九六九年，曼森暗中規劃六起好萊塢凶殺，幾週前，阿姆斯壯才登月，踏出人類的那一小步。古巴裔美國人亞道夫．德．赫穌斯．康斯坦佐（Adolfo de Jesús Constanzo）自小學習牲祭儀式，長大利用儀式來滿足自身需求，在墨西哥市成為毒梟撒旦教的嗜血頭頭，製造大量恐怖狀態、毒品販運和屠殺。巴關・希瑞・羅傑尼希（Bhagwan Shree Rajneesh）確實天賦異稟，說話頭頭是道，但他以此為煙幕，遮掩自身渴求毒品、性交、多台勞斯萊斯，並掩飾內心的暗黑意圖。在一九八〇年代，他還允許信徒於美國國土施放生化武器。在曼森大開殺戒的「迴旋溜滑梯（Helter Skelter）」凶殺事件後九年，吉姆・瓊斯（Jim Jones）撼動了全世界，他在蓋亞那一處叢林營區（又稱瓊斯鎮）監看教徒集體自殺。

性偏差是常見線索，可能只是自尊、權力和無情融合時無可避免的副產品，也是邪教領袖的正字標記。羅克・泰里奧（Roch Thériault）是加拿大某支歸返大地邪教的領袖，自稱摩西，扮成先知，引領信徒（暱稱螞蟻山之子）走入荒野，避開世界末日，幫助信徒生活變得平等、幸福美滿、從罪中得釋放。實行方式則是殘暴的統治，泰里奧毆打信徒、施行手術且不施打麻醉、用剪線器截斷信徒的腳趾、刻意讓信徒挨餓、剝奪睡眠、奴役、性虐待、火燒生殖器，還有其他野蠻手段。大衛・柯瑞許（David Koresh）同樣是誇大狂，而且性愛成癮。柯瑞許接手基督復臨安息日會的大衛教派，把自己塑造成救世主，讓多位女信徒懷孕生子，接著歷經美國菸酒槍炮及爆裂物管理局五十一天圍攻，在一九九三年四月帶著大家走向死亡。

本書蒐集的邪教領袖生命故事，許多結局也都同樣容易預見——他們死於各種暴力，但也有少數幾位自然死亡，通常令他們聲名狼藉的邪教早已解散多時。基斯・拉尼爾（Keith

Raniere）創立NXIVM教團，操縱一群性奴隸，奴隸身上烙印其姓名縮寫。拉尼爾依然活著，被關在專門囚禁性罪犯和戀童癖的聯邦監獄，六十二歲的拉尼爾會在獄中花上許多時間懊悔內省——確切來說是一百二十年。

若領袖和所有信徒一同死去，正如吉姆．瓊斯所為，此舉看似能夠證明過往的真誠。不過，恢復上帝十誡運動教集體自殺事件雖然是史上最慘絕人寰的悲劇，創教者克莉多尼亞．瑪琳達（Credonia Mwerinde）卻沒有躺在近千名信徒的屍體之中。多數信徒一加入運動教，便即刻奉獻俗世財產。這些死亡的信徒生前被關進木造教堂，出口被釘死，接著遭人放火。瑪琳達（她自稱能看見聖母瑪利亞，依此創立了運動教）消失無蹤，自此之後沒有人再見過她，或是運動教積攢的金錢。

這些邪教領袖的自戀不只一次逐漸膨脹，超越自我中心，成為致命力量，足以顛覆甚至凌駕現實。孤立則讓狀況更加危險。一九九七年三月，在加州聖菲牧場（Rancho Santa Fe）的一座門禁社區，天堂之門創立者馬歇爾．阿普懷特（Marshall Applewhite）啟動複雜的逃跑計畫，要讓教徒登上幽浮，跟隨海爾—波普彗星，引領所有人前往嶄新、下一階段的存在，拋下皮囊（他們穿著成套襯衫、運動長褲、Nike運動鞋）。實際上，這是一場集體自殺，帶走了三十九位天堂之門信徒的生命，阿普懷特也在其中。

仔細閱讀本書記錄的邪教領袖生活和受其啟發的信徒的命運，會發現奇怪卻常見的致命共生：信仰、領袖魅力、性變態。這些邪教領袖可能殺了人，但若沒有信徒的盲目奉獻，一切都不可能發生。信徒渴望打破生活常規，甚至打破常理界線。這些引人注目的支配者緊緊掌控為數不多卻經常是肥羊的信眾，善於贏得信任並且用來做壞事，但是信眾執意實踐信仰，且精力無窮無盡。一頁翻過一頁，每個案件都顯示，邪教領袖點燃了火，身陷領袖掌握的人則悲慘地成為燃料。

Chapter 1

羞恥

查爾斯・曼森和曼森家族

一九六九年八月兩晚，曼森和一小群信徒犯下連續凶殺，手法出奇殘忍，全美上下開始恐懼，邪教成為揮之不去的噩夢。曼森的審判廣受媒體報導，其中少不了晚間新聞的聳動報導，刻劃一名迷失青年的形象，這名青年獻身給自由之愛、迷幻旅程、種族戰爭，還隨機殺害電影明星和好萊塢名人。曼森待過青年青年之家，坐過牢，因此幾乎擁有整套犯罪心理學工具包：操控他人、樂於隨機使用暴力、誤以為自己萬分重要，自命不凡。曼森也曾是詐欺犯，他抵達加州時，正值六〇年代反文化運動高潮。曼森將一群輟學逃家的人變成嗜血邪教，教團大開殺戒標誌「花的力量時代」（Flower Power era）已告終。

洗腦凶殺

比起其他邪教領袖，曼森和曼森家族才是將邪教形象帶入現代人民意識的頭號罪人。這絕非誇大，就像說美國前副總統高爾（Al Gore）發明網路一樣真實。曼森和他暗中規劃、宛如惡夢的泰特－拉比安卡凶殺事件（Tate-LaBianca murders），標示出美國面目全非的時刻。

曼森如何獲致如此惡名？六〇年代是電視的天下，當時只有三家電視台，大多數美國民眾每晚都會準時收看六點新聞。一九六〇年九月二十六日，甘迺迪贏了史上首次電視總統辯論時，電視可以說成了讓歷史轉型的力量。

廣播聽眾認為贏的人應該是甘迺迪的共和黨對手，也就是當時的副總統尼克森。不過，電視觀眾看到完全不同的景象：尼克森緊張得直冒汗，發著低燒，而瀟灑氣派的對手隨著一小時辯論開展，顯得益發自信、盛氣凌人。

甘迺迪賢伉儷和電視兩情相悅的羅曼史，就職後依然持續：風華絕代的賈桂琳透過電視轉播，為全國民眾導覽白宮。之後，像齣難以磨滅的悲劇，電視鏡頭再次讓第一夫人走進美國家庭，鏡頭捕捉到賈桂琳身穿一襲洋裝，血跡斑斑，這時她的先生才剛遇刺。三天後，電視台直播國葬，整整七小時。全國百分之九十三的電視機都鎖定國葬直播。

接下來動盪的十年，電視繼續帶動話題，全美各地抗議不斷，社會動盪不安。越戰又叫「客廳戰爭」，因為每晚新聞節目都在播放美國轟炸越南、處決市民和越南和尚引火自焚的畫面。詹森總統意識到，越戰可能會毀了他的政治生涯，甚至宣稱哥倫比亞廣播公司（CBS）和國家廣播公司（NBC）落入越共掌控。

一九六五年，聯邦和地方警察驅趕阿拉巴馬州伯明罕的民權示威群眾，ＡＢＣ新聞台中斷既定節目表，插播震驚全國的報導：執法單位踐踏、棒擊、用催淚瓦斯對付六百位市民。這是轉捩點，五個月後，詹森政府因此通過投票權利法案。一九六七年，大規模動亂徹底影響了底特律和紐瓦克市中心擁擠的貧民區，一九六八年則深深影響了邁阿密、芝加哥、瓦茨和華盛頓特區，電視畫面上都是城市深陷火海。

一九六九年七月二十日，估計五億三千萬人同時觀看電視直播阿姆斯壯和太空人夥伴執行太空任務，阿波羅十一號先繞月，接著登月小艇成功登月，他們走下梯子，首次踏上地球唯一的天然衛星。這場鼓舞人心的活動被譽為人類能力至高表現的最好證明。

接著沒幾週，八月九日、十日，電視播出人類最黑暗的一面，晚間新聞報導洛杉磯七人凶殺案，殘忍的處決連續兩晚發生。

直到整整兩個月後，十月十二日夜晚，警方才終於逮捕曼森。曼森個頭小，不到五呎三（約一百六十公分），那時體重一百三十磅（約五十九公斤），但依然能震懾身旁圍繞的信徒。曼森指使信徒犯下的凶殺，立刻讓該區陷入一片驚慌。第一起凶殺，受害者是電影明星莎朗蒂（Sharon Tate）和四名友人，地點在比佛利山莊（Beverly Hills）榭洛街（Cielo Drive）一〇〇五〇號寓所；下一起受害者是萊諾・拉比安卡（Leno LaBianca）、羅絲瑪麗・拉比安卡（Rosemary LaBianca）夫妻，時間就在隔晚，地點則在洛斯費利茲（Los Feliz）韋弗利街（Waverly Drive）三三〇一號。

按照曼森謬誤百出的想像，這幾起殺戮引起的恐懼應該會引發全國種族戰爭。曼森指望的末日衝突未曾發生，但此次犯罪的隨機本質，以及殺害過程的殘忍細節，都隨著調查進展一一詳細報導，每晚電視都播報聳動細節，美國大眾看得心生恐懼。

如果說這效果是曼森的目的，那就太抬舉他了。不過，他確實渴求他人目光。凶殺發生時，曼森大半生都待在感化院和監獄。曼森有離奇本領，總能把焦點轉到自身，並藉此獲得想要的東西；在獄友和心理學家的報告中，這是其中一項大家很快就會提及的特質。一九六七年，曼森素行良好，獲釋出獄，他在舊金山的終端島監獄服刑了六年半，原本判十年偽造文書罪，於華盛頓州麥克尼爾島聯邦監獄服刑，一出獄便遇上國家分裂，不同年代的人意見分歧，爭論婦女解放、民權運動和越戰。才幾年前，在加州大學柏克萊分校的言論自由運動時期，學生再三警告同儕不要相信任何年過三十的人。雖然一九六九年的時候曼森才三十四歲，但幾乎所有「家族」成員都是青少年或二十出頭。

曼森凶殺案讓全國看見年少輕狂可以多麼迅速就走偏。所有曼森家族成員都是嬰兒潮世代，七千六百多萬名小孩於一九四六年至一九六四年間誕生，大家一同進入青春叛逆期，凡事都想嘗試。他們也確實極易受曼森影響，曼森交叉使用迷幻藥（LSD）和心理操控。曼森從未踏足案發現場，其實也根本不需要，因為即使他連一根指頭都沒抬，兩次殺戮也都寫滿了他的名字，無庸置疑。

曼森幾乎可說全權掌控了家族，也利用這份權力達到目的，彷彿親自揮動那些凶器：刺刀、菜刀和點二二口徑高標長管左輪手槍，又稱「邦特蘭特裝型」。這就是他的犯案手法——借刀殺人。曼森連續兩晚展示了將孤兒、逃家的人及大學中輟生變成動機強烈的殺人犯有多麼輕而易舉。即使曼森遭到逮補，依然能繼續掌控家族。一群年輕女子身著鄉村洋裝，守在法院外頭，跪在地上告訴所有詢問的記者，她們正在守夜，「等待我們的父親重獲自由。」法庭上，曼森和同謀天天搞新花樣破壞法庭秩序，高歌、狂笑，或在訴訟程序進行時齊聲誦念，甚至恐嚇主審法官查爾斯．H．歐德爾（Charles H. Older），曼森一度大喊：「為了基督公義，真該砍下你的頭！」

多虧凶殺審理有如馬戲團的氣氛，幾百萬名好騙的青少年和年輕人開始有威脅性，他們想要「開啟感官、調整頻率、脫離社會」[1]。凶殺後的幾個月，「cult」一詞立刻有了可怕的新意涵，當然也因為需要全新的字彙才足以形容這些犯罪的駭人本質。直到那時，cult 通常指稱某一教派，其信仰系統與傳統宗教相異，即使該教派的開端與實踐都能追溯回一個或多個傳統宗教。老早就有各式各樣分裂而生的教派，源自佛教、猶太教、伊斯蘭教和基督宗教，世人認

1　譯註：一九六六年 Timothy Leary 所說，指敞開神經知覺、和諧融入萬物、跳脫體制。

為多數都不具危險或惡意。專家愛用新宗教運動（New Religious Movement）一詞來稱呼這些教派，簡稱NRM。

在曼森之前，cult經常跟流行文化元素有點沾邊，形容歌手或電視節目的死忠粉絲。但曼森凶殺之後的幾年，世人開始用來指稱精神科醫生羅伯特・傑伊・利夫頓（Robert Jay Lifton）所謂的「破壞性」教派，這類教團會系統性地傷害、殺害他人或自身。在曼森之前當然也有危險的教派領袖，但這些人物未能吸引美國大眾的注意，直到泰特－拉比安卡凶殺事件才開始引起關注。照利夫頓所言，破壞性教派很好辨識，因為通常具備三項鮮明特色：深具魅力的領袖，隨後成為信徒崇拜的對象；態度不變，讓領袖趁機從信徒身上取得性和／或錢財利益；近乎全面的掌控，可歸因於領袖有能力執行利夫頓所謂「思想改造」，又或者口語一點：洗腦。

腹地有患

曼森的祖母南西・馬多克斯（Nancy Maddox）信奉基督教基要派，亡夫是乞沙比克暨俄亥俄鐵路（Chesapeake & Ohio Railroad）車長，她獨自在肯塔基州東部貧困的亞士蘭鎮（Ashland）拉拔五個孩子：格倫納（Glenna）、艾琳（Aileene）、路德（Luther）、桃樂絲（Dorothy），和么女艾達凱瑟琳（Ada Kathleen）。南西是名嚴母，但也希望自己的黃金法則紀律可以協助孩子擊敗困境，成為有道德又虔誠的人。但是艾達凱瑟琳也想像其他女孩一樣開心玩樂，十五歲開始偷溜出州界，越過艾昂頓大橋（Ironton Bridge），前往俄亥俄河另一側的舞廳。她開始和一位年長八歲又帥氣的沃克・韓德森・史考特（Walker Henderson Scott）上校交往，並且藏起交往的蛛絲馬跡，不讓媽媽知曉分毫。史考特也藏著祕密：他是已婚人士。不過，已婚或單身都沒差，一發現艾達凱瑟琳懷孕，他便溜出鎮外。

南西把艾達凱瑟琳送到辛辛那提（Cincinnati）生孩子，小媽媽（青年孕婦）在那裡不會引起太多地方八卦。艾達凱瑟琳無依無靠，剛好證明她能臨機應變。儘管有孕在身，她依然說服二十五歲的威廉・曼森（William Manson）娶她，一九三四年十一月十二日，十六歲的她生下一名男孩：查爾斯。

只不過，母職並未讓艾達凱瑟琳變得安分。她的家人（包括兒子）日後談及她年少輕狂的日子，都提到其中一件非常不像話的事件：這名年輕媽媽推著嬰兒車到酒吧，用小男孩換了一大罐啤酒。一九八七年，辛辛那提WCPO電視台新聞主播艾爾・肖特爾科特（Al Schottelkotte）挖出另一版本，作為曼森特輯的內容播出：據說，那家酒吧的一名女服務生向凱瑟琳吐露自己很想要小孩，接著凱瑟琳提出這筆交易。服務生覺得凱瑟琳在開玩笑，橫豎送了一大罐啤酒，隨後驚喜地發現，那名客人不但喝完啤酒，還真的隻身離開酒吧，留下小寶寶。艾達凱瑟琳的哥哥路德花了好幾天（這是充滿謎團的細節，暗示那名服務生可能決定好好把握新到手的母職）才找到那間酒吧和那名服務生，帶小寶寶回家。

艾達凱瑟琳很快便開始把曼森丟給親戚照顧，好出去玩樂、參加派對，她經常和路德一起失蹤多天。起初，先生威廉還樂於幫忙養育其他男人的孩子，但這些失蹤立刻讓他感到厭煩，在一九三七年和艾達凱瑟琳離婚。這對艾達凱瑟琳而言根本不痛不癢，那時她正忙著告史考特上校，想讓上校吐出扶養費。官司是打贏了沒錯，但孩子的生父從頭到尾只付了二十五美元。

難道曼森遺傳了媽媽的壞心肝，逐漸憎惡當初拋棄他的生父？不管原因為何，都證據不足。不過，有群業餘偵探密切合作，幾乎調查了曼森生活、人際的所有面向，不管多隱微都不放過。其中許多人認為曼森與其家鄉亞士蘭的凶殺懸案有關，懸案發生時間為泰特－拉比安卡凶殺案四個月前：達爾文·史考特（Darwin Scott）是上校史考特的哥哥、曼森的伯伯，他陳屍家中，有十九處刺傷，身上還插了一把屠刀。比佛利山莊和洛斯費利茲兩起犯罪十分相似，因此許多人認為這三起案件都和曼森脫不了關係，因為曼森痛恨一切到想殺人，並且渴望復仇。不過肯塔基警探的結論更為合理，達爾文·史考特前科累累，長時間坐牢，不幸還是個賭徒，經常積欠黑社會巨額賭債。

年輕心理變態的童年恥辱

能夠證實的是，早在曼森變成精神錯亂、只想復仇的曼森家族領袖之前，就已經經歷問題重重的異常童年。暴力的邪教領袖經常沒有父母悉心養育，而且多數經歷虐待、忽視、犯罪等對待（然而這並非常規，本書接下來要介紹的領袖，許多位都生長在有愛與支持的家庭，享有良好的教育）。即便都是邪教領袖，曼森的童年依然與眾不同，嚴重缺乏關愛——經常與警方發生衝突，傷痕累累；在親戚家流轉，承受大家勉為其難的照顧，這些親戚在他頻繁進出感化院和問題男孩之家之間探訪。麻煩很早就發生了：一九三九年八月，曼森才四歲，他媽媽和舅舅結夥搶劫法蘭克・馬汀（Frank Martin）。

這對兄妹喝了一整晚，誘騙馬汀到加油站，發動攻擊搶錢，他們拿番茄醬瓶抵在馬汀背上，好讓他誤以為是槍。他們根本沒有好好掩飾身分，隔天就被警方找到並逮捕。曼森甚至可能眼睜睜看著媽媽被帶走。幾週後，法官判定艾達凱瑟琳在本案的角色需服刑五年，當地媒體稱本案為番茄醬瓶搶案。

由於媽媽入獄，曼森的近親必須決定該由哪位家族成員照顧男孩才合適。最佳人選看來是他的阿姨格倫納和姨丈比爾・湯瑪斯（Bill Thomas），他們還有個八歲的女兒喬・安（Jo Ann Thomas）。一家三口住得離艾達凱瑟琳的服刑監獄最近，曼森就這樣被送去那裡了——西維吉尼亞洲的麥克梅欽（McMechen）中產階級小鎮，多數鎮民都在礦坑或鐵路局工作。

沒有多久，曼森就開始為新家人製造麻煩。曼森比學校其他男同學都還嬌小，但總想要成為眾人關注的焦點，只要有人視而不見，就會衝動亂來做壞事。那個時代的標準處罰（不乖就打一頓）對曼森毫無效果，而且就算在學校慘遭同學毆打，他似乎一點也不擔心，不管霸凌的人有多高大，曼森都會回嘴。傑夫・桂恩（Jeff Guinn）寫了一本包山包海的傳記《曼森：查爾斯・曼森的一生和時代》（暫譯，*Manson: The Life and Times of Charles Manson*），描述曼森的表姊喬．安聽爸媽的話照顧他，這也是為何曼森和一名身型較大的男孩在操場爭執時，喬．安衝上前保護年幼的曼森，咬了大男孩的手指，把他嚇跑。

老師對喬．安的行為大感意外，沒料到平日溫柔的女孩會大打出手。喬．安解釋是表弟先被騷擾，自己只是想挺身保護。但老師向曼森確認到底發生什麼事時，曼森只回道這和他一點關係也沒有。這不打緊，老師早就發現曼森撒謊成性——他們之前目睹曼森召集同學（尤其是女同學）攻擊他不喜歡的同學。每次老師質問曼森是否事先影響同學的想法，曼森都說自己是無辜的，那些女生只是做她們自己想做的事而已。因此當曼森和喬．安的說詞有出入，老師別無選擇只能相信喬．安。

喬．安現在看清了表弟的真面目：天賦異稟的騙子，毫無忠誠，隨時把自己的惡作劇推到他人頭上，只要不是自己負責就好。儘管如此，喬．安也直接面對表弟深層病態的暴力。

極其可怕的事發生時，喬．安十歲，曼森七歲。那天喬．安的父母白天都不在家，她負責照顧曼森，還要打掃家裡。曼森拒絕幫忙任何家事，這倒也不奇怪，小男孩經常如此。但曼森沒有就此打住，他跑去院子，找到一把鐮刀，重回屋內搞破壞，喬．安根本沒辦法做家事。當曼森拒絕停手，喬．安把他踢出去，鎖上紗門，

曼森尖聲大叫，用鐮刀大砍紗門。曼森的眼神嚇壞了喬．安，她確信表弟打算一進到屋內，就用鐮刀砍她。幸好父母正巧開車回來，在情況變得更糟之前，曼森丟下鐮刀。

童年時期曼森展現的這些人格特質符合心理病態的公認特質，這也是邪教領袖常見的狀

況，最典型的是毫無道德觀、反社會行為、愛無能等多項特徵。臨床心理師羅伯特・D・海爾（Robert D. Hare）開發的檢核表協助諮商師鑑別潛在心理病態者。能推測海爾會幫年幼的曼森勾選哪些特徵嗎？病態說謊？常為自身利益而操控他人？缺乏同理心？花言巧語、表面魅力？儘管有些精神科醫生會遲疑是否要為如此年幼的孩童貼上如此駭人的標籤，這些特徵都屬於海爾所分的兩類：心理病態量表的第一種「因子一」——最常見的診斷，病患特徵為「自私、無情、利用他人毫不手軟」，以及心理病態量表的第二種「攻擊型自戀」。

難道這代表在如此年幼的時期，曼森就已經是心理病態，深受可識別的心理疾病所苦？不盡然。雖然他的行為明顯符合早期預警，但這些同樣的特徵（包含天性愛撒謊）也和高智商及外向人格成正相關。海爾醫生警告，這些所謂的心理病態核心特徵確實非常實用，展示這些特徵的人可以贏得關注、獎勵和人緣。雖然看似是個危險的反差，但心理病態者火藥庫中最厲害的武器之一，便是在他們狀況好的時候，人人都喜愛他們作伴。

不過，曼森最初是怎麼養成這些特質的呢？舉例來說，普遍認為兒童麻木或冷漠可能是出於天生，也可能是後天養成。有些兒童天生富有同理心，但如果成長時受到家長虐待，家庭環境不安穩，那份同理心會在自我防衛之下消磨殆盡，也會反映出四周環境。被這樣養大的孩子，最有機會能夠治好，也比較有機會過上正常、豐饒的人生——前提是諮商師能及時提供協助。

較危險的孩童則是生長在家庭有愛、鄰里友善的環境，卻依然展現和在完全相反環境下長大的孩子同樣的特質。依照心理病態研究權威的看法，哈維．M．克萊克利（Hervey M. Cleckley）於一九四一年出版的影響深遠的著作《理智的面具：試圖闡明所謂心理病態人格的幾項議題》（暫譯，*The Mask of Sanity: An Attempt to Clarify Some Issues on the So-Called Psychopathic Personality*）和海爾醫生制定的海爾心理病態人格檢核表，幾乎都認同有些孩童因天生的遺傳傾向，容易成為心理病態。即便還在蹣跚學步，這些孩童就渴望刺激，不斷說謊，性格殘忍（或者依臨床文獻中的分類，又形容為「無情冷漠（callous unemotional，簡稱CU）」）。如果他們有時看起來既體貼又善解人意，只是因為想從他人身上獲得些什麼，這也就是克萊克利所謂「理智的面具」。這樣的情況也不少見，海爾認為約百分之一的孩童深受其擾，數量大約等同展現自閉症或躁鬱症症狀的孩童。

如果沒有接受治療，可以預見心理病態的孩童到了八九歲時會對他人暴力相向，十四到十六歲時會參與犯罪活動。預防狀況變糟的方法就是持續關注和及時正向介入，充滿愛的家庭確實發揮了防護效果，但依然不是萬無一失，然而那樣的家庭環境正好是曼森童年所缺。

曼森年幼時經歷許多創傷事件。從小，媽媽就丟下曼森去找樂子，可能導致嬰兒曼森難以和媽媽建立恰當的依附情結。看著媽媽被捕，再去探監，幾乎肯定會造成創傷。這些都是慢性病，會改變曼森的本質。不過，曼森也經歷一些毫無疑問差勁透頂的苦痛體驗，可能因此扭曲了他看待自己和世界的視角。曼森五歲時住在阿姨、姨丈家，小學一年級的導師瓦爾納夫人（Varner）在麥克梅欽臭名遠播，不僅言語虐待學生，還威脅恐嚇學生。

瓦爾納夫人在曼森入學首日，花了一整天嘲笑他媽媽坐牢，毫不留情。他跑回家大哭，眼神流露的軟弱令比爾姨丈感到噁心，隔天姨丈強迫曼森穿喬．安的洋裝上學，據說這樣做對外甥有益，會變得堅強，學會怎麼當個男人。數十年後，曼森依然清晰記得那次羞辱。

紐約兒童心智研究所（Child Mind Institute）創辦人哈羅德．科普萊維奇（Harold Koplewicz）認為，在正確的支持環境中，變裝打扮對想要探索自身性別認同的孩童而言，會是正向有益的表達型態。但對於那些被家長角色強迫的孩童而言，變裝會成為嚴重受創的體驗。事實上，雖然我們無法肯定這麼一次事件會成為轉捩點，尤其是曼森的童年早已如此悲慘。值得注意的是，有多名犯下一連串令人毛骨悚然的罪行的連環殺手，全都經歷過相同的體驗。亨瑞．李．盧卡斯（Henry Lee Lucas）撒謊成性，起先自白殺害六十人，接著一百人，隨後又說三千人。在他小時候，他的媽媽賣春，常幫他穿洋裝打扮成女生，也叫他看著自己和客人性交。事後證明媽媽是他的第一位受害者，殺害時間為一九六〇年。奧提斯．圖爾（Otis Toole）也被媽媽強迫變裝（媽媽都叫他蘇珊），五歲時遭父親友人強暴，智商測驗為七十五，似乎都隨機選擇殺害對象：橫跨六歲男童、十八歲的搭便車男性、六十歲老男人，以及從佛羅里達州塔拉哈西夜店誘拐的二十歲女性。多伊爾．藍恩（Doil Lane）小時候被迫打扮得像個女生，長大成人殺了兩名女孩（一位八歲，一位九歲），承認自己喜歡女生內褲。查爾斯．艾爾布萊特（Charles Albright），人稱「德州眼珠殺手」，他媽媽不只把他打扮得像女孩，給他洋娃娃，還幫助他學習剝製標本。儘管如此，媽媽還是不允許他買剝製店陳列的昂貴玻璃眼珠。後來他殺了三名妓女，三次謀殺因一個共通點而串起：三名受害者的眼珠都遭移除，技術精湛宛如外科醫生。

除了這些連結，近日研究近一步發現，特定羞辱事件導致的敏銳羞恥感，與未來出現心理病態行為有關。對許多人而言，反應是內在的，羞恥會導致不同負面結果，可能會貶低自我、自殘，甚至自殺。羞恥感會帶來攻擊行為，在其他人身上或在曼森身上看來都是真的。這通常會引發一連串危險行為：羞恥源頭總是藏匿無蹤，卻能反覆觸發無明火，事件跟著發生。儘管無法斷定一年級事件就是引爆曼森羞恥的源頭，曼森縝密的復仇計畫和自身的不穩定（無法控制的怒火從未承認的汙名中燃起）對研究邪教模式的人而言相當耳熟。

犯罪失敗

一九三〇年代晚期四〇年代初，曼森與湯瑪斯一家生活了兩年半，同住期間培養了三樣興趣，熱情一生未減：音樂（湯瑪斯家有台鋼琴，大家開心的時候會一起彈琴唱歌）、刀和槍。一九四二年末，差不多曼森八歲生日的時候，媽媽獲得假釋，曼森開始和媽媽一起生活。一開始曼森興奮到不行，可以重回媽媽身邊，不用再和嚴格的親戚同住。在人生晚年，曼森在《曼森自述》（暫譯，*Manson in His Own Words*）一書的訪問中告訴諾爾．艾蒙斯（Nuel Emmons），媽媽一出獄給他的擁抱，是他童年裡唯一的開心記憶。

艾達凱瑟琳和兒子在麥克梅欽待了八週，艾達在酒吧當外場服務生；接著他們搬往南方，住在西維吉尼亞州查理頓，艾達在馮記二十四小時超市當收銀員。沒多久，艾達就發現兒子令人不安的行為，例如不停蹺課、大部分時間都在女人身邊打轉，說些甜言蜜語，騙錢買糖果。當曼森開始偷東西並怪罪他人時，艾達向媽媽南西求助，教導曼森是非對錯。但這些道德教誨對孫子的影響不大，沒有比當初對自己女兒的教育好上多少。

最後艾達完全不知道該如何和兒子溝通。而她在酒吧下班又回去狂飲的行為只是雪上加霜，她聘請一堆不能全然信任的保姆看顧曼森。艾達終於用上最後一招，決定找間專收問題少年的機構，看看是否能教導曼森一些紀律和道德品行。

因此，在一九四七年，曼森十二歲時，艾達把他送去印第安納州，就讀由聖十字兄弟會（Brothers of Holy Cross）經營的特勒荷特校區（Terre Haute）吉爾特男孩矯正學校（Gibault School for Boys）。

突然之間，飽受折磨的男孩時期變成充滿小罪小惡的青春期。第一年的聖誕節假期，曼森獲得許可，能和阿姨、姨丈、寬宏大量的表姊喬．安一起過節，還是喬．安邀請他的。但是曼森立刻就利用了湯瑪斯一家的熱情好客。曼森打算趁大人上教會時偷拿姨丈的槍，他打開蓮蓬頭讓水聲蓋過翻找槍袋的聲響。喬．安這時其實有點怕曼森，並沒有出手阻止，不過當父母回家問道為何水開著時，她叫父母去問表弟。父母轉身去找曼森，當場抓到他偷拿槍。

回學校才幾個月，曼森就逃跑了，開始搶當地商店。但當時曼森十三歲，毫無耐心，不會小心翼翼，也不會提高警覺，無法不留痕跡地逃離現場，於是很快又被捕補了。一九四八年，一名法官因為曼森之前就讀吉爾特男校，以為曼森是天主教徒，把他送至內布拉斯加州男孩之城，一個收留問題男孩的地方，此處還是十多年前史賓塞．屈賽（Spencer Tracy）那部電影《孤兒樂園》的靈感來源。晚報《印第安納波利斯新聞》（*Indianapolis News*）那時有一篇特別報導刊登了曼森的照片，寫到曼森告訴法庭審判人員：「我覺得我在牛和馬附近工作會很開心。我喜歡動物。」

曼森沒有在男孩之城久待。才第四天，他就和另一位青少年布萊基．尼爾森（Blackie Nielson）偷了一輛車，開往伊利諾伊州。兩名少年不知從哪裡搞來一把槍，毫不遲疑用在兩次持槍搶劫中。隨後他們打算繼續犯罪實習，在布萊基的職業竊賊叔叔手下做事，糊口飯吃。

一九四九年早期，警方又再次逮捕曼森，這回送往普蘭非（Plainfield）的印第安納少年矯正學校（Indiana Boys School），這間機構更加嚴格。曼森現在十四歲，必須在成群凶神惡煞的少年中存活，他們的罪行五花八門，有些人搶劫，有些人攻擊過度而失手殺人。那裡的工作人員和較年長的矯正生經常虐待像曼森一樣體型嬌小、年紀較輕的男孩。照曼森所言，他才剛入校就慘遭強暴。

曼森日後的描述可當作心理學家所謂「解離狀態」的教科書範例：「你知道的，被強暴，他們可以就這樣抹掉……我不覺得有人受到侵犯，這才是可怕的事。我只是想著：『清乾淨，就這樣。』」這樣的自我遠離機制，如果極端一點會變成多重人格障礙，但也可能協助曼森處理這些創傷記憶和應付周遭環境。這也為曼森令人不安的特質提供了早期證據：他能麻痺自己不受疼痛影響，脫離客觀現實。曼森後來說明，那時是他第一次發明所謂的「發瘋遊戲」。由於他實在太嬌小，根本無法威嚇其他學生，所以就裝瘋賣傻，嚇唬同學，讓他們以為他真的瘋了，他會揮舞雙臂、尖叫、做可怕的鬼臉。後來電視台記者到監獄採訪時，曼森也實際展示了一遍。

無庸置疑，曼森在印第安納少年矯正學校痛苦不堪，一直活在恐懼之中，害怕受到肢體暴力和性虐待，他在一九四九年企圖獨自逃跑四次。那年十月，他和六名男孩發動學校史上最大型集體逃跑，但警方很快就在加油站抓到打算闖入的曼森。兩年後，曼森和兩名男孩再次逃跑，這次最遠逃到了猶他州比佛（距印第安納州一千六百哩），開著偷來的車，沿途從加油站偷補給品。儘管這回成功脫逃，還偷車兜風了好遠，但沒隔幾天又被警方逮捕。

曼森後來被送往華盛頓特區的國家男童訓練學校（National Training School for Boys），做了一堆檢測，發現他是半文盲，讀寫能力不足，但也發現他的智商是一零九，比普通人稍微高一些。當然，曼森也證明自己可以操控校內心理學家，入學沒多久就讓其中一位相信，他的人生要轉變只差一件事，那就是提升自信，而要發揮所有潛力，就只能轉到自然橋榮譽營（Natural Bridge Honor Camp），一家位於維吉尼亞州洛克布里奇郡（Rockbridge County）安全等級最低的監獄。就是在那裡，湯瑪斯一家又重新進入曼森的人生。他們向管理員保證，如果曼森提早出獄，他們會提供住處，幫他找工作。假釋聽證日排在一九五二年二月，與此同時，曼森需要做的就是低調行事，遵守規則。然而，差一個月就要聽證了，曼森卻被抓到和一位囚犯肛交，還拿刮鬍刀抵著對方的脖子。曼森被轉送到維吉尼亞州彼得斯堡（Petersburg）的聯邦感化院（Federal Reformatory），紀錄顯示他繼續犯下「八次違規，其中三次是同性戀性交」。

精進個人監獄技能

又過了五年，此時曼森變成會危害一般大眾的那種囚犯。下一站，他來到契利科什（Chillicothe）一間安全等級最高的感化院，毫無假釋的可能，直到二十一歲生日：一九五五年十一月十二日。

頗出乎意料的是，儘管契利科什院方認為就連在那裡曼森都「完全不適合」矯正，但此時曼森終於竭盡全力要改變自身。曼森學習閱讀，不僅可以看七年級等級的讀物，還成為模範囚犯，也認真在院內的汽車修理廠工作，傑出的服務還贏得嘉獎，終於因此獲釋。他十九歲的時候，實際上已經花了七年待在六間不同的感化院。

這麼長一段拘禁期間，各個指導員、個案社工師和精神科獄醫都會定期為曼森評估。這些觀察是理解那時期的曼森的最佳材料，因為他們最早看到曼森即將成為操控大師的那一面。第一站，吉爾特男孩矯正學校，一名指導員描述曼森表現平平、討人喜歡、情緒不穩，有一點迫害妄想症。等到了印第安納少年矯正學校（專門接收「無法管教的孩童」），老師逐漸起疑，抱怨曼森「只會看對象乖乖做事，都是他認為可以從中獲得些什麼的人」。下一站國家男童訓練學校，評估結果差異加劇，曼森說服一位個案社工師他有侵略型反社會人格，成因就是「不愉快的家庭生活，假使那樣還能稱得上是家庭生活的話」。另一位個案社工師就更沒耐心了，觀察曼森後說道：「這名男孩試圖留下他認真想要矯正的印象，但事實上他一點都不努力。」不過同一名社工師也發現曼森渴望權力，說道：「我覺得時候到了，曼森會力圖在小團體中成為重要人物。」曼森欺瞞的技巧似乎被自然橋榮譽營最後一份報告證明，一名輕信曼森的精神科醫生寫道：「大家會覺得所有謊言背後有位極度敏感的男孩，這名男孩還沒放棄向這世界索求情和愛。」仔細讀完這份檔案，很難不發現心理遊戲和精心算計是曼森最主要的自我防衛方式。

一九五四年四月假釋之後，曼森離開俄亥俄州契利科什，回到西維吉尼亞州，居無定所，起先和湯瑪斯一家住，接著去惠靈（Wheeling）和媽媽住，後來到了最喜愛的地方，和南西外婆一起住，外婆十分溺愛曼森。但外婆開了一個條件，如果想要一起住就得遵守：曼森每週日都得上對街的拿撒勒人教會（the Church of the Nazarene），祖孫倆一起做禮拜。於是乎，曼森盛裝打扮、聽道，吸收聖經教條，未來某日這些都會用在他的信徒身上，例如女人要服從男人，訓誡那些拋棄自身認同而事奉更高權柄的人。

曼森無法融入這般成聖的圈子，一點都不令人意外。曼森試著和其他主日學小孩做朋友，但很快又拉遠了距離，他聊起矯正學校的暴力過往，還宣稱自己吸過了哪些毒品。曼森發現炫耀其實不容易，特別是對方根本聽不懂「走水路[2]」是什麼意思。後來某次曼森拜訪表姊喬·安（現在嫁給一名當地牧師），表姊發現他在巴結一位青少女，少女來表姊家尋求協談，曼森把少女捧上了天，喬·安立刻明白表弟要勾引這位易感的訪客。結果喬·安讓少女立刻遠離曼森並且離開他們家。

但如果從其他方面來看，曼森似乎誠摯地想努力過正常中產階級的生活。曼森在當地賽馬場惠靈賽場遇到一位離婚爸爸查理·威利斯（Charlie Willis），威利斯把他介紹給人人喜愛的小女兒羅莎莉（Rosalie）。小倆口約會時看來不會長久，但在一九五五年一月十三日，兩人結婚了，許多人猜測一定是女生懷孕了。但就算真的先有後婚，那名孩子也從未出世。曼森找到工作，交了朋友，開始玩音樂（一如往常想要拿出最佳表現時一樣）並且開始學習吉他和弦。只不過，他想過安穩生活的心沒能維持多久。新婚幾個月後，羅莎莉真的懷孕了，曼森卻開始偷車賺外快。一九五五年夏天，曼森決定帶著羅莎莉離開麥克梅欽，探望目前住在洛杉磯的媽媽。兩人也算是倒楣，在洛杉磯就被捕了，他們駕駛偷來的車越過州界，構成聯邦犯罪。

在判刑之前，法官請精神科醫生艾德溫·麥克尼爾（Edwin McNeil）評估曼森。曼森聲稱不知道該怎麼賺錢維生，都怪自己從小在感化院長大。麥克尼爾醫生建議法官從輕量刑，因為

2 譯註：注射毒品。

曼森的太太懷有身孕。法官採納麥克尼爾的建議，判了曼森五年緩刑。但是一九五六年二月還有另一個審判期日，於是曼森決定和羅莎莉一起跑路。

躲避警方只是一時，一九五六年四月二十三日，兒子小查爾斯．曼森（Charles Manson Jr.）剛出生不久，二十一歲的曼森就被送往聖佩德羅（San Pedro）的終端島監獄（Terminal Island）。一進監獄，曼森繼續他真正的教育，向皮條客學習如何揀選容易下手的女生，她們通常缺少強大的家長角色，可以輕易利用這些心理弱點。皮條客教導曼森家暴施虐者的慣用手法：首先孤立一位女性，說服她你才是唯一真正愛她的人，接著毆打她，好叫她擔驚受怕、乖乖聽話。

曼森結合監獄操場私人課程和卡內基（Dale Carnegie）規劃的監獄官方課程，卡內基著有《如何贏取友誼與影響他人》[3]（*How to Win Friends and Influence People*）一書。監獄體系普遍都有教學課程，有時是大學學分班，有時是為了積極利用時間。不過這倒是曼森人生第一堂心無旁鶩上的課。回到牢房，曼森會練習卡內基建議好好運用的台詞。書中列出的許多想法呼應曼森自己的操控傾向，而且其中一條建議，曼森謹記在心：「讓對方覺得這是他的主意。」

一九五八年九月三十日，曼森終於出獄，那時羅莎莉和兩人的兒子卻與另一名男人同住，曼森只好搬回去和媽媽住。曼森並沒有離開監獄太久。

3 二〇一五年龍齡出版，黑幼龍編、詹麗茹譯。

曼森實際運用在終端島學到的暗黑課程，一九五九年年中開始替一名年輕女性拉皮條，這名女性叫雷歐娜（Leona）。當曼森偷了一張社會安全補助金支票，偽造簽名，打算在勞爾夫超市（Ralph's）兌現，雷歐娜成功懇求法官放過曼森一馬。為了增加收入，曼森還偷了信用卡和車子，再一次開著贓車往北越過州界，這回載著雷歐娜及其一名友人去賣春，這違反了《曼恩法》，一九六一年六月，二十六歲的曼森再次坐牢，這回被關在華盛頓州麥克尼爾島矯正中心（McNeil Island federal penitentiary）。

這回入監，曼森開始研究山達基（Scientology），學習更多操控技巧。他認為山達基信仰的前世和靈魂不朽概念，可以用來瞄準陷入困境的女性，還能以個人名義提供解脫之道，讓女性擺脫過去的創傷。曼森也精進了吉他技巧，滿足早些年在湯瑪斯家耳濡目染的音樂愛好。他還認識了披頭四，那時披頭四已經紅遍大街小巷，就算在聯邦監獄也可以輕易聽到他們的歌。儘管有前科和個人往事，曼森依然幻想自己會深受世人仰慕愛戴，正如披頭四一般。曼森開始寫歌，規劃職涯，打算運用自己厚實的男中音、天生的藍調韻律感以及寫歌的才能，成為知名的音樂人。就連監獄員工都認為，等曼森獲釋，他有望靠音樂賺錢。

等到可以申請假釋，已經是一九六七年，曼森三十二歲了，大半輩子都待在監獄。不過，相關當局都沒料到，曼森要求繼續待在監獄。難道他已經有預感將來自己會成為什麼人物了嗎？雖然不可能有答案，但也很容易明白他為何想留在監獄。監獄是充滿規則的世界，他知道如何在這樣的世界生存、茁壯。如同早些年曼森向精神科醫生麥克尼爾坦承，對於要如何在外面的世界生存，他毫無頭緒。要求在獄內繼續生活，其實展現了曼森少有的自我覺察。

自由之身踏入愛之夏

美國監獄系統並非設計來留宿任何人，更不用說想要花納稅錢持續入住的人。一九六七年，曼森獲得假釋，法院允許重新將其安置到舊金山。那年是愛之夏[4]的時期，舊金山聚集了數十萬名年輕男女。目前為止，曼森只待過監獄，童年也住在與世隔絕的小鎮，簡直是雙倍無防備就踏入瞬息萬變的世界。反戰人士和嬉皮舉目皆是，其中有一群人特別醒目：黑豹黨（the Black Panthers），全國總部設在奧克蘭（Oakland）灣區，成為舊金山街頭最受矚目的存在。曼森多年來在監獄操場嚴格的種族隔離之下戰兢而行，對黑豹黨感到膽怯，這群新進果斷、不吐不快的非裔男女隨時準備好為自身權利而戰。

儘管如此，在舊金山一片反威權氣氛之下，監獄經驗反而讓曼森輕易融入街頭，享受大家的親切歡迎，和在家鄉麥克梅欽體驗過的各種拒絕天差地別。曼森在加州大學柏克萊校園花了幾天，混入抗議人士和嬉皮群體，研究這群年輕人，模仿他們，很快就意識到在反文化潮流之

4 The Summer of Love，以舊金山為中心的社會運動，倡導愛與和平，並熱衷於音樂、迷幻藥物、嬉皮文化。

下，其實可以把前科累累的歷史重塑為勇氣徽章，象徵著獨立、不可抑制的革命精神。曼森察覺這時期民眾態度的根本轉變，瞭解到自己不用努力，也確實可以在這個世界出人頭地。

運用所有新發現的花招，曼森魅惑了在柏克萊認識的瑪麗・布魯納（Mary Brunner），假裝有相同的理念和社會觀，立即操控了瑪麗。瑪麗不僅和他上床，還給他地方住。曼森說服瑪麗只有他會傾聽、重視瑪麗，在他眼中瑪麗既特別又美麗，接著曼森不停灌輸瑪麗他所詮釋的「自由之愛」新概念，他在海特－艾許伯里（Haight- Ashbury）跟嬉皮學來的概念。瑪麗在柏克萊圖書館工作的時候，曼森會站在街角，隨口說說不同宗師的名言佳句，彈彈吉他，吸引那些易感的人，他們通常想到舊金山參與這場革命。

其中一位還借車給他，曼森往南開到洛杉磯，在那裡遇到了琳內特・弗洛姆（Lynette Fromme）。琳內特父母離異，逃家來此，深深陷入曼森的滔滔不絕，跟著他回到嬉皮大本營海特－艾許伯里。

曼森告訴琳內特：「大家都叫我園丁，因為我照料所有花之子（flower children，嬉皮的別稱）。」

曼森立刻將他對自由之愛的看法付諸實行，和琳內特做愛，瑪麗在一旁看，再和瑪麗做愛，換琳內特看。隔了幾週，他多了第三名信徒——一位男人邀請曼森吃晚餐，讓曼森搭便車，他十幾歲的女兒露絲・安・摩爾豪斯（Ruth Ann Moorehouse）對曼森一見傾心，曼森甚至成功帶走一架老鋼琴，以物易物，換來一輛福斯麵包車，代表現在可以和越來越多信眾一起旅行。

曼森的發展策略就像標準的龐氏騙局。曼森的話術其實沒有多特別，同時代其他許多人的推銷方式也差不多：曼森告訴大家拋棄顧忌和財產，愛所有人，東摻一些山達基和卡內基，西混一些披頭四歌詞和聖經金句。即便台詞很常見，曼森現場即興，舉手投足充滿魅力，出人意表：唱歌、細語、布道、蠱惑輪流來，不論是在一群人之中，還是一對一親密對談，人們都深深受其吸引。這一風格成功招致信徒，而為了試探他們是否虔誠，曼森會叫信徒輪流出去，帶更多人進來。曼森沒多久便擁有一大群虔誠的女信徒，即使其中一條教誨是女人要服從男人，聽命行事。短短幾個月，曼森就可以命令嬉皮後宮的任一位跟任何人做愛，還開始稱呼信徒為「家族」。

曼森接下來著手擴張版圖，開著七彩手繪麵包車，在舊金山附近四處旅行，招募更多女性加入他的邪教，許多都是青少女，學校一放假就跑來舊金山。一九六七年秋天，家族成員越來越多，曼森舉家搬遷至洛杉磯。沒多久，曼森就發現餵養這麼一大家族著實不容易，大家必須翻垃圾桶覓食才活得下去。幾個月後，一九六八年，曼森和盲眼老人喬治・史潘（George Spahn）談好，家族可以免費待在五百英畝大的史潘片場（Spahn Ranch），又叫史潘電影牧場（Spahn Movie Ranch），位在洛杉磯北方的聖蘇珊娜山脈（the Santa Susana Mountains）。為了換宿，家族成員得做家事，幫忙騎馬穿過那裡的老舊電影場景。

雖然自由之愛這一共有氛圍讓這群人立刻融入六〇年代的精神，但事實上除了相信曼森，這群人沒有更高的信仰，曼森的話語和心血來潮團結起這群失根的人，這代表他們越過了那條劃分公社和邪教的隱形界線。舉例來說，其中一種不明羈絆是曼森說服信徒去做的活動，雖然看似無害，實則違法，曼森稱之為「驚悚潛行」：夜闖民宅，重新擺設，完全隨機入侵，且沒有東西失竊，益發令屋主不安。犯罪、冒險精神和陌生信仰在那些日子並不奇特，因為嬰兒潮年輕人在全國落地開花變成花之子，聚在一起做社會實驗，結果經常很糟糕。然而，沒有什麼比曼森家族的所作所為更加引人注目了。

在洛杉磯，這無人起疑的教派開始急速成長。曼森擅於發現性格堅忍的問題青年，說服他們曼森本人就是生活中所有問題的解答。其中一個在史潘片場上持續執行的儀式是團體迷幻旅程，這倒也無傷大雅，而曼森會全程把大家的迷幻感知編入他逐漸茁壯的個人啟示中。出名信徒中有一位琳達．卡薩比安（Linda Kasabian），她在曼森的凶殺審判扮演關鍵角色（卡薩比安無罪開釋）。她之後談起第一次遇見曼森時，她心想：「這就是我一直尋覓的人。」當然，曼森技巧純熟，依照潛在家族成員所想而示人，謎一般的開示也得以隨人闡釋（例如曼森告訴琳內特．弗洛姆（綽號唧唧）：「走出房間的方式不是穿過門，只要不想著出去，就自由了。」）。招募如此成功，曼森開始教導查爾斯．華森（Charles "Tex" Watson，綽號德克斯）等男信徒如何拉人入教派。在全盛時期，家族人數超過百名。

失意音樂家預知末日

曼森還是希望自己的音樂能名揚四海，一九六八年春天，他以為機會終於來敲門，海灘男孩（Beach Boys）樂團的鼓手丹尼斯．威爾森（Dennis Wilson）順道載了兩名搭便車的曼森家族成員。曼森趁機自我介紹，還向威爾森兜售毒品，並強迫家族女成員獻身。不久，曼森利用威爾森對音樂的不安，成功讓威爾森買下曼森創作歌曲〈停止存在〉（Cease to Exist）的版權，後來威爾森和海灘男孩據此重新製作。因為威爾森，曼森才認識了厲害的唱片製作人泰瑞．梅爾徹（Terry Melcher），他是女星桃樂絲黛（Doris Day）的兒子。曼森、威爾森和梅爾徹三人常常在梅爾徹家碰面，榭洛街一〇〇五〇號，正是未來曼森凶殺案最慘無人道的案發現場。

梅爾徹手上握有許多暢銷歌曲，他為飛鳥樂團（the Byrds）和海灘男孩製作專輯。曼森希望梅爾徹可以為自己施展同樣法力，獲致名氣，踏上心心念念的音樂之路。無奈事與願違，梅爾徹在史潘片場目睹曼森和特技演員大打出手，便斷然拒絕同曼森簽署任何唱片合約。

一九六八年十二月二日，海灘男孩版本的〈停止存在〉發行，是〈青鳥飛越山丘〉（*Bluebirds over the Mountain*）的B面歌曲，也收錄在《二〇／二〇》專輯。威爾森改動多處，曼森十分惱怒：歌名變成〈永遠學不會停止愛人〉（*Never Learn Not to Love*），歌詞也改了，威爾森重寫黑暗心靈邀請「停止存在」，成為較符合海灘男孩風格的誘人邀約「停止抗拒」。

火上添油的是，威爾森並沒有為曼森掛名，認為自己應得所有功勞，可能是因為他多少想拿回曼森家族時不時揩油累積的大筆開銷，還有曼森家族在史潘片場玩壞一台他的法拉利。威爾森說自己確實支付曼森十萬美金，還給了一輛摩托車，答謝他對這首歌的付出。但最令曼森惱火的可能是那張單曲銷售不佳，期待能聲名大噪的歌曲發行也草草結束，只是一首B面歌曲，排行榜名次就卡在六十一，這份恥辱終結了曼森根本不算起步的音樂職涯。

威爾森不知道惹怒曼森會有多危險。曼森發現自己沒有掛名時，威脅要殺了威爾森。現在換成威爾森要和曼森一刀兩斷。如此尷尬的結局導致曼森和音樂界再也沒有任何瓜葛，他的星夢也跟著消逝。現在曼森僅有的，就是家族對教派不滅的忠誠。

曼森開始預言詭譎幻象：未來末日降臨前，肯定會爆發一場血腥無比的種族戰爭，非裔美國激進派和白人自由派會與反對黑人民權的種族主義者為敵。甚至預示在戰爭存活的少數幾名白人會立刻遭激進派屠殺。依曼森所言，家族成員得在暴動時尋找庇護，逃到死亡谷（Death Valley）的一處「無底洞」才會安全，通過無底洞就能前往一座祕密地下城，安穩生活，遠離地上的戰爭。戰爭一落幕，曼森家族就會重回大地，成為美國唯一倖存的白人。暴動消滅老舊腐敗的文明，換曼森家族統治非裔美國人（曼森認為非裔美國人沒有自治能力）。最後，曼森還有一個毫無邏輯的念頭，莫名認為這場種族戰爭的激進派勝利人士會真心歡迎曼森統治。

一九六八年十一月二十二日，披頭四推出《白色專輯》（*White Album*），隨即空降排行榜，曼森更加確信這張當紅專輯為即將爆發的種族戰爭帶來了神祕象徵。曼森開始稱那場戰爭為「迴旋溜滑梯」（*Helter Skelter*），就是使用披頭四的同名歌曲。在曼森已然扭曲的心中，披頭四也知曉逐漸逼近的末日；曼森說服信徒，很可能只有他一人理解披頭四真正的訊息。這當然是一派胡言，披頭四的歌名源自英國孩童喜愛的遊樂設施。

曼森瘋狂的情節明顯是妄想，有些觀察員據此認定曼森患有妄想型精神分裂症。曼森的假釋委員會中，至少一位支持該理論，即使一名FBI探員在監獄訊問，發現曼森神智清醒——儘管他聰明伶俐、心狠手辣、麻木無感、操控人心、說謊成性。打從一開始，這樣的性格混合注定不安穩。然而現在再加上不受約束的復仇慾、近日怒火的導線、幾近全權掌控一大群嗑藥昏頭的信眾（全員認為曼森蒙神眷顧，可預示將來），零星小火便可引燃這一桶混合燃料。

一九六九年八月九日，在燠熱濕悶的史潘片場，曼森正處在這樣的心理狀態，但要處理的可不只是熱氣。信徒鮑比・薄索列（Bobby Beausoleil）英俊異常，年紀輕輕，是名來自海特－艾許伯里的演員，近期殺害了音樂教師兼藥頭蓋瑞・欣曼（Gary Hinman）而鋃鐺入獄。薄索列待在獄中越久，就越可能洩密於警方，曼森擔心不已。

才兩週前，七月二十五日，曼森和薄索列人在欣曼家討債，欣曼因毒品買賣未果欠錢。欣曼不服，曼森便拿刀砍了他一隻耳朵。這下顯然欣曼成了危害，輕易就能領著警方找到家族。曼森告訴薄索列：「你知道該怎麼辦。」

曼森究竟是像黑幫老大，或只是依循卡內基原則行事？不論如何，看來這是曼森恪守的準則：讓對方覺得這是自己的主意。曼森從來不需要下達追殺令。果然，兩天後的七月二十七日，薄索列刺殺欣曼，還有兩名曼森最虔誠的信徒蘇珊・阿特金斯（Susan Atkins）和瑪麗・布魯納從旁協助，兩人輪流用枕頭壓住欣曼的臉。薄索列用死者的血在牆上寫下「政治豬頭

（political piggy）」，蓋上豹掌印——黑豹黨的記號。家族成員打算誤導警方，期待警方分析犯罪現場證據，會認定犯人為黑豹黨。

不過薄索列在現場留下指紋，八月六日，警方在路邊的死者車上發現熟睡的薄索列。一經扣押，立刻發現薄索列的指紋與欣曼住處的一致。薄索列不是特例，日後曼森家族還有更多做事馬虎、粗心大意的人。

薄索列入獄，曼森斷定現在需要犯下一連串隨機卻相仿的殺人事件，好分散警方注意力。有人得死，有錢又有名的人，慘遭殺害會登上頭條的人。曼森煩惱的不只是可能再次入獄，還得持續控制信徒，相信他不是凡人，而是等同耶穌的存在——畢竟，曼森老愛跟信徒說他的名字「Manson」也就是「Son of Man」，意思是「人子[5]」——也是魔鬼。

精神科醫師利夫頓指出邪教領袖此一特殊手法，經常向信徒展示自己的地位等同強大的神靈角色，優於全人類，利夫頓稱之為「神祕操控」。按照利夫頓所言，這是思想改造（又稱洗腦）的關鍵步驟之一。神祕操控流派的另一面包括邪教領袖佯裝能預見未來，在信眾面前自詡為先知。伎倆要成功，邪教領袖得重新詮釋日常巧合，每個巧合都像預言成真。這正是曼森的目的，為此試圖把六〇年代晚期再平凡不過的活動——披頭四發行新專輯——解讀成末日種族

5 聖經用語，專指耶穌，也就是彌賽亞、救世主。

戰爭即將到來的徵兆。

曼森家族早就深信領袖是當代先知。只不過夏日將盡，黑人卻還沒有起身反抗白人的壓迫。曼森絕不能讓信徒覺得他錯了，也不能認為他可能沒那神聖。於是曼森一邊分散警方注意力，不再調查他和欣曼凶殺的關聯，一邊規劃、煽動種族戰爭：曼森家族會犯下新的連續凶殺案，並且再一次把黑豹黨牽扯進來。

這全仰賴一連串事件激起非裔美國人，面對這些凶殺引發的種族主義復仇，群起反抗——而這麼暴力混亂的情境會迫使警方停止調查欣曼死因。混亂之中，曼森家族會救出薄索列，曼森會避開刑期，家族全員會撤退到兩百五十哩開外，到死亡谷的地下應許聖所。如同曼森多數計畫，這次也純屬幻想，部分基於該區實際存在的廢墟和廢棄礦坑，部分則基於他對披頭四《白色專輯》的妄想詮釋，這張專輯應該預示了種族戰爭（〈黑鳥〉（*Blackbird*）一曲）和革命，此外，曼森家族相信啟示錄裡預言的基督正是曼森。

由於這則預言本就詭異，也很難認定曼森僅憑神祕操控之力便掌控了信徒。家族成員攜手踏上的數百場迷幻旅程，毫無疑問讓成員相信種族戰爭無可避免，而不用多久，這一小群根本餵不飽自己的嬉皮和中輟生會開始統治世界。先不管為什麼會得出如此詭異的結論，一九六九年八月九日，夜幕低垂，虔誠的曼森家族已準備好回應曼森的每道指令。

魔鬼行事

曼森選擇德克斯為首領導本次重大任務。會有德克斯這綽號是因為他來自德克薩斯州某小鎮，小時候在家鄉看似前途大好：田徑選手、橄欖球新星、班上票選人氣王、優等生，還是「應援團長」——頗負盛名的職務，被指派炒熱現場，帶領全體學生歡呼加油。德克斯就算跟了所有迷幻旅程，依然認為自己是天生領袖，甚至可能是曼森的敵手。曼森利用這一野心，說動現在腦袋不靈光的德克斯起身行動。

曼森也提醒德克斯該負的個人責任：一九六九年六月，德克斯搞砸了一場詐騙，本來會幫家族賺到搬去死亡谷所需的兩千五百美元，因此曼森開槍射了一名藥頭伯納德．克洛（Bernard "Lotsapoppa" Crowe，綽號：洛薩波帕）。洛薩波帕撿回一條命，卻從來沒有找警方。不管如何，曼森已經展露自己為了捍衛德克斯，殺人也無妨。德克斯能同樣挺身而出嗎？

要啟動種族戰爭計畫，曼森並不認為需要殺掉特定人士，只需要夠有錢又有名，名聲響亮到其死足以撼動社會，觸發戰爭。因此曼森把手伸進自己怨恨的深井中，挖出唱片製作人梅爾徹的地址，誰叫他拒絕和曼森簽約。曼森知道梅爾徹已經不住在比佛利山莊榭洛街一〇〇五〇號，但不管現在是誰住，想必一樣有錢又有名。曼森把這主意偷渡進入德克斯腦海之後，德克斯提議說其實他打算去梅爾徹之前住的地方。曼森操控德克斯的手法簡直出神入化，就連多年後FBI調查時，德克斯依然相信慘絕人寰的殺戮是他自己的主意。

曼森想要三名女信徒陪同德克斯。在曼森家族，女人必須遵守男人的命令，不得質疑——

這是曼森之前在拿撒勒人教會學到的信條，再依己意扭曲了一下。曼森沒有告訴三名女信徒當晚的殺人計畫，因為他明白不管德克斯的命令為何，她們絕對會照做。

為了好好協助凶殺，曼森從頭號大信徒中選了派翠西亞．克倫溫科（Patricia Krenwinkel）。光看外表，這名中產階級又上教會的甜姐兒可能不太適合曼森家族，但她十五歲時父母離異，中學就轉校了三次。因此，克倫溫科從小到大都覺得孤獨。只有一半血緣關係的姊姊帶她認識了迷幻藥，後來兩位女生都企圖自殺。克倫溫科初遇曼森之時，曼森立刻注意到這名十九歲女生既孤獨又缺乏自信。克倫溫科是曼森家族的理想人選——有些微故障但沒有全壞，即使曼森年紀大了十三歲，還是決定誘惑克倫溫科，成功讓她相信曼森就是她的靈魂伴侶。曼森可以給她所需的愛和關注。曼森可以為她的生命帶來意義。當她意識到曼森並非自稱的那般，曼森已經確保她因為太過恐懼而無法逃離。曼森會叫克倫溫科站著別動，再朝她擲刀。曼森會在其他女人面前羞辱克倫溫科，叫她裸體站在信徒中央，嫌她又醜又笨。曼森讓她生活在地獄，卻又給予恰到好處的關注，讓她覺得永遠都沒辦法逃離。

阿特金斯是這場殺戮的另一名共犯，也是一九六七年曼森假釋出獄遇到的頭幾名女性之一。阿特金斯十五歲時，媽媽癌症逝世，接下來的青年時期都在恐怖情人、酗酒和嗑藥之間循環。阿特金斯早就試過自殺，還因搶劫坐了幾個月牢，也當過撒旦教徒活動的「吸血鬼」上空舞者。看過曼森彈吉他，阿特金斯就成了手到擒來的獵物。兩人上床時，曼森注意到阿特金斯的不安，命令她假裝現在在和父親做愛。曼森爐火純青的洗腦效力強大：阿特金斯已經幫薄索列殺了欣曼，曼森知道時候到了，可以再次倚賴阿特金斯。

此外，曼森還需要一名把風司機。卡薩比安最近剛離婚，離開美國迷幻馬戲（the American Psychedelic Circus）公社，和德克斯上床，最終來到史潘片場加入曼森家族，還貢獻了五千美金（從前夫朋友那裡偷來的錢）。不同於多數曼森信徒，卡薩比安擁有正規合法的加州駕照，這也是為何當初即使其他家族成員反對，曼森依然執意要她加入。

一九六九年八月八日，德克斯開著一輛黃色一九五九福特，因為全車四人只有他知道目的地。德克斯準備了一條繩子、一把刀、一把．二二口徑高標「邦特蘭特製型」左輪手槍（.22-caliber Hi Standard “Buntline Special” revolver）。黃色福特蜿蜒開過班乃狄克肯楊區（Benedict Canyon）抵達榭洛街一〇〇五〇號，三名女人開口詢問今晚究竟要做什麼。要偷車嗎？還是要再做另一回「驚悚潛行」？三人完全不知道他們即將要犯下美國史上最惡名昭彰、最慘絕人寰的凶殺案。

德克斯把車停在那間住處的大門。從這裡看不到主屋，但圍欄有閃閃發光的聖誕樹燈飾，這些燈飾是梅爾徹和演員女友甘蒂絲．柏根（Candice Bergen）兩人搬出去之前柏根佈置的，沒多久，導演羅曼．波蘭斯基（Roman Polanski）在一九六九年二月中租下這裡。德克斯用大鐵剪剪斷房屋電話線，接著倒車下山坡，停在眾人視線範圍外。

四人翻過圍欄，德克斯說明現在他們要進入屋內，殺光所有人。但說時遲，那時快，正要動身之際，一輛車開下車道，停在大門口。四人躲在灌木叢，看到駕駛座坐著十八歲的史蒂芬・帕倫特（Steven Parent），帕倫特剛好來拜訪朋友威廉・加瑞森（William Garretson），加瑞森是榭洛街宅邸的年輕管理員，就住在主屋外的小客房。

當帕倫特搖下車窗，打算開啟電動大門時，德克斯步出灌木叢，揮刀砍他。帕倫特只求饒命，但德克斯接著近距離射了他四槍。四周峽谷吞噬了槍聲，使得近處槍聲聽起來像從遠方傳來。主屋沒有人察覺異狀，而加瑞森待在管理員小屋，已經調高了音樂音量，什麼動靜都沒聽到。

這四位家族成員現在朝主屋前進。德克斯派卡薩比安去檢查是否能從窗戶進屋，然後注意到一扇半開的走廊窗。德克斯剪開紗窗，往上推開窗，派卡薩比安回大門口把風。隨後，德克斯、阿特金斯、克倫溫科三人進入主屋。

這幢宅邸占地廣大，那晚有四人居住其間：社工師艾比蓋兒・福爵（Abigail Folger），也是福爵咖啡（the Folger's Coffee）的財產繼承人；波蘭演員沃伊塞赫・弗里科夫斯基（Wojciech Frykowski），福爵的男朋友；電影明星莎朗蒂，老公正是羅曼・波蘭斯基，她肚中懷有八個月大的胎兒；好萊塢髮型師傑・塞布林（Jay Sebring），莎朗蒂的前男友，但兩人依然關係緊密。大家才剛吃完晚餐回來，正準備好好歇息一晚。莎朗蒂剛剛和波蘭斯基通話，波蘭斯基打來說他完成下一部想執導的電影《碧海騰蛟龍》（*The Day of the Dolphin*）的劇本了，再過幾天就會回家。只要先生不在家，莎朗蒂總喜歡請朋友來家中作伴。

莎朗蒂出道時是時尚模特兒，後來成為演員，一路從電視連續劇《靈馬艾德》（*Mister*

Ed）、《打擊魔鬼》（*The Man from U.N.C.L.E.*，後來電影《紳士密令》的原型）的小配角，到電影主演，例如《天師捉妖》（*The Fearless Vampire Killers*），她拍攝這部片時認識了波蘭斯基；以及一九六七年的《娃娃谷》（*The Valley of the Dolls*），這部片讓她獲得金球獎提名。曼森家族沒有人知道目前是誰住在樹洛街，不過只是運氣好，剛好遇到期望中的名人，這些人的死會震撼全國。

午夜過後，宅邸的居民全都準備入睡了。福爵在客房讀書，弗里科夫斯基在客廳沙發上熟睡，塞布林和莎朗蒂在主臥聊得起勁。

家族成員發現的第一人是弗里科夫斯基。德克斯叫阿特金斯查看四周是否有人，而他的聲音吵醒了弗里科夫斯基。德克斯踹了弗里科夫斯基的頭，說道：「我是魔鬼，來這裡幹一番魔鬼行事。」並且警告弗里科夫斯基，要是敢通知其他人或求救，他就死定了。

阿特金斯路過福爵敞開的房門，朝她揮手；福爵習慣莎朗蒂家隨時都有訪客，早已見怪不怪，也朝阿特金斯揮了揮手，回頭繼續讀書。走廊盡頭，阿特金斯看到莎朗蒂和塞布林在聊天，但是他們沒有看到她。

與此同時，克倫溫科發現自己把刀忘在車上，於是跑回大門想要向卡薩比安借刀，反正把風用不到刀子。

德克斯聽完阿特金斯匯報，命令阿特金斯和克倫溫科把所有人帶到客廳。兩人持刀威脅之下，所有人都照做了。德克斯綁住塞布林的雙手，接著把繩子繞過他的脖子，套在天花板梁上，再把另一端繞在莎朗蒂的脖子上。莎朗蒂開始哭泣，塞布林抱怨他們對待孕婦太過粗魯。德克斯回以一槍，射中塞布林的腹部。髮型師不支倒地，血汩汩流出，染透客廳地毯。

德克斯下令交錢，但大家都沒有多少，除了福爵錢包裡的七十元美金。

德克斯本來預期會帶著一大筆錢離開，作為曼森家族遷徙基金，所以當事情變糟時，他非常生氣——畢竟這戶可是電影明星住的豪宅，但屋裡沒人身上的錢比他還多。

緊接著第一階段攻擊之後的殺戮非常醜惡、暴力、完全無法預測。但即使這幾位受害者手無寸鐵，而且殺手既冷血又被洗腦，行為毫無良心，殺人依然不是件容易的事。當天稍早，德克斯和阿特金斯都嗑了迷幻藥和安非他命，這又讓緊接而來的浴血之亂更加無法捉摸。

德克斯開始反覆刺傷塞布林。塞布林倒地奄奄一息之際，弗里科夫斯基掙扎著想幫自己鬆綁。阿特金斯試圖刺他，但弗里科夫斯基掙脫了，兩人在地板上扭打，阿特金斯的拳頭幾乎都落在弗里科夫斯基的大腿和脛骨上，痛得他放聲大叫，卡薩比安因此從大門口跑來。一抵達就剛好看見弗里科夫斯基踉蹌步出前門，整個人倒在灌木叢中。卡薩比安打算向弗里科夫斯基道歉，卻被跟著出來的德克斯打斷，然後看著德克斯留弗里科夫斯基一人在草坪等死。驗屍官後來報告，弗里科夫斯基被槍擊了兩次，刺傷五十一處，頭部遭那把邦特蘭特製型左輪的槍托重擊，倒地不起。

四人中只有卡薩比安在這場大屠殺流露出一絲道德愧疚，她著實嚇壞了，朝主屋大喊有人來了、有人來了。希望這個謊可以嚇到其他家族成員，結束這場血腥活動。但沒有人理她。

福爵打算衝出前門逃跑，但是克倫溫科緊跟在後，把福爵推倒在地，掄起刀反覆刺她，血液浸濕了福爵的白色睡袍。克倫溫科可能對自己剛才的作為感到過於震驚，跟德克斯說她不確定福爵是否真的死了。德克斯過來幫忙，殺死福爵，接著指派克倫溫科去檢查小客房是否還有人該殺。

克倫溫科陷入兩難，既害怕違背指令，卻又不想把更多人捲入這場大屠殺。因此她朝小客房走去直到德克斯看不見為止，等了一分鐘才轉身，回頭告訴德克斯沒有其他人了——這也是為何管理員加瑞森成為唯一倖存者。

到這個時間點，只剩下莎朗蒂還活著。她懇求殺手饒她的孩子一命，他們可以拿她當人質，再等兩個禮拜，寶寶就會出生，到時候她自己任憑他們處置。但是阿特金斯和德克斯嗑藥嗑到茫了，殺人殺上了癮，而且對曼森家族的教條深信不疑，根本不會在意腹中胎兒的幸福。阿特金斯刺了莎朗蒂十六刀，之後阿特金斯在法庭作證時還說那時她告訴莎朗蒂：「女人，我對你沒有任何憐憫！」而莎朗蒂則哭求著找媽媽。

不到半小時，每一位待在榭洛街一〇〇五〇號主屋的人都死了。

後來，阿特金斯記起曼森要他們在現場做一些「巫術般的」事情，於是她拿原本綁住弗里科夫斯基雙手的毛巾，在前門用血潦草地寫下「PIG（豬）」，模仿之前她看到薄索列殺了欣曼之後寫在牆上的字。接著德克斯、阿特金斯、克倫溫科離開犯罪現場，走下車道和卡薩比安碰頭，她正在黃色福特旁等著他們。他們拿出事先帶來的另一套乾淨衣物換上，除了卡薩比安，因為她一直等在車旁，身上一點血跡都沒有。

開回史潘片場的路上，卡薩比安把染血的衣物綑起來，並且按照德克斯的指令，全丟下一座陡峭的山丘，之後再往前開一點，也把武器都丟在同一條路上。他們繞遠路開上一條小巷，看到某戶人家的草坪上有澆花水管，便打開水龍頭洗掉手上和臉上的血。屋主出來要制止他們，但他們跳上車，揚長而去——只是速度不夠快，屋主已經記下車牌號碼。

等他們回到史潘片場時，曼森非常不開心。他們幾乎沒有帶回多少錢，而且依他們的描述，曼森認為他們並沒有製造一個夠有看頭的犯罪現場。曼森隨後自己開回榭洛街一〇〇五〇號，擦掉所有成員的指紋，還在莎朗蒂屍體旁的沙發上鋪了一面美國國旗，希望在這個年輕人燒國旗抗議越戰的時代，把國旗和死掉的孕婦並列會獲得莫大關注。但即便莎朗蒂的死訊占據了八月十日的所有媒體版面，曼森仍舊不滿足。依他所知，警方還沒發現任何線索，會把這起犯罪現場連結到黑豹黨或欣曼凶殺案。很明顯，這些殺害還不足以挑起他所預示的種族戰爭。得死更多人。

迴旋溜滑梯

八月十日晚上，曼森召集德克斯、阿特金斯、克倫溫科、卡薩比安以及另外兩名虔誠信徒史蒂夫·葛洛根（Steve "Clem" Grogan，綽號：克萊姆）和萊絲莉·范·豪頓（Leslie Van

Houten）。十九歲的范．豪頓是這群人裡面年紀最小的，在洛杉磯郊區蒙羅維亞（Monrovia）中產階級家庭出生長大，但隨著五年前雙親離婚，范．豪頓開始吸毒、逃家。一年前，她因為薄索列而認識曼森。范．豪頓最終變得只想討曼森歡心，為此無所不用其極。她在家族中最主要的功能是解讀曼森即興創作時潦草寫下的歌詞。不過那晚曼森幫她安排了更重要的任務。

曼森說明，因為八月九日早晨，他所有顧慮的事幾乎同時被搞砸，所以他想要大家至少再犯兩起模仿凶殺案，而且他會親自跟在一旁確保沒有差池。曼森下令車上要塞七個人，一道開往拉比安卡夫婦位在洛斯費利茲韋弗利街的住處。曼森之前參加過隔壁鄰居的派對，相信每位住在韋弗利街的人都富到流油，殺死他們絕對能轟動全國。他完全不知道他所選定的受害者只開了間雜貨店，而太太羅絲瑪麗也只是另外投資了一間精品店。

拉比安卡家的後門沒鎖，曼森和德克斯發現先生萊諾睡在沙發上，便把他叫醒，拿槍指著他。曼森總是想要受人喜愛，就連他打算殺掉的人也不例外，曼森向萊諾保證這只是搶劫，然後詢問屋內還有誰。萊諾告訴他們還有太太在臥室裡，並且指出臥室的位置。曼森把羅絲瑪麗帶到客廳，拿走了她的錢包，再把克倫溫科和范．豪頓帶到屋內。曼森命令眾人把羅絲瑪麗帶回臥房，並且要大家保證「每個人都有事情做」，這樣大家才能背負一樣的責任。隨後，曼森和卡薩比安、阿特金斯、葛洛根一起回到車上，開車離去，找尋下一個凶殺目標。

同一時間，德克斯拿了枕套蓋住萊諾的頭，堵住他的嘴，再用檯燈電線綁住他。接著他走回臥室對羅絲瑪麗做了一樣的事。克倫溫科抓了一把菜刀來料理羅絲瑪麗。自昨晚殺了艾比蓋兒・福爵後，克倫溫科不想要雙手再染血，但是違背曼森這一念頭則更嚇人。克倫溫科和范・豪頓在臥室乖乖等待，此時德克斯正拿著曼森給的刺刀，一刀一刀捅死萊諾。羅絲瑪麗聽到先生的哀嚎便起身掙扎，克倫溫科開始拿刀刺她。德克斯走進臥房給上致命的一擊，結束羅絲瑪麗的性命。

這一次，家族成員特別留心，為曼森獻上他指定的吸睛現場。德克斯在萊諾的肚子刻下「WAR（戰爭）」一字。克倫溫科現在克服了起初的猶豫不決，在萊諾的肚子上插了一把叉子，在喉嚨刺進一把刀。范・豪頓用拉比安卡夫婦的血，在牆上寫下「DEATH TO PIGS（豬該死）」和「RISE（起身而戰）」。到目前為止范・豪頓沒有參與多少實際凶殺，因此德克斯叫她掀起羅絲瑪麗的洋裝多刺幾刀，即便羅絲瑪麗早已死透。范・豪頓還看到一袋硬幣，立刻拿走。現在夫妻都死了，三人才意識到其實他們餓了，於是翻找冰箱，吃掉西瓜，也喝了巧克力牛奶。待凶殺後點心時間結束，他們還把西瓜皮整齊地放在洗碗槽，同時克倫溫科用血在冰箱門上寫下「HELTER SKELTER（迴旋溜滑梯）」。

就是這段選自披頭四《白色專輯》的致命文字，日後成為解鎖曼森家族祕密的關鍵，也是最終能將曼森依凶殺定罪的主要證據。曼森以為警方看到牆上的文字，就真的照字面理解，認為非裔美國社群犯下這一連串模仿凶殺案。然而，打從一開始，警方就沒有認為欣曼、莎朗蒂及其友人、拉比安卡夫妻這三起凶殺有任何關聯。事實就是，八月十二日時，警方告訴記者，莎朗蒂命案和拉比安卡凶殺案兩者毫無關聯。警方真的覺得手上的案子毫不相關。

曼森想要啟動「迴旋溜滑梯」，卻失敗了。而且即便警方忙著調查他竊車一事，看來他好像還能躲開這幾場大屠殺的後果。曼森家族一直都在偷車，偷來改造成沙丘越野車，為了之後躲在地下城等待種族戰爭結束時使用。這些小罪也是一開始警方在八月十六日下午四點來到史潘片場逮捕曼森家族成員的原因。曼森和另外二十五名信徒被圍補，遭警方拘押。其中一名遭到拘押的成員是阿特金斯。阿特金斯渴望注目和肯定，這種性格也是她一開始就認同曼森家族的原因。一關進監獄，他就開始向獄友維吉尼亞・葛瑞漢（Virginia Graham）炫耀她在前一週的殺戮中做了哪些事。葛瑞漢問她有什麼感受，阿特金斯回道，殺害身懷六甲的莎朗蒂時，一部分的自己也被殺掉了，現在她覺得既疲憊又興奮，還感到平靜，因為她知道這只是「迴旋溜滑梯」的開端。

這正是調查員需要的轉機，可以證明莎朗蒂和拉比安卡兩回凶殺案是同一個團體所為，而那個團體不會是黑豹黨。曼森、德克斯、阿特金斯、范・豪頓、克倫溫科五人全都被控凶殺。卡薩比安最初就不大願意參與殺戮，後來又被這些殘殺嚇到，早已駕著家族旅行車離開史潘片場，因此當曼森以為她出門跑腿，她實則逃到了新墨西哥州（New Mexico）主動成為汙點證人。自願提供的證詞為卡薩比安換來檢察官授予豁免，縱使她從未要求過。她坐在證人席上七天，檢察官文森・巴格魯希（Vincent Bugliosi）讚賞她始終一致的證詞，才能讓所有涉案人士都被判有罪。

曼森兀自作繭自縛。信徒越來越多，曼森不能只憑一己之能，必須仰賴其他人事物來挖掘不為人知的弱點，加以剝削，從中獲取個人利益。曼森重蹈覆轍落入相同陷阱——只要是邪教領袖都必定會跌落的陷阱——那就是打造自身成為神一般的存在，而且是可以預示未來的神明。曼森從未想過自己會因此進退兩難，也未曾懷疑自己臨時應允的種族戰爭、地下聖所，以及未來由他這幫格格不入的家族成員所統治的國家（當然可能會是全世界）都只不過是拖延時間的伎倆。曼森早就忘了自身那段不幸的犯罪過往：數度逃離感化院卻都失敗、多次漫不經心入室行竊、那張偷來打算在勞爾夫超市兌現的支票、想當皮條客卻又再次入監。就算曼森再怎麼善於利用人心不安，天賦也有派不上用場的時候，例如謀劃執行一場駭人犯罪，又或者一連串錯誤創造的鐵證：阿特金斯莫名其妙地把刀留在榭洛街住宅的沙發靠枕下方；德克斯拿邦特蘭左輪槍托重擊弗里科夫斯基，用力過猛，那把便宜槍枝的握把碎裂，散落在屍體旁讓警方得以追查；德克斯用沾血的手指按下那扇電動大門的按鈕。曼森家族是犯罪組織，但家族具備的技巧都在常人預料之中——就是隨機挑選來自破碎家庭、既失業又嗑迷幻藥的人能具備的技巧。有這麼忠誠的信徒，還有他的救世目標，曼森最遠大的計畫卻直接帶領他重回此生待最久的地方——監獄。而且，儘管現在曼森臭名昭彰，這計畫也帶他重回幾乎自出生就注定的初始羞恥。

封鎖家族

一九七〇年六月十五日，曼森的審判開始，每晚電視都會播報訴訟程序，但審判迅速變調，成為擄獲全美目光的鬧劇。曼森視每次出庭如一場表演。他不停擾亂法庭秩序。他會攻擊法官，直接衝向法官，拿鉛筆當武器，同時大喊：「你真該被砍頭，老頭！」他對阿特金斯、克倫溫科和范・豪頓下達明確指令，告訴她們怎麼一起搞破壞。某一次這些女信徒就在法庭唱起歌來，下次曼森會要求她們鬼吼鬼叫，再下次則是剃成光頭。曼森在額頭刻了「X」，還面對鏡頭發言。現場聽這段發言的人可以聽出裡面混雜著真實、前後不一、自怨自艾、操弄人心的言論，曼森說：「我是你們造就的。我曾經良善，但現在我什麼都分不清。你們這群人對我毫無憐憫。我頭上的記號象徵著拒絕、反基督教會、墜落十字、魔鬼符號、死亡、恐怖、懼怕的黑印記。我想過要當個好人，但你們不讓我當好人。」

許多女信徒沒有受審，她們就駐紮在法院外圍，打扮得像花之子，接受媒體訪問。有些女信徒則企圖威脅，甚至毒害目擊證人。家族成員芭芭拉・霍伊特（Barbara Hoyt）同意和檢察小組談話，便收到數名曼森信徒提供她夏威夷旅行作為交換，希望她拒絕作證。露絲・安・摩爾豪斯和芭芭拉・霍伊特兩人一同前往夏威夷，就在出發前，摩爾豪斯把可能致死的十劑迷幻藥藏在漢堡裡，哄騙霍伊特吃下。霍伊特有幸活了下來，更加渴望成為檢方的目擊證人。

曼森使出最終險招，打算叫他的共同被告向法庭宣布他們自行犯下這些凶殺，曼森並未涉入其中。但是，范・豪頓的辯護律師羅納德・休斯（Ronald Hughes）察覺曼森正在操控他的客戶。休斯律師對曼森的觀察可能過於敏銳了——就在結案陳詞開始前的週末，休斯去露營，卻失蹤了。發現他遺體的那天，正是每一位被告都被判死刑的那天。

雖然羅納德・休斯的死因至今未解，但有一位匿名的曼森家族成員聯絡首席檢查官文森・巴格魯希，告訴檢察官曼森家族該為此負責。巴格魯希檢察官在結案陳詞時，把曼森描述成「馬屁精奴隸部落的獨裁主人」。就算領袖身陷囹圄，家族依然樂意為他殺人。

儘管被判刑，曼森家族成員都只做了一會兒死刑犯。一九七二年二月，加州廢除死刑，曼森、德克斯、克倫溫科、阿特金斯、范・豪頓以及加州系統判定的每一位殺人犯（包括刺殺甘迺迪（Robert F. Kennedy）的瑟漢（Sirhan Sirhan）），刑期都變成終身監禁得以假釋。

曼森信徒一進監獄，遠離了曼森的影響，終於開始動腦思考。許多信徒嘗試新的身分認同。德克斯轉信基督教，成為獄中大受歡迎的牧師，雖然一位ＦＢＩ探員調查德克斯時懷疑其真誠，認為這名主謀殺手只是為了假釋而裝模作樣。儘管德克斯盡一切心力改過向善，加州假釋委員會依然認為他會危害社會，最近一次拒絕他的假釋是在二〇二一年十月。

阿特金斯也重生為基督徒，換一種方式奉獻。他於二〇〇九年癌逝，創下全國女囚最長坐牢紀錄——現在克倫溫科和范・豪頓雙雙打破該紀錄。

薄索列也轉向另一種狂熱，加入標榜白人至上的雅利安兄弟會（Aryan Brotherhood）。二〇二二年，薄索列同樣被拒絕假釋。

克倫溫科一變成死刑犯，就開始獨立思考。尤其在得知曼森輸了牌局，把她賣給另一名囚犯後，更加確認自己一定要獨立思考。克倫溫科宣稱她已經重拾自我，而且有了自我意識，當初所有刻意不讓自己感受到的懊悔也回來了。她現在是模範囚犯，同時身兼不同監獄團體的導師。儘管她宣稱自己已經逃離曼森的影響，在二〇二二年，她第十五次提出假釋，依然被駁回。

這群殺人犯裡面年紀最小的范．豪頓也成為模範囚犯，並且公開表達她對曼森家族殺戮的懊悔。父母離婚鬧得雞犬不寧，導致她在十四歲反覆吞食迷幻藥，後來她認為多虧父親重返她的生命中，幫助她能重新認識真實人生，並且重拾自己的道德規範。在二〇一三年的假釋聽證，她的律師指出她的價值觀已經完全改變了，迥異於一九六九年遭受曼森操控的狀態。二〇一六年和二〇一七年，她受到兩次推薦，得以申請假釋，但加州州長傑瑞．布朗（Jerry Brown）為了尊重受害者家屬的要求，連續拒絕兩次假釋。家屬還是認為范．豪頓是毫無悔意的殺人凶手。她最近一次在二〇二二年二月向加州最高法院（the California Supreme Court）申請假釋遭拒。

對某些人而言，曼森的洗腦效力過強，無法擺脫影響。頭號信徒瑪麗．布魯納因為信用卡詐欺而坐牢，所以錯過這幾場凶殺，但後來出庭作證反對曼森家族，她對此依然感到愧疚。她和其他多位家族成員後來因偷竊槍支都被關進監獄。他們希望用偷來的槍在一九七一年劫機，好讓曼森自由。布魯納在一九七七年獲釋，之後便從大眾眼前消失。

綽號唧唧的琳內特．弗洛姆企圖刺殺總統福特（Gerald Ford）失敗，因此在一九七五年坐牢。但獄中生活不足以動搖她的信念。她和家族成員好朋友珊卓．古德（Sandra Good）關在一起，兩人自稱「曼森教會的姊妹」。弗洛姆對曼森的信念堅如磐石，聽聞他診斷出癌症之後，在一九八七年十二月二十三日逃獄去見他一面。弗洛姆再次被捕，最終在二〇〇九年十二月假釋出獄。在ABC新聞台特別節目《曼森女孩》（*Manson Girls*）訪問中，弗洛姆坦承，即便殘暴的殺人事件都過了五十年，她還是深深愛著曼森。她說：「今生能遇見他，是我莫大的榮幸。我知道對那些認為他邪惡至極的人而言，這句話在他們耳裡聽起來是什麼感受。」

卡薩比安出席了十七天審判，作證曼森家族成員的罪行，讓所有成員都判了死刑，後來的人生少有人知。最後一次有人看到她時，她住在華盛頓州塔科馬（Tacoma）的低收入戶住宅，使用第二個化名，住得其實離她四位兒女中最小的孩子家不遠。自從那些判刑以來，這些年她一直活在罪惡感中，痛苦掙扎，也在嗑藥、酗酒中受苦。她也說過，她沒有一天不想起那些無謂的凶殺。她的逮捕紀錄證實這些年來她的自我控管出了問題：一九七六年，在新罕布夏州（New Hampshire）納舒厄（Nashua），她阻止消防員撲滅篝火，此舉妨害治安被開罰一百美元；一九八二年，在新罕布夏州拉科尼亞（Laconia），她在一場摩托車集會裸露胸部，被控妨害風化暴露罪；一九八七年，在佛羅里達州卡納維拉角（Cape Canaveral）被控酒駕；一九九六年最後一次被捕是因為持有快克古柯鹼和安非他命，那段時光如他所述為「毒品、酒精、自我毀滅時期」。曼森過世後不久，她接受《每日郵報》（*Daily Mail*）採訪，告訴記者，直到今天，她依然可以聽見受害者的尖叫聲——「如果我讓自己重返當時的話。」

Chapter 2

羞恥

亞道夫·德·赫穌斯·康斯坦佐與毒梟撒旦信徒

亞道夫・德・赫蘇斯・康斯坦佐（Adolfo de Jesús Constanzo）在邁阿密土生土長，是一名古巴難民的美國籍兒子，毫無爭議是本書描述的所有邪教領袖之中，最殘虐而且最系統性嗜血的領袖。康斯坦佐在一九八〇年代晚期短暫的恐怖統治期間，從墨西哥市（Mexico City）到德州布朗斯維（Brownsville）附近的邊境城鎮，經查證至少犯下十六條人命的祭祀凶殺，不過他無法饜足的殺人慾望、殘暴無道和四處恐嚇的行徑都顯示死亡人數可能更多。康斯坦佐從小住的家充斥著動物排泄物以及腐敗的祭祀牲口遺骸（業餘手法處理的牲禮），這成為他無法磨滅的羞恥感，驅使他終其一生殘暴無情。康斯坦佐生命中少數秩序的來源之一，來自薩泰里阿教（Santería）一支名為帕羅瑪尤貝教（Palo Mayombe）暗黑教派的訓練，學習在儀式中如何不顧活物的痛楚依然殺生獻祭。他在邁阿密進行精心策劃的入會儀式後，在墨西哥開立店面，從算命師搖身一變成為「教父」——毒梟撒旦教（Narcosatanists）的主腦。毒梟撒旦教是暴力犯罪集團和邪教，運作機制得仰賴他例行殺人血祭帶來的超自然保護。活人獻祭極端殘虐，祭祀期間，他會活剝受害者的皮，一塊一塊肢解，將器官一個個摘除，或拿開山刀朝頭顱迅速一擊斃命，或者拿刀往受害者脖子一抹。

注定殺生

是什麼讓十五歲單親媽媽生下的小男孩，長成青年之後開始淡定放血、切除性器、取出受害者大腦，丟入他的恩甘加（nganga）靈鍋之中？他的童年究竟發生了什麼事，導致他變成如此冷酷無情的系統性殺手？

一九六二年十一月一日，在邁阿密小哈瓦那區（Little Havana），迪莉婭．奧羅拉．岡薩雷斯．德爾．瓦萊（Delia Aurora Gonzalez del Valle）生下康斯坦佐。十幾歲的迪莉婭才剛逃離革命後的古巴。我們對康斯坦佐的養育過程知之甚少，因為在承受他一連串殺戮的波及之後，還在世的家人都避免任何曝光報導，這倒也不難理解。我們只能從信徒的證詞得知男孩的童年，其中一位信徒在他小時候是迪莉婭的朋友。

這些消息來源指出，迪莉婭是虔誠的薩泰里阿教信徒。薩泰里阿教在古巴已經有四百多年的發展歷史，當初由被迫為奴的西非、中非人帶來，後來成為後裔的信仰。薩泰里阿教最純粹的形式以約魯巴（Yoruba）習俗為基礎：催眠儀式、擊鼓儀式、動物祭祀和神祕的先祖通靈儀式。在古巴，有人改良了儀式，把家鄉神祇的超能力加諸於天主教聖人身上，藏起原本崇拜的真實本質。這樣簡單的偽裝讓他們幾乎得以繼續公開執行祖先的儀式。

迪莉婭有回宣稱自己向來都是會做禮拜的天主教徒，從來就不信什麼薩泰里阿教（此說無從驗證，因為她在一九八九年五月為兒子認屍之後不久，便完全從大眾眼前消失，從此杳無消息）。她的說法和康斯坦佐信徒以及她自己前鄰居的說法有出入。邁阿密警方也搜集到大量證據，證明她確實參與薩泰里阿教和帕羅瑪尤貝教（Palo Mayombe），後者是古巴和海地的主要宗教，也是康斯坦佐和毒梟撒旦教信徒（後來大家都如此稱呼）宣稱他們所信奉的宗教。這些描述提供了基礎，讓我們可以更全面了解康斯坦佐早年的生活。

薩泰里阿教和大多數其他亞伯拉罕諸教（主要為猶太教、基督教、伊斯蘭教）最大的差異在於，薩泰里阿教沒有明確的誡律。對薩泰里阿教來說，法術可以行善也能作惡，咒語可以助人也能害人。不論幫助人還是傷害人，端看作法唸咒的人而定。現在，加勒比地區和拉丁美洲境內，各行各業都有人信奉薩泰里阿教。如同許多古巴難民，康斯坦佐的媽媽也帶著這一傳統抵達美國。

康斯坦佐才剛出生，他的爸爸就拋家棄子，沒有證據顯示康斯坦佐是否知道爸爸是誰。那些年裡，迪莉婭至少嫁了三位男人，但在康斯坦佐生命中最具影響力的男性角色則是教導他帕羅瑪尤貝教儀式的人。

帕羅瑪尤貝教是另一種古巴宗教，很多方面都和薩泰里阿教十分相似。帕羅瑪尤貝教結合了天主教和源自西非的宗教儀式（跟著強迫為奴的人一併來到美國，而本次案例則是來自剛果河沿岸）。瑪尤貝（Mayombe）是非洲一區域，也是這一宗教的發祥地，而帕羅（palo）是西班牙語的「木棍」或「樹枝」的意思，施咒前會用帕羅攪拌靈鍋。

雖然薩泰里阿教和帕羅瑪尤貝教沒有正式的連結，但雙方信仰與信徒重疊的部分很大。比起請求神明幫忙，帕羅瑪尤貝教的信眾反而想要透過咒語控制亡靈。如同薩泰里阿教，這些都能用在好事和壞事上。大家經常將帕羅瑪尤貝教描述成薩泰里阿教的病態表親，這是因為它和死者的關聯，但值得注意的是，它本身並不具有邪惡的本質。

康斯坦佐六個月大時，迪莉婭帶他和一名海地籍的帕雷洛祭司（palero）見面，帕雷洛祭司相當於帕羅瑪尤貝教的牧師。迪莉婭希望對方可以提供一些育兒指引。帕雷洛祭司檢視這名男嬰，宣告他是天選之人，並且預言男嬰將來會成就一番事業，甚至自願擔任康斯坦佐的精神導師，也就是他的教父（padrino），迪莉婭對此欣然允諾。

生於貧窮髒亂

但這段師徒關係根本還無法開始，迪莉婭就已經離開邁阿密，跟隨即將要嫁的男人搬到波多黎各聖胡安（San Juan），那時康斯坦佐還在蹣跚學步。沒有人知道他們家在波多黎各的生活，也沒人知道康斯坦佐的第一位繼父是誰（總共有三位）。依迪莉婭所言，他的兒子是模範孩童，是當地天主教教會的輔祭男孩，還打得一手好網球。

即便康斯坦佐在公開場合看起來過著模範生活，但一回到家裡，他就被拋進大不相同的世界。他的童年有兩個關鍵的形塑環境，兩者似乎都預示了日後他對待陌生人、敵人和親信一貫出名的暴行和殘忍：生長環境的貧困髒亂以及早年的牲祭教育。

生長在毫無衛生的環境，幾乎無時無刻不亂成一團，似乎深深影響著康斯坦佐。迪莉婭是囤積狂，幾乎無法丟棄任何東西，這個強迫行為會立刻把他們生活的每個地方都變成一團混亂。而且更糟的是，家裡很快就充斥著排泄物和動物遺骸，擺明了迪莉婭雖然是門外漢，還是企圖執行薩泰里阿教的儀式。

薩泰里阿教祭司又名薩泰羅（santero），經常獻祭求告神明。若是處理小事，祭品通常都是小東西，例如食物或花朵。雖然牲祭並不少見，但殺生總是得嚴肅以對。只有至關重要的情況才會需要殺生。甚至在犧牲一頭動物時，殺生儀式常伴隨著祈禱，儀式末了，與會者會烹煮並分食該頭牲禮的肉。

不過，迪莉婭看來一直活在幾乎連續不止的緊急狀態，時時刻刻都在警惕。或許是為了對抗想像中的威脅，她也囤了一大群動物，只要一有機會就殺來獻祭。鄰居說他們在自家門口發現遭到斬首的雞隻。其中一位宣稱自己看到迪莉婭把無頭的山羊屍體直接丟在路上。雖然這些事件都有回報到警局，但沒有人提出指控。

幾乎可以肯定康斯坦佐看到周遭的混亂而產生的反應，他年紀輕輕便發展出不尋常的特質：他變得過度乾淨整潔。他差不多在會自己穿衣服時，就開始在每晚睡前悉心擺放好衣服，而且他也將自己的房間以及其他任何可以視為己物的東西維持得乾乾淨淨，一塵不染。

康斯坦佐就像查爾斯．曼森，似乎從最易受影響的年紀開始就一直在對抗深深的羞恥感。曼森的羞恥似乎來自特定事件（目睹媽媽被逮捕入獄、被迫打扮得像個女孩去上學），康斯坦佐的羞恥則來自家中未曾停歇的貧困髒亂。邁阿密的鄰居覺得他家很可疑，有小朋友的家庭都不會邀康斯坦佐去玩，而且所有鄰居都知道迪莉婭永遠不會應門，這與警方的報告一致。拒絕應門可能和囤積狂常見的行為有關，一般稱之為「門鈴恐懼」，有些人會讓生活空間亂到無法見客，這樣就能避免外人進入屋內，相當清楚任何外人一看到他們的生活環境都無法苟同。這經常導致諸多不健康的結果，例如嚴重的社會疏離。這一套行為模式有時又稱戴奧基尼氏症候群（Diogenes syndrome），常伴隨老年失智而發生，但難以確認一般人是否也常患此症，因為幾乎半數的囤積狂都有高智商，成功躲過任何偵查，也避免除了最低限度之外與世界的一切互動。也因此迪莉婭能在兒子過世後消聲匿跡，一點也不令人意外——兒子在世的時候，迪莉婭都一直在練習這個伎倆。

康斯坦佐在這種骯髒貧困的環境中長大，他內化的羞恥感有各種名稱：有毒羞恥感（toxic shame）、核心羞恥感（core shame）。但這些持續不斷的感受，不論壓抑得多深，都會造成同樣的行動，而似乎就是這些行動觸發了曼森的危險怒火。或許康斯坦佐認為自己的強迫症（他堅持掌控中的所有事物都要維持一定秩序）證明他遠離了掌管媽媽宇宙的混亂。但從迪莉婭似乎偏好家中毫無秩序的狀態，也不難推測，康斯坦佐長大成人後還是保有小時候無法克制的潔癖，後來潔癖不僅轉變成對所有環境都要求一塵不染，還轉而追求極端的秩序，下令所有信徒發血誓完全服從於他，違者就會被處死。帕羅瑪尤貝教從旁加強這一狂熱，因為它將世界一分為二：信者和不信教的人。就如同他的教父一般，康斯坦佐開始相信，作為一名帕雷洛祭司，他對不信教的人沒有任何責任。他們全都該死。

凶殺見習

康斯坦佐一家在聖胡安住到一九七二年，那年迪莉婭的丈夫被診斷出癌症。就像康斯坦佐媽媽的其他許多方面一樣，沒人知道為何她選擇現在這一刻搬回邁阿密，獨留配偶在聖胡安直到隔年過世。

此趟重返邁阿密，十歲的康斯坦佐和海地籍教父重新聚首，就是那位在康斯坦佐仍是男嬰時查看他，並認為他是天選之人的男人。這男人的來歷鮮為人知，幾則相關消息都是由康斯坦佐信徒的二手陳述，但報告都同意這男人能有穩定收入，都是靠其為邁阿密毒品交易的黑社會巨頭施展帕羅瑪尤貝教咒術而來。

帕羅瑪尤貝教分為兩個教派，一是帕羅基督派（Palo Cristiano），一是帕羅猶太派（Palo Judio）。所謂基督派的帕雷洛祭司和神一同做工，鄙視惡靈；而另一派的帕雷洛祭司則和惡靈共事——這派會叫「猶太派」和猶太教沒有任何關係，純粹因為這群信徒拒絕遵守奴役他們的人所信奉的教條。這派認同某種神靈般的存在，他們稱之為卡迪恩龐貝（Kadiempembe），意思是「噬魂者」——其實有點類似西方概念的魔鬼。康斯坦佐的教父就是這一傳統的帕雷洛祭司，敬畏卡迪恩龐貝的力量。

康斯坦佐第一次站在教父的恩甘加（剛果語，廣義指死者或超自然力量）面前時，他才剛滿十一歲。在帕羅瑪尤貝教，這個字也代表之前提到的靈鍋，就是帕雷洛祭司和靈界互動的主要工具。恩甘加是帕雷洛祭司的力量來源，也是帕羅瑪尤貝教獨有的文化。

恩甘加必須盛滿牲祭的屍骸、神聖藥草、昆蟲、一段樹枝（又叫帕羅）。但最主要的材料是人體殘骸。恩甘加需要人的頭顱、大腦，以及同一具屍體的多根骨頭，才能好好地代表死者的靈魂。只要殘骸適當地擺放在恩甘加內，該名亡靈便會執行帕雷洛祭司的命令。

帕雷洛祭司相信恩甘加會服從自己的擁有者，絕不二心。與惡靈合作的帕雷洛祭司盛讚暴力罪犯和瘋子的殘骸，因為這群帕雷洛祭司相信這種人執行任何命令都不會遲疑，不管命令有多邪惡。大家相信亡靈不需要事先給祭品，雖然帕雷洛祭司通常都會在命令達成之後，拿公雞血作為獎勵祭祀亡靈。

學習帕羅瑪尤貝教是康斯坦佐唯一最像接受真正教育的時候。他在家會參與媽媽不專業的殺生，而在教父的監督下，他學習複雜精細的死亡儀式。這是第二個形塑環境，他有系統地學習如何殺生，造就他出名的殘忍作風。

這一發展至關重要。自一九七〇年起，FBI罪犯剖繪專家報告顯示，最冷酷無情的連環殺手有一項共通的常見特質，那就是童年幾乎都參與刻意的動物虐待和惡意對待。康斯坦佐的教父幫他上的正是這種行為的大師課，訓練這名男孩相信為了成為強大的帕雷洛祭司，即使殺了另一個人類（尤其是不信教的人），也是合情合理。就這樣循序漸進認識各種儀式，康斯坦佐開始相信，只要帕雷洛祭司下令，任何系統性殺生都合情合理。

如果目標是要能殘忍地殺生，他所接受的訓練超乎尋常地嚴格、有架構且有效。姑且不論是否刻意而為，這訓練完美地逐步灌輸康斯坦佐心理學家所稱的暗黑三角人格：心理病態，或缺乏懊悔之心；惡性自戀，或施虐性自大狂；馬基維利主義，或自私自利剝削他人者。

餵食恩甘加

康斯坦佐的教父把自己的恩甘加放置在花園小棚屋，四面無窗。在此區，教父都會要康斯坦佐矇住雙眼，教父宣稱他的恩甘加會殺掉還未準備好見它的人。在康斯坦佐訓練全程，教父都堅守這項原則，只要發現康斯坦佐打算偷窺，教父就會毆打他。

教父教導他如何為藥頭施展保護咒——藥頭是穩定客源，而且出手大方。教父態度明確，雙方僅僅只是生意往來。教父看不起藥頭本身，而不信教的人用毒品摧毀自身，對他而言無關緊要。就像迪莉婭，教父也認為不相信薩泰里阿教和帕羅瑪尤貝教的人，其性命就沒什麼價值。利夫頓醫生稱此種態度為「生存支配」（dispensing of existence）。此一概念解釋了邪教成員的共通行為，成員通常都相信只有他們自己活在神聖的圈子裡，因為他們知曉更高階的真相，而能夠擁有這種真相，則代表其他不知道真相的人，即便是自己的家人，也都活在神聖圈子之外。邪教成員利用此一關鍵信念來合理化對外人（例如不信教的人）犯的所有罪行，就連凶殺也變得合理。

教父教導許多原則，其中一項是禁用毒品。

教父告訴康斯坦佐，假使某天有靈附身，卻發現這具身體受非法藥物汙染，靈體會殺了康斯坦佐。康斯坦佐終其一生都遵守這條原則。一九七七年，康斯坦佐十四歲，開始宣稱自己能通靈——這一特異能力成為他邪教的基石，不僅是其違法事業的必要元素，還是導致其最終垮臺的關鍵因素。第一項通靈證據似乎瞄準了他的媽媽。迪莉婭是瑪麗蓮夢露的粉絲，康斯坦佐告訴媽媽他做了夢，夢中得知這名電影巨星其實不是自殺身亡，雖然這剛好和許多人懷疑的一樣，不過媽媽聽了這個新結局，心情大好。

三年後，他媽媽發誓康斯坦佐早就預言了雷根總統會遇刺。當地薩泰里阿教社群的其他成員則駁斥她的炫耀吹噓。前房東爾內斯托．皮察多（Ernesto Pichardo），同時也是邁阿密知名的薩泰羅祭司說道：「光聽迪莉婭說的，你會以為康斯坦佐是另一位救世主。」

邪教領袖經常自視甚高，並且認為一切天經地義，這些感受肯定其來有自。迪莉婭和教父對待康斯坦佐，宛如他是「天選之人」，教導他：其他人都比不上他。但對他的老師而言，他肯定不是另一位救世主。一九八一年，康斯坦佐自高中畢業，成績很差。他到當地社區大學上了一學期，之後就輟學了。隨後他開始偷東西，但徹頭徹尾失敗，第一次被捕是因為偷衣服，後來又因偷電鋸而被捕。

揀選虔誠信徒

一九八三年早期，康斯坦佐的教父督促他搬到墨西哥市，開始經營施咒事業。薩泰里阿教和帕羅瑪尤貝教並非墨西哥隨處可見的儀式，但另一支名為庫蘭德洛派（Curanderismo）的民俗法術頗受歡迎，此派使用藥草來療癒精神。康斯坦佐和他的教父認為，會相信民俗法術的人可能也比較容易相信其他的民俗法術。沒多久，康斯坦佐和媽媽開始定期前往墨西哥市，媽媽介紹他認識瑪麗亞．德爾．羅西奧．奎瓦斯．葛拉（María del Rocío Cuevas Guerra），這位熟人後來成了開業初期的客戶，在康斯坦佐死後，則提供了當局有關康斯坦佐謎樣童年最多的細節。

在墨西哥市，康斯坦佐在玫瑰區（Zona Rosa）路邊一家咖啡廳開店，開始幫顧客讀牌，加上他，這個一級觀光區就有十四位算命師了。多虧童年的訓練，他非常擅於為顧客施展看似法力滿滿的咒語。一九八三年四月，康斯坦佐正是在那裡遇到年輕的商學院學生馬丁．昆塔納．羅德里格斯（Martin Quintana Rodriguez），馬丁剛結束一段認真的關係，而康斯坦佐假裝自己不需要馬丁說出口就知道這件事。兩人開啟一段他們自己的戀情，康斯坦佐對認識的人都只說馬丁是保鑣。保鑣這一稱呼並非只是掩護的說詞：馬丁身體健壯、肌肉發達，站在康斯坦佐旁邊，確實令人望而生畏。

開業才幾週，康斯坦佐精準到嚇人的算命師名聲便傳開了。他究竟如何獲致神算的名聲，原因不得而知，但他似乎有天賦，這種神力不是解釋成操控人心就說得過去。每當他需要吸引更多信徒，又或者增加盈餘時，他也會毫不遲疑使用這項天賦。

康斯坦佐才和馬丁相遇沒幾週，豪爾赫．蒙特斯（Jorge Montes）和奧馬．奧雷亞．奧察亞（Omar Orea Ochoa）就在他的占卜桌旁找到這位算命師。豪爾赫五十一歲，之前是一名模特兒，本身也是讀牌師，平常在玫瑰區同志酒吧幫人算命，這也是他和奧馬相遇的場景。奧馬十八歲，是新聞系大學生，自從有位算命師告訴他，某天他會遇見一名強大的男人，而那人會改變他的一生，奧馬就深深著迷於各種超自然神力，無法自拔。因此當豪爾赫聽聞謠言說有一位新來的神準讀牌師，奧馬迫不及待親眼瞧瞧。

幾乎就在這兩個男人坐下要求算命的同時，康斯坦佐就告訴奧馬，他年輕時的預言就要實現了，並且宣稱自己其實就是多年前奧馬算命時得知的那位強大男人。奧馬立刻就迷上了康斯坦佐。

算命結束，三人聊了好幾個小時。康斯坦佐告訴他們他是薩泰羅祭司，還炫耀他的宗教有多強大。當康斯坦佐提到他計劃搬到墨西哥市生活，豪爾赫立刻說要幫他找模特兒工作。康斯坦佐確實長相帥氣，而且豪爾赫說他可以利用自己事業的人脈來幫忙。豪爾赫也說他還可以幫忙宣傳康斯坦佐的讀牌服務。

康斯坦佐本來就計劃在墨西哥市多待兩週，之後就要回邁阿密，但這次相遇之後，他幾乎把所有時間花在奧馬身上，向奧馬解釋薩泰里阿教和帕羅瑪尤貝教，以及他能施展這兩個宗教的不同法術。他會擲螺貝（薩泰羅祭司常用此法預測未來），告訴奧馬貝殼預言前途一片光明。再一次，他似乎又知道從來沒有人跟他說過的事，他一一唱名奧馬兄弟姐妹的名字，證明他的神力。

康斯坦佐開始引誘奧馬，在玫瑰區請他吃飯，送他昂貴的禮物，這些招式顯然很有效。第一次相遇以來才幾天，兩人就睡在一起了。而且幾乎從一開始，康斯坦佐就開始制定規範，彼此之間迅速變成凌虐關係，他還讓奧馬知道馬丁也是他的情人。康斯坦佐打算讓馬丁成為他的「男人」，而奧馬則會成為他的「女人」。奧馬不喜歡與人共享康斯坦佐，但也同意了這樣的安排。

情緒勒索的模式很早就浮現。康斯坦佐拒絕做一些簡單的事，即便奧馬求他，他也不願意施展任何法術，除了一些簡單的伎倆，例如擲螺貝。接著康斯坦佐也吝於展現情意。他對奧馬非常慷慨，但也維持距離。康斯坦佐似乎從未感到滿意或開心，即便奧馬和馬丁做的都是他想要的事。這是「間歇獎勵」（intermittent reward）系統：長時間不展現情意或不認同情緒，隨後短暫給予激情熱愛。間歇獎勵系統是凌虐關係的正字標記，只會讓奧馬更加渴望取悅康斯坦佐，而康斯坦佐很快就能情緒勒索奧馬。情緒勒索則為更加公然暴力的剝削行為鋪路。

儀式型附身

一九八三年，康斯坦佐多次往返邁阿密和墨西哥市，在玫瑰區建立聲望，並向教父回報日後展望。他長時間向教父學習帕羅瑪尤貝教，從旁協助教父，終於在一九八三年尾聲，二十一歲的康斯坦佐準備好出師了。如職業術士所說，他即將「拉亞多恩帕羅（rayado en palo）」，意思是「晉級帕羅」。

我們所知的出師過程都是根據多名信徒所言，這些信徒跟在康斯坦佐身旁，參與所有儀式，而康斯坦佐告訴他們，這些儀式都是仿自他的個人經驗。信徒的說法與學術文章講述的帕羅瑪尤貝教出師儀式一致。

儀式執行前好幾個禮拜，康斯坦佐就必須開始準備。他在美洲木棉樹下睡了七夜，美洲木棉樹是帕羅瑪尤貝教的聖樹，原生於加勒比地區、熱帶和亞熱帶美洲地區，以及西非部分區域。美洲木棉十分高大，樹形常令觀者不安，樹冠巨大且向外伸展，巨大的板根有時甚至高於成人，輕易就能庇蔭在此過夜的薩泰羅祭司。接著康斯坦佐把整套新衣帶到一處墓地，把衣服埋在一座墳裡，放置不管，歷時三個星期五。這段期間，他用神聖藥草做了一系列淨化沐浴，還得每天向卡迪恩龐貝獻祭。正如他的教父，康斯坦佐選擇了卡迪恩龐貝作為他的個人引路靈。

出師儀式當天，康斯坦佐到墓地挖出之前埋下的衣服，整套換上，再前往教父存放恩甘加的小棚屋。走進小棚屋，教父矇住康斯坦佐雙眼，隨後開始問一連串問題，確認康斯坦佐正確執行每一道儀式準備步驟。帕雷洛祭司相信，在一場宗教儀式裡，靈會附身在人身上，因此如果身體沒有做好準備，附身過程會變得危險。

終於來到最後一道問題，教父問康斯坦佐是否真的想要走完儀式。教父說明，儀式一旦完成，康斯坦佐的靈魂就會消亡，無法改變。康斯坦佐答道，他的靈魂早已死去，他不信任何神。於是教父用木棉樹枝刷他，接著拿一隻活雞刷遍他全身上下。根據帕羅瑪尤貝教教義，那隻雞是用來吸走不屬於他的任何東西，用來淨化他的身體以供靈體附身。淨化之後，教父劃開雞脖子，把血倒入恩甘加。接下來，教父在刀子上燃燒一小堆火藥，開始在康斯坦佐肩膀上刻下一連串符號。此舉叫做「拉亞多」，意思是「切入、晉級」，這也是儀式名稱的由來。這些符號專屬於康斯坦佐，就像他的識別符號，可以區分他和其他帕雷洛祭司。

教父刻完康斯坦佐的符號，便立刻為他拿下矇眼布條。現在屋內點燃了蠟燭，燭光閃耀。十年訓練，康斯坦佐第一次看到照亮的恩甘加，充滿血液、動物遺骸，以及一顆人類頭顱。教父頒授一只布做的黑袋。袋中裝著帕雷洛祭司稱之為基森格（kisengue）的骨頭：人類脛骨。這枝陰森森、類似文憑管的骨頭，是剛出師的康斯坦佐用來攪拌恩甘加、召喚亡者、執行法術的權杖。出師儀式圓滿結束，康斯坦佐相信現在自己可以自行操控那股力量，他是獨當一面的帕雷洛祭司。

情人、施虐者、神在地上

一九八四年，康斯坦佐決定長期移居墨西哥市，在玫瑰區租了一間公寓。豪爾赫依約前來幫忙康斯坦佐的生意，把他的服務形容成通靈兼療癒，四處推薦，甚至還幫他安排了一些模特兒兼差。那年年底，奧馬和馬丁都搬進康斯坦佐的公寓，雖然兩人的家人都反對。康斯坦佐的反應預示了之後的暴力行為：當奧馬的姊姊打算報警逮捕康斯坦佐，指控他染指未成年的時候（雖然奧馬那時已經成年），康斯坦佐威脅要殺了姊姊。馬丁的哥哥阿爾弗雷多（Alfredo）在一次家庭聚會上悄悄把康斯坦佐拉到一旁，告訴康斯坦佐他反對這段關係，而康斯坦佐則輕聲細語道，如果哥哥敢介入這段關係，他會挖出哥哥的心臟。

康斯坦佐非常維護自己新的多角戀關係，但並不妨礙他在情人搬進公寓後暴力對待。康斯坦佐叫馬丁和奧馬替他擦鞋、擺放他的衣服，並且要他們在屋內像打雜僕役般行事。有時他會堅持兩人穿上女裝，這時倔強的馬丁通常都會成為目標，康斯坦佐會毆打馬丁，就算馬丁身強力壯，還是任他爆發，默默接受不還手。每次康斯坦佐爆發，奧馬就會躲起來。

每隔一段時間馬丁就會受不了，跑去和阿爾弗雷多住。然後康斯坦佐會開車到阿爾弗雷多家，朝馬丁大喊他愛他，他需要他回來，不然他會自殺或乾脆把兩人都殺了。每隔幾個月，這個惡性循環就會重複一次。

虐待關係專家薇菈・E・穆拉迪安（Vera E. Mouradian）已經洋洋灑灑列出多種暴力行為，雖然穆拉迪安沒有直接評論過本案，康斯坦佐確實展現許多穆拉迪安記載的操控手法，例如吝於展現情意、把伴侶當僕人使喚、叫伴侶做一些丟臉的事。這兩人還繼續待在他身邊，你可能覺得奇怪，但其實這在虐待關係之中相當普遍。施虐者時常展現迷人的一面，讓受害者時時抱有希望，覺得總有一天痛苦的循環會結束，但這一天永遠不會來。研究員茲拉特卡・拉科維奇－費爾瑟（Zlatka Rakovec-Felser）寫到，這種「蜜月期」最典型的就是施虐者道歉請求原諒，並且淡化自己的暴力行為。康斯坦佐是三人之中最重要的經濟支柱，揮霍大筆金錢在馬丁和奧馬身上，用自己的財力重新贏得他們的信任與忠誠，帶他們血拼購物，買車給他們開。

儘管康斯坦佐的家庭生活因為自己的專橫跋扈而疲於奔命，但他的生意相當不錯，業績蒸蒸日上。英俊的外貌加上準到離奇的算命，讓他在玫瑰區大受歡迎，也幫他賺進大筆鈔票。豪爾赫、奧馬和馬丁協助他處理日漸龐大的客戶群。康斯坦佐開始覺得有必要讓三位成為正式助手。他決定把他們納入帕羅瑪尤貝教，就像他教父對他做的一樣。他使用豪爾赫公寓的一間空房，塗黑窗戶，打造薩泰里阿教和帕羅瑪尤貝教的神殿。他一位一位領著他們進來，為每一位殺雞祭祀，在每一位背上刻下獨一無二的符號。三位現在是拉亞多——「晉級」了。每一位都起誓絕對服從康斯坦佐，而這一忠心誓言是康斯坦佐另外加的要求，並非儀式原有。沒有證據顯示康斯坦佐的教父提過任何類似的要求。

這些入教儀式來得倉促，而對薩泰里阿教和帕羅瑪尤貝教傳統有些認識的人開始懷疑康斯坦佐的身分地位。他自己出師未滿一年，不會有人認為他經驗足夠；相較之下，他自己的訓練持續了快十年。但是康斯坦佐對傳統方式不屑一顧，只想要用自己的方式：快速。康斯坦佐透過快速聚集信徒，讓每位信徒都發誓聽命行事，創造以他為中心的獨有宗教形式，從帕羅瑪尤貝教邁出重要的一步，朝邪教的方向發展。

超自然力量的證據

一九八五年，康斯坦佐越來越出名，多虧一連串事件幫他打響名號，都是他有如神助的神祕力量。訴說第一起事件的人是瑪麗亞．葛拉，康斯坦佐媽媽的朋友，兩人在康斯坦佐初訪墨西哥時碰過面。

瑪麗亞宣稱她親眼目睹康斯坦佐在雨天想通靈算命，他藉故離開公寓，去外頭喘口氣——然後就踏出三樓的陽台墜落。

瑪麗亞一陣驚慌衝到大廳，卻發現康斯坦佐走回建築物內，毫髮無傷。路人在街上聚集，圍住一台車頂凹陷的車子。他們說有個男人跌落車上，人不僅沒死，全身上下一點傷也沒有。

只要有人提起這件軼事，康斯坦佐只是笑笑，卻也從未否認過。瑪麗亞堅稱康斯坦佐在可能致命的墜落後活了下來，安然無恙，她的堅稱完全就是神祕操控。神祕操控的其中一種正字標記是所謂計劃好的自發事件，操控者把預先安排好的事件說成某種更強大力量的操控。很難說瑪麗亞的故事究竟哪個面向是真的，如果有任何真實可言的話。此一事件也可能是康斯坦佐設計好的，說服媽媽他真的有強大神力，並且希望當地傳說可以就此產生。又或者，康斯坦佐意外跌落陽台，人沒事純粹只是因為運氣好。

一九八五年中，這起事件發生後沒多久，一位時運不濟的房仲弗朗西斯科（Francisco）尋求康斯坦佐的協助。房仲業績一落千丈，不知道要找誰幫忙。康斯坦佐答應了，並收取四千五百美元的費用。他擲了螺貝，也做了林匹亞（limpia，意思是淨化儀式），並指示房仲買下墨西哥市市中心一處殘破房產。弗朗西斯科投入所有財產買下房產，花了兩萬美元。

一九八五年九月十九日，墨西哥市發生規模八．一的強震，是史上最嚴重的地震，死亡人數達上千名。但因為城市重建，弗朗西斯科得以用二十五萬美元脫手。發生了這麼一個自然、無法預測的事件，看來好像康斯坦佐參照自身神祕的預知能力，真的給了弗朗西斯科如預言般的關鍵建議。

下一次狡猾的巧合讓康斯坦佐第一次沾染上死亡和神祕力量，而他樂見精心雕琢的傳說多了新賣點。一切都要從玫瑰區的一名歌手說起。這名歌手找上康斯坦佐，希望把債討回來，一間夜店老闆拒絕支付積欠的薪水，還叫保鑣痛揍歌手一頓。歌手想要報仇，希望康斯坦佐能幫上忙。康斯坦佐做了一場儀式，他把人偶浸入雞血，接著告訴歌手找一座新墳，把人偶擺上去。

歌手按照指示做了。接下來，康斯坦佐把那隻死雞留在夜店老闆家門口，連續寄給老闆好幾封恐嚇信件，這些信也都浸過血。夜店老闆相信神祕儀式的力量，相當鄭重其事處理這些恐嚇，不過他開始酗酒，一個月後就死於心臟病。

在此之前，康斯坦佐執行的咒術包括淨化和算命，但現在有人死了。奧馬、豪爾赫和馬丁只在乎他們所有人都插了一手，表達了他們對這起事件感到不安和懊悔，但康斯坦佐向他們保證，他們做的根本不是壞事，因為神明總是尋求正義。如果更強大的力量決定給這名男人心臟病，那就只因為他罪有應得。

康斯坦佐的心理操縱奏效了。康斯坦佐之前向三人組信徒說明過生存支配，這一概念指的是邪教成員有權決定誰能活命，依據的基準則是他們如何解釋神明的隨心所欲。康斯坦佐精心營造環境，讓奧馬、馬丁和豪爾赫可以代他殺人，卻依然覺得事事合理。

夜店老闆之死代表了雄心壯志的發展方向，康斯坦佐採取了合乎邏輯的下一步，致力打造自己的帕羅瑪尤貝教團。他和三位新帕雷洛祭司著手製作他們自己的恩甘加。這段過程的描述是根據康斯坦佐信徒後來的說詞。

一九八五年某夜，康斯坦佐說服奧馬、豪爾赫和馬丁去墓地，他們抵達事先選定的墳墓，挖到看見棺材才停手。他們拿走了棺內屍體的頭顱、脛骨、肋骨、手指頭和腳指頭。這些部位並非隨機挑選。一旦亡靈受到帕雷洛祭司控制，每一部位都對應不同的能力。靈魂安住在頭顱和肋骨內，亡靈因此得以思考，脛骨和腳趾頭讓亡靈可以走動，手指頭則讓亡靈能使用雙手。

康斯坦佐和另外三人很可能知道屍體的身分，而且相較之下，這座墳大概非常新。他們是帕雷洛祭司，所以想要一顆有完整大腦的頭顱，相信如果有大腦，亡靈會更聰明。他們把遺骸帶回豪爾赫的公寓，隨後展開另一儀式，決定亡靈是否適合大家。待遺骸擺放在臨時祭壇上之後，康斯坦佐斜躺在地，蓋著床單。有把刀子被拿到燭火上方，刀子上擺了七小堆火藥。很快地，康斯坦佐開始全身動彈不得，臉部肌肉扭曲變形，表情嚇人。

康斯坦佐模仿帕羅瑪尤貝教的靈體附身，又是另一個神祕操控的例子。在場的三人打從心底相信這就是亡靈占據了康斯坦佐身體的徵兆，而不光只是裝模作樣。奧馬詢問靈體是否會為康斯坦佐工作，康斯坦佐以亡靈的身分說話，答道他會。刀上的火藥都燒盡了，儀式圓滿結束。要是那名據稱亡靈的傢伙拒絕為他們工作，又或者七堆火藥沒有同時點燃，他們就必須把遺骸埋回去，找一副新的頭顱和骨頭，重頭來過。

四人把頭顱和骨頭放進靈鍋。接著把其他不同動物的屍體和血液也加入釜中，還有昆蟲、硬幣、鹿角、鐵路釘、藥草和樹枝——全都是儀式所需。恩甘加準備好之後，康斯坦佐告訴這三名信徒，現在大家都是一家人了。他祈求恩甘加保護他的家人，無病無災，大富大貴，信徒滿門。

初次見血和一樁新生意

康斯坦佐的生意持續成長。從一九八零年代中期的生意記錄看來，他有三十一名常客，許多位支付的單次費用都是好幾千美元。有一位毒販三年來總共支付了四萬美元。康斯坦佐的服務大受歡迎，他甚至製作價目表，分發給客戶。

差不多在製作自己的恩甘加時，康斯坦佐遇見了弗洛倫第諾・文圖拉・古提雷茲

（Florentino Ventura Gutierrez）。從康斯坦佐帳本找到的所有客戶中，文圖拉可能最令人感到震驚——他曾經擔任墨西哥聯邦司法警察（the Mexican Federal Judicial Police，此機構相當於墨西哥的ＦＢＩ）局長，沒多久便升官成為國際刑警組織墨西哥分部部長。

文圖拉是墨國數一數二位高權重的警察，私底下非常迷信。在販毒集團興起的時期，作為一名高階警官，他的敵人數不清有多少，因此尋求法術來保護自己。最後，他聽聞了康斯坦佐的大名，願意付一大筆錢，請求這位當地名人協助。墨西哥市助理檢察官費德里科・龐塞・羅哈斯（Federico Ponce Rojas）懷疑文圖拉也提供緝毒內線消息給康斯坦佐。和文圖拉合作的美國緝毒局幹員質疑這些指控，但康斯坦佐就是在和文圖拉變熟稔時涉足毒品生意，甚至偶爾會向潛在客戶說自己是文圖拉的外甥。

文圖拉並非康斯坦佐在警界唯一的人脈。一九八五年，豪爾赫介紹他和薩爾瓦多・維達爾・加西亞・阿拉爾孔（Salvador Vidal García Alarcón）認識，這名緝毒警察遭人持刀攻擊，臉中間留下一道疤。猙獰的傷口讓他每個角度看起來都長得不一樣。豪爾赫告訴康斯坦佐，維達爾漸漸相信自己的「三張臉」——左側、右側、正面是被三位不同靈體附身：蘇族印第安人、古巴藥頭和非洲巫師。康斯坦佐知道緝毒警察欠他人情可能會有許多好處，便急切地要豪爾赫安排兩人會面。康斯坦佐先展示一番擲螺貝算命，接著複述豪爾赫告知的資訊：很可能有三位靈體附在維達爾身上。再一次，康斯坦佐成功利用神祕操控欺瞞顧客。

維達爾相信這些靈體控制了自己，並對此感到憤怒，這讓他成為最佳目標。經驗老到的算命師如康斯坦佐知道如何利用維達爾的弱點，沒多久維達爾就加入康斯坦佐逐漸茁壯的邪教。

維達爾入教對康斯坦佐想達成的野心至關重要。他不只是一名警官，還身型魁梧、面目猙獰。在入教儀式期間，康斯坦佐告知維達爾他要成為教團專屬的執法人員，康斯坦佐命令他代表教團做任何事時，他都必須遵守，不得有異議。維達爾同意了。

康斯坦佐提出如此激進的要求，可能確實增強了維達爾投入教團的程度。心理學家奈吉爾．巴伯（Nigel Barber）寫到，邪教如果要求越多，會運作得越長久。這不適用於非宗教團體，但對嚴格的宗教而言，要求越多，信徒越加篤信。只要以更遠大的使命為名，信徒都樂意自我犧牲，即便很過分也能忍受。

社會學家理查．索西斯（Richard Sosis）和艾瑞克．R．布雷思勒（Eric R. Bressler）研究十九世紀的八十三個宗教社區，發現如果社區只有兩項代價高昂的要求，例如終生獨身或上繳所有財產，平均運作時間不到十年。社區如果有六到八項要求，運作期達五十年；而超過十一條要求的社區，則長達一甲子。要求越多，運作越長久。

現在緝毒警官完全成了核心成員，康斯坦佐可以得知毒品戰爭兩邊人馬的消息。據說維達爾早就貪汙腐敗，和許多黑社會人物有關係，他開始把康斯坦佐介紹給一大家子迷信的毒品走私犯。不過就像他教父先前做的一樣，康斯坦佐開始建議這些走私犯何時該運貨。現在維達爾也進入教團的羊圈[6]，康斯坦佐可以告訴走私犯繼續運毒，再指示維達爾確保沒有人干涉這些毒販。這樣一來，康斯坦佐把賄賂轉成徹頭徹尾的神靈保佑，賄賂本是毒品交易生意相當常見的成本開銷。

現在康斯坦佐接手貪腐警官一度扮演的關鍵角色，開始累積過去難以想像的大筆錢財。但是他在努力經營新生意的同時，也帶領信徒一起走上肆意殺人之路。一九八六年秋天，康斯坦佐認識了古柯鹼走私販吉列爾莫．卡爾薩達（Guillermo Calzada），開始建議對方運送毒品的良辰吉時。維達爾確保沒有警方插手卡爾薩達的貨運，這當然提高了卡爾薩達的利潤。到了一九八七年早期，康斯坦佐因毒品販運賺進的錢財高達好幾萬美元，就是在這段時期，信徒注意到一項變化：他對於金錢追求越來越無法自拔，人也變得越來越殘暴無情。

根據加州大學柏克萊分校（the University of California, Berkeley）心理學教授雪莉．強森（Sheri Johnson）所言，自戀狂總想要追求權力和財富，無法克制自己。雖然無法對康斯坦佐做臨床診斷，但他很明顯缺乏同理心，總是急於利用他人，還有膨脹的自負感，三者都是自戀的經典標記；依照信徒證詞，他經常展現這些特徵。

卡爾薩達每個月都會支付康斯坦佐好幾千美元的服務費。但康斯坦佐越來越瞭解毒品生

6 聖經用語，成為一分子。

意，企圖提高收費標準。他向奧馬和馬丁抱怨，要不是有他的法力和人脈，卡爾薩達不可能賺那麼多。他還說，利潤的一半應該歸他。奧馬和馬丁當然同意他的說法。一九八七年四月，某次在卡爾薩達家的會面，康斯坦佐大膽提出均分利潤一事。卡爾薩達拒絕了，康斯坦佐陷入暴怒，東道主下逐客令。

康斯坦佐怒不可遏。但幾天後，他還是向卡爾薩達道歉，把自己的行為怪罪到邪惡咒術頭上。為了表示自己的懺悔，康斯坦佐堅持要為卡爾薩達全家免費舉行儀式。

一九八七年四月三十日，康斯坦佐造訪卡爾薩達家。參加儀式的有卡爾薩達自己，還有他的太太、媽媽、女僕、生意夥伴、祕書和個人保鏢。

康斯坦佐開始帕羅瑪尤貝教誦念，接著兩個男人持槍衝進來。其中一位是馬丁，信徒宣稱另一名持槍男子是維達爾，但他從未被起訴。兩人屠殺了卡爾薩達以及他的核心成員。他們除去各個身體部位，手指、脊椎、心臟、性器，甚至兩顆腦袋。隨後把屍體丟棄在附近的一條河裡。警察找到屍體的時候已經過了一週，當時只剩三具能夠辨認，其他四具太過支離破碎，無法與任何紀錄匹配。

據奧馬所言，馬丁拒絕進一步談論那起事件。但是康斯坦佐向信徒重述這幾起凶殺時，都說犯罪現場帶回的那些身體部位，之後會在恩甘加裡繼續活下去。康斯坦佐娓娓道來凶殺的恐怖細節，毫無疑問讓信徒都成了幫兇，但大家都太過忠誠，根本不會離開教團。

司法心理學教授凱薩琳．瑞姆斯藍（Katherine Ramsland）寫道，會在犯案時尋找同伴的殺手尤其擅長分辨、誘騙那些可能參與犯罪的人。前獄中心理學家阿爾．卡利斯勒（Al Carlisle）如此描述這種關係：「主導人需要跟隨者忠誠不二，才能顯示他真金不怕火煉。恭順的跟隨者需要主導人的神力和權力，因此他們試圖成為主導人的影子，反映其信念和行為準則。每一位都彼此需要，一切才合情合理。」看來到了卡爾薩達凶殺發生時，康斯坦佐和信徒的互動早就成為這種模式了。

事隔一個月，一九八七年六月，維達爾調職到馬塔莫羅斯市（Matamoros），就在德州邊境南方。康斯坦佐抓住這次機會，擴展運作範圍，要維達爾找尋當地夥伴合作。奪走卡爾薩達的毒品生意之後，康斯坦佐還是不滿足，想要持續擴大自己的影響範圍。

勾引教母

一九八七年七月中，維達爾搭上當地一位毒品走私犯艾里歐．艾爾南德斯（Elio Hernández）。艾里歐二十幾歲，才剛擔起家族事業七個月；之前是由艾里歐的其中一名兄弟掌管事業，但那名兄弟六個月前在一家餐廳外慘遭槍擊。艾里歐生性暴力又衝動，而他是否能領導事業也還是未知數，他的生意頭腦也不太靈光。艾爾南提斯家族簡直走投無路，而這正是康斯坦佐提供服務的理想對象。

為了打入艾爾南提斯家族世界，康斯坦佐祕密設計了非常迂迴的計畫。首要目標是艾里歐的前女友，二十二歲的莎拉．阿德瑞特（Sara Aldrete）。維達爾告訴康斯坦佐，雖然艾里歐現在結婚了，但他依然愛著莎拉。按照康斯坦佐扭曲的思考模式，莎拉成了他進入艾里歐世界的門票。

一九八七年七月三十日，康斯坦佐和馬丁在車道上硬插車，停在莎拉正前方，拒絕移開。後方排了越來越多車，康斯坦佐走向莎拉的車，堅持要碰面。莎拉只想要他趕快離開，不要擋路，但不管是出於惱怒，還是恐懼，她最終還是下了車碰面。

康斯坦佐告訴莎拉他是邁阿密來的律師。莎拉問起康斯坦佐戴的串珠項鍊，康斯坦佐說：「這是我宗教信仰的一部分。」接著就不願意多說。利用莎拉的好奇心，康斯坦佐建議他們隔天一起吃吃午餐，再碰一次面。整個八月，康斯坦佐開始籠絡莎拉。他和莎拉提起薩泰里阿教和帕羅瑪尤貝教，不過沒有提到任何活人獻祭。他宣稱自己是「薩泰羅祭司基督派」（Santero Cristiano），或許是為了不要嚇跑她。他告訴莎拉他可以讀牌預知未來，莎拉請他幫自己算命。康斯坦佐拒絕了，說莎拉現在還沒準備好。

康斯坦佐努力追求時，莎拉同時和另一名男人約會。但是那位名叫奧古斯特（August）的男人接到一通匿名電話，告訴他莎拉腳踏兩條船，導致兩人吵架、分手。下次康斯坦佐和莎拉談話時，康斯坦佐宣稱早已預測到這回分手。接著他終於答應幫莎拉讀牌，並預測了三件事：下學期會收到學校給的錢、會接到一通老友打來的電話、曾經交往過的對象會帶著問題出現。

接著就像他對許多人做的一樣，康斯坦佐逐一說出莎拉人生的細節。雖然三項預測的背景資料細節都是依據維達爾的調查，但對莎拉而言，一切好似神奇法力。前兩項預測迅速成真：莎拉獲得學校獎學金，有位老友真的打來。當莎拉告訴康斯坦佐的時候，康斯坦佐的反應像是一切正如他所料。他充滿自信又精準的預言說服了莎拉他真的能看見未來。莎拉成為信徒，而且很快就成了女友。康斯坦佐為了接近艾里歐而訂的計畫如他所願一一展開。

康斯坦佐和莎拉的關係就如同他與奧馬、馬丁的關係——慷慨大方但冷淡無情。那年十月，他終於告訴莎拉他是雙性戀，莎拉非常生氣，但也沒有離開。除了莎拉和艾爾南提斯家族的聯繫，康斯坦佐對莎拉一點興趣都沒有，但仍繼續操控莎拉待在自己身邊，盼望艾里歐來找莎拉。對康斯坦佐來說，莎拉只不過是拿來剝削的資源。對權力無止盡的追求主導了康斯坦佐的每個決定，就連親密關係也不例外。

到了十一月，花在莎拉身上的時間和金錢都值得了。莎拉在塔可餅攤前聽到有男人喊他的名字，轉身就看到艾里歐．艾爾南提斯越過街朝她跑來。艾里歐告訴她，最近家運走下坡，他哥哥十個月前被人殺死，他害怕自己的性命也要不保。經營毒品事業的緊張感和責任開始讓艾里歐感到痛苦了。莎拉跟他說有位男人可以幫忙。康斯坦佐的第三個莎拉預言成真了，現在計畫的下一步已經到位。

就像康斯坦佐身旁發生的諸多類神祕事件一樣，這次相遇也令人難以置信。康斯坦佐可以非常有耐心，但不會只是聽天由命，至今我們還是無法證實他是否出手促成這次相遇，雖然他肯定早就看出艾里歐極可能成為神祕操控的目標。背負著領導壓力、面對競爭藥頭的恐懼、哥哥死後家族陷入困境，艾里歐變得極為脆弱，不堪一擊，而這正是康斯坦佐需要的狀態。莎拉一人攸關計畫成敗：艾里歐對莎拉的喜愛是康斯坦佐可以利用的弱點。康斯坦佐計劃先讓艾里歐入教，再利用莎拉控制他。

但首先，康斯坦佐需要先讓莎拉晉級，坐上教團高位。他希望莎拉成為教母，就如同他自己是教父那般。一九八八年三月二十三日，莎拉到墨西哥市參加自己的晉級儀式，墨西哥市是康斯坦佐過冬的地方。當晚奧馬為莎拉矇眼，領他進去一間昏暗的房間。康斯坦佐開始用嘶聲誦念起來，這是平常恩甘加靈體附身時他使用的聲音。有人拿了一隻雞刷遍莎拉全身，再把雞頭砍下，一切都跟康斯坦佐和其他人的晉級儀式一樣。有一頭更大隻的動物刷過她身上，聽到叫聲，莎拉才意識到是一頭山羊。有人砍了山羊頭，接著莎拉感受到血潑灑在身上。康斯坦佐撕開她的上衣，在她背上刻下數個符號，如同他對其他所有正式入教的信徒一般。康斯坦佐應答道。莎拉的靈魂已死，現在所有信徒都是她的家人了，她是他們的教母，他們會服從教母。從此刻開始，莎拉就是拉雅達（rayada）。

莎拉現在可以擔任康斯坦佐的中尉，負責馬塔莫羅斯市的運作。只要康斯坦佐不在，信徒得向莎拉呈報，而如果有人違規，是由莎拉決定要不要告訴康斯坦佐。康斯坦佐向信徒發布三道諭令。其一，完全遵照他的指令。他說神明沒有憐憫，他自己也沒有。其二，信徒必須相信神明和亡靈，特別是噬魂者卡迪恩龐貝。他提醒信徒不能上教會，因為基督徒是牲畜。其三，康斯坦佐禁止吸食任何毒品，就像他的教父以前下的禁令。康斯坦佐跟信徒說，如果不保持純淨，必死無疑。

就在隔天，康斯坦佐為艾里歐舉行拉亞多儀式。儀式過後，康斯坦佐向艾里歐提出一筆交易。他要分一半艾里歐的家族收益，同時也提出退款擔保：如果家族收益沒有增加，他分毫不拿。這樣的安排看似對艾里歐很慷慨，但對康斯坦佐而言，艾里歐只是另一枚可以剝削的棋子，滿足他自私自利的控制慾望。康斯坦佐提出這筆交易是因為他有自信絕對可以拉抬業績。

康斯坦佐把利潤增加歸功於自己的強大法力，但是新生意更應歸功於單純的人際網絡。艾里歐的哥哥被殺之後，艾爾南提斯家族就失去了重要人脈。維達爾運用自己的貪腐警察關係來建立這些連結。一如往常，康斯坦佐的法術比起說是超自然力，更應歸功於貪汙警察的武力和通風報信。

屍體增加

幾週後，他們一回到馬塔莫羅斯市，康斯坦佐便叫艾里歐聚集身邊最忠誠的人馬。艾里歐挑了五位員工。阿爾瓦羅・達里奧・德・萊昂・瓦爾德斯（Alvaro Darío de Léon Valdés），人稱杜比（El Duby），殘忍無情的打手，最出名的事蹟就是酒吧槍戰。卡洛斯・德拉亞塔（Carlos de la Llata）靠著販賣古柯鹼給觀光客來賺錢。奧雷利歐・查韋斯（Aurelio Chavez）是艾爾南提斯家族運作基地聖艾倫納牧場（Rancho Santa Elena）的領班。塞爾吉歐・馬丁內茲・薩利納斯

（Sergio Martínez Salinas）是艾里歐的同輩親戚，有運送毒品的經驗。最後一位馬利歐・法比奧・龐塞・托雷斯（Malio Fabio Ponce Torres），是艾里歐的大學同學，家境殷實。康斯坦佐看出最後一位有潛力成為新生意的領導人。這五人成為康斯坦佐在馬塔莫羅斯市教團運作的核心。一九八八年四月上半期，康斯坦佐和莎拉在莎拉家客廳，教導五人小組薩泰里阿教和帕羅瑪尤貝教。莎拉和艾里歐辦完儀式後三週，五位新進成員也完成了入教儀式。康斯坦佐現在有忠心耿耿的步兵，愛怎麼使喚就怎麼使喚。

同時，維達爾負責確保艾爾南提斯家族事業再次興旺。新成員入教儀式後不久，維達爾安排讓聯邦政府沒收的一噸大麻落入艾里歐手中。康斯坦佐的儀式確保他的神力被視為一切成功的關鍵，他告訴艾里歐他擲了螺貝，宣稱他因此得知買家的名字。這批毒品獲利二十萬美元。

一九八八年四、五月間，他們繼續好幾筆類似交易。康斯坦佐和維達爾掌控生意通路，確保交易成功，但康斯坦佐每每呈現給艾里歐看時，一切都像黑法術的成果。艾爾南提斯家族從沒賺過這麼多錢，但仍不足以滿足康斯坦佐。五月尾聲，康斯坦佐向艾里歐提議，偷走競爭對手的毒品來擴大版圖。

新血奧雷利歐得知有位三流毒販赫克托・德拉富恩特（Hector de la Fuente），他的大麻都存放在奧雷利歐家旁邊的農場。奧雷利歐告訴康斯坦佐這則消息。五月二十八日，這幫人追捕德拉富恩特，在鎮上找到人，綁架他之後，強迫他帶著他們去找藏匿的大麻。五人小組、俘虜、康斯坦佐和馬丁一群人抵達該座農場，發現農場主人五十二歲的兒子莫伊塞斯・卡斯提羅

（Moises Castillo）在那裡幫忙年邁的父親。康斯坦佐的人拿走大麻，也帶走德拉富恩特和卡斯提羅，一路開回聖艾倫納牧場。

他們藏起毒品，帶著兩位俘虜到附近一座果園。康斯坦佐拿著槍，問德拉富恩特和卡斯提羅他們是否是好基督徒。德拉富恩特沉默不語，但卡斯提羅十分肯定。接著康斯坦佐提醒他的人，基督徒是牲畜，而牲畜就要拿來犧牲。他瞄準兩人頭部開槍。他的人儘管有毒品交易的經驗，卻也都愕然不已。康斯坦佐可能非常樂見此景：他用行動明確宣示，為達目的，可以不擇手段。康斯坦佐隨意的處決很可能讓近侍心生不安，但這也讓不少暴力分子開始崇拜他，他的邪教信徒也越來越多。一九八八年六月，康斯坦佐為艾里歐的姪子賽拉分．艾爾南提斯（Serafin Hernández）入教，姪子的綽號是「小賽拉分」，這是為了區別他和他爸爸還有艾里歐的兄弟，大家都叫他們「老賽拉分」。

小賽拉分身為艾里歐的姪子，大家認為他沒有能力，但有項特質讓他成為教團有價值的一員：他是美國公民，比起其他信徒，更容易穿越邊界。小賽拉分熱愛自己的教團成員身分，很快就到處誇耀，還喜歡在馬塔莫羅斯市的酒吧或俱樂部惹人注目。他成為厲害的招募要員，才到了一九八八年六月中，就幫馬塔莫羅斯分支拉進五名信徒，規模從七名信徒變成十二名。同一時間，維達爾繼續安排利潤豐厚的毒品交易，而康斯坦佐把成功都歸給自己，沒有絲毫遲疑。儀式地點從莎拉的公寓改到牧場上一座小屋，隨後他決定，是時候為聖艾倫納牧場設置正式的恩甘加。他開始收集要丟入靈鍋的材料。

一九八八年七月初同時期，康斯坦佐和馬丁回到墨西哥市，莎拉也一起匯合。剛到沒多久，信徒豪爾赫帶著問題來找康斯坦佐：豪爾赫想要擺脫室友拉蒙・帕斯・艾斯基維爾（Ramón Paz Esquivel），拉蒙是古董業務，同時也在墨西哥市紅燈區以藝名「克勞蒂亞」表演。拉蒙是豪爾赫之前的情人。雖然他們還住在一起，關係卻越來越糟。

拉蒙經常醉醺醺回家，還帶著年輕男人，搞得整棟建築鬧哄哄。他偶爾也會拒絕豪爾赫的算命客戶進來公寓。最近，拉蒙指控豪爾赫偷了一件他的古董。他氣呼呼衝進屋裡，翻倒家具，抬腳踢了豪爾赫其中一隻心愛的貴賓狗，豪爾赫終於忍無可忍。康斯坦佐同意幫忙。

一九八八年七月十六日晚上，康斯坦佐、豪爾赫以及馬丁、奧馬等信徒聚在豪爾赫的公寓等待拉蒙。拉蒙一進屋，這群人就把他拖進浴缸。馬丁綁住他雙手，用大力膠帶封住他的嘴。康斯坦佐又再次嚇了眾人一跳，他趁拉蒙還活著的時候，砍下拉蒙的腳趾、手指和性器官。豪爾赫和奧馬逃出房間——康斯坦佐在信徒面前宰殺過許多牲畜，但從來沒有出現如此殘暴的舉動。接著康斯坦佐剝下拉蒙的部分皮膚，最後才殺了他。康斯坦佐用容器盛起拉蒙的血，拿了許多根骨頭，最後從顱內取出大腦。現在康斯坦佐有了他需要的人類靈魂，可以放進聖艾倫納牧場的恩甘加。

康斯坦佐認為拿來獻祭的受害者如果在劇痛和恐懼中死亡，會更適合他的恩甘加。

薩泰里阿教和帕羅瑪尤貝教學者米真・岡薩雷斯－威普勒（Migene González-Wippler）寫道，有些帕雷洛祭司會虐待黑貓，再入滾水烝之，最後才放入恩甘加。依岡薩雷斯－威普勒所言，此類大釜專門設計來做邪惡之事。

看來康斯坦佐把這項哲學應用在他的人類受害者身上。他創造自己的宗教，這宗教不僅接受這般殘忍暴戾和凶殺，還成為儀式執行不可或缺的一部分。

不論是凶殺習慣，還是毒品交易，康斯坦佐都越來越暴虐無道。一九八八年八月，康斯坦佐和維達爾設定了至今最大的毒品交易量。維達爾扣押了七十五公斤的古柯鹼。他們計劃以八十萬美元賣給名叫「鉤子」（El Gancho）的買家。艾里歐和他的兄弟奧維德早就安排好了交易，因為鉤子是他們親家的兄弟。鉤子和艾爾南提斯家族做過一些小交易，但是康斯坦佐沒有打算要把那批毒品給鉤子，他打算把錢偷走，同時自己留下古柯鹼。艾里歐和奧維德不想背叛親戚，但康斯坦佐相當堅持。一如既往，信徒的感受只是次要，遠遠比不上他追求財富重要。

鉤子一抵達，一臉歉意的艾里歐和奧維德就舉槍威脅他，拿走了錢。鉤子說這樣他自己會有麻煩，他的老闆會因為這樣受辱而殺了他。他們不理他的警告，搶錢之後過了三天，鉤子綁架奧維德和他兩歲的兒子。鉤子要奧維德打給艾里歐，告訴他如果隔天不還錢，死的就是奧維德和兒子。但是艾里歐知道康斯坦佐絕對不會心軟。出於絕望，艾里歐和兄弟老賽拉分向警方求助，警方拒絕幫忙，除非兄弟倆承認這樁綁架是毒品交易失敗的後果。

艾里歐無路可走，又去問康斯坦佐的建議。康斯坦佐的答覆就是命令艾里歐去找一名活人獻祭。艾里歐開車在那附近兜轉，直到看見一名流浪漢，就說要載他一程。流浪漢接受了好意。艾里歐開回聖艾倫納牧場，一抵達，信徒就把流浪漢拖進恩甘加小屋。康斯坦佐把所有人趕了出去，只留下艾里歐在一旁幫忙。康斯坦佐活生生切開那個男人，把身體不同部位丟進恩甘加，祈求靈體保護奧維德，才殺了流浪漢。等康斯坦佐把流浪漢的大腦丟入恩甘加，他叫艾里歐切開流浪漢挖出心臟。

隔天，奧維德和兒子平安返家。警方事後調查，也沒能找出原因。但對教團來說，這都是康斯坦佐儀式的直接結果。

最終錯誤

康斯坦佐越來越沉迷殺人的狀態，其實是連環殺手的常見模式。犯罪學家史蒂芬・J・賈南傑洛（Stephen J. Giannangelo）在《連環殺手心理學：暴力理論》（暫譯，*The Psychopathology of Serial Murder: A Theory of Violence*）一書中詳細描述殺手如何隨著時間流逝變得越來越危險。如果殺了人卻免於受罰，殺手的自信就會增加。如果說殺人讓罪犯情緒亢奮，那麼他們的自信會導致更多凶殺。要不了多久，殺手對於殺人越來越自在，甚至迫切渴望殺人。

此一模式和康斯坦佐類似。殺了卡爾薩達一家之後，康斯坦佐等了一年才再次殺人。不過只要他開始定期殺人，就會想要越殺越多。康斯坦佐的殘酷無情，令眾人恐懼，也加強了對教團的掌控，他也會利用恐嚇操控信徒做任何他要求的事。艾里歐抱怨莎拉不跟他睡，康斯坦佐命令莎拉陪睡。莎拉抗議，但康斯坦佐反駁道，只要這麼做，莎拉對艾里歐的控制絕對會比之前更強大。莎拉最終還是同意了。

莎拉的反應正是神祕操控受害者常見的特徵。利夫頓醫師認為，順利被操控的人會把自己的心理狀態與領袖的合併，甚至會開始自己操控他人。即使不情願，莎拉還是和艾里歐上床，正是在進行操控。

從一九八八年晚期到一九八九年早期，康斯坦佐加快殺人的速度。十一月，他殺了一名使用古柯鹼的信徒，吸食毒品違反他之前下的禁藥令。十二月，他處決了維達爾的兩名警察助手，因為維達爾不在信任他們。光是一九八九年二月，這個邪教就殺了五個人。其中一名受害者是十四歲的男孩，在數名教團成員注意到這名青少年步行經過牧場後，康斯坦佐命令艾里歐殺了他。後來艾里歐才發現那名少年其實是他的親戚。

康斯坦佐毫無同情心。他要信徒毫不遲疑殺人，用和他一樣的眼光看待人——不過是達成目的的手段而已。然而，二月大殺戮無法滿足康斯坦佐，他還渴望更多受害者。但在三月中，他執行了某回儀式，並將該儀式視為壞預兆。他其實也不算錯——那件令人髮指的行為最終會毀了毒梟撒旦教。

這時候，康斯坦佐已經成功在墨西哥馬塔莫羅斯市邊境建立起大規模的毒品走私生意。他在凶殘教團的領袖地位也已鞏固。信徒很快就注意到康斯坦佐對儀式法術漸漸不感興趣，而且看來只對殺人有興趣，現在變成每週都會殺人。

一九八九年三月十三日，康斯坦佐又要求另一名受害者。他的那幫信徒在他們自認的地盤邊緣某家酒吧裡找到一名古柯鹼藥頭，便把藥頭拖回牧場。康斯坦佐開始恐怖的例行步驟，在藥頭還活著的時候肢解、剝皮。但神奇的事發生了，藥頭竟然沒有叫。康斯坦佐沒能讓他吭聲，也因此無法展示男人的痛苦和恐懼。康斯坦佐認為，祭品如果沒有展示出明顯的痛苦，儀式就是大失敗。於是他想了個替代方案：需要殺一名比較軟弱的人。康斯坦佐說理想目標是年輕人，例如大學生，那種只說美語的典型白人外國佬。現在時機正好，馬塔莫羅斯市正好迎來春假高峰。

隔天晚上，一九八九年三月十四日，毒梟撒旦教信徒馬利歐和小賽拉分開著小賽拉分的貨卡出門執行命令。凌晨兩點，大學生魚貫橫越大橋，從墨西哥這一側的馬塔莫羅斯市前往德州那一側的布朗斯維，兩人看到一位落單的二十一歲金髮男生馬克・基爾羅伊（Mark Kilroy），一番搏鬥合力把他架上貨卡。他們載著戰利品回到聖艾倫納牧場，綁住他的手腳，留在貨卡後車廂一晚。

隔天早上，牧場管理員多明哥・雷耶斯（Domingo Reyes）發現基爾羅伊，帶了雞蛋給他當早餐。管理員替這名少年感到遺憾，但也知道最好不要幫其他忙。雷耶斯趁著還沒有任何毒梟撒旦信徒發現他幫忙之前，趕緊回到自己家。

一九八九年三月十五日中午，毒梟撒旦教信徒聚集在牧場的恩甘加小屋。信徒在小屋外頭等待，康斯坦佐在屋內虐待、強暴、殺害馬克・基爾羅伊。他命令杜比（馬塔莫羅斯市最初五人小組的一員）挖出基爾羅伊的心臟。基爾羅伊的大腦被丟入恩甘加，康斯坦佐在一旁說受教育的人類會讓恩甘加更聰明，更能夠執行複雜的任務。

但現在教團的暴力信仰跨越了國際邊界，會受到更詳細的檢視：馬克．基爾羅伊的朋友早已發了瘋似的找他，警方開始調查基爾羅伊失蹤一案。基爾羅伊的父親每天站在布朗斯維和馬塔莫羅斯市中間那座橋上，發送印有兒子照片的尋人傳單。案件可能在墨西哥國土發生，屬於墨西哥警方的法律管轄範圍，但是基爾羅伊的叔叔是美國海關督導員，要求美國這一側也要展開調查。

墨西哥警方背後也有強大的誘因要解決這起案件。負面新聞報導會傷害觀光發展，旅遊是當地許多社區賴以為生的產業，馬塔莫羅斯市也不例外。但是三月最後兩週，兩邊都沒有進展。基爾羅伊在深夜消失，那時馬塔莫羅斯市街上到處都是喝醉的大學生。幾乎沒辦法找到可靠的線索。

一九八九年四月一日出現轉機，雖然當時警方還不知道。一台紅色貨卡開過墨西哥聯邦警察毒品臨檢點，完全沒有減速。司機不是別人，正是小賽拉分。現場員警發覺小賽拉分形跡可疑，於是不漏痕跡尾隨紅貨卡。小賽拉分就算知道也不會介意。他加速開過臨檢點，如同他稍後跟警方說的，因為他相信教團法術能隱身，警察真的看不見他。

警方一路尾隨到聖艾倫納牧場，接著在牧場外蹲點監視。小賽拉分一離開，警方就四處查看，推論牧場很可能是毒品運送中繼站，接著開始監視艾爾南提斯一家。警方的直覺很準，一個禮拜後的四月八日，毒梟撒旦教經手了一筆三十萬美元的毒品交易。聯邦警察目睹交易全程，也透過手機竊聽所有對話。警方逮捕了艾里歐、小賽拉分，還有其他兩名教團成員，隔天就以毒品交易起訴四人。

艾爾南提斯家族面對警方並不害怕，依逮捕的警官所言，他們雖然被捕，但看似莫名地不在意，幾乎可以說樂在其中。警方不明白為什麼這些男人如此篤定自己不會有事，但艾爾南提斯一家相信這些靈體還有他們的教父會保護他們免受傷害。

接著警方逮補了牧場管理員多明哥，他不是藥頭，也不是邪教成員。他很快就嚇到什麼都說了，他坦承家族有參與毒品交易，也說偶爾有人進去牧場就沒有離開過，其中一人是金髮美國人。警方立刻興趣大增。警方拿出一張馬克．基爾羅伊的照片給多明哥看，他立刻就指認這名年輕人就是失蹤的那名青年。

同一天，聯邦警察逮捕艾里歐和小賽拉分，奧維德打給康斯坦佐，通知他信徒被警察扣押。結果，四月十日一大清早，康斯坦佐、馬丁、杜比和莎拉便逃離墨西哥市，要去和奧馬碰頭。

四月十一日，小賽拉分為警方指出好幾具屍體的位置。在聖艾倫納牧場的埋藏地點。他們在那裡找到馬克．基爾羅伊的遺骸，還有數不清無法識別、殘缺不全的屍體，每具都失去了手指、耳朵、腳趾、心臟、性器和大腦。

有幾座墳墓被做了記號，地面上露出幾根鐵絲，基爾羅伊的也是。小賽拉分向警方解釋，鐵絲纏繞著每具屍體的脊椎。只要屍體完全分解，就能抓著鐵絲抽出地下的脊柱，這是康斯坦佐跟教父學來的可怕手法，駭人耳目。受害者的脊椎可以加工做成項鍊，康斯坦佐告訴信徒，配戴項鍊會帶來好運。

小賽拉分和其他也被逮捕的信徒什麼都告訴了警方，包括康斯坦佐是誰、他做了哪些慘無人道的事，毫無保留。他們相信警方也無技可施，因為康斯坦佐的法力保護著他們。信徒如此有自信，認為警方拿他們沒轍，都是因為康斯坦佐完美執行的神祕操控。毒梟撒旦教犯下令人髮指的罪行，卻能逍遙法外如此長時間，主要是因為維達爾貪腐的警方作業。但是經過無數次儀式和獻祭，康斯坦佐早已騙倒眾人，信徒都認為是他的法術讓一切長久順利成功。他的手法如此奏效，信徒就連坦白一切後，都仍對康斯坦佐的法術深信不疑，反而不太相信法律可以控告他們。

殺手血淋淋的結局

發現聖艾倫納牧場的屍體後，警方開始追緝康斯坦佐和餘黨。康斯坦佐、馬丁、莎拉、奧馬和杜比藏匿在一名信徒的公寓大廈，看著電視上案情開展。新聞報導都會放一張康斯坦佐的照片。因此，這五人把頭髮又剪又染，好讓人難以認出，準備過逃亡生活。

這群人盡量待在屋內，在同一地點待上數天，就會移動到下一處。這段時間，康斯坦佐變得專橫跋扈，尤其是面對莎拉的時候。與此同時，莎拉每天越來越不安，變得無精打采，懇求著要回自己父母家。康斯坦佐擔心莎拉會背叛教團、向警方通報，於是告訴她永遠不能離開。

康斯坦佐依然確信他們可以逃跑，認為自己的神祕操控和狡詐機巧可以帶大家安然渡過危機。他打給維達爾，命令他到瓜達拉哈拉（Guadalajara）偷一批古柯鹼，這樣就可以金援這群在逃的人，一路賄賂出境墨西哥。康斯坦佐逐漸瘋狂的行為也表現在他重新聯絡上瑪麗亞，希望她可以幫忙安排假護照，還要求她去找整形醫師來改變他們的外貌。瑪麗亞就是康斯坦佐媽媽的那位老友，目睹他從三樓墜落的奇蹟。

整個四月，這幾位毒梟撒旦教徒四處藏匿，而警方每次都晚了一步。但是他們快要沒有時間了。調查員注意到卡爾薩達凶殺、豪爾赫室友拉蒙的凶殺和聖艾倫納牧場的屍體這幾案的相似之處。兩名帕羅瑪尤貝教專家從邁阿密飛來，建議警方燒了聖艾倫納牧場的恩甘加小屋，警方在一九八九年四月二十二日燒毀小屋，希望徹底激怒康斯坦佐，讓他行事出錯。這一招改變了結果。逃匿的教團成員待在遠方一間小房子，看到新聞報導中警方燒毀恩甘加和小屋。康斯坦佐用機關槍射擊電視，一怒之下搗毀房子，爆氣了一個多鐘頭。

四月底，警方搜捕漸漸逼近康斯坦佐。維達爾打電話通知康斯坦佐，他目睹恩甘加被摧毀，相信現在康斯坦佐的法力也消散了，因此他再也不會幫忙。現在沒有收入來源支持這群人繼續逃亡。維達爾原本是忠誠的夥伴，最後也和康斯坦佐保持距離。

康斯坦佐眼睜睜看著結局將近，要他忠誠的信徒答應集體自殺。如果無可避免與警方當面對峙，奧馬應當要先殺康斯坦佐和馬丁，再殺其他人，最後自殺。康斯坦佐接著宣布他們要重回墨西哥市。四月二十七日，他們返抵墨西哥市，躲藏在瑪麗亞找的一間昏暗骯髒的公寓裡。五天後，康斯坦佐派莎拉出門打電話給那名整形醫師，希望醫師能改變他們的外貌，讓他們順利逃出國。莎拉打了電話，但是醫師拒絕幫忙。

在這趟外出，莎拉也打給爸媽的鄰居，確認家人安危。她還放出訊息說康斯坦佐綁架了她，而她想要逃跑。五月六日，警方調查毫無關連的另一件案子時，來到他們躲藏的公寓外頭。但是康斯坦佐一看到警車，就朝其他人大喊逃亡結束了。趁其他人拿起武器時，他開啟窗戶，拿起衝鋒槍瞄準一名員警。照莎拉和杜比所言，康斯坦佐大喊一聲：「媽媽，結束了！」然後開火。

隨後展開凶殘的槍戰。康斯坦佐把他們日漸減少的鈔票一捆捆丟出窗外，希望看熱鬧的人可以把場面搞得一團混亂，他們可以趁機逃跑。他發現這招沒用，便拿槍瞄準附近一個瓦斯桶，希望引爆瓦斯桶，轉移警方注意力。那招也失敗了。康斯坦佐感受到一切都要玩完了，命令奧馬燒了剩下的好幾千美元。如果康斯坦佐不能擁有，其他人也休想。杜比交給康斯坦佐最後兩盒彈匣，康斯坦佐終於爆發了。

康斯坦佐要求所有人立刻集體自殺，但奧馬拒絕殺害康斯坦佐和馬丁。康斯坦佐命令杜比先動手，但也是徒勞。康斯坦佐向杜比保證，他絕對會死而復活。莎拉在一旁尖叫著同意，杜比不情願地執行最後任務。康斯坦佐和馬丁走進臥房衣櫃，杜比朝他們射擊，清空彈匣。他沒有殺其他人。忽然之間，惡夢結束了。

莎拉跑向警方，宣稱自己遭到狹持。杜比後腳跟著出來，遭警方逮捕。警方衝進大樓，找到躲在床下的奧馬。現場一片混亂，警方和困住的毒梟撒旦教信徒斷斷續續開槍戰，因此警方最初回報是他們殺了康斯坦佐。但是稍後彈道證據顯示杜比射殺了自己的教父，就如同他不久後招供的一樣。

警方稍後細細訊問他們三位，這些詳細的審問就是大眾對毒梟撒旦教犯罪的認識來源。莎拉依然堅稱自己清白，說她原本和康斯坦佐離開馬塔莫羅斯市是要去墨西哥市度假。但她沒有打包行李，而且她父母完全不知道這趟旅程。她無法說明為什麼離開公寓打電話時沒有逃跑。此外，她對康斯坦佐的毒品交易和大殺戮知之甚詳，說自己無辜實在說不過去。莎拉堅持她對這些黑暗面知道的不多，都是康斯坦佐分享的隻言片語。儘管如此，她重述了許多康斯坦佐的殘暴罪行，她的身分迅速從關鍵目擊證人變成重大嫌疑犯。

杜比招供時知無不答。起初他對自己是殺了教父的人感到懼怕。但他迅速重拾在教團的自大態度，告訴警方不管他們做什麼都不會影響到他。他似乎還是相信康斯坦佐會死而復生，拯救遭警方扣押的他。康斯坦佐的神祕操控連在他死後都仍然健在。

奧馬就不太願意承認任何事。他堅持自己和教團的任何凶殺都沒關係，他只是康斯坦佐的情人和管家。一部分確實如此：奧馬沒什麼參與毒梟撒旦教的活動，而且他住在墨西哥市，離馬塔莫羅斯市有好幾百哩遠。不過最後他還是認了，豪爾赫室友拉蒙被獻祭時，他人也在場。

杜比被判殺人有罪，受害者為康斯坦佐和馬丁，判刑三十年。豪爾赫殺害拉蒙，判刑三十五年，刑罰較重可能是因為豪爾赫並非按照要求行事，直接殺了拉蒙。

莎拉在康斯坦佐和馬丁凶殺一案被判無罪，但被判共謀罪，得服刑六年。隨後在一九九四年，莎拉被判殺人有罪，因為她參與好幾起教團活人獻祭凶殺，判刑六十二年。同時，艾里歐、小賽拉分、塞爾吉歐以及另一名信徒大衛．塞爾納（David Serna）每人都被判六十七年。

奧馬還沒等到判決，一遭扣押，他的健康就開始急速惡化，被逮捕後一年未滿，一九九〇年二月，他就因愛滋併發症死於心臟病，年方二十四。另外兩名信徒遭起訴，但沒有受審：一位是馬利歐，也是馬塔莫羅斯市最初五人小組的一員；一位是奧維德，就是那位艾里歐的兄弟。兩人就此消失，從未被捕。

維達爾被控持有並販售古柯鹼，但他的罪行沒有涉及凶殺。警方宣稱他是一名好警官，只是被販毒黑錢和巫術信仰蒙了心，變得腐敗。他們也堅持沒有其他警察參與康斯坦佐的邪教。維達爾涉嫌販毒，會面臨七到二十五年刑期，但沒有證據顯示他有服過刑。

聖艾倫納牧場成為墨西哥政府的財產。活人獻祭之後，就閒置在那，畢竟沒有人想要在那裡工作。偶爾，夜間會有人來挖地，很可能都是祭司、術士來找屍體殘骸，於是開始有人謠傳毒梟撒旦邪教還未解散，不過根本沒有當地活動可以證實這則謠言。

可是康斯坦佐確實留下一個謎團，暗示故事還沒結束。警方搜查他在墨西哥市波莫納街（Calle Pomona）的公寓時，發現其中一個房間有祭壇，在那裡辦過許多墨西哥市邪教的入教及其他儀式。

地板上正在腐爛的血液和動物屍骸之中，有一圈又大又乾淨的圓，圓上一點髒汙都沒有。曾經有東西在那裡，但就在警方搜查前被移走——那是恩甘加。康斯坦佐和那幾名信徒逃跑時帶走了它，但他們移動時並沒有帶在身邊。

總歸一句，康斯坦佐最初的恩甘加依然存在於世上，至今還未被人找到。

Chapter 3

剝削

巴爾・希瑞・羅傑尼希

巴關・希瑞・羅傑尼希是一位機智詼諧的哲學教授，他重新包裝自己，變成一名魅力四射的宗師。一九七〇年代，他的靜修處最初開在印度，一九八〇年代在奧勒岡（Oregon）吸引大批信眾，多數是歐美人士，羅傑尼希高舉自由之愛、資本主義，以及透過跳舞開悟。羅傑尼希開始吸笑氣（學名氧化亞氮，短效吸入性全身痲醉劑），並且用他的祝福交換信徒贈送的昂貴禮物，他在全盛期擁有九十三台勞斯萊斯，其中一台擋泥板後方還藏著催淚瓦斯槍。當羅傑尼希陷入毒品吸食的漩渦，他的信眾發動一場生物攻擊，在奧勒岡餐廳的沙拉吧添加沙門氏桿菌（salmonella），試圖操縱當地選舉，想要贏得附近小鎮的選票。最終他們掌控了全區，不僅將公眾裸體設為合法，還騷擾任何反對的居民。只不過當政府蒐證控告他們時，命運已定：羅傑尼希被遣返回印度，他的跟隨者被關進監獄，而多台勞斯萊斯通通被法拍，賣了七百萬美元。

物質宗師

巴關．希瑞．羅傑尼希扮演悟道宗師，無人能出其右。翩翩飛舞的袍子，又長又白的鬍鬚，促狹慧黠的眼睛，定期與橘衣信徒談話，他激起眾人強力奉獻。在他的達顯（darshan，梵語，又譯福德，意思是有機會目睹神，蒙神祝福）期間，羅傑尼希給予信眾新名字作為遁世者（sannyasin）或棄絕者，成為離棄世界的人。在靜修處或寺院時，他總是坐在正中央像極了王座的椅子上，隨心所欲開講，話語充滿神祕、樂觀和寬容，例如：「神並不嚴肅，要不然祂沒辦法創造這麼美麗的世界，無處不是音樂和歡樂，滿溢著愛。」

一九七〇年代，上千名尋求心靈更新的人蜂擁至印度，羅傑尼希打造了狂熱信眾，多數都是西方人，信眾湧入他的靜修處，學習他獨創的招牌「動態靜心」（dynamic meditation），上下彈跳、瘋狂舞蹈，也成為一種冥想形式。羅傑尼希宗師把這視為頓悟心法傳授給信徒。嬰兒潮世代可能缺乏耐心或精力，無法做傳統的靜坐冥想，但在羅傑尼希的指引下，可以自由自在練習這套新的高能量東方靈修形式。為了證明自身虔誠，許多信徒捨棄世俗財產，為羅傑尼希獻上精美禮品，還有定期大筆捐獻，羅傑尼希聖人總能過著極盡奢華的生活。他告訴信眾：「資本主義是一種自由狀態。這正是為什麼我支持資本主義。資本主義允許你享有各種自由。」[1]

但宗師過著上流奢華的生活，信眾則為了他連續工作十二小時，支撐著他奢侈的生活型態。一九八〇年代，健康問題和迫害擔憂致使他棄印度而去，並且叫最虔誠的信徒瑪．阿南德．席拉（Ma Anand Sheela）支付一切費用，在美國建立一座新的靜修處。席拉本身就是位堅忍不拔、魅力滿滿的領袖。席拉埋頭執行任務，在奧勒岡的安特羅普（Antelope）外圍買了六萬四千英畝房地產。才幾個月，在席拉的領導之下，信眾把那塊地變成自給自足的社區，有房子、下水道、一座發電廠，還有好幾千英畝田地，種植社區賴以為生的糧食蔬果。

數以千計的遁世者開始造訪這座位於安特羅普的新靜修之地，又名羅傑尼希社區（Rajneeshpuram），不過他們立刻發現自己和周遭社區格格不入。當地人反對這批定居在遙遠谷地的新住民，也不贊同他們奇怪的生活方式，例如一絲不掛四處亂逛，又或者無時無刻不傳來陣陣做愛聲。小鎮居民威脅要改變那些能讓羅傑尼希社區營運的法律時，席拉動員人馬，招收遊民，安置在社區內，如此一來，他們就能投票否決任何規範。靜修處開始購買安特羅普的房地產，並且接管當地生意。而當州政府和地方政府繼續調查時，席拉改用暴力，策劃凶殺甚至發動生物戰爭。

同時，羅傑尼希找到一名常駐好萊塢的牙醫提供他藥品（他的最愛是笑氣），於是開啟自我放縱的生活。他開始定期舉辦行車達顯：羅傑尼希坐著勞斯萊斯，徐徐經過崇拜的信眾，信眾會在羅傑尼希社區的路上列隊，投擲花朵。但在那段時期，相關當局已經漸漸逼近了。一九八五年，警方突襲社區，發現充足的證據可以提起告訴，包括殺人未遂、電話竊聽和密謀詐騙政府。不到幾個月，奧勒岡靜修處變成一座鬼城。席拉認罪，但只服刑三十九個月就被釋放、驅逐出境，之後宣稱自己無辜無罪。

魔鬼小孩頭上有詛咒

一九三一年十二月十一日，在印度古其瓦達村（Kuchwada）的羅傑尼希祖父母家，引起這番騷動的男人出生了。男孩的耆那教徒父母為他命名為香卓拉．穆罕．耆那（Mohan Chandra Jain）。耆那教是古老的印度宗教，奉行反暴力、誠實、貞潔、一生無長物。耆那教徒相信追求更高的靈魂層次是首要之務，目標為終止業力、透過輪迴轉世讓生命圓滿。此一哲學可以說明許多羅傑尼希的教誨，雖然他很明顯並沒有嚮往身無長物的人生。

出生那時，羅傑尼希寶寶是無法預測的難題。他的祖父委託數一數二的占星師為羅傑尼希排本命盤，本命盤也就是羅傑尼希出生當下各個行星的宮位。大家相信這些命盤和行星宮位的幾何學結果會主導一個人的性格和未來，雖然每個結果都依個人解讀而有所不同。而且因為這位占星師不相信羅傑尼希會活過七歲，拒絕排完他的命盤，除非他活到七歲。因此，死亡威脅一直籠罩著羅傑尼希童年早期。而這則預言似乎相當可信，家人為此更加緊張不安，因為羅傑尼希患有慢性氣喘，後來還感染天花，差點喪命。

羅傑尼希的父母住在加達爾瓦拉（Gadarwara），兩人忙碌異常：他們有十一個小孩，還繼承了羅傑尼希奶奶的生意，必須額外負責經營家族事業。為了減輕重擔，羅傑尼希父母把他送到外婆家。外婆很快成為羅傑尼希童年塑造期裡影響力最大的人。她養育羅傑尼希時，給予他絕對的自由，沒有設下任何限制。

但伴隨著這般無拘無束的獨立自主，羅傑尼希內心有其黑暗的一面，漸漸沉迷於死亡和神祕學。

羅傑尼希好不容易活過七歲，一九三八年，占星師信守諾言，終於排了男孩的命盤。但是占星師的解讀並沒有隨著男孩年紀增長而變好。這次，預言說羅傑尼希不會活超過二十一歲，這則預言接下來會一直籠罩著男孩的年少時期，如影隨形。

知曉悲慘預言的那年，外公也過世了，對羅傑尼希而言是很大的衝擊。關於痛失外公，他形容是「所有依附的消滅」。沒多久，他開始疏遠身邊的人，或許是為了保護自己不受哀痛所苦，又或者因為外公走了之後，他和外婆搬回加達爾瓦拉父母鬧哄哄的家。

在那裡，羅傑尼希開始上學，他擅長學習，但愛和老師爭論，經常挑戰老師的權威。他拒絕按照期待發展，經常叛逆搗蛋，陷入麻煩。他自己說過、認識的人也這麼說：他很莽撞又愛惡作劇。他不僅以打破規則為樂，也喜歡測試自己的體能極限，好似要嘲諷籠罩整個年少時期的死亡陰霾。

戰勝死亡

羅傑尼希十四歲時決定，如果命運乖舛是注定，那麼他會正面迎擊。他說：「假使真如占星師所說，我會早死，那麼最好及早準備……為什麼我不自己往前，在半途會一會死亡呢？如果我就要死了，那麼最好死得明白。」[2]他前往加達爾瓦拉外圍的一間老廟，在廢墟之間冥想了七天。他安排好每天有人送來食物和飲水，除此之外都是獨自一人，等待死亡。

從西方觀點看來，事情會變成這樣無非是羅傑尼希雙親的錯，但並非如此。相反地，這應當視為印度靈性的應證，以及年輕男孩剛毅致志的徵兆。在印度，這般機警的孩童不僅受到認可，還受到鼓勵。即使年紀輕輕，他展示的精力和耐力，許許多多有志者都無法企及。如此一來，在這麼多邪教領袖之中，羅傑尼希如鶴立雞群一般，具備真正的靈性恩賜，日後他也藉此過上自我放縱的生活。

他說：「我等了整整七天……死亡都沒有來……陌生詭異的感受浮現……如果你覺得好像即將死去，人會變得沉默冷靜。任何事都不再令人煩惱了，因為一切煩憂都關乎生命。」[3]羅傑尼希透過說服自己即將去世，似乎創造了經歷瀕死體驗的理想情境，許多人認為瀕死是心靈或宗教體驗。研究人員相信瀕死狀態會增強腦部活動，而且認為瀕死狀態和神祕體驗有許多共通點，例如靈體分離、超脫時空、宇宙一體或天人合一、沁人心脾的正向情緒、神聖感受、直覺啟示。有些人說瀕死時平靜安穩，有些人說自己看到他界神明。這些現象都能說明羅傑尼希在廟裡待了七天、成功活下來之後親自述說的一切。

當然，羅傑尼希沒有死，但根據布魯斯・葛瑞森（Bruce Greyson）的研究，這些瀕死經驗可以讓一個人的生命走上嶄新的方向。葛瑞森醫生曾於維吉尼亞大學（University of Virginia）任教，也是瀕死經驗研究權威學者。轉換生命方向尤其常見於那些生命充滿掙扎、懷疑或恐懼的人，而這正是羅傑尼希經歷的生活。

即使他打敗了命運，活超過預言的死期，但就像他現在相信的一樣，死亡依然籠罩著他。羅傑尼希十五歲時，女朋友莎茜・沙瑪（Shashi Sharma）死於傷寒，但依照耆那教的轉世信仰，女友答應下輩子還會回到他身邊。二十二年後，一名英國女子來參加靜心營，羅傑尼希記起了那則誓言。女子名叫克莉絲汀・吳爾芙・史密斯（Christine Wolf Smith），在沙瑪死後隔年誕生。靜心營是讓新信徒學習動態靜心的地方，而在他的指導底下，克莉絲汀恢復記憶，想起了前世，成為他的長期伴侶。

儘管如此，羅傑尼希依舊是個人啟示的翹楚。十九歲離家，前往賈巴爾普爾（Jabalpur）就讀西卡瑞尼學院（Hitkarini College），離雙親大約九十哩。然後在一九五三年，寺廟廢墟瀕死經驗後七年，他宣布自己成道（enlightenment），來標示自己的二十一歲生日，同時也顯示自己又過了一次七年死亡循環。

成道一詞，不同宗教各有解釋，但就耆那教傳統而言，成道的寓意又更為深遠。通常，成道被視為一種平靜的存在狀態，在這樣的狀態下，靈魂超脫一般煩憂，以及日常生活顧慮。但就耆那教而言，成道指的是正覺（kevala jnana），也代表更多人宣稱的獲得了無上慧（supreme wisdom），甚至是全知。當然，有很多大學生就像羅傑尼希一樣相信自己什麼都知道，其實並非不尋常，但羅傑尼希完全清楚他的宣稱並不尋常。他才剛起步，試圖讓自己與眾不同，甚至說服了自己，比起全人類，他的精神運作完全是不同層次，是更高的層次。

羅傑尼希的教育包含高層意識境界，他完全愛上了喬治・葛吉夫（George Gurdjieff）的教導。葛吉夫是希臘裔亞美尼亞靈性哲學家。葛吉夫聲稱大多數人類體驗只是情感和身體的失聯狀態，因此都活在半意識的催眠狀態。然而，葛吉夫稱這種狀態為「清醒的沉睡」（waking sleep），這種狀態可以克服，也可以超越，越過之後就能往前進，達成高層意識境界的最大個人潛力，同時獲得內在成長。其中特別吸引人的方面，也是後來羅傑尼希邪教失控成長的關鍵，那就是葛吉夫相信覺悟的師父可以看見弟子的個人需求，並且為弟子指出該完成哪些工作才能達到高層意識境界。這成為一條特別實用的原則，幫助羅傑尼希剝削數千名信眾，信眾全都興高采烈在羅傑尼希社區無酬勞動，認為他們藉由挖茅坑、洗豆芽、培養沙門氏桿菌，都正在上升到更高的層次。

挑戰權威

或許因為羅傑尼希已經成道，也可能因為他在獲得全知之前本就聰明又好鬥，所以他在課堂上頂撞師長也不覺得哪裡不對。他老愛妨礙課堂教學，質疑教授的論據和權威，常常大搞破壞，以至於學校最終請他離開。他轉校到同樣位在賈巴爾普爾的迪恩耆那學院（D.N. Jain

College），一九五五年取得哲學學士文憑。

在新學校，羅傑尼希並沒有收斂，依然一樣好辯，但是教職員想了個絕妙辦法：免除他出席課堂的義務，他的胡鬧把戲就無法干擾同儕。他只需要出席考試。他參加了學校辯論社，還贏得全國冠軍，他爭辯不停的性格終於獲得獎勵。

大學一畢業，羅傑尼希就進入薩加爾大學（University of Sagar）就讀，獲得哲學碩士學位。接著到賴布爾（Raipur），獲聘為梵文學院（Sanskrit College）的講師。但據研究羅傑尼希的哲學理念的精神科醫生阿尼爾．庫馬爾．邁索爾．納加拉吉（Anil Kumar Mysore Nagaraj）所言，賴布爾梵文學院校方擔心羅傑尼希對學生造成負面影響，而且削弱學生士氣，帶壞學生性格，於是校方立刻解聘他。下一站，賈巴爾普爾大學（Jabalpur University）。羅傑尼希擔任講師，一九六〇年僅僅二十九歲便升等為哲學系教授。他開始在年度信仰大會（Sarva Dharma Sammelan）演講，各種信仰每年齊聚在賈巴爾普爾舉辦大會。羅傑尼希根據講稿創作了四百多本書，全都是他的所思所想，題材五花八門，例如基督教和哈西迪教派（Hasidism）、瑜伽和禪宗，甚至西格蒙德．佛洛伊德（Sigmund Freud）和亨利．福特（Henry Ford）。他借用許多不同哲學理念，揀選那些吸引他的趣聞，再創造自己的信仰系統。

到了一九六二年，羅傑尼希教導的幾個核心都快要成型了，他開始推出靜心營，帶領三至十天不等的靜修，專門傳授他稱之為「動態靜心」的獨門冥想法。此法獨特，結合了動態與靜態。按照他的指示，這項靜心法需要十分鐘強烈紊亂的呼吸，接著十分鐘爆發自我，同樣隨興亂動身體。（他建議：「發瘋吧！尖叫、大喊、哭泣、跳躍、踢踹、搖動、唱歌、跳舞、大笑，盡情發洩。毫無保留，運動全身。」[4]）第三階段十分鐘，他命令門徒上下跳躍，手臂高舉過頭，重複喊：「呼！呼！呼！」直到氣力放盡。動態靜心的第四階段為十五分鐘：門徒只需要待在原地靜止不動。（他說明：「第四階段會自然發生。事情會發生，但不是源於你的行動。如恩典般降臨：你會變得真空如無物，感受到有東西充滿。當你不在時，某些靈性的東西注入你。你人不在當下，因為你沒有作為；人一沒有作為，自我就消失了。」[5]）第五階段，同樣持續十五分鐘，門徒會歡慶並放縱自我。

這儀式顯然沒有多少紀律可言。多數時候冥想修習要求學生正坐，淨空思緒，但羅傑尼希的靜心法則是告訴門徒先變得緊張，接著耗盡精力，為的就是透過純粹癲狂來切斷思緒。對過度好動、沒有相關訓練，又絲毫不願靜坐的西方人而言，這簡直是毫不無聊的完美冥想形式，因此從一開始就大力支持。

羅傑尼希一邊開發這半鍛鍊、半靜心的法門，同時四處巡迴演講，批評正統印度宗教，讚揚資本主義，這樣雖然引來宗教相關當局的仇恨，但也贏得多位印度富商的愛戴與支持。一九六五年，富商合力建造了生命覺醒中心（Jeevan Jagruti Kendra），資助羅傑尼希的旅行和演講，免去了他原本還得去學校工作的負擔。

他的資本注意哲學後來記錄在著作《來，來，又來了》（暫譯，*Come, Come, Yet Again Come*），他寫道：「資本主義不只是意識形態；資本主義不是強加於社會之上，而是自然發展。不像共產主義、法西斯主義、社會主義，這些都只是意識形態，這些主義都必須強制施加。資本主義是一種自由狀態，這正是為什麼我支持資本主義。」[6]

性與領袖魅力

一九六八年，羅傑尼希三十七歲，他的演講甚至開始著墨更多性與色慾的接納。這些演講最終集結出版，書名為《了解性、超越性：從性到超意識》（*From Sex to Superconsciousness*），印度媒體因此封他為「性愛大師」。保守人士嚇壞了，大力反對他的教導，但他早就瞭解最慷慨的支持者很高比例都來自西方，對他而言這也是厲害的行銷手段，因為西方那時性革命正開始蓬勃發展。他故意煽風點火，甚至在演講時像魔術師一樣催眠聽眾，隨時隨地都能成為目光焦點。

在色慾的神性相關演講中，羅傑尼希宣稱措辭必須如此強烈，才能達到心靈成長並進入高層意識境界。他相信宗教和社會規範把性和羞恥、丟臉劃上等號，限制了所有形式的性表達。他堅稱這些負面意涵阻礙兒童獲得恰當的教育，兒童需要瞭解自身的慾望。此一觀點和保守派領導者衝突，他們不滿意有人公開討論這種事，他們覺得這種事就該私下談論——和執行。

這些看法羅傑尼希根本不屑一顧。他說太多宗教只關心來世的獎勵，而忽略此時此刻的美好、歡樂和神性。他教導大家，通常愛是通往成道的法門，不過性尤其如此。而且他提倡透過性表達來釋放愛，愛常常被困在體內。他說道：「全體人類社會都病得不輕。如果想要改變這個不治的社會，就必須接受性能量是神聖力量，性吸引力的本質就是宗教。」

對西方人而言，這席話也正合時宜：同性戀解放運動、女性主義、開放式婚姻、無責離婚（No-Fault Divorce）攜手顛覆長久以來的社會規範，鼓勵所有形式的性表達。這一運動恰好與羅傑尼希的哲學理論完美吻合。休・B・厄本（Hugh B. Urban）在《左巴佛陀：全球奧修運動的性、靈、資本主義》（暫譯，*Zorba the Buddha: Sex, Spirituality, and Capitalism in the Global Osho Movement*）討論羅傑尼希運動的形成，巧遇第二波女性主義興起。羅傑尼希相信女性優越，宣稱女性能有多次高潮，證明了女性擁有更強大的性能量。這令男人心生懼怕，好幾個世紀以來男人都在壓迫、支配女人。蘇珊・吉恩・帕爾默（Susan Jean Palmer）出書討論邪教中女性的角色，帕爾默寫道，其實這是很吸引人的訊息，特別是對三四十歲的中上階層而言，這些人先前往印度，稍後再到奧勒岡，成為他的門徒。

羅傑尼希相信他可以提供這些女人一條通往自由和成道的路，除了他，沒有任何其他女性主義運動可以提供。他說：「我的願景是下一個時代是女性的時代。男人已經試了五千年，還是不成功。真是夠了！現在女性能量必須被釋放……婦女解放運動之類愚蠢的運動無法為女性帶來自由…… 如果我們在這世上創造幾名女佛陀，那麼女性將不再受制於任何枷鎖與束縛。」[7]

就在這一關鍵轉折時刻，從一九六〇年代晚期到整個一九七〇年代，羅傑尼希開始轉變成社會學家所謂的魅力型領袖（charismatic leader）——政治經濟學家、社會學家馬克斯・韋伯（Max Weber）常用這一字形容許多邪教人物。韋伯在政治社會學著作中提出三種權威型態：魅力、傳統、法定。依韋伯的定義，理性法定權威和傳統權威較為普遍常見。理性法定權威的正當性來自複雜規範，受到國家法律管制。而傳統權威指的通常是世代沿襲的權力。

但魅力型領袖則獨樹一格，大眾認為他們好像有天生超凡能力或特質。羅傑尼希完全符合這點，毫無疑問，他宣稱自己成道，因此最近自封「巴關」，意指「有福的人」。這立刻在孟買（一九九五年前官方名稱是Bombay）迅速發展起來，一九六九到一九七四年間，他在孟買建立總部，住的是昂貴公寓，靠著演講和靜心營賺錢。

孟買時期，羅傑尼希總是讓崇拜的信徒隨侍在側，他們相信他的哲學可以拯救人類。在虔誠信徒之中，他倡導的性覺醒不單單只是理論。一九七〇年代早期，許多女信徒說宗師對她們動手動腳。羅傑尼希利用「私人達顯」時段和門徒性交：他和一名女門徒持續到半夜，接著他會睡到凌晨四點，然後與另一名女門徒開始另一場達顯。有時他會要求門徒自行脫去衣衫，又或者在他面前性交，這樣他才能「治療他們的能量」。

我們漸漸發現，這五年間，羅傑尼希哲學的兩項關鍵開始結合。關鍵一是他極度鄙視機構權威和社會規範，關鍵二是他渴望建立佛土（Buddhafield）——在這片應許之地，他的「新人類」（New Man）可以寧靜安詳地活著。他的信眾相信憑他一己之力，就可以兌現拯救全人類的諾言。

這符合韋伯針對魅力型領袖所列出的最後一條要件，這類型領袖在信徒面前看起來像具有「革命力量」，生來就要「從根本改變現有態度的中央系統……轉而朝向……『世界』。」[8]羅傑尼希的佛土引起許多迴響，大家開始提及前所未見的新世界，在那裡人人皆可成道得救。二十年間，羅傑尼希把自己打造成擁有更強大能量的存在，吸收那些崇拜他，認證該能量的人。他建了一個社區，旨在解放信徒，不再受社會限制和義務所困。他應允的好處（例如性和成道）成了餌，確保門徒全心奉獻，說服多位忠實門徒相信羅傑尼希正是唯一能引領他們前往應許之地的宗師。

虔誠左右手

一九七〇年代早期，大批信徒湧入孟買（一九九五年以後官方名稱改為Mumbai），很快地，羅傑尼希不論到哪裡，都有大批信徒簇擁。他的門徒席拉在書中寫道：「他現在已經不是普通老百姓，而是公眾人物了。大家認為他是聖人、教長、宗師。」[9]該書論及羅傑尼希，以及席拉協助擴張邪教時扮演關鍵卻時常犯法的角色。

席拉一九四九年出生，本名是席拉・安巴爾・帕特爾（Sheela Ambalal Patel）。一九六〇年代晚期，席拉十七歲時離開家鄉巴洛達（Baroda），離開特權中產階級家庭，到海外旅行，最後在紐澤西（New Jersey）就讀蒙特克萊爾州立大學（Montclair College）。就讀大學時，她愛上多金學生馬克・席爾曼（Marc Silverman），席爾曼十八歲時被診斷出何杰金氏淋巴癌（Hodgkin's Lymphoma），醫生說他只剩兩年可活，席爾曼猶豫是否該結婚。但是席拉還年輕，不管命運如何安排，一心只想和馬克在一起。席拉寫道：「我們決定不鳥邏輯，把所剩不多的時間花在歡樂的事物上。我們想要感受愛，不想放棄人生，就算人生有限。」[10]因此，一九六八年，儘管馬克患有絕症，兩人還是結婚了，而且馬克還比醫生預測的再多活了整整十一年。

一九七二年，席拉二十二歲，收到媽媽眼睛要動手術的通知。席拉返回家鄉陪伴，馬克打算一完成學業也前往印度。到醫院探病後，席拉和爸爸一時興起決定去孟買拜訪親戚。他們待的公寓距離眾人喜愛的羅傑尼希只有幾個街區，兩人趁興想去瞧一眼名人。頗出乎意料的是，羅傑尼希的個人祕書瑪・尤佳・拉克西米（Ma Yoga Laxmi）歡迎他們入內，並直接帶他們去見羅傑尼希。祕書身穿門徒的橘色長袍，祕書一職則是羅傑尼希用來稱呼每一位同意為他工作的人。羅傑尼希立刻把席拉擁入懷中，而席拉日後寫道自己則是「看著他，內心完全融化」。

新情緒溢滿心中，席拉報名了羅傑尼希的靜心營，並且立刻改用羅傑尼希給予的名字：瑪・阿南德・席拉。「瑪」是羅傑尼希給予所有女門徒的稱謂，意思是「母親」。「阿南德」代表「喜樂」，而「席拉」本就是她的名字，意思是「性格堅強之人」。她立刻成為門徒。席拉寫道：「我把所有事情都交由教主決定。他說『穿橘色』，那我就會穿橘色。他說『裸體走上大街』，那麼我就會裸體走在大街上。我行事沒有其他理由。要知道為什麼我做這些事，你得詢問我的教主，他已經成為我的絕對指引。」

席拉的先生馬克終於到印度與她匯合時，起先有些懷疑，但沒多久也成為門徒。羅傑尼希急於測試他們的忠誠，派他們到喀什米爾（Kashmir）巴哈爾加姆（Pahalgam）山區，在隆冬嚴寒天氣下待上三週。他們接下任務，在山上待了三週，沒有暖氣，只帶了極少量食物。席拉寫道：「我們下山時感到超級良好——我們高分通過他出的態度測驗。」

心理學家利夫頓在《思想改造和極權主義心理學：中國的洗腦研究》（*Thought Reform and the Psychology of Totalism: A Study of "Brainwashing" in China*）這本討論洗腦的書中，稱這類例子為「藉由純正程度來自我神聖化」。[11]這類測驗，甚至可以稱為試煉，通常都設計得非常困難，注定失敗。但失敗反而激勵虔誠信徒更加努力嘗試，甚至挑戰更不可能的任務，只希望能取悅教主。當然，極罕見的情況下，真的有信徒完成不可能、不舒服、羞辱人的任務，那麼宗師就會知道這群信徒願意付出多少努力來取悅他。

席拉和馬克闖關成功後，羅傑尼希要求他們飛回美國，把事情都安排好。席拉在書中沒有解釋究竟是哪些事，但他們的返美以及最近的信仰轉變，都讓他們和家人有所齟齬，特別是馬克的家人。席拉這樣描述馬克的父親：「他寧願兒子是罪犯，也不要兒子穿著橘色紗籠逛大街。」這又是利夫頓列出的另一類洗腦手法，他稱之為「社會環境改造」（milieu reform），其中一項做法是切斷與外界的溝通，教徒因此得依賴邪教領袖。

但這些洗腦伎倆遠遠不及羅傑尼希下一波提出的要求。一九七四年，羅傑尼希厭倦了孟買的一切，責怪這座城市害他童年的氣喘復發。除了這些託辭，還有許多實際考量：門徒人數大增，他需要更大的地方，才能真的讓入教人數一飛衝天。在祕書拉克西米的協助下，羅傑尼希在孟買旁邊的浦納（Pune）買地，打算建造靜修處，而在一九八〇年之前，此處成為總部，什麼事都在這裡舉行。這一靜修處也登記為非營利慈善機構，機構名為羅傑尼希基金會（Rajneesh Foundation），基金會最終涉入了財金騙局。

掌權女士

浦納靜修處營運資金來自羅傑尼希信徒的捐獻，靜修處隨著富裕門徒占比、捐獻金額雙雙上升，營運狀態也越加良好。依照席拉所說，門徒可以指定用途，有些希望能引起他的注意，專門指定給羅傑尼希使用。席拉寫道：「有些人會更樂意捐錢支應他的個人需求，而不是捐給機構營運。」因為他們盤算著能藉此接近宗師。多虧門徒慷慨解囊，羅傑尼希開始縱情奢靡，特別是勞斯萊斯汽車。在印度，勞斯萊斯常被認為是大君的奢侈品。他的唯物主義行徑招致許多批評。席拉語氣充滿崇敬，宛如羅傑尼希只是聰明伶俐卻愛搗蛋的孩子，寫道：「沒有人認為真正的靈性領袖會過著奢華的生活。一般大眾認知的靈修代表物質生活貧乏。他的生活方式讓許多傳統宗教人士大為不悅。」

一九七四年至一九八〇年間，浦納靜修處迅速擴張，六千名門徒住在社區或社區附近，一起參加療程、演講、靜心工作坊，有時門徒會半裸或以全裸狀態參加。有些工作坊叫會心團體

（Encounter Group）工作坊，提供密集團體療程，鼓勵性探索，據說是為了幫助門徒走出自己的心理或精神舒適圈。其他人單純認為羅傑尼希患有窺淫癖。

身為權力結構的一員，席拉有機會瞭解羅傑尼希願意剝削門徒到什麼程度。羅傑尼希開始收費，不僅參與團體治療、靜心要付費，就連進入靜修處也要入場費。關於浦納靜修處財政結構的改變，席拉寫道：「大家都瘋狂想要成道，滿心熱忱地急於擺脫自我，急於成為靜心之人，為此要他們做什麼都願意。門徒掏空口袋，奉上昂貴禮物，證明自己全心投入。如此這般的剝削，簡直骯髒、醜惡、令人反胃。」羅傑尼希開始強力展現自己的權力。席拉寫道：「他要所有人明白誰才是教主。」

宗師十分清楚自己的矛盾，有回說道：「（一致）對我而言是天方夜譚。我活在當下，不管我現在這一刻說了什麼，都只在這一刻是真的。我沒有參照過往，也完全沒有思考未來。」但事實不盡然如此。他顯然總是留心還有哪些方法可以從大批信眾身上榨取更多奉獻。為了達成此一目標，他還任命靜修處的一小群女信徒來幫忙，下放權力給她們，讓她們負責各種活動，維持靜修處營運順暢。

加拿大社會學家帕爾默撰寫《月亮姐妹、奎師那母親、羅傑尼希愛人：新興宗教的女性角色》（暫譯，*Moon Sisters, Krishna Mothers, Rajneesh Lovers: Women's Roles in New Religions*）一書，研究羅傑尼希和信徒的互動，書寫這一群位居浦納社區高階地位，身處中央核心的女信徒。她們漸漸被稱為「掌權女士」，其中許多位都獲准晉見羅傑尼希本人，這可是不尋常的機會，身分角色也很多元，從個人祕書到個人護士都有。不過也管理日常活動，並且經常負責重大決定。

但羅傑尼希還是會鼓勵她們彼此競爭、起內鬨，來顛覆自己授予她們的權柄，導致這群女人陷入不安，擔心她們在小團體裡的相對地位。席拉寫道：「羅傑尼希刻意營造衝突易起的環境。衝突被認為是深入內心的好工具。按照他的說法，衝突應該要能看得見且摸得到，不應該被壓抑或推到表面底下。唯有如此，衝突才能幫助內在成長。」

據說羅傑尼希比較喜愛和女人共事，因為他覺得女人比較不具攻擊性，更願意接納、承受他的能量。對他而言，接受性（Receptivity）是能否成道的關鍵。不過也代表了概念本身撲朔迷離，定義標準為承受度（Openness），只有羅傑尼希本人才能辨別程度高低。羅傑尼希若認為某些門徒比其他門徒更願意接納，此舉雖然被門徒視為極大讚賞與光榮象徵，但實際上則代表了他成功地讓身旁環繞著焦慮不安的人，多數是女人，每一位都極力爭勝，言行舉止都是要向宗師證明自己。

拉克西米屬於掌權女士小組，曾經擔任羅傑尼希早期的個人祕書，羅傑尼希對她的評價恰好說明他平常如何對待這群女信徒：前一秒才委派任務，下一秒卻毀謗他們。他評價道：「千萬記得拉克西米從來沒有獨力完成任何事。她就是台完美的車子，而這正是為何她雀屏中選負責這份工作……不管我說什麼，她就會去做。」[12]這裡可以發現兩相衝突的概念協助羅傑尼希掌控他的信眾。拉克西米遵循的行為規範，只有羅傑尼希知道標準。但是拉克西米又不夠強大，無法按照自己的意思行動。她做任何事都跟隨羅傑尼希的指導。

神逃跑了

由於缺少規章條例，靜修處發生的行為似乎常常快要失控。一九七九年，一名德國門徒沃爾夫岡．多布羅沃尼（Wolfgang Dobrowolny）造訪靜修處，拍了一部電影《浦納靜修處》（暫譯，*Ashram in Poona*），其中有一段十五分鐘毫不保留的殘暴鏡頭。一場裸體會心團體療程中，原本大家打枕頭戰是為了釋放暴力念頭，卻打到失控。這部電影還在美國上映，《洛杉磯時報》（*Los Angeles Times*）報導：走在成道的路上，少不了「幾處骨折和黑眼圈」[13]。組織運作若依魅力領袖的興致而行，經常會出現這種混亂場面。這是因為魅力型領袖排斥規範，但規範通常會形塑更加井然有序的治理型態。

三十歲的席拉此時正好經歷自己的震盪期。一九八〇年，席拉三十三歲，先生馬克長年對抗何杰金氏淋巴癌，終於不敵病魔離世。席拉寫道：「他過世之後，（羅傑尼希的）教導給我力量，接受他的離去。」馬克過世之後，席拉把自己都奉獻給了羅傑尼希和靜修處，傾其所有，無所保留。但隨著靜修處的暴力越演越烈，她也明白了當初躲進治理細節，希望尋得安穩一舉，不大可能達成了。靜修處還發生各種事件，精神崩潰、身體損傷和性侵害等等。幾乎每起暴力都無人聞問，因為門徒總是被叮囑，要把會心團體工作坊和憤怒療程視為必經過程，協助他們克服顧忌，而西方人十分吃這一套。在此同時，羅傑尼希和鄰近社區的關係漸趨敵對，因為區內過於擁擠，問題嚴重難解。羅傑尼希顯然需要有所行動，讓虔誠的教徒團結一心，同時讓靜修處變得低調一些。

依照溫．麥科馬克（Win McCormack）列的羅傑尼希運動時間表，雖然不清楚確切時間點，但另一次關鍵轉變發生在一九八一年早期。羅傑尼希立誓止語，不再公開演講。他相談的對象只剩貼身幕僚。信徒稱呼此一時期為他工作的「終極階段」，代表他要與信徒「相交越深」。席拉說自己得知羅傑尼希的誓言時，失望透頂。席拉寫道：「我簡直不敢相信自己的耳朵。這代表我愛的男人步向終點。我不明白他為什麼要這麼做。」

或許席拉著實擔心羅傑尼希止語之後，暴力事件會更加激烈，而他們的運動可能忽然瓦解。又或者席拉對突然出現的權力真空感到亢奮。不管那時她的意圖為何，羅傑尼希靜默不語，正好給了她機會奪權。

起誓之後，羅傑尼希的個人祕書拉克西米奉命為門徒另尋新地，大小必須足以容納六千名信徒。羅傑尼希也相信，離開浦納就能免受印度教基本教義派的迫害，隨心所欲實踐他們的宗教。雪上加霜的是，印度總理英迪拉．甘地（Indira Gandhi）視羅傑尼希及其信徒為社會之害。二〇一八年，查普曼．威伊（Chapman Way）和麥克蘭．威伊（Maclain Way）執導 Netflix 紀錄片短影集《異狂國度》（*Wild Wild Country*），片中席拉述說大眾普遍持有的偏見導致了「僵局」，要在印度境內尋找新地點，幾乎不大可能了。

席拉想家了，懷念待在美國的歲月，但她也用羅傑尼希信徒的邏輯思考。多虧美國憲法保障宗教自由，席拉相信所屬的社區可以在美國生存，不受印度教傳統教派人士的迫害。因此，一九八一年春天，席拉建議把社區搬遷至美國，羅傑尼希欣然同意。席拉寫道：「不消說他有病在身，靜修處又如此擁擠，與正統派印度教教徒衝突日益加劇，他最後決定搬離（浦納）。」那可能是席拉自行修正的歷史，因為許多新聞報導都說這趟突如其來的遷徙其實是為了逃稅。

因為在浦納有那麼多人似乎都需要知道羅傑尼希就近在咫尺，所以離境安排的一切事宜都必須保密到家。

一九八一年六月一日，一切終於安排就緒，羅傑尼希、席拉和少數幾名親信坐上其中一部勞斯萊斯，誰都沒說一聲就駛離社區。英國廣播公司第一台（BBC One）播報頭條：「神逃跑了。」該報導提供不少細節說明：「毫無預警，一聲不響，他的信徒完全不知道他的行蹤。」

信徒痛失他們魅力迷人的領袖，身心交瘁。珍・斯托克（Jane Stork）在《打破魔咒：我的羅傑尼希信徒生活，以及漫漫長路重返自由》（暫譯，*Breaking the Spell: My Life as a Rajneeshee, and the Long Journey Back to Freedom*）寫道：「我帶著子女、先生前往印度，一同和他（羅傑尼希）生活。我所知的一切，我的整個世界，都崩潰瓦解了。」羅傑尼希開到靜修處大門時停了一下，接著就直接開走，留下斯托克獨自思索自己的命運。她寫道：「如果他不再回來，那接下來我該怎麼辦？」整個社區陷入恐慌，不只是珍・斯托克，還有其他人都疑惑之後會發生什麼事。

但是羅傑尼希留下明確訊息給那些被拋下的信徒：「你們每一位都要回到自己的家鄉，待上一段時間，日常起居都要實踐我的教導。每一位都要居住在自己舊有的環境，過著靜心的生活。等到新社區準備好，等到我邀請你來佛土生活，大家就可以回來了。」這裡的「佛土」是應許之地，依羅傑尼希的描述，在佛土可以免除社會限制，可以擁抱生命、歡笑，可以真正成道，變成新人類。這片佛土（和羅傑尼希的福祉）是席拉現在唯一關心的事。

羅傑尼希毫無疑問賦予了信徒生命意義，但不論他多麼能說善道，全都被他剝削的才能抵消了。正如稍早提及，他精於利用信徒的仰慕，來餵養自己奢靡的生活。但面對席拉，羅傑尼希需要更加巧妙的剝削手段。他收了一位具備超凡奉獻精神的女人——這點他心知肚明，畢竟他當初派席拉前往喀什米爾山頂，在嚴寒之中冥想——他算準女人最脆弱的時刻，趁機利用如此犧牲奉獻的心神。如同洗腦研究專家瑪格麗特・辛格（Margaret Singer）所說：「任何人處在孤獨的時期，幾乎都脆弱易感，這時可能輕易上當，相信邪教吸收新成員時說的花言巧語、撒的彌天大謊。」[14]上百名信徒之中，羅傑尼希選擇席拉當祕書，而自從先生馬克過世以來，席拉一直感到孤獨，現在宗師的關注和肩負的工作，極可能填滿了她孤寂的心靈。

席拉在回憶錄中憶道：「失敗的恐懼令人難以招架。為他工作一點也不輕鬆。他把我逼到快要發瘋。」席拉在美國四處尋覓，看了田納西州（Tennessee）、科羅拉多州（Colorado）和亞利桑那州（Arizona）的房地產，寫道：「我從拚命尋找，漸漸心灰意冷。我覺得會永遠無法找到當初答應（羅傑尼希）的那塊土地。」她終於打給親兄弟畢彬（Bipin），畢彬在美國生活多年，而且依席拉所言，畢彬的人脈很廣。席拉鬆了一口氣，畢彬不只建議幾句就算了，還告訴席拉他會安排人接席拉到奧勒岡，看看他認為完全符合席拉需求的房地產。

席拉到了奧勒岡，望著大泥濘牧場（Big Muddy Ranch），六萬四千多英畝的荒漠，離波特蘭約四小時車程。席拉非常喜歡這塊地的大小和偏遠的位置，完全符合席拉想像中未來這場運動的願景。

席拉寫道：「那天是一九八一年六月十一日，剛好是我的亡夫馬克第一個忌日。這一連結對我而言意義重大。我告訴那位房仲我要買這塊地。」她在當地居民和地主面前說話輕聲細語，表現得像一位甜美年輕的少婦。為了更徹底融入，她沒有使用羅傑尼希幫她取的名字，而是用法定姓名席拉．席爾曼（Sheela Silverman）。她立刻著手討當地官員和附近牧場主人的歡心，盡其所能贏得他們的好感。她甚至辦了一場舞會，還購買牛隻，儘管羅傑尼希社區茹素。

佛土的自由之愛

羅傑尼希對執行任務都有不切實際的要求，例如要花多長時間蓋五十間A字形房屋來容納門徒，但席拉努力在他要求的日期前完成。只是，把信徒逼到極限還不夠，社區工程一開始，席拉便幾乎遊走在法律邊緣，除了滿足自己的需求，也為了達成羅傑尼希的命令。舉例來說，席拉被告知當地的土地使用分區管制法律強制規定牧場用地只能供一百五十人居住，但立刻忽略不理。

可能是因為擔子又更重了，或者只是拿羅傑尼希的期限當藉口，席拉變本加厲，開始假傳聖旨。這次權力轉移，羅傑尼希社區開始從魅力型領袖轉型成傳統權威。換句話說，韋伯認為在傳統權威結構下，遵從命令的義務來自個人忠誠感，而這份忠誠感「本來無窮盡」。「正向行動中心」（Positive Action Center）位在奧勒岡波特蘭，是一家邪教諮商中心，其共同執行長阿德里安．格里克（Adrian Greek）研究罪惡感如何被當作控制工具。在《新共和》雜誌（*New Republic*）訪談時，他指出邪教領袖（尤其是羅傑尼希和席拉）喜歡「設定完美標準讓信徒遵守，都是常人無法企及的標準，藉此引發罪惡感。然後，如果你覺得很糟糕，這是你自身的錯。」這便開啟了循環，為了要拔除罪惡感，信徒盡全力變得更加投入，也讓自己墮入邪教領袖的陰謀詭計。若失敗再次發生，這下就更加難受了。格里克說：「人在改變時總會覺得壓力太大（還有罪惡感），因為人總是無法『成道』。」[15]

這股扭曲的動力幫助席拉管理工作人員，這群羅傑尼希信徒變成生產力極高的勞動力。他們開始耕種數千英畝的土地，甚至蓋了醫療設施和大眾運輸系統。他們經常一天工作十二個小時，就為了打造羅傑尼希社區。席拉解釋道，他們如此拚命是因為這次靜修之處會是他們的應許之地，所有的努力都是為了打好基礎，讓全社區可以過著靈性生活——或者至少過著羅傑尼希描繪的生活：瘋狂的靜心活動，以及隨處可得的性伴侶。

越來越多門徒來到新社區，工作人員數量迅速來到一百多名。他們使用預製組合屋和預切建材，讓手上的任務提前完工。他們建造新辦公室，還蓋了羅傑尼希專屬的家。光是為了工作人員和即將抵達的信徒，牧場上搭建的A字形房屋數量，沒多久就超過法規許可的建築物上限。

短短幾個月，羅傑尼希社區一片欣欣向榮。門徒的農地足以供應所有住民；電力網供電數百戶；他們還鋪設了管路系統、基礎設施和道路；他們設立銀行操作系統，蓋了靜心廳，甚至蓋了一座自己的私人機場，同時在波特蘭收購了一間飯店，遠道而來的門徒可以先行歇息，再動身前往牧場。那裡還有一家購物中心，除了披薩店，還有專賣紅衣、橘衣的精品店。

一九八一年八月二十九日，購入牧場才過了八十天，羅傑尼希抵達奧勒岡，初次踏足他新的應許之地。佛土已經住了數百名門徒，全都熱烈歡迎他的到來。雖然社區接下來四年持續擴張，但是席拉和他的團隊前兩個月都在和時間競速，打造所有羅傑尼希需要的事物，這樣羅傑尼希才能加入信徒，和他們一起靜心崇拜。

席拉如此描述他的來臨：「他……走向他的新房子。他看起來充滿歡喜，金光煥發。門徒歡欣鼓舞。我沉浸在對他的愛中。我打造了新的家園……他可以在這裡建立最宏偉的社區。」

雖然席拉在宗師身邊言行舉止沉著平靜，但顯然和其他所有人相處都像仇家。她懷疑羅傑尼希的私人裁縫維納（Veena）利用自己的職位圖利，於是調動維納的宿舍到臨時居住區，並且要維納負責照料社區的豆芽。聽到維納非常享受新環境，甚至似乎過得更好，席拉再次調動她到更偏僻、隔熱效果不佳的小屋，還得與其他七名信徒同住。

因為席拉實在太能幹，這種小手段也就隨她去了。信徒一批批湧入這偏僻的靜修處。在那裡，遠離附近安特羅普居民窺視的目光，離臨海的大城市也有好幾小時之遙，信徒過著所謂放縱的生活，也就是幫羅傑尼希贏得「性愛大師」名號的生活方式。宿舍雖然男女有別，但大家被鼓勵在獨立小屋盡情性表達。諾亞・麥斯威爾（Noa Maxwell）的雙親在他六歲時加入這門邪教。他告訴《衛報》（*Guardian*）他還記得走在靜修處裡，「隨時隨地都會聽見性愛高潮的聲音，持續整夜……」[16]

誰要為這樣的自由之愛付出代價？女人，看來也是如此。儘管羅傑尼希理應支持女力，但整體氛圍可以倏地變得惡意滿滿。在溫・麥科馬克的《羅傑尼希編年史》（暫譯，*The Rajneesh Chronicles*）一書中，前信徒蘿莎琳・史密斯（Roselyn Smith）訴說自己目睹的性剝削，她說：「我覺得巴關因此吸引了許多男信徒。大家都驚豔於他吸引來的男人有多優秀，條件有多好，醫生、律師和許多專業人士。性愛就是他最主要的誘餌之一。大家都聽過，如果到浦納，如果到奧勒岡，就有做不完的愛，那裡的女人都非常開放，絕對能滿足需求……任君挑選。你可以走向任何人，問道：『要來一發嗎？要回我家嗎？』如果對方拒絕，還有其他許多人可以問。我想他吸引了這麼多男信徒，正是因為如此。因為在其他地方，男人可無法這麼做。」[17]

同一時期，反羅傑尼希情緒逐漸在安特羅普滋長。羅傑尼希於一九八一年抵達，沒多久，該州的重量級環保團體奧勒岡千友會（1000 Friends of Oregon）便開始抗議羅傑尼希門徒，請願移除牧場上的建築物，並且引證他們認為這些新居民違反了哪些土地使用法規。羅傑尼希信徒需要更多建築，但現有建物已經大幅超越當地建築許可允許的數量。於是乎信徒想了個聰明解方，卻激怒了街坊鄰居。

那時奧勒岡州法律規定，一百五十人以上的任何團體都有權透過投票建立城市，而這正是羅傑尼希門徒所做的。一九八一年十月，他們投票一致同意讓牧場變成城市，後來這座城市又叫羅傑尼希社區。一投完票，席拉立刻告訴記者：「這是一座非常美麗的城市……宇宙間未曾存在過的美麗城市。市民生活和諧，充滿愛……正是宇宙的典範。」達希爾・愛德華・保羅森（Dashiell Edward Paulson）就讀奧勒岡大學，在博士論文〈一九八一年至一九八五年羅傑尼希社區的日常：羅傑尼希運動的魅力與權威〉（*The Routinization of Rajneeshpuram: Charisma and Authority in the Rajneesh Movement, 1981–1985*）中寫道羅氏邪教，使用了韋伯理論來討論席拉在牧場的行為。保羅森寫道：「接到任命後，席拉迅速鞏固自己的權力。」[18]

從來沒有任何信徒質疑席拉的地位，因為照慣例，羅傑尼希的祕書在社區內總是一人之下，萬人之上。保羅森如此形容席拉：「一位傳統統治者，手法也很傳統。但為了處理複雜、費力又昂貴的社區發展，她也建立起龐大的金字塔官僚體系。」韋伯稱席拉金字塔為「官僚行政器官」，而保羅森則說金字塔「撐起她的權力，把她的獨裁與有效率的行動結合在一起」。

持槍信徒

席拉駁斥安特羅普的強烈反對，說這正好證明當地居民愚昧又偏見。媒體瘋狂報導隨之而來。她參與《莫孚・葛里芬秀》（*The Merv Griffin Show*）、《唐納休秀》（*Donahue*）和時事辯論節目《交火》（*Crossfire*）等許多新聞、談話節目，一次又一次理直氣壯地宣告，羅傑尼希社區不會消失。她無所畏懼、兇悍好鬥、言行挑釁，似乎喜愛媒體關注的每分每秒。

世界各地建立了奧勒岡社區的縮小版，遍及義大利、德國、英國、印度、澳洲和日本。每個中心經營多樣事業，打造自給自足的環境。有些提供建設服務，有些則是開餐廳或舞廳。但所有羅傑尼希中心都會派會員去奧勒岡參加世界節（World Festival），活動命名巧妙，為期五天的慶祝會在教主日（Master's Day）達到高潮，這段時間是特別設來榮耀羅傑尼希自己。來自世界各地的門徒，成千上萬人齊聚一堂，在世界節舞動身體、享用美食、做日光浴、靜心修行。慶祝是為了提升士氣，但也正好大賺一筆，支援牧場運作。

席拉相信全世界工作的門徒數量高達三萬人，尚未做桑雅斯（sannyas，意指發誓降服於羅傑尼希）、還不是門徒的信徒數量則多達五十萬人。這宣稱令人半信半疑，但也是席拉自信滿滿的經典範例。成功立市之後，社區不僅成為合法的獨立市，羅傑尼希社區還選出門徒加入安特羅普市議會，到一九八三年一月，門徒占了七名議員席次中的六席。不過，儘管顯然是席拉的勝利，她和外界的紛爭才正要開始。

一九八三年七月的某個深夜，席拉接到電話：波特蘭的羅傑尼希飯店慘遭炸彈攻擊。攻擊造成的損失金額為二十萬美元。《洛杉磯時報》報導該名炸彈客史蒂芬・P・帕斯特（Stephen

P. Paster），沒有說明動機，但他「隸屬一個武裝、穆斯林基本教義派組織」。[19]羅傑尼希社區立刻責怪奧勒岡州食古不化、令人憎恨的社區，認為他們要為爆炸案負責。在回憶錄中，席拉寫道：「我們明白警察和法庭不是站在我們這一方……州政府不僅沒有保護我們的權利，甚至故意侵犯我們的權利。」她相信如果她「沒有採取強硬手段來保護」羅傑尼希社區，就沒有人會做了。

為了回應這些威脅，先不論真的還是想像的，席拉和門徒決定武裝自己，購買大量武器，訓練自己如何正確使用這些武器。因為羅傑尼希社區是一座獨立市，所以可以成立自己的執法單位，他們稱之為和平部隊（Peace Force）。向安特羅普居民說他們只是維持和平，沒有別的打算，但這團體採取威嚇手段，還騷擾城鎮居民，晚上巡邏時會用強光照射民宅。他們甚至逮捕安特羅普當地糾察員，控訴他「威脅」，沃斯科縣立刻撤回告訴。兩個社區間的緊張和不信任迅速惡化。

導致羅傑尼希社區這座城市在一九八三年遭到審查的主要原因很多，但這些強勢手段絕對是其中之一。奧勒岡州檢察長戴夫．弗羅恩梅爾（Dave Frohnmayer）小時候在安特羅普附近長大，和多位當地居民至今依然關係緊密，這些居民可以直接找他。他們認為羅傑尼希社區不是自稱的農業團體，根本就是宗教組織。這讓檢察長開始顧慮，他想知道這座新城市是否符合憲法規定的政教分離（此規定來自美國憲法第一條修正案的「禁止建立宗教條款」（establishment clause），也呼應奧勒岡州憲法的規定），以及相信這支和平部隊遵守的不是州法律，而是他們的靈性教主。

一九八三年十月，因為羅傑尼希社區違反政教分離規定，檢察長裁定取消其獨立市資格。羅傑尼希信徒強烈抨擊這項裁決，認為裁決不僅充滿偏見，還是宗教迫害。但是席拉沒有因此而受到任何動搖。在《異狂國度》中，她問道：「我們要怎麼解決問題呢？我們越有創意，政客反對我們的破壞力道就越強。他們的仇恨迫使我們採取這些方式自保、自衛，繼續生存下去。」

珍．斯托克在《異狂國度》中說道：「那時那些奧勒岡人才知道，我們武裝起自己，我們知道如何使用武器，實在逼不得已，我們才準備好戰鬥。」羅傑尼希逃離浦納靜修處時，珍還住在印度，後來終於搬到牧場，席拉還曾經稱讚珍是技巧特別高超的神射手。這可能是為何珍被拉進席拉的緊密小圈子，珍形容這一小群人是席拉的中尉。他們常常在席拉家相聚。席拉家由牧場上的多間平房組成，大家稱之為耶穌林。羅傑尼希信徒和鄰居的全面戰爭現在開始。

選民壓制和生物攻擊

一九八四年九月，巴士陸續抵達羅傑尼希社區，送來數千名可能成為新居民的人，剛好就選在沃斯科縣十一月選舉前兩個月。別名「共享家園」（Share-a-Home）的倡議計畫是席拉的點子，完全設計來支持兩位參與當地選舉的羅傑尼希信徒，兩位都是以未列名（write-in）候選人身分競選。這群人幾乎都從奧勒岡州的大城市和大鄉鎮過來，起先他們看起來很享受這個新環境。想要繼續待在羅傑尼希社區的話，他們只有一件事要做，那就是登記成為選民，再把票投給他們被告知要填寫的候選人姓名。

只不過，牧場突然湧現一大群新選民，不可能沒有人發現。沒多久，奧勒岡州務卿諾瑪．保拉斯（Norma Paulus）便暫停選民登記程序。這項裁定適用所有新選民，但顯然目的就是為了防止發生奧勒岡政府認為該邪教團體「極可能選舉造假」一事。可想而知，這也帶來了麻煩。一位前街友蹺掉新環境的靜心活動，得知自己沒有權利登記成為選民，氣憤難平，出手攻擊席拉，掐住她的喉嚨，把她抬離地面。席拉和其他幾位門徒化解了衝突，前街友被載離牧場，丟在好幾哩外的公園長椅上。當初為了增加選民數量，席拉無意間吸收許多患有醫療問題和社會功能障礙的人，這些狀況也是為何這群人最初會流浪街頭的原因。很快，越來越多新居民的行徑開始變得難以捉摸，甚至益發暴力。顯而易見，現在共享家園計畫徹底失敗了。席拉幾乎小命不保，那場可怕的衝突之後，牧場供應的啤酒都偷偷添加了好度（Haldol）鎮靜劑。

更加不祥的是，共享家園倡議開始沒多久，羅傑尼希醫療中心（Rajneesh Medical Center）護士瑪．阿南德．普佳（Ma Anand Puja）向西雅圖一家醫療用品供應商購買了幾瓶鼠傷寒沙氏桿菌（Salmonella Typhimurium），接著負責在羅傑尼希社區的實驗室裡監督桿菌的培養。溫．麥科馬克根據在普佳手下工作的人最終總算提供給相關當局的資訊寫道，普佳和席拉認真討論哪一種病原體適合帶進牧場，最後終於決定用沙門氏桿菌，食物中毒常見的成因。誤以為食物中毒爆發，就能讓足夠多的人身體不適，進而影響投票結果，但人數也不至於多到引起公權力的注意。

小藥瓶一送達牧場，他們便開始培養沙門氏桿菌，希望能夠灑在沃斯科縣的各個沙拉吧。事件發生後數年，《時代》雜誌（*Time*）菲力普・埃爾默－德威特（Philip Elmer-DeWitt）於報導中寫道，羅傑尼希信徒「在十家當地餐廳和超市，把桿菌灑在藍紋起司沙拉醬、桌上的奶精裡和馬鈴薯沙拉上」，[20]都是為了讓選民身體不適，如此一來，羅傑尼希信徒就能掌管全縣。

隨後席拉宣稱她和沙門氏桿菌下毒事件毫無關係。

當食物中毒真的爆發時，當地居民打從一開始就懷疑羅傑尼希信徒。中毒範圍很廣：七百五十一人出現腸胃炎症狀，超過一百五十人症狀嚴重，最年輕的是新生兒（在子宮內便遭受感染），最年長的是八十七歲的老人家。奧勒岡州第四選區眾議員詹姆斯・韋佛（Jim Weaver）直接指控這群邪教成員規劃這起細菌攻擊，而他選在美國眾議院所有議員面前公開譴責。正如他後來和麥科馬克說明的：「沙門氏桿菌出現在這麼多家餐廳卻沒有共通連結，實在不大可能，除非有人故意為之。」儘管從一開始韋佛就確信羅傑尼希信徒毒害了他的選民，還是花了好幾個月才建立起明確連結。

羅傑尼希社區起初聽到這些指控，認為這是荒唐奇想，全都一笑置之。在爆發後立刻舉辦的一場記者會中，珍・斯托克站上講台捍衛社區道：「我們會感染這個世界——用流行病毒般大量的喜悅和歡笑。」至於投票結果，這次生物攻擊完全就是場災難，食物中毒爆發，社區選民忽然動員起來，投票率之高讓羅傑尼希信徒一席都沒有當選。

藥品、自殺、凶殺

席拉費盡心力打造的一切開始瓦解，速度之快，令人害怕。除了官司纏身，席拉還得知有一群從洛杉磯來的人帶頭作惡，這群人別名好萊塢幫（Hollywood Crowd），喜歡羅傑尼希的教導而來到奧勒岡。瑪・阿南德・哈夏（Ma Anand Hasya）的前夫是《教父》製片艾伯特・S・魯迪（Albert S. Ruddy），席拉認為他在教團的存在值得特別爭論一番。哈夏和魯迪奉獻給羅傑尼希數十萬美元，還有鑽石手錶。哈夏很快就獲得宗師喜愛，待在他屋子的時間越來越長，而且席拉都不在場。

許多門徒形容席拉打翻了醋罈子，珍・斯托克在《異狂國度》仔細說道，席拉很快就發現好萊塢幫拿藥品給羅傑尼希，特別是笑氣和煩寧（Valium，抗焦慮症藥物）。這消息讓席拉羞愧不堪，覺得慘遭背叛——她為了這男人犧牲奉獻，為了這男人完成這麼多任務，教主竟然沒有遵守自己的教導過著純淨生活，甚至還對她說謊。席拉生氣地向羅傑尼希對質，求他停止這一切，想想他的教導，想想他的信徒。但是羅傑尼希只叫席拉不要插手。席拉怒火中燒，因為非法用藥會讓羅傑尼希和警方的關係更危險。他的簽證狀態已經遭到審查，奧勒岡官員除了控訴該社區假結婚真移民的詐欺，還展開另一起調查。

奧勒岡州聯邦檢察官查爾斯・特納（Charles Turner）成為席拉的頭號敵人，因為特納能決定要不要審羅傑尼希的案子。一份二〇〇五年的司法部新聞稿寫道：「一九八五年五月，席拉和最緊密的親信開會，密謀殺害該名美國檢察官。會議中，珍・斯托克同意參與這場陰謀，並且接著執行許多行動來促成此事，包括自願擔任殺手、購買武器，以及監視他們以為是檢察官透納工作地點的車庫。」

羅傑尼希依然遵守止語誓言，而且偏愛與世隔絕，再加上用藥日增，他對這些密謀都不知情。但這不表示他無辜——是他創立了這個體系，他鼓勵信徒（尤其是最親近的信徒）採取極端手段，競逐他的目光。同樣也是他為信徒設下目標，要他們超越任何型態的壓抑，不論是性、心理、社會層面，雜七雜八加總起來，信徒開始覺得不管做什麼，只要是為了羅傑尼希，即便是最不像樣的行徑，都能合理解釋成奉獻。一連串事件因此發生，毒害大眾、謀殺未遂等事件紛紛出籠。

但是謀殺政府官員此等重罪不是席拉心中唯一的念頭。依照斯托克所言，席拉為了要贏過好萊塢幫，搶占先機，在羅傑尼希的住處安裝竊聽器，也因此聽到羅傑尼希詢問醫生史瓦米・達瓦拉吉（Swami Devaraj，又叫喬治・梅瑞迪（George Meredith）醫生）如何能有尊嚴地死去，沒有痛苦。達瓦拉吉開立了不同藥物來達到目的，羅傑尼希命令他去把藥買來。

稍後羅傑尼希告訴席拉，他規劃在一九八五年七月六日死去，那天正是教主日，世界節最盛大的慶祝日。斯托克記得那場在耶穌林的會議，席拉在會中轉告小圈子的人這則消息。為了拯救宗師性命，席拉告訴大家，他們必須殺了達瓦拉吉。斯托克描述當下全場肅靜，沒有人自願，直到她自己舉起手。

席拉拿出凶器：裝滿毒的注射器。斯托克接著說道一九八五年世界節事發當天：「我靠近達瓦拉吉……他靠向我，我把注射器刺進他身體裡……他掙扎了一下，我把注射器拔出來丟掉……他搖搖擺擺地走遠，我轉身離開……我想要一個人獨處。一部分的我覺得自己拯救了羅傑尼希的性命，我完成了該做的事。但內心深處，我傷痛欲絕。從小到大我都清楚明白不可殺人，而現在我卻試圖殺人。」不過，達瓦拉吉躲過一劫，雖然被暗算，還是活下來了。

失樂園

一九八五年九月，席拉和珍．斯托克一起離開社區，隨後還有幾名親信。他們飛到德國，待在當地一家羅傑尼希中心，但熱情歡迎並不長久。同一個月，羅傑尼希打破止語，打算摧毀席拉，或許因為有關當局顯然步步逼近，他想要和打著他名號做的事情保持距離。他毀謗席拉，控訴席拉行為凶殘還犯罪，包括謀殺未遂，以及牧場內外的毒害事件。席拉反駁，說羅傑尼希是個騙子，剝削、操控他的信徒。羅傑尼希反擊道：「要嘛她得自殺，才能從這一切犯罪的重擔中解放，不然她就得坐牢，一輩子痛苦。」[21]

那時整座社區擠滿了記者，許多門徒形容那場景太過令人震驚，就像看到父母在大庭廣眾之下爭吵，也告訴記者他們覺得憂鬱、遭到背叛、困惑不安、痛苦難耐。由於羅傑尼希的指控，聯邦政府現在有充分理由可以帶著搜索令進入社區，就在十月，他們突擊了牧場。席拉急忙逃離羅傑尼希社區，通常細心謹慎的祕書，這回難得粗心：政府幹員搜集各項證據，錯綜複雜的竊聽線路、違法藥物，還有牧場上的實驗室，直接證明沙門氏桿菌一案和羅傑尼希社區有關。他們還找到美國檢察官特納刺殺計畫、精心規劃的包辦婚姻，進而以「共謀詐欺政府」罪提出告訴。

姑且不論羅傑尼希宣稱一切違法活動都是席拉所為，其他人根本不知情，聯邦政府和州政府起訴羅傑尼希和他的首席顧問，證據無懈可擊。起訴的案件包括一級傷害罪、謀殺未遂、非法竊聽，以及詐騙美國政府三十五次。沒多久，席拉、珍・斯托克、瑪・阿南德・普佳三人在德國被捕，引渡回美，全都獲判有罪。席拉涉及毒害大眾和竊聽，被判三次二十年徒刑，但因表現良好，坐牢二十九個月之後就獲釋。

瑪・阿南德・普佳服刑三十九個月。珍・斯托克因為涉入該教團多起犯罪，服刑將近三年。

一九八五年十月，羅傑尼希被捕，遭起訴違反移民法、共謀犯罪。起訴判決的時候，他承認自己向移民局說了兩次謊，被罰四十萬美金，被迫離開美國國土。律師費都是拍賣勞斯萊斯和其他牧場上的財產而來，社區開始解體了。到了一九八六年底，另有十八位羅傑尼希信徒被判有罪，犯下違反州法的罪行。

就算暴力、不法行為和失能情況嚴重，還是有信徒回憶起牧場時光，認為那是此生最美好的幾年，也依然認為羅傑尼希的教導不同凡響、啟發人心。他們聲稱教主遭人誤解了。許多信徒繼續遵守他的教導直至今日。

羅傑尼希自己稱呼安特羅普山谷（Antelope Valley）社區為「失敗的美好實驗」。他終於返回印度浦納靜修處，在那裡生活直到一九九〇年，心臟衰竭帶走他的性命，享年五十八歲。羅傑尼希去世之後，人稱喬治・梅瑞迪醫生，又叫史瓦米・達瓦拉吉的史瓦米・艾瑞圖（Swami Amrito）朗讀一則宗師為信徒準備的訊息：「我離開這副飽受折磨的身軀，身軀早成了我的地獄。別為我哭泣。我永遠與你們同在。」[22]

直到今日，浦納靜修處還是持續歡迎訪客，只是改名叫奧修國際靜心度假中心（Osho International Meditation Resort），又或者偶爾稱為靜心俱樂部（Club Meditation）。靜修處對「巴修水療中心」（Basho Spa）引以為豪，不僅有網球場、游泳池、三溫暖，還有其他「靈性成長設施」。線上論壇第一手參訪評論都強調，訪客都年輕漂亮，好似羅傑尼希宣揚過的自由之愛依然是水療區經營章程的一部分。除了美麗的人、各種誘惑，還會時不時加收日間或夜間靜心長袍費用，靜修處就是隻金雞母。信徒分成兩派，一是奧修之友基金會（Osho Friends Foundation），一是奧修國際基金會（Osho International Foundation），兩派上法院互告，都想掌控靜修處的房地產和資產。度假中心也設立「多元大學」（Multiversity）提供課程，例如「奧修無念禪」（Osho No-Mind）教導學生學習透過亂語來清除雜念，又或者「內在譚崔體驗」（Inner

Tantra Taster）[23]指引整班男女學生「感受性極巔的能量」。不易得知這樣的發展是否為羅傑尼希的意思，不過他在地球上的最後一句話是：「我把我的夢留給你們。」

Chapter 4

剝削

吉姆·瓊斯和人民聖殿教

吉姆．瓊斯是最滑頭的邪教領袖，在鎂光燈下魅力四射、富有同情心，暗地裡卻愛虐待、操控他人。他為自己的教會取得資格，成為基督門徒會底下的人民聖殿教會（Peoples Temple of the Disciples of Christ），特別之處在於提供信仰療法，同時提倡政治行動主義，憑藉二十世紀中期逐漸成長的非裔美國人民權運動（Civil Rights Movement），吸引貧困社區、非裔社區的信徒。他把虔誠信徒視為提款機和性愛樂園的綜合，在洛杉磯和舊金山布道，每週入帳三萬美金，並且與任一位信徒都能發生性關係，不分男女，只要他們需要「治療」。隨著權力增長，他變成加州政治圈的一股勢力。但當議會調查教會的行動逐漸逼近，他和近千名信徒逃離美國。瓊斯承諾大家一處社會主義樂園，但當信徒被迫與世隔離，瓊斯又掌控了全局，他的偏執和癮頭則反過來掌控他，帶給信徒和全世界一場噩夢。

赤道謀殺

一九七八年十一月十八日，九百零八名美國市民集體死亡，事發地點在蓋亞那（Guyana）茂密的叢林深處一處偏遠的林間空地，大小為三千英畝，名為人民聖殿農業計畫（Peoples Temple Agricultural Project），又稱瓊斯鎮（Jonestown）。這場殘殺從第一位死者到最後一位，前後只花了四個多小時。人民聖殿的長年領袖吉姆．瓊斯，信徒都喚他為「父親」或「爹地」，也與信徒一同死去，參與了近代史上規模最大的集體自殺。幾乎所有信徒看來都是親手了結自己的性命，但有三百名嬰兒與未滿十歲的兒童則是死於父母之手。

多虧有現場錄音檔（FBI暱稱為「死亡卡帶」），世人可以清楚得知當天究竟發生了什麼事。幾乎所有喪命的人都吞服混合了風味愛飲品（Flavor Aid，南美版的酷愛飲品（Kool-Aid））和氰化物的毒液。卡帶錄到吉姆．瓊斯勸信徒快快喝下。二十四年前，瓊斯創立了人民聖殿，起初只是在印第安納波利斯（Indianapolis）最貧困區域的小教會。卡帶錄到他說：「很簡單，很簡單。身體不會抽搐。」語帶熱血激情布道者的抑揚頓挫，歌唱般說道：「這真的很簡單。拜託快點喝了吧，免得來不及。」

大家都看到喝下毒液會全身抽搐。小孩先喝了。有些嬰兒還太小或太害怕，無法自己喝，得由護士拿著注射器，裝著致命混合液，注入孩子張大的嘴中。他們立刻就死了，通常只需要幾分鐘。成人需要比較多時間，要二十分鐘或更久。氰化物中毒無法安穩死去，有些受害者會嘔吐或口吐白沫，有些會吐血。儘管目睹淒慘死狀，信徒還是不停歡呼讚嘆，「感謝爹地」，聲音之大幾乎蓋過兒童嚎啕大哭。

全程都有武裝守衛監視，守衛包圍分發混合毒液的集會大棚。

守衛要大家排隊，確保沒有人逃進叢林。守衛看著所有人民聖殿教信徒喝下毒藥，再吞下自己的份。只有兩名死者不是死於氰化物中毒：瓊斯本人和情婦安妮・摩爾（Annie Moore）。經推定，雙雙死於自行造成的槍擊傷口。

當地氣候炎熱，屍體腐敗速度之快令人震驚。等到隔天早晨，當地青年常備兵帶領蓋亞那國防部隊（Guyana Defense Force）軍人到瓊斯鎮，才發現這齣慘劇，多具屍體都已經變形，動物吃掉大半屍體，有些甚至膨脹隆起。有些屍體長了膿腫，或許是在注射部位四周形成，這是那些不願意自行服毒的信徒得面對的命運。瓊斯不想留任何活口，他相信此舉會名留青史，贏得英勇果敢之名。卡帶中，瓊斯吟詠道：「這不是自我毀滅，而是革命之舉。」依他所想，這場集體自殺會顯得很高尚，其崇高程度不亞於歷史意義重大的馬薩達（Masada）猶太人集體自殺，他們寧願燒毀家屋，自我了結性命，也不願向羅馬大軍投降。為了達到相同的歷史影響力，吉姆・瓊斯想要所有信徒一齊死去。

儘管他煞費苦心，還是有好幾位成功逃過一劫。蓋亞那國防部隊成員抵達時，看到一名七十六歲的非裔美國老奶奶海亞辛斯・斯拉什（Hyacinth Thrash）走出平日睡覺的地方，四處找尋她妹妹希波菈・愛德華茲（Zipporah Edwards），但遍尋不著。信徒被召集到大棚服用氰化物毒液時，斯拉什老奶奶躲在床底下。她喊道：「噢，天啊！他們來到這裡殺光所有人，只剩我還活著！」兩名律師馬克・連恩（Mark Lane）和查爾斯・蓋瑞（Charles Garry）來此造訪，神奇地說服了守衛放他們走。他們替人民聖殿教辯護，美國那邊還有針對瓊斯和人民聖殿的訴訟要打。兩人在叢林中東拐西繞，一直沿著瓊斯鎮外圍，直到找到路通往最近的城鎮，但也遠在十三哩外。

另一名信徒葛洛弗・戴維斯（Grover Davis）是個七十八歲非裔美國老爺爺，他走向守衛，守衛問道：「你以為你要去哪裡？」戴維斯簡單答道：「我不想死。」守衛讓他走入叢林，說道：「祝你活得快樂。」戴維斯躲在溝渠裡，直到瓊斯鎮不再傳出聲響。

一個人要怎麼吸引這麼多信徒，而且信徒還如此盲目、瘋狂地相信，就算自殺也在所不惜？怎麼會等到為時已晚，才有人意識到他造成的危害？邪教領袖名冊中，吉姆・瓊斯代表了極端，不只是造成的死亡人數居冠，他也完美符合利夫頓的洗腦或稱思想改造的條件清單：神祕操控、社會環境控制和生存支配等等。這是窮盡一生的事業。但打從一開始，瓊斯就學會隱藏自己對權力的欲求和才能，打著良善、正直、革命的花俏名號，招搖撞騙。

鄉村牧師

一九三一年五月十三日，幾近經濟大蕭條谷底，詹姆士・「吉米」・瓊斯（James "Jimmy"

Jones）在印第安納州（Indiana）鄉間克里特（Crete）誕生了，雙親為萊奈塔（Lynetta）和詹姆斯・蘇曼・瓊斯（James Thurman Jones）。一個月前，明尼亞波利斯（Minneapolis）發生搶食暴動。美國失業率攀升到百分之十六，躁動不安的人民朝雜貨店玻璃窗丟磚頭，希望能搶到足夠的食物，至少家人有一兩夜能飽餐幾頓。國家某些區域也出現自給農業農夫棄耕，這些無人耕種的「塵暴區」[7]難民聽聞華盛頓特區有工作機會，動身前往，住在臨時營地胡佛村（Hoovervilles）。這對當時的胡佛總統（Herbert Hoover）而言是天大的諷刺。胡佛本是名成功商人，後來在一九二八年競選美國總統，承諾要是當選一定讓「家家戶戶有雞可吃，有車可開」。

到了吉米誕生的時候，瓊斯家已經無法過上好日子了。在印第安納州東部，瓊斯家族在當地頗負盛名：吉米的爺爺約翰・亨利・瓊斯（John Henry Jones）在蘭道夫縣（Randolph County）鄉村地方擁有大片農地，同時是地方政治中的有力人士。爺爺還幫兒子詹姆士・T・瓊斯（James T. Jones）和媳婦萊奈塔（Lyneta）付了頭期款，在克里特買下小片農地，當地只有二十八位居民。萊奈塔很倔將，這是她的第三段婚姻，而第二段婚姻只維持了三天。萊奈塔

7 Dust Bowl，一九三〇年代，奧克拉荷馬度過長時間的巨大沙塵暴，沙塵暴是長期乾旱及土地過度使用的結果。一九三〇年時，這個區域先是發生了旱災。到了一九三四年時，旱災將大平原（Great Plains）變成沙漠，這個沙漠就是塵暴區。

也很有野心，在少有女性接受高等教育的時代，她讀了農業與商業學院。那時女人若能獨立思考，生活並不是那麼容易，這次萊奈塔嫁給年長十五歲的人，以為終於能安穩生活。

但詹姆士不是身心健全的人。早在入伍參加第一次世界大戰之前，他就不太可靠，戰後返家又身心破碎，戰爭如影隨形，揮之不去，還因為芥氣攻擊而一直受呼吸道問題所苦。他無法負荷體力活，沒辦法繼續靠種田維生。一家人迫於鄉村的貧困生活，前往印第安納州林恩（Lynn）環境較差的區域居住。林恩有九百五十位居民。

瓊斯一家在鐵路機廠附近租屋，屋內只有幾件從農村帶來的物品，房子後方還有一間車庫，上半部是乾草倉。詹姆士負責道路工程，但戰爭的傷還有後遺症，再加上他是老菸槍，根本不可能保住飯碗。萊奈塔在附近的玻璃工廠上班，薪水支付大部分家用，但老公卻在當地俱樂部打牌，賭掉自己的失能給付。

萊奈塔在工廠的工時長，工作又辛苦，她會一邊抽菸一邊和同事聊政治，髒話連篇。同樣，詹姆士也不是用心的家長，平日不是做做清潔工，就是在牌桌流連。由於萊奈塔嚴格執行的不尋常規定——除非她在家，否則兒子不能待在家（或許她不想讓兒子和老公獨處，因為她不在家就沒辦法保護兒子），吉米童年多數時光都在林恩街道東轉西晃。

這其實沒有聽起來那麼殘忍，附近親戚有值班表，會輪流看顧吉米。當然，這條嚴苛規定形塑了他的性格。多數時候，他必須連續照顧自己好幾個小時甚至好幾天。在某種程度上，他學到如何與陌生人交流，憑直覺得知對方的需求，改變自己的行為，迎合對方喜好。

在街坊鄰居之間，吉姆．瓊斯成了無家可歸的孩子，拉著小紅托車在林恩四處轉，聚集流浪動物，帶回乾草倉飼養。一群野狗開始跟著他打轉，他在當地變得惡名昭彰。他很快就發現這幫狗帶來的力量：只要野狗在身邊，爸媽要管教他就難了。

許多鄰居阿姨可憐他，邀他來吃飯，小吉米變得精於找出她們的興趣，愛她們熱愛的事物。這也是他交朋友的起手式，而影響最深的莫過於梅特爾．甘迺迪（Myrtle Kennedy），就是那間拿撒勒人教會（Church of the Nazarene）的牧師娘。梅特爾很高，六呎二吋（約一百八十五公分），瘦巴巴的，慈悲為懷。無業游民都聽聞過梅特爾，遊民通常搭著貨運列車經過城鎮，梅特爾會烤好幾打派，一份一份放在窗台上，給路過的遊民吃。

不一會兒工夫，梅特爾就帶著小吉米上教會了。吉米可以把聖經的長篇經句都記下來。他會披上袍子，在車庫對著他的小小動物園布道：雞、山羊、蜥蜴、蛇，還有那幫跟著他的野狗。他最後愛上了教會，熱愛到一場主日崇拜不夠過癮。與梅特爾做完禮拜，吉米會拜訪其他禮拜堂，一週接著一週，走遍鎮上所有教會：基督教衛理公會、貴格會和基督門徒會。

但儘管參加這麼多基督教聚會，他卻開始不滿家裡沒有這樣的溫柔關懷。他父很明顯母一點都不愛彼此。集體自殺約一年前，在瓊斯鎮某夜的卡帶錄音裡，他回想當年道：「我準備好在三年級尾聲殺人。我真是他媽的見人就想揍，充滿敵意。沒有人愛我，沒有人理解我。那時家長應當參加學校活動。如果家長沒來，孩子就會遭到排擠。我歌唱得還不錯。某次學校有表演活動，每個人該死的家長都來了，就只剩我的沒來。我一個人站在那，我總是一個人。

大家的家人和親戚都來了，但瓊斯家一個人都沒有來。」

小吉米相貌帥氣，但深色的頭髮和眼睛、寬大的臉，以及分明的五官，看起來一點也不像其他金髮碧眼的中西部當地人。然而，放學後，小朋友自由奔跑、三五成群時，他發現自己的組織天份，他像以前吸引那群流浪狗一般聚集附近孩童。他甚至組了棒球聯盟，組球隊、管理球隊，從鄰近鄉鎮招募球員，規劃賽程。後來，在他於瓊斯鎮回顧童年的那一晚，卡帶錄下他說：「總是我把小朋友召集起來，可是我不喜歡這樣，因為我必須負責規劃好多事。所以我很早就不再需要別人的支持。我甚至想不起我何時曾經需要別人……大家感不感激不是重點，他們利用完你，就把你像榨完汁的乾癟橘子一樣扔到一邊。我知道這無可避免，但又想：『我可能沒有才華洋溢，但有顆忠誠的心。』」[1]

他可以一直講自我犧牲的英勇主題直到他在世的最後一日。

瓊斯會帶不同群孩子到穀倉，他也把動物關在穀倉，試著布道，對那群孩子闡述道德、宗教和科學。他會穿上袍子，用蠟燭和許多疊書來裝飾穀倉。炎炎夏日，他會做檸檬水誘引男孩步入他的臨時神廟；秋高氣爽時，他的演講和授課可以延長到一天六小時。

瓊斯飼養信鴿，他喜歡在臨時聚集的會眾面前，送信鴿出穀倉，鴿腿上都綁著神祕訊息。他從來沒跟那些男孩說過是什麼訊息，同時注意到他越強調謎團，他們就更加感興趣。這些斯文加利式控制[8]的早期跡象逐漸頻繁出現，日後成為人民聖殿教領袖時，他也用同樣方式控制信徒。如同傑夫．桂恩在大獲好評的書《邁向瓊斯鎮：吉姆．瓊斯和人民聖殿教》（暫譯，*The Road to Jonestown: Jim Jones and the Peoples Temple*）中寫到，這年輕人早早就顯示了令人擔憂的傾向。瓊斯一度領著男孩走過車庫頂的椽木，這條路極其危險，挑高十呎的地方只有一根梁，其中一位男孩卻步了，想要退回安全的地方，卻發現瓊斯不讓他過去。瓊斯說道：「我動不了。死亡天使抓著我。」他最後還是有讓路，但另一名兒時友人目擊了那一刻，日後向桂恩承認道：「我那時才六歲，但已經覺得他根本就瘋了。」

瓊斯的陰險狡詐，無可否認。鎮民認為只要他媽媽在身邊，他就會受到影響，變得狡猾。吉姆有許多自娛活動，其中一項是聚集同輩親戚或同儕，教導他們什麼是性行為，以及如何生孩子，細節明確清楚。他也可以很殘忍，他曾經在朋友面前，從車庫內十呎高的地方朝地板丟下小狗。再後來，他拿小小動物園的動物來實驗隨興自創的手術流程，如此冷酷無情是心理病態早期經典徵兆之一。

他對死亡的迷戀亦發茁壯。他帶著一幫男孩闖入當地一家棺材店，命令他們躺進棺材，體驗死掉的感覺。還有其他類似的陰森娛樂，某次他辦了一場模擬喪禮，追思車禍過世的人。有

---

8 斯文加利（Svengali）為莫里哀小說《三人行》的角色，以心理操控能力著名。

時他根本不需要觀眾，獨自在遊戲場主持喪禮，其他孩子遠遠地看著。二次大戰爆發後，他對納粹很入迷，喜歡他們的制服，也喜歡看他們踢正步。他研究希特勒，十分欣賞希特勒吸引群眾的方式，後來希特勒在元首地堡自盡，這般好強之舉令瓊斯讚嘆不已。

梅特爾全心事奉拿撒勒人教會，瓊斯也繼續跟著她一起上教會。但他也繼續去其他禮拜堂，承認自己喜歡觀察偏離真實信仰的場合——梅特爾很滿意這個理由。郊外臨街開了一家使徒教會，開始積極招募悔改歸主的人。瓊斯著迷於他們週六晚間的崇拜，形式鬆散，所有信徒感到聖靈充滿，完整搬演聖靈滾動派（Holy Roller）的戲碼，個個扭動身軀、「說方言」、狂喜甚至暈厥、跳躍、吶喊。他開始一有機會就參加大帳篷復興聚會，沉浸在毫無拘束的聖靈法喜、音樂、魅力滿滿的布道風格。沒多久，他便隨身攜帶聖經，在中學時離譜到換教室上課也要帶著。

他對禮拜煽動人心的一面如此熱愛，也確實為他帶來一時片刻的公眾光環與人氣。有一回中學的校隊準備參加大賽，同學請他在賽前誓師大會為對手主持一場模擬喪禮。他用心主持，運用記憶中所有布道者的技巧。他請集會同學起身，並在全體學生高聲歡呼之下，讓敵隊所有選手躺進大墳墓安息。依照桂恩的敘述，其中一位參加的同學說道：「他很擅長營造戲劇化的場面。」

社會主義者的誕生

二次大戰甫結束，萊奈塔離婚了。在吉姆於誓師大會上取得成功的隔年，他即將升上中學三年級，母子兩人搬離林恩，來到里奇蒙（Richmond），一個印第安納州接壤俄亥俄州的小城市，人口約四萬兩千名。搬家要付出代價：現在沒了前夫的失能給付，也沒有前夫家時不時的接濟，單親媽媽萊奈塔手頭很緊。吉姆在新學校表現良好，提早一學期畢業，在附近的里德紀念醫院（Reid Memorial Hospital）找了份差事，擔任夜班勤務員。

在醫院，年輕的吉姆如魚得水。在林恩的童年時光顯示他操控人心和危險的一面，而新環境讓他有機會好好利用另一項才能，也就是能立刻知道他人的需求和喜好。他記得所有人的名字，而且不嫌麻煩地另外幫病患和家屬帶他們需要的東西。他不僅攬下艱困、不愉快的工作（幫病患擦澡或清理大小便失禁），還把場面變得好玩又友善。他值大夜班，工作時間長。等到他可以自由進出醫院各處，越來越熟悉醫院內部運作，他藉此確保慈善病房的貧困病患也能和付費樓層的富裕病患一樣，享有同樣的醫療。

相較於之前的生活，里奇蒙是瓊斯住過最大的城市。克里特和林恩幾乎所有人口都是白人。三K黨（Ku Klux Klan）成員公開活動，好似他們只不過是另一個社會團體或公民團體。一九三〇年代，印第安納州一度有三分之一的白人男性加入三K黨。萊奈塔向來無法接受此事，總是仗義直言，公開討論貧困、不公不義、工會和歧視等議題。吉米也和媽媽一樣對社會議題直言不諱。吉米會在週末拜訪里奇蒙北邊的貧困鄰里，站在街角布道，每次都能吸引一群聽眾，大多是黑人居民，他們不常聽到矮小健壯、整潔體面的白人男性提倡種族平等。

這是吉姆．瓊斯接下來職涯的中心矛盾：他致力解決社會議題，尤其大力反對種族歧視，極力支持市民權利，看來真誠懇切，甚至鼓舞人心。他打動了階級兩端，受壓迫者和坐擁權力的人紛紛與他聯手。成為社運人士的他提出各式深具開創性的政策，在接下來數十年的種族動盪，這些政策為他贏得多數人的信任。吉米從小缺乏愛和肯定，也沒享受過好生活，貧富戰爭對瓊斯而言意義特殊，他也理直氣壯成為社會主義者。

布道的時候，他會採取政治煽動者和社區組織者的態度，再加上之前向聖靈滾動信眾學來的狂熱風格。私底下的他處理生活務實面時，則會採取萊奈塔明智、符合常理的態度。社會正義是連結，社會主義是行銷策略。這成為瓊斯日後說服人轉信的基礎，直到最後一刻依然如此。

對多數邪教領袖而言，政治和他們八竿子打不著。要是遇到一名奸巧甚至兇殘的心理病態要求跟隨者絕對忠誠，人們腦中冒出的第一個問題鮮少會是「他支持哪個政黨？」邪教領袖的根本動機就是規模宏大的自我滿足。不論他們宣揚的是哪一派政治、宗教、種族，也不論他們聽起來有多慷慨激昂，都只是為了達成目的所採取的手段而已。魅力型邪教領袖會依自己方便與否廢止核心原則。自始至終，原則都只有一條：不論我想要什麼，一定會到手，只要能成功，不管得說什麼都可以。

但對瓊斯而言，政治想法是他吸引信徒的法寶，信徒深受啟發，這也是他的魅力泉源。此外，他一生好幾次醜惡地展示出他的牛脾氣、虐待癖，還有除了自己定下的規矩之外，漠視一切法條的態度。舉例來說，他離開林恩前，最好的朋友唐・福爾曼（Don Foreman）替他餞別，吃完飯後福爾曼先行離席，瓊斯的回應是掏出槍，朝福爾曼離去的身影開槍。大學時期，有一次他的室友在上鋪睡得正熟，卻覺得被猛刺了一下，往下看才發現瓊斯伸長手硬把帽針刺穿床墊。他確實很殘暴，但只在暗中偷來。

光天化日之下，瓊斯總是大聲宣揚自己對種族正義和經濟平等的奉獻。事實上，正是他對這些議題嚴肅以待，才引起瑪瑟琳・鮑德溫（Marceline Baldwin）的注意。瑪瑟琳心地善良，美豔動人，是基督教徒，比他年長五歲，住在里德紀念醫院院區護士宿舍。瓊斯告訴瑪瑟琳，他退出中學籃球隊是因為教練有種族歧視，這句話贏得了芳心，只不過從頭到尾都是捏造的。吉姆・瓊斯天生就知道人們愛聽什麼，而他也一直說人們愛聽的話。他告訴瑪瑟琳他在理髮店頭理到一半，理髮師卻開始說一些貶低黑人的話，他就憤而離席。不久後，瑪瑟琳告訴室友，她覺得自己愛上了那位熱情帥氣的年少男子。

醫院同仁完全不知道這名勤務員在街頭布道，也不知道他的政治傾向。在醫院，大家認為他彬彬有禮又朝氣蓬勃，對病患善良到幾近聖人。他態度認真，對醫療流程越來越熟悉，贏得了醫生尊敬，而他的熱心幫忙則獲得了護士的信任。他會熬夜陪伴年長病患，為他們朗讀，或長時間工作，完成沒人要做的苦差事，即便他同時還兼顧中學學業。

這名年輕人的工作倫理和抱負打動了瑪瑟琳雙親。一九四九年，兩人結為連理。那年瓊斯十八歲，瑪瑟琳二十三歲。稍後瓊斯進入印第安納大學（Indiana University）就讀，不確定將來要當醫生、律師，還是教師。新婚頭幾天充滿了緊張，因為他無預警地和瑪瑟琳雙親爭論小倆口應享的權利和特權，他憤憤地告訴瑪瑟琳他不再信上帝了——如果上帝真的存在，這世界就不會有如此多苦難。

瑪瑟琳對此十分苦惱，但她嘗試不同方法來維繫婚姻，也採納父母的建議。後來，一九五二年，這對夫妻在教堂做禮拜，意外發現布告欄的一張傳單——衛理聯合工會社會行動（Methodist Federation for Social Action）頒布新的社會信條。這是瓊斯能相信的那種宣言，傳單明確指出教會應該承諾為了「減少貧窮、老年安全、團體協商、言論自由、監獄改革以及種族權利」盡力而為。跟平常一樣，吉姆行動迅速，告訴親家現在他打算成為全職牧師。到了一九五三年，夫妻倆已經搬到印第安納波利斯，瓊斯在鎮上貧困區的薩莫塞特衛理公會（Somerset Methodist Church）擔任學青傳道，同時在巴特勒大學（Butler University）上課。現在他找到了可以運用自己所有熱情的畢生職志：公眾演講、政治行動與聚集群眾。

賣猴子蓋教堂

瓊斯開始從一個牧職跳往下一個，每一站他都努力擴展會眾、建立聲譽，從其他教會揀選信徒，多虧他還兼差當客座牧師。他換教派的頻率似乎就和換衣服一樣頻繁。起初，瓊斯的願景明確又理想：他想要在印第安納波利斯發起第一次不分種族的衛理公會禮拜。但他很快就拋棄任何特定派別的嚴苛限制，反而選擇支持末後春雨新秩序（New Order of the Latter Rain），或至少從中吸收信徒。這一派別在禮拜總是狂喜無拘束，是五旬節運動（Pentecostalism）其中一支更加狂熱的派別，他們還會提供信仰療法。

瓊斯在貧困區域租了一間店面，並為他的小教會命名「社區團結」（Community Unity）。對他而言，布道要選擇哪一派別不重要，他真正想要大家關注的是在台上布道的他本人。就像之前在醫院工作一樣，只要能成功，他不介意賣力工作。正是這股工作拚勁，而不是他改變人信仰的才華，讓他在一九五四年四月十日初次登上《印第安納波利斯星報》（*Indianapolis Star*）頭版。他也開始挨家挨戶賣猴子，在鎮上騎著單車載著猴子，每隻售價二十九美元，經常在客戶討價還價時，順便傳一些福音。

瓊斯拒絕支付印度公司運費，理由是大部分動物在運送途中死亡，這本來只是鄰里奇聞，後來變成頭版新聞，至少《星報》編輯認為如此。

多虧了他偶爾在復興聚會露面，越來越多人認為他是講道壇上深具魅力的牧師。末後春雨的巡迴布道之中，牧師既要娛樂大眾，還要能讀心，偶爾兼當江湖郎中，猜中陌生信眾的名字、車牌號碼和社會安全號碼。瓊斯在林恩和里奇蒙已經見習過聖靈滾動派，認為自己可以青出於藍。

瑪瑟琳第一次看瓊斯以牧師身分上台時，害怕他會在信眾面前嚇到動彈不得。但當他閉上雙眼，克服最初的焦慮，立刻進入了只能稱之為催眠的狀態。他全速講道，毫無間斷，將近一小時整，會場全體五旬節信眾都為之深深著迷。他看似隨口就能援引聖經經句。他請觀眾上台，信眾在他碰觸之下倒地。瑪瑟琳不敢相信他擁有如此強大的吸引力。

月桂街會幕堂（Laurel Street Tabernacle）的當地牧師羅素・溫伯格（Russell Winberg）當時也在場，會後邀請瓊斯當客座牧師，定期來他的講道壇布道。在月桂街會幕堂，瓊斯開始例行展示信仰療法。他會籲請疾病纏身的信徒自願上台，在眾目睽睽下治好病痛。瓊斯為這派信仰療法注入嶄新層次的戲劇效果和「真實感」。如果會眾罹癌，瓊斯會站在他們身邊，運用靈活手法和動物內臟，活生生扯出一坨鮮血淋漓的肉塊。瓊斯和他的治癒奇蹟成了末後春雨教派巡迴布道的熱門資產。牧師可以在巡迴布道時互邀彼此講道，也可以建立自己在地方或全國布道會的名聲。

當然，瓊斯的治癒奇蹟只是早年鄉村布道的延伸；他用玻璃罐存放「惡性腫瘤」擺在後台。不論是否真的有任何醫療益處，他的療法某種程度也算成功：瓊斯開始從月桂街會幕堂會眾中，挑選一隊極其忠誠的擁護者。他迅速組織自己的大型布道會，邀請全國知名牧師威廉・伯蘭罕（William Branham）主講。伯蘭罕總是在體育場館布道，歡迎所有種族的信眾。瓊斯在那場活動籌募到的奉獻，加上賣猴子存下來的錢，足夠在印第安納波利斯種族混住區買一間自己的教會。那是幢新房子，飾有尖塔和彩繪玻璃，這對瓊斯而言簡直邁進了一大步。教會正式名稱為「救贖之翼」（Wings of Deliverance），但他幾乎立刻就稱之為「人民聖殿教」。

這時期的新信眾，有幾位最終跟隨著他一路前往蓋亞那，例如建築工人傑克・阿諾德・

畢姆（Jack Arnold Beam）、活潑愛八卦的帕蒂・卡特梅爾（Patty Cartmell），以及非裔美國青年吉姆・考伯（Jim Cobb），他的病痛似乎在瓊斯的碰觸下立刻痊癒。瓊斯預言考伯將來會成為人民聖殿的好領導者；不過幾句鼓勵的話，青年的命運就注定了。

或許瓊斯在救贖之翼教會挑到最重要的皈依者是黑人男子阿奇・伊詹姆斯（Archie Ijames），之前伊詹姆斯為了親眼看看傳說中的伯蘭罕，參加了瓊斯、伯蘭罕聯合布道大會。沒想到結束後，他印象最深刻的卻是瓊斯全力尋求社會正義。看著吉姆・瓊斯，伊詹姆斯看到了同道中人。而瓊斯則是告訴伊詹姆斯他注定要擔任神職。不過真相是瓊斯的教會需要一位黑人領導者，向大家證明他言行一致。

有了專屬的空間，事情都不一樣了。之前收到的捐獻都要繳一部分給月桂街會幕堂等會場的管理者，現在瓊斯可以直接收到信眾的每一分錢。雖然到了一九五八年，信眾僅將近百名，但瓊斯聚集的人氣已經足夠讓市府承認他是一名社會正義領袖。他開設了一間慈善廚房和兩家養老院。這幾間機構有兩項功能：首先，確立人民聖殿教使命在大眾眼中的模樣；其次，提供瓊斯絕佳的招募場所，因為來的人都絕望又孤單。

瓊斯也為自己的家庭戴上假面，反映出他的社會使命，但瑪瑟琳最想要的就只是擁有自己的家庭。他們從領養開始，先是一名十歲女孩艾格妮絲（Agnes），女孩遭親母拋棄，遺棄在人民聖殿教會。接著是兩名韓國孩童，瑪瑟琳和瓊斯替他們取名：史蒂芬妮（Stephanie）和路（Lew）。不到兩年，史蒂芬妮死於車禍，瑪瑟琳和瓊斯聯絡韓國的那家孤兒院，領養了史蒂芬妮的朋友蘇珊娜（Suzanne），成為家庭的一份子。一九五九年，瑪瑟琳生下獨子史蒂芬．甘地．瓊斯（Stephan Gandhi Jones）。同時，他們領養了一位黑人孤兒，為他取名小吉姆．瓊斯（Jim Jones Jr.）。他們打算把史蒂芬和小吉姆當作雙胞胎養育，以此比擬人民聖殿教的宏大使命。

尋覓應許之地

到了六〇年代初期，瓊斯開始與人氣爆棚、富到流油的神聖滌凡牧師（Father Divine）結盟，滌凡牧師在哈林區（Harlem）布道，宣稱自己是神本身，也是基督再臨。如同瓊斯，滌凡牧師利用教會舉辦活動，積極支持市民權利、婦女權利、廢除種族隔離。瓊斯想獲得一樣的成就，決定模仿神聖滌凡，開始宣講反物質主義，鼓勵信徒捨棄所有地上的財富，全心奉獻、致力服務人民聖殿教的不同社會分支。神聖滌凡讓信徒為他做工、餵養飢餓的人、在總部工作。瓊斯頻繁造訪哈林區，後來八十歲的神聖滌凡基於稅賦問題把總部遷到費城（Philadelphia），瓊斯便跟著造訪費城，覬覦國際和平宣教運動（Peace Mission movement），打算等老牧師離世就要接手。

不論是否受到神聖滌凡的啟發，瓊斯漸漸開始宣稱自己的神性。大約這時期，信徒開始稱呼他為「父親」。瓊斯開始安排活動，讓大家認為他有神祕能力，某種程度上也是在地上的神。就算有人因此離開，瓊斯也不介意，他只想要那些全心相信「瓊斯個人崇拜」的信徒。瓊斯派卡特梅爾在會眾中臥底，有了卡特梅爾的協助，瓊斯開始在能量滿滿的布道上假裝能讀心，還能預測未來。他盡力打探會眾資訊，好利用他們的恐懼。他會特別為那些信念動搖的人編造預言，預告厄運將至，而為了獎勵他的領導階層，總會預見他們家庭享有成功的未來。

集結虔誠信徒的同時，他也確保善行繼續。人民聖殿教繼續提供飲食和衣物，關懷社區老年人。會眾人數急速增長，黑人與白人信徒的人數相差無幾。在一名理想化的青年傳道敦促下，瓊斯手腕高明地加入新教派基督門徒會，人民聖殿教成了附屬教會，此舉帶來正當性（同時大大地減稅），他只需要每年繳交低廉會費。更棒的是，基督門徒會給會眾極大自由。

人民聖殿教雖然成立不久，但幾乎占據主流地位，加上瓊斯越來越響亮的社會正義名聲，引起了印第安納波利斯民主黨市長查爾斯・H・鮑斯威爾（Charles H. Boswell）的注意。鮑斯威爾也在貧困區域推廣新建設，為了表彰瓊斯，欽點其擔任名譽職位印第安納波利斯人權委員會會長（Indianapolis Human Rights Commission），此職長年以來幾乎沒有實際作為，但是瓊斯嚴肅以待，開始發起全市運動，鼓吹餐廳取消種族隔離。他與教會的黑人嘉賓一同前往餐廳，卻被告知沒有事先訂位，雖然之前從來不需要訂位。接著他會回訪，有時候不只一次，客氣有禮地向餐廳老闆解釋新客戶會對生意有多好，而且他多麼不想採取其他行動：帶著所有教區居民，黑人和白人同胞一齊在餐廳外抗議。

一間走過一間，在他的威脅下，當地餐廳屈服了。對瓊斯而言，他證明了自己說到做到，帶著多種族聯合的信徒光顧每家迫於壓力臣服的餐廳。他們會在離峰時段光顧，這樣餐廳還能在熱門用餐時段維持原本的門面。他們的餐費都由人民聖殿教支付，總是有許多信徒渴望加入用餐行列。

瓊斯擴展行動範圍，打算讓印第安納波利斯警察局取消種族隔離。他啟動就業服務，提供企業可以勝任工作的應徵者，告誡信徒要做一位理想員工，否則不只會賠上他的聲譽，還有人民聖殿教的名聲。

瓊斯嚴重腹痛，他到印第安納大學醫療體系衛理公會醫院（Indiana University Health Methodist Hospital）就醫，經醫生診斷為胃腸潰瘍出血。他雖然痛苦難耐，但還是拒絕了治療，除非醫院廢除種族隔離，重新將黑人病患安置到保留給白人病患的病房。他出院之後不停推動，努力終於有了結果，醫院取消了種族隔離制度。瓊斯短時間內達成這些具體公民實績，成功躋身社運人士圈子，一路待了十餘年。

但並不是每個人都欣賞他的努力。這段時期，在美國支持種族融合，很可能會被貼上共產黨的標籤。市長勸瓊斯低調一點，但瓊斯不同意，就連有人憤怒地在人民聖殿教堂畫上納粹卍字符，在地上煤堆中插了根炸藥，他也不屈服。瓊斯接著又成功搏了幾個版面，主動聯絡「伊斯蘭國度」（Nation of Islam）和三K黨，但都吃了閉門羹。他大量印刷兩個組織的回絕信，在市政府派發，似乎是在說：「看吧！我多麼溫和中立啊！兩邊極端都回絕了我。」

後來，他自己的恐懼終於迫使他放棄這項積極進攻的政治行動計畫。甘迺迪總統警告國人核戰危機的可怕，已經開始經營預言事業的瓊斯接著預測即將會有災難：他不知道何時，因為還沒有獲得完整的畫面，但災難會在三點零九分爆發，但他說不準是凌晨三點，還是下午三點。前陣子住院，他的醫生正色告誡他出了院得好好休息一段時間。有了這兩樁緊急事件，瓊斯告訴人民聖殿教信徒他會去尋覓安全的地方，大家都可以一起撤退，遠離即將降臨的災難。《君子雜誌》（*Esquire*）一九六二年一月號刊登一篇報導，題為〈世界九處躲藏地點〉（*9 Places in the World to Hide*），瓊斯打定主意，能造訪幾處就盡量去看看，務必找到最佳躲藏地。

隨後他和家人來了趟長時間遠距離旅行，到巴西、英屬蓋亞那（British Guiana，蓋亞那在一九六六年獨立）和夏威夷，調查哪處最能接納一大群遵守社會主義原則的基督教會眾。他們離開美國本土逾一年，人民聖殿教只提供極少的資金援助，瓊斯靠著教英文糊口，或到資金充裕的傳教機構找工作。

在里約熱內盧（Rio de Janeiro）的時候，某位外交官太太原本應該要接待他。後來幾年，他常常講述這則故事，雖然事實如何無從考察。那時，瓊斯一家正在替當地一間孤兒院籌錢，那位外交官太太願意捐五千美元，前提是瓊斯和他做愛。瓊斯說自己陷入了道德困境，畢竟如果把自己的道德標準看得比孤兒迫切的需求還重要，不是太自私了嗎。他向瑪瑟琳尋求建議，瑪瑟琳同意比起任何個人考量，孤兒的需求才是當務之急。這段插曲總能讓布道變得特別精彩刺激，但想要傳遞的訊息和性一點關係也沒有：對瓊斯而言，只要能達到目的，用什麼方法都可以；他期許信徒也能為達目的不擇手段。

瓊斯指派溫伯格代替他管理人民聖殿教，但是伊詹姆斯通知他溫伯格看來想把教會帶往不同方向，溫伯格打算把自己立為首領，同時不再強調行動主義，儘管大批信徒都是受行動主義吸引而入教。信徒漸漸喪失興趣，教會活動的收入再也不足以支援人民聖殿教獲得當地盛讚的善行。瓊斯在巴西認識了愛德華．馬爾敏牧師大人（Reverend Edward Malmin），說服這名年長傳道士代他回美國，重整人民聖殿教的財務。但等到馬爾敏牧師大人抵達，溫伯格早已公開造反。伊詹姆斯警告瓊斯如果再不回來領導，所有心血會毀於一旦，瓊斯和家人匆忙返美。甘迺迪總統遇刺身亡不久，溫伯格已經帶著數十名白人信徒離開人民聖殿教。

神就在我們之中

到了一九六四年底，瓊斯告訴印第安納波利斯信眾他接獲了新預言：印第安納波利斯一九六七年會毀於原子攻擊。他們必須趁早離開。於是人民聖殿教打包行囊，前往加州紅木谷（Redwood Valley）重新安頓。紅木谷位在猶凱亞（Ukiah）外圍，在舊金山往北一百三十哩處。瓊斯說明此處會中選是因為遠離核微粒降落區，隔了一段安全距離。許多信眾不相信新預言，繼續留在印第安納州，這些心存疑慮的人非常幸運，避免捲入十四年之後瓊斯鎮的真實災難。

但還有一百四十名瓊斯信徒真的搬到了加州。這代表人民聖殿教信徒得大幅改變生活模式。現在每戶家庭要共同生活，一起住在破爛的社區公屋，禁止一切物質享受，動物和孩童在院子四處閒晃。瓊斯在新環境承擔了更多責任，但方方面面的組織都依然高效。在瑪瑟琳的協助下，他也在附近的州立門多西諾醫院（Mendocino State Hospital）幫多名信徒找到工作。接著他著手開設兩家療養機構、一家慈善廚房、一間勒戒中心和三間類似大學宿舍的住處，希望能在之前印第安納波利斯時期開創的形式上擴大規模。

他也用更直接的方式成功宣傳了教會的正當性。人民聖殿教成功申請為基督門徒會全國體制下的完整會員。基督門徒會是新教改革宗派的主流，現在又稱基督會（Christian Church）。從那時起，人民聖殿教獲得了其他新興宗教運動所缺乏的官方認證。但這得付出代價：人民聖殿教必須支付大筆年費。門徒會領導階層非常開心，因為人民聖殿教的會員到了一九六九年增為三百名，一九七三年則來到三千名，那時每年支付超過一萬五千美元給門徒會組織。

信徒極速增長多數來自瓊斯深入接觸的舊金山和洛杉磯黑人鄰里。到了一九七〇年，瓊斯在舊金山新開的人民聖殿教分會主持崇拜會，兩年後，另一分會在洛杉磯開幕。在七〇年代的人民聖殿教巔峰時期，會員高達八成都是黑人。瓊斯立刻將信徒的虔誠奉獻視為政治實力，因為他能要求信徒把票投給他選中的自由派政治人物，人民聖殿教變成加州政治的強大勢力。一九七五年的舊金山市長選舉選情緊張，不到五千張票就能左右結果。舉例來說，人民聖殿教派出的選民幾乎就能確保喬治．摩斯康（George Moscone）勝選，最後他也確實當選了市長。在關鍵選區，瓊斯和信徒會挨家挨戶掃街拜票。摩斯康以十二票對一票的差距勝選。

摩斯康心懷感激，指派瓊斯擔任舊金山房屋管理局（San Francisco Housing Authority）局長。瓊斯崛起於政治掮客圈，第一夫人羅莎琳・卡特（Rosalynn Carter）和副總統華特・孟代爾（Walter Mondale）等民主黨的全國知名人物，似乎都把他當成受人敬重的政治人物來對待。

根據提姆・萊特曼（Tim Reiterman）的《烏鴉：吉姆・瓊斯牧師大人與人民聖殿教的驚人故事》（暫譯，*Raven: The Untold Story of the Rev. Jim Jones and His People*）一書，瓊斯也在其他方面運用他的人民聖殿勢力，為地方勢力和警察長官烤蛋糕，發送粉絲信給各黨派的政治人物，稱讚保守派提倡的法律與秩序，讚賞自由派的越戰立場。人民聖殿報刊登了州長雷根（Ronald Reagan）和FBI局長J・埃德加・胡佛（J. Edgar Hoover）等執法單位和當選官員客氣有禮的回應。

在政治拓展工作中，信徒提莫西・史托恩（Timothy Stoen）扮演了關鍵角色。依傳統價值來看，他是最成功的信徒，擁有史丹佛法學院學位，律師事業也是生意興隆。他和太太葛雷絲（Grace）獲得特殊待遇，瓊斯允許他們享有傳統中產階級便利舒適的生活（收音機、洗碗機和電視），他們在自己的公寓生活，其他信徒卻得拋棄房地產。史托恩過著傳統富裕的生活，但很快就讓自己變得不可或缺。他收起私人事務所，到地方檢察官辦公室工作，出手干涉法律問題，在真正問題到來前先處理掉。太太葛雷絲對這一安排不大熱衷，但她愛先生，所以還是跟在史托恩身邊。

瓊斯忙著透過政治參與打響自己名聲的同時，加快腳步舉辦高度操控的信仰療法聚會，安插虔誠的信徒當暗樁，與會群眾根本身處大型詐騙現場。暗樁會化老妝，甚至塗黑臉，假裝癲癇發作或假死。虔誠信徒已經領會瓊斯的訊息，相信只要達到目的，任何手段都是正當手段。信徒適時演出安排好的橋段，覺得把更多人帶進教會是在拯救生命。

與此同時，瓊斯握緊了對信徒的鐵腕統治。大部分信徒都長時間工作，上繳百分之十或二十五的收入給教會。在社區內，他受到蘇聯政府的特權階層影響，借用蘇聯政府結構方式，成立了所謂的規劃委員會（Planning Commission）。表面上，規劃委員會是個領導委員會，但初期成員人數迅速擴增，涵蓋超過一半的信徒。瓊斯一直分割過於龐大的規劃委員會，切成多個小規模菁英單位，創造階層獎勵制度，讓信徒彼此敵對，因為信徒都爭相競逐瓊斯的目光，想要搶得親近他的機會和位置。

一九七一年，瓊斯在某場糟糕的布道會上，微妙地混合了政治議題、個人故事、厚顏無恥的裝腔作勢，還帶著惡性自戀者典型的自信，自作主張宣告他本人正是社會主義之神（God Socialist）。他宣布：「我會終結貧窮！」開始向會眾侃侃而談他將如何大有所為。他說道：「我在這裡就是你們的榜樣和典範，你們只要努力就能成功。你可以成為自己的神！不是降卑，而是復活，走出任何你必須忍受的經濟狀況、不公義、種族歧視或奴役境況，遠走高飛。你體內沉睡著救贖之鑰！」方方面面看來，這就只是典型的正向思考力訊息，再加上一些左派思想。但沒多久，這股力量似乎轉移到他身上，他成為信徒崇拜的對象。瓊斯留了一頭波浪般烏黑亮麗的秀髮，如同舊約預言的基督二次降臨。他又說：「我活在肉身裡面，正如之前行走在地上……索羅門形容道祂的頭髮黑如烏鴉，如以賽亞書七章二十節說，用剃刀剃淨。我是用剃刀剃淨才來到地上！我頂著黑如烏鴉的髮來到地上！我作為社會主義之神來到地上！」

到了一九七二年，洛杉磯和舊金山兩邊人民聖殿的奉獻合計每週超過三萬美元。瓊斯從基金中拿了一點，購入十一輛遊覽車，挑選一群榮譽信徒參加他的福音宣道巡迴，沿途停靠丹佛（Denver）、休士頓（Houston）、紐約和華盛頓特區，全程一直盡可能販售宣傳材料，例如治療油（其實只是廉價橄欖油）或瓊斯舊袍子的碎布（慈善商店買的袍子剪成碎片）。最差勁的是，新信徒得支付更高的金額來買小尺寸的瓊斯個人照。

返回加州的路上，瓊斯認為順道回一趟家鄉印第安納州是很好的主意。帶頭的遊覽車靠邊停下，詢問先知老家的方向，純屬巧合，他們攔下來問路的人正是唐·福爾曼，瓊斯曾經拿槍射擊的那位兒時友人。福爾曼詢問先知的姓名，聽到是吉姆·瓊斯，他一點也不意外，笑出聲，搖搖頭，指著一條塵土飛揚的老路。

相較之下，瓊斯的心靈導師梅特爾很開心看到以前她負責帶著上教會的孩子返鄉。瓊斯和遊覽車停留在林恩一晚，向全鎮展示當初村裡的小伙子現在有多成功。

性愛、毒品、社會主義

回到加州，瓊斯發現想要成功沒那麼容易。紅木谷的新址受到地方官員審查。雖然人民聖殿教的孩童可以上公立國小，但是老師疑心漸起。這些小孩很早熟，在教室公開談馬克思主義和辯證唯物論。一位老師發現某群人民聖殿教孩子互相比較瘀青擦傷。小朋友群聚時，他們都說是在週末玩的時候受傷，但老師關切地又再一一單獨詢問，每個孩子都坦承人民聖殿教領導者強迫他們所有人獨自攀登險山，沒有人陪同監管。

還沒踏上回程往西，瑪瑟琳早就懷疑先生外遇多次，但一回到加州，先生藏都不想藏了。正值七〇年代，自由之愛和換妻概念占據了全國想像，瓊斯鼓勵會眾經營更加開放的關係，性共享成為人民聖殿教的核心特色，恰好符合瓊斯雜亂無章的共產宣言。瓊斯認為和男女信徒做愛是對他的羊群提供額外事工。他幫助男人克服障礙，終於不用在壓抑內心的同性情慾；他協助苦惱的女人，因為他看得出來這些女人需要談一段備受尊重的感情。

他也有許多萍水相逢的豔遇。信徒頻繁詢問是否能和他做愛，讓他備感挫敗，為此曾經宣布任何想要他的人都可以獲得他，總計和男男女女發生十六次性行為。他有許多短暫的情婦，但和卡洛琳．雷頓（Carolyn Layton）維持一段長期、開放的關係，雷頓的先生也是人民聖殿教信徒。雷頓認為這段外遇是祝福，好似成為天選之人，一頭栽進熱戀，毫不在意向任何人公開他們的關係，就連對瓊斯的孩子也不隱瞞。雷頓後來懷孕，在一九七三年產下吉姆．瓊（Jim Jon），又名基莫（Kimo）。

除了外遇不斷，瓊斯的藥癮也變大了——安眠酮（Quaaludes）和煩寧注射劑都是心頭好，他開始無法控制自己。這可能是因為他瘋狂想要控制其他所有面向。他幻想有人要行刺，他的反應是增強個人維安等級，提供彪形大漢信徒槍枝，這些信徒隨侍在側，如影隨形保護瓊斯，也會巡邏社區。在這段控制狂熱之中，他走了一步日後會纏身的棋，不只對他，還有其他九百零八位瓊斯鎮居民。一九七二年，史托恩夫妻有了兒子約翰．維克多．史托恩（John Victor Stoen）。但瓊斯瞞著葛雷絲，要求提莫西簽署宣誓書，聲明其實吉姆．瓊斯才是孩子真正生父。這當然是謊言，但瓊斯知道，日後這對夫妻要是動搖信念，這樣的忠誠證據就可以當成籌碼。

瓊斯掌握信徒的危險手法開始引起媒體關注。瓊斯在林恩上演一場返鄉滑稽劇碼之後，《印第安納波利斯星報》記者想起了人民聖殿教在六〇年代還只是靠著信仰療法噱頭經營的教會，並告訴《舊金山觀察家報》（*San Francisco Examiner*）宗教版編輯萊斯特．金索文（Lester Kinsolving）。金索文開始和《印第安納波利斯星報》首席記者卡洛琳．匹克凌（Carolyn Pickering）合作。他們在紅木谷社區外蹲點監視，訪問當地居民，蒐集各種故事，例如涉嫌兒童虐待和其他謠言。他們也上人民聖殿教做主日崇拜，親眼看見武裝巡邏隊。

他們的系列報導開始刊登之後，舊金山市民讀到荒唐的主張（瓊斯據說已經讓四十人起死回生），人民聖殿教宛如嘉年華會的信仰療法現場；以及信徒稱呼瓊斯為神，他也只是意思意思否認一下，說道：「如果你說『他是神』，有些人會認為你瘋了。他們沒辦法體悟。我很高興你痊癒了，但我只不過是神的使者……我擁有超凡的療癒能力。」一篇報導揭露一名青少女是如何被指婚嫁給另一名信徒，登記編入門多細諾（Mendocino）的福利名單，然後被迫簽字將自己的福利支票轉讓給人民聖殿教。

瓊斯的回應是派出忠誠信徒到《觀察家報》總部外頭舉牌抗議。《觀察家報》想要訪問瓊斯本人，瓊斯乘著加長禮車登場，安排記者在後座與他的武裝維安人員並坐。依照萊特曼所言，記者會被載到危險區域轉轉，經過空蕩蕩的內河碼頭（Embarcadero）。提莫西坐在前座，不透露自己身分，威脅要採取法律行動。《觀察家報》砍了這系列後續的報導。

一九七三年，除了從救贖之翼時期起就最為忠誠的吉姆・考伯之外，還有另外七名人民聖殿教青年領導者，共八人一齊叛逃出走。這些叛教者後來又被稱為「八大革命家」（Eight Revolutionaries），是第一批起身反抗瓊斯的信徒，甚至在其中一個聖殿社區拿槍指著瓊斯。他們要瓊斯放他們走，而且不會懲罰還留在教會的家人。迫於情勢，瓊斯答應了。作為交換，他們不會插手，瓊斯和人民聖殿教可以繼續順利運行。

那年十二月十三日，瓊斯趁著在洛杉磯有幾次講道，順便去麥克阿瑟公園（MacArthur Park）附近的電影院看正在上映的克林伊斯威特（Clint Eastwood）電影《緊急搜捕令》（*Dirty Harry*）。那座公園是知名的同志尋歡幽會地，臥底警察則在電影院釣魚。因此，當瓊斯邀請一名年輕男子做愛，接著露出下半身開始自慰，他立刻被逮補。提莫西編了荒謬至極的說詞，還附上醫生診斷，說明瓊斯最近動了泌尿手術，需要定期刺激陰莖。洛杉磯警察局（LAPD）放了瓊斯，但為建了他的檔案，並且開始監視人民聖殿教。

瓊斯明白自己正在耗盡政治人脈的情意，於是召集教會高層領導者瑪瑟琳、伊詹姆斯、畢姆和卡特梅爾。他說，是時候找尋他們的避難所了。畢姆和伊詹姆斯飛到蓋亞那（中選原因是蓋亞那和美國之間沒有引渡條約），打點當地政府，承諾捐贈兩百萬美金給首都喬治城（Georgetown）的發展委員會。他們買了一塊三千兩百英畝的地，到了一九七四年尾聲，他們的社會主義烏托邦開始興建。

死亡與疑心

剛從柏克萊畢業的音樂家鮑伯．休士頓（Bob Houston）和太太菲莉絲（Phyllis）在一九六九年加入人民聖殿教。社區的共同生活讓兩人漸行漸遠，在其中一場聖殿教的「淨化昇華」（catharsis）會，或者又叫批鬥大會，大群教友質問夫妻倆的性生活。那年代坦率精神當道，瓊斯似乎也很喜歡這種精神，夫妻倆也同樣坦率，全部據實以告：他們超過一年沒有做愛了。瓊斯問他們是否想各自找新對象，菲莉絲答道只要低調謹慎，她不介意。於是瓊斯允許鮑伯和初來乍到紅木谷社區的喬伊絲．蕭（Joyce Shaw）發展新戀情。

喬伊絲和鮑伯一起編輯人民聖殿教雜誌《永活常存的道》（*Living Word*），但兩人關係才開始變得親密，瓊斯就命令鮑伯回到菲莉絲身邊。這般專斷獨行的權力令他流連忘返，頻繁使用，強迫信徒順從，逐漸剝奪他們的自我。不過，瓊斯很快又軟化態度，允許喬伊絲和鮑伯重回彼此懷抱。兩人甚至獲得許可，得以在外面工作，在舊金山租房子住。兩人都被指派加入規劃委員會，委員會成員都是經營人民聖殿教事業的核心成員。他們晉身菁英階層，享有自由也有責任，必須賣力工作，開放舊金山的家給其他成員過夜。

鮑伯幾乎總是有兩份工作，一邊當音樂治療師、老師或輔導員，一邊在舊金山的南太平洋（Southern Pacific）鐵路機廠當夜班轉轍工，這是安穩的工會職位。鮑伯深思熟慮又好學，綽號是「小教授」（Little Professor）。他也認真看待人民聖殿教的政治理念，研讀馬克思主義和世界時事。瓊斯講道時詢問大家是否有問題，鮑伯誤以為瓊斯是認真的，積極發問時事問題，但瓊斯並沒有答案。沒多久，教會弟兄姊妹開始調侃鮑伯的華麗辭藻和資產階級重理智的態度；若沒有瓊斯鼓吹，調侃不可能發生。同時，喬伊絲越來越忿忿不平，想要離開教會。麥克阿瑟公園事件之後，人民聖殿教受到益發嚴密的監視，開始要求信徒奉獻所有薪資。瓊斯計畫帶所有人前往蓋亞那，只要一安排好就行動。信徒會收到小筆零用金來應付日常開銷。三餐則由人民聖殿教的設施共同提供。儘管配給有嚴格管制，喬伊絲仍設法存到足夠的錢，買了張公車票回到父母家。雖然她還愛著鮑伯，但某晚下班後她就這麼離開了，不辭而別，回到俄亥俄州。拋棄人民聖殿教的理由不勝枚舉：瓊斯採行體罰，實施杖刑，用的是木頭「教育大板」（Board of Education），甚至安排拳擊賽。要被體罰的信徒和體型更健壯的對手配成一組。若是意外防守得當，避開一場痛毆，他們會被迫再打一次，對手會比上一次更強壯高大。瓊斯給予殘暴體罰正當理由：這世界本來就是殘酷的地方。瓊斯告訴信眾，人民聖殿教總是有愛的地方，教會的矯正總是仁慈寬大。外面的世界不會這麼良善。

自不用說，還有更加狡詐極端的手段可以保證瓊斯掌控一切。父母們被要求簽署告解書，承認他們和自己的孩子發生性關係，這些文件都留在教會檔案庫，以免有人想要離開教會或向媒體爆料。有時人民聖殿教信徒會直接被要求在空白紙張上簽名，之後才會補上內容。就是這些粗暴蠻橫的對待和殘酷無情的心理虐待，導致八大革命家起身叛教。喬伊絲開始打長途電話給鮑伯，說服他離開教會。

鮑伯是忠實信徒，但他也會接喬伊絲的電話。一九七六年十月四日，他寄出喬伊絲上一通電話提到的衣服，接著去上那天的第一份工作，在少年感化院擔任緩刑輔導員。那天值第二個班時，他在鐵路機廠被碾死。機廠的火車速度和步行一樣慢，而且鮑伯生性謹慎細心，也在那裡工作多年了。喬伊絲相信鮑伯不可能犯下任何致命過失，更不可能像人民聖殿教宣稱的那樣自殺。他們的對話充滿希望，雖然鮑伯使用人民聖殿教的電話時態度不是很明確。喬伊絲懷疑鮑伯遭瓊斯的劊子手謀殺，不是被迫躺在鐵軌上，就是被敲暈了放在危險的地方。

喬伊絲寄信給鮑伯的父母，說明自己的懷疑。多年來，喬伊絲照顧鮑伯和前妻菲莉絲的兩名孩子，現在還是非常愛這兩個孩子（喬伊絲和鮑伯在瓊斯和規劃委員會的要求下結婚）。喬伊絲警告鮑伯的父母千萬不能讓他們的孫子去蓋亞那。她說信徒在那裡都慘遭虐待，生活很苦。在最近一次的舊金山聚會裡，瓊斯宣布他在飲料中下毒，當暗樁在會眾之間假裝嗆到並死亡時，瓊斯只是靜靜看著大家的反應。接著他揭曉事實上根本沒下毒，恭喜一點都不驚慌的信徒，因為他們已經準備好執行他所謂的「革命自殺」（revolutionary suicide）。這些活動可能是在演練將來在蓋亞那採取的緊急措施。

鮑伯之死餘波未平，人民聖殿教稱那天早上鮑伯打了封辭呈交給瓊斯。喬伊絲在和鮑伯雙親的通信中，解釋為何那是扯謊：她先生從來沒有打過字。那封信對其他教會信徒而言顯然是個警告。對相信瓊斯是神的信徒而言，這只是證明叛徒自會遭受慘烈報應。而對那些失去信仰、開始質疑人民聖殿教當前路線的信徒而言，這記警鐘更加清晰響亮：這顯示瓊斯為了懲罰離教的人，願意採取多麼暴力的手段。

鮑伯的父親山姆・休士頓（Sam Houston）是美聯社（Associated Press）舊金山分社的攝影師，他把喬伊絲的信拿給《舊金山觀察家報》數一數二的調查記者萊特曼。沒隔多久，喬伊絲就回到加州，也開始和萊特曼聊聊。鮑伯的父親沒有就此打住，他聯絡上美國議員里奧・瑞恩（Leo Ryan）。瑞恩議員很特殊，喜歡親身參與自己關心的議題。一九六五年瓦特暴亂[9]爆發後，瑞恩去當代理教師，想要了解在那裡生活究竟是什麼樣子。瑞恩之前擔任加州眾議院議員時，負責監獄改革委員會，讓自己匿名入監服刑，調查監獄狀況，獲得第一手資訊。有人認為這些都只是宣傳噱頭，但是瑞恩開始組織記者聯盟、召集國會調查員，為人民聖殿教招來多餘

---

9 Watts riots，警察盤查兩名非裔男子引發瓦特暴亂。

的目光。

不可思議的是，在瓊斯心中，這項活動只是次要，另外還有更大的危機：一九七六年，瓊斯政治代理人的太太葛雷絲．史托恩離開了人民聖殿教。但因為提莫西簽了那份文件，聲明瓊斯才是他們兒子約翰．維克多．史托恩的爸爸，因此現在約翰在法律上是葛雷絲和瓊斯的孩子，即使兩人從未同床共枕。葛雷絲離開的時候，沒辦法帶約翰一起走。她知道兒子會被用來與她作對，提告要求監護權。打官司期間，她威脅要抖出人民聖殿教的祕密，畢竟她是教會斡旋律師的太太，幾乎知道所有事情。

一九七六年末，《舊金山紀事報》（*San Francisco Chrcnicle*）記者馬歇爾．基爾達夫（Marshall Kilduff）開始報導瓊斯和房屋管理局。事情不太對勁，局長出席令人昏昏欲睡的官僚聽證會時，竟然帶著一隊武裝維安人員。基爾達夫認為瓊斯是個有趣的人，安排了一趟紅木谷參訪，教會為他和其他訪客帶了一場導覽。導覽到一半，他發現身旁其他訪客都是暗樁，他們在那裡只負責提問，好讓導遊可以談論人民聖殿教的奇事、療養院、青年之家和勒戒中心。基爾達夫跟編輯討論要寫一篇相關報導，但他們跟他說行不通，幾年前萊斯特．金索文已經寫過這個主題了，迴響不大。

但基爾達夫沒有卻步。他決定把這當作自由撰稿文章，四處推銷，《紐約》雜誌（*New York*）的姐妹出版品《新西部》週刊（*New West*）欣然接受這個機會。有了這本風格犀利的週刊邀稿，基爾達夫開始深入調查。他花了好幾個月才找到報導的來源：大部分是八大革命家，還有少部分來自其他叛教者，例如葛雷絲．史托恩。一九七七年最後一週，報導終於見報，內容駭人聽聞。《新西部》週刊提供第一手報導，詳述五花八門的虐待行為。埃爾莫．梅特爾（Elmer Mertle）回想他女兒被皮帶鞭打七十五下的那天，告訴基爾達夫：「她被打得那麼慘，其他孩子都說她的屁股看起來像漢堡排。」柏蒂．馬拉勃（Birdie Marable）詳細描述重返東部的回程，遊覽車上擠滿了人，大家被迫擠在走道上、卡在行李架間，有時甚至得塞進下方的行李廂。韋恩．皮地拉（Wayne Pietila）和吉姆．考伯談到他們看守雞胗的責任，在瓊斯的信仰療法詐騙現場，雞胗會是他摘除的「癌」。沃特．瓊斯（Walt Jones，和吉姆．瓊斯沒有任何親戚關係）揭露，經營情緒障礙青年之家、照顧問題孩童可以獲得州補助，而青年之家必須簽字轉讓超過百分之五十的補助金給人民聖殿教。米奇．圖榭特（Micki Touchette）述說每次在洛杉磯和舊金山的禮拜都可以收到一萬五千美元的奉獻。這則驚人消息驅策瑞恩議員徹底調查人民聖殿教。

就在《新西部》報導刊登前，瓊斯的間諜網路早已拿到報導影本。瓊斯打給舊金山聯合衛理公會格萊德紀念教堂（Glide Memorial United Methodist Church）的塞西爾．威廉斯（Cecil Williams）牧師。他詢問既是朋友也是支持者的威廉斯牧師，如果要打官司，他的勝算有多大，也告訴牧師他人在洛杉磯國際機場（Los Angeles International Airport），準備前往蓋亞那。威廉斯是個鬥士，是頭幾位願意涉入同志權利運動的非裔美國籍牧師，他督促瓊斯待在美國，起身為自己也為人民聖殿教作戰。離去只會被視同認罪。

但瓊斯沒有聽從友人的建議，當天就前往蓋亞那，之後再也沒有踏上美國故土。

黑暗之心

每個邪教的編年史上，尤其是那些直接遭暴力了結的邪教，總是有那麼一刻，邪教越了線，變得殘酷無情。在某些案例中，魅力領袖轉變成嗜血瘋子的過程是如此漸進，難以明確指出究竟何時發生。在紅木谷，瓊斯已經開始使用所有洗腦經典技巧：分離信徒及其家人、處罰異議分子、下達絕對指令（例如安排婚姻、拆散家長與孩子）、配給食物、金錢、睡眠等珍稀資源，以及建立「我們與他們」的敵對心態，進一步分化忠實信徒和世界上其他所有人。但一抵達瓊斯鎮，他的所有行徑都更為囂張、暴力且瘋癲。人民聖殿教的樂園時光轉瞬即逝，從抵達蓋亞那到集體自殺發生，不過短短一年又四個月。

在舊金山，瓊斯的支持者一頭霧水。摩斯康市長覺得被背叛了，但在公開場合還是支持瓊斯，依然相信他被告知的消息，認為他的房屋管理局局長很快就會回來，直到一九七七年八月二日，瓊斯透過廣播唸出了辭呈。大批信徒遷徙至蓋亞那，包機每次載四十到五十人，加重蓋亞那移民署官員的工作，畢竟他們比較習慣人民逃離蓋亞那。許多信徒不但沒有向家人道別，還趁著深夜離開；有些信徒首先前往東岸，希望能藉由多趟短程航班輾轉到蓋亞那來掩飾行蹤。到了九月，近千民信徒來到叢林深處的臨時社區。

信徒從凱圖馬港（Port Kaituma）越過崎嶇殘破的道路抵達瓊斯鎮大門，第一個映入眼簾的是路口處的木製牌坊上懸吊著招牌宣告：

歡迎！

人民聖殿農業計畫

接著，他們經過安檢哨。原本運送機械的箱子設有假底，武器藏在裡面，偷渡至蓋亞那。瓊斯使用暗語，透過廣播命令運送更多軍火，例如「聖經」代表槍枝。《烏鴉》一書中，提姆．萊特曼引用一段廣播訊息，瓊斯說：「我要你們去第二大道（Second）與使命街（Mission）路口的聖經交易所（Bible Exchange）。」指的就是位在那個路口的舊金山槍枝交易所（Gun Exchange）。他繼續道：「他們有賣手電筒，黑色金屬的那款，二十四吋長。聽明白了嗎？」

終於，新來的信徒抵達一片林間空地，可以看到瓊斯鎮全貌。有好幾英畝大蕉園和好幾塊農地。生活區域先是一棟作為食堂的大建築，旁邊有淋浴間，闢了一條路徑通往這區域最活躍的空間：集會大棚，是個金屬遮雨棚覆蓋的大型聚會空間。五間宿舍，男女分住，旁邊有一區共居家庭專用的小屋聚落，另外還有幼兒園、學校和電台廣播室。生活區域後方是一大片休閒區域。相較之下兩幢最氣派的建築（畢竟這裡可是叢林深處）是設計給權貴嘉賓和過夜訪客的東屋（East House），和瓊斯一家人居住的西屋（West House）。但就算是瓊斯的起居空間，理應最具備隱私、最沒有公社感，依美國標準而言還是稍嫌貧困，與其說是奢華舒適的度假屋，不如說是堅固粗獷的木屋。他的住處擁有其他人沒有的個人特色，成堆的藥物儲藏在床下的小櫃子、小抽屜裡：配西汀（Demerol）、安眠酮等強效鎮靜劑，還有用來抵抗吸毒到斷片的安非他命等級藥物，協助他再次得體會客。

在蓋亞那，瓊斯再也無法維繫讓他像神一般的說詞。瓊斯鎮不只毫無隱私，根本沒有祕密。從高齡者睡的宿舍望出去就能看到瓊斯的木屋，這意味著他的外遇、特別治療、長時間嗑藥恍神動也不動，全都被看得一清二楚。飢餓、疲憊、勞役繁重的信眾從生活區域就能看到瓊斯享用比基尼女郎端來的大餐。雖然瓊斯用藥恍惚的時間拉長，也沒能阻止他廣播發表口齒不清的演講和其他聽不懂的話。他沉浸在偏執幻想之中，認為敵人威脅著他們的社會主義樂園。瓊斯習慣連續講好幾個鐘頭不停歇，就算早已胡言亂語，依然堅持老習慣，長篇大論講個不停。有次的鎮內廣播特別語無倫次，總部直接關閉麥克風，在田裡做活的每個信徒都享受這寧靜的幸福。有時瓊斯的高級副手會直接拿走麥克風，打斷瓊斯的發言。

在瓊斯滔滔不絕的演講中，頭號公敵竟是提莫西．史托恩。史托恩在一九七七年初來訪，協助緩和人民聖殿教和喬治城國家政府間緊張的政治情勢，當然也來看兒子約翰。自從葛雷絲退教，瓊斯就不讓提莫西參加會議，而提莫西懷疑瓊斯打算讓約翰與父母作對。提莫西知道隨時都有人監視，盡量讓自己行跡不那麼可疑。他並不想在晚上偷偷帶走兒子，但約翰和瓊斯如此親近令他困擾。雖然他和葛雷絲幾年前就離婚了，他還是覺得自己背叛了葛雷絲。

離開瓊斯鎮不久，提莫西加入葛雷絲的加州訴訟，想要搶回約翰的監護權。一九七七年九月七日，瓊斯向信徒宣布史托恩的背叛，告訴大家提莫西與葛雷斯聲請法院來逮捕他，並想奪走約翰。那晚他再次耍了個老戲法：上演西屋襲擊戲碼。這時瓊斯已經因為藥物濫用而持續生病好一陣子，他宣布自己已經治好嚴重傷害，但無可否認的事實還在，瓊斯鎮腹背受敵。接下來六天，他禁止信徒休息，宣布有一組美國臥底打擊部隊奉命來瓊斯鎮，任務可能是逮捕瓊斯，或者殺光所有人。

在這場應該要發生的戰役中場，瓊斯命令所有信徒前往集會大棚。他語帶哀傷告訴大家，目前態勢已經無可挽回了。幾年前經歷舊金山恐怖之夜的信徒可能已經認出接下來的走向——瓊斯公開重提他的革命自殺理念。最終，瓊斯唯一的婚生子史蒂芬成功勸退父親，讓他忘卻一九七七年的夏日圍攻幻想。但是瓊斯沒有停止呼籲信徒一起集體自殺。

每次公共廣播系統隨著劈啪聲開啟，信徒就會聚在集會大棚。瓊斯開始講不停，說自殺乃是為了人民聖殿農業計畫，潛力無窮。信徒會提問，瓊斯總是講得天花亂墜，唬得信徒改變想法，告訴他們沒有比集體自殺氣魄非凡的姿態了。這類活動在瓊斯鎮被稱為「白夜」（White Nights）。一次長達數小時，直到日落，瓊斯會恐嚇信徒，沒多久大家就要走向死亡。但每回白夜結束，就像幻想圍攻一般，彷彿不曾存在。瓊斯會叫信徒重回瓊斯鎮的日常生活，甚至會放假一天讓大家休養復原。

白夜過後，瓊斯會表彰效忠者運用了意志力，因為唯有忠德才能戰勝圍攻勢力。現在心理學課堂經常以這些白夜演練舉例，說明極端的操作制約：瓊斯透過反覆演練和獎勵，逐漸將毫無人性的行為正常化。這麼一來，不正常的革命自殺概念也變得正常，甚至成為日常例行事務。

調查逼近

鏡頭回到美國，提姆．萊特曼和人民聖殿教離教者萊昂．布魯薩德（Leon Broussard）會晤。布魯薩德遊蕩街頭已久才遇見瓊斯，瓊斯向他保證，住在瓊斯鎮可以免受欲求想望所苦。但布魯薩德一踏上蓋亞那，就被迫工作，無法休息，必須修剪草地、從鋸木場搬運木頭到建築工地。這些勞動繁重累人，無法跟上步調的信徒除了遭受公開鞭刑，還被剝奪飲食。布魯薩德扛了太多木頭，肩上生瘡，起先向維安人員要求休息，但被特別挑出來處罰。接著他向瓊斯要求休息，瓊斯威脅要把他丟入長寬深各九呎的大坑，布魯薩德之前看過據說工作不夠賣力的信徒被迫在坑裡盲目地整天挖土。隨後布魯薩德逃出瓊斯鎮，一路逃到凱圖馬港，向人抱怨他遭受虐待，這引起美國大使館的注意。為了避開多餘的注目，人民聖殿教依布魯薩德要求，付錢讓他飛回家。這一互動代表瓊斯鎮現在正式成為美國政府警戒的觀察對象，正好為瓊斯的偏執預言（即將爆發的攻擊）火上添油。

萊昂．布魯薩德的說詞無法驗證，但提供了提姆．萊特曼調查小組足以攻擊瓊斯的材料。調查小組除了向擔憂的家屬收集證詞，也自行展開調查。瑞恩議員辦公室開始相信在蓋亞那確實有暴行。瑞恩開始檢視各種選項。他的第一直覺總是想辦法親自介入，於是開始前置作業，打算來一趟官方拜訪，實際走訪瓊斯鎮。

同時，瓊斯現在脫離了文明世界，行為舉止處處展示瓊斯鎮是主權國家。他邀請古巴、南斯拉夫、甚至北韓等共產主義集團國家的代表團來訪。他寫了封八千五百字的信，標記著「急件！急件！急件！」，寄給卡特總統（Jimmy Carter），副件寄給國務卿范錫（Cyrus Vance）和美國駐蓋亞那大使館，解釋人民聖殿教盡心盡力，協助毒蟲和長期失業的人成為堂堂正正好市民，但美國政府的態度卻如此不公不義（例如不轉寄人民聖殿教徒的社會安全福利支票到蓋亞那）。這封信讀起來非常像嗑藥過頭的男人顛三倒四的話語經由高薪請來的法律顧問增刪潤飾的結果——而實際情況正是如此。

在瓊斯鎮，許多信徒都在學俄文，因為瓊斯提議，如果沒辦法在蓋亞那建立自己的社會主義樂園，他們可以再次移動，目的地是蘇聯。蘇聯駐蓋亞那大使館官員費奧多．紀莫費耶夫（Feodor Timofeyev）甚至參訪了瓊斯鎮，欣賞共產主義歌曲的表演。參訪期間，紀莫費耶夫附和幾聲，贊同人民聖殿教遷址到他的國家，但是當消息傳回美國大使館，美國領事理查．麥考伊（Richard McCoy）調侃這個前景，紀莫費耶夫強調他沒有著手安排任何事，之後也沒有打算。

同時，瓊斯漸漸無法隨時隨地掌控所有信眾，為此他建立了既是診療所又是精神病房，也算是懲罰室的「全天護理單位」（Extended Care Unit）。在全天護理單位裡，瓊斯和他的醫生拉里・沙赫特（Larry Schacht）儲放大量強效藥物：超過一萬劑抗精神病藥物托拉靈（Thorazine）以及上千顆藥丸。設施內還存放兩萬劑配西汀和五千劑煩寧，還有好幾批嗎啡、安眠酮、易致癮的安眠藥和手術鎮靜劑。瓊斯和沙赫特醫生會替異議分子和危險分子施打藥物，基本上就是把忠誠信徒變成活死人。

瓊斯的兒子史蒂芬發現全天護理單位內部的真實情況，已經準備好要弒父了。瓊斯鎮曾經為他的人生帶來意義，但父親的人格只是越來越低劣，史蒂芬現在相信父親想要全面掌控瓊斯鎮生活的瘋狂行徑已經徹底升級成精神疾病。但瓊斯鎮最不需要的就是暴力政變。史蒂芬將希望寄託於活得比父親長久，瓊斯現在看起來一天比一天虛弱。他的藥物濫用已經失控，一部分也是為了掩蓋藥癮真相，他罹癌的謠言開始在鎮上散布。事實上，卡爾頓・古德萊特（Carlton Goodlett）醫生診斷瓊斯有肺部感染，隨後告訴某家報社，瓊斯狀態極糟，如果沒有和信徒一起死，頂多也只能再活個幾週。

瓊斯的情婦卡洛琳・雷頓和瓊斯一家住在西屋，甚至告訴史蒂芬，如果瓊斯，她會支持史蒂芬。史蒂芬把所有時間用來讓大家活得開心和健康，包括同父異母的弟弟小吉姆。他們組了籃球隊，最終獲得資格參加喬治城的錦標賽。瓊斯不想讓他們去參加，但是瑪瑟琳設法說服了他讓兒子去參賽，轉述史蒂芬需要一些休息時間。因此，一九七八年十一月初，史蒂芬和球隊打包行李離開瓊斯鎮，打定主意不再回來。

與此同時，首次也是唯一一次的瓊斯鎮官方考察開始行前準備。為了回應選民的抱怨，瑞恩議員組織了事實查證代表團，查看目前住在蓋亞那的美國公民生活福祉狀態。瓊斯明白，除非他願意指揮信徒起身反抗，否則沒有方法可以防止瑞恩踏進瓊斯鎮。瑞恩背後有美國政府撐腰，瓊斯鎮的居民依然是美國公民。瓊斯告訴信徒這是另一次信心考驗，希望能讓越少人離教越好。

陪同議員前來的還有一群記者，包括提姆．萊特曼和NBC新聞團隊。另外還有一小群擔憂的家屬，包括鮑伯．休士頓的家人，他們希望能查看鮑伯孫女的狀況，以及前人民聖殿教青年領導吉姆．考伯。連提莫西都來了，雖然他只留在喬治城，遠遠避開瓊斯的怒火。

十一月十五日，他們的飛機在喬治城著陸。兩天後，十一月十七日傍晚，他們飛到凱圖馬港，人民聖殿教派人護送他們進入人民聖殿農業計畫。經過哨兵室，目光所及處沒有武器。他們後來會發現瓊斯命令將所有槍枝鎖在別處。

集會大棚舉辦了歡迎派對，訪客一邊吃東西，一邊聽音樂。就連瓊斯都出席了，看著他的人民載歌載舞，與白夜迥然不同。集會大棚掛著布條寫道：凡是遺忘過去的人注定要重蹈覆轍[10]。

聖殿成員團團圍住瑞恩議員和記者，對瓊斯鎮和他們敬愛的領袖讚頌不已。輪到瑞恩站在

10 出自西班牙裔美國哲學家桑塔亞那（George Santayana）。

大群信徒面前演講，也不得不承認這和他預期的不一樣。他告訴大家，瓊斯鎮看起來對許多信徒而言是非常美好的地方，為此他十分開心。提姆．萊特曼和記者同仁可沒這麼輕易相信。他們找到機會堵住瓊斯，在大家面前臨時開始訪談，這樣瓊斯就不能太過防備。記者窮追猛打想知道瓊斯鎮是如何體現社會主義價值，瓊斯承認：「瓊斯鎮反映出我認為最好的人事物。」

瓊斯接著要年紀輕輕的約翰出來告訴在場所有人，他在這裡很開心，一點都不想回家，也不想回到葛雷絲和提莫西身邊。瓊斯聳聳肩表達遺憾，瓊斯鎮是這麼平和的地方，美國代表團不但白跑一趟，還浪費納稅人的錢。但當瑞恩議員一行人詢問是否能留宿，才不用大老遠開回凱圖馬港，瓊斯回絕了，告訴他們房間不夠。即使他們堅持，說他們可以睡在集會大棚，瓊斯依然不為所動——瓊斯鎮顯然不再歡迎他們了。

一群人坐上車，信徒護送他們回到凱圖馬港。安全離開瓊斯鎮之後，NBC記者唐．哈里斯（Don Harris）擠到萊特曼身邊，掏出一張有人在派對中塞給他的紙條，給萊特曼看。紙條上寫道：「弗農．葛斯尼（Vernon Gosney）和莫妮卡．巴格比（Monica Bagby）。請救救我們脫離瓊斯鎮。」整趟旅程下來，這還是第一個顯示事情終究沒那麼完美的跡象。結果，整晚有超過十多名信徒偷偷接近他們團隊中的不同人。瑞恩現在重新抖擻精神。

十一月十八日早上，他們重回瓊斯鎮，瑞恩議員告訴瓊斯他們至少要帶回十二名想退教的人離開。遺憾的是，鮑伯．休士頓的兩位女兒都不在內，她們和瑞恩以及家人合照，臉上掛著微笑，但想要繼續留在蓋亞那。最後幾次訪問中，瓊斯似乎喪失所有精力與熱情。他試圖表現得毫不在意，告訴離教者，如果他們想回來，瓊斯鎮永遠歡迎他們。

瑞恩在聚集清點要帶走的離教者時，萊特曼和其他人注意到，瓊斯整個下午都在一旁和畢姆、卡特梅爾等高級副手交頭接耳。接著卡特梅爾慷慨激昂說道：「從來沒有人離開過。瓊斯自始至今表達的都是愛。我在他身邊二十一個年頭，來到這裡也已經十八個月。對我而言，這就是全世界。他代表天底下所有良善與慈愛。他的缺點只有一個：他的心地太過寬大。」

瓊斯對在場媒體說的最後幾句話帶著一絲隱約的不安：「很遺憾我們的毀壞來自內部。我們想要的僅僅是不受打擾。我會繼續努力。時間會判定我是否成功。」接著離教者被帶到瑞恩代表團的護送車隊。最後，瑞恩議員在集會大棚裡再走一圈，滿懷信心，希望能帶走任何想脫隊的人，瓊斯則遠遠看著他。

突然之間，人民聖殿教成員唐．斯萊（Don Sly）衝出人群，亮出自製刀片，跑到瑞恩背後，劃開他的襯衫，將刀片架在瑞恩的喉嚨上。瑞恩的同伴衝向集會大棚時，較為冷靜的信徒齊力把斯萊拖離大棚。瓊斯雖然也在裡面目睹事發經過，卻依然毫無反應、沉著冷靜、一聲不吭。

瑞恩命令代表團帶著離教者立刻登上護航車隊。瓊斯輕聲向瑞恩保證斯萊會被送到蓋亞那相關當局接受懲處。此時此刻，代表團只想帶著離教者離開。有幾位離教者哭了起來，回頭指著瓊斯鎮，說還有其他人想離開，但太害怕了不敢出聲。在最後一刻，卡洛琳的前夫賴瑞．雷頓（Larry Layton）跳上貨卡，告訴他們他也想要離開。吉姆．考伯悄聲告訴萊特曼，雷頓是位死忠信徒，不可能背叛人民聖殿教。兩人做好準備：事有蹊蹺。這顯然是個陷阱。

失樂園

在人民聖殿教喬治城總部，史蒂芬晚起了。瓊斯鎮隊打輸了籃球比賽，雖然他們謊報獲勝。接著，事情突然爆發了。人民聖殿教喬治城辦公室的首席接線生莎朗．阿莫斯（Sharon Amos）接到瓊斯鎮傳來的密碼訊息：他們要盡可能收集大量武器「報復敵人」。阿莫斯和其他成員認為這則訊息是命令他們找出待在喬治城飯店的提莫西並殺掉他。

阿莫斯已經準備好要行動，但史蒂芬阻止了她。瓊斯之前也曾下達強硬指令，例如在白夜的時候。史蒂芬想要在採取任何行動前，先調查清楚美國事實查核團發生了什麼事。

同時，議員代表團也還在凱圖馬港。他們叫來第二架飛機載多出來的人。飛機終於抵達，代表團分成兩組登機。氣氛緊張，情緒緊繃。賴瑞．雷頓舉止一直鬼鬼祟祟，開車護送他們到飛機跑道的人民聖殿教守衛怒目瞪視叛教者。當他們準備登機時，地平線上出現另外兩台貨卡，朝著跑道駛來。其中一名記者注意到前來的人帶著武器。

瓊斯當了好幾個月自己國度的首領，現在採取了終極手段：他要發動裡應外合戰。

人民聖殿教貨卡開火，賴瑞．雷頓同時衝向第二台飛機的機長，試圖壓制他。瓊斯不打算留任何活口。幾分鐘內，瑞恩議員、記者唐．哈里斯、攝影師鮑伯．布朗（Bob Brown）和葛雷格・羅賓遜（Greg Robinson）、叛教者派翠西亞・帕克斯（Patricia Parks）都被殺死。幾乎其他所有人都受了傷，包括萊特曼，他們驚慌失措，跑進飛機跑道周圍的高大草叢和軟爛濕地裡，另一名記者史提夫・宋（Steve Sung）裝死倒在其中一架飛機旁。

幾分鐘後，槍聲停歇，蓋亞那警力出現，人民聖殿教貨卡立刻駛離現場。機長反擊，賴瑞・雷頓失去意識，被留在現場。倖存者爬出藏匿地點，眼睜睜看著大屠殺後的殘酷場景。凱圖馬港居民詢問萊特曼和那些還能說話的人，同時其他傷患接受最好的包紮。代表團的倖存者向蓋亞那人保證他們不是人民聖殿教之前警告的ＣＩＡ間諜。當地居民早就懷疑瓊斯和他的教團，不需多費唇舌，他們就明白人民聖殿教才是真正的威脅。冒著小命不保的風險，當地居民幫助倖存者藏匿在凱圖馬港。

瓊斯鎮的噩夢才剛開始。耳熟能詳的公共廣播系統警鈴迴盪在整個社區。瓊斯和畢姆、卡特梅爾、卡洛琳・雷頓、瑪瑟琳等人民聖殿教領導者開完私人會議，直接出來宣告他預卜即將發生暴力行動。時至今日，他還要靠超能力重塑已道德破產的計畫。他告知大家，幾名信徒自作主張帶著武器追捕瑞恩議員。瓊斯聲稱賴瑞・雷頓打算劫機，迫使飛機墜毀在叢林。這很可能是瓊斯最初的計畫，但沒想到機長反擊，毀了計畫。

就是這一刻，所有還留在瓊斯鎮的信徒最後一次聚集在大棚，而廣播訊息傳到喬治城總部時，大家都提高了警覺。瓊斯開始向瓊斯鎮的信徒宣講，人在喬治城的阿莫斯則對廣播延遲感到不耐煩。訊息終於傳來了：白夜指令。阿莫斯相信虔誠的信徒已經走在「去見『弗雷澤夫人』」的路上，這個人民聖殿教廣播代碼指的是死亡。革命自殺開始了。阿莫斯抓著他的三個孩子克莉絲塔（Christa）、馬丁（Martin）和里昂（Liane），把他們押進浴室。所有人都進了浴室後，阿莫斯先割開克莉絲塔的喉嚨，再來割馬丁的，最後她要大女兒里昂殺死她，可是里昂力氣不夠，刀子只插到一半就進不去了，阿莫斯自己完成剩下的工作。接著里昂割喉自盡。這是瓊斯鎮大屠殺的第一輪自殺。

籃球隊隊員李．英格朗（Lee Ingram）那時在樓下的客廳。聽到樓上傳來一陣騷動，跑上樓發現大家的屍體。克莉絲塔奄奄一息，正在垂死掙扎。接著，史蒂芬從喬治城回來，目睹了殘暴的場景。英格朗立刻打到舊金山，告訴他們不要理會稍早的廣播訊息。史蒂芬試著聯絡瓊斯鎮，但沒有任何回音。很無奈地，已經太遲了。

在瓊斯鎮，瓊斯早已登台演出。他看似重拾舊日精力，口齒清晰，行走自如，不需協助。他看起來幾乎鬆了一口氣，重述他稍早告訴聖殿領導者的話，表達他堅信他們已經來到了終點。他明白地告訴大家：「如果我們不能活得安穩，還可以死得平靜。」此話一出，博得滿堂彩。光聽卡帶，無從得知是這是真摯情感，還是信徒預期這次只不過是另一場排演，他們的回應會受到評斷。不管真相為何，白夜排演得到成效：信徒都知道流程了。

現場有三百多名孩童，瓊斯溫柔說道：如果父母現在不先拯救自己的孩子，之後美國政府會來宰殺所有孩子。瑞恩議員遇害的消息一傳來，瓊斯認為提莫西．史托恩也已經遇害，現在瓊斯鎮所有人都得死。小孩能在父母身旁安靜地死去，可以稱得上是萬幸。但在這個恐怖日子留存下來的錄音裡，真相清晰可聞：整個下午孩子驚嚇不止，個個放聲大哭。

瓊斯在最後的時刻重新包裝所有事情，好似要赦免自己的罪，免除所有責任。他不是要殺了他們，他依然是大家的父親，在這漫漫長夜裡哄大家入睡。瓊斯問到是否有異議時，只有一位信徒克莉斯汀・米勒（Christine Miller）出聲，說道如果大家之前可以一起來到蓋亞那，肯定能再做一次：「好，大家一起飛到俄羅斯吧，這是我的提議。我認為只要相信，沒有不可能。」但瓊斯駁回：「你以為現在我們汙名滿天下，俄羅斯還會歡迎我們？」點明他們曾有的價值已蕩然無存。

米勒回道：「嗯哼，我可不這麼認為。我覺得只要還有一口氣，就有希望。這就是我的信仰。」

瓊斯的回應提供了一絲安慰。他說：「嗯，我們總有一天會死，那裡不會有希望。因為人遲早都會死。」瓊斯語帶溫柔，告訴米勒活在恐懼之中不值得。在這群信眾面前，所有人都站在瓊斯那邊，米勒打了退堂鼓。

這場對話發生的當下，沙赫特醫生領了一群人走出全天護理單位，抬著大桶大桶液體，混合了風味愛飲品和氰化鉀（早在一九七六年就走私進瓊斯鎮了）。舞台旁邊設了一個分發站，準備好杯子和注射器，沙赫特醫生的護士會替那些害怕到不敢喝的人直接將溶液注射到口中。接著瓊斯叫大家去排隊，下令先給小孩致命混合液，於是隊伍開始形成。他監督一場勉強稱得上氣氛嚴肅的慶典，邀請支持者和信徒上台發言，推翻克莉斯汀・米勒的挑戰。

一位接著一位，追隨者感謝瓊斯、祝福瓊斯。他們鼓勵大家團結一心。第一批喝下毒藥的孩子紛紛開始咳嗽、呼吸困難。當孩子都死光了，盛讚瓊斯的言語還在繼續。大人要較年長的孩子幫忙安慰年幼的孩子。可以聽到有人宣稱毒藥不會引發苦痛，只是嚐起來怪怪的。

在離大棚遠遠的地方，瓊斯的情婦瑪麗亞．卡薩里斯（Maria Katsaris）交給麥可．普羅克斯（Michael Prokes）和另外兩名兄弟一只手提箱。箱子裝著價值將近七十萬美元的各國貨幣，還有收件人為費奧多．紀莫費耶夫的信件，安排從境外戶口匯款近一千萬美元至蘇聯。

他們收到的指令是把行李箱帶到喬治城的俄羅斯大使館。事實上，他們走了一哩路後，把大部分的錢埋起來，丟掉行李箱，口袋塞了四萬八千美元，一走了之。他們稍後都被蓋亞那警方逮捕。

瓊斯的律師馬克．連恩和查爾斯．蓋瑞逃進叢林，向守衛保證會告訴全世界人民聖殿教故事的終章。信徒史丹利．克萊頓（Stanley Clayton）假裝要去哨兵室查看，但實際上害怕得躲進大蕉園。他離開的時候，看到武裝維安人員圍繞著某一區域，拒絕喝下毒藥的人（其中一名是克莉斯汀．米勒）就在裡面，被迫被用注射器餵毒。

就算死亡就在眼前發生，讚嘆瓊斯的證詞仍然不輟。男男女女逐一倒地。有人大喊，有人尖叫。沙赫特醫生站在舞台上，告訴還活著的人他們即將踏入另一端：不是結束，是新的開始。一小時的大屠殺進展到一半，瓊斯收到消息，瑞恩議員死了。他告訴信徒：「都結束了。都結束了。多麼好的遺產呀。」他督促大家加緊腳步。時間不多了。敵人要來了。

幾小時後，夜晚籠罩瓊斯鎮，平日九百位居民的聲響，現在一絲都聽不到。悄然無聲之中，史丹利．克萊頓聽到槍響，接著冒險走出藏匿地點。集會大棚靜悄悄的。他走過大片信眾，屍體成堆成堆疊在大棚下，排滿走道，在毒液大桶邊擠成一團。每個人都死了：畢姆、卡特梅爾和瑪瑟琳。舞台上，瓊斯就倒在平日發表幾小時演說的椅子下，死於頭部槍擊。日後，驗屍報告顯示高濃度戊巴比妥（pentobarbital，在協助人類善終合法的國家及地區的麻醉藥首選）。肝臟中有致死濃度，但腦部的濃度較低，此一差別清楚表示瓊斯當天下午先為自己施打了鎮靜劑，而且他的巴比妥藥癮讓他可以容忍超大劑量。

西屋裡還有十三具屍體，包括約翰和卡洛琳。卡洛琳的姊妹安妮．摩爾死於槍擊，只有她和瓊斯一樣死於槍擊。她的日記就攤在身旁地上，傷感地懷想蓋亞那早期生活，寫道：「似乎這世間所有美好的事物都一直受到攻擊。」最後一行，墨水換了顏色，寫道：「我們會死是因為你不讓我們活下去。」

電視不會報導這場革命

瓊斯並沒有猜錯：人民聖殿教留下了無可抹滅的遺產。但他的革命自殺沒有獲得任何同情。現在沒有人會認為瓊斯有任何遠見或願景。隨後賴瑞．雷頓便成為法院起訴的唯一一位人民聖殿教成員，指控的罪行是共謀殺害美國議員。

少數幾位倖存者過著不一樣的生活。集體自殺那時，葛洛弗・戴維斯七十八歲，回到美國領著社會安全福利支票過活，直到一九九三年過世。海亞辛斯・斯拉什搬回印第安納波利斯，和當地作家瑪莉安・K・湯恩（Marian K. Towne）合寫《唯一活口：蓋亞那瓊斯鎮的倖存者》（暫譯，*The Onliest One Alive: Surviving Jonestown, Guyana*）。她在一九九五年過世。

麥可・普羅克斯（Michael Prokes）長期擔任人民聖殿教媒體宣傳專員，就是奉命帶那箱錢離開瓊斯鎮的一員。他回到美國召開記者會，保證有天大頭條。記者會上，他讚美瓊斯，接著走進浴室，朝頭部開了一槍。

有些人彼此尋求慰藉。阿奇・伊詹姆斯仍是忠誠的普世主義者（universalist），常常聯絡前聖殿教成員，試圖建立倖存者聯盟。提姆・萊特曼出版了最權威的人民聖殿教歷史書。瓊斯鎮停駐在歷史裡，在語言中被紀念：「喝酷愛飲料」已成為盲目、無條件忠誠的代名詞。

事發後幾年，史蒂芬做了不容易的決定，與父親的遺產和解。他能放下憤怒，單純是因為這樣才能忍受剩下的人生。他的兄弟小吉姆和提摩西都是瓊斯鎮籃球隊員，也都倖存下來，他們的韓裔美國籍姊妹蘇珊娜（瓊斯後來領養的孩子）也順利活著。蘇珊娜嫁給麥可・卡特梅爾（Mike Cartmell），兩人在大家前往蓋亞那之前就離教了。麥可的媽媽就是人民聖殿教領導者帕蒂・卡特梅爾。瓊斯與瑪瑟琳領養的另外兩名孩子路和艾格妮絲，還有瓊斯與卡洛琳・雷頓的兒子基莫，全都和家人一同死在瓊斯鎮。

從瓊斯鎮移動九百多具屍體，還有喬治城總部與凱圖馬港飛機跑道的屍體，實是運輸物流的噩夢。奧克蘭陸軍基地（Oakland Army base）會是將屍體運回美國的入境首選，因為大部分死者的家屬都住在北加州，但是如同艾瑞卡．梅爾曼（Erika Mailman）在《滾石》雜誌（Rolling Stone）的報導，那裡的官員不想要處理大批傷痛欲絕的民眾，因此屍體被送到德拉瓦州（Delaware）的多佛空軍基地（Dover Air Force Base）。過了初期的人潮，還有超過四百多具屍體無人認領，要嘛沒有家人，要嘛去一趟德拉瓦的旅費太昂貴。瓊斯畢生專挑窮人和受到迫害的人傳教。未認領屍體中至少一半是孩童，雨水刷去名牌上的文字，他們的年齡還太小，沒有任何可供識別的特徵紀錄，例如指紋或牙齒印模。

加州奧克蘭長青墓園自願提供這些遊魂最後的安息之地。墓園老闆巴克．坎普豪森（Buck Kamphausen）認為亡者值得好好安葬。他有一座小丘，以階梯架構來埋葬，不會占太多空間。多虧小吉姆和史蒂芬籌募的基金，坎普豪森可以設立小小的紀念碑，四塊花崗岩石板刻寫有人指認的亡者。

正式揭碑因為官司拖了好幾年，有些家屬不希望吉姆．瓊斯也在名單上。但是最終他的姓名也在上面，依字母順序排列。每年十一月十八日，倖存者和遺族會聚在墓園，懷念那日逝去的生命，並為他們祈禱。

Chapter 5

施虐

羅克・泰里奧和螞蟻山之子

螞蟻山之子是末日邪教團體，一九七〇、八〇年代在加拿大偏遠區域建立共同生活社區，初期是烏托邦孤立主義，宣揚純淨和健康的生活方式，但結束的方式正是虐待失控的駭人實例。要決定邪教領袖羅克．泰里奧（Roch Thériault）的殘酷無情和受害者的消極順從究竟哪一樣更令人不可置信，實在不簡單，信徒多年來都容許他的殘暴和無情。法裔加拿大人泰里奧英俊帥氣，還有推銷天賦，和信徒一同住在荒郊野外，對多位妻子暴力相向，還在自宅手術中虐殺了其中一位妻子，死後還繼續褻瀆遺體。他粗暴對待許多自己的孩子，行徑之殘忍，一位媽媽寧願讓嬰兒凍死在雪地裡，也不願把孩子交給泰里奧。泰里奧截斷其中一位太太的一隻手臂後，終於被關進監獄，但他依然能高度掌握忠心耿耿的信徒，甚至在配偶探訪期間，讓三位妻子懷孕，又多了三名孩子。但他的邪惡魅力對其他囚犯一點影響都沒有——他總算在六十三歲死去，一位獄友帶著偷渡進牢裡的刀子結束了他的性命。

惡性自戀者

幾乎每一位邪教領袖都算是自戀型人格，但自戀型人格障礙症（narcissistic personality disorder）涵蓋的行為很廣泛，在某些極端例子下，特定個人會變成截然不同的生物。馬歇爾・阿普爾懷特（Marshall Applewhite，本書稍後會詳加介紹）被多數評論認為是親切的人，激起他人無條件的奉獻，也因此能夠掌控並帶著三十八個人走向死亡。但沒有人會用親切來形容羅克・泰里奧，尤其是那些認識他的人。

泰里奧粗暴、巧妙地殘酷、冷血又無情。他一次又一次表現自己喜愛暴力，尤其是否決人類天性的暴力。他殘害自己的孩子，並且對造成的痛苦毫不在意，病態地冷漠，幾乎可以肯定他患有心理疾病，而他患的疾病自成一類：惡性自戀，綜合了強烈的反社會、妄想和虐待人格障礙症。比起尋常自戀者，惡性自戀者在海爾精神病態人格檢核表（Hare Psychopathy Checklist）都會獲得較高分，因為他們殘忍無情利用他人，以及他們的反應型憤怒和衝動暴力行為。但是泰里奧和亞道夫・康斯坦佐不一樣，他並沒有整個童年都在為暴力未來受訓，不用學習如何在儀式上宰殺動物，也沒有學習過他人的痛苦可以用來達成自己的邪惡目的。泰里奧的動機似乎只是單純喜愛觀賞他人痛苦掙扎。他對虐待他人毫無悔意，甚至進展成「愉悅殘酷」（joyful cruelty，心理學家史考特・派克（M. Scott Peck）以此建議將「邪惡」納入有用的診斷術語）。

一九四七年五月十六日，泰里奧在魁北克薩格奈（Saguenay）出生。薩格奈河（Saguenay River）畔的薩格奈是座工業小鎮。他的父親亞桑特．泰里奧（Hyacinthe Thériault）是名油漆師傅，因為工作考量，一家人不久後搬往魁北克南部的塞特福德礦城（Thetford Mines），那裡是世界主要石綿來源地之一。或許泰里奧發現心理學家經常尋找童年受虐中最嚴重的虐待證據（相信虐待會製造更多虐待），因此他很喜歡告訴大家他自己在還是男孩時慘遭暴力對待。他發現這個故事可以幫他得到想要的東西，不管是從他想要引誘的女人身上，還是從法院指派的心理學家身上獲得免罪診斷。他告訴其中一位妻子吉賽兒（婚前姓拉弗朗斯）（Gisèle (née Lafrance)），父親毆打他的胃部，造成了嚴重傷害。其他時候，他宣稱在九到十二歲大的時候，父親把他推下樓梯。（長大之後他飽受胃潰瘍之苦，將此歸咎於他宣稱自己遭受的暴力，雖然重擊從未是胃潰瘍的成因。）

泰里奧四處散播童年受虐導致他做出不可饒恕的行為，例如他殘暴地殺了一名女子，再寫信給該女子的姊妹，信中寫道：「從兩歲開始，我在家一直受到虐待，打得有夠慘，簡直不如一條狗，直到十四歲，父親揍完我，就把我丟到屋外，叫我永遠不要再踏進家門。」但是，泰里奧從來沒有被趕出家門。他是好學不倦、天資聰穎的學生，喜歡閱讀，而且據他所言，也會到戶外玩。雖然泰里奧長大後常在卑微的道歉裡將父親形容成暴力酒鬼，不過鄰居和親友的印象並非如此。他父親駁斥自己打過兒子，但承認：「必要時我會處罰他。」那時候體罰很常見，而泰里奧又擅於說謊，每次犯錯就怪罪兄弟姐妹，這似乎給了爸爸管教的機會。只不過，有位鄰居記得亞桑特和兒子們會玩叫作「骨頭」的遊戲：大家坐在廚房餐桌，穿著厚重工作靴，互踢雙腿，直到有人投降。

難道比起同時代家教一樣嚴格的孩子，泰里奧受到更多傷害嗎？又或者他本就具備心理病態常見的冷血無情？心理學家肯特．基爾（Kent A. Kiehl）在《精神病患的低語：無良心的科學》（暫譯，*The Psychopath Whisperer: The Science of Those Without Conscience*）書中提到，這些孩童的邊緣系統本來就發育不良，那是大腦專門處理情緒的部位。他們可能會有不活躍的杏仁核（大腦的恐懼中心），無法辨別他人或自身的沮喪與恐懼，於是容易犯下暴力罪行。他們或許能夠學習理解情緒，但因為他們無法感受情緒，這只會加強他們操控他人的能力。他們傷害他人時，心跳不會加快。恐怖、屈服、沮喪等極端情緒，反而只會引起他們的好奇。

患有這程度心理病態的人也沒有辦法控制自己，年紀稍長之後容易追求性愛、毒品和酒精等事物，並且總是過度或過量，就為了尋找任何形式的刺激，創造內在火花，取代他們平日無法體驗的各種情緒。他們無法感受到恐懼，又無法理解合理言行的邊界，兩項特質正好混合成易燃體質，也因此想透過造成他人痛苦來尋求滿足，透過物質濫用來提振精神，享受每一次他們能威脅、傷害他人的時刻。

這就是導致泰里奧在加拿大荒郊野外折磨、閹割、殘害、肢解、謀殺、糟蹋信徒的原因嗎？現在仍難下定論。但按照家人和鄰居所言，他的童年處境和他日後享受傷害他人的程度，根本不成比例。其實可以推測他天生就能從他人的痛苦中找到歡愉。

自學成醫

泰里奧在學校表現優異，老師都覺得他聰明、勤奮又有創意，成績名列前茅，將來大有可為。他自學英文，不僅因為實用，還能炫耀自己比只會說法語的家人聰明得多。但他的父母是勞工階級，泰里奧和兄弟姊妹都一樣，在地方教育的最高年級七年級之後，就沒有再接受正式教育了。比起到附近城鎮繼續教育，他選擇待在家，修剪草坪、做些工作來幫助家計。

他繼續靠著讀聖經自學（聖經是他手邊少數的幾本書之一），憑藉現在能長時間流利宣道的能力，他看似認真研讀了經文。他的家人是天主教「大天使米迦勒朝聖者」（Pilgrims of Saint Michael）一派的信徒，這個派別在經濟大蕭條之後的魁北克信徒眾多。這一群體融合了社會保守立場和所謂社會信用（social credit）的經濟民粹主義，推廣將利潤重新分配給勞動人民。如果這些理念真的施行，就會十分類似二〇二〇年總統大選民主黨初選候選人楊安澤（Andrew Yang）提議的全民基本收入（universal basic income）。

大天使米迦勒朝聖者有另一個名字——白貝雷帽（White Berets），因為他們頭戴白色貝雷帽，挨家挨戶發送文宣冊子，努力說明社會信用的好處，希望能增加信徒。信徒戴著顯眼的白貝雷帽很引人注目，尤其當泰里奧先生帶著所有孩子一起齊步走在鄰里，一邊宣傳白貝雷帽教派一邊募款，非常醒目。

泰里奧的朋友嘲笑他參與如此公開的活動，重擊了他脆弱不堪的自尊。自戀狂雖然有膨脹的自我，但又執著於獲得他人認可，無法容忍別人開玩笑。泰里奧厭惡被迫和父親一起宣教，此外也漸漸討厭天主教。他無法將雙親的信仰和鄰里朋友的嘲笑分開看待，泰里奧一家在鎮上推廣白貝雷帽教派時，大家都會笑他們。

但就算這情況把他折磨得如此淒慘，他還是獲得了寶貴的經驗：在出了名困難的銷售情境下騙取陌生人的認可。後來泰里奧變得格外擅長這種陌生推銷的說服方式。我們不禁想，雖然他這麼討厭白貝雷帽，卻還是從痛苦中獲益不少。

一九六一年，泰里奧離開校園，之後開始打零工賺錢，還喜歡在舞廳打發時間。現在，他不再需要拿高分來顯示自己高人一等，他發現自己可以展現另一特殊優勢。保羅·凱赫拉（Paul Kaihla）和羅斯·拉弗（Ross Laver）合著的《野蠻救世主駭人故事：邪教領袖羅克·泰里奧和深愛他的女人》（暫譯，*Savage Messiah: The Shocking Story of Cult Leader Rock [sic] Thériault and the Women Who Loved Him*）談論泰里奧和螞蟻山之子，書中提到，泰里奧後來寫道：「從小我總是相信自己不一樣，不是普通人。這不單只是心理感覺，還是解剖學上的差異。」他漸漸確信自己天賦異稟，過人甚遠，還吹噓自己的性愛技巧人人稱羨。就算不考慮這些，他本來就又高又帥，有著一雙深邃藍眼睛配上濃密眉毛。他可以天南地北高談闊論，立刻吸引挖礦小鎮上的女性。

青年時期，泰里奧大把時間都花在當地舞廳，周旋在女人之間，但現在二十歲的他想要試試傳統生活，在一九六七年和十七歲的法蘭欣·格雷尼爾（Francine Grenier）結婚。他蓋了新家，離父母家只有半哩——實在不像自稱十四歲就被踢出家門的人會做的事。但是泰里奧和法蘭欣沒在新家待多久，未滿一年就搬到蒙特婁（Montreal），泰里奧應徵成為煙囪檢查員。酒精開始在他的生活中出沒，但他是開心的醉漢。小倆口看來很滿意目前的生活。一九六九年一月，他們迎來大兒子羅克－希爾萬（Roch-Sylvain）；一九七一年四月，迎來老二法蘭索瓦（François）。

但法蘭欣還懷著法蘭索瓦時，泰里奧的健康突然亮紅燈，抱怨嚴重胃痛（這胃痛日後會折磨他一輩子）。醫生診斷為胃潰瘍，在現代，吃藥就能有效控制胃酸分泌，但在一九七一年，泰里奧被迫忍受迷走神經切除手術，現在幾乎不採取這種減酸手術了。迷走神經是體內最長的單一神經系統，手術過程中會切斷迷走神經穿過胃部的部分。那時大家還不太清楚迷走神經的功能，但現在已知迷走神經深深影響身體的副交感神經系統。副交感神經系統主要負責讓身體靜下來，休息復原；交感神經正好相反，主管壓力回應，也就是所謂戰鬥或逃跑反應。

這次不成功的手術真的造成泰里奧情緒劇變嗎？手術後法蘭欣立刻注意到這一轉變，從此她先生的性格越來越暴虐。醫生切除泰里奧一大部分的胃，難道他也失去了控制新陳代謝和情緒的關鍵元素嗎？一九七一年秋天，他接受第二次手術，術後好幾個月都在嘔吐，伴隨著無法承受的劇痛。後來他才知道，因為手術沒做好，所以患了傾食症候群，影響消化功能，食物直接從胃跑到小腸，身體根本沒時間好好吸收任何營養。現在，動過胃繞道手術的病患大約只有百分之十五會被傾食症候群影響，而且通常可以直接透過飲食調整解決。只有不嚴格遵守飲食調整的人才會承受可怕的症狀，包括疲倦、嘔吐、胃痛——泰里奧開始被這幾種症狀折騰。

慢性疼痛影響了他的心理健康。本來隨遇而安的愛家男子，現在被痛苦吞噬，沉迷邪惡黑暗的嗜好。泰里奧具備學者般的能力，可以迅速吸收複雜資訊，總是推動他廣泛閱讀。但現在他只專注閱讀醫學教科書，一本接一本，向朋友長篇大論講解人體解剖和傾食症候群。法蘭欣後來提到，那次手術是他們美滿婚姻的轉捩點，泰里奧完全變了一個人，原本寵溺妻小的貼心好老公變成只在乎自己的男人。更糟的是，自學讓他自信過剩，覺得自己可以行醫，危險萬分。

慢性疼痛、沉迷醫學和喝酒自療，全都導致泰里奧接下來的崩壞。他辭掉工作，強迫逐漸擴大的家庭重回塞特福德礦城。他開始告訴大家自己的內臟都變成塑膠，活不久了。偶爾，他不談論自己的健康，會開始針對法蘭欣的外表。多年來，他堅持要法蘭欣穿長裙，避免其他男人覬覦，現在態度突然反轉，要求法蘭欣穿短裙露大腿。

這些看似隨意控制法蘭欣的衝動，以及強加自己的意志在他人身上，都是虐待傾向的早期警訊。沒多久，他的虐待行為就會籠罩他們的婚姻，隨後更成了他整體存在的展示。突然對法蘭欣的衣著打扮有興趣不是他出現新的病態性執著的唯一證據。他開始說要創建裸體社區，甚至問了姻親是否能使用他們的一小塊地。這是他第一次顯露想要創造另一個世界的慾望，在那裡他可以主宰一切。最初未能實現的這一股衝動，似乎最簡單、最根本地表達了他的慾望，因為裸體社區的會員可以讓他二十四小時運用自己的解剖知識。

業務變先知

回到塞特福德礦城，泰里奧獲得許可，在爸媽的地開了一間木工店。泰里奧看來繼承了爸爸的務實才幹，不僅適合當一位好商人，後來也因此成功存活下來。起初店面生意興隆，他砍樹做傢俱和木製廚具。有些鄰居對當地政府建照的核發程序不滿時，泰里奧似乎記得所有市政法規，代表鄰居四處辯論，指出法規不一致之處，正好能讓那些不贊成的鄰居好好利用。他成功為大家發聲，於一九七五年被選進市議會。他還加入當地天主教社會組織阿拉米斯俱樂部（Aramis Club），俱樂部平時資助慈善活動和地方舞會。

他被指派為發展推動小組組長，直到他沒完沒了的各種不妥建議迫使俱樂部撤銷他的會員資格，例如強烈希望會員穿上背面印著撒旦圖片的衣服。他在市議會也經歷類似遭遇。他其實很少出席市府會議，少數出席的那幾次卻霸凌議會同僚，只因大家反對他不完善、財政負擔不合理的建築提案。不久後，大家也把他踢出市議會。

在此期間，他在婚姻中的表現也沒有比公民參與好多少。他把時間都浪費在外地偷情（他總是佯稱自己要「出差」），傢俱生意的營收逐漸減少，夫妻倆被迫另謀生計，以免入不敷出。法蘭欣當起服務生，泰里奧開始到附近的大城鎮經營木工生意。沒多久後，法蘭欣明白，比起花時間做生意，先生花更多時間帶女人到飯店幽會。線索之一：他出差回來後總是確信胃藥影響了他的性表現。按照醫生建議，他停止服藥，但又開始喝酒舒緩疼痛。那次停藥後不久，法蘭欣在一九七六年與先生離婚，帶著兩個小孩離開。

是什麼讓泰里奧忘卻婚姻和生意的失敗呢？二十六歲、美麗誘人的吉賽兒·拉弗朗斯（Gisèle Lafrance）成了他宣洩強迫症的新對象。正如法蘭欣懷疑的，泰里奧出差時在某家假日飯店（Holiday Inn）的舞會上認識了吉賽兒。過沒多久，吉賽兒在塞特福德礦城租了房子，泰里奧週末都會去那裡，甚至讓吉賽兒和兒子碰面。他被踢出阿拉米斯俱樂部之後，個人信仰也再次轉變：他決定成為基督復臨安息日會教徒，開始披著修道士長袍在鎮上走動。他成功轉變信仰的第一人是吉賽兒，而這也開啟了他結合性交與福音的傳道模式，日後仍是自由之身時依然繼續這種傳道模式。

基督復臨安息日會引起泰里奧注意的許多理由都和大衛·柯瑞許相同（本書第六章會詳細介紹柯瑞許）：安息日會的架構允許大家尋求個人啟示，而這種認證很可能會被無恥之徒當作墊腳石，藉此獲得如神之力，掌控他人。安息日會沒有特別鼓勵在世先知，但傳統上一直都接受這個概念。泰里奧還沒利用基督復臨安息日會來主宰他人，先栽入了源自利未記的嚴格飲食規定，還有安息日會對健康的關注，帶著新歸信主的熱忱恪遵一切。皮埃爾·齊塔牧師（Pierre Zita）看到這股熱情，派他去吸引其他可能因他的轉變深受感動而加入安息日會的人。

泰里奧又一次成功運用自己的銷售長才，大放異彩。沒多久，不論是在教會還是地方有機食物運動之中，他都幫自己建立了領導地位，這兩個群體的人大多重疊。泰里奧可以拿自己的健康當例子，說明安息日會成就的奇事：他戒了酒，開始過著茹素的健康生活，並相信審判日和基督復臨近了。他的推銷話術吸引了幾十名新信徒。

到了一九七七年夏天，齊塔牧師安排泰里奧到新牧區，打算吸引鄰近城鎮普列西維勒（Plessisville）的法國天主教徒。就像之前一樣，泰里奧召集年輕信徒，包括一名男子雅克・菲塞特（Jacques Fiset）和他的三名女性友人：十九歲的香塔爾・勒布里（Chantal Labrie）、十八歲的法蘭辛・拉芙蘭姆（Francine Laflamme）和二十一歲的索朗吉・博伊拉（Solange Boilard）。出師告捷，泰里奧明白自己想要親自領導這群新教徒。他花了將近一年才執行自己的計畫，但才幾個月就著手規劃另立門戶。

推銷成功的關鍵是泰里奧善於辨別可以利用的弱點。年輕的索朗吉・博伊拉聰明又叛逆，家庭生活一點也不美滿。她痛恨父親，他是個酒鬼，還會虐待家裡所有人，她因此極易墮入所謂暴力代間循環——此一循環始於家長，孩子長大落入類似小時候和父母的關係漩渦，循環就會繼續，不是成為受害者，就是施虐者。姑且不論泰里奧是否曾經受虐，他倒是能迅速辨認出他人弱點。

和索朗吉不同，香塔爾・勒布里家庭美滿。她愛做白日夢，有一對溺愛又保護她的父母。儘管如此，她容易悶悶不樂，很可能患有憂鬱症。法蘭辛・拉芙蘭姆則是活潑外向，但對外表極度不安。男性邪教領袖常常利用年輕女性的自我懷疑，讓她們覺得自己有人愛，認為自己是漂亮、特別的存在。泰里奧也利用自己強大的魅力，迷倒任一位相遇的人。

要不了多久，他就邀請雅克、索朗吉、香塔爾、法蘭辛，週末到吉賽兒公寓外紮營，他們可以聽他長時間講述末日近了，內容卻是雜亂無章。邪教領袖可以採取的最有效步驟之一是孤立信徒。這是利夫頓思想改造八原則之一：社會環境控制。把所有人聚集在吉賽兒家只是第一步，泰里奧還利用各種策略切斷信徒與他人的連結，增強團體動力。

對泰里奧而言，這是多產的夏天，立刻增加六名新核心成員：年輕女子妮可．魯埃爾（Nicole Ruel），她擺脫不了媽媽死亡的陰影，一認識泰里奧就立刻搬進吉賽兒的公寓。二十四歲的克洛德．歐萊特（Claude Ouelette），本來想要去驗光師學校，但沒多久便定期參加泰里奧的讀經聚會。接著，兩名法蘭辛的中學友人也開始出席：十八歲的「瑪麗莎」（Marise，大家提到時都只稱呼他瑪麗莎）和二十歲的荷賽．佩爾蒂埃（Josée Pelletier）。後來，在羅索湖（Lake Rosseau）附近舉辦的安息日會靈修講座中，泰里奧又招募了兩名年輕美麗的皈信者：加布里艾爾．拉瓦萊（Gabrielle Lavallée）和朋友尤蘭德．吉尼伯特（Yolande Guinnebert）。兩人都是法裔加拿大人，在度假區工作。

到了一九七七年秋天，泰里奧說服新信徒中輟大學學業，幫他的戒菸計畫準備素食餐點。不久後，他們每週都可以從安息日會信徒奉獻收到好幾千美元，而泰里奧全都拿來中飽私囊。他還沒對信徒下手，但已經開始無恥地剝削他們，把他們變成可以獲利的勞動力。

九月又有兩名新信徒：二十四歲的建築工人亞克．吉蓋爾（Jacques Giguere）和太太，二十三歲的蛋糕廠作業員瑪麗斯．格雷尼爾（Maryse Grenier）。亞克．吉蓋爾受到泰里奧提倡簡單生活的演講所吸引，但瑪麗斯不過是跟隨自己深愛的男人而已。她和大家保持距離，而且是唯一沒有淪為泰里奧發洩性慾對象的女性。

那年秋天，泰里奧信徒的父母發現孩子竟然沒有回大學上課，試圖從他們認為的危險詐欺男（猜得也沒錯）手上拯救自己的孩子。香塔爾的父母特別堅持，急著把女兒送去看心理學家，據說這樣可以協助香塔爾掙脫泰里奧的掌控。香塔爾同意待在精神科醫院一個月。

香塔爾突然出走，泰里奧記取的教訓是，他的信徒還不夠與世隔絕。他需要把信徒帶到絲毫不受外界影響的地方，尤其得遠離想要干涉他計畫的父母或心理治療師。這也說明了為何一九七七年十月他在波司的聖瑪麗（Sainte-Marie-de-Beauce）租了一間兩層樓房，離塞特福德礦城大概只有一小時車程。泰里奧接著開到那家精神醫院，哄騙香塔爾重回團體，最後大家終於一同開到集體生活的新家。

泰里奧的信徒開始散發傳單，招募更多人參加他們新總部舉辦的戒菸計畫，新總部也被稱為「健康生活診所」（Healthy Living Clinic）。法蘭辛的朋友荷賽與瑪麗莎也退學了，永久加入團體，就連泰里奧開始加強控制信徒，要求他們穿長袍、放棄俗世財產，他們依然繼續在泰里奧的診所工作。

雖然許多人喜愛大家庭氛圍，吉賽兒事後想想才覺得，所有新加入的女信徒似乎都愛上了她的男友。但是她提醒泰里奧，她才是真正睡在泰里奧身邊的人。一九七八年一月八日，泰里奧和吉賽兒結婚，他同意結婚是對傳統道德的最後讓步。但是他從來沒有認真看待自己的結婚誓言。沒有蜜月，團體活動如常舉行，和之前沒有兩樣。雖然吉賽兒沒多久便懷上第一胎，但泰里奧看來似乎對她不再感興趣。吉賽兒威脅要離開，肢體暴力隨之而來。

螞蟻山之子

波司聖瑪麗居民開始對泰里奧起了疑心。泰里奧沒有支付帳單。聖瑪麗居民對他首次對吉賽兒施暴一事毫不知情，但關於他言行舉止的傳聞越來越令人恐懼不安。一九七八年三月，他說服患有白血病的潔菈爾丁．加涅．奧克萊爾（Geraldine Gagné Auclair）出院，保證他可以用葡萄汁治癒白血病。但奇蹟並沒有發生，在他的照顧下，潔菈爾丁沒多久就過世了。泰里奧立刻用有利於他的角度向信徒說明這場悲劇，告訴他們他的親吻讓潔菈爾丁起死回生，但接著他明白了神的旨意，再次把潔菈爾丁的靈魂送入來生。

警察找不到理由判定泰里奧需要為潔菈爾丁的死負責，但持續監視他和健康生活診所。雖然泰里奧試圖向信徒講述自己的神祕親吻，展示自己的屬靈能力，但顯然這只展示了他如何誘拐一名患有絕症的脆弱女子，把她當成洩慾對象，毫不在乎對方命在旦夕。這只是開頭，日後還有許多冷血無情的例子。

基督復臨安息日會對他改變信仰的骯髒手段和他那群烏合信眾感到十分困擾，在一九七八年春天把他趕出組織。泰里奧開始意識到，他實在引起太多人注目，包含教會、警方、鎮上債主，以及信徒的直系親屬。如果要持續控制信徒，繼續默默營運，不引人注目，他就必須帶他們到更遙遠的地方孤立起來。因此，一九七八年六月五日一早，他調度了貨卡、巴士、汽車和露營車，載著十幾名信徒離開波司的聖瑪麗，永遠不回來。家長向警方通報小孩失蹤，相關當局搜尋泰里奧之前落腳的地方，只發現一只棄置的包包，裝著那幾名失蹤人口的身分證。

現在他們都到了荒郊野外，地點如此偏僻，根本不需要使用身分證；如此遙遠，泰里奧和虔誠信徒之間不再有人能插手。現在他只缺一樣東西：一個他能稱為自己這一派的宗教。遷徙後一個月，一九七八年七月六日，他宣布：不到短短幾個月，末日就會來臨——就在隔年二月十七日。這策略深具創意。基督復臨安息日會本就有多位先知預測過末日時間。事實上，安息日會的建立就是始於其中一次失敗的預言，那次失敗後來被稱為「大失望」（Great Disappointment），威廉・米勒（William Miller）牧師預測基督會在一八四三年三月二十一日到一八四四年三月二十一日之間復臨，但預測失敗，導致信徒大失所望。

泰里奧增添了新細節，把末日將近的預言變成他個人的版本。他聲稱冰雹風暴會摧毀大地，只有他和他的信徒蒙神憐憫，因為神把泰里奧當作個人先知。直到最後，他們會在荒野平安無事過日子。世界一崩毀，他們會開始重建社會。

泰里奧和信徒立刻開始找尋紮營的理想地點，等待末日來臨。他們在魁北克加斯佩半島（Gaspé Peninsula）的偏遠地區，距離聖喬格鎮（Saint-Jogues）十三哩半處找到一片林間空地。泰里奧的信徒遠離文明世界，幾乎不會和外界聯絡。他成功隔離了信徒（確保他對信徒的完全掌控，這一步不可或缺），稱呼新家為「永久的山」（Eternal Mountain）。賦予新家園神聖氛圍，泰里奧盤算的是為他的剝削打上神祕之光。

事實上，如預期般，光是在荒野存活就得做許多活。泰里奧的信徒多數為大學生年紀的青年，他叫信徒砍樹、掘井，從頭開始蓋木屋。他自己則一點力都沒出，因此除了他之外，每個人始終感到精疲力盡。信徒實在太高效多產，他為他們取了綽號：螞蟻山之子。這是另一招專門摧殘信徒意志的手段。把新兵操到幾乎不支倒地，隔天大清早再把他們叫醒重來一次，就是新兵訓練的重點，泰里奧在此之上新增經典洗腦手法的獨家荒野版本。面對末日鬼影幢幢，泰里奧說服大家配給糧需要省著吃，必要的睡眠也得略過。絕對的疲憊讓大家事事依賴他，就連最基本的決定也是。

大家在營地上蓋東西，泰里奧忙著依據聖經重新為大家命名。吉賽兒變成以斯帖（Esther）、索朗吉現在是拉結（Rachel）、加布里艾爾則是得撒（Thirsta），他自己是摩西（Moses），摩西是先知，也是頒賜律法者。但事情並沒有就此打住。每當大家能休息一下，他就叫大家說出自己的祕密，剝奪他們的隱私，暴露他們的弱點。他提醒大家，他們離開的社會充滿邪惡——整個社會就是邪惡，他們的家人也是邪惡。他重新為社區命名「死者世界」(World of the Dead)。

泰里奧建立自己版本的邪教領袖手段：讓信徒和社會對立；強迫信徒公開祕密、暴露弱點；實際孤立信徒；疏遠信徒與他們的家人。在這個完全依照他的言語運轉的世界，他意識到自己不需要繼續對吉賽兒忠誠。

但出於扭曲心理，他讓一切變成好像都是吉賽兒的主意。吉賽兒告訴他其他女生越來越寂寞，他開始和那些女生睡覺，好像他只是應吉賽兒要求而展露善意。沒多久，他就把自己與每位女人的連結說成神聖之事，並利用自己的先知權威和每一位都結了婚。他引述聖經的大衛王（King David）為例，大衛王有許多妻子，列名的只有八位，實際上有更多。或許他也想要和大衛王一樣擁有二十名孩子，沒過多久，除了瑪麗斯之外，他讓死者世界的所有女人都懷了孕，只有瑪麗斯對他毫不在意，依然對丈夫亞克忠誠。

林中生活艱困，泰里奧立刻捨棄之前身為安息日會教徒遵守的健康生活。他要求信徒長途跋涉，到最近的城鎮聖喬格買垃圾食物和酒。他開始掄起木棒毆打惹他生氣的人。泰里奧也開始形塑他的末日邪教中最使人疏離的原則之一：他的三名親生子被視為神選之人，但其他人的後代都該當成奴隸或動物對待，例如亞克和瑪麗斯的女兒耶薛（Jeziel）和剛出生的兒子撒母耳（Samuel）。

同一時間，那些憂心忡忡的父母團結起來，說服了相關當局逮捕泰里奧，開七小時車帶他回魁北克市（Quebec City）做心理評估。其中一位精神科醫生相信泰里奧患有思覺失調症。只不過，善於迂迴說服人的泰里奧成功讓醫療專業人士相信他人畜無害——這場景會再三重複。泰里奧因為某些犯罪行為、暴力行徑、暴行嫌疑而受到調查，但他都靠著聰明才智和友善態度，成功騙過法院官員。

事實上，此時泰里奧早已開始粗暴對待違抗命令的信徒，強迫他們為自己的錯誤行為書寫自我貶低的信件給他們的「爺爺」（Papy，法文）看，例如偷吃額外的食物。一封信寫道：「這個過錯導致我的肥胖。我不想當又胖又肥的僕人。站在像您這樣的男人身邊，那實在太醜陋了。」在其他情境下，這可能很滑稽，但對螞蟻山之子而言，如果他們踰越泰里奧時常獨斷的界線，遭受的懲罰經常來得很突然，而且比起行為，懲罰總是不成比例地重。一九七八年秋天，又發生一起類似的食物事件，泰里奧打斷瑪麗斯兩根肋骨。瑪麗斯太過飢餓，懷有身孕的她沒有事先徵得泰里奧的許可，就吃了兩塊鬆餅。當瑪麗斯考慮離開時，泰里奧脅迫亞克，要他砍下瑪麗斯一根腳趾當作懲罰。

摩西坐牢

一九七九年二月十七日，泰里奧預言世界末日的那天，日出日落，一天就這麼過了，世界逃過一劫，泰里奧說這是因為神理解時間的方式和人類的大不相同，自然有所混淆。他的信徒不知為何接受如此荒唐的解釋，或許因為此時他已經徹底摧毀信徒的意志，徹底到他隨便說什麼他們都接受。一九七九年四月，雅克．菲塞特逃離了一段時間，他也拒絕跟記者說任何泰里奧的壞話。

香塔爾的父母持續想拯救女兒，透過與女兒的短暫聯絡，終於弄明白邪教的位置，他們再一次拚了命想救出女兒。雅克．菲塞特的報紙訪問刊登後沒幾天（根據他的描述，他們聽起來像假日出門遠足的教會團體），警方帶走香塔爾和泰里奧做精神狀況評估。

泰里奧又成功騙過精神科醫生，評估結果是精神狀況良好。香塔爾也沒有引起醫生的疑心。但事實上，泰里奧繼續沉淪到極度殘暴的狀態。他最常針對的是撒母耳，瑪麗斯和亞克剛出生的兒子。依照泰里奧的指示，亞克會「處罰」哭泣的撒母耳，脫光兒子的衣服，把還是嬰兒的兒子丟在雪地裡滾動。

比起對信徒的限制，泰里奧允許自己更加自由地來去營地和城鎮，安排了第一段婚姻的兒子來荒郊野外看他。一九八一年三月，十二歲的羅克－希爾萬和十歲的法蘭索瓦終於來訪，泰里奧要信徒舉辦歡迎派對。派對上，患有心理疾病的年輕新信徒蓋・維爾（Guy Veer）無法忍受兩歲大的撒母耳嚎啕大哭，狠狠毆打還是幼兒的撒母耳。撒母耳後來無法排尿，泰里奧回應當務之急的做法並不是送小孩到醫院，而是用刮鬍刀片幫撒母耳的生殖器動手術，強迫撒母耳喝下兩盎司純乙醇，姑且當作術前麻醉。當晚，撒母耳離開人世，他媽媽瑪麗斯早就被毆打到精神受創，飢餓和睡眠不足耗盡她的精力，聽到兒子的死訊時根本沒有反應，只是繼續做家事。泰里奧吩咐信徒，只要有人問起，一律回答撒母耳是被馬踐踏而死，這個理由反映出男嬰受到多大的傷害。

其餘信徒也同樣麻木。沒有人出聲責備。泰里奧決定找替罪羔羊。一九八一年秋天，泰里奧公開審判蓋・維爾。雖然大家都認為維爾無罪，畢竟他患有心理疾病，泰里奧還是說服了大家維爾犯了罪，應該接受閹割作為處罰。他要維爾躺在餐桌上，接著用刮鬍刀片割除維爾的睪丸。手術期間，維爾積極配合，甚至相信泰里奧說的，閹割會減輕頻繁的頭痛。但這樣還是不能滿足泰里奧，又或許他感受到謀殺孩童應該接受更加嚴厲的懲罰。因此他開始模擬蓋．維爾的死刑，恐怖活動嚇得維爾逃跑，向警方供認撒母耳的死亡。雖然維爾的心理狀態不足以完整說明泰里奧就是害撒母耳死亡的凶手，但也給予警方足夠的資訊，逮捕該名邪教領袖。一九八一年十二月十九日，泰里奧被控應為撒母耳的死亡負起刑事責任，判處兩年有期徒刑。

但就算泰里奧被強制分隔，也無法動搖他對信徒的掌控。信徒離開荒野，搬到南方十六哩的新卡來爾（New Carlisle），就為了離監獄進一點。教團裡的關係益發複雜，許多女信徒現在都生了泰里奧的孩子，在他服刑期間又有更多女性懷孕生子。幾年前他替索朗吉．博伊拉和同為信徒的克洛德．歐萊特主持了婚禮。儘管如此，他還是把索朗吉變成後宮一員，從一九八〇到八六年間，索朗吉為他生了三個孩子。泰里奧身陷囹圄，索朗吉離開了，但吉賽兒誘騙她回來，保證泰里奧一出獄，索朗吉就會成為正宮。她決定留下，最終賠上了性命。

林中血祭

一九八四年二月，泰里奧服刑完，不到兩個月就說服忠誠信徒拋棄都市生活，和他一起重回樹林深處，這次搬到位於永久的山以西九百哩、多倫多東北邊一百哩處的英語區安大略省，距離僅有五百人口的金芒特（Kinmount）十三哩的一塊林地，就在人口只有兩百五十人的柏恩特河小鎮（Burnt River）郊外。孤立信徒的不只是荒郊野外，還有語言隔閡。所有信徒都是法裔加拿大人，只有泰里奧可以說一口流利英語。他們從未這麼依賴他。

這次搬家讓信徒住得越來越偏遠，開啟一段更為暴力的時期。泰里奧採取許多邪教全權控制的經典策略，例如立下看似隨意的規則，迫使螞蟻山之子幾乎一直處在恐懼之中，害怕無法達到他的標準。恐怖氛圍會被性愛打斷，大家可以喘口氣，即便只因為性愛似乎讓泰里奧放鬆。

重回荒郊野外，泰里奧再次袖手旁觀，看著信徒建造新房舍和教團需要的其他設施。螞蟻山之子此時已經成為技巧純熟的木工，他們蓋的木屋、穀倉和儲藏小屋粗糙堅固，簡直就是拓荒時期經典的鄉村社區。但是教團人數越來越多，因為女信徒一直為泰里奧生更多孩子。（加布里艾爾憶道：「羅克總是告訴我們必須那麼做，因為他就像聖經的亞伯拉罕，他必須妻妾成群，兒孫滿堂，他的血脈才能延續。」）食物不足的話，泰里奧會叫信徒去金芒特商店偷東西，一九八五年一月，偷竊行為告終，大家在店內行竊被捕。

即便環境如此偏僻，泰里奧依然與外界保持聯絡，足以注意到艾力克斯．喬瑟夫（Alex Joseph）的新聞，這名人見人愛的猶他州比格沃特市（Big Water）自由黨市長公開和八位妻子生活，某種形式的重婚。泰里奧似乎受到了激勵，八〇年代至少前往比格沃特市拜訪了喬瑟夫三次。這幾次拜訪，喬瑟夫辦了輕鬆愉快的歡迎儀式，介紹頭戴金冠的泰里奧，稱呼他為以色列人的王（King of the Israelites）。泰里奧回到加拿大荒郊，開始穿長袍、戴金冠，舉手投足彷彿自己就是王族。

儘管現在多了新的傲慢，他依然得餵養自己的孩子。他想到新點子：在路邊販售農產品，由信徒販售新鮮現採農作物。蔬果攤生意出乎意料地好，泰里奧擴展生意，多開一家麵包店，店名就是以他那群刻苦耐勞的信徒命名：「螞蟻山之子」。他借了烘焙設備，教團有好幾年都能真的賺到錢。

但就算獲得這些世俗成就，泰里奧還是與自己內心的魔鬼搏鬥，同時造成他人的痛苦和敗壞。他又開始喝酒。他一次帶兩名或兩名以上的妻子上床，要妻子互相競爭誰先達到高潮。他會主持好幾小時的戰鬥，信徒必須打得你死我活，每擊中一拳，他會給予獎勵積分。他也會精心規劃集體性狂歡聚會，尿在每位參與的信徒嘴中。他會鞭打信徒，或者用斧頭握柄痛毆他們。某次他為雅克．菲塞特施行了成人割禮，或至少是宣判他要接受割禮，命令信徒切下雅克陰莖尖端。

根據信徒寫的懺悔信，顯然每個人都恐懼於泰里奧的力量，不論是屬靈還是實體，又或者兩者合併。索朗吉又名拉結，寫信給他（「我的主人摩西您好」），為了一陣「透過您行使」的憤怒而道歉，如同索朗吉的說法，好似泰里奧只在「強大神靈」附身時才會虐待他人，而這可憐的男人只不過是天上報仇的器皿。索朗吉寫道：「我真的相信您所做的不是出自於您，而是來自更高的存在。」事情扭曲到了極點，甚至連泰里奧朝索朗吉投擲刀子、拿步槍射擊，對「奶奶」（Mamy，對索朗吉的稱呼）造成不明傷害，每次索朗吉都會向泰里奧道歉。

對理性的一般人而言，這些毫無緣由的虐待暴行實在難以理解。為什麼沒有信徒離開？因為他把信徒操到徹底枯竭，根本無力思考逃跑？還是信徒害怕身體會立刻受到傷害？泰里奧稱自己是神的先知，如果信徒背棄他，主會擊殺所有人。聽了這些話語，信徒依然相信他嗎？又或者信徒察覺暴行的步調和規模逐漸上升，但又無能為力，只能祈禱泰里奧再次引起相關當局的關注，而這次相關當局終於能干涉，永遠解決他。

信徒向地方政府申報小撒母耳的死亡，此舉確實引起社工關注，開始找尋泰里奧虐童的證據。但到了一九八五年一月二十六日，社工的懷疑才加深。那天，加布里艾爾把自己的男寶寶留在寒冷的野外過夜，比起把孩子交給泰里奧接受懲罰，加布里艾爾寧願讓寶寶暴露在加拿大隆冬的天氣之下。他的寶寶凍死了，儘管致命虐待如此明顯，驗屍官寫下的死因只是嬰兒猝死症候群（sudden infant death syndrome）。

第二個嬰兒死亡了，社工懷疑自己的努力是否只是徒勞。他們諮詢非營利組織精神虐待協會（Council on Mind Abuse），協會確認泰里奧領導的確實是邪教組織。面對官方監視，泰里奧看似舉手投降，但在同年三月，社工訪視邪教裡的孩子時，他大發雷霆。到了六月，泰里奧酒越喝越多，他告訴忠誠信徒，末日終於來臨。他會用業餘無線電對飛過上空的飛機發送假的求救訊號。

來訪的社工把孩童消極被動的態度解釋為可能遭受虐待的證據。但不論社工認為這一切多麼明顯，都沒辦法收集到足夠明確的證據，保證可以一次移動所有人。直到瑪麗斯終於在一九八五年十月帶著尚在稚齡的兒子耶薛和多馬（Thomas）一起離開，後來社工訪問瑪麗斯多次，才聽到持續的可怕性虐待與身體虐待。泰里奧會做出令人髮指的行為，例如把孩子的衣服釘在樹幹上，讓孩子懸掛在樹上，接著催促信徒拿石頭丟孩子當作懲罰，然後在任何人真的丟出石頭的前一刻喊停。泰里奧生氣時會亂摔嬰兒，或把嬰兒丟進湖中。嬰兒哭泣時，泰里奧會壓碎他們的手指，讓他們安靜。他會在火堆上方同時舉起兩名寶寶，強迫信徒祈求他饒過寶寶。他從來沒有真的把孩子丟進火裡燒，但他的暴力行徑來得無常，而且所有信徒都知道他可以極度殘忍。他要孩子在一旁看著大人性狂歡，鼓勵自己的孩子虐待動物。一九八五年十二月六日，社工在安大略省警察局（Ontario Provincial Police）協助下，終於突襲泰里奧的社區，帶走全體十三名孩童。

或許有人會認為相關當局終於取得足夠證據，能把泰里奧關進監獄。可惜，他再次利用了兒童心理學家靈敏易感的性格。他相信心理學家同為法裔加拿大人，會明白他在多數人都說英語的政府機構手中受到不公的對待。法院指派的心理學家不僅認為泰里奧無罪，還表達他相當欽佩泰里奧費盡心力讓孩子重返自然。法院文件顯示泰里奧被形容成偏見社會下的受害者，建議法院讓孩子重返他們的大家庭，所幸法官申飭了所有相關人士，在一九八七年十月二十七日，裁決所有孩童都應永久遠離該邪教。但就算泰里奧在一九八六年八月向警方坦承自己傷害了太太與孩子，助理檢控官艾力克斯·史密斯（Alex Smith）還是決定不起訴他。泰里奧還是自由之身，真是令人難以想像。接下來兩年，信徒又為他生下九名孩子。所有孩子在出生幾天內都被裁定交由國家監護。

小孩不在了，螞蟻山之子的大人行為逐漸墮落，只剩性狂歡和暴行互相循環，泰里奧變得更加危險莫測。他運用精神和生理威脅讓信徒繼續對他唯命是從，同時他也利用這樣的掌控施加更多難以言喻的虐待行為。他用焊槍灼燒多名信徒的肌膚。他毆打一名孕妻的肚子，打得太用力，太太隨後流產。他看似趁興拿起鉗子，隨手拔除信徒的牙齒。他毆打信徒、刺傷信徒、射擊信徒。在不同情境下，泰里奧強迫克洛德．歐萊特吃下人糞和死老鼠。恐怖循環似乎還包括泰里奧大哭崩潰，尋求原諒，乞求神收手，不要再命令他做出如此敗壞的行為。信徒看來似乎從沒學到教訓，每次都原諒泰里奧。信徒堅信泰里奧要嘛本身就是天主，不然就是為神說話的代言人。

到了一九八八年，睡眠不足、糧食短缺、時時警戒與恐懼被虐待的生活型態，造成的身體殘害開始反映在信徒糟糕的健康狀態上。但是泰里奧毫無憐憫之情。他注意到克洛德．歐萊特一些微不足道的踰矩行為，反應是處罰歐萊特，拿橡皮筋纏住其睪丸直至腫脹、發紅。歐萊特抱怨疼痛時，泰里奧切開陰囊，移除一粒睪丸，再用炙熱的鐵燒灼傷口。

索朗吉尤其無法承擔龐大的壓力。或許有人認為泰里奧會同情那些胃部劇痛的人，然而，他再次增強了酷刑的程度，即興發明了醫療程序，宣布他要幫索朗吉的肝動手術。滿屋子人嚇到悶聲不吭，每位信徒都怕一出聲就吸引了他的注意。

他先混合了糖蜜、油和水，幫索朗吉灌腸，接著切開她的胃，徒手扯出腸子，弄得腹腔到處是胃酸，最終導致索朗吉喪命。沒有人出手阻止。途中他決定插管，強迫數名忠誠信徒把管子插入索朗吉喉嚨。他交給加布里艾爾針線，叫他縫合索朗吉。索朗吉整晚在地上扭動，直到隔天早晨不可避免與世長辭。

光看就知道泰里奧被無法控制的瘋狂附身。埋葬索朗吉之後，他開始確信自己能讓索朗吉死而復生。他強迫信徒挖出索朗吉，切開她的頭顱，要信徒和他一起射精到正在腐爛的腦上。接著他切下幾根肋骨，像項鍊一樣戴著，深信有天他可以像亞當一樣，透過聖經描述的手法重造索朗吉。為了他們自身，也是為了索朗吉，信徒焚化了索朗吉，希望這樣總算能阻止泰里奧的褻瀆行為。

救命之刀

一九八八年過去了，外界對索朗吉的失蹤沒有任何官方反應。社工注意到索朗吉不在，收到令人不盡滿意的回覆，警方依舊沒有展開正式調查。一直等到一九八九年夏天，泰里奧砍斷加布里艾爾的手臂，這般可怕行徑總算讓安大略省警察局再次有所行動。

自從加布里艾爾成為信徒，泰里奧似乎都會特別針對他。有次泰里奧拔除了加布里艾爾的八顆牙齒，打算治好他的牙痛。但一九八九年七月二十九日，他犯下的虐待行為總算讓他親眼看到自己的邪教灰飛煙滅。泰里奧採取另一種過度殘暴的手法，打算治療加布里艾爾抱怨的手指僵硬，他拿獵刀刺穿加布里艾爾的右手，釘在餐桌上。加布里艾爾就這樣被晾在那邊一個多小時，大量出血，還得拚命讓自己不要因為疼痛而暈厥，害怕自己就像失去意識的索朗吉一樣，立刻被泰里奧殺死。

最後泰里奧總算還他自由——用美工刀切斷他的上手臂。

加布里艾爾活了下來，逃跑到女性庇護所，但是雅克・菲塞特（那名在泰里奧堅持之下，陰莖尖端遭切除的男子）找到他，成功說服他回到營地。一九八九年八月十一日晚上，泰里奧要信徒壓制住加布里艾爾，用乙炔焊槍加熱傳動軸，再去燙加布里艾爾的殘肢。三天後，加布里艾爾再次逃跑，重新回到那間女性庇護所，終於被送往醫院。

警方詢問加布里艾爾的傷，但他無法坦承泰里奧罪行的真相，反而說自己出車禍，卡住無法動彈，泰里奧才切斷他的手臂。

他的說詞太不連貫，警方決定逮捕泰里奧。八月十九日，警方來到營地搜查，只發現人去樓空。加布里艾爾的截肢事件讓大多數信徒忍無可忍，終於鼓起勇氣永遠離開泰里奧。而泰里奧則和雅克、妮可及香塔爾一起逃亡。他們成功躲避警方追捕將近兩個月，住在營地附近隱蔽在林間的臨時小屋。一九八九年十月六日，泰里奧終於遭到逮捕。在泰里奧被扣押以後，吉賽兒才敢坦承索朗吉死亡的全部真相，相關當局因此得以正式控告泰里奧犯下殺人罪。

十二月十八日，法官下令羅克・泰里奧要依二級謀殺罪受審。即使面對如此沉重的指控，泰里奧還是可以騙倒精神科醫生，讓醫生認為他是好人。一九九一年十二月的法院文件顯示，醫生描述他為「聰明、好奇、敏感的男人」，同時對於醫學「知之甚廣」，醫師認為泰里奧的問題源自他是「在平凡世界生活的非凡人類」。

但法院可不這麼認為，一九九三年一月十八日，泰里奧被判十年有期徒刑，表現良好或許可以假釋。歷經十年的折磨與殘害，活下來的多數信徒就像加布里艾爾和吉賽兒一樣，成功逃離泰里奧的掌控。加布里艾爾接受必要的治療，開始認出泰里奧操控受害者的危險心理學手段。之後他出版了回憶錄《羊盟》（暫譯，法文版原名：*L'alliance de la brebis*；英文版書名：*The Alliance of the Sheep*）。

泰里奧的大兒子也出版了一本回憶錄，描述他們經歷的噩夢。但大多數泰里奧的受害者都隱姓埋名，尋求親朋好友的庇護，掙扎著想過好自己的生活。許多人接受安大略犯罪受害補償會（Criminal Injuries Compensation Board）的幫助，補償會提供暴力犯罪受害者金錢援助。

但是，依然有幾位泰里奧信徒永遠無法逃離他的掌控。法蘭辛、香塔爾和妮可為了離泰里奧近一點，在他服刑的監獄附近租了房，繼續探訪、為他生孩子。她們甚至一起開了麵包店，利用螞蟻山之子時期錘鍊的技能。這次創業的生意很好，直到客人發現這間可愛小麵包店的女老闆都是瘋子殺人犯的太太，而且她們都希望先生獲釋後可以重回他身邊。

雖然泰里奧可以把精神科醫生和多位信徒耍得團團轉，但有一群人絲毫不為所動：他的獄友。一九九三年四月，為了他自身的安全考量，泰里奧被轉到金斯頓監獄（Kingston Penitentiary）——就算身處鐵石心腸的殺人犯之中，他可怕的罪行依然超乎尋常地令人憎惡和冷血。

他的魅力攻勢沒法拯救他自己。二〇一一年二月二十三日，監獄監視器錄到泰里奧的獄友馬修・傑拉德・麥克唐納（Matthew Gerrard MacDonald）用偷渡的自製刀具刺進泰里奧的脖子，接著拿著凶器走向守衛，說道：「那坨屎倒在裡面。刀子在這。我幹掉他了。」

羅克・泰里奧的一生結束了，但他的案子依然令人心生不安，他犯下的暴行太過極端，他可以行使的權力太過迷人，有時候逃跑的人還會回心轉意，自己決定回到他身邊。很難知道泰里奧死後究竟對生還者還有哪些影響，以及他的信徒或在他的殘酷暴行下長大的人，是否有天真能逃離他所帶來的恐懼。

Chapter 6

誇大狂

大衛・柯瑞許和大衛教派分支

大衛・柯瑞許（David Koresh）預言自己會在一片榮耀火焰中死亡，他帶著邪教大衛教派分支和美國聯邦政府決一死戰，每天媒體都在報導這場血淋淋的對決。他的一小幫信徒抵禦了整整五十一天，FBI包圍他們位在德克薩斯州威科（Waco）外的據點。一九九三年四月十九日，最終對決張力滿滿——七十六名大衛教派信徒死亡，其中二十一名未滿十六歲。「威科」成為生存主義者（survivalist）圈的熱門字，也讓準備者（prepper）團體相信，如果政府準備那麼多武力攻擊人民，人民必須準備更多火力，才能頂住或制伏那樣的攻擊。在卡梅爾山（Mount Carmel）據點，柯瑞許大量囤積機關槍、手榴彈等武器。他也堂而皇之採行多妻制：認為據點裡的每一位女性都是他的妻子，就連十一歲大的女孩也是。但如同許多邪教領袖自稱神的使者，他似乎也一樣寧願死去，也不願放棄自稱的神權，不願承認自己只是凡人。總的來說，他的末日預言成真了，至少是關於他自己的那部分——據點在身邊起火時，他死於頭部槍擊，下手的可能是信徒。

一百五十年預言失準

看到魅力型領袖保證末日將近，身旁圍繞狂熱追隨的信徒時，千萬當心。這提議非常危險，死亡成了祝福，變成應該迎接的命運，甚至鼓勵信徒赴死，因為對忠實信徒而言，死亡不是結束，而是踏進專屬樂園的第一步。這是據點內多數人的心態。卡梅爾山據點位在德州阿克斯特爾（Axtell），距威科十二哩遠。一九九三年四月血腥大混戰後，即便到了現在，還是有倖存信徒依然認為領袖柯瑞許的預言萬分神聖。

大衛教派信徒在卡梅爾山過著集體生活。基督復臨安息日會在一八六三年成立，而大衛教派是從安息日會三度分離出來的支派。安息日會相信基督即將復臨，這也是為何名稱裡有復臨二字。但就連這個根本信念也是源自一則失準的預言：威廉·米勒牧師預言基督會在一八四四年十月二十二日復臨。但復臨運動和預言日期都失準了，後來被稱為「大失望」。南北戰爭前，社會改革和宗教動盪為期多年，米勒派的餘民齊心協力修改信仰教條，重新說明十月預言的那天其實開啟了「調查審判」（investigative judgment）時期，這段時期上帝和基督會審判忠實信徒的信仰和行為。

不過，因大失望而建立的新信仰無法指出這段調查時期何時會結束。忠實信徒十分重視上帝啟示，這不光是信仰的原則，他們還能明確指出近代歷史的啟示。信徒也相信即將來臨的決定性時刻，而透過地上的信仰和善行，信徒會被領到天上會幕。雖然審判日沒有明確的日期，但大家總覺得即將來臨，深信末日將近，為信仰增添了一些緊迫。

有血有肉的神聖使者與即將來臨的最終審判，這兩個信仰讓信徒特別容易受到柯瑞許等魅力型領袖的影響。一般來說，基督復臨安息日會和分支派別都樂於接受先知，而且如果有人聲稱上帝為自己受膏，信徒基於信仰都覺得至少必須聽對方把話說完。事實上，目前卡梅爾山據點依然是大衛教派的合法財產，據點入口矗立了兩面紀念牆，高度及腰，由一塊塊刻有姓名的石板組成，總共八十二名死者（一九九三年二月二十八日首次突襲，六名信徒喪命，接著開啟長時間包圍，其餘七十六名在四月十九日過世）。牆上立了一塊墓碑刻道：「復臨運動七牧人」。七牧人代表各個啟示，從源頭的懷愛倫（Ellen G. White，在南北戰期間創立基督復臨安息日運動）開始，一直列到維農・韋恩・豪威爾（Vernon Wayne Howell，大衛・柯瑞許的本名，柯瑞許創立了「基督復臨安息日運動大衛分支教派」）。

威科慘案是美國史上死傷最慘烈的ＦＢＩ對峙，乘載著一百五十年的預言、期盼和啟示。魅力型領袖可以利用這些好材料培養狂熱的信眾。柯瑞許利用人性本有、同時也是信仰包含的慾望，一步一步轉變愛起內鬨的分派餘民，成為一群翹首期盼末日來臨的狂熱信徒。

就在突襲前，近百名柯瑞許的學生住在卡梅爾山。這團體涵蓋多種族、多民族：除了四十五名黑人，還有英國人、澳洲人及加拿大人。根據當時對突襲的描述，該團體名為大衛教派分支，但到了一九九三年，他們改口自稱「七印學生」（Students of the Seven Seals），他們相信柯瑞許已經蒙神膏立，得以解釋啟示錄記載的七印。他們全副心神都放在柯瑞許應許的天國，願意忽視柯瑞許和還沒進入青春期的少女性交，他殘暴虐待這些性交下誕生的嬰兒，也沒人作聲。柯瑞許做的所有卑劣行為，他們都視而不見，因為他們相信只有柯瑞許可以揭開七印，迎來末日。他們對他不像樣的行為視若無睹，他們認為，反正末日近了。對柯瑞許和他的八十一名信徒而言，末日將以一團烈焰降臨。

棄兒見異象

抵擋美國政府近兩個月、惡名昭彰的男人，在一九五九年八月十七日德州休士頓誕生，媽媽是十五歲的邦妮・蘇・克拉克（Bonnie Sue Clark），爸爸是十九歲的鮑比・韋恩・豪威爾（Bobby Wayne Howell）。維農還沒出生，鮑比就已經和另一名女人遠走高飛了。邦妮以缺席的爸爸為兒子命名，或許還希望鮑比有天會回來。

邦妮是冷漠的家長，維農才四歲，她就隻身前往加州，把維農丟給外婆爾琳（Earline）和患有心理疾病的阿姨貝芙利（Beverly）照顧。隔了三年，邦妮才回家，還帶著暴力酒鬼洛伊・哈德曼（Roy Haldeman）一起回來。虐待和遺棄這兩個極端的家長行徑，對維農將來心理的健全發展不是好徵兆。就心理層面而言，在最早期的塑形年齡階段就發生嚴重的忽視對待，孩子會沒有健康的典範學習如何建立和善連結，也沒有親密模式可以參考，甚至沒有重要基本親子連結的記憶，這些孩子長大會發現自己難以和他人建立連結。這些孩子會操控他人，首先激起他人情感，再重現自己承受過的遺棄，看來是要懲罰那些居然蠢到感受到愛的人。尤其是對那些極度渴望情感的人而言，身體虐待會造成創傷後壓力症狀，一輩子糾纏不清。

經歷嚴重的童年創傷，長大的維農創立邪教，說服許多人他是上帝化身，最終引領大家在烈焰中死亡，原因為何？他肯定在早期就出現顯而易見、令人擔憂的跡象。在基督復臨安息日會的達拉斯中學（Dallas Junior Academy），小維農過得非常痛苦。他被診斷出讀寫障礙，從一般教室轉到特殊指導教室。再一次，維農又失去機會，被迫無法與他人建立有意義的連結。這些重複上演的放逐，非常有可能導致他開始用扭曲的方式思考，認為自己一直被單獨挑出來針對。或許是因為他被驅逐出人群，所以才想要支配人群。

可以明白為什麼他會對全權掌控抱有幻想。等他終於可以回到遊戲場和同學一起玩，大家總是捉弄他，欺負他。維農也指控，在八歲那年，一群年紀比較大的男孩打算強暴他。而在媽媽回來之後，儘管維農強力反對，她還是帶著維農搬到達拉斯的新家。在達拉斯，維農加入了媽媽的教會：基督復臨安息日會。討厭被針對的青年開始練舉重，對自己的身體越來越自豪。

基督復臨安息日會教徒非常看重先知講道之能，安息日會創立者懷愛倫也寫下多頁文字詮釋自己的夢兆，維農深受吸引。或許因為讀寫障礙，維農幾乎無法有好的傳統學術表現，但他決定盡全力記下聖經的大量章節，特別是啟示錄富有詩意和神祕氣息的預言。啟示錄預言充滿象徵、數字，以及其他聖經經句和古代先知等參照。維農特別喜愛這種黑暗的內容，指涉巨大能力攻擊充滿罪孽的世界。

放學後，他會花好幾個小時研讀聖經和禱告。說不定他把一些熱情分給學校課業就能讀完高中？誰也說不準。許多有讀寫障礙的孩子在班上成績總是吊車尾，但這些不尋常的宗教狂熱行為還伴隨著證據，顯示維農深受神經和心理問題所苦。一九九三年三月四日，圍攻期間，柯瑞許和FBI談判時說：「人們……說上帝對他們說話。嗯，我也有同樣的問題，你知道的，但我的問題從小就出現了。」接著他繼續描述一九七〇年左右時，他十二歲的體驗：「有天晚上……我直視著，嗯哼，北方宇宙，並且祈求。有顆非常美麗的星星，像是宇宙中一團爆炸。這顆星星……成了我的一部分，伴著我長大。我跟媽媽提起這顆星星，她說，喔……去睡覺，星星就會離開。」

這些童年體驗可能是狂喜幻覺，成因很可能是顳葉癲癇（temporal lobe seizure）。神經內科醫生奧立佛・薩克斯（Oliver Sacks）參加世界科學節（World Science Festival），接受記者約翰・哈肯柏利（John Hockenberry）訪問，形容這些幻覺是「突然一股至福或極度歡愉傳來，覺得自己被送到天堂……狂喜幻覺幾乎總是有些神祕、宗教、性愛的傾向」。[1]這只是猜想，但包含薩克斯在內的許多神經內科醫生都寫過狂喜癲癇在宗教中扮演的角色。他們的理論是，某些影響力最深遠的宗教人士，不論是聖經的保羅，還是聖女貞德（Joan of Arc），又或者是早期基督復臨安息日會的懷愛倫，都很可能是因為患有顳葉癲癇而看見異象。這並不是說宗教體驗都不能相信，也不是要把維農和歷史或聖經人物相提並論，只是提供現代醫學推測——顳葉癲癇和許多人聲稱的神祕體驗有關聯。

性、啟示、搖滾樂

維農總是說自己在一九七〇年代早期輟學，沒有讀完九年級，雖然位在達拉斯外圍的加蘭中學（Garland High School）紀錄顯示他唸完十年級，後來兩年偶爾還會回學校上課。他打零工，也做雜活。除了努力背聖經經句，也一直在練吉他，參加當地樂團徵選。他的中學朋友賴瑞・麥克唐納（Larry MacDonald）記得大家就像一般青少年，平日樂趣就是玩音樂、喝啤酒和吸毒。但如果和女生相處，「小維有點安靜，但如果他真的想要，他可以切成開機狀態，就像登台表演。小維可以變得魅力十足。」

維農的吉他技巧日益進步，開始參加一些小型演出，更加引人注目。他在十八歲有了初次性體驗，對象是青少女琳達（Linda），後來他說到琳達時都用「未成年美少女」來形容——這在維農的性生活中是特別明顯的主軸。他說琳達非常漂亮，後來告訴ＦＢＩ談判專家，琳達看起來比實際年齡成熟。但維農因為狂熱的宗教信仰，內心很衝突，還無法接受自己發生性行為，所以離琳達遠遠的。只不過，他肯定沒有離多遠，因為琳達找到了他，說自己懷孕（雖然維農主張自己不孕），但墮胎了。這場騷動好似重燃兩人的情感，維農最後和琳達一起住在琳達爸爸家。

在這段同居時期，維農的行為揭露了他個人與宗教倫理都存在矛盾的一面。一方面，對於墮胎或沒有結婚卻和性伴侶同居，他絲毫沒有感到不安。相反來說，基於宗教理由，這對情侶拒絕使用保險套，這也驗證了維農其實沒有不孕，因為琳達又懷孕了。此時，琳達爸爸展示了同樣令人困惑的優先順序：他把維農踢出家門。

維農思索：為什麼琳達爸爸允許青少年女兒發生婚前性行為，卻在女兒一懷孕時，就禁止維農再見他女兒呢？這次，琳達把小孩生下來了，但她爸爸不讓維農參與小寶寶的人生。除此之外，就沒有琳達的消息了。看來她爸爸成功切斷了她和未來的大衛・柯瑞許的連結。維農沒有任何插手的餘地，其中一個原因很可能是琳達真的未成年。

我們無法知道這名二十歲的年輕爸爸是否真的想成為家庭的一份子，但面對喪失所有探視權，維農的反應是更加投入宗教，只要遇到牧師，他就會針對自己的經文詮釋徵詢意見。他的媽媽邦妮為憶道：「他針對啟示錄和創世紀提及的大胸部女人製作了圖表，可惜不太成功。」

維農上德州泰勒（Tyler）的基督復臨安息日教會，開始對牧師的女兒有意思，不久就改變原本隨意的聖經詮釋，朝另一方向發展。他向上帝禱告，訴說自己喜歡十五歲的珊蒂・柏林（Sandy Berlin），接著告訴珊蒂的父親L・哈特利・柏林（L. Hartley Berlin），自己禱告結束，一抬頭，就看到聖經翻到以賽亞書第三十四章十六節：「這都無一缺少，無一沒有伴偶。」維農認為這段經文是上帝給的個人訊息，告訴維農他應該和這名年輕女子在一起。或者如他所說：「上帝告訴我，我可以擁有你的女兒。」

柏林牧師不同意他對聖經的詮釋，維農繼續堅持，隨後就被趕出教會了。這是維農第一次宣稱上帝告訴他應該和哪位特定女子在一起，但不會是最後一次。儘管第一次出師不利，但想必他明白自己找到了重要戰略。

維農依自己可疑的動機解讀經文，初嚐了操控的滋味，同時放縱自己的另一熱情。被趕出教會之後，在一九八〇年左右，二十一歲的維農搬到洛杉磯發展音樂職涯。他甚至錄了幾首歌，多虧珊蒂．柏林把這幾首歌拿給一家唱片公司，最終才能在一九九四年發行。維農一直與珊蒂保持聯絡，認為珊蒂和他長期交往。他不是第一個經歷性拒絕後錄製專輯回應的音樂人，也不是第一位想透過發專輯滿足對他人目光渴望的邪教領袖。但就像查爾斯．曼森，維農發現市場不太需要嗓音單薄、自信過剩的歌手。

上帝要維農娶青少女

因為參與基督復臨安息日會，維農理解安息日會信徒一直以來都嚮往有血肉之軀的先知。他讀了啟示錄，知道如果末日近了（同時新一輪千禧年即將到來，許多人都相信末日即將來臨），地球上絕對會出現一位先知說上帝的話語。媽媽邦妮告訴維農，她有聽過這號人物。因此在一九八一年，維農啟程前往德州威科十二哩外的卡梅爾山，尋找大衛教派的據點。

大衛教派分支仍相信安息日會的許多概念，透過一系列先知，可以追溯系譜直接到懷愛倫。一九二九年，維克多．霍特夫（Victor Houteff）帶著一群人離開基督復臨安息日會，成立安息日會大衛教派（Davidian Seventh-day Adventists），相信聖經中提到的大衛王後代會回來統治巴勒斯坦（Palestine）。一九三五年，他們在卡梅爾山建立據點。

接著，在一九六〇年代，安息日會大衛教派再次分裂，新一派繼承卡梅爾山據點的人自稱大衛教派分支。最初領導者為班傑明．羅登（Benjamin Roden），這群人相信末日近了，聖經描述的末日序曲已經一一發生。他們懷抱對永生的期待生活，強烈恐懼在地上的一切都將終結。他們是福音教派，打算建立自己的會眾，拯救靈魂。

羅登過世，遺孀洛伊絲（Lois）接管，宣布她看到異象——有關女性聖靈（Holy Spirit）的預言。正值洛伊絲掌權時期，維農抵達卡梅爾山。一九八一年維農初抵卡梅爾山，聽了六十一歲的洛伊絲講道。洛伊絲的訊息和他自己的互相呼應，接著他迅速成為左右手，跟著洛伊絲四處巡迴招募信徒。有人認為他們結婚了，大家普遍認為他們倆在一起。

維農告訴洛伊絲，聖靈會照耀著他們，維農會讓洛伊絲懷上上帝揀選之人（the Chosen One）。維農引述聖經中年邁父母得子的前例：施洗約翰（John the Baptist）之父九十九歲、其母八十八歲才生下他；亞伯拉罕的妻子撒拉（Sarah）九十歲才生下以撒（Isaac）。這是維農第一次展現自戀，想像自己蒙上帝揀選，要讓六十幾歲的女人懷孕——對方既是女性聖靈的化身，也是父神（God the Father）的母系相對存在。

一九八三年，洛伊絲指定維農接班，維農開始對教友布道，最初在逾越節（Passover）講道，他稱之為「蛇的根」（The Serpent's Root）。早在中學，他就開始磨練記住聖經經文的技巧，布道時一邊引述經文，一邊詮釋，完全得心應手。他可以即席布道，經文隨手捻來，令其他大衛教派分支信徒大為折服，他們知道維農的正式教育只唸到初中。他們認為，這個男人沒受過什麼教育，卻能即興布道連續好幾小時，引聖經據典毫不費力，證明他被神觸碰著。大衛教派分支不在乎維農和洛伊絲的年齡差距。但維農一獲得信眾的注意力，就開始搶奪洛伊絲的權力，說服大家洛伊絲已經失去了「預言之靈」（spirit of prophecy），甚至燒毀一部分出版業建築，阻止洛伊絲的雜誌《SHEkinah》[11]繼續印行，這本雜誌探索神聖的女性本質。

到了一九八四年，維農已經確認教派中有自己的追隨者，切斷了與洛伊絲的關係，下一步則顯露自己究竟累積了多少權力。此時也是轉捩點，他開始偏離合理的行為標準，相信自己的先知地位逐漸高昇，可以開始掌控虔誠信徒。他在二十四歲宣布，神要他娶十四歲大的瑞秋．瓊斯（Rachel Jones），瑞秋的父母都是大衛教派分支信徒。

在德州，十四歲需要家長同意才可以結婚，否則就是違法，但瑞秋雙親相信維農和他講述的訊息。維農也知道如何恰到好處地傳遞訊息，適度懷疑神的旨意，但強調自己非常渴望遵循主的道。瑞秋的父親忠實盡責，前往威科法院，簽名同意十四歲的女兒嫁給維農。比起普通法和個人道德觀，瑞秋的父母相信某些更為宏大的人事物：他們相信維農．豪威爾。

11 由 She 及 Shekinah 組合而成，Shekinah 為希伯來文，意思為神榮耀的同在。

靈性戰場

透過與洛伊絲．羅登交好，維農在大衛教派分支內贏得強而有力的基礎盤，但樹立了至少一位敵人：洛伊絲的兒子喬治．羅登（George Roden）。喬治相信自己能優先獲得領導卡梅爾山會眾的權力，也是合情合理。喬治性情不穩，經常動粗，爸爸死後由媽媽接管時，他已經覺得自己被忽略。現在喬治眼睜睜看著維農在虔誠信徒心中地位節節高升，把挫敗都怪罪到這名半路殺出的程咬金身上。一九八五年，媽媽日漸衰弱，喬治終於出手維護自己的控制權，把維農及其追隨者趕出集會。

大衛教派分支的多數信徒，包括他那位還是青少女的妻子以及親家，現在都效忠維農。他和多數信徒一起往東遷了九十哩，抵達德州的帕勒斯坦。當地環境原始，大部分會眾都過著拓荒者一般的生活。悶熱的夏日裡，他們在夾板建造的斜屋頂棚屋裡，聽著維農讀經講道，每次都可能長達數小時。

搬到帕勒斯坦沒多久，維農又娶了兩名靈性妻子，正式成為一夫多妻者。一九八六年，他娶了十四歲的凱倫．道爾（Karen Doyle）和十二歲的蜜雪兒．瓊斯（Michele Jones）。戀童癖的臨床定義是受到前青春期兒童的吸引，維農聲稱多妻的目的是這些女孩可以為他生孩子，所以其實他是戀青少年癖（hebephilia），認為青春期最早幾階段的未成年人有性吸引力。但不論用什麼正式名稱為這種行徑貼標籤，這行為本身顯然就是性虐待。

一九八六年，洛伊絲去世，喬治接管卡梅爾山。少了母親的引導，喬治在講道壇胡亂揮舞雙手，更多信徒紛紛離開，跑去跟隨維農。羅登重新將卡梅爾山命名為「羅登村」（Rodenville），在一段介紹影片中，他背著一把半自動步槍，展示了他對維農近乎犯罪的執迷。羅登說道：「基本上這就是聖戰，何梅尼（Khomeini）對上以色列，就是維農．豪威爾和我的關係。」

與此同時，維農前往不同安息日會社區招募信徒，包括加州、加拿大、英格蘭、澳洲和以色列。維農修長帥氣，說話溫柔，容易親近。遇到任何人，他通常會露一手音樂才華，幾乎都能找到彼此的共通點，而且任何表現出一點興趣的人，他都知道如何立刻引誘。就像世上許多患有自戀型人格障礙症的人一樣，維農明白如何即刻反映他人的興趣和關心，讓他人覺得被看見和受關心。他運用自己的關注，就像盞聚光燈，只要照在信徒身上，他們就會覺得受到特別青睞。

維農的核心成員變多，部分原因是他繼續娶更多女人。第二波婚姻的配偶年齡稍微大了一點：十七歲的羅賓．邦茲（Robyn Bunds）、十六歲的妮可．根特（Nicole Gent）和二十歲的達娜．沖本（Dana Okimoto）。他告訴這幾位女人，他體內帶著神的種子（God's seed），她們都是被特別揀選出來的，要懷上神的孩子。

當維農身邊漸漸培養出類似邪教的信徒時，卡梅爾山在喬治的治理下變成一座鬼城。喬治搞錯方向，試圖重新奪回他認為自己應得的地位——大衛教派分支領導。於是喬治向維農下戰帖，展開一場詭異的靈性大對決。喬治挖出二十年前葬在卡梅爾山公墓的安娜．休斯（Anna Hughes），裝進無蓋棺材，放在據點的禮拜堂祭壇上，披上以色列國旗（大衛教派分支的其中一個核心記號）。接下來的行為顯示了喬治的心理狀態有多不穩定：他向維農下戰帖，挑戰究竟誰能讓安娜．修斯起死回生。

兩人都挑戰失敗，也不知道到底有多少位信徒觀看這場駭人比賽。維農強調自己希望有證據，想看看羅登挖出那具屍體的照片。但這藉口也無法說明為何他與七名信徒深夜抵達卡梅爾山，帶著五把半自動突擊步槍、兩把步槍、兩把一二口徑散彈槍，以及將近四百發子彈。羅登顯然只有一把烏茲衝鋒槍，在隨之而來的戰鬥中，他躲在樹後方，樹幹上布滿了彈頭，羅登的手受了小傷，逃離現場。

交火過後，維農和參與突襲的信徒都以謀殺未遂遭到起訴。經過威科法院為期兩週的審判，法官宣判他們無罪，維農的案子則以無效審判作結。維農的四太太羅賓．邦茲後來說，在法院訴訟程序中，維農哭泣的樣子令她十分感動。這讓羅賓覺得與維農更加親近，維農是她唯一看過哭泣的男性。

幾個月後，喬治因為另一起不相干的謀殺未遂被起訴、逮捕。卡梅爾山無人掌管，財務狀況落入一團混亂。到了一九八八年，維農湊齊了錢，繳清該地的欠稅。完成最後這步，他現在擁有卡梅爾山全部領地，並且完成大衛教派分支內部晉升的所有步驟，成為最高領導人。

另一方面，喬治已經完全失去他認為自己可以直接繼承的土地和地位，心理健康急速惡化。他成功出獄回歸社會生活了一陣子，但在一九八九年，他開始想像維農雇用他的室友來殺害他，於是決定先採取行動，開槍射殺無辜的室友。羅登因精神錯亂被判無罪，被送到德州奧德薩（Odessa）以西三百哩的大泉州立醫院（Big Spring State Hospital）。說來諷刺，喬治活得比死對頭多幾年，但再也沒有清醒過。一九九七年，喬治六十歲，被人發現倒在大泉醫院地板上，死於心臟病發。

新亮光

美國精神醫學學會（American Psychiatric Association）官方手冊《精神疾病診斷與統計手冊》（*Diagnostic and Statistical Manual of Mental Disorders*，簡稱DSM）尚未認為誇大狂是一種症狀，但文獻卻又頻繁提及，好似誇大狂是自戀型人格障礙症（NPD）的同義詞。某種意義上來說，NPD和誇大狂在維農一例顯示的只是程度差異。自戀狂受到過度膨脹的自負感所苦，誇大狂除了告訴他人自己是神子，還會說服整群信眾相信這個主張，並且執行他的誡命。自戀者要求他人過多的欽佩，而這個誇大狂相信自己有權讓教派的所有女信徒受孕，如此一來她們就能懷上神之子。

教團回到卡梅爾山之後，維農直接奪走前面幾位先知的所有服飾與權利。值得特別一提的是，維農並沒有發明先知體系，只不過是利用既存的傳統而已。基督復臨安息日會原本就已經極度相信神的啟示。即使有些時期人間沒有神的先知直接向神的民訴說神的道，安息日會信徒依然相信這麼一位先知隨時都會降臨，傳遞聖言。

虔誠信徒因此隱隱興奮，持續不斷。在大衛教派分支這種派別又特別如此，一脈相承的體系，部分可以循著「預言之靈」，一路追溯至創立者懷愛倫。創立多年的中央教會由一個總會、多個分支與地方聯合教會共同管治。

中央教會認為先知講道之能是「二十八條基本信仰」（28 Fundamental Beliefs）其中一條，多數人認為「預言的恩賜」（the Gift of Prophecy）是專屬懷愛倫的教導。但是大衛教派分支相信這一恩賜是活傳統（living tradition），一路從維克多．霍特夫傳到班傑明．羅登，再到洛伊絲．羅登，最後傳至維農．韋恩．豪威爾。相信先知直接與神連結是信仰的傳統中心元素。信徒現在擁有直接與神溝通的代表，行為往往會走向極端。道成了肉身，在他們中間，他們深受啟發。想要遵守先知的話，因此輕易武裝起自己，突襲敵手據點，又或者在先知帶領之下連續讀經十五小時，他們也能輕鬆聆聽、一同祈禱。

一九八九年八月五日晚上，維農傳遞的訊息徹底改變了他們教派的本質，但他們似乎因為要遵守先知的話，所以繼續支持，一言不發。出乎所有人意料，他稱呼自己的啟示為「新亮光」（New Light）。那時他人在加州波摩納（Pomona），住在一間團體家屋裡，屋主是西岸大衛教派的信徒（維克多．霍特夫在加州也發布了啟示，因此加州也有一群為數不少的信徒，至今依然住在那裡）。依照一名前信徒所言，那晚他目睹維農中斷連續四小時的讀經，抬頭望著天花板，側耳好像在傾聽些什麼。接著他突然宣告自己的婚姻以及所有信徒的婚姻都無效。

從那時刻起，所有女信徒都任他挑選。他可以擁有一百四十名妻子，而且他是她們的完美佳偶。他告訴她們，他有權——而且是天賦職責——讓她們所有人懷孕，因為要為樂園準備好許多人。男信徒（立刻成了這些女人的前夫）也會被給予各自的完美佳偶，但要等到天堂才會執行，就像夏娃是用亞當的一部分所造成，他們的佳偶也會由他們的肋骨造成。在那之前，男信徒會維持單身，放棄地上的妻子，獻給維農，現在他們全都隸屬「神的軍兵」（God's Army）。這是大衛教派分支轉變的關鍵時刻，似乎毫無預兆就轉變成性瘋狂邪教，全心崇拜一位終將死亡的凡人（mortal man）。這人不過就是一夫多妻高手，以及法院判定的強暴犯。

在宣布新教義之前，維農依法改名為大衛，以聖經的大衛王命名。在一九三四年，維克多．霍特夫當初創立安息日會大衛教派，也是以大衛王來命名。維農多加了「柯瑞許」（Koresh），以波斯王居魯士（Persian king Cyrus）命名，他被先知以賽亞稱為受膏者，希伯來文叫柯瑞許。

現在要理解為何有人想和這樣的人有任何關係，其實有點困難。這個男人早就對未成年女孩下手，允許自己和任何女信徒性交。然而，在FBI好幾個小時的談判中，卡帶清楚錄到，柯瑞許自己詮釋他和信徒相信聖經記載著令人信服的證據，證明這樣的行為是上帝的諭令。

現在我們可以明白這條啟示多麼迅速招致一九九三年四月十九日的事件。只不過大衛教派無法預知未來的現實事件。柯瑞許說服信徒他是神的羔羊（Lamb of God），他有權揭開啟示錄記載的七印。如同啟示錄第四章四節所述，在新王國，他就是坐在寶座上的那位，被二十四位長老圍繞。顯然，柯瑞許一提到聖經就會非常有魅力——他說服了大衛教派分支，那二十四位長老都是他的孩子。事實上，那些長老就是他目前負責要帶到世上的孩子——依照神的諭令，不是只和他的妻子，而是和教內任一名女信徒。

柯瑞許宣布「新亮光」教導時，四太太羅賓．邦茲正好在波摩納的房子，那時已經為柯瑞許生了一個兒子。事實上，邦茲的雙親擁有那棟房子。離開教團之後，邦茲說早期她聽聞柯瑞許的其中一名妻子會生下彌賽亞。後來邦茲告訴《威科論壇先驅報》（*Waco Tribune-Herald*）記者：「就像選美比賽。大家爭奇鬥豔，想成為神眼中最好的女人，像童話故事一般。那時候我還是一廂情願，還沒看清現實。」

這條新諭令創造許多緊繃。柯瑞許懲罰那些不願交出妻子的男信徒。他分開男女信徒，只有讀經的時候才可以共處一室，一舉一動都在他警惕的眼皮底下。為了證明他沒有錯，他曾經叫一位女信徒在男信徒面前撩起裙子，露出內褲。接著他詢問是否有人感到興奮，所有男信徒都舉起了手。柯瑞許說這證明了他是對的：男人無可避免地想和女人做愛，這就是為什麼男女必須分開。

但是羅賓．邦茲從來就不喜歡和其他女人共享柯瑞許，生下兒子之後，她自己也逐漸成熟，意識到這一切是多麼荒唐。柯瑞許開始和羅賓的媽媽上床，那是最後一根稻草，接著羅賓就離開大衛教派分支。出於報復，柯瑞許綁架他們的兒子，希望帶兒子回到威科據點。執法單位堅持要柯瑞許歸還兒子，於是他把兒子送到地方警局，獨自離開加州。

即便離開大衛邪教，羅賓還是覺得在那次啟示宣告之前，柯瑞許不是一直都這麼壞，說道：「他完全變了一個人。他以前人很好，很謙虛，彬彬有禮。但是這些年來，他越來越沒有以前的樣子了。因為現在他有權掌控他人，他變得討人厭、滿嘴髒話、咄咄逼人。」

當心巴比倫淫婦

哈佛神經內科醫生伊凡．莫瑞（Evan Murray）研究歷史宗教人物的精神病症。冷靜地討論了亞伯拉罕、摩西、耶穌（Jesus）和聖保羅（Saint Paul）等受神光照的人，其理論發人深省。總結討論分析的限制時，他提出一個有趣的觀察面向：如果按照人在社會中無法好好溝通或運作來看，患有精神病症或精神錯亂的人，總是無法和神智清醒的人清楚區別。他主張大衛．柯瑞許依然可以吸引追隨者，建立高強度的社會連結，即使柯瑞許顯然罹患多疑症、自大妄想症、救世主妄想，還有其他許多思覺失調症狀。簡單來說，柯瑞許的瘋狂並不妨礙信徒喜愛他，享受有他作伴。

柯瑞許經常說明自己透過異象獲得聖經上的頓悟，或許他並沒有捏造，或許他只是患有顳葉癲癇，伴隨葛斯文症候群（Geschwind syndrome）臨床表現。此一失調症的典型特徵包括說話過多、強迫書寫症、強烈宗教感、強烈道德感，一般而言還有強化的精神生活。這些症狀會隨著時間越演越烈。柯瑞許看來符合這一描述：他可以一次帶領長達十九小時的讀經，也會一連寫上好幾天預言。雖然這些症狀經常伴隨低落性慾，但在某些個案中也會伴隨性慾亢進（hypersexuality），柯瑞許看來就是這類案例。

他曾經告訴一名信徒他在一九八五年造訪以色列時看見的異象：「站在錫安山上，我遇見天使，純粹由光組成的存在。他們是圍繞那天上寶座，也就是神祕寶座（Merkabah）的戰士，安坐火驥，佩戴烈焰寶劍。他們只允許那些可以揭開七印的人進入更高的界域，進入那與我們的世界同時並存的千千萬萬個世界。」柯瑞許說明天使帶他到天上見神，神授予他解經鑰匙。他說：「那時我知道了自己的命運就是揭開七印，為我們的社區開啟道路。」

關於顳葉癲癇，神經內科醫生薩克斯寫道：「這類癲癇幻覺和一般精神性幻覺有許多相似之處，即使癲癇病患可能沒有任何精神病史。性格堅定、時時懷疑的人才能抗拒得了這類幻覺，進而不去相信或遵從幻覺，尤其是如果這類幻覺帶有一絲啟示或主顯的性質，同時似乎指向一個特別甚至崇高的命運。」[2]很明顯大衛．柯瑞許不是那麼堅定且多疑。童年飽受忽視和虐待造成他渴求關注、需要依賴，更難抗拒這些異象。

這些據稱的異象再加上他那看似特別有毒的自戀型人格障礙症，柯瑞許手握利器剝削自己的信徒，且心中無情，可以忽略行為帶來的後果。即將來臨的災難所需的形成因素都齊備了。但要創造威科那般規模的慘劇，還需要其他許多強化因素。

柯瑞許總是在尋找末日將近的神蹟、奇事。他特別相信極右派陰謀理論的「新世界秩序」（New World Order）。這套理論相信超級富豪、世界主義菁英和暗黑人物組了一個陰謀集團，想要逐漸瓦解所有民族國家，奪走人類的自由，並依照他們的原則來掌控世界。

這些理論深受美國基本教義派和愛國者運動民兵成員支持，也加深一般大眾對政府的不信任。槍枝展會出現新世界秩序的文宣，而在一九八〇年代晚期、九〇年代初期，新世界秩序開始吸引其中一派陰謀論者，他們認為自己可以點出所有邊緣現象（幽浮、亞特蘭提斯失落的大陸）和瘋狂理論（甘迺迪暗殺理論、撒旦儀式）背後的運作方式。對柯瑞許而言，新世界理論非常有吸引力。他構思了一切，說自己和大衛教派分支信徒應該前往以色列，代替以色列作戰，擊退聯合國。

但美國在一九九一年參與波斯灣戰爭，柯瑞許轉變信念，改而相信美國政府才是啟示錄提及的巴比倫——「奧祕哉！大巴比倫，作世上的淫婦和一切可憎之物的母。」這對許多基督復臨安息日會教徒而言並不是太過跳躍的思考邏輯。閱讀懷愛倫一八五八年《善惡之爭》（*The Great Controversy*）書中的教導，信徒早就相信啟示錄預言有間假教會掌控了地上每一位王。柯瑞許只不過稍微更新懷愛倫的預言，變成現代版本。

開始玩弄這些點子沒多久，柯瑞許相信最終會有一場衝突。在這將近的末日，他會領導一切，因為如同經文所示，他是揭開七印的那位。他告訴大衛教派分支信徒這場大戰正好會發生在神安置他們的地方：就在德州。大衛教派分支信徒開始儲備食物、武器、彈藥和丙烷等用品，出於宗教義務備戰。他們也在柯瑞許的讀經會受到鼓舞。這正是為何富含生動詩意圖像的啟示錄對柯瑞許這類自行膏立的先知那麼重要：末日總是近了，永生危在旦夕，啟示錄激起浮想聯翩的畫面，總是歡迎不同詮釋。

這份新任務讓大衛教派分支有理由參加槍枝展。事實上，他們開展了買賣武器的副業，改造武器盈利或自用。一九九〇年，他們在會展現場開設一家槍枝店舖叫「磁袋子」（Mag Bag），買賣武器和零件。柯瑞許和信徒保羅．法塔（Paul Fatta）、麥可．舒德（Michael Schroeder）參加德州所有槍枝展，利用現在開始獲利的新創事業募款支援大衛教派分支。對這些屬靈戰士而言，他們相信巴比倫攻擊迫在眉睫，而槍枝交易不僅是獲利來源，還是明智決定。柯瑞許採用新世界秩序的常見觀點，認為政府已經準備好沒收市民的槍枝。他告訴信徒，武裝起來才能保護自己不受威權統治。

偏執預言轉變成自我應驗預言的時間點就在一九九二年夏大，一名UPS司機到卡梅爾山送一只破爛包裹，盒子裡是武器和手榴彈空殼。將空殼交給技術高超的人，就可以變成實彈。司機遵守公司政策，向相關當局回報這起可能的違法武器交易。

武器日漸增加的情報標記了大衛．柯瑞許和大衛教派分支走上盡頭的起點。頃刻之間，壓力大增。一九九三年一月，美國菸酒槍炮及爆裂物管理局（Bureau of Alcohol, Tobacco and Firearms，簡稱ATF，本政府單位的責任之一為管制武器的非法使用與交易）派遣臥底幹員住在卡梅爾山旁邊。一名威科突襲倖存者後來宣稱，柯瑞許一開始就知道新鄰居是聯邦政府幹員。他甚至歡迎新鄰居來參加讀經。柯瑞許的態度是，或許這些人來到這裡是因為任務在身，但神給予他的訊息或許也會感動他們。柯瑞許真的認為他可以轉變這些幹員的信仰——此心態事後證明是致命的過度自信。

先知帶著一把克拉克手槍

ATF最近展現的興趣似乎確認了教團內原本就已高漲的懷疑，他們被視為外來者，政府是他們神聖使命的敵人。同時，柯瑞許也助長這樣的說法。現在的活動似乎證實了他的預言，最終戰役即將展開，聖經中的大巴比倫以美國政府的形式顯現，準備好隨時發動攻擊。

新亮光啟示之後，羅賓・邦茲等一群理性信徒逃離大衛教派分支，前往較為清明理智的地方，但ATF突襲當日（一九九三年二月二十八日），根據預測，還有將近百名大衛教派分支信徒在卡梅爾山據點，全員備戰，集體對峙。這早已超越他們平日訓練時觀看的暴力戰爭電影了（柯瑞許在一九九〇年強制實行）。

柯瑞許的二十名高級副手和個人保鑣（又稱「勇士」（Mighty Men），源自聖經中保護大衛王的戰士）全都熟習AK四七步槍和M十六步槍的操作。勇士負責訓練其他信徒如何開槍殺人，就連兒童也要學習。最高階的勇士是四十二歲的道格拉斯・韋恩・馬丁（Douglas Wayne Martin）和四十八歲的史蒂夫・施奈德（Steve Schneider）。馬丁是柯瑞許最信任的顧問，也是最令人驚訝的信徒之一。他和太太席菈（Sheila）生了七名孩子，幾名已成年，一名重度殘障，至少四名和爸媽一同住在卡梅爾山。馬丁有大學學位，是哈佛法學院出身的律師，之前在達蘭（Durham）的北卡羅來納中央大學（North Carolina Central University）任教。

馬丁一邊住在偏遠的卡梅爾山，過著沒有自來水的艱苦生活，一邊在鎮上職業，賺來的每分錢都上繳給柯瑞許。地方法官都知道他是誰，一名威科市議員甚至是他的摯友，市議員說他是午餐良伴。馬丁是名盡責的律師，在與ＦＢＩ對峙期間，他還安排把錢送出據點，賠償那些他再也無法代表的客戶。《華盛頓郵報》（*Washington Post*）報導：「教團的代表律師蓋瑞．寇克（Gary Coker）也在ＦＢＩ剪斷電話線之前和馬丁講到話。他說馬丁看來『幾近冷靜』，形容馬丁很善良，還是個很愛孩子的好爸爸。」[3]

史蒂夫．施奈德正好相反，這名高級副手脾氣糟糕，無預警就會爆發一陣狂怒。一九八九年新亮光啟示之後，他的太太是首批成為柯瑞許女人的其中之一，名字變成朱蒂．施奈德．柯瑞許（Judy Schneider Koresh）。雖然婚姻被拆散，史蒂夫非常憤怒，但還是對柯瑞許忠心不二。史蒂夫告訴聯邦談判專家他的領袖擁有「我們不知道的能力，要是我們膽敢挑戰，便是犯了大錯」。史蒂夫受過一些大學教育，曾經希望成為牧師，但後來遇到了柯瑞許。史蒂夫管理教團其中一個事業：音樂製作公司「彌賽亞居魯士製作公司」（Messiah Cyrus Productions），專門錄製柯瑞許的音樂，在地方社區發行，例如〈威科狂人〉（*Mad Man in Waco*）等歌曲。

勇士中有兩位也是音樂人：麥可．舒德和大衛．希伯杜（David Thibodeau）兩人都是鼓手。史蒂夫・施奈德在一九八〇年代晚期和舒德及他太太凱西（Kathy）相遇，告訴他們大衛・柯瑞許這號人物。不久後，舒德夫婦不辭千里開車拜訪，聽柯瑞許講道，兩人都很喜歡。一九八九年，他們開著破爛的廂型車，載著小孩前往卡梅爾山，加入教團。麥可的媽媽說那台廂型車實在太破爛，她不確定他們是否可以成功抵達。但不知為何，她知道自己不會再見到他們了。

麥可和凱西都是慕道友，雖然柯瑞許強加的規則很嚴苛，但他的訊息深深吸引著他們。那時凱西懷孕了，但在新亮光教義下，一抵達還是被分開了。柯瑞許在澳洲和加州都有手下，如果有需要，也會利用信徒的家作為臨時作戰基地。他派麥可去加州的其中一個據點，協助招募更多「神的軍兵」。麥可沒有辦法見上新生兒子一面，直到兒子六個月大。他回到德州後經常睡在工作的修車廠。就像其他先生一樣，麥可幾乎都只在讀經時看到太太，當然一舉一動都在柯瑞許的監視下。

柯瑞許不想讓大衛教派分支的男信徒花時間和他們自己的孩子相處，因為親子連結會直接威脅他的掌控。但是凱西偷偷安排麥可和兒子碰面。就像馬丁和施奈德一樣，即便槍枝交易是他們獲利最高的事業，許多信徒在社區裡都有工作，把薪水上繳給柯瑞許。

柯瑞許開始感到來自外界的壓力。他知道自己的觀點和做法被視為犯罪行為，尤其是一夫多妻制，以及和十一歲的兒童發生性關係。一九九二年二月，ＡＴＦ發動突襲一年前，德州衛生及公共服務部（Texas Department of Health and Human Services）和兩名警員造訪了卡梅爾山據點，調查監護權官司中提及的柯瑞許性侵年輕女生事件。這次調查毫無成效，因為教團中沒有人要說柯瑞許的壞話。柯瑞許更加確信他會為自己殉道。但他打算在倒下之前奮戰一番。他為武裝衝突做好準備，總是隨身帶著一把克拉克手槍，就連睡覺也不例外。

勇士備戰

同年，據說幾名勇士前往加州波摩納的邦茲家拿機關槍轉換工具包。邦茲一家在大衛教派分支信徒離開之後發現工具包，把工具包存放在車庫，就怕柯瑞許回來拿。即使羅賓和媽媽都已經離開了教團，還是無法完全逃離柯瑞許的掌控。某種程度而言，他們看似還是在保護柯瑞許。如果真的切斷所有聯絡了，他們大可直接丟棄違法武器，但是柯瑞許對邦茲一家還是有足夠的影響力，他們默默地儲藏他的財產。

一九九三年ATF突襲令寫道，那名UPS司機抵達卡梅爾山送包裹時，「他看到許多觀測站都有部署人員，而且相信這些人員帶著武器」。一名員工回報，他和一個大約八歲大的男孩聊了一會兒：「那名男孩說他等不及長大當個男人。兒童保護服務單位（Child Protective Service）的喬伊絲．斯帕克斯（Joyce Sparks）詢問他為什麼這麼急著想長大，他回答，長大了就會拿到一把『長槍』，像那裡的所有男人一樣。」

雖然ATF突襲卡梅爾山的主要動機是懷疑他們持有非法武器，但突襲令也引述了多人提出性虐待指控。有些是第一手消息，基於一年前觀察到的可疑行徑，同時羅賓．邦茲的造訪也提供許多資訊，她告訴警方柯瑞許「在那裡都定期和年輕女孩發生性關係，女孩年紀從十一歲到成年不等」。有時候柯瑞許似乎對他犯下的罪行十分自豪，臉不紅氣不喘。前大衛教派分支信徒馬克．布勞特（Marc Breault）在監護權官司出庭作證，發表不利於柯瑞許的證詞，說柯瑞許曾經問過他覺得柯瑞許真正最喜歡的妻子是誰。布勞特猜是第一位妻子，也是唯一一位合法的妻子瑞秋。但是柯瑞許告訴他其實是瑞秋的妹妹。柯瑞許說：「馬克，你能相信嗎？她十二歲的時候就跟我在一起了！」

ATF臥底幹員勞勃．羅里葛茲（Robert Rodriguez）開始造訪該教團，表面上是要聽柯瑞許講道，實際上打算取得更多教團的武器資訊。大衛．希伯杜經歷首次突襲和最終對峙，兩次都倖免於難，後來宣稱柯瑞許一開始就知道羅里葛茲是臥底，但還是歡迎他踏上他們的土地，參加讀經。不需要神聖預言，也看得出來他們受到監視——據點上方有直升機盤旋，而頻繁的航班則向柯瑞許洩露了消息。

羅里葛茲待在據點的那幾週和柯瑞許成了朋友。包圍期間，柯瑞許和FBI對話，說道羅里葛茲人很好，他很喜歡羅里葛茲。儘管這麼親密，羅里葛茲還是無法確認大衛教派分支是否真的擁有非法武器。但他查清了卡梅爾山據點確實有大量武器，也有能力把半自動武器改造成全自動武器。這代表ATF有相當理由可以搜查。之後軍方文件揭露疑點，ATF可能在申請突襲令時知情說謊，宣稱當地有安非他命製毒工廠，這是少數幾個可以允許他們從軍方獲得額外戰術訓練的情境。

一九九三年二月二十五日，美國聯邦地區法院（United States District Court）德州西區地院（Western District of Texas）發出搜索令，逮捕大衛．柯瑞許並且搜索據點。授權幹員搜出重軍火和自製手榴彈、炸彈製作原料、突擊步槍，以及改造成全自動所需的零件。

四月圍攻之後，大衛教派分支倖存者提出訴訟，德州巡警（Texas Ranger）出庭作證：「建築燒毀幾小時後，從燒焦的據點殘骸中找到大約三百支突擊步槍、手槍。」——大量武器。預期會有重裝武備反抗是ATF突襲全程的重要因素。

二月二十八日，ATF準備好向卡梅爾山發出「不敲門搜索令」（no-knock warrant），幹員不用先宣布，就可以直接進場搜查，嚇嚇大家。但在前幾天，超過幾百名幹員和後勤人員湧入威科，住滿當地汽車旅館，準備搜查卡梅爾山據點。媒體得知有大批聯邦幹員，各自準備了報導團隊。突襲當天早上，一名KWTX-TV記者迷路了，不知道怎麼去卡梅爾山，向當地一名郵差問路。結果那名郵差認識住在那裡的一位居民：他的小舅子大衛．柯瑞許。

臥底幹員羅里葛茲人在據點讀經時，柯瑞許接到警告電話，羅里葛茲明白自己的底被掀了，但兩人之間沒有摩擦。柯瑞許一點都不驚慌，他很高興地讓羅里葛茲離開。就某些方面而言，接獲這則消息讓柯瑞許很開心：他預言的對峙即將成真。

勇士和大衛教派分支的其他戰鬥人員立刻起身行動，穿戴前妻縫製的黑色鎧甲防護衣。後來一份聯邦宣誓書聲稱道格拉斯．韋恩．馬丁戴著一串手榴彈項鍊。女人和小孩躲了起來，雖然目前知道至少有一名女性（是位前警官）參戰，那天早上死了六名大衛教派分支信徒，其中一名就是她。

羅里葛茲通知ＡＴＦ，大衛教派分支信徒知道突襲要發生了。聯邦幹員的計畫取決於出乎意料的造訪，然而行動沒有延後，ＡＴＦ指揮高層決定繼續突襲。他們想要衝入據點，希望殺得大衛教派分支猝不及防——回想起來是個極為愚蠢的決定。大衛教派分支信徒有大型據點作為保護。據點裡還有數十輛交通工具，連巴士都有。建築物排列緊密，成矩形陣仗，左後方有一座大水池，還有重要的戰略建物，例如水塔和瞭望台。還有集體居住區、柯瑞許生活區、儲藏用體育館和數個水泥碉堡。擊潰卡梅爾山據點不是簡單的任務。

早上九點半，ＡＴＦ試圖執行突襲令和搜索令。一百名探員爬上兩台家畜搬運車，藏在大型防水油布下。直升機載著槍手盤旋上方。但ＡＴＦ人員的內部通訊極糟。有些人聽過完整簡報，知道卡梅爾山據點的武器數量，但有些人則是毫無頭緒。大家依然在爭論，無法確認究竟是哪一方先開火，但隨之而來的大規模槍戰持續了兩小時。倖存者凱西．舒德在訪問中說道：「我看到家畜搬運車停在前面，有人下了車。他們全身黑，帶著槍。接著我聽到槍響，就蹲下來找掩護。」凱西和孩子把床下所有東西拖出來，才能躲進去。子彈射穿房內的水瓶，瓶內液體在地上流乾。小孩大哭，但槍擊持續太久，其中一位還真的在床底睡著了。

交火期間，柯瑞許撥打九一一，宣布：「我是大衛．柯瑞許。我們全體遭人射殺。」那通電話一開始似乎是合理的緊急求助，但結束時卻成了ＦＢＩ談判專家所謂的「聖經胡謅」，柯瑞許一逕講著七印。胡謅之間，他還轉述現場死傷人數，情緒毫無波動，著實令人驚訝。這位男人（為人父許多次）展現的既不是同理，也不是懊悔。然後，他忽然變得興奮：「我已經教導這件事四年了！我們知道你們會來，也知道你們要做的一切！」他講的並非稍早得知的突襲事實，他炫耀的是自己解讀啟示錄徵兆的能力，啟示錄早在多年前就告知他，政府鎖定了他。他想要所有人知道，他不僅預測到這項突襲，向信徒宣講此事，還採取步驟確保大衛教派分支做好萬全準備面對他預言的對峙。

深入了解邪教領袖生活的迷人之處，其中之一是他們代表人類行為中令人無法想像的極端。或許，我們觀察他們的童年，可以看到塑造怪物的力量，因此可能深感同情：查爾斯．曼森目睹心愛的媽媽被警方拖走；亞道夫．德．赫穌斯．康斯坦佐活在羞恥之中，試圖在髒亂不堪的家中保留一小塊有秩序的角落；維農．韋恩．豪威爾，也就是現在的大衛．柯瑞許，從小被遺棄、遭毒打、罹患讀寫障礙、飽受異象困擾，試圖在生活中尋找可以協助他融入的事物。

然而我們現在看到的是邪教領袖生活中的典型時刻——一個關鍵轉捩點，即便命運逼近，他們依然還有選擇，可以避開血流成河，轉身離開自找的災難；但一次又一次，邪教領袖看到即將發生的未來，卻無法或不願意改變路線。就是在這個時刻，他們會處在最黑暗、最極端的狀態，繼續選擇暴力，甚至凌駕他們自己的求生直覺，因為他們就是這樣的人。

對大衛．柯瑞許這般的邪教領袖而言，未來似乎沒有回頭路，木已成舟。從那時刻起，他只會越來越堅定投入自己一手推動的命運，整個人像活過來一樣，做出衝動的決定、忽視危險、激勵信徒、在他的預言書中瘋狂書寫。他知道死亡即將來到所有人身邊，但這對他來說不是那麼重要，重要的是所有混亂與暴力都完全依照他斷定的方式降臨，全世界都會看見。

柯瑞許性侵未成年人和青春期前的兒童，其實不太需要預知異象，也能猜到總有一天政府會來逮捕他。但在此之前，他的信徒早已不再用這種方式思考柯瑞許這個人。突襲和後續的圍攻反而增強了信徒的忠誠，他們現在相信柯瑞許總是對的；他提及的戰爭來了，代表他一直以來都是對的。如果末日在據點外步步逼近，信徒顯然不想離開柯瑞許的身邊，也不想放棄永生。這是認知失調的實例：大衛教派分支回應明顯對他們不利的證據時，把一切都轉變成恰好相反的證據。不利證據包括他們的領袖會動用暴力，實行一夫多妻，還是個戀童癖，同時政府團團圍住他們，結局想必不會幸福美滿。但他們覺得自己是對的，不論是相信天國近了，還是相信柯瑞許是啟示錄預言的那名使者。

但是，美國政府也沒有好到哪裡去。ＡＴＦ面對強力的反擊，被攻得猝不及防——這實在太愚蠢，因為搜索令授權他們搜查的正是大批武器和炸藥。那天ＡＴＦ的行動顯示他們不僅沒有做好準備，還嚴重自信過剩。槍戰持續了兩個多小時，將近尾聲，政府卻彈藥用罄——這指出他們低估了風險，或證明他們缺少火力管制，又或者兩者皆是。他們也沒有想過要帶急救醫護人員。知道對方已經預期他們會來，還是倉促決定進攻、部隊之間溝通不良、執法單位之間協調不足、彈藥有限、現場缺乏急救支援——全部加總起來，成了ＡＴＦ的一場噩夢。

幾小時內，FBI接管本次行動。

FBI成功讓雙方停火，ATF幹員才得以進入戰區，清出傷亡人員。當地警方鼓勵柯瑞許接受政府提供大衛教派分支的醫療協助，但柯瑞許說沒有人受重傷，而且他們一點也不想要任何幫忙。他沒有吐露大衛教派分支已經有六人死亡，自己的左手腕和臀部都受了傷。其中一名受傷的信徒後來在電話上告訴談判專家，他不想出來接受治療，寧死也要支持他們教團。

鎂光燈焦點威科

為期五十一天的僵局於焉展開。主要由三方談判——大衛教派分支的領導群柯瑞許、馬丁、舒德；麥克倫南郡警局（McLennan County Sheriff's Office）警督賴瑞．林區（Larry Lynch）。以及ATF的吉姆．卡瓦諾（Jim Cavanaugh）。同時，FBI訂定好行動和策略。他們允許柯瑞許在達拉斯一家廣播電台播送他的宗教教導，並且和CNN來一場電話訪問。一九九三年當時，二十四小時有線新聞台還是相對較新的媒體。儘管如此，相較於十年前只有不到兩千萬戶安裝有線電視，那時已經超過五千萬戶安裝了。幾乎一夜之間，大衛教派分支從邊緣團體搖身一變，成為全世界新聞播送的頭條故事。對柯瑞許而言，要嘛這是一場夢，要嘛很可能是顛葉癲癇的幻覺成真。

三月一日，突襲隔天，柯林頓總統（Bill Clinton）接獲最新消息。他贊同FBI局長威廉・塞申斯（William Sessions）的方法：盡量拖延與柯瑞許的談判，希望藉此換來和平解方。日子一天拖過一天，十名兒童被送出據點，還有一小群成人，其中包括兩名年長女性也離開據點。凱凱・舒德後來告訴《坦帕灣時報》（*Tampa Bay Times*）：「那些小孩都不想離開。但是大衛說未滿十二歲的孩童不算。是大衛決定要小孩離開的。我可不這麼認為。就算我們要死，也應該一起死，永遠在一起。我覺得『我把小孩推入大巴比倫』，一個充滿邪惡的世界。」[4]

才停火幾小時，凱西・舒德聽到外面傳來槍響。她有不祥的預感，認為是先生麥可試圖回到據點時被ＦＢＩ擊殺。她沒猜錯：麥可的屍體就躺在雙方中間的地上，過了四天才被移開。

到了三月一日當地時間傍晚五點，ＦＢＩ已經全面掌控局勢。同時柯瑞許越來越焦慮不安，裝甲車逐漸接近戰術位置，據點的電話線被切斷，只剩一條通往談判專家的線路。

ＦＢＩ謹記十五年前的瓊斯鎮大屠殺，想要確認柯瑞許和他的教團沒有計劃集體自殺，尤其還有這麼多小孩在據點裡。柯瑞許向他們保證，大衛教派分支完全沒有想要自殺。對話持續到三月三日清晨，柯瑞許保證只要他的一卷宗教教導卡帶在全國電視台播放，他一定會投降。我們再一次看到他優先考量的不是大衛教派分支信徒的安危，他唯一關心的是自己思想理念的曝光。

中午一點半，他的卡帶透過視博恩電視台（Christian Broadcasting Network，又名基督教廣播網）全國播出。然而，到了晚上六點，柯瑞許食言，違反與ＦＢＩ的協定，說神告訴他只管安坐等候。聯邦幹員陷入困境。他們全都熟習如何在人質談判中與嫌犯打交道，但這次場景很獨特：沒有人拿槍指著任何人的頭。這些人質是意識形態的俘虜。除此之外，大衛教派分支相信比起害怕柯瑞許，ＦＢＩ才是他們應該感到害怕的對象。如果熊熊大火吞噬一切，就像柯瑞許多年來講道的那樣，他們保證會獲得永生。但要是他們離開了據點，就會被丟進火湖（Lake of Fire）和其他非信徒一起受苦——這才是阻礙大家逃出來的最大因素。

柯瑞許似乎很享受他的新觀眾。ＦＢＩ親自體驗到柯瑞許詮釋聖經的旺盛精力。三月三日過去，整天都是沒完沒了的宗教教導。談判專家聽到柯瑞許是如何蒙神膏立揭開七印。依倖存者所言，柯瑞許也告訴信徒，現在正在第五印，大迫害來臨時他們都會殉道。他們準備好迎接死亡，接著等待其他殉道者，才能一起在新王國永生不死。七印的循環完成後，他們都會進入新王國。

柯瑞許也描述了一下自己的槍傷，說道：「我想要坐起身，但我做不到，還有我的屁股……被射傷真的很戲劇化。全身都受到驚嚇，而且……子彈好像碎裂還是刺穿，或者不知道對我屁股上方的骨頭做了什麼，因為子彈從肚子射進來，穿過去打到屁股，接著好像從我背後繞了出去。」[5]

然而這一切都沒有令他放慢腳步。三月四日，他花了將近八小時和談判專家通話。在ＦＢＩ這一側，氣氛則是逐漸緊繃。催淚瓦斯也變成選項之一，但之前從來沒有對兒童使用過，尤其建議用量非常大。

三月五日，九歲的希絲．瓊斯（Heather Jones）離開據點，夾克上別著媽媽寫的字條。寫道一旦小孩全都離開據點，大人就會死。這則訊息確認了ＦＢＩ最大的擔憂，也是據點裡大家討論的事情之一。柯瑞許和施奈德不斷否認教團有計畫集體自殺，但ＦＢＩ不相信他們，尤其在對峙前，有些已經離開教團的信徒通報，柯瑞許一直宣導信徒他們都要為了他而死。被釋放的孩童告訴心理學家的話讓ＦＢＩ更加相信集體自殺很可能會發生。

這個可能性看起來像正在形成的災難。據點裡的建築物內還有數十名孩童。ＦＢＩ已經知道大衛教派分支有大量武器和彈藥，也有充足的食物，可以餵飽信徒一整年。從三月六日到八日，對話毫無進展。ＦＢＩ遞送鮮奶給小孩，柯瑞許寄出一卷錄影帶，拍攝小孩，說他們很快樂，他也指出幾名自己的孩子，其中一名是和第一位青少女太太瑞秋所生的八歲大兒子居魯士。

三月九日，ＦＢＩ切斷送往卡梅爾山的電力，繼續重新安排據點四周的裝甲車，調整成更具威脅性的位置。這激怒了柯瑞許和施奈德，但兩人不知道的是，所有行動都是ＦＢＩ談判專家和作戰小組之間緊繃的徵兆。對ＦＢＩ而言，他們看到勇士用夾板遮住窗戶，開了幾個供槍管使用的孔洞，認為這手段不是那麼堂堂正正。柯瑞許拒絕談判，直到電力恢復，圍攻期間一直都是這種貓捉老鼠的談判方式。

三月十日和十一日來了又走，還是沒有進展。柯林頓總統的司法部長珍妮特．李諾（Janet Reno，首位女司法部長，三月十二日宣誓就任）被總統交辦要加速解決威科情勢。同一天，凱西・舒德離開據點，再一次向ＦＢＩ保證沒有集體自殺的計畫。多年後，凱西告訴《坦帕灣時報》的蘇珊・阿紹夫（Susan Aschoff），她離開的原因是被抓到偷抽菸。柯瑞許說她「毫不在意神的律法，肯定會阻礙全體教團發展」。[6]

ＦＢＩ開始利用酷刑技巧想要強迫教團投降。刻意挑在寒流期間斷電。朝據點內照射強光中斷他們的睡眠，大聲播放音樂，有時甚至播放兔子垂死的尖叫聲。另外兩名男人離開據點，接下來幾天，七名成人陸續離開。其中一人麗塔・里德爾（Rita Riddle）說，她看到聯邦小組的人在坦克車上朝著大衛教派分支信徒比中指，還露屁股。那樣的行為讓教團覺得自己更加正當。這些粗魯無禮、對神不敬的男人是誰？

關於據點內的感受，我們只能推測。信徒驚恐嗎？還是他們相信等待的天命將來，樂在其中？孩子覺得心煩嗎？ＦＢＩ小組肯定顯露了他們壓力很大。談判專家依然認為可能有和平解方，但攻擊性更強的戰術選項逐漸獲得更多人支持。三月二十三日，另一名英國籍成人信徒李文斯頓・費根（Livingstone Fagan）離開據點，遭到逮捕。圍攻二十年後，費根依然深信柯瑞許，在《電訊報》（*Telegraph*）科爾・莫頓（Cole Moreton）的訪問中坦承：「我不想要離開，但大衛要我這麼做。如果我們都被殺死了，需要有人在外面發聲，從我們的觀點訴說故事。」[7]

從三月二十九日到三十一日，司法部長李諾的代表和ＦＢＩ官員開會討論聯邦幹員的內鬨。柯瑞許第一次獲准和他的律師講話。沒多久，他同意接下來過完逾越節，全部信徒都會離開據點。逾越節是大衛教派分支重要的節日之一。四月四日，在和律師的對談之中，他重複了這個保證。逾越節在四月五日，但到了四月七日，柯瑞許再次拒絕告知究竟哪天會離開。催淚瓦斯重回考慮中，柯瑞許多次毀約的前例更加堅定催淚瓦斯這一選項。

威科對峙大事件在邪教史上如此引人注目，是因為有許多公開的原始資料可用。柯瑞許聽起來不像高超的聖經詮釋者，比較像運用一些圖像再宣稱幾則預知異象的人，確實比起他的經文分析還更具說服力。下方列出這類對話的其中一例，本對話經過編輯，長度適中，更為清晰好懂：[8]

柯瑞許：「所以每位宣稱對神的認識有所領悟的人，只會有一次決定性試驗——展示七印——如果無法，那他們就只能等待，直到有人可以。」

FBI：「大衛，你是如何變成可以詮釋七印的人呢？」

柯瑞許：「嗯……一九八五年的時候，我人在以色列。那時有幾位俄羅斯太空人——為什麼我會告訴你這些，是因為我們有兩名目擊者。那些俄羅斯太空人回報，他們看到七名天使般的生物朝地球飛去，翅膀的大小就像巨無霸客機。嗯。所以到底發生了什麼事呢？一九八五年我在以色列，我和這些人碰面。說真的。」

FBI：「你和誰碰面？」

柯瑞許：「不、不、不。聽好——俄羅斯太空人在他們的太空站。」

FBI：「好的。」

柯瑞許：「然後他們朝下發送無線電給總部。他們嚇壞了。」

FBI：「好的。我明白了。」

柯瑞許：「說他們看到七名天使般的生物朝地球移動。」

FBI：「好的。然後你和這七名天使般的生物碰面。」

柯瑞許：「沒錯。」

FBI：「哪裡？」

柯瑞許：「以色列。」

FBI：「好的。但以色列哪裡？」

柯瑞許：「錫安山上。」

FBI：「噢，好的。」

柯瑞許：「對。我跟你說。這真的很讚。天使其實沒有翅膀，但他們有梅爾卡瓦（Merkhavah）[12]。」

FBI：「有什麼？」

柯瑞許：「梅爾卡瓦。」

FBI：「那是？」

---

12 Merkhavah 是 Merkabah 的另一種拼法，指神的寶座或神的戰車。

柯瑞許：「那是——那是艘太空船。」

ＦＢＩ：「太空船？」

柯瑞許：「那是交通工具。可以……可以乘著光移動。光的折射是——」

ＦＢＩ：「噢，好的。」

柯瑞許：「你知道彩虹怎麼形成的嗎？」

在四月九日耶穌受難日（Good Friday），對峙進入第四十一天，柯瑞許寄了四封信給ＦＢＩ。一封寫道：「天上召喚你們接受審判。」當然，ＦＢＩ找了心理學專家來分析證物，他們斷定柯瑞許正處於高度妄想狀態，很可能罹患精神病。他們也同意柯瑞許根本沒有打算離開據點。務必謹記，柯瑞許受了嚴重槍傷，還在康復中。即便子彈已經不在他體內，還是會有感染的風險，很可能直到最後都一直承受著劇痛。

火焰在等候

四月十一日復活節過去，他們沒有投降。逾越節在四月十三日結束，依舊沒有進展。但在四月十四日，柯瑞許宣布他正在埋頭書寫啟示錄七印詮釋。他保證手稿一完成，就會離開據點。兩天後，他通知ＦＢＩ，現在已經寫到第一印，他們可以粗略估計要等多久：按照目前速度，還要再十二天。對峙第五十天，大衛教派分支信徒爬上一座電塔，FBI告訴他們已經斷電了，因為電塔可以當成射擊塔使用。他們高舉孩子給聯邦幹員看。這是緊張的一刻，政府不知道他們是否想請政府饒過孩子，或只是把孩子當成人肉盾牌，掩蔽這場對戰中初次的攻擊行動。大衛教派分支信徒明白自己遭人監視，舉起牌子上面寫道：「火焰等候著。以賽亞書第十三章。」

隔天日出前一小時，四月十九日，被指派的FBI談判專家用大聲公告訴大衛教派分支，對峙即將結束，FBI會使用催淚瓦斯，教團所有人都會被逮補。早上六點零一分，FBI派出兩台戰鬥工兵車，在木造建築上敲出大洞，車子曲臂裝上噴頭把催淚瓦斯投入據點。FBI知道教團購置防毒面具，但就算是防毒面具，配戴八小時之後也會失去效能。最新戰略是按程度增加施壓力道。

圍攻初期，竊聽裝置被藏在傷口縫合工具包或裝小朋友鮮奶的冰箱內，偷渡進據點，FBI透過那些裝置竊聽，聽到據點內的信徒討論在不同區域潑灑燃料，還聽到信徒計劃射擊聯邦幹員。FBI暫且不開火，但繼續用戰鬥工兵車的曲臂破牆，朝室內灌入更多催淚瓦斯。到了早上九點半，其中一台工兵車在建築物底部打出一個洞，大小足以讓人逃跑。

早上十一點，局勢看似對FBI有利，但四十五分鐘之後，據點其中一面後牆倒塌。

正午過後，卡梅爾山據點內好幾處冒出火舌。FBI偷偷緊急派遣消防車，但消防車停放在遙遠的待命區。聯邦幹員呼籲柯瑞許投降，領著大衛教派分支信徒到安全的地方。九位信徒離開據點，立刻遭到逮捕，其中一名是大衛．希伯杜。他們是最後一批離開的，而且是少數幾位逃過接下來災難的信徒。

這一小群信徒被帶離建築物後，沒多久據點內部就傳來槍響。由於看起來不像瞄準FBI，聯邦幹員相信建築物內的人不是自殺，就是射殺彼此。撲滅大火之後，發現七十六名大衛教派分支信徒死亡。二十二名兒童和幾名女信徒一直藏匿在水泥碉堡內，直到碉堡倒塌，壓在他們身上，無一倖免。有些人因重擊創傷立刻死亡，有些人因吸入煙霧稍晚才死。所有屍體都殘缺不全，找到的部分也焦黑到無法辨識。解剖報告列出的任何識別特徵都是根據有限的牙醫紀錄或DNA樣本推測。小朋友從未滿一歲到十三歲大都有，其中兩名死者才幾個小時大，很可能是在攻擊時出生的，隨著戰況開展刺激了孕婦的生產。其中四名是韋恩．馬丁的孩子。DNA檢驗證明其中十四名都是柯瑞許的親生孩子，包含兩名新生兒。

其中有七名青年，年紀介於十四到十九歲之間。據點的成人來自世界各地：二十三名美國人；二十名英國人，多數是牙買加裔；也有個別來自加拿大、以色列、澳洲和紐西蘭。二十個人都有槍傷，包括六名孩童，還有三歲的小孩被刺傷——很可能是大火燒起來威脅到大家時，對他們進行安樂死。

柯瑞許死在史蒂夫．施奈德旁邊，兩個人都死於致命的槍傷，無法確認死亡順序，但一致認為是施奈德先朝柯瑞許的頭射擊再自殺。

事後數月、甚至數年，還有一項懸而未決的爭議——究竟是哪個人或單位需要為引燃大火、吞噬卡梅爾山負責。據點裡偷偷設置了十一台監聽器，錄了多卷卡帶。FBI監控這些竊聽器，但主張平常有太多背景噪音，什麼都聽不清楚。事件過後許久，卡帶音質得以改善，錄音變得比較清晰，似乎顯示據點裡的大衛教派分支信徒在走廊潑灑燃料，還提醒彼此要依照柯瑞許的指示，只能在「他們第一次開著坦克進來時」點燃，「就在他們進來的時候」。

但FBI也確實根據竊聽器獲得的資訊即時行動。早上七點四十分，就在戰鬥工兵車開始打牆一個小時又四十分鐘之後，FBI總部得知柯瑞許告訴據點裡的人還不要點火，「留到最後一刻再點」。現場FBI戰術指揮官下令小組可以輪番使用軍事催淚瓦斯，這一步在運作計畫時沒有人同意，因為展開催淚瓦斯攻擊會引發火災。在這中間，竊聽器錄到施奈德說：「手稿快要完成了。」

FBI大約花了三個半小時商討行動，疑惑為何沒有人從據點現身時，被告知竊聽器似乎傳來大量信徒躲在塔下方的水泥碉堡的消息，催淚瓦斯無法影響到那裡。九十分鐘後，戰鬥工兵車輾進塔的一部分，屋頂塌陷。稍後，在聽到信徒討論要點燃火炬、現在是否該點火之後，FBI打落最後一個運作中的監視裝置。中午過後，人質救援隊察看局勢，等待協助任何出現的人，那時獲得回報，說看到有人在潑灑東西，看起來好像要點火。幾分鐘後，FBI裝有紅外線攝影機的飛機回報了第一起火災。

兩邊的情況完全亂成一片。和施奈德與柯瑞許通話已經不在選項內了——五點五十九分，在賽基（Sage）通話後不久，一台戰鬥工兵車不小心切斷了對話線。單位之間的回報令人困惑，現場單位看似在妨礙彼此，又或者只是獨立運作，同時也在努力彼此協調。可以責怪的地方很多，但後來的調查終於還給聯邦政府清白，政府不需要為火災負責任。柯瑞許需要為當日所有死者負責，大衛教派分支信徒也同樣有錯，他們沒有按照每日指令離開據點。

但陪審團的裁決並沒有說服新世界秩序陰謀論者，也無法說服反政府愛國者運動成員。慘劇發生之後，民眾開始呼籲「勿忘威科」（Remember Waco），其實就是籲請極右派團體提供武力，專注儲存武器，反抗未來可能發生的政府攻擊。威科大火第二週年，美國國內恐怖分子提莫西．麥克維（Timothy McVeigh）和泰瑞．尼可斯（Terry Nichols）在奧克拉荷馬市（Oklahoma City）的艾佛瑞德．P．默拉聯邦大樓（Alfred P. Murrah Federal Building）前引爆一卡車炸藥，這是一場被誤導的復仇，因為他們相信政府過度干涉了卡梅爾山。有一百六十八人死亡，大部分是聯邦人員，還有大樓托兒中心的孩子。曾參與威科對峙的許多FBI幹員也調查這件新的可怕罪行。

幾年後，凱西．舒德說：「我從來都不想去那裡。」但她認為，如果柯瑞許真的是先知，自己就能去天堂；如果他不是先知，就要為凱西的死負責。對凱西而言，這似乎值得冒險。另一位在對峙早期就離開據點的母親在一小時的訪問中提到柯瑞許二十四次。該名母親從未提及自己的先生或小孩，他們後來都死於威科大火。柯瑞許的力量讓許多人都上癮，無法自拔。

大火燒起來的時候，倖存者李文斯頓．費根人在監獄裡。他的孩子比他還要早離開據點，但他在監獄電視上看到烈火，那場大火後來燒死了他太太和母親。他因自願過失殺人和武器罪服刑十四年。他的孩子由在英國的哥哥扶養。對他而言，錄音卡帶的內容都是「精心策劃」和「虛假、捏造」的。二十五年後，他在《電訊報》的訪問中說自己根本認不出任何人的聲音。他還是相信柯瑞許預言的末日近了：「時機點還是由上帝掌控。柯瑞許告訴我們要看著事情的發展。」[9]

許多倖存者每年會在四月十九日舉行聚會，但不是在卡梅爾山，回去那裡太痛苦了。卡梅爾山還是教團的合法財產，現在據點已看不太到建築物的痕跡。訪客中心裡有寫著「威科：預言應驗」的橫幅，圍繞著「復臨運動七牧人」的照片，包括大衛・柯瑞許和懷愛倫。走在據點裡，如果沒有地圖，不大可能指出起火地點。在德州曾為據點的平坦土地上，有些地方現在看起來較為蒼白，部分地面讓人可以想像據點大致的平面圖。但德州這一區酷熱異常，在四月那一天的大火吞噬了七十六名信徒的土地上，重新長出的草已在豔陽下褪色了。

Chapter 7

施虐

基斯・拉尼爾和 NXIVM 教團

基斯．拉尼爾在任何邪教領袖中都顯得特出，不是因為他比其他人惡劣，而是因為他的邪教發生時代最接近當代。NXIVM 的成長顯而易見，一開始在柯林頓執政後期只是「高階人士成功課程」（Executive Success Program），直到二〇二〇年大選前幾天，法院針對拉尼爾的敲詐勒索和性販運罪行裁決最後量刑。打從一開始，他就利用平凡無奇的商業模式（多層次傳銷）吸引女人。他的追隨者包括烈酒大亨的女兒、有線電視劇配角演員、資深專業配音員，大家只是參加自我協助課程，最後卻捲入他的性奴後宮（共六人）。但即便情況如此異常，如果拉尼爾沒有在 NXIVM 內另外建立特別團體的話，還是可能掩人耳目。這個特別團體集結了幾名性奴隸，身上都烙印著他的姓名縮寫，並且被迫臣服於他們的終極「主人」。烙印的消息一傳到《紐約時報》（*New York Times*），整個金字塔型騙局倒塌了，拉尼爾逃亡到墨西哥，被美國聯邦政府逮捕，判處一百六十年刑期。

討女人愛的萬人迷

打從一開始，拉尼爾就受到弱女子吸引，始於一九六〇年在紐約布魯克林（Brooklyn）出生那一刻。佛洛伊德和當代心理學家都認為，男孩和母親的關係是影響許多終生行為模式的主因。年幼的基斯要照顧母親，母親是舞蹈老師，但酗酒又有心臟疾病[1]，培養出日後他賴以維生一輩子的才能。他後來創立了 NXIVM 提供自助課程，NXIVM 其中一條主要原則是，學員需要正視自己的情緒創傷，部分原因是他天生擅長發掘出那些弱點。隨著自助課程發展，整個團體也逐漸變成只崇拜基斯個人魅力的邪教組織，他也變得更會利用這些弱點，通常都是進行虐待行為，獲得自己的利益和扭曲的歡愉。

和女人培養關係時，拉尼爾後來都會向她們解釋，他的媽媽薇拉（Vera）常喝酒又獨自一人，身邊只有兒子陪伴。《浮華世界》（*Vanity Fair*）雜誌記者蘇珊娜．安德魯斯（Suzanna Andrews）訪問那些最了解拉尼爾的女人時，發現他聲稱自己必須時常照顧媽媽。[2]他的父親詹姆士（James）是廣告主管，駁斥兒子的言論，雖然他在基斯八歲時和太太離了婚。依詹姆士所言，薇拉雖然可能喝了太多酒，實在不應該，但從來不會多到成為問題。

儘管有些出入，但所有言論都同意這對父母讓獨子接受良好教育。約莫在六〇年代晚期，薇拉和詹姆士的婚姻開始觸礁時，他們發現兒子智力可能高於常人，帶他接受了測試。[3]測試結果確認兒子天資聰穎。拉尼爾長期交往的女友之一芭芭拉．鮑契（Barbara Bouchey）上加拿大廣播公司（CBC/Radio-Canada）的《揭祕：逃亡的NXIVM》（*Uncover: Escaping NXIVM*）Podcast告訴主持人喬許．布洛克（Josh Bloch），拉尼爾的父親說：「幾乎就像切換開關一樣。突然一夜之間，他變得像耶穌基督，比所有人都優越、都好，就像神明一樣。」[4]研究人員艾迪．布魯莫曼（Eddie Brummelman）、布萊德．J．布許曼（Brad J. Bushman）和同事一起研究自戀型人格障礙症的發展，發現「如果父母認為，比起其他兒童，自己的孩子非常特殊，孩子會內化「自己優於其他個體」此一觀念，這正是自戀的核心觀點。」[5]這說法似乎適用年幼的拉尼爾，他開始相信自己比其他人都還要好。

沒多久，他在學校的行為顯現自傲的新態度。《大紀元時報》（*Epoch Times*）記者博文．蕭（Bowen Xiao）訪問基斯的同學L. M.，他們都在一間華德福[13]的學校讀書。L. M.說道：「他總是炫耀自己有多聰明、數學有多好，四處走動，彷彿自己是迷你版教授……他的聲音非常溫柔，彬彬有禮。但只是假象。假象可以吸引人。」[6]自傲的行為可能是男孩相信自己才智過人的副產品。

他也可能是在學校發洩，藉以回應家裡的問題。[7]雙親離異之後，基斯和媽媽住，媽媽酒越喝越多了。在喬許．布洛克的訪問中，芭芭拉．鮑契說：「拉尼爾學會當個夜貓子，都是

---

13 Waldorf，以自然教育為主的人性化教育方式。

媽媽的緣故。他要一直在晚上看護媽媽，因為晚上媽媽偶爾會同時吃藥、吸毒又喝酒。」[8]

或許這種不健康的生活方式導致基斯在學校的叛逆行為日益嚴重。到了十歲左右，光是自誇已經無法讓他滿足了。他開始嘲諷甚至折磨同學。L. M.提到一則令人不安的故事，稱之為「小瓶子事件」。有天搭公車，他坐在基斯旁邊，不小心把姊姊嚴守的祕密說溜嘴。起初L. M.毫不擔心，畢竟大家都是朋友，他沒想過基斯會拿這個祕密做不利於他的事。但當天稍晚，拉尼爾走向L. M.說道：「你知道，這就像手上有這麼一個裝了毒的小瓶子，我可以拿著瓶子在你頭上晃。」[9]

出於困惑，L. M.問他這是什麼意思，他說明：「嗯，我不知道。我只是覺得如果我告訴你爸媽和姊姊，他們不會太開心。」

L. M.想不透他要什麼。他沒有要求任何事，也不是勒索，只是看起來好像很享受握有祕密，有威脅L. M.的權力。他沒有就此罷休。他偶爾會出其不意打給L. M.，說著同樣的三個字：「小瓶子」。就這樣一再重複，像私人暗號般，表示他會利用自己知道的祕密對L. M.不利。他持續騷擾L. M.，直到L. M.終於崩潰，告訴媽媽。在媽媽介入之後，沒完沒了的電話終於結束。

顯然拉尼爾還有另一面，在深夜照顧媽媽時就會浮現。進入青春期，他開始顯露自己較為柔軟的一面。鮑契告訴喬許．布洛克，薇拉坐在第一排觀看兒子和班上女生的互動。無意間聽到的對話讓她十分擔心，甚至有天晚上驚慌失措地打給前夫。鮑契回想詹姆士告訴她：「打電話到家裡的女生有數十位，薇拉不小心聽到兒子和女生的對話，兒子告訴每一位女生同樣的話：『你是特別的。你是我生命中重要的人，我愛你。』薇拉說：『他對不同女生都這麼說。』」他們不可能全都是特別的存在，薇拉的結論是兒子在騙人，而且手段高明。

羅伯特．海爾的「心理病態人格檢核表」也列出病態說謊、操弄、過度膨脹的自我，這幾項特質都在這名年輕的學生身上逐漸浮現。得知智力測驗結果而展現的自傲行為，顯示膨脹的自我。這起「小瓶子」事件暗示他喜歡玩影響力遊戲。他對數十名青少女說謊，證明他有能力欺騙大家。回頭看，這些令人擔憂的徵兆很顯眼，而且只會隨著長大成人更加明顯。但要是當初斷言十三歲大的基斯是心理病態，仍言之過早。

一九八四年，二十四歲的拉尼爾依然繼續勾搭青少女。《奧爾巴尼聯合時報》（*Albany Times Union*）記者詹姆士．M．奧達托（James M. Odato）和珍妮佛．吉許（Jennifer Gish）報導，那年拉尼爾在紐約特洛伊（Troy）社區劇團巡迴遇見十五歲的吉娜．梅利塔（Gina Melita）。和劇團朋友一起出遊時，拉尼爾對吉娜毛手毛腳，不僅摸了臉還有腿。吉娜涉世未深，就算被成年男性吃豆腐也不覺得怎麼樣。事實上，奧達托和吉許寫道：「她認為和年長聰明的男性在一起很酷……」[10]兩人開始出去玩，到遊戲機台中心打電動（後來據說拉尼爾從八〇年代的機台遊戲獲得靈感，要 NXIVM 成員稱呼他先鋒 Vanguard）。但拉尼爾和吉娜約會不是為了可以一起參與適齡活動：他想要更多成人面向的活動。不久他們開始交往，他在昏暗的房間裡鎖死房門，和吉娜做愛。[11]

那次的經驗很痛苦，吉娜嚇到了，她原本不知道會這麼痛。[12]儘管吉娜很不舒服，拉尼爾堅持再做一次。即便紐約州的法定同意性交年齡為十七歲，拉尼爾已經犯下強暴罪，還是不停纏著吉娜要再來一次。吉娜感受到這段關係不對勁[13]，提出分手，拉尼爾還建議兩人可以繼續維持性愛關係。吉娜告訴吉許和奧達托，那時她才明白拉尼爾根本不在乎她的感受。[14]

差不多同一時期，拉尼爾開始在安麗（Amway）工作。[15]他的工作夥伴艾瑞克・J・盧迪（Eric J. Roode）上喬許・布洛克的 Podcast 節目時提到，拉尼爾開始談論要成立公司，模仿安麗的多層次傳銷結構。也是在這時期，拉尼爾的首波邪教追隨想法開始萌芽。在他紐約克利夫頓公園（Clifton Park）的連棟住宅中，他開始引起眾人注目。他會主持長時間的哲學討論會，而且把討論會轉變成某種嬉皮聚會。數十名妙齡女子會聚在他家，聽他談論哲學。

除了想要改善安麗行銷結構的模糊夢想之外，拉尼爾也談及人類潛力運動，想要藉由解鎖隱藏在每個人體內尚未開發的資源，為世界帶來正向改變。懷疑論者認為這顯然就是自我沉迷的運動，參與者可以想像自己只要專注提升自我，就能改善全球狀況。但拉尼爾沒有這類遲疑，或許只是因為這種主題特別吸引在座聆聽的女性。拉尼爾用溫柔的聲音、看似全神貫注的態度，告訴在座所有人他看見每一個人的巨大潛力。依那時的一位友人所言，拉尼爾經常在聚會鼓勵大家「去實現他們擁有的各種天賦和才能」。[16]

很快地，參加聚會的人開始認為基斯・拉尼爾是某種宗師。沒多久，他開始談論要如何混合這些元素（創立多層次傳銷機構的目標，以及協助人類實現潛力的想望），他想要要結合兩者。最終大家達成共識，同意基斯應該再做一次智力測驗。他們很可能相信，如此一來可以增加可信度，讓他得以開啟夢想事業。[17]《奧爾巴尼聯合時報》記者艾琳・加德納・基尼（Irene Gardner Keeney）報導，於是在一九八七年，二十七歲的拉尼爾做了超高智商學會（Mega Society）贊助的四十八題智力測驗[18]，答對四十六題。這代表他的智商高達一七八，令人欣羨[19]。為了對照，大家相信愛因斯坦的智商為一六零。有了最新聰明才智名次，拉尼爾開始宣傳自己比發明相對論的男人「更聰明」。

光這樣還不夠，他立刻開始主張他的智商是落在極高的二四〇範圍。[20]雖然無人能重現事發經過，但拉尼爾和他的新高智商被登記在澳洲版《金氏世界紀錄》（*Guinness Book of World Records*），該紀錄宣傳他是地球上擁有最高智商的男人。[21]奧爾巴尼（Albany）當地報紙報導了這令人料想不到的發現：地方居民之中竟然有天才，並稱他為「千萬人才有一位」的天才。[22]記者似乎不太在意那份應該能證明拉尼爾智商的測驗，其實是不計時、帶回家做的測試。也沒有人想要費心檢查這位應該是天才的男子在大學的成績。

如果有人查過，就會發現基斯・拉尼爾從任色列理工學院（Rensselaer Polytechnic Institute）畢業，成績平均績點（GPA）為二・二六。[23]但是在一九八八年，都沒有人報導這些事實。拉尼爾現在二十八歲了，不停宣傳超高智商學會的測驗結果，還有他被收錄在《金氏世界紀錄》裡，將自己形塑成「全世界最聰明的男人」。就是這番聲稱協助他在兩年後建立了第一家以剝削為目的的企業：一間名叫「消費者購買線股份有限公司」（Consumers' Buyline, Inc.，簡稱CBI）的嶄新公司。如同安麗，CBI也是多層次傳銷公司，保證提供會員折扣較高的居家用品。就像在此之前的其他多層次傳銷公司一樣，CBI也說要提供會員金錢獎勵，招募更多新人加入。

CBI幾年下來營運得很不錯，大批訂購居家用品，幫會員省錢，就像好市多（Costco），只是沒有巨大方形建築。

拉尼爾有能力同時販售概念和他自身特殊的地位，也帶動了新會員的擴展業務。但打從一開始，拉尼爾也利用四處宣傳的機會，達成許多不道德的目的。CBI的運行方式比較像龐氏騙局，專門設計將會員費一路往上移動，最終交到公司總部。CBI也吸引了他可以剝削的女人。CBI是他經營的第一家公司，之後還開了許多家，但每一家似乎都給予他更多機會接觸脆弱的女性，走向有毒、日漸虐待的關係。

一九九一年：托妮・娜塔莉

一九九一年，托妮・娜塔莉（Toni Natalie）三十三歲，已婚，在紐約羅徹斯特（Rochester）的CBI推廣說明會遇見基斯・拉尼爾。托妮的先生拉斯第（Rusty）經營日曬機生意，現在生意陷入麻煩，想要兩人一起去看看CBI會員推廣說明會。拉斯第說：「他從任色列理工學院畢業，擁有三主修。是鋼琴家、單車選手、美國東海岸柔道冠軍，而且還想要拯救世界。」

托妮的第一印象是：這男的實際上好矮，而且有點愛吹噓，聲稱自己智商二四〇，十分可疑。但當他一開始推銷，托妮的疑慮漸漸消失。

說明會結束，托妮走向拉尼爾，如同她在回憶錄《課程：潛入基斯．拉尼爾的內心，NXIVM的起落浮沉》（暫譯，*The Program: Inside the Mind of Keith Raniere and the Rise and Fall of NXIVM*）中寫的，她問拉尼爾：「你智商二四〇，為什麼不去治療癌症或改變世界，把世界變成更美好的地方？」[24]拉尼爾答道：「我正在改變世界。我正在把世界變成更美好的地方。你不想一起來嗎？」[25]托妮報名了，才過了幾個月，就成為CBI最佳業務之一。到了一九九一年夏天，她和先生做得太成功，販售許多CBI會員資格，拉尼爾頒發公司的區域最高獎給他們，獎項包括一萬六千美元獎金。拉尼爾想要托妮搬到奧爾巴尼，在CBI全職工作，一直在電話中重複提這件事。托妮總是拒絕，自己還有兒子、先生和家庭要顧。拉尼爾再次表現得像擁有更崇高的使命，他告訴托妮：「我覺得你生來注定要協助我改變世界。」

托妮提醒拉尼爾，她已婚而且中學肄業。拉尼爾回應道，聰明才智分很多類型，托妮是情緒感情方面的聰明。經過這次推銷，她和先生拉斯第決定開三個半小時的車，到克利夫頓公園親眼瞧瞧這家公司。CBI大樓外觀毫不起眼，但辦公室則是另一番風景。一九九一年，個人電腦還不是標準辦公設備，但辦公室有數十位CBI員工在多台電腦前敲打鍵盤。

托妮也很驚喜地看到女性員工比男性多。其中一名女員工還帶著女兒一起上班，女兒十二歲，戴著牙套。托妮看著那位十二歲的女孩翻閱代數課本。拉尼爾甚至特別撥出時間教女孩數學。第一次造訪，托妮印象最深的是公司執行長花時間教女孩數學。多年後，托妮才明白「師生關係」無法解釋三十一歲男子和十幾歲少女之間的複雜關係。法院文件顯示，到了一九九一年，拉尼爾已經強暴該名女孩整整一年，誘騙手法就是提供代數和拉丁文指導。私底下，他真正教女孩的是如何「像成年人一樣擁抱，骨盆要對著骨盆」，接著奪走女孩的童貞。[26]《奧爾巴尼聯合時報》後續報導指出，拉尼爾持續性侵該名女孩，「不單在自己家中，也會在空蕩蕩的辦公室和電梯裡，還有該商辦區的CBI清潔雜物間……」[27]

托妮造訪了CBI總部之後，拉尼爾開始不停打給她，試圖說服她搬家。拉斯第覺得拉尼爾這麼頻繁打電話很有趣，不覺得受到威脅，畢竟拉尼爾不是大帥哥。但拉尼爾靠著強大的傾聽能力來彌補外貌。而且托妮似乎未曾遇到任何人如此關心自己。托妮自然而然開始分享，告訴這位陌生人，先生已經兩年沒和她做愛了。她承認中輟生的身分讓自己非常不安。她如此信任拉尼爾，甚至說出自己四歲時慘遭性虐待。[28]

拉尼爾日後會把他用來安撫個人情緒觸發點的手法命名為「整合」。但面對托妮時，他還在試圖解決這手法帶來的各種問題。托妮起初拒絕了拉尼爾，一直到幾個月後，她先生的日曬機生意終於跨掉，於是她向拉尼爾尋求金援。拉尼爾聲稱自己沒辦法簽任何貸款，除非托妮和公司經理喬治．魏斯（George Weiss）見面。於是，托妮非得再次前往位於克利夫頓公園的CBI總部不可。

這次，她和其他與拉尼爾比較親近的人即將發現：拉尼爾遠比表面上還要糟糕。兩人在拉尼爾家中碰面，家裡非常髒亂：垃圾桶的垃圾都滿出來、洗碗槽堆疊的碗盤殘留著結塊的食物、地板上四處都是衣物，就像青少年的房間。顯然，世界上最聰明的男人不知道怎麼打掃和保持居家環境整潔。她注意到的第二件事是拉尼爾的同居夥伴。拉尼爾還是和大學時期的前女友凱倫・溫特萊納（Karen Unterreiner），以及前女友的兩位朋友帕姆・卡弗里茨（Pam Cafritz，她的父親剛好是有影響力的華盛頓特區社交名流）和克莉絲汀・基夫（Kristin Keeffe）同住。這組合很奇怪，事業成功的高階人士和三名女性同住，好像他們是大學室友。

拉尼爾說到做到，帶托妮去見喬治・魏斯。但比起同意貸款，拉尼爾反而提供托妮工作機會，拉尼爾想為CBI新增一門肌膚保養生意，希望交由托妮負責。雖然薪水非常好，足以償付先生的債，但有個大缺點：托妮必須搬到克利夫頓公園。雖然拉尼爾之前已經拜託托妮搬家很多次，但現在托妮對高薪很心動，而且覺得可以說服先生和兒子一起搬家，畢竟都是為了全家好。但是托妮也會被迫斷了與家鄉手足、朋友及生意夥伴的聯絡。托妮答應一定會考慮這份工作機會。會面結束後，拉尼爾要求送托妮回飯店，跟著她走去貝斯特韋斯特飯店（Best Western）的房間……頓時一切變得再明顯不過，這才是拉尼爾邀請托妮來克利夫頓公園的原因。現在兩人獨處，拉尼爾執行日後的招牌策略：表面上假裝表達同情，藉此瓦解甚至擊潰脆弱的女人，讓女人持續說上十二到十三個小時，要對方說出來童年遭遇的性虐待。托妮就這麼說了：四歲的時候，某位親戚性侵她，接著持續了好幾年。但每一次托妮講完痛苦的回憶，拉尼爾就有另一個問題。托妮要求喘口氣，想要闔眼休息。拉尼爾向托妮保證，他想要的只是減輕托妮的痛苦。接著要求托妮再說一次這個經歷，從頭說起。托妮終於崩潰了，不停哭泣，身心俱疲。

看到托妮流淚，拉尼爾說自己可以幫助她，只是她需要接受這份工作，搬到克利夫頓公園。托妮被打敗了，最終讓步，但到了那一刻，更主要是因為想要鬆一口氣，而不是真的想要讓步。她屈服於拉尼爾有點直接又持續不斷的精神攻擊，毫不令人意外。研究員帕姆・洛（Pam Lowe）、凱西・漢弗瑞斯（Cathy Humphreys）和賽門・J・威廉斯（Simon J. Williams）做了一系列研究，發現「睡眠剝奪和建立權力體系、控制他人之間有所關聯」，此外也發現「睡眠剝奪很顯然是惡人用於虐待的直接策略……為了削弱女性的心理和身體復原力。」[29]即便拉尼爾沒有虐待托妮的身體，但他不停歇的情緒勒索同樣有害。托妮中計了，認為他富有同理心，但其實只是控制策略。

接下來五個月，托妮開始在CBI全職工作，和拉尼爾一同出差，出席各地說明會，介紹公司可以改變人的美德。有次出差，拉尼爾告訴托妮他愛上托妮了。托妮格外高興，世界上最聰明的男人竟會愛上她這個中學中輟生。她和拉斯第離婚，帶著兒子離開，說服自己這是早就該做的必要改變。

一九九二年春天，拉尼爾的室友克莉絲汀・基夫幫托妮找房子，找到的房子離拉尼爾家很近，近到拉尼爾只要走過自家後院就能到托妮家。托妮搬進新家的第一晚，拉尼爾馬上善用地利之便，前往拜訪，兩人做了愛。不像拉斯第（很久沒有碰托妮），拉尼爾精力無窮。經歷幾年冷宮生活後，現在托妮證明自己還有慾望，感到重建自我了。

拉尼爾也反覆告訴托妮，他愛托妮，托妮讓他成為完整的人，托妮是特別的存在。但隔天早上在CBI停車場，托妮還沉浸在餘溫中，傾身親了拉尼爾臉頰，他卻立刻推開她，說他們的關係必須保密，不然大家會認為托妮的工作都是睡來的。拉尼爾說，這樣講還比較可信，畢竟事實上托妮連中學都沒畢業。

他的話語像是直接賞了托妮一個耳光。私底下，拉尼爾說托妮沒有學歷證明根本沒關係；但托妮一旦在拉尼爾的感情中找到安全感，拉尼爾態度說變就變，變得冷淡又粗魯，像績效評估會遇到的糟糕經理。從那時起，拉尼爾就堅持不能公開他們的戀情。

另一方面，公司本身也沒發展得多好。一九九二年五月二十一日，《奧爾巴尼聯合時報》揭露CBI是金字塔型騙局，詳細報導紐約、緬因（Maine）對CBI的深入調查，還有阿肯色（Arkansas）檢察長持續採取的法律行動。[30]

拉尼爾回應這篇報導的方式，好似將報導視為個人攻擊，而根據精神分析師海因茲·寇哈特（Heinz Kohut）所言，拉尼爾用寇哈特形容為「自戀暴怒」（narcissistic rage）的順序揭露自身缺點。拉尼爾為記者貼上騙子的標籤，阿肯色州檢察長則是挾恨報復他。他錄了五分鐘的訊息給CBI員工，頌揚自己的高道德標準。起初大家相信他，但拉尼爾對員工的支持不滿意，於是越來越疑神疑鬼，控訴政府監視他、開他信箱、竊聽電話。[31]

在辦公室外，他的情緒甚至更不穩定，常常在晚上去托妮家，不分時段，搖醒托妮，命令托妮做他稱之為「身體安慰」（physical comfort）的事。托妮漸漸受不了他的情緒爆發，也受不了他既要性愛，又拒絕公開承認兩人的關係。托妮威脅要離開，但拉尼爾說服托妮再次接受他，同意公開承認兩人關係，接著搬去和托妮及其子一同生活。要是早幾個月，托妮可能會很興奮，但現在托妮覺得自己犯了大錯，而且鑄下大錯的感覺揮之不去。

表面上看來，托妮似乎做了正確的決定。九〇年代中期，多起法律訴訟纏身，CBI 終於撐不住而倒閉，拉尼爾和托妮開創新事業「全國健康網路」（National Health Network），生意興隆，用七折價販售保健食品。他們在紐約沃特福（Waterford）買房，美輪美奐，兩人一起養育托妮的兒子麥可（Michael）。

但三年後，托妮終於承認拉尼爾有些地方非常不對勁。首先，拉尼爾拒絕做任何家事。他就在家裡四處坐著，預期托妮會無微不至服侍他。再來，他就是個夜貓子，整晚不睡，白天才睡。和前夫不同，拉尼爾每天都要做愛。如果托妮性致缺缺，他會硬上。托妮開始習慣把自己關在衣櫃裡，就為了擺脫拉尼爾。

托妮覺得被困住了，對於拉尼爾暴力又操縱的行為，她無人可傾訴，直到遇見了南希・薩爾茲曼（Nancy Salzman）。薩爾茲曼光顧保健食品公司，想看看有沒有產品可以解決嚴重便祕。托妮提供多樣保健食品，解決她的便祕問題，薩爾茲曼想要提供自己領域的專業治療，報答恩情。托妮告訴《The Cut》報導網站的凱蒂・希尼（Katie Heaney），那時薩爾茲曼說自己是全球第二位神經語言程式學（Neurolinguistic Programming，簡稱 NLP）專家，雖然這完全是想像出來的資格，但薩爾茲曼表現得像擁有這份資格的人可以當心理師。[32]看起來薩爾茲曼這項伎倆還頗為成功，說服了聯合愛迪生（Con Edison）和美國運通（American Express）這類大公司的高階主管付錢請她定期諮商。托妮也一樣相信了薩爾茲曼的資格，同意碰面。第一次諮商，她告訴薩爾茲曼，自己雖然忽略拉尼爾沒完沒了的性攻擊，還是試圖理解拉尼爾難以預測的行為，以及為何拉尼爾拒絕在晚上睡覺。薩爾茲曼微笑回應道：「噢，這很簡單！我可以幫你，他是反社會人格者。」[33]

接著，薩爾茲曼建議托妮定期接受諮商療程。但奇怪的是，到了下一次諮商，薩爾茲曼改變口徑，不再談論拉尼爾的心理病態行為。取而代之，薩爾茲曼告訴托妮，她的親密關係問題源自兒時遭遇的性虐待。這真的非常離奇，因為關於童年創傷，托妮一個字都沒提。儘管如此，托妮還是很感激有人看出自己的困境，沒想到要詢問薩爾茲曼的消息來源，又或者為什麼薩爾茲曼徹底轉變對拉尼爾的意見。要是托妮進一步詢問，或許會明白心態轉變的原因就是拉尼爾本人。

到了托妮坐下來做第二次諮商時，拉尼爾早就影響了薩爾茲曼。第一次諮商結束，薩爾茲曼違反心理師保密原則，告訴了拉尼爾。隨後，拉尼爾向神經語言程式學家訂了一堂自己的諮商。托妮接著連續參與了三次諮商，等到她拼湊起所有拼圖，為時已晚。托妮已經落入煤氣燈效應的圈套，被一位佯裝心理師的女人操縱了心理。

同時，拉尼爾和薩爾茲曼開始規劃創立新公司，這次的產品價值會非常高，而且幾乎無法定義。托妮在書中寫道：「基斯認為幫消費者省錢這招太平凡了，配不上他的奇才——此外，還充斥著法律風險。人生教練、高階主管成功心法和自我提升等空話噱頭……才是該走的方向。」[34]就像前一家公司CBI，拉尼爾決定新創事業「高階人士成功課程」（簡稱ESP）也會採取多層次傳銷結構。然而這次，會員不需要哄騙大眾購買居家用品，而是販售自我提升「技術」（Technology）。所謂的科學課程，和薩爾茲曼提供給企業客戶的神經語言程式學諮商大同小異。

兩人展開高階人士成功課程前置作業時，拉尼爾和托妮的關係開始瓦解。一九九九年四月某個午後，拉尼爾和托妮大吵一架。托妮唯一一次說服拉尼爾幫忙洗衣服，告訴拉尼爾不要烘新毛衣，但拉尼爾還是放進烘衣機，結果毛衣縮水，拉尼爾拒絕負責，反而朝托妮大喊，要托妮承認自己錯了。托妮不願意照做，拉尼爾便要求托妮向自己乞求「重申權威」。拉尼爾甚至寄了封分手信，附上一本米爾頓（Milton）《失樂園》（*Paradise Lost*）影本，指出幾處路西法悖逆神的地方和托妮的行為簡直毫無二致。這次吵架突顯出托妮覺得拉尼爾在這段關係中有問題的地方：他瞧不起人、自我中心、冷酷無情。終於，托妮永遠離開了拉尼爾。拉尼爾展開報復，提出幾次站不住腳的訴訟，沒有一次成功，繼續騷擾了托妮好幾年，要他的跟隨者闖入托妮家、偷走信件、打給當地電力、電信公司，終止托妮的電信服務和電力供應。

儘管這些報復持續不斷，拉尼爾最終還是放手了。拉尼爾心中有不同以往的商業規劃，還有一位全新的女性合作夥伴：南希・薩爾茲曼。有了薩爾茲曼的參與，拉尼爾開啟ESP

事業……經過多年發展，ESP幾乎不知不覺地從拉尼爾熟悉的多層次傳銷模式（托妮告訴《The Cut》：「因為基斯本質上是個挫敗的安麗業務。」），轉變成NXIVM愉虐SM邪教團體。

一九九九年：南希・薩爾茲曼

幾乎從初次碰面，薩爾茲曼和拉尼爾就開始規劃自我提升技術。筆記本寫滿模糊的術語，例如「理性探究」（rational inquiry），對他們而言這就像邏輯一致性或高效洞察力，實際上，這指的是只有參加高階人士成功課程才能習得的技術，課程最低費用為七千五百美元。另一個術語是「意義探索」（Explorations of Meaning，簡稱EM）——密集互動課程，學員接受指引，探索負面經驗，再重新整合、減輕心理負擔。另一個主要概念包括辨認「解構」（Disintegration）：生活中發生的心理失能狀態限制了個人，但其實可以治療，而且完全解決，只要參加更高昂的密集課程即可。拉尼爾日後宣稱，他的高階人士成功課程可以幫大家成為奧斯卡得獎女演員、奧林匹克奪牌運動員或成功的新創公司老闆，儘管他從來無法舉出任何一位ESP畢業生作為實例。

打從一開始，ESP吸引的盡是聰明、能幹的人，口袋有閒錢，又剛好走到人生路口，尋尋覓覓某種知識，拉尼爾和薩爾茲曼似乎正好擁有該知識。ESP承諾「更多、更好」，用字模稜兩可：更成功、更幸福、更確信，代表忠實的學員（自稱 ESP 人）走在正確的道路上。ESP 人真心誠意渴望提升自我。因為自我提升密集課程費用貴得離譜，也就代表任何報名的人都可能遇到其他同樣聰明又成功的人——遇到這種人，他們就能確認自己加入的是精選的菁英團體。拉尼爾利用這種互相吹捧的社群，高效賺取金錢。

從一開始，拉尼爾和薩爾茲曼就打算利用曖昧的承諾和高昂的註冊費，打造一頭龐大的吸金怪物。但他們幾乎無意識地設立了第二項推力，也就是嚴格階級制度，多年後還轉變成病態機制，專門用來展示支配關係。一開始可說毫無惡意，單純只是分級系統（拉尼爾的靈感源自山達基的「運作中的希坦」（Operating Thetan）八等級），每位註冊 ESP 的學員都會看到分級標示，後來 NXIVM 則使用榮譽綬帶，每位 ESP 人都要披在肩上。榮譽綬帶系統決定了每位成員的階級位置。新生是白色，接著是黃色、橘色、綠色、藍色、紫色。金色專屬薩爾茲曼，而只有拉尼爾一人達到「超凡」（Ethereal）階級，代表他應該是無色——雖然他的榮譽綬帶看起來是白色。要晉升到下一個色彩階級，可以繳費，也可以招募新人，只不過價格驚人。綠色大概一百萬美元，藍色五百萬美元左右。最終，提供拉尼爾性服務也算。唯三位紫色會員都是他的女友或前女友，像是托妮。除了視覺分級之外，學員都應該稱呼薩爾茲曼「長官」（Prefect），拉尼爾「先鋒」（Vanguard）。

為了朝下一個顏色邁進，學員首先必須為自己的榮譽綬帶獲得四道階級條紋。這也是參考山達基的「通往完全自由之橋」，一種提供學員一條畢業途徑的方法。ESP／NXIVM 版本提供階級條紋途徑，供學員朝超凡前進。榮譽綬帶階級也提供團體內衡量相對等級的方法。學員總是知道自己位在團體的哪階，這是重要的劃分，因為位階較低的學員需要遵守更多規則，學習如何和較高階的學員互動。首先是特別的握手方式。一般的握手只會發生在同階級之間，手在相同高度，但較低階的ESP人必須把手放在較高階學員的手下方。ESP人也得起立以示對高階學員的尊重，而且必須互相鞠躬，此外所有人都得向先鋒鞠躬。[35]

所有拉尼爾強制的儀式規矩、等級分階綬帶都具備重要功能。按照荷蘭格羅寧根大學（University of Gronigen in the Netherlands）的社會心理學家南吉・庫登堡（Namkje Koudenburg）和同僚所言：「眾多個人聚在一起做些特定的行動，一種『我們』的感受就會默默浮現。」[36]透過給予ESP人穿著打扮、互動方式的統一系統，拉尼爾打造了新的團體認同。

ESP人也學到世界其實分成兩群：寄生群體和生產群體。雖然所有人類生來就是寄生群體，相互依賴才能生存，但有些人類一直處在這樣停滯不前的狀態，長年利用他人的成功來維持自身存活。等號的另一端是自給自足的生產群體——那些獨立自強、動力足夠獲取成功的人類。拉尼爾自稱先鋒，佩戴超凡榮譽綬帶，抬舉自己為終極生產者。

如果學員想要離開寄生狀態，必須全身心投入ESP課程。每堂EM課程都由一名教練帶領一名學員，透過分析思緒或行為，來尋找學員負面情緒反應的根源。一找到根源（通常是某個童年記憶或其他成長經驗），教練會詢問學員特定問題，來幫助學員削弱該童年時刻的影響力。這樣一來，「意義探索」其實與山達基的「聽析期間」（auditing session）十分相似。

ESP保證只要做了足夠的意義探索，學員就可以摧毀負面情緒的終極根源，負面情緒只會妨礙學員獲得成功。換句話說，EM會給予學員類似生產者的有效「心態」。越多人達到這種心態，世界就會越美好。為了幫助世界提高生產者數量，拉尼爾強烈鼓勵學員招募新人。事實上，白色學員在階級條紋途徑前進的主要機制，不是透過意義探索的自我提升，而是招募新人（帶來的收入可以計算）。

很可能沒有學員比得上「長官」的招募速度。薩爾茲曼在一本ESP手冊寫道：「自從人類發明文字書寫以來，對人類而言，應該沒有任何發明比拉尼爾先生的技術來得重要。」一九九九年，薩爾茲曼鼓勵女兒——十九歲的蜜雪兒（Michelle）[37]和二十二歲的蘿倫（Lauren）[38]參加課程。法庭證詞顯示（由《滾石》雜誌記者EJ・迪克森（EJ Dickson）重述），加入ESP幾年後，蘿倫開始和拉尼爾發生性關係[39]。不太清楚她媽媽是否立刻就知情，因為媽媽自己也是加入沒多久就和拉尼爾上床。把家人都拖進ESP之後，長官繼續招募朋友，在二〇〇〇年說服友人芭芭拉・鮑契加入。鮑契後來寫到第一次參加工作坊時的狀況：「我那時四十歲，管理九千萬美元，每年成長金額為九十萬美元，客戶最小規模為一百萬美元，而我的存款高達一百五十萬美元。」[40]儘管人人稱羨，鮑契就像許多ESP早期招募的學員，人生正好經歷難關，第二段婚姻快要瓦解了。鮑契也是立刻愛上拉尼爾，借他錢、在他描述不實的文件上簽字。後來拉尼爾要鮑契也在商品期貨交易單上一併簽名，但押錯了寶，於是鮑契為了支付追加保證金，花光畢生積蓄。一直要到好幾年之後，鮑契才能重新累積近似當年積蓄的金額——她開始管理NXIVM兩位明星招募員莎拉．布朗夫曼（Sara Bronfman）和克萊兒．布朗夫曼（Clare Bronfman）的數百萬繼承財產，兩人是施格蘭（Seagram）酒業集團資產的共同繼承人，也都和拉尼爾發生性關係。

二〇〇二年：丹妮拉

二〇〇二年初，十六歲的丹妮拉（Daniela，姓氏不透露）富有的墨西哥籍雙親赫克托（Hector）和亞莉安娜（Adriana）轉信踏上ESP人道路，送女兒一堂ESP的十六天密集課程。課堂教導引起兩人共鳴，也希望女兒可以從中獲益。丹妮拉打算去瑞士的高級寄宿學校，夢想是到哈佛讀醫學。根據CNN調查員伊曼紐拉．格林伯格（Emanuella Grinberg）和索尼婭．莫格（Sonia Moghe）的報導，在密集課程期間，丹妮拉學到一個數學方程式，該方程式證明世界會在十到十五年內結束。課程尾聲，丹妮拉開始認為自己的目標毫無意義，改變世界的唯一方法就是參加ESP。[41]

蘿倫．薩爾茲曼在母親不斷要求招募新學員的壓力下，開口要丹妮拉放棄之後的正規教育，改由「先鋒」指導課業，丹妮拉同意了。[42]但是丹妮拉抵達奧爾巴尼之後，拉尼爾沒有指導太多課業。事實上，根據後來法庭上的證詞，拉尼爾克意忽略丹妮拉，因此丹妮拉不得不打掃辦公室，就為了有事可做。[43]一年後，拉尼爾終於開始關心他，但那時他也無意指導課業。

四十二歲的拉尼爾親吻才十幾歲的丹妮拉，提議兩人可以做愛，接著又說丹妮拉還太年輕了。

但丹妮拉十八歲生日後沒幾天，拉尼爾態度丕變，不覺得二十五歲的差距有任何問題，還在無人辦公室骯髒的床墊上替丹妮拉口交。[44]根據托妮《課程》一書所寫，丹妮拉接下來幾年都在「輸入資料、打掃高階 ESP 人的房屋、按照指令替先鋒口交」。[45]儘管這麼頻繁替拉尼爾口交，丹妮拉後來在法庭證詞中形容自己「有點性冷感」，一直要到二十五歲左右，才發覺自己有喜歡的情感，對象是年齡相仿的班．邁爾斯（Ben Myers），他在 ESP 的 IT 部門工作。某晚，一群 ESP 人聚在一起看《星際爭霸戰》，丹妮拉和邁爾斯剛好坐比較近，丹妮拉發現坐在邁爾斯身邊讓身體產生強烈的感受，而丹妮拉很喜歡這些感受。第二次團體活動，兩人接吻。丹妮拉告訴拉尼爾戀愛帶來的美好新感受，拉尼爾的反應不是一般人想像所謂全世界最聰明的男人會做的事：拉尼爾大發雷霆，反鎖在浴室內，拒絕出來，告訴丹妮拉他和性伴侶之間有神祕結合，如果性伴侶做錯事，他的身體就會感到疼痛。

丹妮拉自白，但餘波未平，拉尼爾確保現年二十四的丹妮拉被關在自家房內，說服她父母幫忙執行懲罰（兩人相信 ESP 階級所宣稱，視拉尼爾為超凡存在）。起初是限制她使用電話和 iPad，後來改成禁止，還讓她拿不到金錢和移民文件，也不能聯絡外界。丹妮拉的家人成了她自己的獄卒。雖然丹妮拉可以用紙筆，但僅限於寫信給拉尼爾解釋自己犯下的過錯。她被告知必須「重建並增強」對拉尼爾的性慾，拉尼爾寄電子郵件詢問她和邁爾斯實際上對彼此做了哪些事，要求她一一交代清楚細節。根據莎拉．伯曼（Sarah Berman）在《Vice》雜誌的報導，拉尼爾問道：「他伸手進衣服碰觸你的陰部，還是隔著衣服？」另一封電子郵件，他命令道：「我需要成為你人生的全部。事情只能這樣。」[46]

後來丹妮拉告訴法院，在孤立期間，自己實在太孤單、太煩躁，開始抓手臂，無法停止。拉尼爾愛好瘦巴巴的女人，這也是強加於丹妮拉的其中一項條件，如果丹妮拉想要重獲自由，體重必須低於一百磅（約四十五公斤）。如果丹妮拉偏離減重飲食，例如多吃一點或喝咖啡加糖，就會被控「傲慢」。這成為拉尼爾標準的責備：任何違背他心中想望的意志展現，都是所謂「傲慢」缺陷的證據。丹妮拉無事可做，於是開始想像要去附近沃爾瑪（Walmart）會做的每一步驟，沿路會看到的每樣事物，店內會播放什麼音樂。房門從來沒有上鎖，但是她沒有錢、沒有工作許可證、沒有護照。丹妮拉告訴法院，蘿倫是唯一的訪客，但很殘酷地，每次都是告知最新的規定，規定一直在變，丹妮拉必須遵守這些規定，才能贏回拉尼爾的青睞。今天丹妮拉要寫情書給拉尼爾，隔天要懇求拉尼爾放她自由。丹妮拉把信都交給蘿倫傳遞，但拉尼爾一封都沒打開。丹妮拉被監禁在房內長達兩年，這期間收集了清潔用品想要結束生命。

丹妮拉知道如果沒有解決自己的「道德瑕疵」就離開，父母就會拋棄她。[47]最後，她實在無法忍受，走出房門當面對質唯一的超凡存在。拉尼爾看到丹妮拉，轉身就跑，左閃右躲穿過一群會員（Nxian）想找地方藏身。丹妮拉還沒能當面對質，一群階級較低的學員就先抓住她。她的父親和另一位較高階的學員開車載她回墨西哥，留她獨自一人。[48]

她的自由有苦有樂。丹妮拉的姊妹，年紀較大的瑪里亞娜（Mariana）和年紀較小的卡蜜拉

（Camila）依然是拉尼爾的性伴侶。拉尼爾不避孕，主張避孕措施會干擾性伴侶應該獲得的靈性益處，姊妹倆都懷孕了，也都墮胎了。依檢察官所言，拉尼爾稱呼丹妮拉最小的妹妹「童貞卡蜜拉」（Virgin Camila）[49]或簡稱 VC。二〇〇五年，卡蜜拉十五歲時，拉尼爾就開始和她發生性關係。此外，卡蜜拉還是薩爾茲曼的女僕。這些都無法讓她獲得特別待遇。拉尼爾也會騷擾卡蜜拉，要她減重到一百磅以下，強迫她在 Tinder 上「找其他處女」來和拉尼爾做愛。[50]

兩人的 WhatsApp 訊息也顯示卡蜜拉可能是拉尼爾的第一位奴隸。二〇一五年，拉尼爾強迫已經二十五歲的卡蜜拉也去找個像她自己一樣的奴隸，傳訊息對她說：「我覺得你擁有自己的奴隸會很好，替我誘騙訓練，當成取悅我的工具。」[51]另一則訊息中，拉尼爾詢問卡蜜拉是否同意被烙印。卡蜜拉語帶驚恐回覆：「什麼意思？像牛隻一樣被烙印？」丹尼爾輕挑地回道：「難道你不想為我燃燒嗎？」[52]卡蜜拉不想，之後也不必承受拉尼爾姓名字首烙印在肌膚上的痛苦。[53]

二〇〇二年：布朗夫曼姊妹，莎拉和克萊兒

儘管拉尼爾定期精神錯亂，還習慣把青少女變成性獵物，ESP最初十年仍舊有體面的公眾形象，足以引誘施格蘭酒業帝國數十億美元的繼承人莎拉·布朗夫曼（Sara Bronfman）。二〇〇二年，二十五歲的莎拉和愛爾蘭職業賽馬騎師羅南·克拉克（Ronan Clarke）四個月的

婚姻岌岌可危[54]，這一失敗讓平常快樂的派對女孩開始尋找人生答案。家族友人蘇珊・懷特（Susan White）推薦ESP，莎拉決定試試看自我提升課程。就像許多二十幾歲的年輕人，莎拉的工作經驗有好有壞。她離開紐約大學（New York University），在英屬土克凱可群島（Turks and Caicos）經營跳傘事業，和歐洲賽馬圈人士交流，在比利時參加派對，她第一次聽聞ESP就是在比利時。她在墨西哥市參加密集課程，拉尼爾在那裡有一群富裕的學員，據稱前墨西哥總統卡洛斯・薩利納斯・德戈塔里（Carlos Salinas de Gortari）的兒子埃米利亞諾（Emiliano）和時任總統比森特・福克斯・克薩達（Vicente Fox Quesada）的女兒安娜・克莉絲蒂娜（Ana Cristina）都是學員。莎拉迅速感受到ESP課程的好處，甚至鼓勵妹妹克萊兒（Clare）參加看看。[55]

這對姊妹正好相反：莎拉是快樂有活力的金髮社交名流，克萊兒則是幹勁十足的黑髮奧運明日之星。莎拉環遊世界，追逐不同熱情，但只有三分鐘熱度；克萊兒是馬術大獎賽獲勝選手。蘇珊娜・安德魯斯在《浮華世界》報導布朗夫曼姊妹，寫道克萊兒「有一股叛逆的氣質……她會告訴大家自己已經決定接下來的人生只會與馬匹為伍，因為她不喜歡人類。」[56]

拉尼爾鎖定這對姊妹，因為兩人都富到流油。父親埃德加・布朗夫曼（Edgar Bronfman）身價估計淨值為二十六億美元，兩位女兒的信託基金據說都高達九位數。[57]她們在ESP的初期，拉尼爾就特地為性格倔強的克萊兒開設一系列一對一課程。[58]兩人在這些私人課程的討論沒有留下任何紀錄或在法庭證詞中提及，但是到了課堂尾聲，克萊兒變得跟姊姊一樣對

ESP 充滿熱忱。事實上，據多名前學員所言，克萊兒徹底愛上拉尼爾。[59]

現在姊妹倆都上鉤，四十三歲的拉尼爾已經做好準備，布朗夫曼家族現在逃不出手掌心了。二〇〇三年，拉尼爾逼使莎拉、克萊兒和父親會面。這要求很艱難。埃德加和她們的媽媽（第二任或第三任妻子，因為他的第二段婚姻被法院宣布無效）離婚之後，這對姊妹童年多數時間都沒有和布朗夫曼家族同住，和父親第一段婚姻的孩子沒有任何接觸。他們也不像埃德加較年長的孩子一樣在紐約上流社交圈長大，也沒有上高級昂貴的私校。雖然兩人可以動用個人的鉅額信託基金，但各自都和父親有些問題，基本上父親都缺席她們的童年。

拉尼爾要她們聯絡父親時，姊妹倆遲疑不決，但還是主動聯絡了，而且發現父親也希望修復和女兒的緊張關係。埃德加報名一堂 ESP 專為 VIP 規劃的個人課程，由薩爾茲曼親自指導，費用十萬美元。起初，埃德加愛死了這些課程，開始派專機接送薩爾茲曼到他在維吉尼亞州的家，這麼一來薩爾茲曼就能提供一對一 ESP 療程。[60]

現在拉尼爾的客戶有一名億萬富翁，會員人數三千七百名，所有會員都支付高昂的密集課程費用，令人意外的是，許多會員都會定期上課。不幸的是，這樣的成功總是伴隨一樣缺點：稅務人員隨時可能會來電。依托妮所言，ESP 的利潤之高，拉尼爾突然獲得大筆財富，開始運用策略避稅。拉尼爾切割 ESP 事業旗下多家公司，全部歸入一家雨傘公司名下。在這樣的結構下，拉尼爾沒有直接持有公司，但可以動用公司收入。拉尼爾為新事業命名：NXIVM。[61]

大約此時，拉尼爾收到一則他覺得很幸運的消息：《富比士》（*Forbes*）雜誌想要專訪拉尼爾。他急欲合作，相信全國曝光會迎來更大規模的成功，但他錯得離譜。《富比士》記者邁克爾．弗里德曼（Michael Freedman）仔細調查，找到幾位有勇氣描述課程的內部運作方式的 NXIVM 脫團人士，例如托妮等人。弗里德曼寫道：

> 他的教導總是神祕，充斥著自私自利和無法理解的術語，談的都是倫理和價值，定義基準則是盲目宏大的道德信條，和安．蘭德（Ayn Rand）小說中幹勁滿滿的角色十分類似。他的拿手好戲是把自身利益視為至高無上，不要因他人所想而驅動，避開「寄生群體」（他用來稱呼那些需要協助的人）；唯有如此，才可以忠於自我，真正合乎「倫理道德」。當然，反過來說就是，此一世界觀不支持慈善、團隊合作、同理等其他美德，但或許只是我們還不太明白。[62]

這篇文章毀壞力道極大，威脅到他和克萊兒的關係，整個 NXIVM 帝國陷入危急存亡之秋。此刻，拉尼爾和多位女學員發生激烈性愛，同時也有情感關係。不過，他光是沉浸在《富比士》專訪的興奮之情，忘了照顧克萊兒的感受，關心不足。托妮暗示道，克萊兒總是活在姊姊比較漂亮的陰影下。NXIVM 理應與眾不同。在 NXIVM 競技場上，克萊兒覺得自己值得站上顯要的位置。畢竟她放棄馬術冠軍的夢想，全副心思轉到 NXIVM，全職投入。[63]但就克萊兒看來，拉尼爾和薩爾茲曼似乎在《富比士》雜誌記者面前說更多姊姊的好話。[64]事實上，那篇報導引用莎拉的話作結。莎拉告訴弗里德曼：「我不知道你有多瞭解我的家族，但我生在一個從來不需要努力就能獲得一切的家族，對我而言被授予這條黃色榮譽綬帶，實在非常、非常感動。這是第一件僅憑著我自身努力而獲得的事物。」[65]

克萊兒覺得遭人忽略又無人賞識，決定要報復，打電話給爸爸說拉尼爾跟自己借了兩百萬美元（後來《富比士》聯絡克萊兒確認這筆借款時，她卻否認了）。埃德加現在明白拉尼爾不是自稱的道德宗師，只是想奪走女兒錢財的騙子。《富比士》報導中也有埃德加的評論：「我覺得這是邪教團體。」[66]

拉尼爾沒有解釋報導中質疑他自身行為的段落，反而集中火力攻擊埃德加的評論。拉尼爾的怒火直沖克萊兒，不滿她透露消息給父親。報導刊登之後，克萊兒和莎拉都不再和父親說話，很可能是拉尼爾從旁慫恿。但拉尼爾還不滿足。就像他對丹妮拉做的一樣，拉尼爾控告克萊兒犯了「道德瑕疵」，違反 NXIVM 的使命宣言。克萊兒亟欲重獲拉尼爾的青睞，雙手捧上自己的資產，交由 NXIVM 會員芭芭拉．鮑契管理，莎拉也一樣。

鮑契是成功的財務規劃專家，有能力管理兩姊妹的帳戶。但鮑契沒有告訴任何人，她自己其實也和拉尼爾發生性關係。或許因為這層利益衝突，所以拉尼爾開始說服兩姊妹大手筆揮霍資產時，鮑契並沒有對姊妹倆提出警告。姊妹倆付了一百萬美元幫薩爾茲曼重新裝修住家，為 NXIVM 在紐約奧爾巴尼的總部花了一百七十萬美元，另外為總部附近的 NXIVM 馬場支付了兩百三十萬美元。她們還買了一架私人飛機，供 NXIVM 的高層使用，此外還捐了兩千萬美元的巨額款項給 NXIVM 操控的某基金會。

不只如此，克萊兒另外還為拉尼爾的許多官司做保，控告對象都是拉尼爾視為敵人的存在。任何人叛逃NXIVM，克萊兒都會提起訴訟，也會霸凌那些批評NXIVM課程的人，以昂貴的訴訟威脅。克萊兒資助「敵情研究」（opposition research），調查分析拉尼爾的敵人，從地方政治人物到托妮等對象，蒐集每個人詳盡、有時看似有罪的資料。[67]但拉尼爾還是不滿足。二〇〇五年初，拉尼爾告訴布朗夫曼姊妹，他創造了一條獨步天下的數學公式，絕對能在商品市場大賺幾百萬美元。因為他理應是全世界最聰明的男人，她們同意拉尼爾可以使用她們的基金來買賣。

隨之而來的便是一場財務浴血大戰。拉尼爾在未經深思熟慮的極端商品買賣上燒了六千五百萬美元。[68]最後克萊兒和莎拉總算詢問他這些損失，拉尼爾則宣稱是她們的父親勾結商品結算公司來偷走女兒的基金。

兩姊妹還真的相信這幾筆大損失都是埃德加造成的，而不是拉尼爾。根據羅伯特・海爾的檢核表，心理病態需要持續的刺激，無法好好控制自身行為，此外還非常衝動。[69]這三項因素可以解釋為何心理病態和賭博上癮互有關聯。[70]海爾檢核表上還有心理病態共有的另一項特徵：眾所周知，他們過著寄生蟲般的生活。[71]他們實際上也感受不到愧疚，只要能得到想要的人事物，就算撒謊或發表任何言論都沒關係。

比較一下拉尼爾的行為（靠追隨者的信託基金生活）和ESP課程（警告學員這世界由生產群體和寄生群體所組成）吧！拉尼爾教導Nxian，他是生產群體的完美個案研究，行為卻顯然是寄生群體。他和核心成員的多名女性做愛，不論對方同意還是不同意；他要求全心奉獻的詳盡證明；他一眨眼就花光數百萬美元，起初是鮑契的錢，接著是布朗夫曼的錢。就算拉尼爾的行為都在掠奪，但 NXIVM 的女性學員太迷戀他了，根本看不出來他的寄生蟲真面目。實際狀況是，就連拉尼爾因為在 NXIVM 的所作所為而獲判多次無期徒刑之後，克萊兒依然不離不棄，寫信告知法官她心中的量刑：「我大多數的家人，還有其他許多人都相信我應該否認與基斯和 Nxivm 的任何關聯，而我卻沒有這麼做，大家都難以理解，也無法接受。然而，對我而言，Nxivm 和基斯大幅改變了我的人生，人生變得更加美好。」[72]

二〇〇三年：克莉絲汀．史奈德

拉尼爾開始擴大 NXIVM 服務範圍，創立了一個新課程：JNESS——透過一項運動「促進持續不停地探索身為女性的意義」，來保證女性賦權。學員可以透過十一天密集課程來達成這項模糊的目標，要價五千美元。顯然，不少人聽信他的話術，而成了受害者，其中一名是阿拉斯加原住民克莉絲汀．史奈德（Kristin Snyder）。密集課程期間，史奈德開始出現詭異的行為，說一些和現場討論毫不相關的奇怪言論，打斷課堂進行。參與密集課程的期間，她開始不睡覺，因為她相信拉尼爾也沒有睡覺，想要仿效他。某次她打給爸媽，說自己要為哥倫比亞號太空梭爆炸事件負責。[73]

她持續如此不穩定的言行幾天後，一名 ESP 學監下了決定。他們不會再試圖治療史奈德的情緒爆發，也不會透過理應獲得專利的意義探索技術，整合史奈德的負面情緒反應。他們只想趕走她。按照阿拉斯加州警方所言，史奈德划著獨木舟抵達位於安克拉治（Anchorage）以南一百哩的復活灣（Resurrection Bay）中央，雖然她是獨木舟好手，但根據推斷，她溺水身亡。她的豐田 Tacoma 皮卡停在蘇厄德（Seward）附近的米勒斯蘭丁（Miller's Landing）起點處，警方在座位上發現自殺遺書。遺書寫道：「我參加一堂叫做高階人士成功的課程……我腦袋裡的情緒中心被關閉。外在肌膚還是有感受，但內在器官正在腐敗。如果你找到我或這張紙條……請聯絡我爸媽。我很抱歉，人生，我不知道我已經死了。」[74]翻倒的獨木舟被沖上了岸，但史奈德的遺體至今下落不明——這結果也算合理，因為復活灣有些地方深達九百七十呎。

這張自殺遺書證明了她如果不是罹患臨床精神疾病，就是處在極端的情緒危機中。極度的睡眠剝奪可能也是導致她如此痛苦的原因之一。研究員弗拉維．沃特斯（Flavie Waters）、薇薇安．邱（Vivian Chiu）和同僚發現：「睡眠剝奪會造成精神病症狀……從單純的視覺／體感錯覺，到幻覺，甚至妄想，最終會進入類似急性精神病的狀態。」[75]如果睡眠不足，目前的狀態可能會惡化。不過，另一名二十八歲的女性在墨西哥參加同樣的 ESP 課程，也精神崩潰了一次，但她的案例沒有睡眠剝奪的證據。這表示很可能 ESP 課程還有其他不安定因素，才導致這兩起事件。ESP 的意義探索課程終究是要求學員過度專注在自己最負面的情緒和記憶，藉此移除學員的力量（儘管事實上沒有研究可以指出這樣的治療手法究竟有效或無效）。這課程也很可能在少數 Nxian 身上發揮了相反效果，但也不是能等閒視之的人數。過度專注把負面情緒痛苦地帶到表面，強迫學員達到某個狀態，在該狀態下，情緒痛楚會轉變成精神病。

即便 ESP 的方法可能存在著危險因子，拉尼爾也不會在意，更不會想要移除。依托妮所言，拉尼爾聲稱克莉絲汀・史奈德假裝自殺，因為史奈德是毒品走私集團的一員——這是毫無事實根據的陰謀論。[76]拉尼爾的說詞只是惡性自戀者的經典行為又一範例：惡性自戀者遭遇威脅時，不只更頻繁地撒彌天大謊，還會恐嚇、攻擊他人，全都是為了重新確認他們自身過度膨脹的自我。而事實上，此一行為模式很容易預測，在戒除邪教、迎來康復的相關文獻裡還有特別說法：自戀虐待症候群（Narcissistic Abuse Syndrome）。

二〇一〇年：愛莉森・麥克

在 NXIVM 總部，絲毫看不見墨西哥和阿拉斯加悲劇事件的影響，大家對 JNESS 課程的熱情不減。事實上，這課程宣稱是 NXIVM 的「婦女運動」，只見人數和歡迎程度日益增加。克莉絲汀・史奈德不見蹤影三年後，許多最高階級的 NXIVM 學員前往加拿大溫哥華（Vancouver）一間毫不起眼的飯店會議室參加 JNESS 課程：莎拉和克萊兒搭私人飛機，薩爾茲曼和女兒蘿倫也排好行程出席。這不是巧合。拉尼爾挑選最精明幹練的學員出席。大家到場只為了一人——克莉絲汀・克魯克（Kristin Kreuk）說服來（換句話說就是招募來的）的人，她自己前不久才參加同樣的課程。據說克魯克招募了愛莉森・麥克（Allison Mack），他們一起演出CW電視台（CW TV）強檔連續劇《超人首部曲》（*Smallville*），故事是年輕的克拉克・肯特（Clark Kent）在中學的冒險。麥克才剛獲頒「青少年票選最佳配角獎」（Teen Choice Award for Best Sidekick）。

拉尼爾派出最佳團隊，對麥克施加「愛的轟炸」，目的就是要把她拉進圈內，更常參與課程。NXIVM上層階級幾乎都是才華洋溢的女人。薩爾茲曼幫忙規劃所有課程，布朗夫曼姊妹資助了營運成本的大部分。她們能待在最高階級其實和個人招募技巧無關，主要是為了吸引其他功成名就的女性來上第一堂密集課程。從女性主義的觀點來看，NXIVM的教導幾乎可說是極端守舊派——當時的學員蘇珊・唐斯（Susan Dones）後來告訴《好萊塢報導者》（*Hollywood Reporter*）：「薩爾茲曼談論女人如何被養育成相信一夫一妻制，男人的一般天性就是更加偏好一夫多妻制，為了四處播下種子。」[77]姑且先不論這些教導，拉尼爾希望受歡迎的名人來當發言人，會引誘更多好萊塢類型的學員，他希望愛莉森・麥克會為他做出湯姆克魯斯（Tom Cruise）為山達基所做的一切，並且為紮根在紐約的公司創造獲利豐厚的西岸收益流。

蘿倫全程都坐在麥克旁邊，一起吃飯、聊八卦，課程尾聲兩人看起來就像閨蜜。布朗夫曼姊妹邀請麥克一道飛回紐約，參觀NXIVM總部，親自看看基斯・拉尼爾，麥克答應了，還待了好幾週，深入瞭解課程。抵達總部不久，麥克就有機會參加「先鋒週」（Vanguard Week，或稱V-Week）。原本只是企業會議，旨在推廣NXIVM，但到了二〇〇六年，已經變成為期十天的拉尼爾生日慶祝大會。在V-Week，Nxian從美國各地、歐洲、加拿大和墨西哥前來克利夫頓公園。

前 Nxian 電影製片馬克．維森特（Mark Vicente）喜歡參加一系列暖身活動，最後迎來「大人的夏令營，你可以選擇各種想做的事。打鼓、跳舞、唱歌、（甚至）吟詩。」[78]一切都是向拉尼爾致敬，認可他創造了一場哲學運動。對於拉尼爾的崇敬不限於 V-Week；崇敬拉尼爾已經成為 Nxian 生活的每個面向，女性學員尤其如此。

在總部，拉尼爾幾乎天天打排球，成排成排的女學員會在場邊等待。每到中場休息，女學員就會衝向他，好似他是搖滾巨星，告訴拉尼爾她們內心深層、壓抑的問題。[79]這種程度的奉承說明了為何拉尼爾宣稱自己是獨身主義者，卻還是可以持續和多名女學員發生性關係。根據 FBI 的報告，到了二〇〇〇年初，拉尼爾任一時刻都有十五到二十位性伴侶。[80]或許愛莉森．麥克當初沒辦法發現這一面，但無法忽視的是大家奉先鋒為神明，失控地崇拜他。

社會心理學家羅伯特．扎榮茨（Robert Zajonc）研究發現，幾乎每個人都受到曝光效應的影響。扎榮茨發現物品或概念只要曝光越多次，大家就越可能相信。[81]愛莉森．麥克剛開始認識 NXIVM 的時候，看到拉尼爾在 V-Week 受人吹捧，說他是悟道的存在，排球比賽結束又像披頭四一樣受到瘋狂擁戴。接受這些曝光的結果，麥克很可能也開始用同樣的眼光看待拉尼爾。

但還是有其他原因導致麥克這麼輕易受到拉尼爾這樣救世主角色的影響。麥克童星出道，七歲開始拍攝電影、電視劇，十八歲得到《超人首部曲》一角，卻從來沒機會唸大學。到了二十四歲，人生幾乎都在演藝圈，就像童星友人克莉絲汀．拉金（Christine Lakin）告訴《好萊塢報導者》的：「她太想追求更遠大的事，某種徵兆，顯示生命的目的和意義。」[82]

愛莉森．麥克不是 NXIVM 唯一的知名演員。演出《檀島警騎二．〇》（*Hawaii Five-O*）的朴敏慶（Grace Park）和《星際大爭霸》（*Battlestar Galactica*）的妮基．克萊恩（Nicki Clyne）據說都是學員。演員的某些專業特質可能讓他們更容易相信 NXIVM。表演藝術產業的每個人都得面對拒絕，要嘛是甄選時直接被拒絕，又或者是爭取新角色時卻由其他人獲得這種間接拒絕。演員在工作的兩頭不斷面對負面反應，從拍片前角色競爭到上映後糟糕的評價。面對這一切需要勇氣和毅力。想要存活，每位演員都必須學習如何面對認知失調。這種生存機制是實際的工作要求：有能力協調內心深信的事（我很擅長）和客觀現實（沒人認為我擅長）兩者間的矛盾。這樣持續的自我價值衝擊，或許能說明為何 NXIVM 吸引了那麼多愛莉森．麥克這類好萊塢影星：他們加入時就非常渴望安穩的接納與支持。這可能也說明了為什麼就算法院審理時鐵證如山，麥克後來還是繼續相信拉尼爾和他的課程。

先不提演員的獨特心理，所有參加 JNESS 課程的女性都在尋求自我賦權。拉尼爾可以利用這個矛盾：出席課程就等於暴露了目前她們缺乏力量和自尊。這樣的不安導致她們容易接受灌輸。麥克迅速吸收 NXIVM 關注的負面情緒，二〇〇七年，她在部落格上寫道：「我經常習慣說自己笨。我已經變得太容易達到目標，太輕易就成功，忘了事實上自己根本沒有好好受過教育。」[83]本質上，JNESS 哲學加強了她本來就有的習慣性自我貶低。

愛莉森・麥克一加入 NXIVM，就成了該組織強力支持者，並且躋身頂尖招募員行列。她向親朋好友宣傳拉尼爾的聰明才智，以及 NXIVM 無人能及的倫理道德。[84]此外，她不停帶演員朋友參加課程。史考特・強生（Scott Johnson）和蕾貝卡・孫（Rebecca Sun）在《好萊塢報導者》寫到，麥克打算招募葛萊美獎得獎歌手凱莉・克萊森（Kelly Clarkson）和《第七天堂》（*7th Heaven*）的演員碧佛莉・米契（Beverley Mitchell）。麥克也在 Twitter 上公開招募《哈利波特》（*Harry Potter*）電影演員艾瑪・華森（Emma Watson），懇求艾瑪・華森來試試 NXIVM「獨一無二的人類發展和婦女運動」。[85]

麥克狂熱投入 NXIVM 幾年後，她的健康開始被影響。蘇珊・唐斯在溫哥華的瑜伽教室巧遇麥克，她說到了二〇一〇年，麥克很明顯已經和拉尼爾發生性關係，因為麥克已經「臉色蒼白，就像其他拉尼爾的女人一樣，她們都開始有點病態」。[86]拉尼爾特別喜愛瘦到過度不自然的女性，這些女性都受到操控，心甘情願瘋狂減重。結果就是，他的女人都有相似的嚇人外貌：隨著體重下降，她們的頭就顯得太大，最後看起來都像搖頭娃娃。[87]

二〇〇八年：NXIVM 九人組

到了二〇〇八年，鮑契已經加入 NXIVM 九年了，幾乎全程都與拉尼爾有性關係，還參與了財務運作。知道一些內部運作的方式，鮑契漸漸懷疑起 NXIVM 究竟是如何營運。鮑契不是一個人，另外八名女學員也有類似的顧慮，其中一名是高階 Nxian 蘇珊・唐斯。這九名女學員感受到人多就是力量，當面對質先鋒。弗蘭克・帕拉托（Frank Parlato）的網站專門調查 NXIVM 和基斯・拉尼爾，某篇文章寫到蘇珊・唐斯講述主要論點，批評拉尼爾性剝削女學員、向會員借大筆現金、洗錢、拒絕繳交州政府或聯邦政府稅金。[88]

在奧爾巴尼的當面對質中，拉尼爾既居高臨下，又鄙視眾人，還用溫和卻嚇人的方式威脅大家。對話全程錄製，唐斯後來把影片上傳到 YouTube。影片中，唐斯和鮑契坐在沙發的一端，拉尼爾坐在另一端，但距離近到拉尼爾可以時不時伸手，碰觸兩名女性，強調論點或要她們安靜。錄製到的對質中，拉尼爾的句子既模糊又複雜，意義就像他的聲音一樣難以分辨。回應明確的抱怨時，拉尼爾告訴兩名女性說她們根本沒有資料可以了解真實情況。[89]她們也沒有「領導經驗或拯救人們生活的經驗」，但實際上他究竟在說什麼，一點都不清楚。那時，其中一名女性（畫面上面孔模糊）提醒拉尼爾他其實也沒有負責領導過企業，因為他的公司 CBI「沒幾年就倒閉了」。CBI 的失敗確有其事，因此影響了拉尼爾冷靜的言行。他猛地回嘴：「那才不是……」接著控制下來，不讓情緒影響他。在下一段對話裡，他說的話簡直意義不明：「嗯哼，事情是這樣的。我因為自己的信仰而被攻擊。我得做出決定：應該要請保鑣嗎？要武裝還是不用？而我也害人慘遭殺害，因為我的信仰、因為他們的信仰、因為我說過的話。你可能會說我現在在領導一個組織，而這個組織做了一些很好的事。你知道，這是一句古老的基督教格言（adage）：『光越亮，蟲越多。』因此我認為我們有的是非常亮的光。」如同那些裝腔作勢想要掉書袋的人，他的 adage 一字發音錯誤，變得和 mirage 同韻[14]。

儘管九名女性直接提出擔憂，卻沒有獲得希望中的答覆，或者該說根本沒有得到任何答覆。拉尼爾拒絕處理抱怨，也不肯承認自己做錯任何事。她們受夠了，集體離開。這是至今拉尼爾面對最多高階會員同時刻意退出，而且 NXIVM 那時才剛獲得數一數二最大的勝利：莎拉運用自己的人際關係，聯絡上佛教僧侶丹增東登（Tenzin Dhonden），說服達賴喇嘛（Dalai Lama）來訪奧爾巴尼，與布朗夫曼姊妹一同站上舞台，發表一場題為「困難時期的慈悲倫理」（Compassionate Ethics in Difficult Times）的演講。

演講前言，達賴喇嘛指著兩姊妹，感謝她們，接著朝第一排「受人敬重，我想是領導人」的拉尼爾和薩爾茲曼揮手。多人集體出走之後，拉尼爾希望藉由達賴喇嘛獲得的神聖光環也消失無蹤了（來訪當晚，達賴喇嘛親自為拉尼爾披掛白色榮譽綬帶），而當場所有人談論的都是 Nxian 的叛離，還有傳來傳去的不法行為謠言。多虧有克萊兒的錢，拉尼爾上法院告蘇珊・唐斯。唐斯每起案子都自己擔任辯護律師，打敗那些被請來威嚇她的高薪律師。

二〇一七年：莎拉・艾德蒙森

14 應該要和 damage 同韻。

「NXIVM 九人」出走事件之後，NXIVM 總部營運不當的疑雲越傳越烈，儘管當面質疑拉尼爾會被視為「道德瑕疵」。唐斯和鮑契的投訴都公開存放在網路上（不正當經濟活動，以及越來越大膽地逃稅），這些投訴讓溫哥華、墨西哥、紐約市的 NXIVM 管理階層擔心了起來。考慮到現在烏雲罩頂，明理人不會認為這是開啟全新龐氏騙局的好時機，這次採用性奴隸階級制度，先鋒是在頂端的終極「大主人」（Grand Master）。拉尼爾想像系統中有數千人在他的掌控之下，甚至有天會有當選的政治人物也受到吸引而加入。拉尼爾經常像這樣不經深思熟慮，他為新的團體命名「主宰順從的女性夥伴」（Dominus Obsequious Sororium，簡稱 DOS），據說原本應該是拉丁文的「主宰奴役女性」（Master over Slave Women），但事實上只是胡言亂語——混合了假的詞源、錯誤的詞彙、拼錯的拉丁文，翻譯起來比較像是「姊妹們的精力主人」（Eagery Master of Sisterses）。但是會員已經習慣這些無意義的首字母縮寫詞，例如 JNESS 或假品牌名稱 NXIVM，沒有人敢叫拉尼爾解釋這種學生會犯的錯。出乎意料，新團體僅靠口耳相傳逐漸成長，成員需要接受邀請才能加入。通常都用首字母 DOS 稱呼，跟一九八〇年代的微軟作業系統同名。

名稱並不是這個團體思慮最欠周詳的地方。蘿倫・薩爾茲曼、愛莉森・麥克等最頂尖招募員面對可能入會的人，都對團體的性愛本質避而不談，而是宣傳 DOS 精挑細選 NXIVM 學員（那些急著想要在階級體系內晉升的學員）組成「最高機密女性團體」。這正是二〇一六年蘿倫招募莎拉・艾德蒙森（Sarah Edmondson）時用的話術。蘿倫初次提起時，說道：「你對自身的成長有多投入？」在《傷疤：逃離 NXIVM 的真人真事，束縛我人生的邪教》（暫譯，*Scarred: The True Story of How I Escaped NXIVM, the Cult That Bound My Life*）一書中，艾德蒙森記下自己的回應，現在想想實在單純得太過危險：「當然是不惜一切。」

艾德蒙森在好萊塢打滾多年，是演員也是配音專家，有數十年電影、電視演出經驗，還擁有美國、加拿大娛樂圈名人的聯絡方式。她也被視為 NXIVM 的頂尖招募員，多數時候在溫哥華前哨招募會員，一手規劃溫哥華的課程，也經常帶領課程。蘿倫是她的密友，也是她兒子的乾媽，還是她和同為 Nxian 的安東尼·阿姆斯（Anthony Ames）結婚時的司儀。阿姆斯是布朗大學（Brown University）的四分衛。艾德蒙森徹底信任蘿倫。

儘管如此，她還是對入會的先決條件感到憂心，蘿倫只說是「擔保品」——她必須拍攝影片，控訴雙親、先生根本沒有犯下的暴行。但因為她不相信課程中最親近的朋友會出賣自己，所以同意發誓服從，對先生保密自己的會員身分——蘿倫解釋道：「這是個人選擇，你對自己的成長做的選擇。」更奇怪的是，還得稱呼她的朋友「主人」（Master）。尊稱不單單只是形式而已，不過蘿倫說笑打發道：「不會把你關進籠子裡啦！」[90]

蘿倫的隨意玩笑讓艾德蒙森安心不少。此外，就像其他多位 DOS 新成員一樣，艾德蒙森早就內化先前 JNESS 設下的所有原則：女人既軟弱又不可靠，還缺乏恆心毅力。她很喜歡自己成為姊妹圈的一員，大家會幫她成為最強大、最理想的自己。但一交出擔保品，蘿倫·薩爾茲曼就變了個人。艾德蒙森被要求得在六十秒內回覆主人的訊息，不論時間。蘿倫也纏著艾德蒙森要招募自己的奴隸，這樣艾德蒙森才能跟蘿倫一樣成為主人。

一開始就充滿了警示：蘿倫語氣突變，提出各種居高臨下的要求，推翻她稍早保證「主人」稱謂不過是形式而已。現在蘿倫能命令艾德蒙森做事，似乎也不再如同以往對待朋友那般客氣有禮了。儘管警示越來越多，艾德蒙森還是遭到朋友脅迫，蘿倫手上握有擔保品。[91]那正是為何加入DOS幾個月後，蘿倫邀請艾德蒙森飛到奧爾巴尼，參加入會儀式，正式認定她們的新關係時，艾德蒙森同意了。《傷疤》一書寫到，二〇一七年三月九日，艾德蒙森和其他四名女人的雙眼都被矇住，由蘿倫領著走進房子。蘿倫拒絕洩露地點，艾德蒙森偷看了一下，認出周遭環境，那是愛莉森．麥克的家，雖然沒有看到麥克的身影。

蘿倫接著領大家到另一間房內，叫大家摘下矇眼布條，脫光衣服只留內衣褲。艾德蒙森認出其他女人，過去幾年在NXIVM的活動中見過。她也注意到房內有一張大型按摩床。蘿倫說過入會儀式的其中一個步驟是在身上刺青，圖案很小，只是一個符號，代表四大元素。

艾德蒙森從來不曾刺青，現在也不可能會想要。稍早她提出疑慮時，蘿倫用典型的Nxian風格回答，刻意把她的抗拒描述成需要克服的障礙。蘿倫問道：「你要怎麼思考刺青的意義？如果這代表你的性格、你的力量被刻畫在身上呢？」艾德蒙森吞下異議。丹妮爾．羅伯茲（Danielle Roberts，長期的Nxian，同時也是名醫生）走進房內，艾德蒙森稍稍安心了，至少是醫護專業人員來進行刺青。

不過等到蘿倫拉下牛仔褲，秀自己的骨盆給大家看，上面刺著大家即將要有的符號，那股安心瞬間消逝。艾德蒙森嚇壞了。那不是刺青，而是烙印：凸起新生的傷疤，「就像一塊肉懸在……她身上最脆弱的部位」。[92]艾德蒙森一點都不想要那種東西，但現在她無法離開。蘿倫還握有她的擔保品：她對父母、先生的可怕指控。而且依照蘿倫目前令人不安的優勢地位，艾德蒙森現在相信蘿倫真的會使用擔保品。

她們叫艾德蒙森戴上口罩，說是基於衛生考量，艾德蒙森聽命戴上了。艾德蒙森依照指示幫忙壓住第一位奴隸的雙腿，丹妮爾．羅伯茲醫生在奴隸的肌膚上拖曳著一枝電烙筆。全程耗時四十五分鐘，房內所有人都必須聽著痛苦的叫喊。輪到艾德蒙森了，她爬上按摩床，面朝上躺下。在蘿倫的提示下，她說：「主人，你可以烙印我嗎？那會是多麼大的榮幸。」[93]接著當羅伯茲醫生拿著灼熱的電烙筆畫過肌膚時，她盡力躺著不動。

幾週後，烙印開始癒合，艾德蒙森比較能看出圖案設計。她可沒看見四大元素的符號，反而是四個字母。首先是一組 A、M，接著是一組 K、R，就是基斯．拉尼爾和愛莉森．麥克的首字母縮寫，現在永遠烙印在肌膚上。

從那刻起，艾德蒙森不再感到無能為力，而是越來越憤怒。儘管她們宣稱 DOS 是女性專屬團體，現在她很肯定基斯．拉尼爾也是一份子。擔保品和沒必要的痛苦烙印都是拉尼爾的主意：他們真正擁有他的簽名了。為了確認自己的猜疑，她詢問製片人馬克．維森特，十年前就是維森特招募她加入 NXIVM。維森特和太太已經決定要離開了。艾德蒙森希望能找到盟友，打電話給維森特。

維森特也有相同懷疑。他告訴艾德蒙森，他對於年輕女學員在奧爾巴尼遭遇的事感到很不安。他說：「好像有類似階級條紋途徑的另一套途徑，而愛莉森・麥克是新途徑的頭。」在裡面的女人遭到勒索，不得不和拉尼爾做愛。[94]大家得拍照證明自己有做，還會被事先告知：「你獲得許可，好好享受。」維森特透露的內容顯示 NXIVM 組織結構背後扭曲的邏輯：NXIVM 的每個分支都獨立運作，完全不知道其他分支在做什麼。就算維森特和艾德蒙森兩人都是高階 Nxian，各自也只知道 DOS 謎團的零星資訊，不知全貌。

艾德蒙森現在明白自己被選為其中一員的原因。她的體重超過一百一十磅，所以懷疑自己為何會被選中來服侍拉尼爾。但她是課程數一數二的業務，而且事實上南希・薩爾茲曼一開始還會揀選理想目標交給她招募。薩爾茲曼在溫哥華中心 Facebook 頁面選定的對象都非常有魅力。艾德蒙森感到厭惡，立即決定尋求 FBI 的協助，管他什麼擔保品。

整整一個小時，她告訴聯邦當局自己和其他人都遭遇了什麼對待，全盤托出十二年 NXIVM 的經歷和最新發展：DOS 的掠奪噩夢。接著，《紐約時報》的貝瑞・麥爾（Barry Meier）寫道：「艾德蒙森小姐向紐約州衛生部（New York State Department of Health）投訴丹妮爾・羅伯茲醫生執行烙印，一名領有執照的骨科醫生，同時也是拉尼爾先生的追隨者……」[95]

下一步，艾德蒙森告訴媒體她的故事。她第一個聯絡的是弗蘭克・帕拉托，這位部落格主之前受聘於拉尼爾，工作是挖掘 NXIVM 敵人的黑歷史和糟糕事。工作期間，對於拉尼爾的所作所為，帕拉托越挖越驚嚇，一開始是布朗夫曼姊妹信託基金的數百萬美元，拉尼爾全都投資在糟糕的商品交易上，數百萬美元就這麼泡湯了。拉尼爾聽聞帕拉托詢問布朗夫曼姊妹她們的財務損失，直接開除了他，或許還告上法庭，因此激怒了帕拉托，開始挖掘更多骯髒事，不屈不撓。二〇一七年，艾德蒙森告訴帕拉托，拉尼爾從欺騙繼承人進展到直接殘害人體。帕拉托抓住機會報導了這則故事，看著敵人身陷火海。

帕拉托的報導首度揭露 NXIVM 內部存在烙印女性的團體。消息一出，有些會員退出，但大多數的 NXIVM 會員繼續留著，轉而攻擊艾德蒙森，為她貼上「愛發脾氣的弱女人」標籤。[96]執法單位的回應也同樣令人失望。截至目前為止，她的指控沒有帶來任何重大影響。於是她加倍努力。

Nxian 反駁帕拉托只是一名憤恨不平的前員工，但《紐約時報》就沒這麼好打發了。二〇一七年十月十七日，記者貝瑞・麥爾報導一篇封面故事〈祕密團體烙印女體〉（Inside a Secretive Group Where Women Are Branded）。之後，主流媒體的報導爆炸了。美國廣播公司（ABC）的《二〇／二〇》節目報導了這則故事，訪問馬克・維森特、托妮・娜塔莉和莎拉・艾德蒙森。[97]《Vice》做了弗蘭克・帕拉托的個人報導。[98]《朝代》（*Dynasty*）演員凱瑟琳・奧森伯格（Catherine Oxenberg）參加梅根・凱利（Megyn Kelly）的晨間節目，告訴觀眾她擔心自己的女兒英迪雅（India）成了身上有烙印的 DOS 奴隸。[99]舉國關注，終於刺激了 FBI 正式展開 NXIVM 犯罪調查，請莎拉・艾德蒙森回去，詳細詢問整整兩天半。[100]

二〇一九年：金字塔崩塌

拉尼爾一注意到聯邦執法單位訪問女學員的ＤＯＳ會員資格，就逃亡到墨西哥瓦雅塔港（Puerto Vallarta）的門禁社區，一幢別墅一週花費高達一萬美元。五位ＤＯＳ會員又稱頂尖奴隸，其中包括愛莉森．麥克和蘿倫．薩爾茲曼，也飛去陪伴拉尼爾。她們規劃了「再次承諾典禮」，據說這五名女會員要在典禮上同時發生性行為。後來蘿倫出席法庭作證，回想當時道：「一定還有這個以外的其他方式可以讓我成長。難道我除了集體口交之外沒有其他成長方式嗎？」[101]

薩爾茲曼和其他人還算幸運，那場「典禮」從來沒有舉行。二〇一八年三月二十五日，墨西哥聯邦幹員（Mexican Federales）闖入當年五十八歲的拉尼爾的太平洋避難所，薩爾茲曼正在樓下打蔬果汁。相關當局闖入別墅，薩爾茲曼衝上樓到臥房保護拉尼爾，或協助他藏匿。聯邦幹員重擊房門，最終破門而入，薩爾茲曼蹲伏在地。[102]後來，薩爾茲曼在法庭上說：「從頭到尾我都在想，他們可以直接射擊房門，十分合理地想著：『我現在可能就會被槍擊。』」同時，拉尼爾藏身於衣櫃。官員終於進房，薩爾茲曼本能地呼喊先鋒的名字，洩露他的行蹤。薩爾茲曼在法庭上說：「我選擇相信我們一直以來受到的訓練。我選擇愛。我從來沒有想到，我會選擇基斯，而基斯也選擇了基斯。」

逮捕拉尼爾之後不久，墨西哥當局把拉尼爾引渡回紐約。在紐約，拉尼爾遭指控「敲詐勒索、共謀強迫勞動、共謀電匯詐欺、性販運」。[103]聯邦當局相信拉尼爾創立了DOS（而不是愛莉森・麥克，拉尼爾後來企圖在法庭上怪罪愛莉森・麥克）。聯邦當局控告拉尼爾脅迫女性做愛，再在女性身上烙印自己的姓名首字母。但即便聯邦當局講述本案時，把拉尼爾視為一手創立DOS的主腦，後續案件仍證明，如果沒有共犯員工強大的說服力，他其實沒辦法打造DOS；因此，克萊兒・布朗夫曼、蘿倫・薩爾茲曼、南希・薩爾茲曼、愛莉森・麥克，以及NXIVM會計凱西・羅素（Kathy Russell）都收到類似的控訴。

二〇一九年四月二日，蘿倫認罪一起敲詐勒索，也承認自己協助監禁丹妮拉兩年。法庭上，蘿倫表達懊悔，說道：「我很抱歉，我糟糕的決策、決定導致其他人受傷，不只是本案的受害人，還有社群的幾百名會員，以及他們的親朋好友。」[104]

四月八日，前CW電視台演員愛莉森・麥克認罪一起共謀敲詐勒索和一起敲詐勒索。法庭上，麥克承認強迫女性交出「丟臉的資訊和照片」，也就是會員所謂的「擔保品」，只要拉尼爾要求，她就會將擔保品交給他。[105]麥克獲判三年刑期，罰款兩萬美元。

長官南希・薩爾茲曼、億萬繼承人克萊兒・布朗夫曼和會計凱西・羅素都承認與NXIVM相關的罪行。法庭上，克萊兒表達懊悔，但法官認為她的陳述顯示她還是繼續相信拉尼爾，因此判克萊兒服刑八十一個月，超過檢方提出的刑期。

南西・薩爾茲曼則因共謀敲詐勒索獲判三年半刑期，於二〇二二年二月開始服刑，地點在西維吉尼亞州奧德森（Alderson）的聯邦監獄。蘿倫・薩爾茲曼和凱西・羅素都不用進監服刑，分別獲判五年和兩年緩刑，還要做社區服務。

二〇一九年五月七日，被逮捕後一年多，拉尼爾終於出庭布魯克林聯邦地區法院（Federal District Court in Brooklyn）。托妮．娜塔莉坐在法庭內，一路追蹤進度。托妮之前有幾次和拉尼爾講話的機會，拉尼爾總是預測下次再碰面，托妮不是死了就是在監獄。命運如此巧妙，相反的事發生了：托妮是自由來去的那個人，現在是拉尼爾要在監獄裡待著。

助理檢察官譚雅．哈賈（Tanya Hajar）在開庭陳述說道：「被告自稱心靈導師，但他就是強暴犯。他鎖定年輕女孩，挑出幾人特別關照，但這只是藉口。他說要照顧她們，教導她們，但也只是誘拐她們性交的藉口。」[106]

拉尼爾的律師馬克．艾格尼費羅（Marc Agnifilo）在開庭陳述時把他的顧客和前英國首相邱吉爾相比，敦克爾克大撤退之後，邱吉爾發誓要防衛大英帝國。艾格尼費羅主張拉尼爾的行為都是出於善意。他把DOS形容成「姊妹會」，創立目的是為了讓會員可以「擁有更美好的生活」，還指出超過一萬七千人參加NXIVM課程，其中多數人都獲得美好的體驗。[107]

但是艾格尼費羅的描述在審判中站不住腳。在三天緊鑼密鼓的作證中，蘿倫從頭開始告訴陪審團她與「先鋒」基斯．拉尼爾的關係。她告訴陪審團拉尼爾答應要生寶寶，卻什麼都沒有。蘿倫也試圖解釋她對拉尼爾的深厚情感，說道：「他是我的主人。我最重要的人。我仰慕他。我想要像他一樣。」[108]

蘿倫告訴陪審團她對拉尼爾的愛導致她限制自己的飲食、一週發送三次陰道私密處的近照給拉尼爾、協助拉尼爾撰寫 DOS 的厭女規則。[109]她承認自己依拉尼爾要求做過最糟的事是早些年協助監禁丹妮拉。年紀輕輕的丹妮拉懇求蘿倫給她移民文件，她才能回到墨西哥時，蘿倫拒絕了，相信如此一來拉尼爾會認為她是負責任的媽媽，給她承諾已久的寶寶。[110]

辯護方傳丹妮拉出庭。丹妮拉證實拉尼爾確實監禁她兩年。雖然門沒有上鎖，但她的房間除了不叫籠子之外，事實上就是個籠子，因為父母都是忠誠的 Nxian。拉尼爾告訴兩人，他們的女兒犯了「道德瑕疵」，監禁女兒只是要治癒過剩的傲慢，但拉尼爾奪走她的自由，事實上只是因為她對另一名男人有興趣。

只提供名字的演員妮可（Nicole）被愛莉森・麥克招募進 DOS，告訴陪審團她可怕的體驗：被綁在桌上，陌生人對她口交。[111]抱負遠大的馬術選手希薇（Sylvie）在法庭上哭出來，回憶自己被脅迫要引誘拉尼爾。[112]儘管檢方證人講述拉尼爾種種駭人行徑，辯護律師馬克・艾格尼費羅依然堅持宗師發生的所有關係都是經雙方同意。[113]

二〇一九年六月十九日，經過六週聽證後，陪審團一致裁決七項罪名均成立，包括敲詐勒索、強迫勞動、性販運。[114]托妮諷刺地穿上一件條紋「監獄」上衣聽裁決。一年多之後，二〇二〇年十月二十七日，拉尼爾獲判一百二十年刑期，併科罰金一百七十五萬美元。他在受害者面前朗讀聲明時，依然說自己是無辜的，並且主張多數目擊證人都撒謊，但對於自己引起的痛楚和憤怒，還是宣稱自己「深感抱歉」。他以奇怪的誇張方式表達懊悔，說道：「現在的處境是我自己造成的，全都是我做的。」

二〇二一年一月二十一日，拉尼爾從布魯克林大都會看守所（Metropolitan Detention Center in Brooklyn）轉到亞利桑那州土桑（Tucson）一間安全等級最高的監獄。這所監獄提供性犯罪者管理計畫，因此拉尼爾很可能餘生都會與拉里・納薩爾（Larry Nassar）同類的犯人作伴。曾是美國體操隊隊醫的納薩爾是連續強暴犯，也犯下虐兒罪，目前正在同時服六十年、一百二十五年和一百七十五年的刑期。性犯罪者占了土桑監獄服刑人口的七至八成，拉尼爾在監獄被迫放棄「先鋒」名號，改以新稱號稱呼，完全是他應得的下場：聯邦囚犯編號五七〇〇五－一七七。

Chapter 8

脫逃

克莉多尼亞·瑪琳達和恢復上帝十誡運動教

克莉多尼亞·瑪琳達（Credonia Mwerinde）人生的多數時間都刻意避人耳目，她極可能都設計好了。瑪琳達出生在烏干達（Uganda）西南方偏遠的鄉村，一生都極力避免引人注目，就連創立史上數一數二成功卻嚴厲的末日邪教時也是一樣——「恢復上帝十誡運動教」。她喜愛金錢更勝眾人目光，追求金錢的手段殘酷野蠻：信徒奉獻所有世俗財產給十誡運動，生活在神聖的奴役狀態；她的爪牙會出面脅迫任何擋著她路的人。但在她的末日預言接連三次失準（一九九二年、一九九五年、一九九九年）之後，信徒開始離去。她知道結局將近，下令系統性屠殺每一區的信徒，隨後在木製建築裡聚集了五百多名忠誠的擁護者，鎖上門，放火燒了建築物和裡面所有人，帶著十誡運動的基金逃跑，從此在眾人眼前消失。

神聖異象和大屠殺

在所有聲名狼藉的邪教領袖之中，克莉多尼亞．瑪琳達據稱奪走最多條人命。二〇〇〇年三月十七日，在恢復上帝十誡運動教中央區聖母救援之地（Mary's Place of Rescue）上，她帶領五百多名信徒進入一棟他們稱之為「方舟」的木造建築，方舟已經被潑灑了汽油，整整六十八個德國汽油桶。瑪琳達在新信徒眼中是設計師（Programmer），不過她也是該邪教的女大祭司。接著，瑪琳達走出方舟，滑動門閂，鎖上木門。她心知肚明方舟內無路可逃，放火燒了超級易燃的方舟。火勢太過猛烈，當地法醫起初以為信徒的頭爆炸了，雖然更可能是因為火焰吞噬了軀體，頭顱變得易碎，掉落的椽木砸碎了殘骸，瓦解的木頭結構四散在遺體周遭，甚至直接壓在上方。

不過後來發現，這起駭人事件不是瑪琳達應該負責的唯一大滅絕。第一次大火後沒多久，調查員在另外三處教區挖到亂葬岡，散落在卡農古區（Kanungu District）全境。出土的遺體有勒斃、刺傷、毒害的跡象，而且已經開始腐爛，相關當局認為這些凶殺下手匆忙，最晚也是在聖母救援之地大火前一週發生。加上可怕的新發現，瑪琳達可說是下令奪走超過一千零七十五條人命。

一九五二年，在烏干達美麗的布恩迪森林國家公園（Bwindi Impenetrable Forest），保羅．卡夏庫（Paul Kashaku）和法拉齊亞．卡夏庫（Farazia Kashaku）生下瑪琳達。這名女子究竟為何會在烏干達歷史上留下如此血腥的記號呢？瑪琳達的生活並沒有缺乏任何資源。她的家族持有土地、鳳梨園、甘蔗田，這些經濟作物都是種植在國家肥沃的區域。但是瑪琳達似乎另有志向。她學習釀製香蕉啤酒，默默融入當地酒吧老闆的家庭生活，隨後接管酒吧的經營。她多次使用這一模板：默默融入一個家庭，魅惑一家之主，接著把對方的財富和資源轉為己用。獲得掌控，繼續前進。

一九八九年，酒吧生意失敗，瑪琳達盯上虔信天主教的鄉下地區興起的宗教運動，利用身為女人為數不多的奪權途徑：宣稱聖母瑪利亞出現在她面前，此現象又稱聖母顯靈（Marian Apparition）。教會似乎提供了安全，還有來自信徒隨喜捐獻的穩定收入，這樣的權力結構可以任她剝削利用，卻不會遭受懲罰，而且犯下任何罪時，都可以拿來當成道德掩飾。雖然我們不知道究竟有多少是一開始就計畫好的，但瑪琳達再次找到一名強大的男人展示自己的技術：富有的地主約瑟．基布維特爾（Joseph Kibwetere），他創辦了一所天主教學校尼亞卡津加中學（Nyakazinga Secondary School），還在魯瓦沙邁爾（Rwashamaire）持有廣大土地，學校就蓋在那裡。他後來成為十誡運動的公開代言人，瑪琳達則是全權掌握，如同在王座背後垂簾聽政，自願選擇沒沒無聞，因此贏得了「設計師」的稱號。

接下來十年，瑪琳達會使用據說可以和聖母瑪利亞直接聯絡的方式，將恢復上帝十誡運動教打造成一本萬利的宗教企業，資金全來自信徒的慷慨捐獻，他們承諾捨棄一切世俗財產，且凡事聽從設計師。她的擁護者建造教區、耕田種地，不論有技術還是無技術，都還會在當地找工作，辛勤勞動毫無報酬，只為瑪琳達聖誡大人（Holy Commandments）。這些信眾又向自己的親友宣教，招募更多市民。深得青睞的信徒甚至變成劊子手，對那些膽敢判教的人展開報復，有時甚至會以瑪琳達之名殺人，（如同設計師再三向核心成員保證的）因為聖母瑪利亞告訴瑪琳達十誡運動的信徒獲准殺人。

瑪琳達究竟可以累積多少財富、怎麼把這些世俗財產轉變成帶得走的錢財、到底是如何達成最終脫逃，都是關鍵提問，但至今依然成謎。大屠殺之前，全世界都不知道烏干達小角落裡的瑪琳達十誡運動，等全世界意識到時，瑪琳達已經消失了——目前已知一手建構殘酷末日邪教的唯一設計師逃過死亡，避開懲罰，消失無蹤。

暴力頻仍國家催生的殘酷志向

說來駭人，不過瑪琳達的行為雖然凶殘，但和烏干達的長年派系衝突、宗教戰爭、殖民剝削、經濟掠奪相較之下，也不盡然特別突出。最早期的接觸是阿拉伯商隊和歐洲殖民武力，一部分是因為信仰，一部分是經濟利益，造成一連串衝突摧殘著鄉村地區。雖然十九世紀開始，探險家和傳教士一同抵達，但在此之前的當地歷史顯為人知，傳統上許多王國原本就擁有穩固扎實的政治機構，雖然之後某天會在未來的烏干達境內被殖民。

政權總是享有神權的支持。王國內的無數氏族都崇拜一位創世神，稱呼各有不同，包括卡通達（Katonda）、卡祖巴（Kazooba）、魯杭阿（Ruhanga），同時崇拜許多較小的當地神明、精靈。氏族領袖叫做婀曼德瓦（emandwa），形成當地教派，在這龐大的神話體系下，各族都有自己喜愛的一群神祇。從十九世紀中期開始，這些氏族和他們的文化發現自己面臨歐洲和中東傳教士的挑戰。最早到來的是伊斯蘭傳教士，接著在一八七七年，英國聖公會（Anglican）傳教士跟著英國東非公司（British East Africa Company）一同到來，英國東非公司本身跟著阿拉伯商隊，沿著尼羅河（Nile River）上行到位在烏干達金加（Jinja）的源頭——維多利亞湖（Lake Victoria）湖濱。兩年後，一八七九年，英國聖公會傳教士後方跟著出現第二波法國天主教（French Catholic）傳教士。

部落和殖民帝國紛紛簽訂貿易協定時，看似因宗教而起的戰爭四處爆發，一開始是穆斯林和基督徒，接著是新教徒和天主教徒。這些衝突通常極有可能變成浴血衝突，尤其是某一邊得到支援，拿到先進的西方武器時。

這些發生時間重疊的衝突帶來一片混亂，導致一八九四年成立了英國保護國（British Protectorate），目的不在實現該地和平，而是縮小暴力造成的經濟損失。保護國統治全國直到一九六二年殖民統治結束為止。一九六二年，剛解放的國家換下英國保護國的行政首長，改由阿波羅．米爾頓．奧博特（Apollo Milton Obote）執政。這位新總理（之後會成為總統）鞏固自己的權力，受到英軍最高階級的非洲裔將領支持——殘暴的伊迪．阿敏．達達（Idi Amin Dada）。阿敏之前協助英國鎮壓肯亞（Kenya）的茅茅叛亂（Mau Rebellion），協助消滅索馬利亞

（Somalia）的叛亂威脅。

奧博特和他的嗜血軍事行動長結成令人不安的聯盟。奧博特知道阿敏曾經把簡單的任務——控制奈洛比（Nairobi）的偷牛賊——變成殘酷的恐怖活動，阿敏的軍力折磨偷牛賊至死，有時候還直接把人活埋。儘管如此，奧博特立阿敏做武裝軍隊的頭頭，與阿敏合作一起打造走私生意賺大錢。

後來，剛果反叛分子用價值三十五萬美元的黃金和象牙向阿敏換取武器，卻被阿敏侵吞。東窗事發之後，一連串政治操弄隨之而來。布干達王國（Buganda）國王穆德沙二世（Mutesa II）憤恨不平地抱怨這起侵吞竊盜（英國扶植穆德沙二世來瓜分奧博特的權力），但是阿敏和奧博特把穆德沙二世驅逐出境，再度鞏固了手中的權力。奧博特掌權期間，殘酷暴戾人盡皆知，最終阿敏在一九七一年奪權，奧博特時期才結束。

伊迪．阿敏有六位妻子，在各地鄉村都有情婦，多達三十位，原本大家讚揚阿敏是烏干達的救星，而且魅力十足。當他取消流放、允許穆德沙二世返國，以及特赦奧博特的政治犯時，看起來又更加仁慈了。

他因慈悲為懷登上報紙頭條，卻暗中安排多組殺手，殺害阿喬利和蘭戈部族的士兵，大家都以為這兩個部落對奧博特非常忠誠。很快，阿敏開始挑起戰爭，不顧會造成多大的經濟影響。即使富裕的亞洲人和以色列人負責多家公司、大片農園、建設計畫，促進烏干達的經濟，阿敏還是通通都驅逐出境。

即使阿敏鞏固了實權，他的殺手小隊並沒有停止行動。相反地，殺手小隊還增加了活動，種族清洗、肅清政治敵手，暗殺企業利益的競爭者和膽敢調查阿敏罪行的記者。據估計，死亡人數介於三十萬到五十萬之間。阿敏還宣稱自己吃過人肉。阿敏允許巴勒斯坦人劫持一架法國航空空中巴士（Air France Airbus），脅迫機長迫降恩特貝國際機場（Entebbe Airport），從此烏干達的聲譽跌到谷底；好險以色列軍方大膽發動突襲，成功救出幾乎所有人質。儘管在國際間受到這次打擊，阿敏還是給予自己越來越華麗的頭銜：無限期總統閣下（His Excellency President for Life）、伊迪・阿敏博士阿爾・哈吉陸軍元帥（Field Marshal Al Hadji Doctor Idi Amin）、維多利亞十字勳章受獎人（VC）、傑出服務勳章受獎人（DSO）、軍功十字勳章受獎人（MC）、大英帝國司令勳章受獎人（CBE）、地上萬獸與海中魚族之主（Lord of all the Beasts of the Earth and Fishes of the Sea）以及普遍在非洲、尤其在烏干達的大英帝國征服者（Conqueror of the British Empire in Africa in General and Uganda in Particular）。有人說他沒接受梅毒治療，於是陷入瘋狂。阿敏的大多數軍隊都逃亡到坦尚尼亞（Tanzania），和坦尚尼亞軍隊合作，終於在一九七九年四月擊敗阿敏，把他從權力寶座上永遠移除。

廢黜阿敏之後，烏干達歷經十年無人能遏止的內戰。一九八〇年到一九八六年，烏干達政府（又再次回到奧博特手中，現在還受到美國支持）反擊國家抵抗軍（National Resistance Army），抵抗軍由利比亞（Libya）和沙烏地阿拉伯（Saudi Arabia）支持。抵抗軍贏了，安插穆塞維尼（Yoweri Museveni）成為總統，至今依然是烏干達總統。他的統治同樣也發生人權虐待事件和暴力叢林戰爭，在烏干達北部尤其嚴重，約瑟夫．科尼（Joseph Kony）領導聖主抵抗軍（Lord's Resistance Army）在該區域活動。

軍閥科尼既殘暴又嚇人，直到最近才被國際單位緝拿，指控他犯下大屠殺、性暴力、肢解屍體、系統性竊盜等罪行。但從許多方面看來，科尼主張領導的權力是來自靈性動力，這一動力與克莉多尼亞．瑪琳達據稱的神聖起源相同。科尼說自己被召喚要當神的口，被賜予屬靈能力，神祕力量也曾經拜訪他。他甚至還說戰爭是為了十誡而打，就像恢復上帝十誡運動教宣稱要做的一樣。姑且不論這些似是而非的宗教開端，科尼的惡名全世界都知道：殘暴行為、突擊鄉村，強迫男孩成為童兵、女孩成為性奴。

沒有人可以確認瑪琳達早年生活的真實樣貌，但她展現的許多特質都是成年時期烏干達的常態，尤其是披著神聖外皮的暴力，以及剝削他人攫取財富。單單考慮行為本身，瑪琳達的行為特別突出，過度殘忍無情，但在一個歷史浸滿血的國家，她的殘酷其實早有許多先例。

脾氣火爆

一九五二年，瑪琳達在鄉村小鎮卡農古出生，後來該區以小鎮名字命名。她的雙親是保羅．卡夏庫和法拉齊亞．卡夏庫，據推論只知道他們是地主，後來瑪琳達設法從兄弟姊妹手上成功奪取家族土地的控制權。她的童年則顯為人知，只能假設瑪琳達在天主教會接受廣泛的教育，學習到的知識後來也為她的信仰操縱派上用場。我們只知道瑪琳達沒有遵從家族對於良善行為的想法，行為舉止也不像虔誠的天主教女孩。她脾氣易怒，長大之後，性關係開始隨便。

不過對於在嚴格宗教社群長大的年輕人而言，這些行為一點也不奇怪。類似修道院的教養自然而然會造成叛逆的反應，因為家長與老師都會不停發號施令。每當這種環境下的小朋友年紀大到可以自由探索生活，都很可能會開始嘗試以前師長禁止的活動。但是瑪琳達開始走偏時，父母敦促她重回天主教會。傳教士保證，宗教是設計來拯救人類的靈魂。

瑪琳達二十出頭歲時，踢到了人生第一塊鐵板，愛上一名地方衛生官員。或許瑪琳達認為自己終於找到一條離開嚴格家庭生活的道路。但據說那名男性拋棄了她，又或者至少瑪琳達覺得對方都對她不理不睬。她的反應相當暴力：闖入該名男子家中，放火燒了家中物品。

這是第一回事例，充分展示了瑪琳達畢生對火的癡迷。縱火傾向是麥克唐納三要素（Macdonald triad）之一，這三要素指的是三種可以預測日後出現犯罪行為的活動，社會病態的前兆（另外兩項要素是虐待動物和尿床。虐待動物被視為殺人的預演，尿床本身並不暴力，但和另外兩項習慣同時出現的話，則會強化預測）。許多人相信這些行為是殺人傾向的早期徵兆。

瑪琳達的家人和地方當局想必都嚇到了，瑪琳達被送進了精神病院。但她沒有接受任何有意義的問診，診斷結果也不精確。待了幾個月，瑪琳達被簡單貼上「精神障礙」的標籤，就出院回家了。住院經驗顯然沒有幫瑪琳達建立起人生方向，因為她再也沒有尋求心理醫師解決自己的心魔。至於她的診斷，「精神障礙」這個名稱其實沒有多大用處。雖然「邊緣型人格障礙疾患（Borderline Personality Disorder）」一詞依舊意義模糊，僅用以泛稱那些比較難診斷的疾病，但大致描述都與瑪琳達相符：情緒不穩，導致衝動或危險行為的模式，也會出現氣憤難平導致焦慮或憂鬱的時期。這開啟了有害循環，行為和情緒相互滋長。

瑪琳達也出現了社會病態的徵兆。她總是自我中心，同時似乎傾向戲劇化和暴力的決定。許多社會病態會展現自戀特質：渴求關注，並且認為自己的幸福快樂比起生命中其他人的來得重要。如果遭人輕忽（如同該名官員忽視瑪琳達），社會病態者可能會認為是天搖地動的大事，於是也回以同樣劇烈的反應。不論事發經過為何，都是場悲劇。瑪琳達不可能在那個年代獲得適當的診斷，就算在美國也無法獲得。因為她無法接受純粹的天主教教育，還去了一趟精神病院，身邊許多人開始瞧不起她。

惹是生非

瑪琳達決定掌握自己的命運，不只利用天生的魅力，還誘惑他人一同加入她更黑暗、更危險的追尋。剛好在這個時期（一九八〇年代初期），烏干達落入阿敏的殘暴統治，國家經濟大幅衰退，基礎建設陷入混亂。瑪琳達需要一個安全的港口。她透過和艾瑞克．馬齊瑪（Eric Mazima）組成事實婚姻關係尋求臨時庇護。馬齊瑪又是一名卡農古在地人，瑪琳達說服馬齊瑪把第二任妻子趕出妻子的家，投資瑪琳達在卡農古開設酒吧。關於瑪琳達的早期生活描述，其中一個事實是：她擁有釀香蕉啤酒的技術，這是東非的一種主食。

即使忙著經營酒吧，顯然她還能擠出足夠時間培養自己的暴力舉止。瑪琳達的家人收到一封信，寫到一則未經證實的故事，寄件人警告家人說他相信瑪琳達殺了人，細節不多但很嚇人：某晚，酒吧來了一位疲憊的旅人，瑪琳達引誘他上樓。早上，老客人來喝杯瑪琳達馳名當地的香蕉啤酒，發現瑪琳達在刷洗水泥地板上看來像血的汙漬。瑪琳達看到他們來了，依然鎮定如常，詢問大家要喝什麼。

這又再次指出瑪琳達社會病態的一面：不論身處任何危機情況，即使被人發現，她的回應僅是微笑或玩笑，幾乎是嬉鬧的防衛機制。據稱在凶殺發生的同時期，一九八一年三月，瑪琳達決定運用個人魅力滲透神的國。儘管瑪琳達寫下的日期顯然不大可信，但這應是她第一次遇見聖母顯靈的日期。聖母瑪利亞叫瑪琳達悔改自己的罪孽，重回她拋棄的群羊。不論瑪琳達是否相信這次顯靈真的發生了，但是下個十年間，阿敏搖搖欲墜的獨裁專政四分五裂，國家陷入內戰的時候，瑪琳達決定自己值得聖徒的身分。

瑪琳達見證國家墮入混亂，看著軍閥興起財富大增，可能因此認為她也可以走出自己的路，享有名正言順的地位、權力和廣泛的影響力。據稱的聖母顯靈後不久，瑪琳達開始四處宣傳自己淫亂的形象，散播自己不只是酒保還是性工作者的謠言。雖然這宣稱很可能有些真實依據，但她先生馬齊瑪否認了。這絲毫不打緊，馬齊瑪也只做了她先生一陣子。不論是事實還是想像，性工作是她新職涯的初期證據：瑪琳達想要被視為現代版抹大拉的馬利亞（Mary Magdalene），才會盡力想要展現自己深陷罪孽。她的救贖即將來臨。此舉是瑪琳達最巧妙的一步，也是她一邊控制十誡運動，一邊編造敘事主軸的第一項證據。

一九八九年，瑪琳達的酒吧倒閉，婚姻破裂，適逢穆塞維尼鞏固了權力，擊潰北方的約瑟夫．科尼。然而，瑪琳達早就心生他計。一九八九年六月十四日，瑪琳達標記為第二次遇見聖母顯靈的日子。這次異象在卡農古外側的洞穴顯現，聖母瑪利亞向瑪琳達傳達的話語之後會成為十誡運動經典的開場白：《從天上來的及時訊息》（A Timely Message from Heaven）。她的使命清楚明白：罪人悔改的日子近了，罪人需要有人帶領。萬福聖母瑪利亞賦予瑪琳達使命，尋找共同努力的夥伴。

認同瑪琳達異象的富豪

約瑟．基布維特爾是烏干達人、天主教徒，比瑪琳達年長二十歲。依照烏干達的標準，基布維特爾是名家道殷實的地主，家族十分虔誠。宗教領袖在信仰和財富兩方面都特別優待基布維特爾家族。基布維特爾認為自己是天生的領袖。不為貧賤的出身所囿，基布維特爾設想的未來遠大不凡，引領忠信的人走進真理，而天主教會認為必須確保他應得的豐盛。

一九六〇年，他娶了泰瑞莎（Theresa）為妻，並在他那區域的天主教學校系統擔任助理校監。隨著基布維特爾聲望漸長，能繼承的遺產也不停增加：他的家族持續取得大片土地。他奮發向上，不久便負責監督該區政府的農業與建設計畫。但當他試圖邁出下一步，走上自認注定的升遷，成為政務官時，卻未獲提名。他沒有心平氣和接受挫敗，就像許多不習慣自身特權遭到質疑的人一樣。他倉促撤退，回到魯瓦沙邁爾鎮（Rwashamaire）的家族土地上。現在掌管多處大型房地產、數百頭牛隻和生意興隆的工廠，他可以彌補自己沒能進入政府擔任編制內職務的失敗。基布維特爾和瑪琳達很像，也決定如果之前那樣做無法獲得權力，那麼就撤退回他生命中權力的源頭：宗教。

基布維特爾出資建造新的天主教學校「尼亞卡津加中學」。他在自己的學校任教，身兼校長與宗教老師。如同自己接受的教育，教義嚴明。這應該滿足了基布維特爾人生幹勁的所有條件。這不是多大規模的會眾，但基布維特爾成了社群中最受敬重的人，也因此可以直接影響年輕心靈。基布維特爾人生的意義在於影響他人，而沒有比孩童更輕易下手的目標了。正統的宗教學校給師長特別機會，不只教育，還能真正塑造學生。

天主教會提供基布維特爾影響力和資金建造學校。作為回報，基布維特爾的機構供應天主教會飢渴慕義的新會眾，最終不是成為終身教友，就是終身捐獻，甚至成為修士、修女或神父。經營學校期間，基布維特爾的心理發生頗巨大的改變，變得更愛勸人改信，更想尋求權力，而宗教——尤其是宗教的神話編造（mythmaking）——可以給予他任何想望的事物。瑪琳達看出該策略的良性循環——他的聲望隨著創立宗教學校而水漲船高，天主教會派遣學生到他的學校，而接受良好信仰教育的學生會再回到天主教會服務。瑪琳達可能也基於其他原因聽聞基布維特爾：一九八四年四月二十五日，基布維特爾宣稱自己看到聖母顯靈。

毫無疑問，對經營天主教學校的人而言，蒙受聖母瑪利亞的祝福好處多多。瑪琳達可能注意到了聖母顯靈對當地居民的影響，他們看似沒有尋找太多證據來證明他聲稱的神聖異象。當地居民似乎相當歡迎這個概念，好像大家一起蒙聖母瑪利亞的福。如果聖母瑪利亞賜福基布維特爾，肯定也間接賜福整個社群。基布維特爾總是一再提及這件事：他們是天選之人，正如從前的以色列人。

但時光流逝，基布維特爾思索著自己獲得的聲望是否足夠。正當此一信仰危機開始扎根，瑪琳達帶著兩名女性來找他。那兩名女性是瑪琳達的家族親戚或親姊妹。瑪琳達說，聖母瑪利亞要她們尋找基布維特爾，兩名女性為瑪琳達的發言作證。這正是基布維特爾需要的變動：現在不用懷疑萬福聖母瑪利亞選中他的理由了。

瑪琳達站在他面前，請求他協助執行神聖使命。但有件事很快就清楚明白：瑪琳達掌握所有權力。基布維特爾是受人敬仰又有錢的社區領袖，但在這位因聖母顯靈得蒙救贖的陌生人面前，基布維特爾輸得一敗塗地。基布維特爾很聰明，但顯然缺乏不可或缺的火花，無法真正聲名遠播。他是古板的官員，但沒有待在真正的官僚體系。另一方面，瑪琳達的宗教魅力閃耀四方，而且早已卸下所有自我懷疑，就連埋在深處的都清得一乾二淨。

從那天起，除了瑪琳達的家族成員之外，基布維特爾成了瑪琳達真正的頭號皈依信徒。

一九八九年，恢復上帝十誡運動教正式成立。瑪琳達鼓勵基布維特爾相信他自己遇到的聖母顯靈，甚至提供瑪琳達自己的詮釋：聖母瑪利亞特別挑選基布維特爾，成為烏干達新天主教信徒的領袖，新信徒就是他們故鄉那群天選之人。瑪琳達按照自己的需求，形塑基布維特爾的想望。基布維特爾現在成了學生，而且只要隱身幕後代表可以達成更宏大的命運，十分樂意就此退居幕後。

在烏干達社會裡，女人的影響力遠遠比不上男人。瑪琳達成功獲得基布維特爾的忠誠，可說是第一個奇蹟，也是最難發生的奇蹟。瑪琳達成功讓自視為最高權威的男人臣服，不僅視瑪琳達為更高權威的存在，甚至還相信這一切都是自己的主意。基布維特爾的兒子尤維納爾·魯甘布瓦（Juvenal Rugambwa）說明：「接著我們知道的是，她住進我們家，他們決議要在家裡創立邪教。沒多久，她就打敗我們所有人。我父親對她心懷敬畏，只要她下令，什麼都願意做。」

瑪琳達究竟決定要如何打造她的新宗教呢？她告訴基布維特爾，基督打造自己宗教時採取的做法，就是他們形塑運動的方法。信徒放棄地上的財產，財富都要奉獻給他們的使命。基布維特爾開始變賣過剩的財產，這是必要的交換。透過他扭曲的內在邏輯，他終於明白真正的權力和任何世俗財產都沒有連結，真正的權力乃是靈。他賣掉工廠設備，還讓自己的小孩離開自己建立的學校，從此在瑪琳達監督下接受嚴格的自學，會成為日後事業的新規範。

基礎奠定好了。瑪琳達會擔任「設計師」，設立規則，確立十誡運動的路線。基布維特爾會是魁儡領袖，很快就獲頒主教頭銜。但如果想自稱為合法正規的天主教組織，就必須有真的權威人物：神職人員。十誡運動神聖經文後來的內容寫到，基布維特爾跟隨聖母瑪利亞的指引往北，聖母叫他尋找一名「宗教委身者」（religious devotee）修拉斯蒂加・卡馬嘎拉（Scholastica Kamagara）。現實生活中，修拉斯蒂加只是基布維特爾太太泰瑞莎認識的人：木巴拉拉區（Mbarara District）中央教區的喉舌，常為地方發聲。她有家庭關係，而其中一位男子看起來特別完美，他無法實踐抱負，非常挫敗，但他可以讓十誡運動的架構變得完整……可以讓他們加速朝共同想像的末日啟示前進。

方舟船長

在烏干達，大家很重視地位、安全和延續性，可能是因為全體人民的生活處處匱乏。內戰期間的動盪不安，天主教會機構提供人民避難之處，也給予人民高於戰事的權威，而教區的主責神父深諳此理。伯納德・阿圖海爾教授（Bernard Atuhaire）在關於十誡運動的《烏干達邪教悲劇：私人調查》（暫譯，*The Uganda Cult Tragedy: A Private Investigation*）一書中寫道：「宗教教導作為傳教工作的產物，總是用撲朔迷離的手法描述司鐸，太過令人困惑以至於產生『教宗無謬誤』（infallibility of the Pope）這樣的宣稱，一般來說神父和會士都確實如此相信。」[1]多年來神父強制執行這種階級制度，確立自己為社會的仲裁者——能夠同時懲罰或祝福民眾。這一切能運作，全都仰賴大家相信要尊敬神父，尊敬教宗。

這也是為何輕聲細語的多明尼克・卡塔里巴博（Dominic Kataribaabo）神父能夠鶴立雞群。身為神父，他沒有強硬行使神權。多明尼克親切友好、口齒清晰，似乎比較符合典型的和藹神父形象。一九七〇年代晚期到八〇年代早期，他擔任烏干達西部布樹尼區（Bushenyi District，就在木巴拉拉區的正西邊）基塔比神學院（Kitabi Seminary）院長。他就像許多學生的代理父親，多明尼克神父透過同理而非訓誡贏得人心。

最佳範例就是他與學生約瑟夫・卡薩普拉里（Joseph Kasapuraari）的關係。多明尼克神父從這名安靜的學生身上看到與自己的相似之處。每年，神學院都會票選學生領導的各種職務。卡薩普拉里想要擔任其中一職，但自知本身沒有魅力，大家不會投票給他。儘管如此，多明尼克神父宣布當選者時，用戲劇化的節奏、驚喜的語調，告訴神學院師生卡薩普拉里贏得學生會幹部一席。操控這場選舉是深思熟慮的善意，還是贏得忠誠信徒的工具？宗教和這些男人的生命糾纏不清，盤根錯節，難以區分行為是出於慷慨抑或另有所圖。若賦予宗教機構在社會中如此大的權力，就會發生上述那般危險的情事。但對卡薩普拉里而言，別有所圖還是慷慨大方沒有實質差異，多明尼克神父的行為贏得卡薩普拉里一輩子的感激。

有這麼一群追隨者，多明尼克神父確保在魯加齊教區（Rugazi Parish），大家都視他為一名真正神聖的人，同時也是神謙卑的跟隨者。這令他非常開心，其實在溫柔的表象下，這名聖人和基布維特爾、瑪琳達有相同的狀況：人生在世，總想渴求更多。阿圖海爾教授的紀事有一條紀錄：多明尼克神父強迫神學院工作人員在校區後方的沼澤挖掘無數個小時，一切只因他發現一顆閃閃發亮的石頭。在每週的勵志自助會議上，他經常跟學生談及世俗和野心勃勃的事，例如「你很有價值」、「目標永遠都要設高一點」、「如果想要和平，就得準備好作戰」等言論。對多明尼克神父而言，價值不僅限於對神的信仰，也同等地根植於自身。他教導學生要努力過上理想化、自由的獨立生活。

不過，就如同基布維特爾和瑪琳達所認為，未盡事宜的目標會變成危險的燃料。多明尼克神父貪求進步。在木巴拉拉全境教區內，大家都知道魯加齊教區信徒自以為高人一等，認為自己是這一區宗教命運的真正領導人。魯加齊教區的傑出人才一直以來都在烏干達教區擔任深具影響力的職位，例如財務、天主教會報紙編輯，還有人人欣羨的副主教一職。

多明尼克神父總是幻想，在鄰近的卡瑟瑟教區（Kasese Diocese），自己肯定能出任主教一職。但幻想很快就破滅，另一位神職人員獲得該職，多明尼克神父依舊待在小小的神學院。天主教會許多人認為多明尼克太過涉入政治，企圖為自己贏得地方支持，還有人看穿他的文靜舉止，看到潛藏的篡位者。

就在最低潮的時刻，一位朋友前來拯救多明尼克．卡塔里巴博神父：學生卡薩普拉里。卡薩普拉里正好就是修拉斯蒂加的兒子，而修拉斯蒂加則是基布維特爾受到聖母瑪利亞的指示要去尋找的女人，又或者他太太泰瑞莎的指示。修拉斯蒂加和先生約翰．卡瑪嘎拉（John Kamagara，卡薩普拉里的父親）都非常虔誠。他們以前想當修女和神父，但基於種種原因，總是無法當成。他們一直把卡薩普拉里形塑成他們的宗教戰士，覺得兒子有機會救贖他們脫離失敗。他們確保兒子每天參加彌撒，要求兒子努力追求自己在基塔比神學院立的目標。

卡薩普拉里得知多明尼克神父的挫敗之後，修拉斯蒂加接著告訴泰瑞莎這則消息。如此一來，泰瑞莎注意到天主教會有名聖職權威對信仰起疑了。多明尼克神父、卡薩普拉里和他的雙親在瑪琳達眼中都是轉信的最佳人選——既受過高等教育，在天主教世界又都是傑出人士，最重要的是，他們的關係十分緊密。這很快就會成為瑪琳達招募信徒的主導策略——目標是在單次彌撒集會一次招募所有家庭成員，甚至是整個鄰里。這是最有效率的方法，在最短時間內為恢復上帝十誡運動教打造可運作的基層結構。

十誡運動發展期間，多明尼克神父的抱負和瑪琳達、基布維特爾的一致。假使天主教會不想給予他更高的權力，他就親自出手抓住權力。此外，他還可以證明天主教會錯了。打從一開始，十誡運動就沒想過要與天主教切割。十誡運動從來沒有想要說自己是反叛團體——他們只是一群關心的信徒，擔心天主教會偏離了軌道，並且視自己為少數天選之人，要來拯救天主教會。

卡薩普拉里和父母加入十誡運動，成為宗徒，為組織帶來動能。瑪琳達確保多明尼克獲得獎勵，指派他擔任副主教。天主教會沒能給予他自認應得的榮譽，但是瑪琳達提供機會讓他獲得一切他覺得理應屬於他的事物。

然而，在真正可以迎來更多轉信的人之前，還需要一份必備文件。雖然十誡運動依然把聖經列入聖品，但敘事的基本概念是聖母瑪利亞帶著天上的新指令重返人間。任何宗教運動的基礎都奠定在宗教典籍上。典籍包含規則，規則是宗教運動的律法和文化。宗教運動需要用最清晰明白的方式傳達自身宏大的使命。因此瑪琳達需要定稿，編寫《從天上來的及時訊息》最終版。

下方引用一小段《從天上來的及時訊息》，神的語氣讀起來充滿復仇之心：「全人類將面臨巨大苦難，創世以來還沒有人體驗過的苦難。人類會變得心不在焉，而且會在行為中發展出自己的精神，獨立於天主之外，這種精神會讓造物主憤怒。造物主回應的方式是在世界降下刑罰，許多國家會因此血流成河……」[2]

換句話說，末日近了。天主向人類發怒了，這需要聖母瑪利亞和基督本身雙雙使出最後一擊，才能遏止造物主的震怒。瑪琳達及其追隨者是地球信徒的最後防線，十誡運動則是得救的最後機會。為了增加信徒間的緊急感，現在任何行動都是與時間競賽：聖母和聖子交付十誡運動重要任務，在末日之前，盡可能拯救越多人類越好。

末日看似就在不遠的未來。一九八〇年代，烏干達因為內戰而四分五裂，市民覺得脆弱不堪，他們的希望受挫，迫切渴求任何支持。快要來不及了。除了內戰耗損國家，新的千禧年也即將來臨。為了對抗不理性的恐懼和猜疑，恢復上帝十誡運動教可以提供天主教會無法提供的：能到天上的方舟和其他東西。不幸的是，對希望登上那東西的人而言，其實更加靠近的是可怕結局。

利用家人打造邪教

一九七九年阿敏倒台，隨之而來一連串選舉。但是穆塞維尼（最近從流亡回歸）輸給奧博特之後，宣布選舉過程造假，建立了國家抵抗軍，武裝反抗奧博特政府的「烏干達民族解放軍」（Uganda National Liberation Army）和反叛組織「烏干達民族解救前線」（Uganda National Rescue Front）。結果烏干達內戰（Ugandan Civil War）爆發，又叫烏干達叢林戰爭（Ugandan Bush War），從一九八一年到一九八六年，游擊戰衝突肆虐整個國家。

即使國家抵抗軍打贏了，穆塞維尼也在一九八六年就任總統，國家依然嚴重分裂。雙方都犯下滔天暴行，有些罪行甚至堪比阿敏的殘暴。一九六二年獨立以來，烏干達不停尋求安穩，但至今尚未達成。面對持續的動盪不安，人民深切渴慕希望、安全、屬靈的確實。烏干達因此成為末日神學滋生的完美溫床。

在地方社區現場，聲望地位才能獲得安全和繁榮，獲得地位的管道掌握在教育守門人手上，但在十誡運動發祥地等區域，則是掌握在天主教會權威手上。雖然烏干達的新興上流社會針對教育價值會說一些應酬話，但其實新富階級不想給窮人任何真正獲得權力的途徑，因此切斷所有通往知識和向上流動的大道。瑪琳達和基布維特爾看到向上移動的唯一方法：獲得宗教影響力，以及承諾他人也可以獲得影響力。

基布維特爾是十誡運動的主教，多明尼克神父現在擔任副主教一職，瑪琳達掌握了運動的權力結構，也掌握了吸引後來所有轉信之人的模板。烏干達社群由緊密的群體組成，鄉村鄰里交織成一片，家庭關係錯綜複雜。只要能辨認出弱點，切入群體，十誡運動就能夠迅速傳播。阿圖海爾寫道：「訊息的傳播是鏈式方法。先攻陷直系親屬和身旁友人，接著就是他們的親朋好友。」

這是一種分配系統，也是一種行銷計畫。首先，一位家族成員聽聞十誡運動，受到說服轉信入教。宗教狂熱分子會說服最親近的人跟隨他們的引導。接下來，鄰居會看到許多家庭集體搬到十誡運動的教區居住。不久，全村都會知道十誡運動，任何好奇的鄰居只要展現出興趣，招募者很快就會上門。

瑪琳達確保十誡運動盡可能對所有外人開放。在恢復上帝十誡運動教的主要經文《從天上來的及時訊息》中，瑪琳達寫道：「我們的主耶穌基督……說道：『我死在十字架上，我為世上呼吸的人流血；不管是新教徒或是天主教徒，不管是穆斯林還是從來沒有擁抱任何宗教的異教徒，抑或擁抱任何信仰的活人，都讓他來聽這些訊息吧！』」

《從天上來的及時訊息》經文有許多發放方式。首先，鄉鎮間會有信徒口頭引述經文。接著，當地報紙會刊登經文。事有先例，羅馬天主教會（Roman Catholic Church）經常在報紙上刊登通告、祈禱經文、宗教訊息。地方當局和市民已經習慣在新聞事件和社區公告旁邊看到宗教文字。最後，《從天上來的及時訊息》登上當地電波，在烏干達西南方放送，加上多明尼克神父與基布維特爾兩人和廣播電台的關係，從旁協助播送過程。

雖然沒有太多可靠的會員等級紀錄，但很顯然到了一九九三年，十誡運動已經吸收不少信徒。烏干達當局授予非政府組織合法貿易的地位，他們因此可以在不太接受其信仰的鄉鎮設立一些據點。十誡運動明顯的正當合法性，讓他們吸引了新的追隨者。

這一切都是有意為之。瑪琳達的成長環境籠罩在天主教會施加的沉重期待之下，明白同儕壓力和社群期待對虔誠信徒的影響力。潛在追隨者對招募感到退縮，而且宣稱在十誡運動之外也可以擁有好的信仰，他們會被告知這個世界充滿罪孽，除非他們參與十誡運動，否則無法遵守聖母瑪利亞的新戒律。他們被迫加入十誡運動，不然罪孽就會太重，祈禱不足以消除。他們必須加入運動，否則末日來的時候就是站錯邊。

新招募來的信徒一旦進入團體，很快就會發現所有出路都被關閉。只要有人想退出，瑪琳達就會集結那人的虔誠家庭成員，允許他們可以暫時離開，這樣才能回到家鄉，親自阻止那位即將叛教的人回家，實際阻擋他步入家門。在烏干達，家庭的重要性再怎麼強調都不為過，因此背棄十誡運動代表即將面臨無家可歸的風險。通常，離開的人最後會直接回到十誡運動，只為了有個地方可以住。

還有另一因素為十誡運動召來許多信徒：愛滋病在八〇年代晚期和九〇年代初期大肆流行。流通的《從天上來的及時訊息》片段新增了一條訊息，直接指涉：「造物主降下的刑罰，這世界稱之為愛滋病，但是對天主而言，這是懲罰。『我不會給予任何藥物，治療愛滋病的藥物只有悔改，以及恢復我的十誡。做到這兩件事的人會被饒恕，還會獲得治療。只有當所有人都悔改，都向我哀求，恢復十誡，我才會收走懲罰。』」據稱，愛滋病是末日將近的徵兆，只有十誡運動可以解決——同時，這也是行銷機會，幫助信徒招募更多轉信者，加入烏干達西南部多個教區，進入安靜無語和異常殘暴的新世界。

十誡運動教區生活

十誡運動控制了數個衛星城鎮的貿易職位，還有村際間的多個住宅單位，但大部分會員都住在十誡運動營運中心的四個大型教區。聖母形塑之地（Mary's Place of Molding）位在魯瓦沙邁爾鎮基布維特爾的土地上，在恩通加莫區（Ntungamo District）卡甘巴（Kagamba）的外圍。聖母告解之地（Mary's Place of Confession）蓋在米托馬區（Mitooma District）首都米托馬鎮（Town of Mitooma）附近。另外，魯加齊教區的聖母施捨之地（Mary's Place of Giving）建造在多明尼克神父的土地上，就在他的舊神學院外側。多明尼克為十誡運動贏得一大片可耕地。這教區位在最中心，旁邊就是非洲橫貫公路（Trans-African Highway）。

最後是聖母救援之地（Mary's Place of Rescue），位在卡農古與世隔絕的山麓，距瑪琳達出生地不遠。雙親過世之後，瑪琳達繼承了土地，稱自己的父母為「信徒的祖父母」（Grandparents of the Believers），在父母墳墓上方，起造聖母救援之地的祭壇。雖然瑪琳達的父母和十誡運動一點關係都沒有，但他們兩人成了運動的聖像，也就是十誡運動初代的亞伯拉罕和撒拉。

這些大型教區的資金來源是會員奉獻，也是由會員出力維護。只要潛在新人決定入教，就必須拋棄所有物質財產，直接交給十誡運動的領袖。富人被告知要遵守聖經的警戒，變賣所有財產，全心投入十誡運動。窮人要付出更高的代價：他們沒有土地或有價值的東西可以上繳，被迫列出自己的罪孽，接著依照罪孽罰款。

瑪琳達的教導透過基布維特爾的口說得明白，十誡運動期待追隨者過著幾近貧困的生活。基布維特爾解釋：「我們送給你們耶穌和瑪利亞的聖像。我們沒有送給你們牛隻或汽車，因為我們的身體貧乏，但我們的心靈富足，而且在任何通往永生的事物上都是富足。」

不過，領袖私底下過著非常不一樣的生活。舉例來說，卡薩普拉里把個人財產委託給十誡運動外其他地位更高的朋友。信徒對運動中的兩套規則毫不在意，因為他們被教導，天主已經寬容了領袖。

新進受訓者還需要努力贏得一席之地。第一項指派給他們的任務是聽完所有《從天上來的及時訊息》的卡帶，內容據說是聖母瑪利亞直接講述給瑪琳達和基布維特爾聽的。招募期間，他們只看到片段，現在他們獲得完整訊息，顯示他們決定加入十誡運動已經值回票價了。訊息接著鋪陳：世界拒絕遵守十誡，獻身於撒旦。天主準備好毀滅人類，但聖母瑪利亞和耶穌基督為人類贏得最後一次機會。

瑪琳達的藻飾與渲染保證驚人的末日即將來臨。她解釋每個國家會付出不同代價。《從天上來的及時訊息》摘錄描述道：

俄羅斯會遭受不同類型的蝗災。莫三比克（Mozambique）會被自己國家的機械毀滅。日本會一直下雨，除非天父決定停止。法國，懶惰令你們無法承受即將來臨的懲罰，最終只能哀嘆國家滅亡。

如果十誡沒有在全球恢復，這些預言都會成真。瑪琳達顯然讀過啟示錄。就像新約聖經的最後一卷，《從天上來的及時訊息》沒有推翻聖經內文，只是續集——不僅有賣座特效，還預定好了高潮。

十誡運動沒有保證會拯救世界，他們只想要聚集夠多忠實信徒，展開新世界。現在和即將來臨的末日中間這段時光，他們稱作「篩子」（Sieve，命名選擇充分顯露原因，因為篩子有時候也會用來淘金）。篩子時光會結束，根據他們的信仰，結束時間大約在千禧年之前，那年懲罰會降到地上，接著會一連三日暗無天光。

此時輪到十誡運動教區出場了。多艘堅固的方舟裝滿食物和其他日常必需品，維繫信徒生命安然渡過末日時間。受訓者被告知他們剛剛為自己爭取避難的機會，避開世界毀滅的危害。

受訓者先熟記瑪琳達版本的末日，接著開始學習聖母傳達給基布維特爾的訊息，基布維特爾的異象是為了治理十誡運動教區內的每日生活與儀式。平均一天的七分之四要用來禱告，七分之二拿來工作，一天只有七分之一的時間可以休息。當然，身體健全的人通常被迫做超過每日身體能負荷的量，工作項目五花八門，蓋教區建築、耕地種植餵養所有會員、鑿井、照顧牲畜、維護設施，無所不包。這樣一來，領袖可以讓每一教區變成自給自足的單位。

在多明尼克的土地上，聖母施捨之地還開了一間學校。孩童每天都到那間學校上課，生命都耗在十誡運動上。孩童被迫遠離家人，還被告知新家人就是十誡運動，十誡運動的使命就是他們所有的存在，壓力肯定大到不可置信。孩童可以非常有韌性，但是在易受影響的早期階段，灌輸的模式日後會很難根除。瑪琳達和領袖就指望這件事。如果下一代毫不依戀物質財貨或教區外的任何現實知識，持續掌控只會越來越容易。

大致上，社群內所有家族連結都被切斷。追隨者被剝奪出生姓名。所有未婚的女孩男孩都是兄弟姊妹，所有結婚的女人男人都是叔叔阿姨。年長的會員有自己的頭銜。世俗公證婚姻和父母身分不受認可。個人身分不再重要，只有十誡運動是第一優先。

教區不鼓勵說話，除非禱告，或偶爾在集會期間，在領袖委員面前，會員才被允許大吐苦水。手語和文字是唯一可行的溝通方式。訪客參觀十誡運動教區時，只有領袖和挑選出來的信徒才能獲得許可打破靜默，開口說話。外界看起來，就是一群高度順從、犧牲奉獻的信徒。

一九九三到一九九五年間，這個邪教進入極盛，權力最大，凝聚力最強。受訓者在這樣的生活方式下自我維持，期待完成「學習過程」（learning process）的課程，結束後就可以獲得十誡運動的完整認證。早上八點到下午四點，受訓者禁食、聽恩圖姆瓦（Entumwa，盧干達語，意指神聖的使者。通常也叫「領袖的宗徒」）講道。十誡運動的早期營運階段，宗徒非常暴力，為了趕走體內的惡魔，真的會毆打受訓者。但恩圖姆瓦在每堂課尾聲會替每一位會員禱告。

教區任一堂課開始時，要是基布維特爾人也在，主教本人也會親自主持。顯眼的存在讓基布維特爾成為十誡運動的門面，而不是瑪琳達。不過，這都是計畫：瑪琳達或許掌控了追隨者的心靈，但是只有男性現身才可以喚起恐懼和尊敬，兩者都是販售課程給一般大眾不可或缺的因素。這些學習過程課堂每次都連續上五週，結束後，宗徒會挑選幾名他們覺得已經準備好從事十誡運動全職服侍的受訓者，這代表他們完成訓練了。自那一刻起，他們會被稱為阿巴藤得夸（Abatendekwa，盧干達語，意指得救者），或叫門徒。

但就連門徒都認為領袖的階梯很陡峭，或許根本不可企及。最高的領袖被賜予阿貝本貝齊（Abebembezi，盧干達語，意指領導者）之名。除了無所不能的基布維特爾主教和影響力強大的多明尼克副主教，其他人都是「信徒的祖父母」後裔：設計師瑪琳達的姪女烏蘇拉（Ursula）又名「聖女馬利亞．葛萊蒂和天使的揚聲器」（Loudspeaker of Saint Maria Goretti and the Angels）；妹妹安潔莉納（Angelina）又名「聖母瑪利亞的衣飾」（Dress of the Virgin Mary）；以及她的弟弟亨利（Henry）又名「聖若瑟的建築」（Building of Saint Joseph）。[3]

這些頭銜清楚區分最高等級的領袖和一般基層門徒。卡薩普拉里神父的家人又稱為神視者（Visionary）。媽媽修拉斯蒂加是能影響瑪琳達的謀士，自稱「天上揚聲器」（Heaven's Loudspeaker），宣稱可以隨意接收聖人和天使的聲音。還有一名隱身暗處的角色，只知道被稱為班帕塔（Bampata，盧干達語，意指「我們的領袖」），由瑪琳達精心挑選，率領領袖委員會的情報班。她和手下監視每一位門徒和受訓者，派幹員到村莊招募、監視那些反抗十誡運動成員的家人。

除了基布維特爾和多明尼克以外，大部分領袖不是瑪琳達的親戚，就是原本的宗徒，再不然就是來自所謂罪孽世界高社經地位的個人。他們是十誡運動中最有用的會員，可以協助運動擴張版圖，聚在一起定期和負責規劃的領袖開會。討論的議程總是對其他十誡運動的成員保密。然而，高調的叛教者保羅．伊卡齊爾神父（Paul Ikazire）取得一則重要資訊。伊卡齊爾神父在一九九三年後離教，因為末日沒有降臨。根據這名前阿貝本貝齊會員的說法，他離開前沒多久，瑪琳達告訴其領袖自己收到天上來的新訊息：如有必要，十誡運動可以殺人。

消失不是逃脫

在一九九三年，凶殺看起來是可能性渺茫，也是無法想像的極端，即便是由聖母瑪利亞莫名其妙下令的凶殺也一樣。十誡運動多虧了多明尼克副主教的才華，才剛達到巔峰狀態。教宗若望保祿二世（Pope John Paul II）抵達烏干達做「教務訪問」（pastoral visit），多明尼克可以代表十誡運動出席當觀眾。

多明尼克向教宗呈上十誡運動清楚的使命宣言，據說教宗還獻上祝福。事實上，教宗極可能只是善意地揮揮手，祝福前來請願的人。但事實如何都不重要。對於十誡運動的門徒來說，教宗的認許才是關鍵：他們的使命受到天主在地上最高權威的祝福。

在這一方面，副主教對瑪琳達而言是無價的存在。副主教是十誡運動的頭號募款人，擅長說服地方官僚和當局捐款並提供保護。十誡運動教區所在區域的地方行政監察專員經常收到門徒奉上的食物、讚揚和服務。

再一次，瑪琳達打造十誡運動的速度非常驚人。才五年，領袖聲稱十誡運動擁有三千名會員。但不久後，裂縫便開始顯現。

頭一批麻煩來自十誡運動本身的創造神話。全部神學只依賴末日說，起初領袖說在一九九二年。末日沒有來臨，日期被改到一九九五年，但末日依然沒有降臨。第三個版本則是一九九九年十二月三十一日，卻讓讀過書的信徒開始起疑心。宗徒只能說這是「依照天上」的決定。

下一批麻煩源自教區間的合併。基布維特爾的家人起身反抗，奪回土地，於是失去了聖母形塑之地——到了一九九八年，就連基布維特爾的太太泰瑞莎也無法容忍他的遠大抱負。伊卡齊爾神父叛教之後，領袖都認為如果有大批會員離教，聖母施捨之地所在的多明尼克土地太有價值，丟失不起。他們在高點賣掉了可耕地。

許多全職門徒被永久調至卡農古的聖母救援之地。卡農古是瑪琳達與世隔絕的家鄉。小組別還是留在其他開放的教區和貿易村莊，才能維持穩定的假象。順利渡過最後幾年的叛教者提到，門徒和宗徒之間的抱怨及敵意越來越深。某些門徒出聲反對領袖，質疑每件事，例如上繳世俗財產和末日時間不停延後，宗徒會叫他們安靜，態度隨意又輕蔑。

不久，同一群門徒開始消失。宗徒說聖母瑪利亞提早帶走這些信徒，由於他們信仰的力量。接著在一九九八年，政府調查顯示校內有虐待事件，聖母施捨之地的學校關閉了。學童被轉送到聖母避難之地（Mary's Place of Refuge），避免政府進一步調查。

副主教可能買通當局，幫忙遮掩學校關閉的糟糕結果，阻止任何後續調查。但這還不是最糟的挫敗。基布維特爾診斷出躁鬱症，住院接受治療。然而他不吃精神穩定藥品，自行離開醫護設施，回到十誡運動教區，之後外界再也沒有人看過他。

到了二〇〇〇年一月一日太陽升起，顯然天主對人類的復仇非常溫和，幾乎察覺不到。安德魯．圖穆西梅（Andrew Tumusiime）是少數倖存到最終的十誡運動成員，他說這些情況的結果是瑪琳達的安危受到威脅。門徒越來越煩躁不安。二〇〇〇年，日子繼續過，十誡運動投入奇怪的活動。瑪琳達派遣情報頭子班帕塔前往烏干達西南部，班帕塔帶著劊子手圍剿逃到他方或拋棄運動的門徒。

有些幹員打算騙叛教者回來，告訴他們還留在教區的摯愛病了。班帕塔有時候會採取更直接的的方式：綁架。他甚至去自己孩子的學校，到班上打算帶走他們，或許想藉此告訴叛教者，他也是聽從瑪琳達的呼召，聚集自己的家人。老師出面阻止，班帕塔留給幹員一袋鈔票，拋下大家獨自離開。他的孩子沒再見過他。

聖母避難之地邀請那些受十誡運動掌控的地方當局人員以及門徒的家族成員參加二〇〇〇年三月十八日的慶祝大會。這是瑪琳達最後的詭計，掩飾她為真正活動所做的準備，驚人的末日終於要來臨了。

三月十七日早晨，聖母避難之地的門徒得知篩子的尾聲終於來了。不久，這個罪孽的世界會遭受天主早已準備好的懲罰。無人生還，沒有人可以重述究竟瑪琳達做了哪些承諾，但門徒列隊走進自己打造的方舟，出口被封死時，他們還在唱著歌。

方舟內，領袖再三保證這是為了大家的安全。暗無天光的三日，沒有人該冒險外出。但在方舟內，暗無天光的時日並不長，每艘方舟木構都浸透易燃液體，接著瑪琳達放火，烈焰熊熊燃燒，像是末日真的來臨。

超過五百多人慘死。沒人料想到十誡運動慢慢為此刻做準備。主流媒體毫無頭緒，尤其是西方媒體，根本不知道世界上有恢復上帝十誡運動教。不過，調查員前往其他教區和據點，發現情形比原本以為的更加糟糕。首先，在多明尼克神父的教區聖母施捨之地，調查員另外發現一百五十五具屍體。在其他據點也發現相似慘況，有超過四百名受害者。

後來，報導描述卡薩普拉里神父如何四處尋覓，購買大量電瓶補充液，幾可確定被用來毒害其他據點的門徒。總計至少一千零五十五人在十誡運動的最後一週遭到殺害，十誡運動成為當代世界史上死亡人數最多的邪教。

調查員沒有太多進展。卡農古大火開始燃燒時，有人路過，回報看到一台貨卡駛離現場，瞄到後座坐著瑪琳達。後來謠傳烏干達搜救小隊在北方，依基布維特爾的目擊證詞搜索。這些目擊沒能帶來一絲人影，但幾乎可以確定的是，至少瑪琳達成功逃跑了，她可是一手策劃這場毀滅的主腦。

更加仔細搜索聖母救援之地，發現領袖總部下方另外埋了六具屍體，很可能就是阿貝本貝齊。其中一具戴著神職人員的白色小硬領，推測應該是多明尼克副主教本人或是卡薩普拉里，又或者只是參訪教區的其他神職人員。

後續調查沒能發現更多線索。瑪琳達、基布維特爾以及其他倖存的十誡運動領袖的搜查行動完全停擺了。一九九八年，十誡運動學校關閉之後，位在康培拉（Kampala）的烏干達中央政府開始明白該組織的真面目。如果康培拉官員當初早點展開調查，地方行政監察專員和政府人員就無法協助十誡運動的領袖避開刑事控訴。這些都是多明尼克提供及時貢獻，慢慢培養起來的樁腳。地方當局或許真的開始對阿貝本貝齊施加壓力，尤其是對基布維特爾、多明尼克、瑪琳達三人組。

施加外部壓力到已經張力滿滿的邪教本身，瑪琳達知道自己麻煩大了。而且她已經無法再輕易說服許多領袖了。十誡運動的設計就是迅速擴展，為阿貝本貝齊贏得影響力，以及滿足控制人心所需的特定黑暗衝動。但除了擴展版圖、聚集追隨者、累積財富的衝動之外，她從來沒有思考過其他發展。再也無法維繫聖母顯靈的神話之後，一切都開始崩解。

作為經濟企業，十誡運動仰賴轉信者願意捨棄所有世俗財產，轉而支持擴張。但作為宗教機構，十誡運動仰賴最極端的信仰行為、貧困和靜默誓言、和家人分離、相信世界充滿罪孽、相信天主和聖母瑪利亞的蒙福保護、相信他們的代表以肉身顯現，並在世人中間來往。

細節不多，因此不知道究竟是何者先傾頹：經濟模式？還是宗教模式？但是就像查爾斯．

曼森、吉姆．瓊斯等先前許多邪教領袖一樣，瑪琳達發現，只有在編造根本性的欺瞞，推動邪教加速成長的期間，自身的魅力才有效果。謊言一一被拆穿之後，所有魅力都無效。成為邪教領袖是場危險遊戲，但從歷史論述來評斷，只要發現操作手法，就很難遏止邪教領袖操弄人心，全權掌控信徒的此生、來生，直到一切為時已晚。

決定最好燒了五百多人而非承認過錯，或許不是教科書對於「自戀」的定義，但是這般歹毒的舉動清楚明白展示視一切為理所當然的態度以及缺乏同理心這兩項自戀人格障礙症的關鍵特質。但瑪琳達獨樹一幟的原因並非上述行為。

瑪琳達的故事走向開始與他人不同是她的逃脫。最後的報告引人遐思，敘述瑪琳達搭車逃離燃燒的教區，駛進晦暗朦朧，無人知曉。很容易猜想瑪琳達對於大屠殺也是一笑置之，就像那天早上老顧客來喝香蕉啤酒，她停下手邊洗刷地板血跡的動作，抬頭和大家說笑一樣。但是搭車逃離的版本和烏干達或剛果民主共和國（Democratic Republic of the Congo）的現實生活不相符，剛果與烏干達國境相鄰，也是瑪琳達最有可能脫逃的地方。就算瑪琳達真的成功把邪教的土地和房產等非流動性資產換成可以攜帶進入叢林深處逃亡的財產，她獨身一名女性可以自衛多久？除了抵禦潛在攻擊，更不用說要保護四散林中的爪牙（在邪教的腐敗被揭露之後依然忠心耿耿追隨瑪琳達）。

瑪琳達只逃離了歷史而已。她進入的那片叢林藏著反叛武力，凶殘的軍閥一直都留心任何獲取錢財的機會，或任何可以施展暴行或滿足性慾的機會。即使瑪琳達真的獨自躲過偵查，又或者身旁還有一小群支持者，事實上再也沒有人聽到她的消息，現在比較可信的理論是她最終的命運就在叢林裡。沒有人——就連瑪琳達本人——都無法逃脫自己犯下的罪行，結果恐怖，令人難以想像。

Chapter 9

否認現實

馬歇爾・阿普爾懷特和天堂之門

一九九七年三月，在加州蘭喬聖菲，馬歇爾．阿普爾懷特（Marshall Applewhite）連續三天領著三十八名信徒走向死亡。最後一天，雙眼圓睜的天堂之門主腦加入信徒的行列，服用混合了苯巴比妥（Phenobarbital，一種安眠藥）和蘋果醬的伏特加，因此他也能夠「離開自己的運具」——又或者就像世界上其他人所認定的「自殺」。這是美國歷史上規模最大的集體自殺。眾多信徒（年齡從二十五到七十二歲不等）相信阿普爾懷特說的：他們發送射線到海爾—波普彗星後方航行的太空船上，加入「高於人類等級的王國」，王國的先進生物文明就像天堂，這些先進生物介於外星人和天使之間。集體自殺後，每位信徒錄製的最後影片「離開宣言」浮上檯面。影片顯示一群信徒完全同意自殺，既冷靜又謙遜，甚至對於即將發生的大活動感到興奮——故意否認現實的誇張程度，史上無人能及。馬歇爾．阿普爾懷特實際上編織出故事又欺騙眾人，當然是場悲劇，但也是洗腦課的大師班。

靈魂採集人

一九七〇年代早期，馬歇爾・赫夫・阿普爾懷特（Marshall Herff Applewhite）和邦妮・盧・內特斯（Bonnie Lu Nettles）建立了一個類宗教——他們稱之為天堂之門的新創宗教，混合了熱門科幻小說元素、阿普爾懷特童年的長老教會（Presbyterianism）教義和內特斯的新時代靈性觀（New Age spirituality）心靈大雜燴。基本前提很簡單：他們聲稱只有他們知道「高於人類的進化等級」（The Evolutionary Level Above Human，簡稱TELAH），這是只有在幽浮掠過夜空，無知人類偶爾才能瞥見的高等智慧。兩人說他們知道如何前往該等級，甚至可以帶上幾名選定的人，前提是願意證明自己已經準備好接受兩人的教導。

這個新興宗教的原則離奇古怪，無從驗證，而且非常模糊，兩人得以一邊發展，一邊修改公式，甚至制定新的核心信仰。隨著參與人數變多（巔峰時期吸引了將近兩百名信徒），平常大家用類似幼稚園綽號的「波和比」（Bo and Peep）或音高唱名「Do和Ti」（Do and Ti）稱呼兩位創辦人，隨著參與人數變多（巔峰時期吸引了將近兩百名信徒），他們從來沒有改變熱情友好的軟銷手法，初來乍到的人都非常喜愛他們。他們說自己是神放置在地上的天使，來救贖人類，知道自己的規則有時候很嚴格，但他們也能體諒大家，任何加入他們遊牧生活的人隨時都能離開。

表面的善良和接納掩飾了他們對信徒的牢牢掌控，以及他們對信徒日常生活最細微的控制。每日行程，例如誰負責削蔬果、誰和誰同住、誰開車載誰去工作，全都事先安排好。鬆餅有明定的圓周大小。飲食規定相當極端：只吃義大利麵，或只吃蔬菜，或連續三週不進食只喝水、檸檬汁、卡宴辣椒粉或楓糖漿。信徒都穿同樣的服裝，定期交換（甚至連內衣褲都要交換），才不會有人對某件上衣、某個外型、某種自我個體感受產生依戀。信徒都集體行動，例如一起看《星際爭霸戰》（*Star Trek*）和《X檔案》（*The X-Files*），一同前往連鎖餐廳瑪麗・凱蘭德派餅（Marie Callender's）享用最後一餐。他們有自己的詞彙：共處一室的人叫「星團」（Star Clusters）、強制靜默期間叫「墳墓時間」（tomb time）、不是信徒的人叫「路西法派」（Luciferian）或者就叫「路西」（Luci）。

一九五〇年代，學者首次提出洗腦概念，想要藉此理解為何這麼多美國戰俘會和俘虜他們的中國或北韓人合作，後來在一九七〇年代用來解釋為何許多良善、大學年紀的美國青年會相信查爾斯・曼森和共生解放軍（Symbionese Liberation Army）的胡言亂語。有些邪教研究人員，例如《膜拜團體就在我們當中：與其隱性威脅作持續鬥爭》（蘭臺網路出版）作者瑪格麗特・泰勒・辛格經常出庭作證，說明這些原則的運作可以如何應用在特定個人身上。這個概念作為法律辯護，沒能成功說服法官，而且雖然「洗腦」一詞被選入第五版《精神疾病診斷與統計手冊》（*DSM-5*，精神科醫生的專業手冊），背後的概念就連治療圈的醫生也不喜愛。法律和精神病學都只相信自由意志。

但是，羅伯特・傑伊・利夫頓身為第一波治療美國戰俘的精神科醫生，一點一點建構了更廣的論證，說明捕捉者如何成功洗腦俘虜。他的用詞是「思想改造」，強調整段過程的漸進本質。他列出的思想改造原則（社會環境控制、神祕操控、強制坦白、藉由純正程度來自我神聖化、煽動言詞、神聖的科學氛圍、生存支配），幾乎每一條都可以在天堂之門邪教的運作中發現。

一次又一次，天堂之門行經任何鄉鎮，四處張貼傳單，孤獨又軟弱的人都會看到——

幽浮

- 為何出現。
- 為誰而來。
- 何時離去。

——接著，孤獨的人一時興起，就會去參加阿普爾懷特的公眾演講。幾乎都在不經意之間，他們就已經完全顛覆自己的生活、切斷所有家族聯絡、捐獻自己擁有的一切給天堂之門。他們搬到天堂之門的生活區域，開始遵守複雜的生活規則。一九九七年自殺的許多信徒，至少都已經加入二十年了。他們的歸屬感，以及與一個全然認同自己作為的教派之間的連結，一路陪著他們直到嚥下最後一口氣。

天堂之門的信徒都有特定的背景經歷。一方面，他們似乎代表了美國的橫斷面。他們來自不同的州，年齡不同，職業也不同，但大多數長期相信天堂之門光怪陸離信仰和每日嚴格儀式的信徒，都曾遭受嚴重的損失或遺棄。一名蚵農失去了生意、婚姻、兄弟，接著開始吸毒。

還有一名養子親眼目睹摯友溺死；一名失業公車司機無法接受母親死亡；一名樂天女子為自身的不孕哀悼。這群迷失方向的靈魂渴求一些日常生活無法提供的事物。許多信徒都非常聰明。其中一名受害者的前女友說道：「多數人不會想把事情搞得比原本更複雜，但這群人太聰明了，認為這世界肯定比表面來得複雜。」[1]

天堂之門死者的訃聞和紀念碑上，經常重複出現這些特徵。他們太聰明了，沒辦法只為自己好。他們傷透了心。他們失去家人。他們無法融入大學生活、從前的工作或人際關係。他們拋棄所有。他們很友善。他們很奇怪。他們很疏離。這是奇怪的人格特質組合。或許也不意外，所有特質共通之處是——全都可以用來形容天堂之門邪教主腦馬歇爾．赫夫．阿普爾懷特本人。

神聖的格格不入

一九三一年五月十七日，在德州斯普爾（Spur），小馬歇爾．赫夫．阿普爾懷特（Marshall Herff Applewhite Jr.）誕生了。斯普爾是經濟大蕭條時期受到最劇烈、最長久衝擊的地方。阿普爾懷特的父親是牧師，四處傳教，擅長在德州一座城鎮接著一座創立教會，但通常都待不久。他們家有牧師和牧師娘，以及兩名年紀較大的女兒、兩名年紀較小的兒子（馬歇爾和嚴重智能障礙的弟弟，後來弟弟被送到州立照護之家），一家六口經常四處搬遷。他們並沒有特別特立獨行，因為塵暴區的當地環境，許多人時運不濟，都在四處找工作。但這樣的遊牧生活方式確實形塑了馬歇爾的童年，灌輸他終身不信任定居在任何一處。

馬歇爾的姊姊露易絲（Louise）在集體自殺之後，成為阿普爾懷特家族倖存者的發言人，形容弟弟小時候聰明、外向又英俊，還很愛唱歌。馬歇爾天生活潑外向，不管一家人在何處落腳，都可以交到朋友，但是父親太過頻繁舉家搬遷，馬歇爾無法獲得長久友誼才能建立的強大個人羈絆。童年早期如果能有這樣的關係，可以培養兒童同理的能力。青少年時期若能有類似的羈絆（這些羈絆對成人的日常人際互動而言至關重要），可以培養青少年在成年之後繼續茁壯。如同許多突然死亡或死狀慘烈的邪教領袖，任何精神失調的診斷都純屬推測。但是馬歇爾・阿普爾懷特確實展示了童年奔波的經典後遺症，甚至在他有機會帶領自己的家人（包括血緣家人和邪教家人）時，重現了這樣的漂泊無根。如同父親，馬歇爾不停移動天堂之門，一處搬過一處，所有信徒都不受長期依存關係的羈絆。不穩定變成一種靈性挑戰，可以協助追隨者上升到下一等級。

在青春期，這種動盪不安造成馬歇爾只剩一種長久的連結：宗教。他一直都是爸爸的最佳信徒，也是唱詩班的明星，一路從年少的男高音唱到日後迷人的男中音。科帕克利士提電力公司（Corpus Christi Electric Company）董事長佛洛伊德・查普曼（Floyd Chapman）和馬歇爾的爸爸共同創立公園大道長老教會（Parkway Presbyterian Church），記得他父親不僅人見人愛，還非常擅於組織規劃，立刻就為教會吸引了大批信徒。露易絲說弟弟馬歇爾很像父親：「弟弟一直都是天生的領袖，非常有魅力。他可以讓大家相信任何事。」[2]

一九四八年，馬歇爾從科帕克利士提中學（Corpus Christi High School）畢業，接著到德州達拉斯以北六十五哩舍曼（Sherman）的奧斯汀學院（Austin College）註冊入學，夢想成為牧師。一開始主修哲學，參加多個校園組織，十分活躍。大一室友約翰．亞歷山大（John Alexander）說「赫夫」（以前同學都這麼稱呼他）熱愛團體活動，又有個人魅力，參加了無伴奏合唱團、司法委員會，以及長老教會未來牧師的校園協會。

一九五二年，阿普爾懷特畢業，和太太安．皮爾斯（Ann Pearce）相遇，後來到維吉尼亞州里奇蒙的聯合長老教會神學院（Union Presbyterian Seminary）註冊，學習神學。但他很快就發現引領會眾的未來比不上他畢生熱愛的音樂，第一學期還沒結束就退學了，和安一起搬到北卡羅來納州（North Carolina）加斯托尼亞（Gastonia），擔任第一長老教會（First Presbyterian Church）的音樂總監。這對新人生了兩個寶寶，但馬歇爾在一九五四年被徵召入伍，兩人又再次搬家。

接下來兩年他們住在奧地利（Austria）薩爾茲堡（Salzburg），隨後搬到新墨西哥州白沙（White Sands），擔任陸軍通信兵（Signal Corps）的教官。一九五六年，馬歇爾光榮退役，接著註冊科羅拉多大學（University of Colorado）獲得音樂碩士學位，主要研究音樂劇，也在《南太平洋》（*South Pacific*）和《奧克拉荷馬！》（*Oklahoma!*）領銜主演。阿普爾懷特在科羅拉多大學相熟的教授查爾斯．拜爾斯（Charles Byers）形容他「樂天知命」、「很受同學歡迎」。

從科羅拉多大學畢業之後，阿普爾懷特展開吟遊詩人般的生活，只要和音樂有一些關聯，任何工作都接。他也試著在紐約演出，希望能打入歌劇圈，但到了一九六〇年，他發現自己既有才華又生得俊俏，還是更適合音樂劇。然而，僅管試鏡了一次又一次，還是失敗了，沒辦法成為舞台演員，於是一九六一年，他帶著家人到阿拉巴馬大學（University of Alabama）任教。

他在音樂系擔任助理教授四年，同時是合唱團指揮，直到一九六五年遭到學校開除，家人得知理由時相當震驚：他有外遇，對象是一位男同學。這方面的身分認同在後來集體自殺的報導中占據要角——知道自己喜愛同性、對此反覆掙扎，就好像大家要為這些無謂喪失的性命找尋線索和隱藏的動機。他的性認同肯定被視為與他從小接受的傳統宗教養育及希望在教會中發展職涯相互衝突。但在同性戀解放運動（Gay Liberation movement）早期那段日子，阿普爾懷特並非唯一因為性認同而苦苦掙扎的人，很難想像如此常見的內心掙扎（儘管折磨程度因人而異）會導致他帶著三十八人步上死亡。

性傾向顯然對阿普爾懷特而言是重大議題，但要把這件事當成自殺悲劇的關鍵因素，或許一九九〇年代晚期的文化戰爭能說明得更多，而不是天堂之門邪教。背景迥異的宗教人士如《七〇〇俱樂部》（700 Club）的派特・羅伯森牧師大人（Reverend Pat Robertson）和同志友善教會「大都會社區教會普世團契」（Universal Fellowship of Metropolitan Community Churches）創辦人杜羅・培禮（Troy Perry）試圖理解這場集體自殺時，雙雙指向這個偶發事件。羅伯森想要推廣「另類性傾向」（offbeat sexuality，一位嘉賓在他的電視節目上如此稱呼）幾乎總是邪教偏頗靈性的一部分，而培禮則看到「一個男人否認、壓抑神賜的性傾向，以及從更高的層面看來，整體社會拒絕、敵視男女同性戀」的運作。[3]

阿普爾懷特的性傾向問題毫無疑問相當深層，甚至有些極端，後來可以從他對團體內任何性關係都下達嚴格禁令看出來。他希望信徒克服所謂世俗情慾，才能準備好迎接上升到高於人類的下一等級。他頒布嚴格諭令「禁止性愛、禁止人類等級關係、禁止社交」[4]，雖然實際施行比較像是社會控制：阿普爾懷特不想要他的子民除了依賴他之外，還在其他地方找到生命的意義。

他最終的獨身承諾非常徹底：就在集體自殺前，他和六位男信徒宣布放棄世俗情慾，自行化學去勢，這是清空世俗情慾「容器」的第一步。有些人推論去勢會更接近他們雌雄同體的理想，相信到了下一等級，就會成為雌雄同體。在「最終離開」之前，信徒錄製了訪問影片，其中一個最詭異的坦白時刻是，一名信徒談到去勢及隨之而來的通體舒爽，就是他準備好要自殺的關鍵一步。他說：「我無法告訴你這讓我感到多麼自由。我已經準備好要執行下一步，我非常期待能執行，而這下一步不過就是一次臨床手術。」[5]

阿普爾懷特身為天堂之門的「神聖領袖」（divine leader），異常堅持要打造另一個現實（並且控制那些願意接受這個現實的人及其生活），以至於他應該沒有時間思考性傾向在他從合唱團指揮轉變成自殺邪教領袖的過程中扮演了什麼角色。但能確定的是，阿普爾懷特一九六五年與男學生的外遇拆散了他的家庭。遭到阿拉巴馬大學開除之後，他搬到休士頓，但到了一九六八年，和安的分居試驗成了永久分居：安訴請離婚，贏得孩子的完全監護權。

有那麼一小段期間，阿普爾懷特看似一個人過得很好。在休士頓，他在聖湯瑪斯大學（University of St. Thomas）找到音樂教職。聖湯瑪斯大學是一所小型私立天主教大學。他也擔任一間聖公會（Episcopal）教會的唱詩班指揮，同時在休士頓大歌劇院（Houston Grand Opera）演出十五個角色。

在那個時期認識他的許多人，都記得阿普爾懷特穿著講究、魅力十足，喜歡成為關注的焦點，不僅相當慷慨，還可以無止盡地與人社交。儘管他常常現身休士頓社交圈，私底下他卻開始崩解。他想念孩子，很想見他們，無法平衡顯而易見的雙面生活，在公開場合做一位女士的護花使者，同時和公開出櫃的愛人一起住在休士頓的蒙特羅斯（Montrose）區。更添混亂的是，他也和休士頓豪門之女共譜認真的關係，休士頓豪門反對這段關係，還威脅要取他性命。

半世紀之後再來辨別哪件事情先發生，實在別具挑戰：阿普爾懷特忙亂且顯然相互衝突的性追求，或是他的精神崩潰。大學校長湯瑪斯．布拉登（Thomas Braden）神父記得，阿普爾懷特顯露出一些解離狀態的跡象：「那時他的舉止有些詭異。只要跟他說話，他就會提到其他事，跟剛才的對話毫不相關。」休士頓音樂劇團成員、演員派屈克．史威茲（Patrick Swayze）的母親派希．史威茲（Patsy Swayze）告訴《紐約時報》，她記得阿普爾懷特舉止突然變得詭異，成了演員和工作人員八卦的話題：阿普爾懷特「談論幽浮，還宣揚奇怪的宗教」。

阿普爾懷特的病史少有人知，披露的病史都是二手資訊，零碎不完整且多有矛盾。但這樣模糊破碎的病史，卻也讓他成為學者喜愛的研究對象，學者認為用自己喜愛的理論可以輕易解釋這個人。這也是為何邪教領袖都會是最佳個案研究：他們傳記的可怕事實人盡皆知，又容易連接到心理疾病。舉例來說，亞伯達（Alberta）艾德蒙頓（Edmonton）麥科文大學（MacEwan University）的社會學家蘇珊・雷恩（Susan Raine）指出，阿普爾懷特強烈的宗教養育背景和出櫃同志生活實驗的兩相衝突，就是他內在痛苦衝突的來源之一。雷恩主張，這個內在掙扎不僅主宰他的生活多年，還讓他覺得自己的身體就像戰場，導致妄想型思覺失調症發作。該診斷和阿普爾懷特的宣稱一致。阿普爾懷特說自己會聽到他認定為神聖的聲音，而幻聽是妄想型思覺失調症的招牌症狀之一。

雖然無法確認這個診斷，但這個理論看似有理。思覺失調症一點也不罕見，超過百分之一的人口都受其影響。該診斷已可以說明阿普爾懷特說的那些天上造訪以及與他界生物的接觸。診斷也符合另一項出了名的行為：為每一位天堂之門成員安排的高度嚴謹日程表（嚴格的飲食控制、仔細控管行為）都與「治療計畫」一致，治療計畫的目的在於控制那些造成思覺失調症發作的事件。也就是說，阿普爾懷特持續不斷修改規則，就連最微不足道的日常活動也不放過，或許為自己創造了精細完整的自我照護計畫，專門設計來防止他脫離現實的症狀，以及把這些事件的嚴重程度降到最低。他為邪教中每一位成員制定嚴格日程其實也不令人意外——不管還有哪些症狀，幾乎每一位邪教領袖或多或少都罹患自戀型人格障礙症。

可以驗證的是，不久後阿普爾懷特也遭聖湯瑪斯大學解聘。他宣稱因為情緒問題而遭解聘，雖然似乎還有另一種解釋：聖湯瑪斯大學是第二所因為阿普爾懷特與男學生發生關係而開除他的大學。根據相互衝突的說法，焦慮不安的阿普爾懷特入住精神病院，「治療」他的同性戀衝動，此舉今日看來備受爭議。沒人知道他確切的住院時間，但時間長到足以讓他與護士建立長遠的連結，這位護士幫助他一同創立天堂之門，最終導致三十九人（包括阿普爾懷特本人）自殺身亡。

命中相遇

一九七〇年代早期，阿普爾懷特的生活不太順遂。雖然自己外遇導致婚姻破碎，但他依然非常思念孩子。他的父親過世了，雪上加霜，他一直以來都非常崇拜父親。阿普爾懷特更加憂鬱，經濟也陷入困難，被迫四處借錢，奇怪的舉止則讓他與曾經非常歡迎他的富裕圈子漸行漸遠。

就在這喪父之痛和心理困擾的時刻，他遇見即將改變他一生的女人邦妮．盧．內特斯。阿普爾懷特漂白了自己的歷史，他似乎認為精神疾病的大量治療不符合他自己升等存在的地位。依照他的版本，兩人相遇的時機是阿普爾懷特教導兒童音樂劇，而邦妮的兒子也在班上。另一說法是朋友生病，他去探病，就遇見了內特斯。第三種說法是阿普爾懷特自己就是那位生病的人，接受精神疾病觀察，會相遇是因為內特斯是療程的全責護士。如果阿普爾懷特真的自行入院接受同性戀「治療」，就能說明為何兩人的相遇有這麼多版本，以及為何內特斯明明知道兩人初次相遇的經過，卻故意打迷糊仗。

阿普爾懷特的姊姊露易絲提供了另外一種理論：她的弟弟因為爸爸的死，承受嚴重情緒困擾，於是到醫院動心臟手術，雖然心臟阻塞的問題解決了，他卻經歷了瀕死體驗。瀕死體驗依強度不同，會帶來玄奧體驗，徹底深刻地改變一個人。宣稱有瀕死體驗的人常提及漂浮的感受，又或者與身體分離，好似靈魂或意識上升到高於身體架構的地方。其他感受包括溫暖、寧靜、安全、徹底瓦解，以及強烈、閃亮光芒的存在。

只有露易絲宣稱弟弟經歷了可能改變人生的類神祕體驗，因為這次體驗非常符合妄想型思覺失調發作狀態，她很可能只是說了一個漂白過的版本，又或者把真相重新用去汙名化的方式再說一次。但所有不同版本都肯定：在一九七〇年代早期，馬歇爾・阿普爾懷特遇到了邦妮・盧・內特斯。內特斯是護士，也是業餘占星家。兩人都剛好四十幾歲，內特斯比阿普爾懷特年長四歲。跟阿普爾懷特很像，內特斯的家庭宗教信仰非常虔誠，她十一歲時在基督裡重生，到了二十二歲，她嫁給務實的商人約瑟夫・塞格爾・內特斯（Joseph Segal Nettles），生了四名孩子。

但是，到了和阿普爾懷特相遇的時候，她的婚姻正在步入滅亡，全因為她令人不安的新嗜好：熱衷新時代靈性觀和轉世信仰。那時，每週三晚上，內特斯開始在家裡客廳舉行降靈會，幫助鄰居和已不在人間的摯愛取得聯絡。她甚至開始相信一位她認為是十九世紀「方濟弟兄」（Brother Francis）的修士直接與她溝通，下達指令，告訴她該怎麼生活才恰當。

大約在阿普爾懷特和內特斯相遇的時候，很顯然正因如此，她古板耿直的先生開始辦理離婚。她開始覺得自己離周遭越來越遠。她女兒泰芮（Terrie）記得那時生活的氣氛，在集體自殺一個月後，告訴《紐約時報》記者貝里・巴瑞克（Barry Bearak）：「媽媽和我似乎與周遭格格不入。我們會到野外，望向星空，發誓自己剛剛看到飛碟了。我們會想：『如果飛碟帶走我們，不是很有趣嗎？』」

內特斯初次會與阿普爾懷特相遇，很可能是在醫院負責照顧阿普爾懷特，她那時情緒脆弱，而且正在失去二十年來親密的連結。此外，還有其他原因造成她過度解讀這次巧遇，且輕易受其影響。一九七二年，她曾經去算命，對方說她會遇到高挑、神祕的男子，滿頭白髮，膚色白皙。她深陷焦慮，還越來越孤立，把這則模糊的預言放在心上，當成一項承諾，甚至是神聖預言。

阿普爾懷特的確很高，滿頭銀白髮絲。而且就像內特斯，他習慣把自己的歷史講得有如神話，因此毫不意外，按照他的描述，兩人初遇就立刻對上眼，因為他們「共享對神祕祕密的認知」。兩人初次見面後不久，內特斯為阿普爾懷特排了本命星盤，幫他解讀。本命星盤是出生那刻天上星星的定位地圖，對占星師而言作用有點像指紋。內特斯告訴他，兩人的星盤非常相似，簡直不可思議（即使實際上兩人沒有多少共同的重要宮位），而且外星生命還預告了兩人的相遇。阿普爾懷特相信了，相信自己要承擔神聖使命。兩人幾乎立刻就變得極為親近，在同樣需要被視為神聖和強大的基礎上發展出共生依賴關係，而且馬上下了結論，兩人早就透過前世因緣認識彼此好長一段時間。

蘇珊・雷恩相信，差不多那時，阿普爾懷特才剛從思覺失調症發作中復原，因此主張內特斯「害阿普爾懷特初浮現的妄想信念增強了」。阿普爾懷特的姊姊露易絲告訴美國廣播公司新聞頻道（ABC News），和邦妮・內特斯的相遇雖然是單一事件，但「徹底改變了他，從正常人變成後來那樣」。

阿普爾懷特和內特斯的立即連結是天堂之門建立過程的重要時刻，也因此幾乎不大可能分別哪些是真實發生，哪些是日後為了重塑他們自己是天人（celestial beings）才宣稱的細節。內特斯天生聰慧、廣泛閱讀、自信滿滿，似乎主導了兩人關係發展的方向。首先，她創造了情感強烈的氛圍，將巧遇轉變成命中注定、宇宙規模的相會，不論星星還是外星生命都預言了他們的相遇。後來，她說服阿普爾懷特，兩人不只命中一定會相遇，而且還注定要創立自己的宗教。

她從新時代信仰之中採取各家所長加以混合，阿普爾懷特感到特別能共鳴。此時阿普爾懷特已經遭受一連串人生危機的衝擊，開始轉向非傳統靈性尋找答案，不論是占星學、沙漠密宗經文，又或是羅伯特・A・海萊因（Robert A. Heinlein）和亞瑟・C・克拉克（Arthur C. Clarke）的科幻小說。他甚至開始告訴大家自己在德州加耳維斯敦（Galveston）沙灘遇見一位「存在」（presence）。「存在」告訴他人類的起源，以及人類未來的去向。他的藝術家好友海斯・帕克（Hayes Parker）說道：「你聽到會想笑，但赫夫沒在開玩笑，而且他看起來不像發瘋。」

大家逐漸有共識，認為阿普爾懷特真的罹患某種類型的妄想疾病。有些人認為，在那樣的狀態，內特斯鼓勵阿普爾懷特相信自己是某類神聖使者，可能發揮了療癒效果。舉例來說，利夫頓相信內特斯的影響或許幫助預防近一步的精神退化。兩位似乎助長了彼此的自大狂症狀，這一動態強化且認可了彼此最古怪的衝動。[6]

阿普爾懷特和內特斯沒多久就一起住，還共同經營「基督藝術中心」（Christian Arts Center）新時代書店，專賣靈性書籍。後來店名改成「知情地」（Know Place），他們拓展使命，開班授課，教導神祕主義和神智學（theosophy）。一九七二年末幾月，內特斯的離婚通過法院判決。他們兩人口徑一致，彼此之間毫無性吸引力——或許阿普爾懷特因此鬆了一口氣，畢竟他可是因自己的慾望而身心煎熬。真正結合他們的是共同的使命或妄想，最早反覆出現的提議成了後來的天堂之門。內特斯失去了孩子的監護權。沒多久，兩人便結束書店等事業（或說經營不善），決定離開休士頓，拋下兩人經歷的一切心痛和恥辱，踏上阿普爾懷特熟悉的生活：公路之旅。

他界靈感

一九七三年一月一日，阿普爾懷特和內特斯趕著到拉斯維加斯（Las Vegas）。一路上，他們在營地或便宜旅店過夜。自大狂的方便之處在於他們相信自己身為神聖生物，沒有付錢的義務，他們走進餐廳，點菜吃飯，把店家招待的麵包塞進口袋，跑出餐廳，一毛錢都沒付。在內華達（Nevada）和太平洋西北地區（Pacific Northwest），他們打了一堆零工，從抽化糞池到賣血都做，同時只要有任何空閒，就會埋頭撰寫宣言。他們開遍美國和加拿大，在奧勒岡（Oregon）羅格谷（Rogue Valley）接收神對他們「重大使命」的教導。就像那時公路旅行的人經常寫信，他們也寫信向朋友坦承：「不管從社會、醫學、精神病學還是宗教的標準來看，我們和你早就已經失去理智了。」[7]

最終，這段期間兩人討論出他們的信仰大綱。第一條宗旨原則，同時也是他們關係的基礎：阿普爾懷特和內特斯是神派遣的先知，要實現神聖預言。第二條宗旨原則：比起世上所有人，神賜予阿普爾懷特和內特斯較高等級的心智。第三條宗旨原則：阿普爾懷特和內特斯會因信仰而遭殺害，其後所有人都會看到他們從墳墓死裡復活，被傳送到一台太空船上，世人會稱本事蹟為「最終示範」（Demonstration）。為了建立先知的權威，他們引用啟示錄十一章三到五節：

> 我要使我那兩個見證人，穿著毛衣，傳道一千二百六十天。他們就是那兩棵橄欖樹，兩個燈臺，立在世界之主面前的。若有人想要害他們，就有火從他們口中出來，燒滅仇敵。

當然，經過他們的詮釋，阿普爾懷特和內特斯就是聖經預言的那兩個見證人。

公路之旅過了一年半左右，他們獲得第一位皈依信徒：之前在休士頓認識的莎朗．摩根

（Sharon Morgan），如同身處一九七〇年代的許多人，摩根漸漸覺得被困在問題婚姻裡。她離開家人，加入阿普爾懷特和內特斯下一階段的公路之旅，幫忙招募更多信徒，雖然目前還是不大成功。摩根悄悄靠近路人，詢問他們是否有興趣「和兩位能告訴你該如何帶著軀體一起離開這顆星球的人聊聊天」。《紐約時報》的貝里．巴瑞克報導，三人行經休士頓時，摩根的先生威脅要把她關進精神病院，還帶著兩名女兒一起說服她回家。兩歲的女兒跑上前擁抱摩根，六歲的女兒站在後方，流著淚問道：「你會留下來嗎？」面對這麼強大的情緒壓力，摩根拋下了阿普爾懷特和內特斯。

摩根的先生控告阿普爾懷特和內特斯信用卡詐欺，舉出太太在靈性兜風期間收到的所有卡費款項。雖然案件沒有繼續追查，但警方確實拘捕了兩人，例行背景調查時，發現阿普爾懷特的逮捕令。兩人現在開的車是九個月前租的一台水星彗星（Mercury Comet），至今未歸還，在密蘇里（Missouri）獲報失竊。

阿普爾懷特被遣送回密蘇里候審，他為自己辯護，告訴檢察官：「超越地球的一股力量讓我不得不留住這台車。」法官嚴懲不手軟，判刑六個月監禁。此時，阿普爾懷特似乎已經準備好面對現實，打算收手不做天人。他更新工作履歷，才可以找新的教職。但由於在監獄內申請工作大不容易，他大多數時間都花在深思自己採取的方法上，他回想自己和內特斯、摩根共處的時光，仔細思考許多朋友提出的道德質疑，例如他們宣稱擁有神聖使命，卻沒錢付旅店費用。他也開始修改新宗教的主要靈性原則，捨棄神智學和密宗，使用科幻小說，並且建立一個脫離實體且以太空為本的演化論述。

理性的外皮（科幻小說中科學的部分）最終會在未來二十年陸續吸引好幾百名信徒。許多人加入一陣子就離開了，每天例行公事實在太花功夫，但有些人覺得很吸引人，於是採取激進手段，拋家棄子，放棄所有塵世享樂，就為了加入阿普爾懷特和內特斯的使命。阿普爾懷特在監獄牆上釘了一張紙，寫著「信仰大綱」，描繪他們新的宇宙異象，用毛毛蟲打比喻，傳達一條中心教導，寫道良善的生命會上天堂這概念實在太蠢了，就跟相信「如果毛毛蟲死時是條良善的毛毛蟲，就會神祕地在盛開的玫瑰花上醒來，和蝴蝶國王永遠生活在一起」一樣蠢。相反地，他說人類必須經歷結蛹階段，克服自身的人性，才能為下一等級（Next Level）的生活做準備。

要完成這個轉變需要一名老師。在古代，耶穌是神派來的老師。但兩千年之後的一九七四年，神另外派了兩位老師（也就是他和內特斯）。如同許多自封先知的人（和騙子），阿普爾懷特預言「幾個月內」會出現他們真理的「示範」，明顯可見。他寫道，他們兩人會被殺害，接著會和耶穌一樣，在「光雲」（cloud of light）中死裡復活。那光雲是「人類所稱的幽浮」。

許多社會學家現在認為戰後的幽浮狂熱反映了對冷戰的偏執多疑，而不是大家經常以為的，是出自外星生命實際造訪卻被壓下證據。瑞士精神科醫生卡爾・榮格（Carl Jung）詮釋對這類現象的新執著為古代原型的修訂版。在《飛碟：空中可見物體的現代神話》（暫譯，*Flying Saucers: A Modern Myth of Things Seen in the Sky*）一書中，榮格引用十六世紀的小冊子，以及西元前一五〇〇年「火盤」飛越埃及上空的古代報導，指出幽浮根本算不上新現象，只不過是公認的神話意象（mythological image）又一次的嶄新化身。榮格在書中寫道：「我們現在有大好的機會見證傳說的形成，在人類的艱困黑暗時代，外星生命運用『天上』力量試圖介入而產生的神奇故事——又這麼剛好發生在人類認真幻想，思考太空旅行可能性的時代……」[8]

邪教若能結合幽浮和隱晦的宗教儀式與概念，肯定會大受歡迎。一九五五年，英格蘭（England）的喬治・金（George King）創立了空靈協會（Aetherius Society），結合神智學原則、瑜伽，以及他聲稱來自金星的心電感應訊息。更早以前，一九三〇年代左右，亞瑟・貝爾（Arthur Bell）開啟名為人類聯合（Mankind United）的邪教組織，蓬勃發展多年，基本概念是只要組織壯大起來，一群祕密贊助人就會從藏身之處現身，對抗「暗地掌權者」（Hidden Ruler），創立全球週休三日制度。這些都是好萊塢式胡言亂語，但都吸引許多跟隨者，部分原因是人類聯合教保證，強大的贊助人擁有先進科技，可以讓死者在遙遠星球上復活。

阿普爾懷特則是有能力將基督教元素移植到這些基本概念上。許多受其教義吸引的人都出身於傳統基督教家庭，但又非常喜歡阿普爾懷特和內特斯把科學新知融入他們從小就熟悉的信仰中。音樂家麥克・柯尼爾斯（Michael Conyers）在一九七五年到一九八八年參加天堂之門，證實「阿普爾懷特和內特斯的訊息能夠與我的基督傳承對話，但又是現代、新穎的內容。」[9]舉例來說，他覺得迷人之處在於阿普爾懷特解釋耶穌的母親聖母瑪利亞在太空船上受胎。柯尼爾斯說：「聽好，雖然聽起來不可置信，但比起平淡無奇的童女生子（聖靈感孕）好多了。這個答案包含技術知識。聽起來還充滿生命活力。」[10]

蓬勃發展

一九七五年五月，麥克．柯尼爾斯在天堂之門第一場招募會認識阿普爾懷特和內特斯，那時阿普爾懷特才出獄幾個月。兩人在加州歐海鎮（Ojai）飯店房內，寄出傳單給好幾十間教會，阿普爾懷特的監獄書寫則是他們教義的精選內容。少數幾人想多了解他們，其中一名是七十二歲的形上學老師克拉倫斯．克魯格（Clarence Klug）。克魯格安排學生前往瓊．卡爾佩珀（Joan Culpepper）位在洛杉磯的家，和天堂之門碰面。卡爾佩珀以前是廣告總監，兼職物理學家，座右銘是「光怪陸離令我激動」。

柯尼爾斯說這群八十人左右的團體對「幽浮二人組」深深「著迷」，二人組現在再次更新話術：他們是來自下一等級星球的外星生命，下一等級星球也就是傳統基督教徒長久以來認為的「天堂」。他們維持這個主軸，混合偽科學和偽宗教。一會兒說他們正在打造太空船要回下一等級星球，一會兒又說自己是啟示錄第十一章三節提到的「兩個見證人」。旅途中某次他們創造了外星人邪惡種族，名為「路西法族」（Luciferian）——他們版本的撒旦和魔鬼。路西法族會假裝自己是「神」，藉此把人們固定在人類等級。他們保證只要追隨阿普爾懷特（基督復臨的化身）和內特斯（不知為何自稱天父，令人困惑），自此之後都會步在救贖之路上——或依照他們的說法，前往「進化的更高等級」。

阿普爾懷特和內特斯是高等存在的活生生代表，也是地球上唯一可以指引追隨者抵達下一等級星球的存在，讓潛在門徒完全依賴他們。他們現在又自稱Do和Ti。阿普爾懷特說這名字沒有意義，但儘管如此，還是讓二人組成為基本原則，就像音階的頭兩個音。阿普爾懷特和內特斯已經開始的做法算是利夫頓思想改造原則的小型範例：使用煽動言詞或「扼殺思考的格言論證」（thought-terminating cliché）。舉例來說，兩人稱呼前往下一等級星球的必要步驟為「人類個體變形」（Human Individual Metamorphosis，或簡稱HIM）。要達成HIM並不容易，也不便宜，畢竟，這可是天堂。想要獲得入場券，潛在信徒要放棄所有人類依附。他們列出一長串模糊不清但必須拋棄的事物，包括職業、家人、朋友、性關係、物質財產，甚至性別。

他們也擅於使用利夫頓所謂神聖的科學氛圍。透過毛毛蟲的隱喻，阿普爾懷特爭辯大家的變形會是實際生理轉變，大家在下一等級星球會成為不同的物種。這話聽起來既合理又科學還無關乎宗教，但其實非常幼稚，因為這段話只是用最字面的詞彙來說明天堂這一概念。儘管如此，在第一次招募會上，沒有人提出這番異議，而到了說明尾聲，阿普爾懷特和內特斯招募了將近二十四名新信徒。

為什麼有人會相信？神經科學家山姆．哈里斯（Sam Harris）認為，或許是因為阿普爾懷特異常毫不眨眼的凝視目光，這股目光很可能有催眠效果，畢竟顯然不是他的科學可信度或論證有說服力。哈里斯在自己的《說人話》（Making Sense）Podcast中說道：「他不是很聰明的人，也不是憑藉概念連結或字句運用的能力令人大為折服。」[11]但如同影片所示，阿普爾懷特持續睜大雙眼，視線穿透鏡頭。這或許是內特斯女兒泰芮指的「一股難以置信的力量」。哈里斯在《醒醒：不談宗教的靈性指南》（暫譯，*Waking Up: A Guide to Spirituality Without Religion*）一書中寫道：

因此，宗師經常堅持不尋常的眼神接觸並非意外。最好的情境是該行為源自在其他人面前感到真誠的舒適，並且發自內心關切大家的幸福快樂，基於這樣的心態，沒理由移開目光。但是保持眼神接觸也會變成「靈性展演」（acting spiritual），進而成為擾人的裝模作樣。也有人維持直盯盯的眼神接觸，不是因為開放的心態和對他人感興趣，也不是因為想要看起來開放或充滿興趣，而只是想要用攻擊力強又自戀的方式展示自身的主宰地位。心理病態通常可以與人保持特別好的眼神接觸。[12]

哈里斯指出，不論是邏輯思考、表達能力，還是個人魅力，阿普爾懷特都非常平庸，因此必定有其他因素吸引信徒加入。在他出獄之後，民眾開始放棄一切，就為了追隨Do和Ti。最終參與集體自殺的許多信徒都是在卡爾佩珀家初次招募會就加入的追隨者。十九歲的李．安．芬頓（Lee Ann Fenton）獲大學獎學金主修生物學，對於他們說的立即救贖概念感到很有共鳴，她說：「我想要在這輩子克服人類狀態，而不是下輩子。」迪克．喬斯林（Dick Joslyn）二十六歲，大學畢業，是前海軍官員，熱愛各種潛意識學說，相當欣賞兩人簡單明瞭的說明，不像其他新時代宣道家的戲劇化表演。喬斯林覺得他們可以領他入門，瞭解深層複雜的人心。

一九七五年，一群死忠信徒加入天堂之門，堅持待到最後一刻：十九歲憤青大衛．摩爾（David Moore）最終成為ＩＴ專員，而且成了天堂之門可依賴的少數幾名信徒，不論教團移動到何方，這幾名信徒都能在外面找到工作、支持天堂之門；樂儀隊鼓手瑪格麗特．艾拉．里克特（Margaret Ella Richter）擁有加州大學洛杉磯分校（ＵＣＬＡ）電腦科學碩士學位，婚姻最近以離婚告終；蘇珊．鮑普（Susan Paup）深受不孕困擾，和先生一同加入，先生離開之後，她繼續留在天堂之門。前模特兒茱蒂絲．羅蘭（Judith Rowland）是個兩名孩子的媽媽，被自己的媽媽招募，只留了張道別字條給先生，單單說道「我想要與主同行」；前選美冠軍喬伊斯．斯卡拉（Joyce Skalla）也留了張字條給先生，說她「放棄」十幾歲雙胞胎女兒，扶養權全都讓給先生。

自從一九六七年無責離婚法生效，只要雙方同意就可以解除婚姻關係，美國離婚率大幅攀升。儘管如此，這些特殊告別（如此多人同意切斷與父母、配偶、子女的聯絡）的最終結局還是極不尋常。天堂之門信徒約翰．克雷格（John Craig）擁有 身經歷，讓他成為科羅拉多州眾議院（Colorado House of Representatives）的明星候選人。他身高六呎六（約一九八公分），從軍校畢業，打過韓戰，和州長一起獵麋鹿，也會和六名小孩一起滑雪。同時，他是科羅拉多州的頂尖房地產開發商，也是信譽良好的杜藍哥商會（Durango Chamber of Commerce）會員。但是一九七五年夏天，他太太和孩子出遠門去參加游泳比賽，他在丹佛機場遇到了阿普爾懷特和內特斯，不久後便簽了轉讓授權書給太太，消失無蹤。

女兒告訴《紐約時報》：「當時，那個團體說幽浮即將在三四個月內降臨帶走他們。我想，好喔，如果沒發生，他就會回家。但他再也沒回家。」克雷格消失後兩個月後確實聯絡了女兒，邀請女兒到丹佛YMCA參加天堂之門大會。她不能坐在父親旁邊，大會結束，克雷格也不能單獨陪女兒走到車子旁。女兒回憶那次碰面，說道：「老爸看起來一點也不像喪屍。我仔細尋找藥物誘發或洗腦的跡象。但是他口齒非常清晰，非常有活力。他就是我老爸。」[13]

家人終於雇用私家偵探，想要找出克雷格的蹤跡。但私家偵探發現，阿普爾懷特和內特斯隱藏行跡的手法簡直是專家等級。就算偵探可以找到克雷格，他也絲毫沒有返家的意思。克雷格有了新的名字「羅根弟兄」（Brother Logan），成為團體內的關鍵組織者，以及天堂之門的副手，僅在阿普爾懷特和內特斯之下。他也成為天堂之門最活躍的招募者，幫忙印製傳單、租借大會場地、在門口歡迎潛在追隨著。他似乎用自己在科羅拉多州房地產圈的顯要地位交換了更加重要的關鍵位置：天堂之門的實質幕僚長，維持邪教營運的組織者。他用個人財富捐了大筆款項，確立自己在天堂之門的領導角色。

幽浮二人組的話語傳了開來，招募人數大幅增加。一九七五年晚期，在奧勒岡沃爾德波特（Waldport），阿普爾懷特和內特斯舉辦一場大會。隨後，與會者中大約三十人變賣了所有俗世財產，和家人道別，跟隨Do和Ti一起消失。但也因此，天堂之門初嚐了全國關注和負面媒體報導的滋味。

阿普爾懷特飽受傷害，決定帶著天堂之門遁入地下。他和內特斯領著現在將近兩百名信徒在全國過著遊牧生活，某種程度上，阿普爾懷特和新家人一起重現自己的童年。信徒會在洛磯山脈區（Rocky Mountains）和德克薩斯州各個公園、營地裡，睡帳篷和睡袋，經常在白天乞討。不知何時起，一名魅力十足的招募員說服許多信徒，沉溺性愛和大麻還是可以同時遵守教義原則，權力鬥爭於焉浮現。這是相當吸引人的訴求，威脅到阿普爾懷特和內特斯的權威。但是二人組迅速趕走那宗另立的派系，並且持續趕走任何不認真看待教義原則的信徒，直到剩下七十名忠誠的狂熱信徒。

阿普爾懷特和內特斯訂定生活規約，不管是改名還是切斷與外界的連結，目的都是要讓信徒完全依賴他們倆。信徒工作和旅行都要兩兩一組。幽浮二人組自行評估，把最不可能產生依附關係的人湊成一組，如此一來，每個人總是有一名柏拉圖式「查核夥伴」（check partner），這系統強調服從，強調只有二人組才是唯一的真理來源。經歷短暫叛亂，阿普爾懷特和內特斯繼續認為親密友誼會威脅凝聚力。阿普爾懷特鼓勵信徒要像寵物一樣待在從屬位置，意思是信徒唯一的責任是服從領袖。阿普爾懷特鼓勵信徒在每次抉擇時自問：「Do會怎麼做？」

這新的威權設定會導致未來新信徒的出走，但是留下來的人都忠心耿耿，甚至覺得阿普爾懷特的態度慵懶隨興就像父親。阿普爾懷特讓信徒保持忙碌，提出專斷無理的活動，測試信徒能力，執行反覆操作的作業，例如一邊單獨工作，一邊仔細聽音叉的聲音——這項活動是設計來讓信徒專注於下一等級星球，同時忽視人類思緒。另一項活動是信徒要戴上眼罩，只能看見正前方的事物。忍受修道院規則的虔誠信徒（又稱「班級」），被告知要為了「畢業」前往

「高於人類等級的王國」做準備。

維持專注在下一等級的方法很多，其一是檢視夜空。在天堂之門的旅途中，他們會在放鬆一下的時刻一起到天文館小旅行。內特斯聲稱下一等級星球的外星生命接觸了她，隨後告知信徒他們非凡的信仰會在一九七六年六月獲得回報。在懷俄明州（Wyoming）東南部麥迪遜鮑國家森林（Medicine Bow National Forest），二人組聚集起門徒，保證那天傍晚幽浮會來訪。不出所料，什麼東西都沒出現。破曉時分，內特斯道歉說：「好吧，我覺得像是臉被砸了雞蛋。」告知大家她收到訊息，造訪取消了。這或許不是有意的信心考驗，但對那些幾乎放棄了所有的人，他們拋棄孩子、離開穩定生活，這次體驗只是讓他們的信仰更加堅定不疑——或許這才是最初的用意。

天人離開運具

一九七〇年代晚期，天堂之門收到一筆鉅款，可能是某位前成員的遺產或捐款。經濟有了保障，現在天堂之門在城鎮間移動時，隨時可以租屋或買房。然而，生活在當地住宅區，周遭盡是窺視的目光，完全不同於遠離文明網路的營地生活，阿普爾懷特神經兮兮的狀態前所未見。他害怕自己會像耶穌一樣，因為傳布福音而遭到政府暗殺，還擔心不管教團到哪裏，執法單位都會鎮壓天堂之門的活動或存在。二人組想了許多精巧的維安措施當作預防。

麥克・柯尼爾斯回想，二人組會派出指定的信徒，這些信徒獲准可以離開房子，採買日常雜貨或跑腿。因此，雖然每次居住人數都超過十五人，從鄰居看來都只有兩三人在走動。通常這個低調的團體可以連續好幾個月、甚至好幾年都不會被人發現，但是只要阿普爾懷特懷疑鄰居有人發現了，整團就會搬到其他地方——可以如此輕易說搬就搬，都要多虧教團本身穩定的財務，還有克雷格談到優惠租約的技巧。嚴格禁止信徒接觸外界一舉完全符合利夫頓說的「社會環境控制」，不只切斷與親朋好友的溝通，甚至切斷每日與外界的聯絡，教團之外的一切都要切斷。這些限制也可以看起來很正面，因為信徒可以更加專心發展人類個體變形。

只不過，雖然天堂之門信徒的典型生活型態日益嚴格，奠基教義則是活文件，可以改變。幾年下來，阿普爾懷特和內特斯大幅更動了實踐方式和信仰原則。他們採用更多新時代的概念，尤其是「靈魂轉換者」（walk-in）。一九七九年，露絲・蒙哥馬利（Ruth Montgomery）《我們之中的陌生人：來自未來世界的開悟生命》（暫譯，*Strangers Among Us: Enlightened Beings from a World to Come*）一書提及這一概念：連著軀體的人類靈魂其實能夠移動到另一副身軀。

天堂之門的新版本提及，靈魂轉換者不是另一個人類靈魂，而是來自下一等級星球的外星生命。阿普爾懷特和內特斯堅持這顆星球不只真實存在，還是地上宗教指稱為天堂的地方。這代表阿普爾懷特和內特斯並不是Do和Ti，這些人格來自下一等級星球，現在棲居在之前和這些塵世名字有關聯的軀體中。這樣的說詞相當方便，他們兩人可以輕易抹除所有個人經歷和過往連結。「Do」從來沒有因為和男學生的關係而遭到開除，而「Ti」則從來沒有捨棄婚姻，他們兩人最近才從外太空來訪。

從某些方面來說，大筆現金收入允許創立者稍微鬆綁了規則。到了一九八〇年代，天堂之門的八十名信徒獲准可以找工作，多數是在電腦產業或汽車修理。到了一九八二年，信徒甚至開始偶爾打電話給拋棄的家人。一九八三年，信徒可以在母親節安排短暫的家庭小聚。但這不是邁向正常化的一步，而是改善天堂之門形象的一步。信徒被告知要和家人說他們在修道院學習電腦，給人的印象是他們依照自由意志選擇留在天堂之門。

但一九八五年，核心信仰發生了最巨大的轉變，邦妮．盧．內特斯——Ti、「天父」——過世了，依照阿普爾懷特所言：上升到下一等級星球了。

從醫學觀點來看，內特斯的死亡早有預期，她在一九八三年罹癌失去了一隻眼睛。那時，醫生告訴她，癌細胞已經開始擴散到身體其他地方。

內特斯原本是護士，肯定明白診斷的嚴重程度。儘管如此，看起來她繼續堅持二人組之前傳道的內容：她和阿普爾懷特，以及他們的信徒，會在目前生理型態下，被幽浮升起、傳送到下一等級星球。然而，到了一九八五年，癌症擴散到肝臟，她被送到達拉斯的帕克蘭紀念醫院（Parkland Memorial Hospital），用假名雪莉．魏斯特（Shelly West）入院。六月中，內特斯過世。阿普爾懷特不僅失去了一同打造事業的夥伴，現在宗教也陷入危機，他變成唯一的發言人。

內特斯過世後，阿普爾懷特陷入憂鬱，多虧虔誠的天堂之門信徒，他才能走出信仰危機。

信徒提醒他還有使命，幫忙他走出懶散疲倦和沮喪的狀態。但是，面對內特斯生理死亡最真實的後果——遺體火化之後，骨灰撒在附近的湖上，阿普爾懷特顯然必須修正幽浮邪教的其中一條核心原則。他開始告訴信徒，Ti耗盡她留在地球上的肉身，Ti的上升是靈性的上升。Do說，事實上太空船帶走了Ti，直達下一等級星球，Ti會直接進入一副新的軀體。

儘管新教義意外地和之前受人嘲笑的那條十分相似——毛毛蟲死後會神祕地在盛開玫瑰上、蝴蝶國王旁重生，這次卻大轉向，只有一名信徒退出。對大多數信徒而言，他們的信念反而增強了。事實上，這個概念意外地與之前突如其來的轉變相符，那時Do和Ti從外星生命轉變成靈魂轉換者。畢竟，如果Ti可以走進這副軀殼，那麼Ti也能走出去，非常合理。

一九八五年時可能還不是那麼明顯，但這個轉變最終帶來致命結果。信徒接受他們活著的軀體不會登上真實的幽浮前往下一等級星球，不會包機以人類的形式前往天堂，雖然一直以來他們都相信二人組告知的這一方式，但現在阿普爾懷特有必要告知大家究竟會用什麼方式抵達下一等級星球。很遺憾，這次轉變鋪下的理論基礎，導致了十二年後集體自殺的悲劇。曾經只是字面意義的觀念，在Ti過世後都徹底改變了。天堂之門立刻開始談論身體就像「暫時容器」，想前往下一等級，現在需要採取「變形最終行動」，其中包括「離開運具」。

外派小隊

內特斯的死亡造成患有偏執傾向的阿普爾懷特變得更加疑神疑鬼，懷疑大家要殺了他。這團人又遁入更深的地底，低調到外人開始以為天堂之門已經不復存在。一九九〇年代早期，信徒逐漸變少，只剩二十六名。隨著人數下降，阿普爾懷特把天堂之門新事業「高等源起」（Higher Source）網站設計公司（當時最新的科技趨勢）的利潤用來支付播放費用，透過衛星播放十二集阿普爾懷特傳福音的影片。

一九九三年五月，阿普爾懷特在《今日美國》（*USA Today*）刊登一則三萬美元的滿版廣告，稱呼自己的教團「完全匿名得勝」（Total Overcomers Anonymous），警告大眾，人類即將迎來災難審判。

廣告一刊登，至少二十名前信徒又重新加入邪教。第二季的系列宣道影片播出後，一九九四年又有更多人加入。這個團體已經從低點增長了兩倍。天堂之門在新興網路次文化中獲得聲譽，而且不少新信徒是透過官網認識Do。

但是，隨著在全球資訊網早期使用者間逐漸高升的能見度，加上沉潛多年後天堂之門重新現身大眾眼前，再一次帶來審視和批評，而阿普爾懷特總是把一切放在心上，無法不去介意。內特斯死後，阿普爾懷特的憂鬱逐漸增強，而且他似乎顯露了耗竭的跡象——這也不難理解，畢竟他過去二十年都帶著一幫亡命之徒留宿營地或違法租屋地點，管理不停更替的核心信徒，更別提信徒常常既貧困又依賴。他抱怨自己的身體狀況，覺得正在惡化，告訴大家他罹患癌症（雖然解剖報告顯示毫無罹癌跡象）。蘇珊．雷恩指出，如此關注自己的身體狀態，同時又頻繁檢視自己的成就，是妄想型思覺失調症的特徵。

關於他的身體和他的事業，阿普爾懷特發現了值得絕望的理由。

他開始公開討論自殺，說為了上升，包括軀體在內任何屬於人類的事物都必須犧牲。他說：「或許有必要親自處理。」如同之前的曼森家族、毒梟撒旦教、大衛教派分支，天堂之門的信徒逐漸成為自己信仰內在邏輯的俘虜。多年來，他們都說人類是被「人性鉤住」了。他們的紀律——詭異的飲食、在教友面前複誦之前被迫坦承的性衝動——似乎都是設計來幫助大家擺脫這副「哺乳綱人類」的軀殼。現在他們開始談論自己是「即將畢業班級」（graduating class），迫切想要轉移到下一等級星球。

在邪教內重塑思想的可能性不大，但是阿普爾懷特或許打算用奇怪的繞道避免可怕的命運。一九九五年，兩名天堂之門信徒在新墨西哥州建築師工作坊的訪客登記簿簽名，該工作坊專門討論能源效率、離網建築，建材為夯土、回收材料，例如舊輪胎（夯土會塞入輪胎，接著堆疊成牆）、玻璃瓶（在水泥牆內作為結構功能、裝飾功能的元素）。或許他們是受到企業的名字吸引：地球號太空船（Spaceship Earth）。他們購入整套建築說明，幾個月過後，天堂之門搬到新墨西哥州偏遠地區，開始建造自己的「地球船」。這是不是他們逐字遵守信仰的規定，住在自己的太空船上，試圖藉此避免集體自殺？有些信徒抱怨建造工程太艱困，或許大家已經習慣有信託基金支援相當舒適的生活了。阿普爾懷特聽進大家的抱怨，中止實驗，大家繼續前進。

大約在那時期，他們到附近天文館參加一場天文學相關見習，學到一顆新標記的彗星即將靠近地球。這顆天體——同一時間分別由兩名業餘天文學家艾倫・海爾（Alan Hale）和湯瑪斯・波普（Thomas Bopp）獨立發現，因此也以他們的名字命名，它非常特別，有淡淡的鈉尾，是從未觀察到的現象。這場閃亮亮的宇宙活動給了許多陰謀論者靈感，有人打到深夜廣播節目《亞特・貝爾作伴的海岸到海岸早晨》（*Coast to Coast AM with Art Bell*，經常和觀眾談論幽浮一同娛樂）宣稱自己看到另一個分離、類土星的物體跟在彗星後方。

據信，阿普爾懷特美化了這些宣稱和異常現象，變成他需要的實際證據。他使用粗淺的科學證據說服信徒，海爾—波普彗星後方跟著一艘太空船，而Ti本人就在那艘船上，那艘船即將前來，傳送他們到超越人類的存在等級。信徒用這段話來確認他們到目前為止所相信的一切。

一九九六十月，在預期彗星路徑最靠近地球的時刻六個月前，天堂之門在聖地牙哥（San Diego）以北二十五英哩的加州蘭喬聖菲租了占地九千兩百平方英呎的七房物件。該物件是法拍屋，租金划算：每月七千美元，光是他們的網站設計利潤就能支付。房子位在保全社區的尾端，非常隱密，他們甚至移除了路燈，讓他人難以偵察。

電腦專家瑪格麗特・里克特、大衛・摩爾和技術文編蘇珊・鮑普等最容易找到工作的信徒繼續工作。依照各方說詞，天堂之門信徒是準時的員工，每天中午十二點十五分準時吃自己帶的午餐。自殺事件過後，一位雇主記得他們的穿著打扮在南加州特別格格不入：「在夏天穿長袖，釦子都扣到頂。深色衣著。短髮。」另一個小企業老闆使用他們的網站服務，記得他們是：「優秀、聰明、善良的人。他們的信仰很堅定，願意為此放棄性命。」他們對自己的意圖頗為公開，但只會模糊提及即將來臨的大活動，至少一位雇主會對他們講有關三月二十五日的玩笑。他說，他們太常討論彗星了，害他在他們每次進來時都看一下呼叫器，開他們玩笑說：「十八天倒數開始。」

信徒在蘭喬聖菲房子外錄製離開地球宣言，背景是鬱鬱蔥蔥的郊區。影片中，他們對即將來臨的上升看起來非常興奮，咯咯笑著討論即將被發送上星球，似乎知道任何觀看這部影片的人會曲解他們的死亡，可能覺得他們太激進或太不理性，因此他們的語調顯得既關切又令人安心。沒有真實證據顯示他們有猶豫或遲疑。最終證詞成了信仰紀錄文件。不難想像他們如同自己宣稱，圍在餐桌邊談笑。可以看到他們的短髮剪得不太好，戴的眼鏡似乎都是最便宜、最不起眼的鏡框。大多數信徒都穿著不合身的衣服，非常寬鬆，好似跟身形大上好幾號的人借了衣服來穿。

隨著命運日子將近，他們集體默默離開職場。一位信徒告訴雇主，他要離開去經營「修道院事業」；另一位信徒送了老闆一條飾有外星人圖像的保羅領帶，老闆總是十分欣賞外星人。

集體自殺的前一天，大家一起外出吃最後一餐，訂位人數三十九人。他們讓瑪麗・凱蘭德派餅餐廳很好做事，預先訂好相同的餐點：番茄油醋招牌沙拉、火雞派、藍莓起司蛋糕、冰紅茶。他們一餐用掉的檸檬片比餐廳平常一天的用量還多。他們留了五十二美元小費，帳單金額三百五十一美元，差一點高達百分之十五，付現。

一名沒有同住的信徒打給九一一留言，調查員前往查看時，發現屋內留下了一本活頁夾，類似最終意圖的戰術手冊。這名信徒收到 FedEx 快遞來的整卷離開宣言影片，前往查看，就發現了屍體。受害者都穿著黑色T恤，風衣的肩章繡著：「天堂之門外派小隊」。腳上穿的是同款黑白 Nike Decades，大量購買的售價為每雙十四美元（病態的意外結果是，現在古著網站上，每雙要價大約五千美元）。天堂之門的活頁夾列出離開的指示：「十五名同學、八名助理，然後再十五名同學和八名助理，接著互相幫忙。」最後兩名全身包裹紫色布料，連臉部都遮住。每一位都注射了致命的苯巴比妥、蘋果醬和伏特加混合物。

有很多理論討論是什麼迫使天堂之門信徒自殺。這是集體自殺，還是一人自殺加上三十八名預謀殺人？阿普爾懷特強大的影響力似乎無法解釋信徒清醒的離開宣言，以及多位前信徒的證詞，那些證詞稱天堂之門信徒隨時隨地都可以離開。天堂之門的官方網站由信徒維護，還在運作。在網路上可以輕易找到離開宣言影片，以及許多阿普爾懷特雙眼圓睜、語無倫次的演講影片。網站內容想要扭轉局勢，列出「反對」自殺的立場陳述：「『自殺』的真實意義是拒絕前往下一等級的機會。」換句話說，任何沒有一起自殺的人，才是真正犯了自殺罪。這不僅是否認現實，而是逆轉現實。不幸的是，三十九名死者似乎都贊同此說。其中一人說道：「我想，比起這段人生能給的，班上所有人都想要更多。」

致謝

給媽媽：從中學短文到 Podcast 腳本，你總是認真閱讀，支持著我。你從來就不允許我停頓下來。從夢想到工作，你都是我每日持續的正能量，鼓勵我發揮所有潛能。你確保我的夢想成真。給爸爸：你對我的影響深遠無比，總是我的北極星，引領我渡過人生。你積極的態度、動力和創意一直是我的靈感來源。爸媽，你們是我的英雄。

給 Spotify 團隊：我要謝謝每一位夥伴。你們說故事的才華洋溢及全心投入，只有你們對 Podcast 的熱情才能相比。我想要特別感謝朱利安・博羅（Julian Boireau）、羅納德・夏皮羅（Ronald Shapiro）、德魯・科爾（Drew Cole）、卡利・馬登（Carleigh Madden）、凡妮莎・理查森（Vanessa Richardson）、格雷格・波辛（Greg Polcyn）和瑪基・鄂得米爾（Maggie Admire）幫助《邪教》（*Cults*）Podcast 誕生。

本書《邪教事典》如果沒有經紀人班・戴維斯（Ben Davis）和伊芙・阿特曼（Eve Attermann）的協助，不可能出版。我還想要感謝編輯艾德・施萊辛格（Ed Schlesinger）和畫廊出版社（Gallery Books）一直以來的協助。

給凱文・康利：謝謝你在與我合著本書時的陪伴與辛勤努力。

注解

如需更多相關資訊，請閱參考文獻。

Chapter 3. 剝削：巴關・希瑞・羅傑尼希

1. Sanjeev Sabhlok, "A Religious Guru Who Strongly Endorses Capitalism," *Sanjeev Sabhlok's blog*, July 26, 2011, https://www.sabhlokcity.com/2011 /07 /a-religious-guru-who-strongly-endorses-capitalism/.
2. Osho, *Autobiography of a Spiritually Incorrect Mystic* (New York: St. Mar- tin's Press, 2001), 56.
3. Ibid.
4. Amy Cooper, "Screaming Meditation Aims to Help Stress Relief," *Canberra Times*, August 4, 2021, https://www.canberratimes.com.au/story/7370998 /shake-it-off-with-screaming-meditation/.
5. Osho, *The Great Challenge* (New Delhi: Diamond Pocket Books, 2003), 30.
6. Sabhlok, "A Religious Guru Who *Strongly* Endorses Capitalism."
7. Hugh B. Urban, *Zorba the Buddha: Sex, Spirituality, and Capitalism in the Global Osho Movement* (Oakland: University of California Press, 2016), 96.
8. Max Weber, *On Charisma and Institution Building: Selected Papers* (Chicago: University of Chicago Press, 1968),

53.

9. Ma Anand Sheela, *Don't Kill Him!: The Story of My Life with Bhagwan Rajneesh* (New Delhi: FiNGERPRINT!, 2012), 139.
10. Ibid., 114.
11. Robert Jay Lifton, *Thought Reform and the Psychology of Totalism: A Study of "Brainwashing" in China* (Chapel Hill: University of North Carolina Press, 1989), 423.
12. Susan J. Palmer, "Charisma and Abdication: A Study of the Leadership of Bhagwan Shree Rajneesh," *Sociological Analysis* 49, no.2 (July 1, 1988): 126, https://doi.org/10.2307/3711009.
13. Hugh B. Urban, *Zorba the Buddha: Sex, Spirituality, and Capitalism in the Global Osho Movement* (Oakland: University of California Press, 2015), 97.
14. Malcolm McConnell, *Stepping Over: Personal Encounters with Young Extremists* (New York: Reader's Digest Press, 1984), 72.
15. Win McCormack, "Bhagwan's Devious Trap," *New Republic*, April 12, 2018, https://newrepublic.com/article/147905/bhagwans-devious-trap.
16. Sam Wollaston, "Growing Up in the Wild Wild Country Cult: 'You Heard People Having Sex All the Time, Like Baboons,' " *Guardian*, April 24, 2018, https://www.theguardian.com/tv-and-radio/2018/apr /24 /wild-wild-country-netflix-cult-sex-noa-maxwell-bhagwan-shree-rajneesh-commune-childhood.
17. Win McCormack, *The Rajneesh Chronicles: The True Story of the Cult That Unleashed the First Act of*

Bioterrorism on U.S. Soil (Portland: Tin House Books, 2010), 233.

18. Dashiell Edward Paulson, "The Routinization of Rajneeshpuram: Charisma and Authority in the Rajneesh Movement, 1981–1985," Thesis, University of Oregon, 2015, 31.
19. "L.A. Resident Gets 20 Years for '83 Bombing of Hotel Rahneesh," *Los Angeles Times*, November 10, 1985, http://www.latimes.com/archives/la-xpm-1985-11-10-mn-3387-story.html.
20. Philip Elmer-DeWitt, "America's First Bioterrorism Attack," me, September 30, 2001, http://content.time.com/time/magazine/article/0,9171,176937,00.html.
21. Marlow Stern, " 'Wild Wild Country': Most Shocking Reveals from the Sex Cult's FBI Informant," *Daily Beast*, April 2, 2018, updated April 4, 2018, https://www.thedailybeast.com/wild-wild-country-the-most-shocking-reveals-from-the-sex-cults-fbi-informant.
22. Harry David, "Guru's Dying Words: 'I Am Leaving This Tortured Body,' " UPI, United Press International, January 19, 1990, https://www.upi.com /Archives /1990 /01 /19 /Gurus-dying-words-I-am-leaving-this-tortured-body /2188632725200/.
23. Noa Jones, "I Charged My Sexual Energies at the Osho Meditation Resort in India," Vice, April 19, 2015, https://www.vice.com/en/article/xd7qp4/sex-robes-and-gurus-299.

Chapter 4. 剝削：吉姆・瓊斯和人民聖殿教

1. James L. Kelley, " 'You Don't Know How Hard It Is to Be God': Rev. Jim Jones' Blueprint for Nurture Failure,"

Alternative Considerations of Jonestown & Peoples Temple, Department of Religious Studies at San Diego State University, October 26, 2017, updated October 17, 2018, https://jonestown.sdsu.edu/?page_id=70768.

Chapter 6. 誇大狂：大衛．柯瑞許和大衛教派分支

1. World Science Festival, "Hallucinations with Oliver Sacks," YouTube, December 8, 2014, http://www.youtube.com/watch?v=8T_XimPe4xU.
2. Oliver Sacks, Hallucinations (New York: Vintage Books, 2012), 159–60.
3. Sue Anne Pressley and Mary Jordan, " 'Mighty Men' of Cult Enforced Koresh's Rules," *Washington Post*, April 22, 1993, https://www.washingtonpost.com/archive/politics /1993/04/22/mighty-men-of-cult-enforced-koreshs-rules /3fc01c4f-0990-44f7-8d52-a64899271cc7/.
4. Susan Aschoff, "After the 'War' at Waco," *Tampa Bay Times*, February 28, 2000, updated September 26, 2005, http://www.tampabay.com/archive/2000/02/28/after-the-war-at-waco/.
5. United States Bureau of Alcohol, Tobacco, and Firearms, and Federal Bureau of Investigation, "Negotiation Transcript—Tape 25," The Ashes of Waco, The Wittliff Collections at Texas State University, March 3, 1993, https://digital.library.txstate.edu/handle/10877/1782.
6. Susan Aschoff, "After the 'War' at Waco."
7. Cole Moreton, "Waco Siege 20 Years On: The Survivor's Tale," *Telegraph*, March 24, 2013, http://www.telegraph.co.uk/news/religion/9950378/Waco-siege-20-years-on-the-survivors-tale.html.

8. Mark Swett, "David Koresh and the Waco Davidians: An Ultimate Act of Faith?," Cult Education Institute, March 2002, https://cult education .com /group/1220-waco-davidians/24283-david-koresh-and-the-waco-davidians-an-ultimate-act-of-faith.html.
9. Corey Charlton, "25 Years On," The Sun, February 28, 2018, www.thesun.co.uk/news/5672656/british-survivor-waco-cult-david-koresh-still-believes-apocalypse-coming/.

Chapter 7. 施虐‧‧基斯 ‧ 拉尼爾和 NXIVM 教團

1. Suzanna Andrews, "The Heiresses and the Cult," *Vanity Fair*, October 13, 2010, https://www.vanityfair.com/culture/2010/11/bronfman-201011.
2. Ibid.
3. CBC, *Uncover: Escaping NXIVM* (podcast), Episode 3: Sex, Money, and Nazis, September 4, 2018 (8:00 mins.).
4. Ibid., (8:25 mins.).
5. Lenny Bernstein, "How Parents Create Narcissistic Children," *Washington Post*, March 9, 2015, https://www.washingtonpost.com/news/to-your-health /wp /2015 /03 /09 /how-parents-create-narcissistic-children/.
6. Bowen Xiao, "EXCLUSIVE: Delving into the Childhood of NXIVM's Leader," Epoch mes, May 28, 2018, updated May 30, 2018, https://www.theepochtimes.com/exclusive-delving-into-the-childhood-of-nxivms-leader _2540043.html.
7. Vanessa Grigoriadis, "Inside Nxivm, the 'Sex Cult' That Preached Empowerment," *New York Times* Magazine, May

30, 2018, https://www.nytimes.com/2018/05/30/magazine /sex-cult-empowerment-nxivm-keith -raniere.html.

8. CBC, *Uncover: Escaping NXIVM* (podcast), Episode 3: Sex, Money, and Nazis, September 4, 2018 (6:30 mins.).
9. Xiao, "EXCLUSIVE: Delving into the Childhood of NXIVM's Leader.
10. James M. Odato and Jennifer Gish, "In Raniere's Shadows," *Albany Times Union*, February 18, 2012, updated February 22, 2012, https://www.timesunion.com/local/article /In-Raniere-s-shadows-3341644.php.
11. "NXIVM: Gina Melita's Story," *Albany Times Union*, YouTube, April 22, 2019, https://youtu.be/YWJcMMgXNmw?t=53.
12. Ibid., https://youtu.be/YWJcMMgXNmw?t=58.
13. Ibid., https://youtu.be/YWJcMMgXNmw?t=108.
14. Odato and Gish, "In Raniere's Shadows."
15. Josh Bloch, Kathleen Goldhar, Anita Elash, and Dave Pizer, "The Making of Vanguard," CBCnews, September 12, 2018, https://newsinteractives.cbc.ca/longform/the-making-of-the-vanguard.
16. CBC, *Uncover: Escaping NXIVM* (podcast), Episode 3: Sex, Money, and Nazis, September 4, 2018 (13:05 mins.).
17. Ibid., (13:40 mins.).
18. Irene Gardner Keeney, "Troy Man Has a Lot on His Mind," *Times Union*, June 26, 1988, https://www.timesunion.com/7dayarchive/article/Troy-man-has-a-lot-on-his-mind-15640272.php.
19. Christopher Ringwald, "Discount Service Target of Probe," *Times Union*, May 12, 1992, https://www.timesunion.com/7dayarchive/article/Discount-service-target-of-probe-15640345.php.

20. Toni Natalie with Chet Hardin, *The Program: Inside the Mind of Keith Raniere and the Rise and Fall of NXIVM* (New York: Grand Central Publishing, 2019), 20.
21. EJ Dickson, "How NXIVM Was the Ultimate Wellness Scam," *Rolling Stone*, October 8, 2019, https://www.rollingstone.com/culture/culture-features/nxivm-keith-raniere-wellness-scam-sex-cult-848439/.
22. Natalie, *The Program*, 19.
23. Barry Meier, "Once Idolized, Guru of Nxivm 'Sex Cult' to Stand Trial Alone," *New York Times*, May 1, 2019, https://www.nytimes.com/2019/05/01/nyregion/nxivm-keith-raniere-trial.html.
24. Natalie, *The Program*, 21.
25. Ibid.
26. Ibid., 27.
27. Odato and Gish, "In Raniere's Shadows."
28. Natalie, *The Program*, 36.
29. Kimberly Lawson, "Sleep Deprivation Can Be a Weapon in the Hands of an Abusive Partner," Vice, July 9, 2019, https://www.vice.com/en_us /article /3k3myn /sleep-deprivation-can-be-a-weapon-in-the-hands-of-an-abusive-partner.
30. Natalie, *The Program*, 59.
31. Ibid., 63.
32. Katie Heaney (as told to), "What It Was Like to Date a Cult Leader," The Cut, November 7, 2019, https://www.

thecut.com/2019/11/what-it-was-like-to-date-a-cult-leader.html.

33. Taste of Reality, *Pinkshade* (podcast), Episode 34: "Exclusive: Toni Natalie, Keith Raniere's Ex-Girlfriend, On NXIVM," May 15, 2018 (17:40 mins.), https://radiopublic.com/pink-shade-6rVXLa/s1!76165..

34. Ibid., 84.

35. Michael Freedman, "Cult of Personality," *Forbes*, October 13, 2003, https://www.forbes.com / forbes/2003/1013/088.html#421b130 41853.

36. Namkje Koudenburg, Tom Postmes, Ernestine H. Gordijn, and Aafke van Mourik Broekman, "Uniform and Complementary Social Interaction: Distinct Pathways to Solidarity," *PLoS One* 10, no. 6 (2015), https://doi.org/10.1371/journal.pone.0129061.

37. Brendan J. Lyons, "Guilty Plea Ends Salzman's Long Allegiance to Raniere," *Albany Times Union*, March 15, 2019, updated March 21, 2019, https://www.timesunion.com/news/article/Guilty-plea-of-Salz man-ends-her-long-allegiance-13691433.php.

38. Kate Prengel, "Lauren Salzman: 5 Fast Facts You Need to Know," Heavy, April 2, 2019, https://heavy.com/news/2019/04/lauren-salz man/.

39. EJ Dickson, " 'I Was in One Mode: Protect Keith': NXIVM Member Testifies About Naked Meetings, Group Sex, Dungeon Paddlings," *Rolling Stone*, May 21, 2019, https://www.rollingstone.com/culture/culture-features/lauren-salzman-nxivm-sex-slave-keith-raniere-testi mony-trial-836547/.

40. Barbara Bouchey, "Barbara Bouchey Breaks Long Silence and Speaks Out About Commodities Losses and Other

Raniere-Bronfman Matters," Frank Report, October 21, 2019, https://frankreport.com/2019/10/21/barbara-bouchey-breaks-long-silence-and-speaks-out-about-commodities-losses-and-other-raniere-bronfman-matters/.

41. Emanuella Grinberg and Sonia Moghe, "She Says She Dropped Out of School and Left Mexico to Join a Cult," CNN, May 24, 2019, https://www.cnn.com /2019/05/24/us/nxivm-trial-mexican-witness/index.html.
42. Natalie, *The Program*, 125.
43. Grinberg and Moghe, "She Says She Dropped Out of School."
44. Ibid.
45. Natalie, *The Program*, 126.
46. Sarah Berman, " 'I Was Gone from the World and Nobody Noticed': One Woman's Story of Being Trapped by NXIVM," Vice, May 30, 2019, https://www.vice.com/en/article/evy58a/i-was-gone-from-the-world-and-nobody-noticed-one-womans-story-of-being-trapped -by-nxivm.
47. Nicole Chavez and Sonia Moghe, "6 Weeks of Testimony in Nxivm Case Reveal Lurid Details of Alleged Sex Cult, Including Branding Women and Holding Them Captive," CNN, updated June 16, 2019, https://www.cnn.com/2019/06/16/us/nxivm-keith-raniere-trial/index.html.
48. Amanda Ottaway, "Witness in Sex-Cult Trial Says She Was Held Captive for 2 Years," Courthouse News Service, May 31, 2019, https://www.courthousenews.com/witness-in-sex-cult-trial-says-she-was-held-captive-for-2-years/.
49. Michael Blackmon, "A Former NXIVM 'Cult' Member Says She and Her Two Sisters Had to Have Abortions After Sex with Keith Raniere," BuzzFeed News, May 28, 2019, https://www.buzzfeednews.com/article/michaelblackmon/

nxivm-sex-cult-abortion-keith-raniere-sisters.

50. Lisette Voytko, "FBI: Nxivm's Leader Had 'Sex Slave' Cruise nder for Him," *Forbes*, June 6, 2019, https://www.forbes.com/sites/lisettevoytko/2019/06/06/fbi-nxivms-leader-had-sex-slave-cruise-tinder-for-him/#4993ded93962.
51. Robert Gavin, "Messages Suggest Girl Was 15 When Raniere Relationship Began," *Albany Times Union*, June 5, 2019, updated June 5, 2019, https://www.timesunion.com/news/article/Messages-suggest-Raniere-s-relationship-with-13939595.php.
52. Ibid.
53. Voytko, "FBI: Nxivm's Leader Had 'Sex Slave' Cruise nder for Him."
54. Andrews, "The Heiresses and the Cult."
55. Ibid.
56. Ibid.
57. Rozina Sabur, "Seagram Whisky Heiress Clare Bronfman Pleads Guilty to Role in Nxivm 'Sex Cult,' " *Telegraph*, April 20, 2019, https://www.telegraph.co.uk/news/2019/04/20/seagram-whisky-heiress-clare-bronfman-pleads-guilty-role-nxivm/.
58. Barry Maier, "The Journey of the 'Sex Cult' Heiress: From Reluctant Recruit to Criminal Defendant," *New York Times*, August 11, 2018, https://www.nytimes.com/2018/08/11/business/clare-bronfman-nxivm.html.
59. Ibid.
60. Andrews, "The Heiresses and the Cult."

61. Natalie, *The Program*, 147.
62. Freedman, "Cult of Personality."
63. Andrews, "The Heiresses and the Cult."
64. Natalie, *The Program*, 156.
65. Freedman, "Cult of Personality."
66. Ibid.
67. Natalie, *The Program*, 159, in ebook search: "dossier."
68. Andrews, "The Heiresses and the Cult."
69. Kent A. Kiehl and Morris B. Hoffman, "The Criminal Psychopath: History, Neuroscience, Treatment, and Economics," *Jurimetrics* 51 (2011): 374, https://www.ncbi.nlm.nih.gov/pmc/articles/PMC4059069/.
70. Beata Pastwa-Wojciechowska, "The Relationship of Pathological Gambling to Criminality Behavior in a Sample of Polish Male Offenders," *Medical Science Monitor* 17, no. 11 (2011), CR669–75, doi: 10.12659/MSM.882054.
71. Kiehl and Hoffman, "The Criminal Psychopath," 374.
72. Nicole Hong, "Clare Bronfman, Facing Sentencing, Refuses to Disavow 'Sex Cult' Leader," *New York Times*, September 28, 2020, updated October 27, 2020, https://www.nytimes.com/2020/09/28/nyregion/clare-bronfman-keith-raniere-nxivm-sentence.html.
73. Natalie, *The Program*, 128.
74. Ibid., 129.

75. Flavie Waters, Vivian Chiu, Amanda Atkinson, and Jan Dirk Blom, "Severe Sleep Deprivation Causes Hallucinations and a Gradual Progression Toward Psychosis with Increasing me Awake," *Frontiers in Psychiatry* 9 (July 10, 2018): 303, https://doi.org/10.3389/fpsyt.2018.00303.

76. Natalie, *The Program*, 129.

77. Scott Johnson and Rebecca Sun, "Her Darkest Role: Actress Allison Mack's Descent from 'Smallville' to Sex Cult," *Hollywood Reporter*, May 16, 2018, https://www.hollywoodreporter.com/features/how-smallvilles-allison-mack-went-actress-sex-cult-slaver-1112107.

78. Robert Gavin, "NXIVM Turned Lake George Resort into Annual Raniere Birthday Jamboree," *Albany Times Union*, May 10, 2019, https://www.timesunion.com/news/article/NXIVM-turned-Lake-George-resort-into-annual-13836853.php.

79. 20/20, S40, E34, "Woman Recalls Being Branded as Part of Joining Secret Society; What Former NXIVM Members Say About Keith Raniere's Attitude Toward Women; How 'Smallville' Actress Became Involved in NXIVM," April 27, 2018, https://abc.com/shows/2020/episode-guide/2018-04/27-042718-nxivm (~20:00 mins.).

80. Josh Bloch, Kathleen Goldhar, Anita Elash, and Dave Pizer, "The Making of Vanguard: The Story of How NXIVM's Keith Raniere Went from Gifted Child to Self-Help Guru to Accused Sex-Cult Leader," CBC News, CBC Radio-Canada, September 12, 2018, https://newsinteractives.cbc.ca/longform/the-making-of-the-vanguard.

81. Elizabeth Hopper, "What Is the Mere Exposure Effect in Psychology?: Why We Like Things We've Seen Before," ThoughtCo, updated December 13, 2019, https://www.thoughtco.com/mere-exposure-effect-4777824.

82. Johnson and Sun, "Her Darkest Role: Actress Allison Mack's Descent from 'Smallville' to Sex Cult."
83. Hopper, "What Is the Mere Exposure Effect in Psychology?"
84. Johnson and Sun, "Her Darkest Role."
85. Zack Sharf, " 'Smallville' Star Allison Mack Reached Out to Emma Watson About Alleged Sex Cult," IndieWire, April 25, 2018, https://www.indiewire.com/2018/04/smallville-allison-mack-emma-watson-sex-cult-1201957138/.
86. Taryn Ryder, "How Did Smallville's Allison Mack Get Involved with a Sex Cult in the First Place?," HuffPost, May 18, 2018, https://www.huffpost.com/entry/how-did-smallvilles-allison-mack-get-involved-with-a-sex-cult-in-the-first-place_n_5aff3222e4b0a046186b7fbb.
87. Ibid.
88. Susan Dones, "Susan Dones on the Record: A Detailed and Provoca-tive Account from an EXpian [sic]," Frank Report, October 23, 2017, https://frankreport.com/2017/10/23/susan-dones-on-the-record-the-most-detailed-and-provocative/.
89. Frank Report, "NXIVM Cult Leader Keith Raniere Claims to Have Had People Killed," YouTube, December 6, 2015, https://youtu.be/HuStK6xg-7g ?t= 212.
90. Sarah Edmondson, *Scarred: The True Story of How I Escaped NXIVM, the Cult That Bound My Life* (San Francisco: Chronicle Prism, 2019).
91. Ibid., 185.
92. Ibid.

93. Ibid.

94. Ibid.

95. Barry Maier, "Inside a Secretive Group Where Women Are Branded," *New York Times*, October 17, 2017, https://www.nytimes.com/2017/10/17/nyregion/nxivm-women-branded-albany.html.

96. Edmondson, *Scarred*.

97. 20/20, "Woman Recalls Being Branded as Part of Joining Secret Society."

98. Vice News, "The Man Who Blew the Whistle on Alleged Sex Cult Inside NXIVM," YouTube, May 17, 2018, https://www.youtube.com/watch?v=wsoUjimrglE.

99. *Today*, "Actress Catherine Oxenberg Talks About Her Fight to Save Her "Hijacked' Daughter," YouTube, November 2, 2017, https://www.youtube.com/watch?v= Gz9YKhZzzUQ.

100. Edmondson, *Scarred*.

101. Michael Blackmon, "The Founder of an Alleged Sex Cult Hid in a Walk-In Closet When Officials Raided His Mexican Villa to Arrest Him," BuzzFeed News, May 21, 2019, updated May 21, 2019, https://www.buzzfeednews.com/article/michaelblackmon/keith-raniere-closet-arrest-mexico-lauren-salzman.

102. Emily Saul and Lea Eustachewich, "Nxivm Leader Keith Raniere Hid in Closet to Avoid Arrest: Testimony," *New York Post*, May 21, 2019, https://nypost.com/2019/05/21/nxivm-leader-keith-raniere-hid-in-closet-to-avoid-arrest-testimony/.

103. EJ Dickson, "Keith Raniere, Head of NXIVM and Alleged Sex Cult, Found Guilty on All Counts," *Rolling*

Stone, June 19, 2019, https://www.rollingstone.com/culture/culture-news/keith-raniere-nxivm-trial-guilty-all-counts-849967/.

104. Amanda Arnold, "NXIVM Member Admits to Enslaving Woman for Two Years," The Cut, April 2, 2019, https://www.thecut.com/2019/04/nxivm-enslaved-woman-locked-room-lauren-salzman.html.

105. Amanda Arnold, "The Most Disturbing Details from the NXIVM Sex- Cult Case," The Cut, May 7, 2019, https://www.thecut.com/2019/05/the-most-disturbing-details-from-the-nxivm-sex-cult-case.html.

106. Will Yakowicz, "Nxivm 'Sex Cult' Leader Posed as a Mentor for Women, But Was a Predatory 'Crime Boss,' Prosecutors Say," *Forbes*, May 7, 2019, https://www.forbes.com/sites/willyakowicz/2019/05/07/nxivm-sex-cult-leader-posed-as-a-mentor-for-women-but-was-a-predatory-crime-boss-prosecutors-say/.

107. Arnold, "The Most Disturbing Details."

108. Emily Saul and Lea Eustachewich, "Nxivm 'Slave Master' Describes Naked Group Meetings with Leader," *New York Post*, May 17, 2019, https://nypost.com/2019/05/17/nxivm-slave-master-describes-naked-group-meetings-with-leader/.

109. Sarah Berman, "The NXIVM 'Sex Cult' Story Keeps Getting More Disturbing," Vice, May 20, 2019, https://www.vice.com/en_ca/article/evyb5j/the-nxivm-sex-cult-story-keeps-getting-more-disturbing.

110. EJ Dickson, " 'I Was in One Mode: Protect Keith.' "

111. EJ Dickson, "Former Slave Describes Allison Mack's Alleged Abusive, Terrifying Behavior in Detail," *Rolling Stone*, June 10, 2019, https://www.rollingstone.com/culture/culture-features/nxivm-trial-allison-mack-sex-slave-nicole-abuse-

keith-raniere-845830/.

112. Emily Saul, Lea Eustachewich, and Kate Sheehy, "Nxivm 'Slave' Gives Jurors Look Inside Alleged Sex Cult," *New York Post*, May 7, 2019, https://nypost.com/2019/05/07/nxivm-slave-gives-jurors-look-inside-alleged-sex-cult/.

113. Chavez and Moghe, "6 Weeks of Testimony in Nxivm Case."

114. Edward Helmore, "Nxivm Trial: Keith Raniere Found Guilty on All Counts in Sex Cult Case," *Guardian*, June 19, 2019, https://www.theguardian.com/us-news/2019/jun/19/nxivm-trial-keith-raniere-ver dict-guilty-allison-mack.

115. Frank Parlato, "Federal Prisoner #57005-177—a/k/a Keith Alan Raniere a/ k/a Vanguard—to Be Arraigned Today; What to Expect," Frank Report, April 13, 2018, https://frankreport.com/2018/04/13/raniere-to-be-arraigned-today-what-to-expect/.

Chapter 8. 脫逃：克莉多尼亞．瑪琳達和恢復上帝十誡運動教

1. Bernard Atuhaire, *The Uganda Cult Tragedy: A Private Investigation* (London: Janus, 2003), xx.

2. Ibid., 6.

3. Ibid., 52.

Chapter 9. 否認現實：馬歇爾．阿普爾懷特和天堂之門

1. Laura Barcella, "Heaven's Gate, 23 Years Later: Remembering 38 Peo- ple Who Died with Cult Leader," *People*, March 26, 2020, https://people.com/crime/heavens-gate-22-years-later-remembering-lives-lost/.

2. John Holliman, "Applewhite: From Young Overachiever to Cult Leader," CNN, March 28, 1997, http://www.cnn.com/US/9703/28/applewhite /index.html.
3. Mubarak Dahir, "Heaven's Scapegoat," *The Advocate*, no. 733, May 13, 1997, 35.
4. Marc Fisher and Sue Ann Pressley, "Crisis of Sexuality Launched Strange Journey," *Washington Post*, March 29, 1997, https://www.washingtonpost.com/archive/politics/1997/03/29/crisis-of-sexuality-launched-strange-journey/3709d9ff-51ee-4f50-a9cd-a45525d7ad8f/.
5. HeavensGateDatabase, "Student Exit Statements," YouTube, April 9, 2013, https://www.youtube.com/watch?v=wHz9it70TdI.
6. Robert Jay Lifton, Destroying the World to Save It (New York: Henry Holt, 1999), 307.
7. James Brooke, "Former Cultists Warn of Believers Now Adrift," *New York Times*, April 2, 1997, https://archive.nytimes.com/www.nytimes.com/library/national/0402mass-suicide-recruit.html.
8. C. G. Jung, *Flying Saucers: A Modern Myth of Things Seen in the Skies, trans.R. F. C. Hull* (Abingdon: Routledge Classics, 2002), 11.
9. Jaclyn Anglis, "The Twisted Story of the Heaven's Gate Cult—and Their Tragic Mass Suicide," All That's Interesting, November 14, 2021, updated December 1, 2021, https://allthatsinteresting.com/heavens-gate-cult.
10. Ibid.
11. Sam Harris, *Making Sense* (podcast), Episode 7: "Through the Eyes of a Cult," March 24, 2015 (10:00 mins), http://www.samharris.org/podcasts/making-sense-episodes/through-the-eyes-of-a-cult.

12. Sam Harris, *Waking Up: A Guide to Spirituality Without Religion* (New York: Simon & Schuster, 2014) 164.
13. James Brooke, "For Cowboy in Cult, Long Ride into Sunset," *New York Times*, March 31, 1997, https://www.nytimes.com/1997/03/31/us/for-cowboy-in-cult-long-ride-into-sunset.html.

參考文獻

Chapter 1. 羞辱：查爾斯．曼森和曼森家族

- Bugliosi, Vincent, with Curt Gentry. *Helter Skelter: The True Story of the Manson Murders*. New York: Norton, 1974.
- Guinn, Jeff. *Manson: The Life and mes of Charles Manson*. London: Simon & Schuster, 2013.
- Heigl, Alex. "The Manson Murders, 45 Years Later." People.com, August 27, 2014. https://people.com/celebrity/manson-family-murders-45-years-later-2/.
- Manson, Charles, and Nuel Emmons. *Manson in His Own Words*. New York: Grove Press, 1986.
- Ng, Christina. "Charles Manson Denied Parole After Saying He Is a 'Very Dangerous Man.' " ABC News, April 10, 2012. https://abcnews.go.com/US/charles-manson-denied-parole-dangerous-man/story?id=16111128.
- *People* v. Manson, 61 Cal.App.3d 102 (Cal. Ct. App. 1976). https://casetext.com/case/people-v-manson-1.
- Sanders, Ed. *The Family*. Berkeley, CA: Da Capo, 2002.
- Statman, Alisa, with Brie Tate. *Restless Souls: The Sharon Tate Family's Account of Stardom, the Manson Murders, and a Crusade for Justice*. New York: It Books, 2013.

Chapter 2. 羞辱：亞道夫．德．赫穌斯．康斯坦佐與毒梟撒日信徒

- Applebome, Peter. "Drugs, Death and the Occult Meet in Grisly Inquiry at Mexico Border." *New York Times*, April 13, 1989, sec. U.S. https://www.nytimes.com/1989/04/13/us/drugs-death-and-the-occult-meet-in-grisly-inquiry-at-mexico-border.html.
- Barber, Nigel. "Why Cults Are Mindless: Mindless Obedience Keeps Religious Cults Together." *Psychology Today*, August 6, 2012. https://www.psychologytoday.com/za/blog/the-human-beast/201208/why-cults-are-mindless.
- Bourbon-Galdiano-Montenegro, Carlos Antonio de. *Palo Mayombe: Spirits— Rituals—Spells*. Morrisville, NC: Lulu Press, 2011 (self-published).
- Bovsun, Mara. "Spring Break Revelry Turns to Horror as Mexican Druglord Kills University of Texas Student in Sicko Human Sacrifice Voodoo Ritual." *New York Daily News*, March 21, 2015. https://www.nydailynews.com/news/crime/mexican-druglord-kills-college-student-sicko-ritual-article-1.2157613.
- Garcia, Guy. "The Believers: Cult Murders in Mexico." *Rolling Stone* (website), June 29, 1989. https://www.rollingstone.com/culture/culture-features/the-believers-cult-murders-in-mexico-53577/.
- Giannangelo, Stephen J. *The Psychopathology of Serial Murder: A Theory of Violence*. Westport: Praeger Publishers, 1996.
- González-Wippler, *Migene. The Complete Book of Spells, Ceremonies, and Magic*. St. Paul: Llewellyn Publications, 1978.
- Gregoire, Carolyne. "The Real Connection Between Ambition and Mental Health." HuffPost, December 16, 2014. https://www.rollingstone.com/culture/culture-features /the-believers-cult-murders-in-mexico-53577/.

- Greig, Charlotte. *Evil Serial Killers: In the Minds of Monsters*. London: Arcturus Publishing, 2006.
- Mouradian, Vera E. "Abuse in Intimate Relationships: Defining the Multiple Dimensions and Terms. " National Violence Against Women Research Center, 2000. https://mainweb-v.musc.edu/vawprevention/research/defining.shtml.
- Rakovec-Felser, Zlatka. "Domestic Violence and Abuse in Intimate Relationship from Public Health Perspective." *Health Psychology Research* 2, no. 3 (2014). https://doi.org/10.4081/hpr.2014.1821.
- Ramsland, Katherine. *Inside the Minds of Serial Killers: Why They Kill*.Westport, CT: Praeger, 2006.
- ——. "Merging into Murder: Emotional Investment and Certain Mind- sets Can Set Up Moral Meltdown." *Psychology Today*, September 3, 2021. https://www.psychologytoday.com/us/blog/shadow-boxing/202109/merging-murder.
- . "Partners in Crime." *Psychology Today*, July 1, 2014. https://www.psychologytoday.com/us/articles/201407/partners-in-crime.
- Schiller, Dane. "Woman Called Priestess of Satanic Cult Says She's Changed: Inmate Has Served 15 Years in Prison in Mexico for Ritual Sacrifices of 13 People." SFGATE, March 28, 2004. https://www.sfgate.com/news/article/Woman-called-priestess-of-satanic-cult-says-she-s-2774267.php.
- Singer, Margaret Thaler, and Marsha Emmer Addis. "Cults, Coercion, and Contumely." In *The Mosaic of Contemporary Psychiatry in Perspective*, edited by Anthony Kales, Chester M. Pierce, and Milton Greenblatt. New York: Springer, 1992, 130–42. https://doi.org/10.1007/978-1-4613-9194-4_13.

Chapter 3. 剝削・・巴關・希瑞・羅傑尼希

- Birnstiel, Sheela, and Ma Anand Sheela. *Don't Kill Him: The Story of My Life with Bhagwan Rajneesh*. New Delhi: Fingerprint Publishing, 2013.
- Carter, Lewis F. *Charisma and Control in Rajneeshpuram: The Role of Shared Values in the Creation of a Community*. Cambridge, UK: Cambridge University Press, 1990.
- Moshakis, Alex. "What Do Near-Death Experiences Mean, and Why Do They Fascinate Us?" *Guardian*, March 7, 2021. https://www.theguardian.com/society/2021/mar/07/the-space-between-life-and-death.
- Nagaraj, Anil Kumar Mysore. "Osho—Insights on Sex." *Indian Journal of Psychiatry* 55, supplement 2 (January 2013): S268–272. https://doi.org/10.4103 /0019-5545.105549.
- *Oregon Experience: Rajneeshpuram*. Season 7, episode 701.PBS.Aired November 19, 2012. https://www.pbs.org/video/oregon-experience-rajaneeshpuram/.
- Osho. *Sex Matters: From Sex to Superconsciousness*. New York: St. Martin's. 2003.
- Palmer, Susan J., and Frederick Bird. "Therapy, Charisma and Social Control in the Rajneesh Movement." *Sociology of Religion* 53, Special Issue (June 1, 1992): S71–S85. https://doi.org/10.2307/3711252.
- Sarasohn, David. "Antelope's Last Stand." *New Republic*, April 12, 2018. https://newrepublic.com /article /147876 / antelopes-last-stand.
- Stork, Jane. *Breaking the Spell: My Life as a Rajneeshee, and the Long Journey Back to Freedom*. Sydney: Macmillan Australia, 2009.

- Urban, Hugh B. *Zorba the Buddha: Sex, Spirituality, and Capitalism in the Global Osho Movement*. Oakland: University of California Press, 2016.
- Way, Maclain, and Chapman Way. *Wild Wild Country*. Netflix docuseries, 2018.
- Wright, Charles. *Oranges & Lemmings: The Story Behind Bhagwan Shree Rajneesh*. Richmond, Victoria, Australia: Greenhouse Publications, 1985.
- Zaitz, Les. "25 Years After Rajneeshee Commune Collapsed, Truth Spills Out—Part 1 of 5." *Oregonian*, April 14, 2011, updated February 5, 2019. https://www.oregonlive.com/rajneesh/2011/04/part_one_it_was_worse_than_we.html.

Chapter 4. 剝削：吉姆・瓊斯和人民聖殿教

- Bebelaar, Judy, and Ron Cabral. *And Then They Were Gone: Teenagers of Peoples Temple from High School to Jonestown*. Berkeley, CA: Minuteman, 2018.
- Mitchell, Dawn, and Michael Jesse. "Retro Indy: Jim Jones and the People's Temple in Indianapolis." *IndyStar*, November 18, 2013, updated November 18, 2019. https://www.indystar.com/story/news/history/retroindy/2013/11/18/peoples-temple /3634925/.
- Fondakowski, Leigh. *Stories from Jonestown*. Minneapolis: University of Minnesota Press, 2013.
- Gritz, Jennie Rothenberg. "Drinking the Kool-Aid: A Survivor Remembers Jim Jones." *Atlantic*, November 18, 2011. https://www.theatlantic.com/national/archive/2011/11/drinking-the-kool-aid-a-survivor-remembers-jim-jones

/248723/.

· Guinn, Jeff. *The Road to Jonestown: Jim Jones and Peoples Temple*. Illustrated edition. New York: Simon & Schuster, 2017.

· "Jonestown." FBI Records: The Vault. Collection, 1978–1979. https://vault.fbi.gov/jonestown/jonestown.

· Kilduff, Marshall, and Phil Tracy, "Inside Peoples Temple," *New West* magazine, August 1, 1977.

· Klineman, George, Sherman Butler, and David Conn. *The Cult That Died: The Tragedy of Jim Jones and the Peoples Temple*. New York: Putnam, 1980.

· Mahaffie, Michael. *605 Adults 304 Children*. Vimeo.com, 2018. https://vimeo.com /373653092.

· Mailman, Erika. "What Happened After Jonestown?" *Rolling Stone*, November 16, 2018.

· Naipaul, Shiva. *Black & White*.London: Hamish Hamilton, 1980. https://www.rollingstone.com/culture/culture-features/jonestown-jim-jones-bodies-memorial-756320/.

· Reiterman, m, with John Jacobs. *Raven: The Untold Story of the Rev. Jim Jones and His People*. New York, Jeremy P. Tarcher/Penguin, 1982.

· Scheeres, Julia. *A Thousand Lives: The Untold Story of Hope, Deception, and Survival at Jonestown*. Reprint edition. New York: Free Press, 2011.

Chapter 5. 施虐‧‧羅克‧泰里奧和螞蟻山之子

· Bovsun, Mara. "Canadian 'Messiah' Formed Horrifying Doomsday Cult After Ulcer Surgery Left Him with Chronic

Pain." *New York Daily News*, February 4, 2018. https://www.nydailynews.com/news/crime/cana dian-messiah-formed-doomsday-cult-ulcer-problems-article-1.3797931.

- "Cult Leader Roch Theriault Killed in N.B. Prison." CTVNews, February 27, 2011. https://www.ctvnews.ca/cult-leader-roch-theriault-killed-in-nb-prison-1.612633.
- Ford, Laura Grace. Savage Messiah. New York: Verso, 2011.
- Kaihla, Paul, Ross Laver, Ann McLaughlin, and Barry Came. "The Ant Hill Kids." *Maclean's*, February 8, 1993. https://archive.macleans.ca/article/1993/2/8/the-ant-hill-kids.
- McPadden, Mike. "Roch Thériault: The Horrifying Savagery (& Home Surgery) of Canada's Most Violent Cult Leader." ID Crimefeed, September 28, 2017. https://www.investigationdiscovery.com/crimefeed/murder/roch-theriault-the-horrifying-savagery-home-surgery-of-cana das-most-violent-cult-leader.
- Palmer, Susan, Martin Geoffroy, and Paul L. Gareau, eds. *The Mystical Geography of Quebec: Catholic Schisms and New Religious Movements*. Cham, Switzerland: Palgrave Macmillan, 2020.
- Thompson, Emily G. *Cults Uncovered: True Stories of Mind Control and Murder*. New York: DK Publishing, 2020.
- Vandonk, Todd. "Behind the Crimes: Murder, Mutilation, Abuse Part of Life at Ant Hill Kids Commune." Toronto.com, September 10, 2020. https://www.toronto.com/news-story/10148519-behind-the-crimes-murder-mutilation-abuse-part-of-life-at-ant-hill-kids-commune/.

Chapter 6. 誇大狂：大衛・柯瑞許和大衛教派分支

- *Activities of Federal Law Enforcement Agencies Toward the Branch Davidians (Part 1)*. Joint Hearings Before the Subcommittee on Crime of the Committee on the Judiciary, House of Representatives, and the Subcommittee on National Security, International Affairs, and Criminal Justice of the Committee on Government Reform and Oversight, One Hundred Fourth Congress, First Session, July 19, 20, 21, and 24, 1995.Committee on the Judiciary Serial No. 72. Washington, DC: U.S. Government Printing Office, 1996, 221. https://books.google.com/books/about/Activities_of_Federal_Law_Enforcement_Ag.html?id=qisLLk7GOQC&printsec=frontcover&source=kp_read_button&hl=en&newbks=1&newbks
- Benson, Eric. "The Branch Davidians: The FBI Agent Who Can't Stop Thinking About Waco." *Texas Monthly*, April 2018. https://www.texasmonthly.com/articles/fbi-agent-cant-stop-thinking-waco/.
- Gladwell, Malcolm. "Sacred and Profane: How Not to Negotiate with Believers." *New Yorker*, March 24, 2014. https://www.newyorker.com/magazine/2014/03/31/sacred-and-profane-4.
- Mitra, Debkumar. "Following Cult Followings: Why the Crisis Is Omnipresent in India." *Economic mes, August* 27, 2017. https://economictimes.indiatimes.com/blogs/et-commentary/following-cult-followings-why-the-crisis-is-omnipresent-in-india/.
- Murray, Evan D., Miles G. Cunningham, and Bruce H. Price. "The Role of Psychotic Disorders in Religious History Considered." *Journal of Neuro psychiatry and Clinical Neurosciences* 24, no. 4 (October 1, 2012). https://doi.org/10.1176/appi.neuropsych.11090214.
- Newport, Kenneth G. C. *The Branch Davidians of Waco: The History and Beliefs of an Apocalyptic Sect.* New York:

Oxford University Press, 2006.
- Thibodeau, David, Leon Whiteson, and Aviva Layton. *Waco: A Survivor's Story*. New York: Hachette Books, 2018.
- "Waco Siege." Britannica.com. Accessed January 14, 2020. https://www.britannica.com/event/Waco-siege.
- Wessinger, Catherine. "The Deaths of 76 Branch Davidians in April 1993 Could Have Been Avoided—So Why Didn't Anyone Care?" The Conversation, April 13, 2018. https://theconversation.com/the-deaths-of-76-branch-davidians-in-april-1993-could-have-been-avoided-so-why-didnt-anyone-care-90816.

Chapter 7. 施虐··基斯 · 拉尼爾和 NXIVM 教團

- Andrews, Suzanna. "The Heiresses and the Cult." *Vanity Fair*, October 13, 2010. https://www.vanityfair.com/culture/2010/11/bronfman-201011.
- Arnold, Amanda. "NXIVM 'Sex Slave' Gives Harrowing Testimony in Court." The Cut, May 9, 2019. https://www.thecut.com/2019/05/nxivm-sex-slave-gives-testimony-in-court-against-raniere.html.
- Berman, Sarah. "The NXIVM 'Sex Cult' Story Keeps Getting More Disturbing." Vice, May 30, 2019. https://www.vice.com/en/article/evyb5j/the-nxivm-sex-cult-story-keeps-getting-more-disturbing.
- "Cult of Personality." *Forbes*, October 13, 2003. https://www.forbes.com/forbes/2003/1013/088.html?sh=483fdd221853.
- Dickson, EJ. "How NXIVM Was the Ultimate Wellness Scam." *Rolling Stone* (website), October 8, 2019. https://www.rollingstone.com/culture/culture -features/nxivm-keith-raniere-wellness-scam-sex-cult-848439/.

- Edmondson, *Sarah. Scarred: The True Story of How I Escaped NXIVM*, the Cult That Bound My Life. San Francisco: Chronicle Prism, 2019.
- Grigoriadis, Vanessa. "Inside Nxivm, the 'Sex Cult' That Preached Empowerment." *New York Times* Magazine, May 30, 2018. https://www.nytimes.com/2018/05/30/magazine/sex-cult-empowerment-nxivm-keith-raniere.html.
- Grinberg, Emanuella, and Sonia Moghe. "She Says She Dropped Out of School and Left Mexico to Join a Cult." CNN, May 24, 2019. https://www.cnn.com/2019/05/24/us/nxivm-trial-mexican-witness /index.html.
- Heaney, Katie (as told to). "What It Was Like to Date a Cult Leader." The Cut, November 7, 2019. https://www.thecut.com/2019/11/what-it-was-like-to-date-a-cult-leader.html.
- Johnson, Scott, and Rebecca Sun. "Her Darkest Role: Actress Allison Mack's Descent from 'Smallville' to Sex Cult." *Hollywood Reporter*, May 16, 2018. https://www.hollywoodreporter.com/tv/tv-features/how-smallvilles-allison-mack-went-actress-sex-cult-slaver-1112107/.
- Meier, Barry. "Once Idolized, Guru of Nxivm 'Sex Cult' to Stand Trial Alone." *New York Times*, May 1, 2019. https://www.nytimes.com/2019/05/01/nyregion/nxivm-keith-raniere-trial.html.
- ——. "The Journey of the 'Sex Cult' Heiress: From Reluctant Recruit to Criminal Defendant." *New York Times*, August 11, 2018. https://www.nytimes.com/2018/08/11/business/clare-bronfman-nxivm.html.
- Natalie, Toni, and Chet Hardin. *The Program*: Inside the Mind of Keith Raniere and the Rise and Fall of NXIVM. New York: Grand Central Publishing, 2019.
- "NXIVM: Gina Melita's Story." *Albany Times Union*, YouTube, April 22, 2019. https://youtu.be/YWJcMMgXNmw

?t= 53.

· Odato, James M. “A Split from NXIVM.” *Albany Times Union*, May 10, 2014, updated May 12, 2014. https://www.timesunion.com/local/article/A-split-from-NXIVM-5468731.php.

· Odato, James M., and Jennifer Gish. “In Raniere's Shadows.” *Albany Times Union*, February 18, 2012, updated February 22, 2012. https://www.timesunion.com/local/article/In-Raniere-s-shadows-3341644.php.

· Ringwald, Christopher. “Discount Service Target of Probe Pyramid Scheme

· Alleged.” *Albany Times Union*, May 21, 1992. https://culteducation.com/group/907-nxivm/6023-discount-service-target-of-probe-pyramid-scheme-alleged.html.

· Waters, Flavie, Vivian Chiu, Amanda Atkinson, and Jan Dirk Blom. “Severe Sleep Deprivation Causes Hallucinations and a Gradual Progression Toward Psychosis with Increasing me Awake.” *Frontiers in Psychiatry* 9 (July 10, 2018): 303. https://doi.org/10.3389/fpsyt.2018.00303.

· Xiao, Bowen. “Delving into the Childhood of NXIVM's Leader.” *Epoch Times*, May 28, 2018, updated May 30, 2018. https://www.theepochtimes.com/exclusive-delving-into-the-childhood-of-nxivms-leader_2540043.html.

Chapter 8. 脫逃‥克莉多尼亞．瑪琳達和恢復上帝十誡運動教

· Atuhaire, Bernard. T*he Uganda Cult Tragedy: A Private Investigation*. London: Janus Publishing, 2003.

· Behrend, Heike. “Salvation and Terror in Western Uganda. The Movement for the Restoration of the Ten Commandments of God.” *Bulletin des Séances/Mededelingen der Zittingen. Royal Academy of Overseas Sciences*

47, supplement (2001): 77–96. http://www.kaowarsom.be/documents/BULLETINS_MEDEDELINGEN/2001-SUPPLEMENT.pdf.

- Borzello, Anna. "The Zealot Who Ran Uganda's Killer Cult." *Guardian*, March 29, 2000. https://www.theguardian.com/world/2000/mar/30/2.
- Cauvin, Henri E. "Fateful Meeting Led to Founding of Cult in Uganda." *New York Times*, March 27, 2000. https://www.nytimes.com/2000/03/27/world/fateful-meeting-led-to-founding-of-cult-in-uganda.html.
- Fisher, Ian. "Exploring the Deadly Mystique Surrounding a Uganda Cult." *New York Times*, April 1, 2000. https://archive.nytimes.com/www.nytimes.com/library/world/africa/040200uganda-cult-deaths.html.
- "Priestess of Death." *Newsweek*, August 13, 2000. https://www.newsweek.com /priestess-death-159015.
- Reid, Richard J. *A History of Modern Uganda*. Cambridge, UK: Cambridge University Press, 2017.
- Ward, Kevin. "A History of Christianity in Uganda." Dictionary of African Christian Biography. https://dacb.org/histories/uganda-history-christianity/.
- Zimdars-Swartz, Sandra L. *Encountering Mary: From La Salette to Medjugorje*. Princeton, NJ: Princeton University Press, 2014.

Chapter 9. 否認現實‥馬歇爾．阿普爾懷特和天堂之門

- Bearak, Barry. "Eyes on Glory: Pied Pipers of Heaven's Gate." *New York Times*, April 28, 1997. https://www.nytimes.com/1997/04/28/us/eyes-on-glory-pied-pipers-of-heaven-s-gate.html.

- Evans, Claire L. "Higher Source: The Immortal Web Design of the Suicide Cult 'Heaven's Gate.' " Vice, April 2, 2014. https://www.vice.com/en/article/pgapzy/heavens-gate-web-designers-higher-source-suicide-cult.
- Galanter, Marc. *Cults: Faith, Healing, and Coercion*. New York: Oxford University Press, 1999.
- "Heaven's Gate Cult Members Found Dead." History, February 9, 2010, updated March 24, 2021. https://www.history.com/this-day-in-history/heavens-gate-cult-members-found-dead.
- Lane, Justin E. "UFO Cults." In *Encyclopedia of Sciences and Religions*, edited by Anne L. C. Runehov and Lluis Oviedo. Dordrecht: Springer Netherlands, 2013, 2317–20. https://doi.org/10.1007/978-1-4020-8265-8_1498.
- Lewis, James R., ed. *Encyclopedic Sourcebook of UFO Religions*. Amherst, NY: Prometheus Books, 2003.
- Lewis, James R., and Jesper Aagaard Petersen, eds. *Controversial New Religions*. New York: Oxford University Press, 2004.
- Raine, Susan. "Reconceptualising the Human Body: Heaven's Gate and the Quest for Divine Transformation." *Religion* 35, no. 2 (April 2005): 98–117. https://doi.org/10.1016/j.religion.2005.06.003.
- Robinson, Wendy Gale. "Heaven's Gate: The End." *Journal of Computer-Mediated Communication* 3, no. 3 (December 1, 1997). https://doi.org/10.1111/j.1083-6101.1997.tb00077.x.
- Wessinger, Catherine. *How the Millennium Comes Violently: From Jonestown to Heaven's Gate*. New York: Seven Bridges Press, 2000.
- Zeller, Ben. "What the Heaven's Gate Suicides Say about American Culture." The Conversation, March 24, 2017, updated December 4, 2020. http://theconversation.com/what-the-heavens-gate-suicides-say-about-ameri can-